Lecture Notes in Mathematics

Edited by A. Dold and B. Eckmann

540

Categorical Topology

Proceedings of the Conference Held at Mannheim, 21–25 July, 1975

Edited by E. Binz and H. Herrlich

Springer-Verlag
Berlin · Heidelberg · New York 1976

Editors
Ernst Binz
Universität Mannheim (WH)
Lehrstuhl für Mathematik I
Schloß
6800 Mannheim/BRD

Horst Herrlich
Universität Bremen
Fachsektion Mathematik
Achterstraße
2800 Bremen/BRD

AMS Subject Classifications (1970): 18AXX, 18BXX, 18CXX, 18DXX, 54AXX, 54BXX, 54CXX, 54DXX, 54EXX, 54FXX, 54GXX, 54HXX, 46EXX, 46HXX, 57DXX, 57EXX, 58C15, 58C20, 58D99.

ISBN 3-540-07859-2 Springer-Verlag Berlin · Heidelberg · New York
ISBN 0-387-07859-2 Springer-Verlag New York · Heidelberg · Berlin

Printed in Germany

Printing and binding: Beltz Offsetdruck, Hemsbach/Bergstr.

FOREWORD

This volume consists of the proceedings of the Conference on Categorical Topology held at Mannheim from the 21st to the 25th of July, 1975.

Financial support for the conference was provided by the Volkswagen-Stiftung, Hannover. The participants would like to express their thanks to them and to the many others whose assistance was invaluable: To the Rector of the University of Mannheim, Professor. Dr.E.Gaugler, for his hospitality, to the University Administration for handling much administrative work and to Mrs.K.Bischoff for her help in organizing the conference and for the typing of some of the papers appearing here.

CONTENTS

Contents

Contents

Address list of authors and speakers:

H.L.Bentley	The University of Toledo, Dept. of Math. 2801 W.Bancroft Street Toledo, Ohio 43606, USA
F.Berquier	B 41 Toison d'Or Centre du Général de Gaulle 59200 Tourcoing, France
E.Binz	Universität Mannheim, Lehrstuhl f.Math.I 68 Mannheim, A5, BRD
M.G.Bourdaud	Université de Paris VII U.E.R. de Mathématiques Tour 45-55 5me Etage 2, Place Jussieu 75005 Paris, France
G.C.L.Brümmer	University of Cape Town, Dept. of Math. Private Bag Rondebosch, Rep.of South Africa
A.Calder	University of Missouri,Dept. of Math.Sc. St. Louis, Missouri 63121, USA
J.Cigler	Mathematisches Institut der Universität Strudlhofgasse 4 1090 Wien, Österreich
J.B.Cooper	Mathematisches Institut der Universität Linz/Donau, Österreich
C.H.Dowker	Birkbeck College, Math.Dept. London WC1E 7HX., England
R.Dyckhoff	University of St.Andrews, Math.Institute North Haugh St.Andrews, KY16 9SS
J.Flachsmeyer	Ernst-Moritz-Arndt-Universität Sektion Mathematik Ludwig-Jahn-Str.15a 22 Greifswald, DDR
A.Frölicher	Faculté des Sciences Section de Mathématiques Université de Genève 2-4, rue du Lièvre 1211 Genève 24, Suisse
H.Herrlich	Universität Bremen, Fachsektion Mathematik 28 Bremen, Achterstrasse, BRD
R.-E.Hoffmann	Universität Bremen, Fachsektion Mathematik 28 Bremen, Achterstrasse, BRD

Address list of authors and speakers

K.H.Hofmann	Tulane University, Dept. of Math. New Orleans, La. 70118, USA
M.Hušek	Matematicky Ustav University Karlovy Sokolovská 83 Praha 8 - Karlin, CSSR
* V.Kannan	Madurai University, Dept.of Math. Madurai, India
* F.E.J.Linton	Wesleyan University, Dept. of Math. Middletown, Connecticut 06457, USA
S.Mardešić	University of Zagreb, Inst. of Math. 41001 Zagreb, p.p.187, Yugoslavia
P.Michor	Mathematisches Institut der Universität Strudlhofgasse 4 1090 Wien, Österreich
L.D.Nel	Carleton University, Dept. of Math. Ottawa, Ontario K1S 5B6, Canada
P.Nyikos	University of Illinois at Urbana-Champaign Department of Mathematics Urbana, Ill.61801, USA
J.R.Porter	The University of Kansas, Dept. of Math. Lawrence, Kansas 66044, USA
M.Rajagopalan	Memphis State University, Dept. of Math. Memphis, Tennessee 38152, USA
H.-Chr.Reichel	Mathematisches Institut der Universität Strudlhofgasse 4 1090 Wien, Österreich
* W.A.Robertson	Carleton University, Dept. of Math. Ottawa, K1S 5B6, Canada
S.Salbany	University of Cape Town, Dept. of Math. Private Bag Rondebosch, Rep. of South Africa
M.Schroder	University of Waikato, Dept. of Math. Hamilton, New Zealand
U.Seip	Instituto de Matematica e Estatistica Universidade de Sao Paulo Cx. Postal 20.570 (Ag.Iguameti) Sao Paulo/Brasil
J.Siegel	University of Missouri,Dept. of Math.Sc. College of Arts and Sciences 8001 Natural Bridge Road St. Louis, Missouri 63121, USA

Address list of authors and speakers

J.van der Slot	Schimmelpenninckstraat 16 Zwijndrecht, Netherlands
D.Strauss	University of Hull, Dept. of Math. Hull, England
G.E.Strecker	Kansas State University, Dept. of Math. Manhattan, Kansas 66502, USA
D.Tanré	Université de Picardie Théorie et Applications des Catégories Faculté des Sciences 33, rue Saint-Leu 80 039 Amiens, France
S.J.R.Vorster	University of South Africa, Dept. of Math. P.O.Box 392 Pretoria, Rep. of South Africa
J.de Vries	Mathematisch Centrum 2e Boerhaavestraat 49 Amsterdam-O., Netherlands
M.B.Wischnewsky	Universität Bremen, Fachsektion Mathematik Achterstrasse 33 28 Bremen, BRD
O.Wyler	Carnegie-Mellon University, Dept. of Math. Pittsburgh, Pa. 15213, USA

* V.Kannan: "Coreflective subcategories in topology"

* F.E.J.Linton: "The Jónnson-Tarski Topos"

* W.A.Robertson: "Cartesian closed categories of nearness structures"

* J.van der Slot: "Categories induced by perfect maps"

(These papers will appear elsewhere)

INTRODUCTION

Categorical topology, i.e. the investigation of topological problems- pure and applied - by categorical methods, is a rather new and expanding field.

Recent investigations have made apparent that a considerable number of seemingly typical topological problems can best be understood and analyzed by means of categorical terms and methods, e.g.

(1) Completions, compactifications, realcompactifications etc. have been classified as solutions of universal problems; topological modifications such as sequential-, locally connected-, and compactly generated-refinements as solutions of dual problems

(2) the importance of factorization structures and the close relations between certain classes of spaces and certain classes of maps have become apparent

(3) the similarities between topological structures such as topologies, uniformities, and proximities have been traced down to common properties of the corresponding forgetful functors, and have led to the important concept of a topological functor

(4) cartesian closedness has been exhibited as one of the crucial properties not shared by any of the categories Top, Unif, Prox of topological, uniform, and proximity spaces respectively

(5) a hierarchy of topological categories - some of them cartesian closed, others not - has been constructed, and it has been demonstrated that certain classical problems from extension theory, dimension theory, homology theory, topological algebra, functional analysis

(espec. duality theory), and differential topology not solvable in Top are solvable in the realm of certain of these more appropriate topological categories.

None of the above ideas and results have appeared in book form yet. The purpose of this conference was to survey the present state of categorical topology in order to stimulate and organize further research in this area.

The papers in this volume may be classified as follows:

I. Internal aspects of topological categories, such as epireflective and monocoreflective subcategories of Top, Haus, and Unif (Hušek, Nyikos, Salbany)

II. Categorical aspects of extension theory (Bentley, Porter), dimension theory (Dyckhoff, Herrlich), and perfectness (Strecker)

III. External aspects of topological categories, such as topological functors (Brümmer, Hoffmann), and cartesian and monoidal closed topological categories and topoi (Nel, Wyler, Wischnewsky)

IV. Concrete alternatives to the classical topological categories, such as nearness structures (Bentley, Herrlich, Reichel) convergence- and limit-structures (Binz, Bourdaud) syntopogeneous structures (Tanré), frames (Strauss-Dowker), and generalized uniform structures (Vorster).

V. Applications in topological algebra (Hofmann, de Vries, Wischnewsky)

VI. Applications in algebraic topology (Bentley, Calder, Mardešić)

Introduction

VII. Applications in functional analysis (Binz, Cigler, Cooper-Michor, Flachsmeyer, Frölicher, Rajagopalan, Schroder)

VIII. Applications in differential topology (Berquier, Michor, Seip).

Ernst Binz Horst Herrlich

The Role of Nearness Spaces in Topology

by

H. L. Bentley

In 1973, H. Herrlich [25] introduced nearness spaces and since that time, these spaces have been used for several different purposes by topologists. The aim in this paper is to survey some of the applications of nearness spaces within topology, namely: unification, extensions, homology and connectedness. Some topics which would have been included here are dimension theory and function spaces but these topics have been covered in the paper by H. Herrlich [28] which is being presented at this conference and so they will not be discussed here.

1. Unified theories of topology and uniformity.

One of the important roles that nearness spaces can play in topology is that of unification. The idea of unifying the different structures which topologists study by finding a more extensive structure which includes them is an idea which has held wide interest for the last fifteen years. This idea can be made somewhat more precise as follows: Find a concrete category $\underline{A}$ which contains both the category of topological spaces and continuous maps and the category of uniform spaces and uniformly continuous maps as full subcategories and, moreover, the category $\underline{A}$ should share as many mapping properties as possible with the original two categories. Of course, such a vaguely stated problem can have many solutions, some of them trivial. It is a well known fact that the category of uniform spaces already contains a full subcategory which is isomorphic to the completely regular spaces (namely, the fine uniform spaces). However, few topologists would submit to being forbidden to study non completely regular spaces.

Several elegant solutions of the unification problem have been suggested and one or two have been more or less developed. For example, A. Csaszar [10] offered the syntopogenous spaces and M. Katetov [33] the merotopic spaces. Others are D. Doičinov [12], D. Harris [21], and A.K. Steiner and E. F. Steiner [45] (For some of these, unification was a byproduct, not the main objective). The solution which is of interest here, one due to H. Herrlich [25], is the category Near of nearness spaces and nearness preserving maps. This category will now be defined and examined in some detail.

A nearness structure on a set X is a structure given by a set ξ of collections of subsets of X (i.e. $\xi \subset P^2X$) having the following properties (called axioms of nearness structures) ($\mathcal{A}$ and $\mathcal{B}$ denote subsets of PX):

(N1) If $\mathcal{B} \in \xi$ and $\mathcal{A}$ corefines $\mathcal{B}$ then $\mathcal{A} \in \xi$. (1)

(N2) If $\cap\mathcal{A} \neq \phi$ then $\mathcal{A} \in \xi$. (2)

(N3) $\phi \in \xi$ and $\{\phi\} \notin \xi$.

(N4) If $\mathcal{A} \vee \mathcal{B} \in \xi$ then $\mathcal{A} \in \xi$ or $\mathcal{B} \in \xi$. (3)

(N5) If $cl_\xi \mathcal{A} \in \xi$ then $\mathcal{A} \in \xi$.

($cl_\xi A = \{x \in X \mid \{\{x\}, A\} \in \xi\}$ and $cl_\xi \mathcal{A} = \{cl_\xi A \mid A \in \mathcal{A}\}$.)

(N6) If $\{\{x\}, \{y\}\} \in \xi$ then $x=y$.

---

(1) $\mathcal{A}$ corefines $\mathcal{B}$ iff for every $A \in \mathcal{A}$ there exists $B \in \mathcal{B}$ with $B \subset A$.

(2) The convention $\cap\phi \neq \phi$ is adopted.

(3) $\mathcal{A} \vee \mathcal{B} = \{A \cup B \mid A \in \mathcal{A}$ and $B \in \mathcal{B}\}$.

The collections which are members of ξ are called nearness collections of the nearness structure defined by ξ on X. The phrase "$\mathfrak{A}$ is a nearness collection" could be expressed more suggestively as "the sets of $\mathfrak{A}$ are near" and, in fact, the short terminology "$\mathfrak{A}$ is X-near" is customarily used. A nearness space is a pair $X = (S_X, \xi_X)$ consisting of a set S_X and a nearness structure ξ_X on S_X. As is customary in such situations, X is written in place of S_X. The operator cl_ξ appearing in axiom (N5) is usually written cl_X. A mapping $f: X \to Y$ of a nearness space X into a nearness space Y is called a <u>nearness preserving</u> map (or simply <u>nearness</u> map) iff whenever $\mathfrak{A}$ is X-near then $f\mathfrak{A}$ is Y-near.[1]

A nearness space X has an underlying topological space TX whose closure operator is the operator $cl_X = cl_\xi$ which appears in axiom (N5). (Throughout this section, topological space means T_1-space, i.e. finite subsets are always closed.) Also, any topological space X has an associated nearness space NX defined by $\mathfrak{A}$ is NX-near iff $\bigcap cl_X \mathfrak{A} \neq \phi$. These correspondences T: <u>Near</u> → <u>Top</u> and N: <u>Top</u> → <u>Near</u> are functorial and, in fact, N is an embedding of <u>Top</u> in <u>Near</u> as a bicoreflective full subcategory. Henceforth, <u>Top</u> will be identified with its image under N: <u>Top</u> → <u>Near</u>. Thus, a topological space is a nearness space X which satisfies the condition:

(T) $\mathfrak{A}$ is X-near iff $\bigcap cl_X \mathfrak{A} \neq \phi$.

A nearness space X has an underlying uniform space UX which can be described as follows: the underlying set of UX is the same as that of X and an X-cover is a collection $\mathfrak{A}$ for which $\{X-A \mid A \in \mathfrak{A}\}$ is not X-near. A uniform cover in the structure of UX is defined to be those X-covers $\mathfrak{A}$ for which there exists an infinite sequence $\mathfrak{A} = \mathfrak{B}_1, \mathfrak{B}_2, \mathfrak{B}_3, \ldots$

---

[1] $f\mathfrak{A} = \{fA \mid A \in \mathfrak{A}\}$.

of X-covers with $\mathcal{B}_{n+1}$ star refining $\mathcal{B}_n$ for each n. Vice versa, any uniform space X has an associated nearness space NX defined by $\mathcal{A}$ is X-near iff $\{X-A \mid A \in \mathcal{A}\}$ is not a uniform cover. The correspondences U: <u>Near</u> → <u>Unif</u> and N: <u>Unif</u> → <u>Near</u> are functorial and, in fact, N is an embedding of <u>Unif</u> in <u>Near</u> as a bireflective full subcategory. Henceforth, <u>Unif</u> will be identified with its image under N: <u>Unif</u> → <u>Near</u>. Thus, a uniform space is a nearness space X which satisfies the condition:

(U): If $\mathcal{A}$ is an X-cover then there exists an X-cover $\mathcal{B}$ which star refines $\mathcal{A}$.

Another important category which can be embedded in <u>Near</u> is the category <u>Cont</u> of contiguity spaces (V. M. Ivanova and A. A. Ivanov [32] and W. L. Terwilliger [49]).

A contiguity structure on a set X is a structure given by a set ξ of finite collections of subsets of X having the following properties ($\mathcal{A}$ and $\mathcal{B}$ denote finite collections of subsets of X):

(C1) If $\mathcal{B} \in \xi$ and $\mathcal{A}$ corefines $\mathcal{B}$ then $\mathcal{A} \in \xi$.

(C2) If $\cap\mathcal{A} \neq \phi$ then $\mathcal{A} \in \xi$.

(C3) $\phi \in \xi$ and $\{\phi\} \notin \xi$.

(C4) If $\mathcal{A} \vee \mathcal{B} \in \xi$ then $\mathcal{A} \in \xi$ or $\mathcal{B} \in \xi$.

(C5) If $cl_\xi \mathcal{A} \in \xi$ then $\mathcal{A} \in \xi$.
($cl_\xi A = \{x \in X \mid \{\{x\}, A\} \in \xi\}$ and $cl_\xi \mathcal{A} = \{cl_\xi A \mid A \in \mathcal{A}\}$.)

(C6) If $\{\{x\}, \{y\}\} \in \xi$ then $x=y$.

Note that the contiguity axioms are the nearness axioms with a blanket assumption that the collections which are members of the structure ξ must be finite. A contiguity space is a set endowed with a contiguity structure. If X is a contiguity space with contiguity structure ξ then a finite collection $\mathcal{A}$ is said to be X-contigual provided $\mathcal{A} \in \xi$. Contigual maps are defined in the expected way.

A nearness space X has an underlying contiguity space CX which can be described as follows: the underlying set of CX is the same as that of X and a collection $\mathfrak{A}$ is CX-contigual iff $\mathfrak{A}$ is finite and $\mathfrak{A}$ is X-near. Vice versa, any contigual space X has an associated nearness space NX defined by: the underlying set of NX is the same as that of X and a collection $\mathfrak{A}$ is NX-near iff every finite subset of $\mathfrak{A}$ is X-contigual. The correspondences C: Near → Cont and N: Cont → Near are functorial and, in fact, N is an embedding of Cont in Near as a bireflective full subcategory. Henceforth, Cont will be identified with its image under N: Cont → Near. Thus, a contiguity space is a nearness space X which satisfies the condition:

(C) If every finite subset of a collection $\mathfrak{A}$ is X-near then so is $\mathfrak{A}$.

Having embedded these three categories (Top, Unif, and Cont) into Near, a useful new operation is possible: one can take intersections of these subcategories. It turns out that

(1) Top $\cap$ Cont = the category of compact topological spaces.

(2) Top $\cap$ Unif = the category of paracompact (= fully normal) topological spaces. = the category of fine uniform spaces.

(3) Cont $\cap$ Unif = the category of precompact uniform spaces.

(4) Top $\cap$ Unif $\cap$ Cont = the category of compact Hausdorff spaces.

In 1964, O. Frink [17] applied a construction due to H. Wallman [52] to what he called a normal base $\mathcal{G}$ of closed subsets of a completely regular topological space X to obtain a compactification w(X, $\mathcal{G}$) of X and which since has become known as a Wallman-type compactification. Because of a question which Frink raised and which remains unanswered to this day (Is every Hausdorff compactification of X of the form w(X, $\mathcal{G}$) for some $\mathcal{G}$?) these Wallman-type compactifications have attracted considerable attention. E. F. Steiner [46] generalized the construction to allow T_1-compactifications of T_1-spaces by replacing the normal bases of Frink by what Steiner called separating bases. A separating base on a topological space X is a base $\mathcal{L}$ for the closed subsets of X which is closed under finite unions and finite intersections and which satisfies: (S) if $x \notin B \in \mathcal{L}$ then for some $E \in \mathcal{L}$, $x \in E$ and $E \cap B = \phi$. Of course, these ideas can be placed in the nearness space setting by going through contiguities. However, nearness structures allow many more possibilities than do contiguities and arbitrary cardinal restrictions can be made. A separating base $\mathcal{L}$ on a topological space X gives rise to several nearness spaces $N_k(X, \mathcal{L})$ with underlying set X and with k an infinite cardinal number. A collection $\mathcal{A}$ is defined to be not $N_k(X, \mathcal{L})$-near iff for some subset $\mathcal{L}'$ of $\mathcal{L}$ with card $\mathcal{L}' < k$, $\mathcal{L}'$ corefines $\mathcal{A}$ and $\bigcap \mathcal{L}' = \phi$. The existence of these structures is one reason for studying nearness spaces because of the fact that the theory of contiguity spaces is inadequate for treating the Wallman-type realcompactifications (A.K. Steiner and E. F. Steiner [44], R. A. Alo and H. L. Shapiro [1], M. S. Gagrat and S. A. Naimpally [19], H. L. Bentley and S. A. Naimaplly [7]). A nearness space X always has a completion X* (see below). If $\mathcal{L}$ is a separating base on a topological space X and Y denotes the nearness space $N_{\aleph_0}(X, \mathcal{L})$ defined above, then the Wallman-type compactification of X induced by $\mathcal{L}$ is the same as the completion Y* of the nearness space Y. If, in addition, $\mathcal{L}$ is closed under countable intersections and Y denotes the nearness space $N_{\aleph_1}(X, \mathcal{L})$ defined above, then the Wallman-type $\mathcal{L}$-realcompactification of X induced by $\mathcal{L}$ is the same as the completion Y* of the nearness space Y.

There is another line of ideas which can be unified in the setting of nearness spaces and which involve the extension of continuous maps from dense subspaces. These will be examined at the end of the next section.

2. Extensions of topological spaces.

Nearness spaces are a most natural tool for studying extensions of topological spaces; in fact, it was in this context that nearness spaces orginally arose. This point of view has been developed in a recent paper by H. L. Bentley and H. Herrlich [4].

An extension e: X → Y is a dense embedding of a topological space X into a topological space Y (for technical simplicity, one usually assumes that the map e is an inclusion). Every extension e: X → Y induces various structures on X.

1. e: X →Y induces the nearness structure

$$\xi = \{\mathfrak{A} \subset PX \mid \bigcap cl_Y \mathfrak{A} \neq \phi \}.$$

2. e: X →Y induces the contiguity structure

$$\varepsilon = \{\mathfrak{A} \subset PX \mid \mathfrak{A} \text{ is finite and } \bigcap cl_Y \mathfrak{A} \neq \phi\}.$$

3. e: X →Y induces the generalized proximity relation

$$\delta = \{(A,B) \in (PX)^2 \mid cl_Y A \cap cl_Y B \neq \phi \}.$$

It has recently been shown that not all nearness spaces are induced by an extension (H. L. Bentley [3], S.A. Naimpally and J. H. M. Whitfield [38]). Contiguity spaces were introduced and axiomatized by V. M. Ivanova and A. A. Ivanov [32] who proved that every contiguity structure on X is induced by some T_1- compactification. Proximity relations were defined by V. A. Efremovič [14] and then Y. M. Smirmov [42] proved that every proximity relation is induced by a Hausdorff compactification. M. W. Lodato [35] axiomatized generalized proximity relations. W. J. Thron [51] (using a characterization given earlier by M. S. Gagrat and S. A. Naimpally [18]) showed that every generalized proximity relation is induced by some

T_1-compactification. All of these results can be expressed in terms of nearness spaces and thereby become just different aspects of a single general theorem.

A formalization of the idea of a nearness space being induced by some extension is embodied in the following definition.

Definition: A nearness space is called <u>subtopological</u> iff it is a nearness subspace of some topological nearness space.

Definition (G. Choquet [9]): $\mathcal{G}$ is called a <u>grill</u> on X iff $\phi \notin \mathcal{G} \subset PX$ and

(G) $A \cup B \in \mathcal{G}$ iff $A \in \mathcal{G}$ or $B \in \mathcal{G}$.

Definition: If X is a nearness space then $\mathcal{G}$ is called an <u>X-grill</u> iff $\mathcal{G}$ is a grill on the underlying set of X and $\mathcal{G}$ is a nearness collection on X.

Theorem: A nearness space X is subtopological iff each nearness collection on X is a subset of some X-grill.

Theorem: Any extension $e: X \to Y$ induces a subtopological nearness structure on X. If Z denotes the nearness space with the same underlying set of points as X and with the nearness structure induced by the extension $e: X \to Y$ then $TZ = X$. Vice versa, if Z is any subtopological nearness space then there exists an extension $e: X \to Y$ with $TZ = X$ and where the nearness structure of Z is the one induced by the extension $e: X \to Y$ on X.

A more fruitful line of ideas comes from considering the concrete nearness spaces.

Definition: If X is a nearness space then an X-cluster is a non-empty maximal (with respect to set inclusion) nearness collection.

Definition: A nearness space is called concrete iff each nearness collection is a subset of some cluster.

Every cluster is a grill so every concrete nearness space is subtopological. Any extension e: $X \to Y$ which induces a concrete nearness structure on X can be recovered by a completion process. A collection $\mathcal{A}$ of subsets of a topological space X is said to converge (with respect to X) to a point x of X provided every neighborhood of x contains some member of $\mathcal{A}$. A nearness space X is said to be complete provided every X-cluster converges (with respect to the underlying topological space TX of X) to some point of X. Every nearness space X has a completion X*. There is an embedding e: $X \to X^*$ of X as a dense nearness subspace of X* where X* is a complete nearness space.

Theorem: If X is a nearness space, then X is concrete iff its completion X* is topological.

Definition (M.H. Stone [47]: An extension e: $X \to Y$ is called strict iff $\{cl_Y A \mid A \subset X\}$ is a base for closed sets of Y.

Theorem: Any strict extension e: $X \to Y$ induces a concrete nearness structure on X. If Z denotes the nearness space with the same underlying set of points as X and with the nearness structure induced by the strict extension e: $X \to Y$, then TZ = X and e: $X \to Z^*$ is equivalent to e: $X \to Y$. Thus, two strict extensions e_1: $X \to Y_1$ and e_2: $X \to Y_2$ are equivalent iff they induce the same nearness structure on X.

Since every contiguity space is concrete, the structure of a contiguity space X is always induced by some topological extension (e.g. $e: TX \to X^*$).

The question of how generalized proximity structures fit into the nearness space setting remains. There is more than one natural way to embed the category of generalized proximity spaces into Near. The details are as follows.

A generalized proximity relation on a set X is a relation $\delta \subset (PX)^2$ having the following properties:

(P0) If $A \,\delta\, B$ then $B \,\delta\, A$.

(P1) If $A \subset B$ and $A \,\delta\, C$ then $B \,\delta\, C$.

(P2) If $A \cap B \neq \phi$ then $A \,\delta\, B$.

(P3) If $A \,\delta\, B$ then $A \neq \phi$.

(P4) If $A \,\delta\, (B \cup C)$ then $A \,\delta\, B$ or $A \,\delta\, C$.

(P5) If $A \,\delta\, B$ and $B \subset cl_\delta C$ then $A \,\delta\, C$.

$(cl_\delta A = \{x \in X \mid \{x\} \,\delta\, A\}.)$

(P6) If $\{x\} \,\delta\, \{y\}$ then $x = y$.

A generalized proximity space is a set endowed with a generalized proximity relation. Proximal maps are defined in the expected way and the category Prox results.

Let U: Cont $\to$ Prox denote the obvious forgetful functor which associates with any contiguity space X the generalized proximity space UX with the same underlying set as X and with generalized proximity relation δ defined by $A \,\delta\, B$ iff $\{A,B\}$ is X-contigual. The following theorem (which was proved in [4]) gives the basis for discovering many important facts about generalized proximity spaces.

Theorem: The forgetful functor U: Cont $\to$ Prox is topological, i.e. for any generalized proximity space X any family $(X_i)_{i\in I}$ of contiguity spaces and any family $(f_i: X \to UX_i)_{i\in I}$ of proximal maps, there exists a contiguity structure on X, giving rise to a contiguity space Y with UY = X, which is initial with respect to the given data, i.e. such that for any contiguity space Z and any proximal map g: UZ → X, the following conditions are equivalent:

(a) g: Z → Y is a contigual map.

(b) For each $i \in I$, $f_i \circ g: Z \to X_i$ is a contigual map.

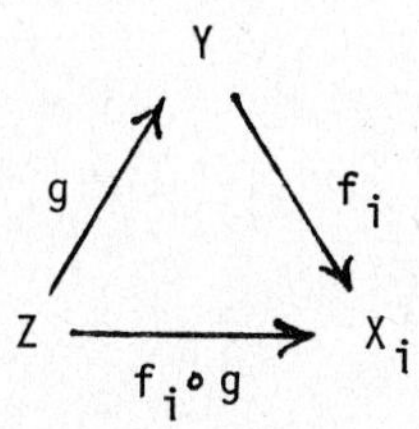

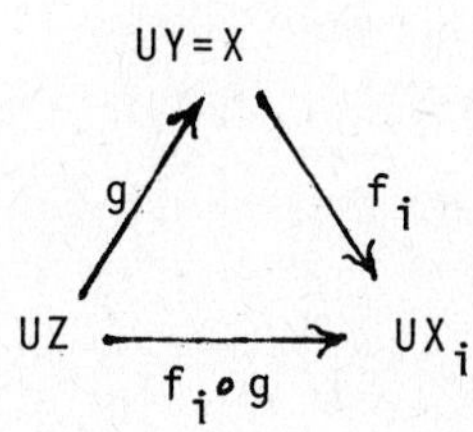

The importance in a functor being topological is well known (e.g. see H. Herrlich [24]). In particular, the fibres of U form complete lattices and so there exist discrete and indiscrete contiguity structures on each generalized proximity space. Thus, there exist at least two ways to define a functor E: <u>Prox</u> → <u>Cont</u> to be a right inverse for U: <u>Cont</u> → <u>Prox</u>, take E to be either the indiscrete contiguity or the discrete contiguity. Thus, let L: <u>Prox</u> → <u>Cont</u> denote the left adjoint, right inverse of U which associates with each generalized proximity space the corresponding discrete (smallest) contiguity and R: <u>Prox</u> → <u>Cont</u> the right adjoint, right inverse of U which associates with each generalized proximity space the corresponding indiscrete (largest) contiguity. Using either of the two functors R or L (which are embeddings as full subcategories) it follows that, since every contiguity structure is induced by some extension, then so is every generalized proximity structure also induced by some extension. Although either R or L can be used in establishing this result, R and not L is the most interesting of thest two functors. In fact, it turns out that if a generalized proximity space X with generalized proximity relation δ satisfies the additional axiom of Efremovič [14]

(E) If not A δ B then there exists $C \subset X$ with not A δ C and not B $\delta(X-C)$,

so that X becomes a proximity space, then the Smirnov compactification of X is the same thing as the completion of the contiguity space RX and these are usually different from the completion of the contiguity space LX. Thus, R is to be favored over L.

An interesting line of inquiry arises by asking what nearness theoretic property of a nearness structure is equivalent to it being induced by an extension e: $X \to Y$ with Y having a certain topological property. It is worth rephrasing this statement in category theoretic terms. Let <u>Ext</u> denote the category of strict extensions, i.e. objects of <u>Ext</u> are strict extensions $e:X \to Y$ and morphisms of <u>Ext</u> are pairs of continuous maps (f,g): $(e:X \to Y) \to (e':X' \to Y')$ for which the following diagram is commutative:

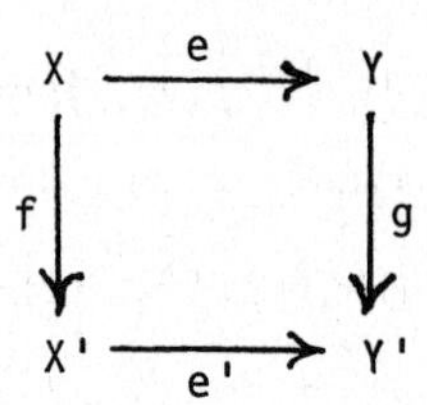

Then there is the "inducing" functor N: <u>Ext</u> $\to$ <u>Near</u> where $N(e:X \to Y)$ is the nearness space whose underlying set is the same as that of X and whose nearness structure is the (concrete) one induced by the strict extension $e:X \to Y$.

For a morphism, N(f,g)=f. Thus, the program which presents itself in this context is the following one: Given a topological property P, what nearness theoretic property P' is such that a nearness space Z has the property P' iff $Z=N(e:X\to Y)$ for some strict extension $e:X\to Y$ where Y has property P. If "property" is understood to mean "full subcategory", then this program can be rephrased as follows. (In the following, subcategory always means isomorphism closed, full subcategory). If B is a subcategory of Top then give an "internal" description of the subcategory B' of Near whose objects are nearness spaces of the form $N(e:X\to Y)$ where Y is in B.

There is a semioperational approach to a solution of this problem. If B is a subcategory of Top and B' is a subcategory of Near, say that B' is a pendant of B iff $B'\cap Top = B$. If B' is a pendant of B which also satisfies the three conditions

(1) Each object of B' is a concrete nearness space.

(2) B' is hereditary (i.e. if Y is a B' object and Z is a nearness subspace of Y then Z is also in B').

(3) B' is completion closed (i.e. if Y is in B' then the completion Y* is also).

then the objects of B' are precisely those nearness spaces of the form $N(e:X\to Y)$ where Y is in B.

Given B it is sometimes possible to find a pendant B' of B so that (1), (2), and (3) are satisfied. A solution of this type of problem with B any of the following topological properties appears in the paper by H. L. Bentley and H. Herrlich [4]:

(a) Compact.

(b) Hausdorff.

(c) Compact Hausdorff.

(d) Regular.

(e) Paracompact.

(f) Realcompact.

A solution for normality has not yet been found.

It was mentioned at the end of the last section that several theorems about the extension of continuous maps from dense subspaces can be unified in the setting of nearness spaces. These are: the extension theorem of Taǐmanov and several generalizations of Taǐmanov's theorem due to R. Engelking [16] and H. Herrlich [23]. This unification has been developed by H. Herrlich in [26] and by H. L. Bentley and H. Herrlich in [4]. Another type of generalization of Taǐmanov's theorem has been given by S. A. Naimpally [38].

The crucial property for the existence of extensions of maps is regularity of the range. If X is a nearness space then for A, $B \subset X$, $A <_X B$ means that $\{A, X-B\}$ is not X-near. A nearness space X is called <u>regular</u> iff for any collection $\mathfrak{A}$, if the collection $\{B \subset X \mid \text{for some } A \in \mathfrak{A},\ A <_X B\}$ is X-near then so is $\mathfrak{A}$. This concept of regularity is a pendant of the topological concept of regularity. (In fact, if <u>Reg</u> denotes the full subcategory of <u>Near</u> whose objects are the regular nearness spaces, then <u>Reg</u> is also hereditary and completion closed.) H. Herrlich's theorem is the following one [26].

Theorem: If Y is a complete, regular nearness space, S is a dense nearness subspace of a nearness space X and $f: S \to Y$ is a function, the following conditions are equivalent:

(1) $f: S \to Y$ is a nearness map.

(2) f can be extended to a nearness map $g: X \to Y$.

Herrlich's theorem has many corollaries, among them the following ones.

Corollary (A. Weil [54]). If Y is a complete uniform space, S is a dense subspace of a uniform space X and $f:S \to Y$ is a function, the following conditions are equivalent:

(1) $f:S \to Y$ is uniformly continuous.

(2) f can be extended to a uniformly continuous function $g:X \to Y$.

Corollary (H. Herrlich [23]). If Y is a regular topological space, $\mathcal{L}$ is a base for the closed sets of Y, S is a dense nearness subspace of a topological space X, and $f:S \to Y$ is a function, the following are equivalent:

(1) $f:S \to Y$ is a nearness map.

(2) f can be extended to a continuous map $g:X \to Y$.

(3) If $\mathcal{A} \subset \mathcal{L}$ with $\bigcap \mathcal{A} = \phi$ then $\bigcap\{cl_X f^{-1}A \mid A \in \mathcal{A}\} = \phi$.

Corollary (R. Engelking [16]). If Y is a realcompact topological space, S is a dense nearness subspace of a topological space X, and $f:S \to Y$ is a function, the following are equivalent:

(1) $f:S \to Y$ is a nearness map.

(2) f can be extended to a continuous map $g:X \to Y$.

(3) For each countable sequence $A_1, A_2, \ldots$ of zero sets of Y, if $\bigcap_{i=1}^{\infty} A_i = \phi$ then $\bigcap_{i=1}^{\infty} cl_X\, f^{-1}\, A_i = \phi$.

Corollary (A. D. Taĭmanov [48]). If Y is a compact Hausdorff topological space, S is a dense nearness subspace of a topological space X, and $f:S \to Y$ is a function, the following are equivalent:

(1) $f:S \to Y$ is a nearness map.

(2) f can be extended to a continuous map $g:X \to Y$.

(3) For each finite sequence $A_1, \ldots, A_n$ of closed sets of Y, if $\bigcap_{i=1}^{n} A_i = \phi$ then $\bigcap_{i=1}^{n} cl_X f^{-1} A_i = \phi$.

(4) For each pair A,B of closed sets of Y, if $A \cap B = \phi$ then $cl_X f^{-1}A \cap cl_X f^{-1}B = \phi$.

3. Nearness spaces and Algebraic Topology.

E. Čech [8] defined a homology theory for general topological spaces. His theory was developed by several topologists, but most notably by C. H. Dowker [13] and E. H. Spanier [43], who showed that the Čech theory satisfies the Eilenberg-Steenrod axioms, and an exposition of this theory appears in the famous book by S. Eilenberg and N. E. Steenrod [15]. Let X be a topological space and let μ be the set of all open covers of X. For each open cover $\mathcal{A}$ of X, let $N_{\mathcal{A}}$ be the nerve of $\mathcal{A}$ and let $H_q(N_{\mathcal{A}})$ be the q-dimensional homology group of the simplicial complex $N_{\mathcal{A}}$. The set μ can be partially ordered by $\mathcal{B} \leq \mathcal{A}$ iff $\mathcal{A}$ refines $\mathcal{B}$ and thus μ becomes a directed set. If $\mathcal{A}, \mathcal{B} \in \mu$ and $\mathcal{A}$ refines $\mathcal{B}$, then there exist "projections" $P_{\mathcal{A},\mathcal{B}} : N_{\mathcal{A}} \to N_{\mathcal{B}}$ which satisfy the condition that for each $A \in \mathcal{A}$, $P_{\mathcal{A},\mathcal{B}}(A) \in \mathcal{B}$ and $A \subset P_{\mathcal{A},\mathcal{B}}(A)$. Using the homomorphisms induced by such projections as bonding maps between groups $H_q(N_{\mathcal{A}})$, a direct spectrum results and the direct limit group, $H_q(X)$, is the q-dimensional Cech homology group of X. C. H. Dowker [13] observed that the above definition makes sense if instead of letting μ denote the set of all open covers of X, μ denotes any set of covers which satisfy the two axioms:

(U1) If $\mathcal{A} \in \mu$ and $\mathcal{A}$ refines a cover $\mathcal{B}$ then $\mathcal{B} \in \mu$.

(U2) If $\mathcal{A} \in \mu$ and $\mathcal{B} \in \mu$ then $\mathcal{A} \wedge \mathcal{B} \in \mu$.(1)

As H. Herrlich [27] has observed, these two axioms (together with the condition that $\{X\} \in \mu$ and $\phi \notin \mu$) give rise to structures which are equivalent to seminearness structures

---

(1) $\mathcal{A} \wedge \mathcal{B} = \{A \cap B \mid A \in \mathcal{A} \text{ and } B \in \mathcal{B}\}$.

(structures which satisfy axioms (N1) - (N4)). Thus, in this terminology, Dowker defined the Čech homology groups for a seminearness space. No one did anything about this for many years, but in 1972, M. Bahaudin and J. Thomas [2] investigated these groups for uniform spaces and then D. Czarcinski [11] studied them for nearness spaces. He showed that these groups satisfy a variant of the Eilenberg-Steenrod axioms for a homology theory and he also showed that a nearness space and its completion have the same homology groups. M. Bahaudin and J. Thomas asked whether a completely regular topological space X and the associated fine uniform space UX have the same homology gropus. This question remains unanswered; however, a partial solution is obvious in the setting of nearness spaces. As nearness spaces, a paracompact topological space X and its associated fine uniform space UX are the same. Therefore, in this case, they have the same homology groups.

4. Connectedness.

The homology theory mentioned above yields, of course, a theory of connectedness for seminearness spaces as a byproduct. It is still interesting to look explicitly at some of the features of this kind of connectedness.

A seminearness space X is called <u>connected</u> iff whenever f: X → Y is a nearness map from X into a discrete nearness space Y, then f is constant. This concept has been studied by S. G. Mrówka and W. J. Pervin [36] in the special case of uniform spaces. Of course, if X is a topological space (i.e. T_1 - space) then this concept is equivalent to the usual topological connectedness (proved by B. Knaster and K. Kuratowski [34] in 1921). One of the primary values in studying connectedness of seminearness spaces, other than the obvious one of unifying the ideas, is that relationships between the various kinds of connectedness is easier to express. Mrówka and Pervin proved that a uniform space is uniformly connected iff its precompact reflection is uniformly connected. This theorem generalizes perfectly to the setting of seminearness spaces and in fact, a quite surprising, albeit simple, further generalization is valid: The precompact reflection can be replaced by an arbitrary bireflection. Let S-Near denote the category of seminearness spaces.

Theorem: Let A be any bireflective subcategory of S-Near such that every discrete seminearness space is an A-object and let F: S-Near → A denote the associated reflector. Then a seminearness space X is connected iff FX is connected.

5. Remarks.

Many topics have been omitted. As was stated earlier, dimension theory and function spaces can be read about in [28] and in general, the paper by H. Herrlich [27] is the basic source of information about nearness spaces.

Bibliography

1. R. A. Aló and H. L. Shapiro, Wallman compact and realcompact spaces, Contributions to Extension Theory of Topological Structures (Proc. Symp. Berlin, 1967), Berlin (1969), 9-14.

2. M. Bahauddin and J. Thomas, The homology of uniform spaces, Canad. J. Math. 25 (1973), 449-455.

3. H. L. Bentley, Nearness spaces and extensions of topological spaces, Studies in Topology, New York (1975), 47-66.

4. H. L. Bentley and H. Herrlich, Extensions of topological spaces, preprint.

5. H. L. Bentley, H. Herrlich and W. A. Robertson, Convenient categories for topologists, preprint.

6. H. L. Bentley and S. A. Naimpally, Wallman T_1-compactifications as epireflections, Gen. Topol. Appl. 4 (1974), 29-41.

7. H. L. Bentley and S. A. Naimpally, L-realcompactifications as epireflections, Proc. Amer. Math. Soc. 44 (1974), 196-202.

8. E. Čech, Theorie generale de l'homologie dans un espace quelconque, Fund. Math. 19 (1932), 149-183.

9. G. Choquet, Sur les notions de filtre et de grille, Comptes Rendus Acad. Sci. Paris 224 (1947), 171-173.

10. A. Császár, Foundations of General Topology, New York (1963).

11. D. Czarcinski, The Čech homology theory for nearness spaces, Thesis, University of Toledo (1975).

12. D. Doĭčinov, On a general theory of topological, proximity and uniform spaces, Dokl. SSSR 156 (1964), 21-24; Soviet Math. 5 (1964), 595-598.

13. C. H. Dowker, Homology groups of relations, Ann. of Math. 56 (1952), 84-95.

14. V. A. Efremovič, Infinitesimal spaces, Dokl. Akad. Nauk SSSR, 156 (1951), 341-343.

15. S. Eilenberg and N. Steenrod, Foundations of Algebraic Topology, Princeton (1952).

16. R. Engelking, Remarks on realcompact spaces, Fund. Math. 55 (1964), 303-308.

17. O. Frink, Compactifications and semi normal spaces, Amer. J. Math. 86 (1964), 602-607.

18. M. S. Gagrat and S. A. Naimpally, Pooximity approach to extension problems, Fund. Math. 71 (1971), 63-76.

19. M. S. Gagrat and S. A. Naimpally, Wallman compactifications and Wallman realcompactifications, J. Austral. Math. Soc. 15 (1973), 417 - 427.

20. M. S. Gagrat and W. J. Thron, Nearness structures and proximity extensions, preprint.

21. D. Harris, Structures in Topology, Memoirs Amer. Math. Soc. 115 (1971).

22. M. Shayegan Hastings, Epireflective Hulls in Near, Thesis, University of Toledo (1975).

23. H. Herrlich, Fortsetzbarkeit stetiger Abbildungen und Kompaktheitsgrad topologischer Räume, Math. Zeit. 96(1967),64-72.

24. H. Herrlich, Topological functors, Gen. Topol. Appl. 4 (1974), 125-142.

25. H. Herrlich, A concept of nearness, Gen. Topol. Appl. 5 (1974), 191-212.

26. H. Herrlich, On the extendibility of continuous functions, Gen. Topol. Appl. 5 (1974), 213-215.

27. H. Herrlich, Topological structures, Math. Centre Tract 52 (1974), 59-122.

28. H. Herrlich, Some topological theorems which fail to be true, preprint.

29. H. Herrlich and G. E. Strecker, Category theory, Boston (1973).

30. W. N. Hunsaker and P. L. Sharma, Nearness structures compatible with a topological space, Archiv der Math. 25 (1974), 172-178.

31. M. Hušek, Categorial connections between generalized proximity spaces and compactifications, Contributions to Extension Theory of Topological Structures (Proc. Symp. Berlin,1967), Berlin (1969), 127-132.

32. V. M. Ivanova and A. A. Ivanov, Contiguity spaces and bicompact extensions of topological spaces, Dok. Akad. Nauk SSSR 127 (1959), 20-22.

33. M. Katětov, On continuity structures and spaces of mappings, Comment. Math. Univ. Carol. 6 (1965), 257-278.

34. B. Knaster and K. Kuratowski, Sur les ensembles connexes, Fund. Math. 2 (1921), 206-255.

35. M. W. Lodato, On topologically induced generalized proximity relations, Proc. Amer. Math. Soc. 15 (1964), 417-422.

36. S. G. Mrówka and W. J. Pervin, On uniform connectedness, Proc. Amer. Math. Soc. 15 (1964), 446-449.

37. S. A. Naimpally, Reflective functors via nearness, Fund. Math. 85 (1974), 245-255.

38. S. A. Naimpally and J. H. M. Whitfield, Not every near family is contained in a clan, Proc. Amer. Math. Soc. 47 (1975), 237-238.

39. L. D. Nel, Initially structured categories and cartesian closedness, preprint.

40. O. Njåstad, A proximity without a smallest compatible nearness, preprint.

41. W. A. Robertson, Convergence as a Nearness Concept, Thesis, Carleton University (1975).

42. Y. M. Smirnov, On proximity spaces, Mat. Sb. (NS) 31 (1952), 453-574; English transl., Amer. Math. Soc. Transl. Ser. 2 38 (1964), 5-36.

43. E. H. Spanier, Cohomology theory for general spaces, Ann. of Math. 49 (1948), 407-427.

44. A. K. Steiner and E. F. Steiner, Nest generated intersection rings in Tychonoff spaces, Trans. Amer. Math Soc. 148 (1970), 589-601.

45. A. K. Steiner and E. F. Steiner, Binding spaces: a unified completion and extension theory, Fund. Math. 76 (1972), 43-61.

46. E. F. Steiner, Wallman spaces and compactifications, Fund. Math. 61 (1968), 295-304.

47. M. H. Stone, Applications of the theory of Boolean rings to general topology, Trans. Amer. Math. Soc. 41 (1937), 375-481.

48. A. D. Taĭmanov, On extension of continuous mappings of topological spaces, Math. Sbornik N. S. 31 (73) (1952), 459-463.

49. W. L. Terwilliger, On Contiguity Spaces, Thesis, Washington State University (1965)

50. W. J. Thron, Proximity structures and grills, Math. Ann. 206 (1973), 35-62.

51. W. J. Thron, On a problem of F. Riesz concerning proximity structures, Proc. Amer. Math. Soc. 40 (1973), 323-326.

52. H. Wallman, Lattices and topological spaces, Ann. of Math. 39 (1938), 112-126.

53. R. H. Warren, Proximities, Lodato Proximities and Proximities of Čech, Thesis, University of Colorado (1971).

54. A. Weil, Sur les Espaces à Structure Uniforme et sur la Topologie Générale, Paris (1938).

Françoise Berquier et Gérard Bourdaud, seuls participants français à la présente Conférence sur la Topologie Catégorique, tiennent à exprimer ici leur reconnaissance envers Andrée Bastiani et Charles Ehresmann. C'est en effet grâce à eux qu'ils participent à ce Colloque. Leur ouverture d'esprit, leur bienveillance et leurs encouragements leur ont été à tous deux précieux dans leurs travaux de recherche.

Cela devait être dit, à un moment où est menée en France une odieuse campagne - ne déshonorant que ses instigateurs - qui tente de discréditer l'équipe "Théorie et Applications des Catégories", et va même jusqu'à interdire administrativement à Andrée Bastiani et Charles Ehresmann de diriger des chercheurs.

UN THEOREME D'INVERSION LOCALE

par Françoise BERQUIER

I. INTRODUCTION.

Soit E un espace de Banach. On a les deux théorèmes suivants :

a) Soit f une application de E dans E, continûment B-différentiable sur un voisinage U d'un point x_o (en d'autres termes, continûment différentiable au sens de Fréchet sur U), i.e. la dérivée de f par rapport à l'ensemble B des bornés de E existe en tout point x de U et l'application Df: $U \to L(E,E)$, $x \mapsto Df(x)$, est continue lorsque L(E,E) est muni de la topologie de la convergence uniforme sur les ensembles de B. Supposons que $Df(x_o)$ soit un homéomorphisme. Alors il existe des voisinages V et W de x_o et $f(x_o)$ respectivement tels que f induise un homéomorphisme de V sur W ; de plus l'application f^{-1}: $W \to V$ est B-différentiable au point $y_o = f(x_o)$ et $(Df^{-1})(y_o) = (Df(x_o))^{-1}$.

b) Soit f une application de E dans E induisant une bijection d'un voisinage V de x_o sur un voisinage W de $f(x_o) = y_o$ et telle que l'application f^{-1} soit continue au point y_o. Supposons que f soit B-différentiable sur V et que $Df(x_o)$ soit un homéomorphisme. Alors l'application f^{-1} est B-différentiable au point y_o et $(Df^{-1})(y_o) = (Df(x_o))^{-1}$.

On sait que les théorèmes a) et b) sont faux en général pour des evt quelconques, comme le montrent les exemples suivants.

1) Désignons par $C(\mathbb{R},\mathbb{R})$ l'espace des applications continues d'une variable réelle à valeurs réelles muni de la topologie de la convergence compacte, et par f_o l'application nulle de $C(\mathbb{R},\mathbb{R})$. Soit exp: $C(\mathbb{R},\mathbb{R}) \to C(\mathbb{R},\mathbb{R})$ l'application exponentielle $f \mapsto exp(f)$, où $exp(f)(x) = exp(f(x))$ pour tout $x \in \mathbb{R}$.

Cette application ne vérifie pas le théorème a) : elle est continûment B-différentiable sur $C(\mathbb{R},\mathbb{R})$ et la dérivée $D(exp)(f_o)$ est l'application identique ; or exp n'est surjective sur aucun voisinage de f_o ([3,5]).

2) Considérons l'espace $\mathbb{R}^{\mathbb{N}}$ des suites numériques muni de la topologie produit. Soit ϕ: $\mathbb{R} \to \mathbb{R}$ une application indéfiniment dérivable à support contenu dans le segment $[1/2,3/2]$ et telle que $\phi(1)=1$. Nous noterons simplement (x_n) l'élément $(x_n)_{n \in \mathbb{N}}$ de $\mathbb{R}^{\mathbb{N}}$.

L'application f: $\mathbb{R}^N \to \mathbb{R}^N$ définie par $f((x_n))=(x_n-\phi(x_n))$ est continûment B-différentiable sur $\mathbb{R}^N$ et la dérivée $Df(f_o)$, où $f_o=(0,0,\ldots)\in\mathbb{R}^N$, est un homéomorphisme ; pourtant f n'est injective dans aucun voisinage de f_o ([1]).

3) L'application f: $\mathbb{R}^N \to \mathbb{R}^N$ définie par $f((x_n))=(x_n-x_n^2)$ vérifie les hypothèses du théorème a) au point f_o (avec les notations de l'exemple précédent) ; or elle n'est injective dans aucun voisinage de f_o ([5]).

4) Considérons l'espace $\mathbb{R}^N$ muni de la topologie précédente. Pour n=1,2,..., posons

$$x_n = (2^{-2^n}, 2^{-2^{n-1}}, \ldots, 2^{-1}, 1, 0, 0, \ldots),$$
$$y_n = (\underbrace{0,0,\ldots,0,1}_{n}, 0, 0, \ldots).$$

Pour $k\in\mathbb{N}^*$, on pose $x_n^k = (2-\frac{1}{k})x_n$ et $y_n^k = (2-\frac{1}{k})y_n$. En particulier $x_n^1 = x_n$ et $y_n^1 = y_n$. Définissons une application f: $\mathbb{R}^N \to \mathbb{R}^N$ en posant

$$f(x_n^k) = x_n^{k-1} \quad (n=1,2,\ldots\ ;\ k=2,3,\ldots\ )\ ;$$

$$f(x_n) = y_n \quad (n=1,2,\ldots\ )\ ;$$

$$f(y_n^k) = y_n^{k+1} \quad (n=1,2,\ldots\ ;\ k=1,2,\ldots\ )\ ;$$

$$f(x) = x \quad \text{si } x \notin \{x_n^k\} \cup \{y_n^k\}.$$

L'application f vérifie les hypothèses du théorème b) au point $f_o=(0,0,\ldots)$; pourtant l'application f^{-1} n'est pas B-différentiable au point $f(f_o)=f_o$ ([1]).

Or on remarque que, si $C(\mathbb{R},\mathbb{R})$ et $\mathbb{R}^N$ ne sont plus munis des topologies précédentes - qui leur confèrent une structure d'evt - mais de topologies plus fines qui font de ces espaces des anneaux topologiques, les applications considérées plus haut vérifient un théorème d'inversion locale analogue aux théorèmes a) et b).

Soit X une variété réelle de dimension finie et $C(\mathbb{R},X)$ l'espace des applications continues sur X à valeurs dans $\mathbb{R}$. Le but de cet article est de donner un théorème d'inversion locale pour des applications

$$\Phi: C(\mathbb{R},X) \to C(\mathbb{R},X)$$

et, dans ce cadre, d'examiner les exemples étudiés plus haut.

II. TOPOLOGIE SUR C(ℝ,X).

Soit $(U_i)_{i\in J}$ un recouvrement ouvert, dénombrable et localement fini de X, formé d'ensembles connexes relativement compacts qui sont des domaines de définition de cartes de X ; soit encore $(\alpha_i)_{i\in J}$ une partition continue de l'unité subordonnée au recouvrement $(U_i)_{i\in J}$. Désignons par $(K_i)_{i\in J}$ l'ensemble des supports des applications α_i ; chacun des K_i est compact et l'ensemble $(K_i)_{i\in J}$ constitue un recouvrement dénombrable localement fini de X. Pour tout $i\in J$, on peut supposer qu'il existe $x\in K_i$ tel que $x\notin K_j$ si $j\neq i$ (sinon on retire K_i du recouvrement).

Soit $C(\mathbb{R},K_i)$ l'espace des applications continues de K_i dans ℝ muni de la topologie de la convergence uniforme.

Si $f\in C(\mathbb{R},X)$, nous poserons $|f|_i = \sup\limits_{x\in K_i} |f(x)|$.

Pour tout $i\in J$, soit ρ_i l'application de restriction

$$\rho_i : C(\mathbb{R},X) \to C(\mathbb{R},K_i)$$

et soit T la topologie sur C(ℝ,X) pour laquelle un système fondamental de voisinages de l'application nulle f_o est formé des ensembles $\bigcap\limits_{i\in J}\rho_i^{-1}(W_i)$, lorsque W_i décrit un système fondamental de voisinages de l'application nulle dans $C(\mathbb{R},K_i)$, les voisinages des autres points de C(ℝ,X) étant définis par translation.

Ainsi, un voisinage de f_o dans C(ℝ,X) contiendra un ensemble $W=\bigcap\limits_{i\in J} W_{K_i,\varepsilon_i}$, où $\varepsilon_i\in\mathbb{R}_+^*$ pour tout $i\in J$ et où

$$W_{K_i,\varepsilon_i} = \{ f\in C(\mathbb{R},X) \ / \ |f|_i \leq \varepsilon_i \}.$$

Dans toute la suite, V désignera le filtre des voisinages de f_o dans T, et κ: C(ℝ,X)xC(ℝ,X) → C(ℝ,X) sera l'application de multiplication. Si F et F' sont des filtres sur C(ℝ,X), nous noterons FF' le filtre engendré par les ensembles $\xi\xi' = \kappa(\xi\times\xi')$, où $\xi\in F$ et $\xi'\in F'$.

Un filtre F sur C(ℝ,X) sera dit *borné* si le filtre VF converge vers f_o dans T ; une partie $M\subset C(\mathbb{R},X)$ sera dite *bornée* si le filtre $[M]$ des sur-ensembles de M est borné.

Notons $\tilde{\mathcal{B}}(T)$ et B(T) respectivement l'ensemble des filtres bornés et des parties bornées de C(ℝ,X).

<u>Proposition II-1</u>: (C(ℝ,X),T) *est un anneau topologique localement borné (i.e.* f_o *possède un voisinage borné dans* T*).*

Δ 1) L'application $(f,g)\mapsto f+g$ est manifestement continue de T×T vers T.

2) L'application κ est continue au point (f_o,f_o) : si $W=\bigcap_{i\in J} W_{K_i,\varepsilon_i}$ est un voisinage de f_o dans T, alors $V=\bigcap_{i\in J} W_{K_i,\sqrt{\varepsilon_i}}$ est un voisinage de f_o dans T et W contient κ(V×V).

3) Soit $f\in C(\mathbb{R},X)$; l'application κ_f: $C(\mathbb{R},X)\to C(\mathbb{R},X)$ telle que $\kappa_f(g)=fg$ est continue en f_o: en effet, si $W=\bigcap_{i\in J} W_{K_i,\varepsilon_i}$ est un voisinage de f_o dans T, alors W contient $\kappa_f(V)$, où $V=\bigcap_{i\in J} W_{K_i,\eta_i}$, avec $\eta_i = \varepsilon_i/\ |f|_i$ si $|f|_i\neq 0$, $\eta_i=1$ sinon.

4) L'ensemble $M_o = \{ f\in C(\mathbb{R},X) \ /\ \sup_{x\in X}|f(x)| \leq 1 \}$ est un voisinage de f_o dans T ; de plus M_o est borné car, si W est un voisinage de f_o dans T, W contient $\kappa(M_o\times W)$. ∇

Proposition II-2: f_o *est adhérent à l'ensemble* Γ *des éléments inversibles de* $C(\mathbb{R},X)$.

Δ Soit $W=\bigcap_{i\in J} W_{K_i,\varepsilon_i}$ un voisinage de f_o dans T. Pour tout $i\in J$, posons

$$\Lambda(i) = \{ \lambda\in J \ /\ K_i\cap K_\lambda \neq \emptyset \}.$$

Il existe (4) un ensemble $(\delta_i)_{i\in J}$ de réels strictement positifs vérifiant :

$$\lambda\in\Lambda(i) \Rightarrow \delta_\lambda \leq \varepsilon_i/\text{card}\,\Lambda(i).$$

L'application $f=\sum_{i\in J}\delta_i\alpha_i$ est inversible et appartient à W. ∇

L'anneau $C(\mathbb{R},X)$ est ordonné par la relation

$$f\leq g \Leftrightarrow f(x)\leq g(x) \quad \forall x\in X.$$

Soit f_1 l'application constante sur $1\in\mathbb{R}$. Posons

$$[f_o,f_1] = \{ \phi\in C(\mathbb{R},X) \ /\ \phi=\sum_{i\in J} a_i\alpha_i,\ a_i\in[0,1]\subset\mathbb{R} \quad \forall i\in J \}.$$

Proposition II-3: *L'ensemble* $[f_o,f_1]$ *a les propriétés suivantes :*

a) Si $\phi\in[f_o,f_1]$, *alors* $f_o\leq\phi\leq f_1$ *et* $f_1-\phi\in[f_o,f_1]$.

b) L'ensemble $[f_o,f_1]\cap\Gamma$ *est complet pour l'ordre induit.*

c) Si $\phi,\psi\in[f_o,f_1]$ *et si* $\phi+\psi\leq f_1$, *alors* $\phi+\psi\in[f_o,f_1]$.

d) Les ensembles $S_\phi = \{ \psi\in[f_o,f_1] \ /\ \psi\leq\phi \}$, *pour* $\phi\in[f_o,f_1]\cap\Gamma$, *engendrent un filtre S qui converge vers* f_o *dans* T.

e) Si $\phi\in[f_o,f_1]\cap\Gamma$ *et* $\psi\in]f_o,f_1] = [f_o,f_1]-\{f_o\}$, *il existe* $\chi\in]f_o,f_1]$ *minorant* ϕ *et* ψ *et telle que* $\chi+f_1-\psi\in[f_o,f_1]\cap\Gamma$.

Δ a) Soit $\phi=\sum_{i\in J} a_i\alpha_i \in [f_o,f_1]$; comme $f_1-\phi=\sum_{i\in J}(1-a_i)\alpha_i$, il vient que $f_1-\phi$ appartient à $[f_o,f_1]$.

b) Si $(\sum_{i\in J} a_{\lambda i}\alpha_i)_{\lambda\in\Lambda}$ est une partie non vide de $[f_o,f_1]\cap\Gamma$, alors $\sum_{i\in J} a_i\alpha_i$ est la borne supérieure de cette partie dans $[f_o,f_1]\cap\Gamma$, où $a_i=\sup_{\lambda\in\Lambda} a_{\lambda i}$.

c) Si $\phi=\sum_{i\in J} a_i\alpha_i$ et $\psi=\sum_{i\in J} b_i\alpha_i$ appartiennent à $[f_o,f_1]$, et si $a_i+b_i \leq 1$ pour tout $i\in J$, alors $\phi+\psi=\sum_{i\in J}(a_i+b_i)\alpha_i \in [f_o,f_1]$.

d) Si $\phi=\sum_{i\in J} a_i\alpha_i$ et $\psi=\sum_{i\in J} b_i\alpha_i$ appartiennent à $[f_o,f_1]\cap\Gamma$, il existe χ dans $[f_o,f_1]\cap\Gamma$ qui minore ϕ et ψ : il suffit en effet de prendre $\chi=\sum_{i\in J} c_i\alpha_i$, où $c_i=\inf(a_i,b_i)>0$ pour tout $i\in J$. Ceci montre que les ensembles S_ϕ, $\phi\in[f_o,f_1]\cap\Gamma$, engendrent un filtre $\mathcal{S}$ sur $C(\mathbb{R},X)$. D'après la prop. II-2, tout voisinage de f_o dans T contient un ensemble S_ϕ ; donc le filtre $\mathcal{S}$ converge vers f_o dans T.

e) Soit $\phi=\sum_{i\in J} a_i\alpha_i \in [f_o,f_1]\cap\Gamma$ et $\psi=\sum_{i\in J} b_i\alpha_i \in]f_o,f_1]$; l'application $\chi=\sum_{i\in J} c_i\alpha_i$, où $c_i=\inf(a_i,b_i)$ pour tout $i\in J$, minore ϕ et ψ , appartient à $]f_o,f_1]$ et $\chi+f_1-\psi$ est inversible. ∇

Définition II-1: Une partie $M\subset C(\mathbb{R},X)$ sera dite *convexe* si $\phi M+\psi M \subset (\phi+\psi)M$, pour tous $\phi,\psi\in[f_o,f_1]$ tels que $\phi+\psi\in[f_o,f_1]\cap\Gamma$.

Il résulte de cette définition qu'une intersection de parties convexes est convexe.

Si A est une partie de $C(\mathbb{R},X)$, l'*enveloppe convexe* de A, notée C(A), est l'intersection de tous les convexes contenant A.

Définition II-2: Une partie $M\subset C(\mathbb{R},X)$ sera dite *absorbante* si, pour tout $f\in C(\mathbb{R},X)$, il existe $\phi\in[f_o,f_1]\cap\Gamma$ tel que l'ensemble $S_\phi.f = \kappa(S_\phi\times\{f\})$ soit contenu dans M.
Autrement dit, la partie M est absorbante si elle appartient à tous les filtres $\mathcal{S}.f = \kappa(\mathcal{S}\times[f])$, pour $f\in C(\mathbb{R},X)$.

De la propriété d), prop.II-3, résulte qu'une intersection finie de parties absorbantes est absorbante.

Définition II-3: Une partie $M\subset C(\mathbb{R},X)$ est dite *symétrique* si $-M = M$.

Proposition II-4: *Dans* $(C(\mathbb{R},X),T)$, *l'application nulle* f_o *possède un système fondamental de voisinages fermés, convexes, absorbants et symétriques.*

Δ Lorsque $(\varepsilon_i)_{i\in J}$ décrit l'ensemble de toutes les suites de nombres réels strictement positifs, les ensembles $W=\bigcap_{i\in J} W_{K_i,\varepsilon_i}$ forment un système fondamental de voisinages symétriques de f_o.

Soit W l'un quelconque d'entre eux.

a) W est fermé : Soit $f\in C(\mathbb{R},X)$ adhérent à W dans T, et soit $\varepsilon>0$. L'ensemble $V=\bigcap_{j\in J} W_{K_j,\eta_j}$, où $\eta_i=\varepsilon$ et $\eta_j=1$ si $j\neq i$, est un voisinage de f_o dans T. Donc $(f+V)\cap W\neq\emptyset$, i.e. il existe $h\in C(\mathbb{R},X)$ tel que $|h|_j\leq\varepsilon_j$ $\forall j\in J$, $|h-f|_i\leq\varepsilon$ et $|h-f|_j\leq 1$ si $j\neq i$. Par suite $|f|_i\leq|f-h|_i+|h|_i\leq\varepsilon+\varepsilon_i$; cette inégalité étant réalisée pour tout $\varepsilon>0$, il vient que $|f|_i\leq\varepsilon_i$. En faisant le raisonnement pour chaque indice i, on obtient que f appartient à W.

b) W est convexe : Soit f et g dans W, ϕ et ψ dans $[f_o,f_1]$ tels que $\phi+\psi$ appartienne à $[f_o,f_1]\cap\Gamma$. L'application $h: X\to\mathbb{R}$ définie par

$$h(x) = \frac{\phi(x)f(x)+\psi(x)g(x)}{\phi(x)+\psi(x)}$$

est continue sur X et, pour tout $x\in K_i$, on a $|h(x)|\leq\varepsilon_i$; donc $|h|_i\leq\varepsilon_i$. En raisonnant pour tous les indices $i\in J$, on obtient que h appartient à W. Comme $\phi f+\psi g=(\phi+\psi)h$, il en résulte que $\phi W+\psi W\subset(\phi+\psi)W$.

c) W est absorbant : Soit $f\in C(\mathbb{R},X)$; choisissons un ensemble $(\delta_i)_{i\in J}$ de réels strictement positifs vérifiant

$$\lambda\in\Lambda(i) \Longrightarrow \delta_\lambda\leq\varepsilon_i/\operatorname{card}\Lambda(i) \quad \text{(cf. prop. II-2)} ;$$

on peut de plus supposer que $\delta_i\leq 1$ pour tout $i\in J$. Posons $\eta_i=\sup_{\lambda\in\Lambda(i)}|f|_\lambda$ et $a_i=\inf(\delta_i,\frac{\delta_i}{\eta_i})$ si $\eta_i\neq 0$, $a_i=\delta_i$ sinon.

L'application $\phi=\sum_{i\in J} a_i\alpha_i$ est inversible et appartient à $[f_o,f_1]$.
Soit $\psi\in S_\phi$ et $x\in K_i$; comme $|\psi(x)f(x)|\leq\phi(x)|f(x)|=|f(x)|\sum_{\lambda\in\Lambda(i)} a_\lambda\alpha_\lambda(x)\leq$
$\leq|f|_i\sum_{\lambda\in\Lambda(i)} a_\lambda\alpha_\lambda(x)$, on obtient que :

1) Si $|f|_i\neq 0$, alors $|\psi(x)f(x)|\leq|f|_i\sum_{\lambda\in\Lambda(i)}\frac{\delta_\lambda}{\eta_\lambda}\alpha_\lambda(x)\leq|f|_i\sum_{\lambda\in\Lambda(i)}\frac{\delta_\lambda}{|f|_i}\alpha_\lambda(x)\leq$
$\leq\sum_{\lambda\in\Lambda(i)}\delta_\lambda\alpha_\lambda(x)\leq\varepsilon_i$. Donc $|\psi f|_i\leq\varepsilon_i$;

2) si $|f|_i=0$, alors $|\psi f|_i=0$.

En faisant le raisonnement pour tous les indices $i\in J$, il vient que ψf appartient à W, pour tout ψ de S_ϕ. Donc W est absorbant. ∇

III. DIFFERENTIABILITE.

Rappelons que M_o est l'ensemble $\{ f \in C(\mathbb{R},X) \ / \ \sup_{x\in X} |f(x)| \leq 1 \}$.

Proposition III-1: *Soit* $r: C(\mathbb{R},X) \to C(\mathbb{R},X)$ *une application telle que* $r(f_o)=f_o$. *Les assertions suivantes sont équivalentes :*

a) $(\forall \mathcal{F} \in \tilde{B}(T))(\forall W \in \mathcal{V})(\exists V \in \mathcal{V})(\exists \xi \in \mathcal{F}) \ / \ (f \in V,\ h \in \xi) \Rightarrow r(fh) \in fW$;

b) $(\forall B \in B(T))(\forall W \in \mathcal{V})(\exists V \in \mathcal{V}) \ / \ (f \in V,\ h \in B) \Rightarrow r(fh) \in fW$;

c) $(\forall W \in \mathcal{V})(\exists V \in \mathcal{V}) \ / \ (f \in V,\ h \in M_o) \Rightarrow r(fh) \in fW$;

d) $(\forall h \in C(\mathbb{R},X))(\forall W \in \mathcal{V})(\exists V \in \mathcal{V}) \ / \ (f \in V) \Rightarrow r(fh) \in fW$;

e) $(\forall W \in \mathcal{V})(\exists V \in \mathcal{V}) \ / \ (f \in V) \Rightarrow r(f) \in fW$.

Δ Il est clair que a)⇒b)⇒c) et que a)⇒d)⇒e). Montrons que c)⇒a). Soit $\mathcal{F} \in \tilde{B}(T)$; le filtre $\mathcal{VF}$ est plus fin que $\mathcal{V}$ et M_o appartient à $\mathcal{V}$, donc il existe $U \in \mathcal{V}$ et $\xi \in \mathcal{F}$ tels que $M_o \supset U\xi$. D'après la prop. II-2, il existe une application γ inversible dans U. Soit $W \in \mathcal{V}$; l'ensemble $W' = \gamma W$ est un voisinage de f_o dans T. Pour ce voisinage, il existe d'après c) un voisinage $U' \in \mathcal{V}$ tel que

$$(f \in U',\ h \in M_o) \Rightarrow r(fh) \in fW' \quad (1).$$

Posons $V = \gamma U' \in \mathcal{V}$; soit $f \in V$ et $h \in \xi$; il existe $k \in U'$ tel que $f = \gamma k$. Comme $\gamma h \in \gamma\xi \subset U\xi \subset M_o$ et que $k \in U'$, on obtient, en utilisant (1),

$$r(fh) = r(\gamma kh) \in kW' = \gamma^{-1} f \gamma W = fW,$$

ce qui établit a).

Montrons enfin que e)⇒a). Soit $\mathcal{F} \in \tilde{B}(T)$ et $W \in \mathcal{V}$. Il existe $U \in \mathcal{V}$ et $\xi \in \mathcal{F}$ tel que $W \supset U\xi$. Pour le voisinage U, il existe d'après e) $U' \in \mathcal{V}$ tel que

$$f \in U' \Rightarrow r(f) \in fU \quad (2).$$

Il existe par ailleurs $U_1 \in \mathcal{V}$ et $\xi_1 \in \mathcal{F}$ tels que $U' \supset U_1\xi_1$. Pour $f \in U_1$ et $h \in \xi \cap \xi_1$, on aura $fh \in U_1\xi_1 \subset U'$, donc, en utilisant (2), $r(fh) \in fhU \subset fU\xi \subset fW$, ce qui établit a). ∇

Définition III-1: L'application r sera dite *petite* si elle vérifie l'une des conditions équivalentes précédentes.

Notons Hom X l'espace des applications $l: C(\mathbb{R},X) \to C(\mathbb{R},X)$ qui sont des homomorphismes continus pour la structure de groupe topologique sous-jacente à $(C(\mathbb{R},X),T)$ et qui vérifient de plus

$$\forall\ f,g \in C(\mathbb{R},X),\ l(fg) = f\,l(g).$$

L'espace Hom X sera muni de la topologie de la convergence uniforme sur les bornés de $C(\mathbb{R},X)$.

Soit $\Phi: C(\mathbb{R},X)\to C(\mathbb{R},X)$ et $\phi_o\in C(\mathbb{R},X)$. S'il existe $l\in \text{Hom}\, X$ tel que l'application $r: C(\mathbb{R},X)\to C(\mathbb{R},X)$ définie par

$$r(h) = \Phi(\phi_o+h)-\Phi(\phi_o)-l(h)$$

soit petite, nous dirons que Φ est *différentiable* en ϕ_o ; l'application l sera appelée *dérivée* de Φ en ϕ_o et notée $D\Phi(\phi_o)$.

La notion de différentiabilité introduite ici est un cas particulier de la notion générale de différentiabilité entre modules quasi-topologiques ([2]) ($C(\mathbb{R},X)$ étant considéré comme module topologique sur lui-même). C'est pourquoi on retrouve avec une telle définition les résultats habituels de calcul différentiel. Mentionnons seulement deux résultats particuliers qui seront utilisés pour démontrer le théorème d'inversion locale.

Proposition III-1: *Si* Φ *est différentiable en* $\phi_o\in C(\mathbb{R},X)$, *alors* Φ *est continue en* ϕ_o.

Δ On a $\Phi(\phi_o+h)-\Phi(\phi_o) = D\Phi(\phi_o).h + r(h)$, où r est petite. Il suffit d'établir que r est continue en f_o. Soit $W\in \mathcal{V}$; il existe $U\in\mathcal{V}$ et $V\in\mathcal{V}$ tels que $W\supset UV$. Utilisant la condition e) pour l'application r, on obtient l'existence d'un voisinage $U'\in\mathcal{V}$ tel que

$$f\in U' \Rightarrow r(f)\in fU.$$

Pour $f\in U'\cap V\in\mathcal{V}$, on aura $r(f)\in fU\subset VU\subset W$, ce qui établit la continuité de r en f_o. ∇

Proposition III-2 (accroissements finis): *Soit* $\Phi: C(\mathbb{R},X)\to C(\mathbb{R},X)$ *une application différentiable en tout point de l'ensemble*

$$[f,g] = \{\ (f_1-\phi)f+\phi g\ /\ \phi\in[f_o,f_1]\ \}.$$

Supposons qu'il existe $B=\bigcap_{i\in J} W_{K_i,n_i}$ *tel que* $\dot{\Phi}(h)=D\Phi(h).f_1$ *appartienne à* B *pour tout* $h\in[f,g]$. *Alors*

$$|\Phi(g)-\Phi(f)|_i \leqslant n_i\,|g-f|_i\ ,\ \textit{pour tout } i\in J.$$

Δ Soit $V=\bigcap_{i\in J} W_{K_i,\varepsilon_i}$ un voisinage de f_o dans T. Pour $M\subset C(\mathbb{R},X)$, désignons par $\overline{C}(M)$ l'enveloppe convexe fermée de M. Posons

$$\Omega = \{\ \phi\in[f_o,f_1]\cap\Gamma\ /\ \Phi((f_1-\phi)f+\phi g) - \Phi(f)\in\phi\overline{C}\{(g-f)(B+V)\}\ \}.$$

Montrons d'abord que Ω est non vide. Comme Φ est différentiable en f, on peut écrire, pour $h\in C(\mathbb{R},X)$:

$$\Phi((f_1-h)f+hg) - \Phi(f) = \Phi(f+h(g-f)) - \Phi(f) = h(g-f)\Phi(f) + r(h(g-f)),$$

où r vérifie (condition e)):

$$(\forall W'\in \mathcal{V})(\exists V'\in \mathcal{V}) \,/\, f'\in V' \Rightarrow r(f')\in f'W' \quad (1).$$

Pour $W'=V$, il existe $V'\in \mathcal{V}$ tel que : $f'\in V' \Rightarrow r(f')\in f'V$.
Comme le filtre $\kappa(\mathcal{S}\times[g-f])$ (notations de la prop. II-3, d)) est plus fin que $\mathcal{V}$, il existe $S_\phi\in\mathcal{S}$ tel que $\kappa(S_\phi\times\{g-f\})\subset V'$. Pour un tel $\phi\in[f_o,f_1]\cap\Gamma$, on aura $\phi(g-f)\in V'$, donc $r(\phi(g-f))\in\phi(g-f)V$; par suite
$\Phi((f_1-\phi)f+\phi g) - \Phi(f) \in \phi(g-f)B + \phi(g-f)V = \phi(g-f)(B+V)\subset \phi\,\overline{C}\{(g-f)(B+V)\}$.
Donc Ω est non vide. Comme $[f_o,f_1]\cap\Gamma$ est complet pour l'ordre induit, il existe $m = \sup\Omega\in[f_o,f_1]\cap\Gamma$.
Comme l'application ι: $[f_o,f_1]\cap\Gamma \to C(\mathbb{R},X)$ définie par

$$\iota(\phi) = \phi^{-1}(\Phi((f_1-\phi)f+\phi g) - \Phi(f)),$$

est continue, et comme $\overline{C}\{(g-f)(B+V)\}$ est fermé, il est facile de voir que $m\in\Omega$. Supposons $m\neq f_1$. Comme Φ est différentiable au point $(f_1-m)f+mg$, on peut écrire, pour $h\in C(\mathbb{R},X)$:
$\Phi((f_1-m)f+mg+h) - \Phi((f_1-m)f+mg) = h\dot{\Phi}((f_1-m)f+mg) + r(h)$,
où r vérifie (1). Pour $W'=V$, il existe $V'\in\mathcal{V}$ tel que : $f'\in V'\Rightarrow r(f')\in f'V$.
Comme précédemment, il existe $S_\phi\in\mathcal{S}$ tel que $\kappa(S_\phi\times\{g-f\})\subset V'$.
Pour $\psi\in S_\phi$, on aura $\psi(g-f)\in V'$, donc $r(\psi(g-f))\in\psi(g-f)V$.
D'après la propriété e), prop. II-3, comme $\phi\in[f_o,f_1]\cap\Gamma$ et $f_1-m\in\,]f_o,f_1]$, il existe $\chi\in\,]f_o,f_1]$ minorant ϕ et f_1-m et tel que $\chi+m\in[f_o,f_1]\cap\Gamma$.
Pour un tel χ , on peut écrire :
$\Phi((f_1-(m+\chi))f+(m+\chi)g) - \Phi(f) = \Phi((f_1-m)f+mg+\chi(g-f)) - \Phi(f) =$
$= \Phi((f_1-m)f+mg+\chi(g-f)) - \Phi((f_1-m)f+mg) + \Phi((f_1-m)f+mg) - \Phi(f) =$
$= \chi(g-f)\dot{\Phi}((f_1-m)f+mg) + r(\chi(g-f)) + \Phi((f_1-m)f+mg) - \Phi(f) \in$
$\in \chi(g-f)B + \chi(g-f)V + m\overline{C}\{(g-f)(B+V)\} = \chi(g-f)(B+V) + m\overline{C}\{(g-f)(B+V)\} \subset$
$\subset \chi\overline{C}\{(g-f)(B+V)\} + m\overline{C}\{(g-f)(B+V)\} \subset (\chi+m)\overline{C}\{(g-f)(B+V)\}$,
en vertu du lemme suivant :

Lemme : Si M est convexe, alors l'adhérence $\overline{M}$ de M est convexe.
Δ Soit ϕ et ψ dans $[f_o,f_1]$ tels que $\phi+\psi\in[f_o,f_1]\cap\Gamma$. Montrons que $\phi\overline{M}+\psi\overline{M}$ est contenu dans $(\phi+\psi)\overline{M}$. Soit $\overline{x}$ et $\overline{y}$ dans $\overline{M}$; il suffit d'établir que $\phi\overline{x}+\psi\overline{y}$ est adhérent à $(\phi+\psi)M$, puisque $(\phi+\psi)\overline{M} = \overline{(\phi+\psi)M}$ (car $\phi+\psi$ est inversible).
Soit V un voisinage de $\phi\overline{x}+\psi\overline{y}$ dans T ; il existe des voisinages V_ϕ de ϕ, V_ψ de ψ, Ω de $\overline{x}$ et Ω' de $\overline{y}$ tels que $V\supset V_\phi\Omega + V_\psi\Omega'$. Par hypothèse, il existe $x\in\Omega\cap M$ et $y\in\Omega'\cap M$. Le point $\phi x+\psi y$ appartient d'une part à $V_\phi\Omega+V_\psi\Omega'$, donc à V, et à $\phi M+\psi M$ d'autre part, donc à $(\phi+\psi)M$, puisque M est convexe. ∇

Nous venons d'établir que $\chi+m$ appartient à Ω ; or $\chi+m > m$ car $\chi \in]f_o, f_1]$; ceci contredit l'hypothèse $m = \sup \Omega$. Donc $m = f_1$ et $\Phi(g)-\Phi(f) \in \overline{C}\{(g-f)(B+V)\}$. Posons $\lambda_i = |g-f|_i$. L'ensemble $(g-f)(B+V)$ étant contenu dans l'ensemble $\bigcap_{i\in J} W_{K_i,\lambda_i(\eta_i+\varepsilon_i)+\varepsilon_i}$, qui est convexe et fermé, il en résulte que $\Phi(g)-\Phi(f)$ appartient à $\bigcap_{i\in J} W_{K_i,\lambda_i(\eta_i+\varepsilon_i)+\varepsilon_i}$, donc que $|\Phi(g)-\Phi(f)|_i \leq |g-f|_i(\eta_i+\varepsilon_i)+\varepsilon_i$ $\forall i \in J$. Cette inégalité étant vérifiée pour tout $\varepsilon_i > 0$, on en déduit que $|\Phi(g)-\Phi(f)|_i \leq |g-f|_i \eta_i$. ∇

IV. STRICTE DIFFERENTIABILITE.

Définition IV-1: Une application Φ: $C(\mathbb{R},X) \to C(\mathbb{R},X)$ sera dite *strictement différentiable au point* $\phi_o \in C(\mathbb{R},X)$ si, dans un voisinage de ϕ_o pour T, on peut mettre Φ sous la forme

$$\Phi(f+h) = \Phi(f) + D\Phi(\phi_o).h + R(f,h),$$

où $D\Phi(\phi_o) \in \mathrm{Hom}\, X$, et où l'application R: $C(\mathbb{R},X) \times C(\mathbb{R},X) \to C(\mathbb{R},X)$, définie par

$$R(f,h) = \Phi(f+h) - \Phi(f) - D\Phi(\phi_o).h,$$

vérifie la condition

(SD) $(\forall B \in B(T))(\forall V \in \mathcal{V})(\exists U \in \mathcal{V})(\exists U' \in \mathcal{V}) \ / \ (f \in \phi_o+U',\ g \in U,\ h \in B) \Longrightarrow R(f,gh) \in gV.$

Si Φ est strictement différentiable en ϕ_o, Φ est différentiable en ϕ_o.

Proposition IV-1: *La condition* (SD) *est équivalente à la condition suivante :*

(SD)' $(\forall V \in \mathcal{V})(\exists U \in \mathcal{V})(\exists U' \in \mathcal{V}) \ / \ (f \in \phi_o+U',\ g \in U,\ h \in M_o) \Rightarrow R(f,gh) \in gV.$

Δ Il est clair que (SD) implique (SD)'. Montrons la réciproque. Soit $B \in B(T)$ et $V \in \mathcal{V}$; comme $M_o \in \mathcal{V}$, il existe $U_1 \in \mathcal{V}$ tel que $M_o \supset U_1 B$, et il existe une application γ inversible dans U_1. L'ensemble γV est un voisinage de f_o dans T. Pour ce voisinage, il existe, d'après (SD)', des voisinages $U \in \mathcal{V}$ et $U' \in \mathcal{V}$ tels que

$$(f \in \phi_o+U',\ g \in U,\ h \in M_o) \Rightarrow R(f,gh) \in g\gamma V.$$

Pour $f \in \phi_o+U'$, $g \in \gamma U \in \mathcal{V}$ et $h \in B$, il existe $g' \in U$ tel que $g = \gamma g'$, et $\gamma h \in \gamma B \subset U_1 B \subset M_o$, donc $R(f,gh) = R(f,g'\gamma h) \in g'\gamma V = g\gamma^{-1}\gamma V = gV$, ce qui établit (SD).∇

Proposition IV-2: *Soit* Φ: $C(\mathbb{R},X) \to C(\mathbb{R},X)$ *une application différentiable sur un voisinage* Ω *d'un point* ϕ_o *et strictement différentiable en* ϕ_o. *L'application* $D\Phi$: $\Omega \to \mathrm{Hom}\, X$, $f \mapsto D\Phi(f)$, *est continue en* ϕ_o.

Δ Il faut montrer que, pour tout $B \in B(T)$ et tout $V \in \mathcal{V}$, il existe $U \in \mathcal{V}$ tel que $\phi_o+U \subset \Omega$ et $(D\Phi(\phi_o+U)-D\Phi(\phi_o)).B \subset V$.

Fixons B et V. Il existe un voisinage $V_1 \in \mathcal{V}$ symétrique et tel que $V \supset V_1+V_1$. Pour V_1 et B, il existe, d'après (SD), des voisinages U_1 et U_1' dans $\mathcal{V}$ tels que :

$$(f \in \phi_o+U_1',\ g \in U_1,\ h \in B) \Rightarrow R(f,gh) \in gV_1 \qquad (1).$$

L'ensemble $(\phi_o+U_1') \cap \Omega$ est un voisinage de ϕ_o dans T, donc il existe $U \in \mathcal{V}$ tel que $\phi_o+U \subset (\phi_o+U_1') \cap \Omega$. Pour $f \in U$, $g \in U_1$ et $h \in B$, on aura, d'après (1),

$$R(\phi_o+f,gh) \in gV_1 \qquad (2).$$

Mais $R(\phi_o+f,gh) = \Phi(\phi_o+f+gh) - \Phi(\phi_o+f) - D\Phi(\phi_o).gh =$

$= D\Phi(\phi_o+f).gh + r(gh) - D\Phi(\phi_o).gh =$

$= (D\Phi(\phi_o+f)-D\Phi(\phi_o)).gh + r(gh).$

Donc $(D\Phi(\phi_o+f)-D\Phi(\phi_o)).gh = R(\phi_o+f,gh) - r(gh)$.

Comme Φ est différentiable au point ϕ_o+f, il existe $U_2 \in \mathcal{V}$ tel que :

$$g \in U_2 \Rightarrow r(gh) \in gV_1.$$

Pour $g \in U_1 \cap U_2$, on aura, en utilisant (2),

$$(D\Phi(\phi_o+f)-D\Phi(\phi_o)).gh \in gV_1+gV_1 \subset gV.$$

En choisissant dans $U_1 \cap U_2$ un élément g inversible, on obtient que $(D\Phi(\phi_o+f)-D\Phi(\phi_o)).h \subset V$, donc que $(D\Phi(\phi_o+U)-D\Phi(\phi_o)).B \subset V$. ∇

V. THEOREME DU POINT FIXE.

Proposition V-1: *Soit* B *un voisinage borné, symétrique et fermé de* f_o *dans* T *et* Φ: B→B *une application vérifiant :*

$$(\forall i \in J)(\exists \eta_i \in]0,1[) \ / \ |\Phi(g)-\Phi(f)|_i \leq \eta_i |g-f|_i \quad \forall f,g \in B.$$

Alors Φ *admet un point fixe unique* $\phi \in B$.

Δ Posons $\phi_{i,n} = \Phi^n(f)_{|K_i} \in C(\mathbb{R},K_i)$; $\phi_{i,0} = f_{|K_i}$, pour une application $f \in B$.

Montrons que la suite $(\phi_{i,n})_{n \in N}$ est de Cauchy dans $C(\mathbb{R},K_i)$. Comme B est borné, il existe pour tout $i \in J$ un réel $c_i>0$ tel que $|\Phi^n(f)-f|_i \leq c_i$. Pour m>n, on a : $\quad |\phi_{i,m-n}-\phi_{i,0}|_i = |\Phi^{m-n}(f)_{|K_i}-f_{|K_i}|_i = |\Phi^{m-n}(f)-f|_i \leq c_i$.

Supposons que $|\phi_{i,m-p-1}-\phi_{i,n-p-1}|_i \leq \eta_i^{n-p-1}c_i$; alors

$|\phi_{i,m-p}-\phi_{i,n-p}|_i = |\Phi^{m-p}(f)_{|K_i}-\Phi^{n-p}(f)_{|K_i}|_i = |\Phi^{m-p}(f)-\Phi^{n-p}(f)|_i \leq$

$\leq \eta_i |\Phi^{m-p-1}(f)-\Phi^{n-p-1}(f)|_i \leq \eta_i^{n-p}c_i$.

Donc la suite $(\phi_{i,n})_{n \in N}$ est de Cauchy dans $C(\mathbb{R},K_i)$. Il existe par conséquent une application $\phi_i \in C(\mathbb{R},K_i)$ telle que $\phi_{i,n} \xrightarrow[n\to\infty]{} \phi_i$ dans $C(\mathbb{R},K_i)$.

Si $x \in K_i \cap K_j$, il est facile de voir que $\phi_i(x)=\phi_j(x)$. Comme B est fermé et que X est une variété de dimension finie, il existe $\phi \in$ B telle que $\phi_{|K_i}=\phi_i$ pour tout $i \in J$. Montrons que $\Phi(\phi)=\phi$. Pour cela, établissons que $\Phi(\phi)_{|K_i}=\phi_{|K_i}=\phi_i$ pour tout $i\in J$. Fixons $i\in J$ et $\varepsilon>0$. Il existe $N_1>0$ tel que $n>N_1 \Rightarrow |\phi_{i,n}-\phi_i|_i \leqslant \varepsilon/2$ (1). D'autre part

$$|\phi_{i,n}-\Phi(\phi)_{|K_i}|_i = |\Phi^n(f)_{|K_i}-\Phi(\phi)_{|K_i}|_i = |\Phi^n(f)-\Phi(\phi)|_i \leqslant \eta_i|\Phi^{n-1}(f)-\phi|_i =$$

$$= \eta_i|\Phi^{n-1}(f)_{|K_i}-\phi_{|K_i}|_i = \eta_i|\phi_{i,n-1}-\phi_i|_i.$$

Il existe $N_2>0$ tel que : $n>N_2 \Rightarrow |\phi_{i,n-1}-\phi_i|_i \leqslant \varepsilon/2\eta_i$ (2).

Pour $n>\sup(N_1,N_2)$, on aura, en utilisant (1) et (2),

$$|\phi_i-\Phi(\phi)_{|K_i}|_i \leqslant |\phi_{i,n}-\phi_i|_i + |\phi_{i,n}-\Phi(\phi)_{|K_i}|_i \leqslant \varepsilon.$$

Donc $\Phi(\phi)_{|K_i} = \phi_i$. Si maintenant $\psi\in B$ est tel que $\Phi(\psi)=\psi$, alors $|\psi-\phi|_i = |\Phi(\psi)-\Phi(\phi)|_i \leqslant \eta_i|\psi-\phi|_i$, pour tout $i\in J$, ce qui est contradictoire. Donc $\psi = \phi$. ∇

VI THEOREME D'INVERSION LOCALE.

Proposition VI-1: *Soit* Φ: $C(\mathbb{R},X)\to C(\mathbb{R},X)$ *une application différentiable en tout point d'un voisinage* Ω *de* $\phi_o \in C(\mathbb{R},X)$. *Supposons que* Φ *soit strictement différentiable en* ϕ_o *et que* $D\Phi(\phi_o)\in$ Hom X *soit un homéomorphisme. Alors il existe un voisinage* V_o *de* ϕ_o *et un voisinage* W_o *de* $\Phi(\phi_o)$ *tel que* Φ *soit un homéomorphisme de* V_o *sur* W_o.

Δ On peut supposer que $\phi_o=f_o$, $\Phi(f_o)=f_o$ et que $D\Phi(f_o)$ est l'application identique Id de Hom X.

Définissons une application Ψ: $\Omega\to C(\mathbb{R},X)$ en posant $\Psi(f) = f-\Phi(f)$.

L'application Ψ est différentiable sur Ω et $D\Psi(f_o)$ est l'application nulle de Hom X (i.e. l'application constante sur f_o). D'après la prop. IV-2, l'application $D\Phi$: $\Omega\to$Hom X est continue en f_o. Donc $D\Psi$ est continue en f_o.

Pour $V=\frac{1}{2}M_o=\bigcap_{i\in J} W_{K_i,1/2}$, il existe $U\in \mathcal{V}$ tel que

$$D\Psi(U).M_o \subset V \qquad (1).$$

Choisissons $U'\in \mathcal{V}$ tel que $U'+U'+U'\subset U$ et $V_o = \bigcap_{i\in J} W_{K_i,\eta_i}$ tel que $V_o\subset\Omega$ et $V_oM_o\subset U'$. Pour $f\in V_o$ et $\phi \in [f_o,f_1]$, on a $\phi f\in M_oV_o$; donc $\phi f\in U'$ et, d'après (1), $\dot{\Psi}(\phi f)\in V$.

Comme $f-\Phi(f) = \Psi(f)-\Psi(f_o)$, on obtient, en appliquant le théorème des accroissements finis,

$$|f-\Phi(f)|_i = |\Psi(f)-\Psi(f_o)|_i \leqslant \frac{1}{2}|f|_i \leqslant \frac{1}{2}\eta_i \quad \forall i \in J.$$

Soit $g \in C(\mathbb{R},X)$; on définit une application $\Phi_g : V_o \to C(\mathbb{R},X)$ en posant

$$\Phi_g(f) = g+f-\Phi(f).$$

Pour $g \in W_o = \frac{1}{2}V_o$, on aura, d'après ce qui précède,

$$|\Phi_g(f)|_i < |g|_i + |f-\Phi(f)|_i \leqslant \eta_i \quad \forall i \in J.$$

Donc, pour $g \in W_o$, l'application Φ_g applique V_o dans V_o. Soit maintenant f' et f" dans V_o et ϕ dans $[f_o,f_1]$. L'élément $f''+\phi(f'-f'')$ appartient à $V_o+(V_o+V_o)M_o \subset M_oV_o+M_oV_o+M_oV_o \subset U'+U'+U' \subset U$, donc, d'après (1),

$$\dot{\Psi}(f''+\phi(f'-f'')) \in V.$$

En appliquant le théorème des accroissements finis, on obtient :

$$|\Phi_g(f')-\Phi_g(f'')|_i = |\Psi(f')-\Psi(f'')|_i \leqslant \frac{1}{2}|f'-f''|_i \quad \forall i \in J.$$

Le théorème du point fixe montre qu'il existe f unique dans V_o tel que $\Phi_g(f) = f \Leftrightarrow g = \Phi(f)$. Donc Φ est une bijection de V_o sur W_o.

Comme Φ est différentiable sur Ω, on sait déjà (prop. III-1) que Φ est continue sur Ω. Montrons que Φ^{-1} est continue sur W_o. Soit pour cela h et k dans W_o ; posons $f=\Phi^{-1}(h)$ et $g=\Phi^{-1}(k)$. De la relation

$$\Phi(f)-\Phi(g) = (f-\Psi(f)) - (g-\Psi(g)) = (f-g) - (\Psi(f)-\Psi(g)),$$

résulte que

$$|(f-g)(x)| \leqslant |(\Psi(f)-\Psi(g))(x)| + |(\Phi(f)-\Phi(g))(x)| \quad \forall x \in X.$$

Fixons $i \in J$; pour $x \in K_i$, on obtient

$$|(f-g)(x)| \leqslant |\Phi(f)-\Phi(g)|_i + |\Psi(f)-\Psi(g)|_i \;;$$

donc $|f-g|_i \leqslant |\Phi(f)-\Phi(g)|_i + |\Psi(f)-\Psi(g)|_i \leqslant |\Phi(f)-\Phi(g)|_i + \frac{1}{2}|f-g|_i$;

il vient finalement :

$$|f-g|_i \leqslant 2|\Phi(f)-\Phi(g)|_i ,$$

soit encore

$$|\Phi^{-1}(h)-\Phi^{-1}(k)|_i \leqslant 2|h-k|_i.$$

Ceci étant valable pour tout $i \in J$, on obtient que Φ^{-1} est continue sur W_o. ∇

Proposition VI-2: *Avec les notations de la proposition précédente, l'application* Φ^{-1} *est strictement différentiable en* $\Phi(\phi_o)$ *et*

$$(D\Phi^{-1})(\Phi(\phi_o)) = (D\Phi(\phi_o))^{-1}.$$

Δ Nous prendrons encore $\phi_o=f_o$, $\Phi(\phi_o)=f_o$ et $D\Phi(\phi_o)=Id$.

Comme Φ est strictement différentiable en ϕ_o, on peut écrire :

$$\Phi(f+h) = \Phi(f) + h + R(f,h),$$

où R vérifie la condition (SD)'. Montrons que Φ^{-1} est strictement différentiable en f_o, avec $D\Phi^{-1}(f_o) = \mathrm{Id}$. Posons

$$\Phi^{-1}(g+k) = \Phi^{-1}(g) + k + S(g,k),$$

et vérifions la condition (SD)' pour l'application S. Soit $V \in \mathcal{V}$; on doit trouver $U \in \mathcal{V}$ et $U' \in \mathcal{V}$ tels que

$$(g \in U',\ g' \in U,\ k \in M_o) \Rightarrow S(g,g'k) \in g'V \qquad (1).$$

Comme l'ensemble $f_1 + \frac{1}{2}M_o$ est borné, il existe un voisinage symétrique V' de f_o tel que V contienne $(f_1+\frac{1}{2}M_o)V'$. Posons $V_1 = V' \cap \frac{1}{3}M_o \in \mathcal{V}$.

Pour le voisinage V_1, il existe $U_1 \in \mathcal{V}$ et $U_1' \in \mathcal{V}$ tels que :

$$(f \in U_1',\ f' \in U_1,\ h \in M_o) \Rightarrow R(f,f'h) \in f'V_1 \qquad (2).$$

Choisissons $W_1 \in \mathcal{V}$ tel que $W_1 + W_1$ soit contenu dans $U_1' \cap U_1$ et que W_1 soit symétrique. Comme Φ^{-1} est continue sur W_o, il existe $W_1^* \in \mathcal{V}$ tel que

$$\Phi^{-1}(W_1^* + W_1^*) \subset W_1.$$

Comme M_o est borné, il existe $U_2 \in \mathcal{V}$ tel que $U_2 M_o \subset W_1^*$.

Soit enfin $U_2^* \in \mathcal{V}$ tel que $(f_1+\frac{1}{2}M_o)U_2^* \subset W_1$. Posons $U = U' = U_2 \cap U_2^*$.

Pour $g \in U'$, $g' \in U$ et $k \in M_o$, on aura $g \in U' \subset U_2 \subset W_1^*$, donc $\Phi^{-1}(g) \in W_1 \subset U_1'$; $g'k \in UM_o \subset U_2M_o \subset W_1^*$, donc $g+g'k \in W_1^* + W_1^*$ et $\Phi^{-1}(g+g'k) \in W_1$.

Enfin, $\Phi^{-1}(g+g'k) - \Phi^{-1}(g) \in W_1 + W_1 \subset U_1' \cap U_1 \subset U_1$.

Il est facile de voir que $S(g,g'k) = -R(\Phi^{-1}(g), \Phi^{-1}(g+g'k) - \Phi^{-1}(g))$; donc, en utilisant (2), on obtient que

$S(g,g'k) \in -(\Phi^{-1}(g+g'k) - \Phi^{-1}(g))V_1$ (en prenant $h = f_1$ dans (2)).

Il existe $\phi \in V_1$ tel que $S(g,g'k) = -\phi(\Phi^{-1}(g+g'k) - \Phi^{-1}(g))$;

or $S(g,g'k) = \Phi^{-1}(g+g'k) - \Phi^{-1}(g) - g'k$;

donc $(f_1+\phi)(\Phi^{-1}(g+g'k) - \Phi^{-1}(g)) = g'k$.

Mais $\phi \in V_1 \subset \frac{1}{3}M_o$; donc $f_1+\phi$ est inversible et $(f_1+\phi)^{-1} \in f_1 + \frac{1}{2}M_o$.

Par suite $\Phi^{-1}(g+g'k) - \Phi^{-1}(g) = (f_1+\phi)^{-1}g'k$, et $S(g,g'k) = -R(\Phi^{-1}(g), (f_1+\phi)^{-1}g'k)$.

Pour $g \in U'$, $g' \in U$ et $k \in M_o$, on a $(f_1+\phi)^{-1}g' \in (f_1+\frac{1}{2}M_o)U \subset (f_1+\frac{1}{2}M_o)U_2^* \subset W_1 \subset U_1$,

donc $S(g,g'k) \in (f_1+\phi)^{-1}g'V_1 \subset g'(f_1+\frac{1}{2}M_o)V_1 \subset g'(f_1+\frac{1}{2}M_o)V' \subset g'V$. ∇

<u>Proposition VI-3</u>: *Dans les conditions de la prop.* VI-1, *l'application* $D\Phi(f) \in \mathrm{Hom}\ X$ *est inversible pour tout* $f \in V_o$.

Δ On a $D\Psi(f)=Id-D\Phi(f) \Leftrightarrow D\Phi(f)=Id-D\Psi(f)$.

Montrons d'abord que $Id-D\Psi(f)$ est injective pour tout $f \in V_o$. Soit $h \in C(\mathbb{R},X)$ tel que

$$(Id-D\Psi(f)).h = f_o \Leftrightarrow h = D\Psi(f).h \qquad (1).$$

Comme $[h]$ est un ultrafiltre borné, il existe β inversible tel que $\beta h \in \frac{1}{2}M_o$. Soit $k \in \frac{1}{2}M_o$ tel que $h = \beta^{-1}k$. En remplaçant dans (1), il vient

$$\beta^{-1}k = D\Psi(f).\beta^{-1}k \Leftrightarrow k = D\Psi(f).k \qquad (2).$$

Posons $\gamma = \inf\{ \phi \in [f_o,f_1] \cap \Gamma \ / \ k \in \phi(\frac{1}{2}M_o) \}$. Comme $\frac{1}{2}M_o$ est fermé, on obtient que $k \in \gamma(\frac{1}{2}M_o)$. Il existe $k' \in \frac{1}{2}M_o$ tel que $k = \gamma k'$. En utilisant (2), il vient : $k = D\Psi(f).k = \gamma D\Psi(f).k' = \frac{\gamma}{2}D\Psi(f).(k'+k') \in \frac{\gamma}{2}(\frac{1}{2}M_o)$, car $D\Psi(f).M_o \subset \frac{1}{2}M_o$.

Donc $\gamma \leqslant \gamma/2$; par suite $\gamma = f_o$ et $k = f_o$. Donc $Id-D\Psi(f)$ est injective.

Montrons maintenant que $Id-D\Psi(f)$ est une application ouverte. Soit $W \in \mathcal{V}$; montrons qu'il existe $V \in \mathcal{V}$ tel que $(Id-D\Psi(f)).W \supset V$.

Il existe μ inversible tel que $\mu M_o + \mu M_o \subset W$. Posons $V=\mu M_o \in \mathcal{V}$. Etant donné $k \in V$, montrons que l'équation

$$(Id-D\Psi(f)).h = k \Leftrightarrow h = k+D\Psi(f).h \qquad (3)$$

possède une solution unique $h \in W$. Recherchons h sous la forme $h=h'+k$. En utilisant (3), le problème se ramène à trouver $h' \in V$ tel que

$$h' = D\Psi(f).(h'+k) \qquad (4).$$

Définissons une application $F: V \to C(\mathbb{R},X)$ en posant $F(h') = D\Psi(f).(h'+k)$.

Pour h' et k dans V, on aura $h'+k = \mu(h_1'+k_1)$, pour h_1' et k_1 dans M_o ; donc $h'+k = 2\mu(h_1'/2 + k_1/2) \in 2\mu M_o$. Par ailleurs

$$F(h') = D\Psi(f).(h'+k) = 2\mu D\Psi(f).(h_1'/2 + k_1/2) \in 2\mu(\tfrac{1}{2}M_o) = \mu M_o.$$

Ceci montre que F est une application de $V=\mu M_o$ dans V, et V est un voisinage borné, symétrique et fermé de f_o dans T.

Soit maintenant h' et h'' des éléments de V ; on a

$F(h')-F(h'') = D\Psi(f).(h'+k) - D\Psi(f).(h''+k) = D\Psi(f).(h'-h'') = (h'-h'')\Psi(f)$.

Pour $f \in V_o$, on a vu que $\dot{\Psi}(f) \in \frac{1}{2}M_o$. Donc, pour tout $i \in J$, il vient

$$|F(h')-F(h'')|_i \leqslant \frac{1}{2}|h'-h''|_i.$$

Le théorème du point fixe assure l'existence d'un unique $h' \in V$ tel que $F(h')=h'$. Cet h' est la solution cherchée de (4).

Montrons enfin que $Id-D\Psi(f)$ est surjective. Soit $k\in C(\mathbb{R},X)$; cherchons $h\in C(\mathbb{R},X)$ tel que $(Id-D\Psi(f)).h = k$. Soit W un voisinage de f_o dans T. On vient de voir qu'il existe $V\in\mathcal{V}$ tel que $(Id-D\Psi(f)).W\supset V$. Soit ν inversible tel que $\nu k\in V$. Il existe $h_1\in W$ tel que $(Id-D\Psi(f)).h_1 = \nu k$. L'application $h = \nu^{-1}h_1\in C(\mathbb{R},X)$ répond à la question.

Nous avons donc établi que $Id-D\Psi(f)$ est un homéomorphisme, pour tout $f\in V_o$. ∇

<u>Corollaire VI-1</u>: *L'application* Φ^{-1} *est différentiable en tout point* $g\in W_o$ *et* $(D\Phi^{-1})(g) = \big((D\Phi)(\Phi^{-1}(g))\big)^{-1}$.

Δ Soit $g\in W_o$ et $f = \Phi^{-1}(g)$. Comme $D\Phi(f)$ est un homéomorphisme, d'après la proposition précédente, on peut supposer que $g=f_o$, $f=f_o$ et $D\Phi(f)=Id$. Définissons une application $s: C(\mathbb{R},X)\to C(\mathbb{R},X)$ en posant $s(k) = \Phi^{-1}(k)-k$, et montrons que s est petite. Soit $V\in\mathcal{V}$; il faut trouver $U\in\mathcal{V}$ tel que

$$(f'\in U,\ k\in M_o)\Rightarrow s(f'k)\in f'V \qquad (1).$$

L'ensemble $f_1+\frac{1}{2}M_o$ étant borné, il existe V' symétrique dans $\mathcal{V}$ tel que $(f_1+\frac{1}{2}M_o)V'\subset V$. Soit $V_1 = V'\cap\frac{1}{3}M_o\in\mathcal{V}$.

Par hypothèse, l'application $r: C(\mathbb{R},X)\to C(\mathbb{R},X)$ définie par $r(h) = \Phi(h)-h$ est petite, donc pour le voisinage V_1 de f_o, il existe $U_1\in\mathcal{V}$ tel que

$$(g'\in U_1,\ h\in M_o)\Rightarrow r(g'h)\in g'V_1 \qquad (2).$$

Comme Φ^{-1} est continue sur W_o, il existe $U'\in\mathcal{V}$ tel que $\Phi^{-1}(U')\subset U_1$. Soit enfin $U_3\in\mathcal{V}$ tel que $(f_1+\frac{1}{2}M_o)U_3\subset U_1$, et $U_2\in\mathcal{V}$ tel que $U_2M_o\subset U'$.

Posons $U = U_2\cap U_3$. Pour $f'\in U$ et $k\in M_o$, on aura $f'k\in UM_o\subset U_2M_o\subset U'$, donc $\Phi^{-1}(f'k)\in U_1$ et, en utilisant (2) avec $h=f_1$, on obtient que

$$r(\Phi^{-1}(f'k))\in\Phi^{-1}(f'k)V_1.$$

Il existe $\phi\in V_1$ tel que $r(\Phi^{-1}(f'k)) = \Phi^{-1}(f'k).\phi$.
Or $s(f'k) = -r(\Phi^{-1}(f'k)) = -\phi\Phi^{-1}(f'k) = \Phi^{-1}(f'k)-f'k$; d'où

$f'k = (f_1+\phi)\Phi^{-1}(f'k)$.
Mais $\phi\in V_1\subset\frac{1}{3}M_o$; donc $f_1+\phi$ est inversible, et $(f_1+\phi)^{-1}\in f_1+\frac{1}{2}M_o$.

On peut donc écrire : $\Phi^{-1}(f'k) = (f_1+\phi)^{-1}f'k$.
Pour $f'\in U$ et $k\in M_o$, on a $(f_1+\phi)^{-1}f'\in(f_1+\frac{1}{2}M_o)U\subset(f_1+\frac{1}{2}M_o)U_3\subset U_1$; donc

$s(f'k) = -r((f_1+\phi)^{-1}f'k)\in(f_1+\phi)^{-1}f'V_1\subset(f_1+\frac{1}{2}M_o)f'V'\subset f'V$, ce qui achève la démonstration. ∇

On peut résumer les résultats précédents dans le théorème :

Théorème VI-1: *Soit* Φ: $C(\mathbb{R},X)\to C(\mathbb{R},X)$ *une application différentiable en tout point d'un voisinage* Ω *d'un point* $\phi_o \in C(\mathbb{R},X)$. *Supposons que* Φ *soit strictement différentiable en* ϕ_o *et que* $D\Phi(\phi_o)$ *soit un homéomorphisme. Alors il existe un voisinage* V_o *de* ϕ_o *tel que* Φ *soit un homéomorphisme de* V_o *sur* $\Phi(V_o)$; *de plus l'application réciproque* Φ^{-1} *est différentiable sur* $\Phi(V_o)$ *et strictement différentiable en* $\Phi(\phi_o)$; *pour tout* $f\in V_o$, *on a*

$$(D\Phi^{-1})(\Phi(f)) = (D\Phi(f))^{-1}.$$

Corollaire VI-2: *Si* Φ *est strictement différentiable en tout point de* Ω , *alors* Φ^{-1} *est strictement différentiable en tout point de* $\Phi(V_o)$ *et* $(D\Phi^{-1})(\Phi(f)) = (D\Phi(f))^{-1}$ *pour tout* $f\in V_o$.

VII. EXEMPLES.

Reprenons les exemples de l'introduction.

1) Cas de l'application exponentielle.

Considérons $\mathbb{R}$ comme une variété de dimension 1, et considérons sur $C(\mathbb{R},\mathbb{R})$ la topologie T définie plus haut.

Proposition VII-1: *L'application exp*: $C(\mathbb{R},\mathbb{R})\to C(\mathbb{R},\mathbb{R})$ *est strictement différentiable en tout point de* $C(\mathbb{R},\mathbb{R})$.

Δ Soit $\phi\in C(\mathbb{R},\mathbb{R})$; définissons une application R: $C(\mathbb{R},\mathbb{R})\times C(\mathbb{R},\mathbb{R})\to C(\mathbb{R},\mathbb{R})$ en posant $R(f,h) = exp(f+h) - exp(f) - h\,exp(\phi)$, et montrons que R vérifie la condition (SD)'. Soit $f'\in C(\mathbb{R},\mathbb{R})$ et $x\in\mathbb{R}$. On a :

$$R(f,f'h).x = exp[f(x)+f'(x)h(x)] - exp(f(x)) - f'(x)h(x)exp(\phi(x)) =$$
$$= exp(\phi(x))\left[exp\big((f(x)-\phi(x)+f'(x)h(x)\big) - exp\big(f(x)-\phi(x)\big) - f'(x)h(x) \right].$$

Il est clair par ailleurs que l'application $R(f,f'h)/f'$ est continue sur $\mathbb{R}$, pour tout $f'\in C(\mathbb{R},\mathbb{R})$.

En appliquant le théorème des accroissements finis sur $\mathbb{R}$, on obtient que

$$\frac{R(f,f'h).x}{f'(x)} = h(x)exp(\phi(x))\left[exp\big(f(x)-\phi(x)+\theta_x f'(x)h(x)\big) - 1 \right] =$$
$$= h(x)exp(\phi(x))\left[f(x)-\phi(x)+\theta_x f'(x)h(x)\right]exp\big(\theta'_x(f(x)-\phi(x)+\theta_x f'(x)h(x))\big),$$

où θ_x et θ'_x appartiennent à $]0,1[\subset\mathbb{R}$.

Soit $W=\bigcap_{i\in J} W_{K_i,\varepsilon_i} \in V$; on cherche U et U' dans V tels que

$$(f-\phi \in U',\ f'\in U,\ h\in M_o) \Rightarrow R(f,f'h)\in f'W \quad (1).$$

Soit $i \in J$; alors $\left|\frac{R(f,f'h)}{f'}\right|_i \leq [|f-\phi|_i + |f'|_i] exp(|\phi|_i + |f-\phi|_i + |f'|_i)$;

posons $\eta_i = |\phi|_i$; on peut toujours trouver deux nombres λ_i et μ_i strictement positifs tels que

$$(\lambda_i+\mu_i) exp(\lambda_i+\mu_i) \leq \varepsilon_i exp(-\eta_i).$$

Alors $U=\bigcap_{i\in J} W_{K_i,\lambda_i}$ et $U'=\bigcap_{i\in J} W_{K_i,\mu_i}$ sont deux voisinages de f_o qui répondent à la question. ∇

2) L'espace $\mathbb{R}^N = C(\mathbb{R},N)$ sera muni de la topologie T définie précédemment. Ainsi, un voisinage de $f_o=(0,0,\ldots)\in \mathbb{R}^N$ contiendra un ensemble $\prod_{i\in N} V_i$, où V_i est un voisinage de 0 dans $\mathbb{R}$ pour tout $i \in J$.

Soit $\phi: \mathbb{R}\to\mathbb{R}$ une application indéfiniment dérivable à support dans $[\frac{1}{2},\frac{3}{2}]$ et telle que $\phi(1)=1$. Soit $f: \mathbb{R}^N \to \mathbb{R}^N$ telle que $f((x_n)) = (x_n-\phi(x_n))$.

Proposition VII-2: *L'application* f *est strictement différentiable sur* $\mathbb{R}^N$.

Δ *Soit* $x_o=(x_{on})\in\mathbb{R}^N$, et définissons une application $R: \mathbb{R}^N \times\mathbb{R}^N \to \mathbb{R}^N$ par

$$R((x_n),(h_n)) = f((x_n)+(h_n)) - f((x_n)) - (h_n-h_n D\phi(x_{on})).$$

Montrons que R vérifie (SD)'. Soit $(a_n)\in \mathbb{R}^N$. On a :

$$R((x_n),(a_n h_n)) = (-\phi(x_n+a_n h_n)+\phi(x_n)+a_n h_n D\phi(x_{on})) =$$

$$= (a_n h_n).(D\phi(x_{on})-D\phi(x_n+\theta_n a_n h_n)), \text{ où } \theta_n\in]0,1[,$$

$$= (a_n h_n).(x_{on}-x_n-\theta_n a_n h_n).(D^2\phi(x_{on}+\theta'_n(x_n-x_{on}+\theta_n a_n h_n))),$$

où $\theta'_n \in]0,1[$. Soit $W=\prod_{i\in J} V_i$, avec $V_i=[-\varepsilon_i,\varepsilon_i]\subset \mathbb{R}$. On cherche deux voisinages $U=\prod_{i\in J} W_i$ et $U'=\prod_{i\in J} W'_i$, avec $W_i=[-\eta_i,\eta_i]$ et $W'_i=[-\eta'_i,\eta'_i]$, et tels que la relation (1) de la prop. précédente soit vérifiée.

Soit $M = \sup_{x\in\mathbb{R}} |D^2\phi(x)|$. On peut toujours choisir deux nombres η_i et η'_i strictement positifs tels que $\eta_i+\eta'_i \leq \varepsilon_i/M$. Ces deux nombres répondent à la question. ∇

3) Soit $f: \mathbb{R}^N \to \mathbb{R}^N$ telle que $f((x_n)) = (x_n-x_n^2)$.

Proposition VII-3: *L'application* f *est strictement différentiable sur* $\mathbb{R}^N$.

Δ Soit $(x_{on})\in \mathbb{R}^N$; définissons une application $R: \mathbb{R}^N\times\mathbb{R}^N \to \mathbb{R}^N$ par

$$R((x_n),(h_n)) = f((x_n)+(h_n)) - f((x_n)) - (h_n-2x_{on}h_n) =$$

$$= (2h_n(x_{on}-x_n)-h_n^2).$$

Soit $(a_n) \in \mathbb{R}^N$; on obtient $R((x_n),(a_n h_n)) = (a_n h_n)(2(x_{on} - x_n) - a_n h_n)$.

Reprenons les notations de la démonstration de la proposition VII-2 ; on peut toujours choisir deux nombres η_i et η_i' tels que $2\eta_i' + \eta_i \leq \varepsilon_i$, et ces nombres sont solution du problème. ∇

4) Reprenons les notations de l'exemple 4), et soit $W = \prod_{i \in J} V_i$, où $V_i =]-1,1[$. Sur ce voisinage, l'application f induit l'identité, donc elle est strictement différentiable.

On peut ainsi appliquer le théorème VI-1 aux quatre exemples précédents.

BIBLIOGRAPHIE

(1) V.I. Averbukh and O.G. Smolianov , The various definitions of the derivative in linear topological spaces, Russian Math. Surveys 23 (1968), n° 4, 67-113.

(2) F. Berquier , Calcul différentiel dans les modules quasi-topologiques. Variétés différentiables (à paraître).

(3) J. Eells , A setting for global analysis, Bull. Amer. M.S. 72 (1966), 751-807.

(4) Vu Xuan Chi, C.R.A.S. Paris 276 (1973).

(5) S. Yamamuro , Differential Calculus in topological Linear Spaces, L.N. 374, Springer Verlag 1974.

Théorie et Applications des Catégories
Faculté des Sciences
33 rue Saint-Leu
80039 AMIENS, FRANCE

Charaktergruppen von Gruppen von S^1-wertigen stetigen Funktionen

E.Binz

Die vorliegende Arbeit ergänzt den in [Bu] und [Bi] beschriebenen Versuch, die Pontrjaginsche Dualität für lokalkompakte Gruppen auf eine größere Klasse von Gruppen auszudehnen. Dies geschieht im folgenden allgemeinen Rahmen.

Es ist wohlbekannt, daß die Kategorie T der topologischen Räume nicht kartesisch abgeschlossen ist. Die Versuche, geeignete kartesisch abgeschlossene Kategorien topologischer Räume zu konstruieren, sind mannigfach.

In [Bi,Ke] wurde T als volle Unterkategorie in eine größere Kategorie, der Kategorie der Limesräume eingebettet. Unter Verwendung der Limitierung der stetigen Konvergenz auf den Funktionenräumen, wird letztere zu einer kartesisch abgeschlossenen Kategorie.

Im Rahmen dieser Kategorie lassen sich bekannte Dualitätstheorien, wie unten kurz skizziert werden soll, ausdehnen. Zu diesem Zweck wiederholen wir aus [Bi] und [Bi,Ke] kurz die Elemente der Terminologie aus der Theorie der Limesräume.

Eine Menge X für die in jedem Punkt $p \in X$ ein System $\Lambda(p)$ von Filtern auf X vorgegeben ist, heißt ein Limesraum, falls für jedes $p \in X$ die folgenden Bedingungen erfüllt sind:

Die Arbeit entstand während eines Aufenthaltes im WS 1973/74 am Forschungsinstitut der ETH, Zürich.

1) $\dot{p}$, der von $\{p\} \subset X$ erzeugte Filter gehört $\Lambda(p)$ an.

2) Mit $\Phi, \Psi \in \Lambda(p)$ gehört das Infimum $\Phi \wedge \Psi$ zu $\Lambda(p)$.

und

3) Mit jedem $\Psi \in \Lambda(p)$ und $\Phi \geqslant \Psi$ ist auch $\Phi \in \Lambda(p)$.

Die Filter in $\Lambda(p)$ heißen die gegen p konvergente Filter. Offenbar ist jeder topologische Raum ein Limesraum. Die Stetigkeit von Abbildungen zwischen topologischen Räumen verallgemeinert sich wie folgt:

Eine Abbildung f vom Limesraum X in einen Limesraum Y heißt stetig, wenn für jeden Punkt $p \in X$ und für jeden gegen p konvergenten Filter Φ, der Filter $f(\Phi)$ gegen $f(p)$ konvergiert.

Auf $C(X,Y)$, der Menge aller stetigen Abbildungen von X nach Y, führen wir nun die Limitierung der stetigen Konvergenz ein. Dabei bezeichnen $\omega : C(X,Y) \times X \longrightarrow Y$ die Evaluationsabbildung, die jedes Paar (f,p) nach $f(p)$ abbildet. In $C(X,Y)$ konvergiert ein Filter Θ gegen $f \in C(X,Y)$, falls für jeden Punkt $p \in X$ und jeden gegen p konvergenten Filter Φ der Filter $\omega(\Theta \times \Phi)$ gegen $f(p)$ konvergiert. Trägt $C(X,Y)$ diese Konvergenzstruktur, so schreiben wir $C_c(X,Y)$.

Wenn X ein lokalkompakter topologischer Raum ist, und Y uniformisierbar ist, dann ist die Limitierung der stetigen Konvergenz mit der Topologie der kompakten Konvergenz identisch. Die erwähnten Dualitätstheorien, wie sie in [Bi] mehr oder weniger ausführlich beschrieben wurden, sind die folgenden:

a) <u>Dualität zwischen Limesräumen und limitierten Funktionenalgebren:</u>

Für jeden Limesraum X ist $C_c(X,\mathbb{R})$, kurz $C_c(X)$, eine Limesalgebra (Operationen sind stetig). Versehen wir die Menge Hom $C_c(X)$ aller stetigen reellwertigen $\mathbb{R}$-Algebrenhomomorphismen mit der Limitierung der stetigen Konvergenz, erhalten wir den Limesraum $\mathrm{Hom}_c C_c(X)$ für den

$$i_X : X \longrightarrow \mathrm{Hom}_c C_c(X) ,$$

definiert durch $i_X(p)(f)=f(p)$ für alle $f \in C_c(X)$ und alle $p \in X$, eine stetige Surjektion ist. Ein Raum X heißt c-einbettbar, wenn i_X bistetig ist. Jeder vollständig reguläre topologische Raum ist c-einbettbar. Für jeden Limesraum X ist insbesondere $\mathrm{Hom}_c C_c(X)$ c-einbettbar. Jede stetige Abbildung f zwischen zwei Limesräumen induziert durch Kompositionen mit f einen stetigen $\mathbb{R}$-Algebrenhomomorphismus und, falls die Räume c-einbettbar sind, wird jeder stetige $\mathbb{R}$-Algebrenhomomorphismus durch eine stetige Abbildung induziert. Zwei c-einbettbare Limesräume X und Y sind deshalb genau dann homöomorph, wenn $C_c(X)$ und $C_c(Y)$ bistetig isomorph sind. Weiter gilt für jeden Limesraum X, daß $C_c(X)$ und $C_c(\mathrm{Hom}_c C_c(X))$ bistetig isomorph sind. Die eben beschriebene Dualität extendiert die Beschreibung kompakter topologischer Räume X mit Hilfe der Banachalgebren $C_c(X)$.

b) Zur universellen Darstellung von Limesalgebren:

Sei A eine kommutative, assoziative Limesalgebra mit Einselelement (über $\mathbb{R}$), für die Hom A, die Menge aller stetigen reellwertigen R-Algebrenhomomorphismen nicht leer sei. Dieser Raum, versehen mit der Limitierung der stetigen Konvergenz, heißt $\mathrm{Hom}_c A$, und wird Träger von A genannt. Der Homomorphismus

$$d : A \longrightarrow C_c(\mathrm{Hom}_c A) ,$$

die universelle Darstellung von A, ist definiert durch $d(a)(h)=h(a)$ für alle $a \in A$ und $h \in \mathrm{Hom}_c A$. Dieser Homomorphismus ist stetig. Er ist genau dann bistetig, wenn A die von d initiale Limitierung trägt, vollständig ist und außerdem A auf Hom A dieselbe schwache Topologie erzeugt wie sie $C(\mathrm{Hom}_c A)$ induziert. Die Theorie der universellen Darstellungen wird durch die Absenz eines genügend starken Stone-Weierstrass'schen Satzes erheblich kompliziert.
Wenn A topologisch ist, ist $\mathrm{Hom}_c A$ in der Kategorie der Limesräume induktiver Limes von kompakten topologischen Räumen, also ein lokalkompaktes Objekt. (Der zu $\mathrm{Hom}_c A$ assoziierte topologische Raum ist dann ein k-Raum.) Die Limitierung auf $C_c(\mathrm{Hom}_c A)$ wird in diesem Falle mit der Topologie der kompakten Konvergenz identisch.
Diese Dualitätstheorie extendiert die Darstellung gewisser topologischer Algebren als Funktionenalgebren, versehen mit der Topologie der kompakten Konvergenz über einem k-Raum als Träger.

c) <u>Die lineare Dualitätstheorie:</u>

Sei E ein $\mathbb{R}$-Limesvektorraum (Limesraum, für den die $\mathbb{R}$-Vektorraumoperationen stetig sind). Der Dualraum E, die Menge aller reellwertigen stetigen Abbildungen, sei mit LE bezeichnet. Dieser Raum, versehen mit der Limitierung der stetigen Konvergenz heißt der c-Dual von E und wird mit L_cE bezeichnet.

Die natürliche Abbildung

$$E \xrightarrow{i_E} L_cL_cE,$$

die durch $i_E(e)(l) = l(e)$ für alle $e \in E$ und alle $l \in L_cE$ definiert ist, ist stetig. Ein Raum E heißt c-reflexiv, wenn i_E ein bistetiger Isomorphismus ist.

Als Beispiele c-reflexiver Limesvektorräume seien $C_c(X)$ für jeden Limesraum X und jeder vollständige lokalkonvexe topologische Vektorraum genannt.

Wenn E topologisch ist, dann ist L_cE induktiver Limes (in der Kategorie der Limesräume) aller abgeschlossenen gleichstetigen Mengen, versehen mit der schwachen Topologie (diese Mengen sind kompakte topologische Räume). L_cL_cE trägt in diesem Falle die Topologie der kompakten Konvergenz, ist also lokalkonvex. Diese Dualitätstheorie verallgemeinert mithin die von Grothendieck zur Vervollständigung lokalkonvexer Räume herangezogene Dualität.

Die Ausarbeitung all dieser Dualitätstheorien ist nicht abgeschlossen. Sie birgt zum Teil erhebliche Schwierigkeiten. Andererseits öffnen die Ausdehnungen ein reiches Feld von Zusammenhängen verschiedener Strukturen vergleichbar denen, die zwischen funktionalanalytischen Eigenschaften von $C_c(X)$ und topologischen Gegebeneheiten in X herrschen (s. [Bi]).

Wir wenden uns nun dem Versuch zu, die klassische Pontrjaginsche Dualität lokalkompakter Gruppen auszudehnen. Dabei beschränken wir uns im Wesentlichen auf den Bereich der Reflexivität und der Beschreibung einiger auftretender Charaktergruppen, lassen aber die Zusammenhänge zwischen Gruppen und Charaktergruppen (Untergruppen, Annihilatoren) der Schwierigkeiten, die bei der Erweiterung von Charakteren auftreten, außer Acht. Herrn Bernd Müller und Herrn Heinz-Peter Butzmann verdanke ich die Form des Beweises von Lemma 8.

Für eine beliebige kommutative Limesgruppe G (kommutative Gruppe, versehen mit einer Limesstruktur für die die Gruppenoperationen stetig sind) bezeichnet $\Gamma_c G$ die mit der Limitierung der stetigen Konvergenz versehene Gruppe aller stetigen Gruppenhomomorphismen von G in S^1. Wir nennen $\Gamma_c G$ die Charaktergruppe von G. Wenn die Limesgruppe G eine lokalkompakte topologische Gruppe ist, dann ist die Limitierung der stetigen Konvergenz auf $\Gamma_c G$ mit der Topologie der kompakten Konvergenz identisch. Für die reichhaltige Theorie der lokalkompakten topologischen Gruppen sei auf [Po] verwiesen. Der natürliche Homomorphismus

$$j_G : G \longrightarrow \Gamma_c \Gamma_c G ,$$

definiert durch $j_G(g)(\gamma) = \gamma(g)$ für jedes $g \in G$ und jedes $\gamma \in \Gamma_c G$

ist für jede Limesgruppe G stetig. Falls nun G eine lokalkompakte topologische Gruppe ist, wird j_G, wie aus der Pontrjaginschen Dualitätstheorie bekannt ist, ein bistetiger Isomorphismus. Wir nennen deshalb eine Limesgruppe G für die j_G ein bistetiger Isomorphismus ist, P_c - reflexiv. Im Zuge der Ausdehnungsversuche der Pontrjaginschen Dualitätstheorie ist für gewisse Typen von nicht-lokalkompakten topologischen Gruppen (ja sogar für Limesgruppen) die P_c-Reflexivität festgestellt worden. So z.B. ist wie in [Bu] gezeigt wurde, für jeden Limesraum X die mit Limitierung der stetigen Konvergenz versehene $\mathbb{R}$-Algebra $C_c(X)$ aller stetiger reellwertiger Funktionen von X, aufgefaßt als Limesgruppe, P_c-reflexiv. Daraus ergibt sich insbesondere, daß ein topologischer lokalkonvexer Limesvektorraum, aufgefaßt als topologische Gruppe, genau dann P_c-reflexiv ist, wenn er vollständig ist (siehe [Bi]). Jede dieser Typen von P_c-reflexiven Limesgruppen trägt aber eine zusätzliche algebraische Struktur, die zum Nachweis der P_c-Reflexivität auch ausgenützt wurde.

Unter $C_c(X,S^1)$ verstehen wir die Gruppe aller S^1-wertigen Funktionen von X, die mit der Limitierung der stetigen Konvergenz versehen sind. Die Gruppenoperationen sind punktweise definiert. In der vorliegenden Note soll nun für gewisse Gruppen der Form $C_c(X,S^1)$, wobei X ein normaler topologischer Raum ist, der eine (einfach zusammenhängende) universelle Überlagerung zuläßt, die P_c-Reflexivität nachgewiesen werden (s.Satz 24). Dabei wird sich am Schluß für gewisse CW-Komplexe eine natürliche funktionalanalytische Interpretation der ersten singulären Homologiegruppen ergeben.

Zum Verständnis des Aufbaus der vorliegenden Arbeit, skizzieren wir kurz den Nachweis der P_c-Reflexivität von $C_c(X,S^1)$.

Der Charakter

$$\varkappa : \mathbb{R} \longrightarrow S^1 ,$$

der jeder reellen Zahl r die komplexe Zahl $e^{2\pi i r}$ zuordnet, induziert (durch Komposition) einen stetigen Homomorphismus

$$\varkappa_X : C_c(X) \longrightarrow C_c(X,S^1),$$

der sogar eine Quotientenabbildung auf das Bild, aufgefaßt als Unterraum von $C_c(X,S^1)$, ist. Die Gruppe

$$C(X,S^1)/\varkappa_X\, C(X) ,$$

die sogenannte Bruschlinskische Gruppe [Hu], die mit $\Pi^1(X)$ bezeichnet wird, wird durch Einführen der Quotientenlimitierung eine Limesgruppe. Sie heißt $\Pi^1_c(X)$. Damit erhält man eine topologische exakte Folge

$$0 \longrightarrow \varkappa_X\, C_c(X) \longrightarrow C_c(X,S^1) \longrightarrow \Pi^1_c(X) \longrightarrow 0$$

Um nun die P_c-Reflexivität von $C_c(X,S^1)$ nachzuweisen, zeigen wir erst, daß $\varkappa_X\, C_c(X)$ und $\Pi^1_c(X)$ (falls vollständig) P_c-reflexiv sind und verwenden dann, um unser Ziel zu erreichen, eine gängige Schlußweise über das Fünferlemma. Dabei beinhaltet ein großer Teil der Arbeit den Nachweis, daß die Duale und die Biduale der obigen kurzen exakten Folge, exakt sind. Leider ist uns nicht bekannt, unter welchen Bedingungen $\Pi^1_c(X)$ vollständig ist. Hinreichend für die Vollständigkeit sind: $\Pi^1_c(X)$ ist diskret (also etwa, wenn $\Pi_1(X)$ oder $\mathrm{Hom}_1(X,\mathbb{Z})$ endlich erzeugt sind) oder $\Pi^1(X) = \mathrm{Hom}(\Pi_1(X),\mathbb{Z})$ gilt. Letzteres ist richtig, falls X ein CW-Komplex ist ([Sp]p.427).

Zum Studium der Charaktergruppe von $\varkappa_X C_c(X)$ machen wir weitgehend von der linearen Dualitätstheorie von $C_c(X)$ Gebrauch. Dabei wird von den Voraussetzungen an X nur der wegeweise Zusammenhang ausgenützt.

Bei den Untersuchungen von $\Gamma_c \Pi_c^1(X)$ verwenden wir die Theorie der Überlagerungen um $\Pi_c^1(X)$ einerseits als Quotient von Gruppen reellwertiger Funktionen von $\tilde{X}$ (der universellen Überlagerung von X) und andererseits als Teilgruppe $\mathrm{Hom}(\Pi_1(X),\mathbb{Z})$ aller $\mathbb{Z}$-wertigen Homomorphismen der Fundamentalgruppe $\Pi_1(X)$ von X (in natürlicher Weise) zu interpretieren. Dabei wird sich herausstellen, daß $\Pi_c^1(X)$ mit einer topologischen Teilgruppe der mit der Topologie der punktweisen Konvergenz versehenen Gruppe $\mathrm{Hom}_s(\Pi_1(X),\mathbb{Z})$ identifiziert werden kann. Diese Tatsache wird hernach beim Nachweis der P_c - Reflexivität von $\Pi_c^1(X)$ wesentlich verwendet.

1) Beziehungen zwischen $C(X)$ und $C(X,S^1)$

Sei X ein Hausdorffscher topologischer Raum. Die Beziehungen zwischen $C(X)$ und $C(X,S^1)$, die im folgenden kurz beschrieben werden, ermöglichen es uns, die P_c - Reflexivität der Limesgruppen vom Typ $C_c(Y)$, wo Y irgendein Limesraum ist, zum Nachweis der P_c - Reflexivität von $C_c(X,S^1)$ für gewisse Typen von topologischen Räumen X, heranzuziehen.

Mit $\varkappa : R \longrightarrow S^1$ bezeichnen wir die Exponentialabbildung, definiert durch $\varkappa(r) = e^{2\pi i r}$ für jedes $r \in \mathbb{R}$. Bezüglich $\varkappa$ ist $\mathbb{R}$ die universelle Überlagerung von S^1. Offenbar ist $\varkappa$ bezüglich der addiven Gruppenstruktur von $\mathbb{R}$ ein stetiger Gruppenhomomorphismus. Dieser induziert den Gruppenhomomorphismus,

$$1) \qquad \varkappa_X : C(X) \longrightarrow C(X,S^1),$$

der jedes $f \in C(X)$ nach $\varkappa \circ f$ abbildet. Der Kern von $\varkappa_X$ besteht aus all denjenigen Funktionen aus $C(X)$, die X nach $\mathbb{Z} \subset \mathbb{R}$ abbilden. Ist X zusammenhängend, gilt $\ker \varkappa_X = \mathbb{Z}$, wo $\mathbb{Z}$ die Gruppe aller konstanten Funktionen aus $C(X)$ deren Werte in $\mathbb{Z}$ liegen bezeichnet. Wir haben also in diesem Fall die exakte Folge

$$2a) \quad o \longrightarrow \mathbb{Z} \xrightarrow{i_1} C(X) \xrightarrow{\varkappa_X} \varkappa_X(C(X)) \longrightarrow o.$$

Dabei bezeichnet i_1 die Inklusionsabbildung. Offenbar ist in (1) die Abbildung $\varkappa_X$ i.a. nicht surjektiv, da z.B. die Identität auf S^1 nicht als $\varkappa_X(g)$ wo $g \in C(S^1)$ darstellbar ist. Um die technischen Auswirkungen dieses Umstandes abzumildern, setzen wir für alles Weitere voraus, daß X ein wegeweise, lokalwegeweise und lokal-halbeinfach-zusammenhängender normaler (Hausdorffscher) topologischer Raum ist. ([Sp](chapter 2). Diese Voraussetzungen garantieren eine einfachzusammenhängende universelle Überlagerung ([Sp](chapter 2). Diese heißt $\tilde{X}$. Die kanonische Projektion von $\tilde{X}$ auf X bezeichnen wir mit u.

Die Faser $u^{-1}(p)$ eines jeden Punktes $p \in X$ kann in natürlicher Weise mit der Fundamentalgruppe $\Pi_1(X,p)$ identifiziert werden. In späteren Ausführungen werden wir oft von der eindeutigen Wegehochhebungseigenschaft Gebrauch machen. Diese Eigenschaft von $\tilde{X}$ besagt, daß es zu jeder stetigen Abbildung $\sigma : [0,1] \longrightarrow X$ nach Festlegung eines Punktes $q \in u^{-1}(\sigma(o))$ eine eindeutig bestimmte stetige Abbildung $\sigma' : [0,1] \longrightarrow \tilde{X}$ gibt, für die $\sigma'(o) = q$ und $u \circ \sigma' = \sigma$ gilt. Diese eindeutige Wegehochhebungseigenschaft von $\tilde{X}$ garantiert einen für das folgende sehr wesentlichen Umstand (siehe [Sp] chapter 2):

Zu jeder stetigen Abbildung s von einem einfach zusammenhängenden wege- und lokal wegezusammenhängenden Raum Y nach X gibt es eine stetige Abbildung $s' : Y \longrightarrow \tilde{X}$ mit $s = u \circ s'$. Setzt man für s'

für einen beliebigen Punkt $p \in X$ einen Wert aus $\varkappa^{-1}(s(u(p)))$ fest, dann ist s' eindeutig bestimmt. Mithin läßt sich jede Abbildung $t \in C(\tilde{X}, S^1)$ zu einer stetigen Funktion f_t, heben. Also ist

$$\text{2b)} \quad 0 \longrightarrow \mathbb{Z} \xrightarrow{i_1} C(\tilde{X}) \xrightarrow{\varkappa_{\tilde{X}}} C(\tilde{X}, S^1) \longrightarrow 0$$

exakt. Neben diesen Sachverhalten werden wir noch für das Bestimmen der Charaktergruppen von $C_c(X, S^1)$ das im folgenden aufzustellende Diagramm (4) verwenden.

Die Projektion $u : \tilde{X} \longrightarrow X$ induziert die beiden injektiven Homomorphismen:

$$\text{3a)} \quad u^* : C(X) \longrightarrow C(\tilde{X})$$

und

$$\text{3b)} \quad {}^*u : C(X, S^1) \longrightarrow C(X, S^1)$$

definiert durch $u^*(f) = f \circ u$ und ${}^*u(t) = t \circ u$ für jedes $f \in C(X)$ resp. $t \in C(X, S^1)$.

Allgemein bedeuten für eine Abbildung a zwischen zwei Mengen A und B die Abbildungen $a^* : \mathcal{F}(B, \mathbb{R}) \longrightarrow \mathcal{F}(A, \mathbb{R})$ und ${}^*a : \mathcal{F}(B, S^1) \longrightarrow \mathcal{F}(A, S^1)$ die zwischen den Räumen aller $\mathbb{R}$-wertigen bzw. S^1- wertigen Abbildungen von A und B induzierten Algebren beziehungsweise Gruppenhomomorphismen definiert durch $a^*(f) = f \circ a$ und ${}^*a(t) = t \circ a$ für jede Abbildung $f : B \longrightarrow \mathbb{R}$ und $t : B \longrightarrow S$ resp..

Offenbar ist

$$\text{4)} \quad \begin{array}{ccc} C(\tilde{X}) & \xrightarrow{\varkappa_{\tilde{X}}} & C(\tilde{X}, S^1) \\ \uparrow u^* & & \uparrow {}^*u \\ C(X) & \xrightarrow{\varkappa_X} & C(X, S^1) \end{array}$$

kommutativ. Dabei ist $\kappa_{\tilde{X}}$ eine Surjektion ([Sp] chapter 2). Diesen Umstand werden wir später wesentlich verwenden. Für die weitere technische Ausnutzung von (4) sollen die Abbildungen in ${}^{*}u(C(X,S^1))$, $u^{*}(C(X))$ und in $\kappa_X(C(X))$ gekennzeichnet werden.

Für jedes $t \in C(X,S^1)$ ist ${}^{*}u(t) : \tilde{X} \longrightarrow S^1$ zu einer stetigen reellwertigen Funktion, sie heiße wieder f_t, hebbar, d.h. es gibt eine Funktion $f_t \in C(\tilde{X})$, für die

$$(5) \qquad \begin{array}{ccc} \tilde{X} & \xrightarrow{\;f_t\;} & \mathbb{R} \\ \big\downarrow u & & \big\downarrow \kappa \\ X & \xrightarrow{\;t\;} & S^1 \end{array}$$

kommutiert. Offenbar ist f_t bis auf eine konstante $\mathbb{Z}$-wertige Abbildung bestimmt. Diese Funktion f_t bildet für jedes $p \in X$ die Faser $u^{-1}(p)$ in die Faser $\kappa^{-1}(t(p)) \subset \mathbb{R}$ ab, d.h. f_t ist fasertreu. Da X bezüglich u die Quotientenlimitierung, d.h. die feinste Limitierung trägt, für die u stetig ist, ist für jede fasertreue Abbildung $f \in C(\tilde{X},S^1)$ die Abbildung $\kappa \circ f$ im Bild von ${}^{*}u$. Also haben wir:

<u>Lemma 1</u>

<u>Für eine Abbildung</u> $t \in C(\tilde{X},S^1)$ <u>ist</u> $\kappa_{\tilde{X}}(f_t)$ <u>genau dann im Bild von</u> ${}^{*}u$ <u>enthalten, wenn</u> f_t <u>fasertreu ist</u>.

Darüber hinaus folgert man weiter:

<u>Lemma 2</u>

<u>Eine Abbildung</u> $t \in C(\tilde{X},S^1)$ <u>ist genau dann in</u> $u^{*}(C(X))$ <u>enthalten, wenn</u> f_t <u>auf jeder Faser konstant ist</u>.

2) Zu den Beziehungen zwischen $C_c(X)$, $C_c(\tilde{X})$, $C_c(X,S^1)$ und $C_c(\tilde{X},S^1)$

Wir versehen $\varkappa_X C(X)$ mit der feinsten Limitierung für die $\varkappa_X : C_c(X) \longrightarrow \varkappa_X C(X)$ stetig ist. Die Gruppe, versehen mit dieser Limitierung, der Quotientenlimitierung bezüglich $\varkappa_X$, heiße $\varkappa_X C_c(X)$. Diese ist eine Limesgruppe. Offenbar ist die Inklusion $i : \varkappa_X C_c(X) \longrightarrow C_c(X,S^1)$ stetig.

Lemma 3

Es ist $\varkappa_X C_c(X)$ ein Unterraum von $C_c(X,S^1)$.

Beweis: Für diesen Beweis fassen wir $\varkappa_X C(X)$ als Unterraum von $C_c(X,S^1)$ auf. Um zu zeigen, daß $\varkappa_X C_c(X)$ ein Unterraum von $C_c(X,S^1)$ ist, genügt es (da $\varkappa_X C(X)$ eine Limesgruppe ist), zu jedem in $\varkappa_X C(X)$ gegen die konstante Abbildung $\underline{1}$ mit Wert $1 \in S^1$ konvergenten Filter Θ einen in $C_c(X)$ konvergenten Filter Θ' zu finden, für den $\varkappa_X(\Theta') \leqslant \Theta$. Sei nun $W \subset S^1$ eine vermöge $\varkappa$ gleichmäßig überlagerte offene Umgebung von 1. Eine Menge U in einem topologischen Raum Y mit Überlagerung $\tilde{Y}$ und Projektion $u : \tilde{Y} \longrightarrow Y$ heißt gleichmäßig überlagert, wenn $u^{-1}(U)$ die disjunkte Vereinigung von Mengen (genannt Blätter) ist, von denen jede vermöge u homöomorph auf U abgebildet wird. Für jeden festen Punkt $p \in X$ gibt es ein $T \in \Theta$ und eine offene Umgebung U von p derart, daß

$$T(U) \subset W .$$

Weil T von p,U und W abhängt, ersetzen wir T durch $T(p,U,W)$. Ohne Verlust der Allgemeinheit können wir annehmen, daß $1 \in T(p,U,W)$. Wir heben nun jedes $t \in T(p,U,W)$ zu einer Funktion $f \in C(X)$ hoch, für die $f(U) \subset W'$, wo W' dasjenige Blatt über W ist, das o enthält. Damit ist die Hochhebung f für jedes t eindeutig bestimmt. Die Gesamtheit all dieser Hochhebungen von Abbildungen aus $T(p,U,W)$ bezeichnen wir mit $T'(p,U,W')$. Das Mengensystem Θ aller Mengen der Form $T'(p,U,W')$ besitzt die endliche Durchschnittseigenschaft, denn sie enthalten alle die konstanten Funktionen $\underline{o}$. (Konstante Funktionen sollen stets mit dem unterstrichenen Symbol, das für den Wert gewählt wird, bezeichnet werden). Offenbar konvergiert der von Θ' in $C_c(X)$ bestimmte Filter in $C_c(X)$ gegen o. Nun bleibt uns noch zu verifizieren, daß $\varkappa_X(\Theta') \leq \Theta$. Sei $\bigcap_{i=1}^{n} T'(p_i,U_i,W_i) \in \Theta$ vorgegeben. Die Punkte $p_1,\ldots\ldots,p_n$ verbinden wir durch einen Weg s. Da die Limitierung der stetigen Konvergenz auf $C(X,S^1)$ feiner als die Topologie der kompakten Konvergenz ist, gibt es eine Menge $T_1 \in \Theta$, die das Bild von s in $\bigcap_{i=1}^{n} W_i$ abbildet. Sei

$$T_o = T_1 \cap \bigcap_{i=1}^{n} T(p_i,U_i,W_i) .$$

Für jedes $t \in T_o$ ist der Weg $f \circ s$ eine Hochhebung von $t \circ s$, falls f eine Hochhebung von t ist. Da die Wegehochhebung eindeutig ist, falls der Anfangspunkt vorgegeben wird, lassen sich alle $t \in T_o$ derart hochheben, daß sie das Bild von s und damit ganz $\bigcup_{i=1}^{n} U_i$ nach $\bigcap_{i=1}^{n} W_i$, abbilden. Folglich gilt

$$\varkappa_X(\bigcap_{i=1}^{n} T'(p_i,U_i,W_i')) \supset T_o ,$$

woraus sich $\varkappa_X(\Theta') \leq \Theta$ ergibt. (Man beachte, daß von X nur der wegeweise Zusammenhang verwendet wurde.)

Da X bezüglich u die Quotientenlimitierung trägt, sind

$$u^* : C_c(X) \longrightarrow C_c(\tilde{X})$$

$$\text{und} \qquad {}^*u : C_c(X,S^1) \longrightarrow C_c(\tilde{X},S^1)$$

Homöomorphismen auf Teilräume.

Diese Tatsache, mit der Aussage von Lemma 3 (X ist wegeweise zusammenhängend!) zusammengefaßt, ergibt:

Lemma 4

Es ist ${}^*u : C_c(X,S^1) \longrightarrow C_c(\tilde{X},S^1)$ ein Homöomorphismus auf einen Teilraum. Im kommutativen Diagramm

6) $$\begin{array}{ccc} C_c(\tilde{X}) & \xrightarrow{\varkappa_X} & C_c(\tilde{X},S^1) \\ \uparrow u^* & & \uparrow {}^*u|\varkappa_X C(X) \\ C_c(X) & \xrightarrow{\varkappa_X} & \varkappa_X C_c(X) \end{array}$$

sind die horizontalen Homomorphismen Quotientenabbildungen (in der Kategorie der Limesräume) auf die Bilder und die vertikalen Homomorphismen Homöomorphismen auf Teilräume.

Bemerkung 5 Die Limesgruppe $C_c(X,S^1)$ ist topologisch, falls X lokalkompakt ist. Sie trägt die Topologie der kompakten Konvergenz.

Die Aussage in Lemma 4 gilt (der Regularität von X wegen) auch dann, wenn die Limitierung der stetigen Konvergenz überall durch die Topologien der kompakten Konvergenz ersetzt wird.

Beweis: Wenn Z ein kompakter Raum ist, dann stimmt auf $C(Z,S^1)$ die Topologie der kompakten Konvergenz mit der Limitierung der stetigen Konvergenz überein. Also ist $C_c(Z,S^1)$ topologisch. Weiter trägt für jeden Raum Y die Limesgruppe $C_{co}(Y,S^1)$ die gröbste Topologie für die die Restriktionshomomorphismen von $C_{co}(Y,S^1)$ und $C_c(K,S^1)$ für alle kompakten Teilmengen $K \subset Y$ stetig sind. Für einen lokalkompakten Raum X ist damit die Evaluationsabbildung $\omega : C_{co}(X,S^1) \times X \longrightarrow S^1$, die jedem Paar (t,P) den Wert $t(p)$ zuordnet, stetig. Mithin ist in diesem Fall $C_{co}(X,S^1)$ mit $C_c(X,S^1)$ identisch. Damit ist der erste Teil der Aussage in Bemerkung 5 bewiesen. Als nächstes zeigen wir, daß jede kompakte Teilmenge $K \subset X$ als Bild unter u einer kompakten Menge $K' \subset \tilde{X}$ dargestellt werden kann, sobald X regulär ist. Sei also $K \subset X$ kompakt. Weil X regulär ist, kann K mit endlich vielen gleichmäßig überlagerten abgeschlossenen Mengen $U_1,\ldots\ldots,U_n$ so überdeckt werden, daß für jedes $i = 1,\ldots,n$ über jedem U_i ein abgeschlossenes Blatt V_i liegt. Offenbar sind $V_i \cap u^{-1}(K)$ kompakt für $i=1,\ldots,n$ und damit $K' = \bigcup_{i=1}^{n} V_i \cap u^{-1}(K)$ kompakte Teilmenge in $\tilde{X}$. Zudem ist $u(K') = K$. Daraus ergibt sich nun sofort, daß

$$u^* : C_{co}(X) \longrightarrow C_{co}(\tilde{X})$$

und

$$^*u : C_{co}(X,S^1) \longrightarrow C_{co}(X,S^1)$$

Homöomorphismen auf Teilräume sind. Zum Beweis, daß $\varkappa_X$ und damit auch $\varkappa_X$ Quotientenabbildungen auf ihre Bilder aufgefaßt als Teilräume von $C_{co}(X,S^1)$ und $C_{co}(\tilde{X},S^1)$ sind, verfahre man wie im Beweis von Lemma 4, ersetze jedoch die U_i durch kompakte Mengen.

Es sei noch hervorgehoben, daß Lemma 4 für jeden wegeweise zusammenhängenden topologischen Raum gilt. Die erste Aussage in Bemerkung 5 gilt für jeden topologischen Raum, während die darauffolgende von der Regularität und dem wegeweisen Zusammenhang Gebrauch macht.

Das Studium der Charaktergruppe von $C_c(X,S^1)$ wird das bis jetzt zusammengestellte Material sowie einige Resultate über die Charaktergruppen von $C_c(X)$ und $\varkappa_X C_c(X)$ benötigen.

3) Die Charaktergruppe von $C_c(X)$

Da zur Bestimmung der Charaktergruppe von $C_c(X,S^1)$ diejenige von $C_c(X)$ herangezogen wird, stellen wir kurz einige Ergebnisse über $\Gamma_c C_c(X)$ zusammen. Diese gelten auch ohne die über X gemachten Voraussetzungen. Sie gelten in der Tat für jeden Limesraum. Mit $Hom_c C_c(X)$ sei die mit der Limitierung der stetigen Konvergenz versehene Menge aller unitären $\mathbb{R}$-Algebrenhomomorphismen bezeichnet. Für jeden Punkt $p \in X$ gehört $i_X(p) : C_c(X) \longrightarrow \mathbb{R}$, der jede Funktion $f \in C(X)$ nach $f(p)$ überführt, dem Limesraum $Hom_c C_c(X)$ an. Die Abbildung

$$i_X : X \longrightarrow Hom_c C_c(X),$$

die jeden Punkt $p \in X$ nach $i_X(p)$ schickt, ist eine stetige Surjektion [Bi].

Der Limesvektorraum $\mathcal{L}_c C_c(X)$, der mit der Limitierung der stetigen Konvergenz versehene Vektorraum aller stetigen reellwertigen linearen Abbildungen, enthält den von $Hom\, C_c(X)$ erzeugte Vektorraum $V(X)$ als dichten Teilraum [Bi] und [Bu]. Das bedeutet, daß

$$7) \qquad a : C_c(X) \longrightarrow \mathcal{L}_c \mathcal{L}_c C_c(X),$$

definiert durch $a(f)(l) = l(f)$ für jedes $f \in C(X)$ und jedes $l \in \mathcal{L}_c C_c(X)$, ein bistetiger Isomorphismus ist [Bu] und [Bi]. Der Homomorphismus $\varkappa : \mathbb{R} \longrightarrow S^1$ induziert einen bistetigen Isomorphismus

$$8) \qquad \bar{\varkappa}_{C(X)} : \mathcal{L}_c C_c(X) \longrightarrow \Gamma_c C_c(X),$$

der jedem $l \in \mathcal{L}_c C_c(X)$ den Charakter $\varkappa \circ l$ zuordnet ([Bu], [Bi]). Für einen einfachen Beweis sei auf [Bi] verwiesen. Die Gruppe

$\bar{\varkappa}_{C(X)}(V(X))$ ist demnach dicht in $\Gamma_c C_c(X)$. Daraus ergibt sich unter Verwendung von (7) und (8), daß

$$9) \qquad j_{C_c(X)} : C_c(X) \longrightarrow \Gamma_c \Gamma_c C_c(X)$$

ein bistetiger Isomorphismus ist.

Diese Beziehungen werden wir gleich im nächsten Paragraphen ausnützen.

4) Die Gruppe $P_c(X)$, ihre Charaktergruppe und die P_c - Reflexivität von $\varkappa_X C_c(X)$

Die Gruppe $P_c(X)$ tritt beim Beschreiben der Charaktere von $\varkappa_X C_c(X)$ auf. Sie ist das Analogon zu $V(X)$, eingeführt im vorigen Paragraphen.

Wir betrachten für jedes $p \in X$ den Charakter $\varkappa \circ i_X(p) : C_c(X) \longrightarrow S^1$. Für jede reelle Zahl r bezeichne $\underline{r}$ die konstante Funktion von X, die den Wert r annimmt. Da für

$$i_X(p)(\underline{n}) = u$$

für jedes $n \in \mathbb{Z}$ gilt, annulliert $\varkappa \circ i_X(p)$ den Kern $\underline{\mathbb{Z}}$ von $\varkappa_X : C_c(X) \longrightarrow \varkappa_X C_c(X)$ und läßt sich deshalb zu einem Charakter, er heiße $j_X(p) : \varkappa_X C_c(X) \longrightarrow S^1$, faktorisieren. Da $\varkappa_X C_c(X)$ bezüglich $\varkappa_X$ die Quotientenlimitierung trägt, ist $j_X(p)$ für jedes $p \in X$ stetig. Die Abbildung

$$10) \qquad j_X : X \longrightarrow \Gamma_c \varkappa_X C_c(X) ,$$

die jedem $p \in X$ den Charakter $j_X(p)$ zugeordnet ist, wie man aus der universellen Eigenschaft der stetigen Konvergenz [Bi] sofort folgert, stetig. Offenbar gilt

$$j_X = \overline{\varkappa}_{C(X)} \circ i_X$$

Für eine Linearkombination

$$\sum_{i=1}^{n} r_i \cdot i_X(p_i) \in V(X)$$

wo $p_i \in X$ und $r_i \in \mathbb{R}$ für $i = 1,\dots,n$, läßt sich

$\varkappa \circ \sum_{i=1}^{n} r_i \cdot i_X(p_i) : C_c(X) \longrightarrow S^1$, genau dann zu einem Charakter von $\varkappa_X C_c(X)$ faktorisieren, wenn $\sum_{i=1}^{n} r_i \in \mathbb{Z}$. Wir bezeichnen mit $P(X)$ die Gruppe aller $\varkappa \circ \sum_{i=1}^{n} r_i \cdot i_X(p_i)$, für die $\sum_{i=1}^{n} r_i \in \mathbb{Z}$.

Diese Teilgruppen von $\Gamma\varkappa_X C_c(X)$ versehen wir mit der Limitierung der stetigen Kovergenz und erhalten so die Limesgruppe $P_c(X)$, ein Unterraum von $\Gamma_c\varkappa_X C(X)$. Unser nächstes Ziel ist nun, die Charaktergruppe von $P_c(X)$ zu studieren.

Zu diesem Zweck halten wir fest:

<u>Lemma 6</u>

Der Homomorphismus

$$^*\varkappa_X : \Gamma_c\varkappa_X C_c(X) \longrightarrow \Gamma_c C_c(X),$$

<u>definiert durch</u> $^*\varkappa_X(\gamma) = \gamma \circ \varkappa_X$ <u>für jedes</u> $\gamma \in \Gamma_c\varkappa_X C_c(X)$, <u>ist ein Homöomorphismus auf einen Teilraum.</u>

Der Beweis ergibt sich unmittelbar aus der Tatsache, daß $\varkappa_X C_c(X)$ bezüglich $\varkappa_X$ die Quotientenlimitierung trägt.
Dieser Homomorphismus $^*\varkappa_X$ bildet $P_c(X)$ homöomorph auf eine Teil-

gruppe von $\Gamma_c C_c(X)$ ab. Diese Teilgruppe heisse $\overline{\varkappa}_{C(X)} P(X)$. Diesen Umstand nützen wir aus, um eine Beziehung zwischen $C_c(X)$ und $\Gamma_c P_c(X)$ herzustellen. Jede Funktion $f \in C(X)$ definiert eine stetige lineare Abbildung $\bar{f} : \mathcal{L}_c C_c(X) \longrightarrow \mathbb{R}$, gegeben durch $\bar{f}(l) = l(f)$ für alle $l \in \mathcal{L}_c C_c(X)$. Durch Komposition mit $\varkappa$ erhalten wir einen Charakter $\varkappa \circ \bar{f} : \Gamma_c C_c(X) \longrightarrow S^1$. Den schränken wir auf ${}^*\varkappa_X(P_c(X))$ ein und erhalten sofort einen Charakter auf $P_c(X)$. Wir bezeichnen diesen Charakter mit $k(f)$. Offenbar gilt $k(f)(\varkappa \circ (\sum_{i=1}^{n} r_i \cdot i_X(p_i)) = \sum_{i=1}^{n} \varkappa \circ (r_i \cdot f(p_i))$ für jedes Element $\varkappa \circ \sum_{i=1}^{n} r_i \cdot i_X(p_i) \in P_c(X)$. Ordnen wir jedem $f \in C(X)$ den Charakter $k(f)$ zu, erhalten wir einen Homomorphismus

$$11) \qquad k : C_c(X) \longrightarrow \Gamma_c P_c(X).$$

Dieser ist, wie man wiederum vermittels der universellen Eigenschaft der Limitierung der stetigen Konvergenz sofort folgert, stetig.

<u>Satz 7:</u> <u>Der Homomorphismus</u>

$$k : C_c(X) \longrightarrow \Gamma_c P_c(X),$$

<u>für den</u> $k(f)(\varkappa \circ \sum_{i=1}^{n} r_i \cdot i_X(p_i)) = \sum_{i=1}^{n} \varkappa(r_i \cdot f(p))$ <u>für jedes Element</u> $\varkappa \circ \sum_{i=1}^{n} r_i \cdot i_X(p_i) \in P_c(X)$ <u>gilt, ist eine stetige Surjektion. Sein Kern ist</u> $\mathbb{Z}$. <u>Weiter faktorisiert</u> k <u>zu einem bistetigen Isomorphismus</u>

$$\bar{k} : \varkappa_X C_c(X) \longrightarrow \Gamma_c P_c(X),$$

<u>dessen Inverser</u>

$${}^*j_X : \Gamma_c P_c(X) \longrightarrow \varkappa_X C_c(X),$$

<u>definiert durch</u> ${}^*j_X(\gamma) = \gamma \circ j_X$ <u>für jedes</u> $\gamma \in \Gamma_c P_c(X)$, <u>ist</u>.

Beweis: Zum Nachweis der Surjektivität von k bilden wir $P_c(X)$ vermöge der Komposition von

$$\Gamma_c \varkappa_X C_c(X) \xrightarrow{\ {}^*\varkappa_X\ } \Gamma_c C_c(X) \xrightarrow{\ \bar{\varkappa}^{-1}_{C_c(X)}\ } \mathcal{L}_c C_c(X)$$

homomorph und homöomorph auf eine Limesgruppe $P_c^1(X) \subset V_c(X)$ ab. Dabei bezeichnet $V_c(X)$ der mit der Limitierung der stetigen Konvergenz versehene Vektorraum $V(X)$. Offenbar besteht $P_c^1(X)$ aus allen Linearkombinationen der Form $\sum_{i=1}^{n} r_i \cdot i_X(p_i)$, wo $r_i \in \mathbb{R}$ und $p_i \in X$ variieren, und für die $\sum_{i=1}^{n} r_i \in \mathbb{Z}$. Zu jedem Charakter $\gamma \in \Gamma_c P_c(X)$ gibt es einen Charakter $\sigma : P_c^1(X) \longrightarrow S^1$ mit

$$12) \qquad \gamma = \sigma \circ \bar{\varkappa}^{-1}_{C_c(X)} \circ {}^*\varkappa_X \, .$$

Man prüft leicht nach, daß σ bestimmt ist durch die Einschränkung auf $M_c(X)$, die mit der Limitierung der stetigen Konvergenz versehene Menge $M(X)$ aller Linearkombinationen $\sum_{i=1}^{n} r_i \cdot i_X(p_i) \in P_c^1(X)$ für die $\sum_{i=1}^{n} r_i = 1$. Jede Strecke, die in $V_c(X)$ irgend zwei Elemente aus $M_c(X)$ verbindet, verläuft ganz in $M_c(X)$, d.h. $M_c(X)$ ist wegeweise zusammenhängend. Wir wollen nun $\sigma | M_c(X) : M_c(X) \longrightarrow S^1$ zu einer reellwertigen Funktion σ' heben. Dazu halten wir in $M_c(X)$ einen Grundpunkt etwa $i_X(p)$ wo $p \in X$ fest und wählen in der Faser über $\sigma(i_X(p))$ eine reelle Zahl r_o aus. Dann verbinden wir jedes Element e aus $M_c(X)$ durch eine Strecke s mit $i_X(p)$. Den Weg $\sigma \circ s$ in S^1 heben wir zu einem Weg in $\mathbb{R}$ mit Anfangspunkt r_o. Diese Hochhebung heiße $(\sigma \circ s)'$. Der Wert von σ' auf e sei dann definiert als der Endpunkt von $(\sigma \circ s)'$. Nun zeigen wir, daß der Wert von $(\sigma \circ s)'$ unabhängig von der Wahl des Grundpunktes ist. Seien $u,v \in M_c(X)$. Je zwei der Elemente $i_X(p)$, u und v verbinden wir derart mit Strecken, daß wir einen in $i_X(p)$ geschlossenen Weg s_1

in $M_c(X)$ haben. Wir verifizieren nun, daß s_1 sich in einem in $r_o \in \mathbb{R}$ geschlossenen Weg hebt. Das von $i_X(p)$, u und r in $M_c(X)$ aufgespannte Simplex Δ liegt in einem endlich dimensionalen Teilraum von $V_c(X)$ und trägt damit nach [Ku] die natürliche Topologie. Offenbar ist s_1 in Δ nullhomotop. Damit hebt sich s_1 zu einem r_o geschlossenen Weg in $\mathbb{R}$. Daraus folgt sofort, daß σ unabhängig von der Wahl des Grundpunktes ist. Zum Beweis der Stetigkeit von σ' betrachten wir einen gegen ein beliebiges Element $u \in M_c(X)$ konvergenten Filter Φ. Ohne Verlust der Allgemeinheit können wir annehmen, daß Φ die Spur eines in $V_c(X)$ gegen u konvergenten Filters ist, der eine Filterbasis aus bezüglich u sternförmig konvexen Mengen besitzt. Also besitzt auch Φ eine Filterbasis von Mengen, die bezüglich u sternförmig konvex sind. In jeder Umgebung von $\sigma'(u)$ gibt es eine offene Umgebung V von $\sigma'(u)$, die als Blatt einer gleichmäßig überlagerten offenen Umgebung U von $\gamma(u)$ auftritt. Weil γ stetig ist, gilt es, eine (bezüglich u) sternförmig konvexe Menge $F \in \Phi$ für die $\gamma(F) \subset U$. Wegen der sternförmigen Konvexität gilt $\sigma'(F) \subset V$. Mithin konvergiert $\sigma'(\Phi)$ gegen $\sigma(u)$, d.h. σ' ist stetig in u. Die Hochhebung $\sigma' : M_c(X) \longrightarrow \mathbb{R}$ ist affin, d.h. es gilt

$$\sigma'\left(\sum_{i=1}^{u} r_i \cdot i_X(p_i)\right) = \sum_{i=1}^{u} r_i \cdot \sigma'(i_X(p_i))$$

für jede Linearkombination $\sum_{i=1}^{u} r_i \cdot i_X(p_i) \in M_c(X)$. Um das einzusehen, verifizieren wir erst, daß σ^1 auf jeder Strecke in $M_c(X)$ affin ist. Für zwei beliebige Elemente $u, v_o \in M_c(X)$ sei $s : [0,1] \longrightarrow M_c(X)$, eine Strecke, die u mit v_o verbindet, d.h. $s(r) = u + r(v-u)$ für jedes $r \in [0,1]$. Die auf $s([0,1]) \subset M_c(X)$ induzierte Limitierung ist die natürliche Topologie. Auf $s([0,1])$ wählen wir eine offene zusammenhängende Umgebung V derart, daß $\sigma'(V)$ in einem Blatt über einer gleichmäßig überlagerten Umgebung U von $\gamma(u)$ liegt. Für jedes von u verschiedene feste Element $v_1 \in V$ gilt:

$$\sigma'(u + r \cdot (v_1 - u)) = \sigma'(u) + r \cdot (\sigma'(v) - \sigma'(u))$$

für alle $r \in [0,1]$. Dies verifiziert man wie folgt: Offenbar ist σ' additiv. Für eine gewisse reelle Zahl r^1 gilt $r^1 \cdot (v_1 - u) = v - u$. Wir definieren $l : s([0,1] \longrightarrow \mathbb{R}$ durch

$$l(v) = \sigma'(u) + n \cdot (\frac{r^1}{n} \cdot (\sigma'(v_1) - \sigma'(u)))$$

(wobei n die kleinste natürliche Zahl mit $\frac{1}{n} \cdot r^1 \leq 1$ ist) für alle $r \in s([0,1])$. Es ist l eine stetige Abbildung für die $\varkappa(l(v)) = \sigma(v)$ für jedes $r \in s([0,1])$ gilt. Somit ist l eine Hochhebung von $\sigma | s([0,1])$, stimmt mithin mit $\sigma' | s([0,1])$ überein. Folglich haben wir

$$\sigma'(u + r \cdot (v-u)) = \sigma'(u) + r \cdot (\sigma'(v) - \sigma'(u)),$$

was wir zeigen wollten. Nun betrachten wir

$$\sum_{i=1}^{n} r_i \cdot i_X(p_i) \in M_c(X)$$

und schreiben diesen Ausdruck in der Form

$$r \cdot \left(\sum_{i=1}^{n-1} r_i^1 \cdot i_X(p_i) \right) + (1-r) \cdot i_X(p_n)$$

wobei $r, r_i^1 \in \mathbb{R}$ und $\sum_{i=1}^{n} r_i^1 = 1$ für $i=1,\ldots,n$.

Wir nehmen nun an, daß für $m \leq n - 1$ für jede Kombination

$$\sum_{i=1}^{n} r_j'' \cdot i_X(p_j) \quad \text{wo} \quad p_j \in X, \quad \text{und} \quad r_j'' \in \mathbb{R} \quad \text{für} \quad j=1,\ldots,m$$

mit $\Sigma r_j'' = 1$ gilt:

$$\sigma'(\sum_{j=1}^{n} r_j'' \cdot i_X(p_j) = \sum_{j=1}^{n} r_j'' \cdot \sigma'(i_X(p_j)).$$

Weil σ' auf jeder Strecke aus $M_c(X)$ affin ist, ergibt sich

$$\sigma'(\sum_{i=1}^{n} r_i \cdot i_X(p_i)) = r \cdot \left(\sigma'(\sum_{i=1}^{n-1} r_i^1 \cdot i_X(p_i)\right) + (1-r) \cdot \sigma'(i_X(p_n))$$

$$= \sum_{i=1}^{n-1} r \cdot r_i^1 \cdot \sigma'(i_X(p_i) + (1-r) \cdot \sigma'(i_X(p_n)$$

$$= \sum_{i=1}^{n} r_i \cdot \sigma'(i_X(p_i)) .$$

Also gilt $\sigma'(\sum_{i=1}^{n} r_i \cdot i_X(p_i)) = \sum_{i=1}^{n} (r_i \cdot \sigma'(i_X(p_i)))$.

Ersetzen wir nun die stetige Abbildung $\sigma' \circ i_X : X \longrightarrow \mathbb{R}$ durch das Symbol f und verwenden die Tatsache, daß σ durch die Werte auf $M_c(X)$ eindeutig bestimmt ist, so erhalten wir

$$\sigma(\sum_{i=1}^{n} r_i \cdot i_X(p)) = \sum_{i=1}^{n} \varkappa(r_i \cdot f(p_i))$$

für alle Kombinationen aus $P_c^1(X)$. Offenbar ist $f \in C(X)$.
Damit gilt aber wegen (12) :

$$k(f)(\varkappa \sum_{i=1}^{n} r_i \cdot i_X(p_i)) = \sum_{i=1}^{n} \varkappa(r_i \cdot f(p_i))$$

für alle Kombinationen aus $P_c(X)$. Also ist k surjektiv und folglich $\bar{k}$ bijektiv. Die Stetigkeit von $\bar{k}^{-1}$ ergibt sich nun wie folgt:
Die Abbildung

$$j_X : X \longrightarrow P_c(X) ,$$

(siehe (10)) induziert einen stetigen Gruppenhomomorphismus

$$^*j_X : \Gamma_c P_c(X) \longrightarrow \varkappa_X C_c(X) ,$$

der jedes $\gamma \in \Gamma_c P_c(X)$ in $\gamma \circ j_X$ überführt. Wie man mühelos verifiziert, gilt:

$$^*j_X \circ \bar{k} = id_{\Gamma_c P_c(X)} .$$

Mithin ist $\bar{k}^{-1} = {}^{*}j_X$, also stetig. Damit haben wir Satz 7 endlich bewiesen.

Lemma 8

Es ist $P(X) \subset \Gamma_c \varkappa_X C_c(X)$ dicht.

Beweis:

Wiederum verwenden wir den zu Beginn des vorangegangenen Beweises eingeführten Homomorphismus

$$\bar{\varkappa}^{-1}_{C_c(X)} \circ {}^{*}\varkappa_X : \Gamma_c \varkappa_X C_c(X) \longrightarrow \mathcal{L}_c C_c(X).$$

Das Bild dieses Homomorphismus besteht aus allen $l \in \mathcal{L}_c C_c(X)$ mit $l(1) \in \mathbb{Z}$. Zu jedem $l \in \mathcal{L}_c C_c(X)$ gibt es, einen gegen l konvergenten Filter Φ, der eine Basis $\mathcal{B}$ in $V(X)$ besitzt [Bu] und [Bi]. Es gelte nun $l(1) \in \mathbb{Z}$. Dann können wir ohne Verlust der Allgemeinheit annehmen, daß jedes Element $\sum_{i=1}^{n} r_i \cdot i_X(p_i)$ aus jeder Menge $N \in \mathcal{B}$ der Bedingung $\sum_{i=1}^{n} r_i \neq 0$ genügt. Wir führen nun „N" definiert als

$$\{ l(1) \cdot \frac{\sum_{i=1}^{n} r_i \cdot i_X(p_i)}{\sum_{i=1}^{n} r_i} \mid \sum_{i=1}^{n} r_i \cdot i_X(p_i) \in N \}$$

ein. Diese Menge liegt in $\bar{\varkappa}^{-1}_{C(X)} \circ {}^{*}\varkappa_X(P(X))$. Der Filter Ψ erzeugt durch $\{ N" \mid N \in \mathcal{B}\}$ konvergiert in $\mathcal{L}_c C_c(X)$ gegen l. Da $\bar{\varkappa}^{-1}_{C(X)} \circ {}^{*}\varkappa_X$ ein Homöomorphismus auf sein Bild ist, ergibt sich, daß $P(X)$ eine dichte Teilmenge von $\Gamma_c \varkappa_c C_c(X)$ ist.

Satz 9 *Die Gruppe* $\varkappa_X C_c(X)$ *ist* P_c-*reflexiv*. *Zudem ist der von der Inklusion*

$$i : P_c(X) \longrightarrow \Gamma_c \varkappa_X C_c(X)$$

induzierte Homomorphismus

$$^*i : \Gamma_c \Gamma_c \varkappa_X C_c(X) \longrightarrow \Gamma_c P_c(X) ,$$

der jeden Charakter $\gamma \in \Gamma_c \Gamma_c \varkappa_X C(X)$ *nach* $\gamma \circ i$ *abbildet ein bistetiger Isomorphismus.*

Beweis:

Aus der Kommutativität von

$$\begin{array}{ccc} \Gamma_c \Gamma_c \varkappa_X C_c(X) & \xrightarrow{\;\;^*i\;\;} & \Gamma_c P_c(X) \\ \uparrow {\scriptstyle j_{\varkappa_X C_c(X)}} & \nearrow {\scriptstyle \bar{k}} & \\ \varkappa_X C_c(X) & & \end{array}$$

folgt unter Verwendung von Lemma 8 und Satz 7 die Richtigkeit der Behauptung des obigen Satzes.

Korollar 10

Für jeden zusammenhängenden und lokal wegeweise zusammenhängenden topologischen Raum Y *mit endlicher Fundamentalgruppe* $\pi_1(Y)$ *ist* $\varkappa_X C_c(Y) = C_c(Y, S^1)$. *Somit ist* $C_c(Y, S^1)$ P_c-*reflexiv*.

Beweis:

Eine Abbildung $f \in C(Y,S^1)$ ist genau dann zu einer reellwertigen Abbildung hebbar, wenn der von f induzierte Homomorphismus

$$f_* : \Pi_1(X,p) \longrightarrow \Pi_1(S^1,f(p))$$

trivial ist ([Sp] chapter 2). Weil $\Pi_1(S^1,f(p)) \cong \mathbb{Z}$ ist f_* für jedes $f \in C(Y,S^1)$ trivial. Mithin ist $C_c(Y,S^1) = \varkappa_X C_c(X)$ (beachte Lemma 3). Nach Satz 9 ist damit $C_c(Y,S^1)$ P_c-reflexiv.

Man beachte, daß die Sätze 7, 9 sowie Lemma 8 für jeden wegeweise zusammenhängenden Raum, also ohne die in §1 generell gemachten Voraussetzungen an X, gelten. Wie in der Einleitung vermerkt, soll zum Beweis der P_c-Reflexivität von $C(X,S^1)$ gezeigt werden, daß die Folge

$$o \longrightarrow \Gamma_c \Pi_c^1(X) \longrightarrow \Gamma_c C_c(X,S^1) \longrightarrow \Gamma_c \varkappa_X C_c(X) \longrightarrow o$$

sowie ihre Duale exakt sind. Zum Nachweis der für diese Dualisierungsoperationen notwendigen Eigenschaften von ${}^*i$ verwenden wir nun die universelle Überlagerung von X. Die gesuchten Eigenschaften (siehe Korollar 13) werden über die Beziehungen zwischen $\Gamma\varkappa_X C_c(X)$ und $\Gamma_c \varkappa_{\tilde{X}} C_c(\tilde{X})$ hergestellt.

5) Zu den Beziehungen zwischen $\Gamma_c \varkappa_X C_c(X)$ und $\Gamma_c \varkappa_{\tilde{X}} C_c(\tilde{X})$

Die Überlagerungsabbildung $u : \tilde{X} \longrightarrow X$ induziert mit (2) folgendes kommutatives Diagramm

13)
$$\begin{array}{ccccccccc}
o & \longrightarrow & \mathbb{Z} & \xrightarrow{i_1} & C_c(\tilde{X}) & \xrightarrow{\varkappa_{\tilde{X}}} & \varkappa_{\tilde{X}} C_c(\tilde{X}) & \longrightarrow & o \\
 & & \uparrow{\scriptstyle id} & & \uparrow{\scriptstyle u^*} & & \uparrow{\scriptstyle \overline{{}^*u}} & & \\
o & \longrightarrow & \mathbb{Z} & \xrightarrow{i_1} & C_c(X) & \xrightarrow{\varkappa_X} & \varkappa_X C_c(X) & \longrightarrow & o
\end{array}$$

Dabei bedeuten u^* den Homomorphismus in (3a) und $\overline{{}^*u}$ die Restriktion von ${}^*u$ in (3b) auf $\varkappa_X C_c(X)$. Wie in (1) vereinbart, soll i_1 als die Inklusion bezeichnet werden. Sowohl u^* als auch $\overline{{}^*u}$ sind Injektionen. Durch "dualisieren" von (2a) erhalten wir

$$\begin{array}{ccccccccc} 0 & \longrightarrow & \Gamma_c\varkappa_{\tilde X}C_c(\tilde X) & \xrightarrow{\;{}^*\varkappa_{\tilde X}\;} & \Gamma_c C_c(\tilde X) & \xrightarrow{\;{}^*i_1\;} & \Gamma_c\underline{\mathbb{Z}} & \longrightarrow & 0 \\ & & \Big\downarrow {}^*\overline{{}^*u} & & \Big\downarrow {}^*(u^*) & & \Big\downarrow \mathrm{id} & & \\ 0 & \longrightarrow & \Gamma_c\varkappa_X C_c(X) & \xrightarrow{\;{}^*\varkappa_X\;} & \Gamma_c C_c(X) & \xrightarrow{\;{}^*i_1\;} & \Gamma_c\underline{\mathbb{Z}} & \longrightarrow & 0 \end{array} \tag{14}$$

Dieses Diagramm kommutiert. Der Zusammenhang von $\Gamma_c\varkappa_{\tilde X}C_c(X)$ und $\Gamma_c\varkappa_X C_c(X)$ soll nun durch die Eigenschaften der in (14) auftretenden Abbildungen beschrieben werden. (Wir identifizieren $\underline{\mathbb{Z}}$ mit $\mathbb{Z}$.)

<u>Proposition 11</u> <u>Es ist</u>

$$0 \longrightarrow \Gamma_c\varkappa_X C_c(X) \xrightarrow{\;{}^*\varkappa_X\;} \Gamma_c C_c(X) \xrightarrow{\;{}^*i_1\;} \Gamma_c\mathbb{Z} \longrightarrow 0$$

<u>exakt</u>. <u>Dabei ist</u> ${}^*\varkappa_X$ <u>ein Homöomorphismus auf einen Teilraum und</u> ${}^*i_1$ <u>eine Quotientenabbildung</u>.

<u>Beweis:</u> Daß ${}^*\varkappa_X$ ein Homöomorphismus auf einen Teilraum ist, besagt Lemma 6. Die Surjektivität von ${}^*i_1$ können wir etwa so einsehen. Zum Charakter $\gamma : \mathbb{Z} \longrightarrow S^1$ findet sich eine lineare Abbildung $l : \mathbb{R} \longrightarrow \mathbb{R}$ mit $\varkappa \circ l|\mathbb{Z} = \gamma$. Offenbar ist $l(r) = r_1 \cdot r$ für ein festes $r_1 \in \mathbb{R}$. Nun identifizieren wir in der offensichtlichen Weise $\mathbb{R}$ mit der Menge der konstanten reellwertigen Funktionen von X. Die Abbildung $l_1 : C_c(X) \longrightarrow C_c(X)$, die jedes $f \in C(X)$ nach

$r_1 \cdot f$ abbildet, ist stetig und linear. Für irgendeinen Punkt $p \in X$ ist dann $i_X(p) \circ l$ eine Ausdehnung von l. Der Charakter $\varkappa \circ i_X(p)$ wird durch ${}^*i_1$ auf Y abgebildet. Um zu zeigen, daß ${}^*i_1$ eine Quotientenabbildung ist, verwenden wir das Diagramm

$$\begin{array}{ccccc} \mathcal{L}_c C_c(X) & \xrightarrow{\;i_2^*\;} & \mathcal{L}_c \mathbb{R} & \longrightarrow & \mathbb{R} \\ \uparrow & & & & \downarrow \\ \Gamma_c C_c(X) & \xrightarrow{\;{}^*i_1\;} & \Gamma_c \mathbb{Z} & \longrightarrow & S^1 \end{array} ,$$

wobei i_2 die Inklusion bezeichnet und die unmarkierten Pfeile die natürlichen Homöomorphismen bedeuten. Das Diagramm kommutiert. Die Komposition der oberen Abbildung ist, da $\mathbb{R}$ nur <u>eine</u> separierte Vektorraumlimitierung, nämlich die natürliche Topologie, tragen kann [Ku], eine Quotientenabbildung. Weil auch $\varkappa$ eine solche ist und überdies $\varkappa^{-1}_{C_c(X)}$ ein Homöomorphismus ist (siehe §3), muß auch ${}^*i_1$ eine Quotientenabbildung sein.

Durch nochmaliges "Dualisieren" des Diagramms 14 erhält man der P_c-Reflexivität der auftretenden Gruppen wegen bis auf eine kanonische Isomorphie das Diagramm 13 zurück.

Der nächste Satz zeigt, daß $\Gamma_c \varkappa_X C_c(X)$ und $\Gamma_c C_c(X)$ innerhalb der Kategorie aller Limesgruppen S für die j_S ein Homöomorphismus auf sein Bild ist (der offenbar alle im Diagramm 14 auftretenden Gruppen angehören) Quotienten von $\Gamma_c \varkappa_{\tilde{X}} C_c(\tilde{X})$ respektive von $\Gamma_c C_c(\tilde{X})$ sind.

Satz 12 Die im Diagramm (14) auftretenden Homomorphismen $^*\overline{{}^*u}$ und $^*(u^*)$ sind Surjektionen mit folgender universeller Eigenschaft: Für jeden Homomorphismus h von $\Gamma_c \varkappa_X C_c(X)$ respektive von $\Gamma_c C_c(X)$ in eine Limesgruppe S, ist $j_S \circ h$ genau dann stetig, wenn $j_S \circ h \circ {}^*\overline{{}^*u}$ respektive $j_S \circ h \circ {}^*(u^*)$ stetig ist.

Beweis: Erst zeigen wir, daß die beiden (stetigen) Homomorphismen $^*\overline{{}^*u}$ und $^*(u^*)$ surjektiv sind. Zum Nachweis der Surjektivität von $^*(u^*)$ halten wir erst fest, daß der Limesvektorraum $C_c(X)$ und der lokalkonvexe topologische Vektorraum $C_{co}(X)$ denselben Dualraum haben [Bu] oder [Bi], also daß $\mathcal{L}C_c(X) = \mathcal{L}C_{co}(X)$ gilt.(Hier machen wir von der Normalität nur insoweit Gebrauch, daß jeder normale Raum c-einbettbar ist [Bi] und mithin $\mathcal{L}C_{co}(X) = \mathcal{L}C_c(X)$ gilt). Nun folgert man leicht unter Verwendung des bistetigen Isomorphismus $\overline{\varkappa}_{C_c(X)}$ (s.(8)) und Bemerkung 5 mit Hilfe des Satzes von Hahn-Banach, daß $^*(u^*)$ surjektiv ist. Um zu zeigen, daß $^*\overline{{}^*u}$ surjektiv ist, sei $\gamma \in \Gamma_c \varkappa_X C_c(X)$. Dann ist $\gamma \circ \varkappa_X \in \Gamma_c C_c(X)$. Wegen der Surjektivität von $^*(u^*)$ findet sich eine "Erweiterung" $\gamma^1 \in \Gamma_c C_c(\tilde{X})$ für die $\gamma^1 = {}^*(u^*)(\gamma \circ \varkappa_X)$. Da γ die Gruppe $\mathbb{Z}$ annulliert, gilt dies auch für γ^1. Also läßt sich γ^1 zu einem Charakter $\gamma'' \in \Gamma_c \varkappa_{\tilde{X}} C_c(\tilde{X})$ faktorisieren. Nun verifiziert man leicht, daß $^*\overline{{}^*u}(\gamma'') = \gamma$. Zur Verifikation der nicht-trivialen Richtung der universellen Eigenschaft von $^*\overline{{}^*u}$ geben wir uns einen Homomorphismus

$$h : \Gamma_c \varkappa_X C_c(X) \longrightarrow S^1$$

für den $j_S \circ h \circ {}^*\overline{{}^*u}$ stetig ist und folgern daraus die Stetigkeit von $j_S \circ h$. Dann bilden wir das Diagramm

$$\begin{array}{ccc}
\tilde{X} & \xrightarrow{\;j_{\tilde{X}}\;} & \Gamma_c \varkappa_{\tilde{X}} C_c(\tilde{X}) \\
\downarrow{\scriptstyle u} & & \downarrow{\scriptstyle *\overline{*u}} \\
X & \xrightarrow{\;j_X\;} & \Gamma_c \varkappa_X C_c(X) \\
& \searrow{\scriptstyle h^1} & \downarrow{\scriptstyle h} \\
& & S
\end{array}$$

worin h^1 die Abbildung $h \circ j_X$ bezeichnet. Daraus lesen wir ab, daß $h^1 \circ u$ stetig ist. Weil u eine Quotientenabbildung ist, muß h^1 stetig sein.

Nun bilden wir

$$\begin{array}{ccccc}
\Gamma_c \varkappa_{\tilde{X}} C_c(\tilde{X}) & \xrightarrow{\;**j_{\tilde{X}}\;} & \Gamma_c \Gamma_c \Gamma_c \varkappa_{\tilde{X}} C_c(\tilde{X}) & \xrightarrow{\;j_{\Gamma_c \varkappa_{\tilde{X}} C_c(\tilde{X})}\;} & \Gamma_c \varkappa_{\tilde{X}} C_c(\tilde{X}) \\
\downarrow{\scriptstyle *\overline{*u}} & & \downarrow{\scriptstyle **(h \circ *\overline{*u})} & & \downarrow{\scriptstyle h \circ *\overline{*u}} \\
\Gamma_c \varkappa_X C_c(X) & \xrightarrow{\;**(h^1)\;} & \Gamma_c \Gamma_c S & \xleftarrow{\;j_S\;} & S
\end{array}$$

Darin ist sicherlich die rechte Hälfte kommutativ. Hier bedeutet $**j_{\tilde{X}}$ den Homomorphismus induziert von

$$*j_{\tilde{X}}^{-1} : \Gamma_c \Gamma_c \varkappa_{\tilde{X}} C_c(\tilde{X}) \longrightarrow \varkappa_{\tilde{X}} C_c(\tilde{X}) ,$$

wobei $*j_{\tilde{X}}$ jedes γ nach $\gamma \circ j_{\tilde{X}}$ abbildet. Man prüft nun nach, daß die linke Hälfte des obigen Diagramms ebenfalls kommutiert und überzeugt sich weiter davon, daß

$$j_{\Gamma_c \varkappa_{\tilde{X}} C_c(\tilde{X})} \circ {}^{**}j_{\tilde{X}} = id_{\Gamma_c \varkappa_{\tilde{X}} C_c(\tilde{X})}$$

gilt. Also ergibt sich

$$^{**}h^1 \circ {}^*\overline{{}^*u} = j_S \circ h \circ {}^*\overline{{}^*u} .$$

Weil ${}^*\overline{{}^*u}$ surjektiv ist folgt

$$^{**}h^1 = j_S \circ h .$$

Da $^{**}h^1$ stetig ist, muß $j_S \circ h$ stetig sein. Weil $^*\varkappa_X$ und $^*\varkappa_{\tilde{X}}$ Homöomorphismen auf Teilräume sind, folgert man nun leicht die analoge universelle Eigenschaft für ${}^*(u^*)$.

<u>Korollar 13</u> <u>Der von der Inklusionsbildung</u>

$$i : \varkappa_X C_c(X) \longrightarrow C_c(X,S^1)$$

<u>induzierte Homomorphismus</u>

$$^*i : \Gamma_c C_c(X,S^1) \longrightarrow \Gamma_c \varkappa_X C_c(X)$$

<u>ist surjektiv und besitzt folgende universelle Eigenschaft:</u>
<u>Wenn für eine Limesgruppe</u> S <u>und einen Homomorphismus</u>
$h : \Gamma_c \varkappa_X C_c(X) \longrightarrow S$ <u>die Komposition</u> $h \circ i^*$ <u>stetig ist, dann ist</u>
<u>auch</u> $j_S \circ h$ <u>stetig.</u>

<u>Beweis:</u> Die Surjektivität von $^{*}i$ und die universelle Eigenschaft folgert man sofort aus Satz 12 und der Kommutativität von

$$\begin{array}{ccc} \Gamma_c \varkappa_{\tilde{X}} C_c(\tilde{X}) & \xrightarrow{\;{}^{**}u\;} & \Gamma_c C_c(X,S^1) \\ \Big\downarrow {\scriptstyle {}^{*}\overline{{}^{*}u}} & \swarrow {\scriptstyle {}^{*}i} & \\ \Gamma_c \varkappa_X C_c(X) & & \end{array}$$

Gemäß der in der Einleitung beschriebenen Konzeption zum Nachweis der P_c-Reflexivität von $C_c(X,S^1)$, wenden wir uns nun dem Nachweis der P_c-Reflexivität von $C_c(X,S^1)/\varkappa_X C_c(X)$ (falls vollständig) zu.

6) <u>Zur Bruschlinski'schen Gruppe</u>

Wie in der Einleitung bemerkt, heißt die Gruppe

$$C(X,S^1)/\varkappa_X C_c(X)$$

die Bruschlinski'sche Gruppe. Wir ersetzen dieses Symbol durch das einfachere $\Pi^1(X)$. Später werden wir auf eine andere Beschreibung von $\Pi^1(X)$ zurückkommen. Den natürlichen Homomorphismus von $C(X,S^1)$ auf $\Pi^1(X)$ heiße b. Die Gruppe $\Pi^1(X)$ versehen wir mit der von b induzierten Finallimitierung und bezeichnen die resultierende Hausdorffsche Limesgruppe mit $\Pi_c^1(X)$.

Unser erstes Ziel ist es, zu zeigen, daß $\Pi_c^1(X)$ topologisch ist. Zur leichteren technischen Handhabung unseres Problems werden wir die Limesgruppe $\Pi_c^1(X)$ als Quotient von Limesgruppen von Funktion aus $C_c(\tilde{X})$ beschreiben. Dazu stellen wir zuerst $\Pi^1(X)$ als Quotient von Gruppen von Funktionen aus $C(\tilde{X})$ dar. Wir fixieren irgend einen Punkt $p \in X$ und führen die Gruppe G_p aller in p verschwindenden Abbildungen aus $C(X,S^1)$ ein. Wir setzen weiter $G_p^o = \varkappa_X C(X) \cap G_p$. Offenbar sind $G_p|G_p^o$ und $\Pi^1(X)$ isomorph.

Nun bilden wir $\varkappa_{\tilde{X}}^{-1}(*u(G_p))$ und $\varkappa_{\tilde{X}}^{-1}(*u(G_p^o))$, die wir kurz mit C_p und C_p^o respektive bezeichnen.

Nach Lemma 1 besteht C_p aus allen Funktionen aus $C(\tilde{X})$, die auf der Faser $u^{-1}(p)$ ihre Werte in $\mathbb{Z}$ annehmen. Eine Charakterisierung der Funktionen in C_p^o erhalten wir über folgende Verschärfung von Lemma 2:

<u>Lemma 14</u>

<u>Eine Funktion</u> $f \in \varkappa_{\tilde{X}}^{-1}(^*u(C(X,S^1)))$ <u>gehört genau dann</u> $u^*(C(X))$ <u>an</u>, wenn sie auf der Faser $u^{-1}(p_1)$ <u>über irgend einem Punkte</u> $p_1 \in X$ <u>konstant ist</u>.

<u>Beweis:</u> Nach Lemma 2 ist $u^*(g)$ für jedes $g \in C(X)$ auf jeder Faser konstant. Sei umgekehrt $f \in \varkappa_{\tilde{X}}^{-1}(*u(C(X,S^1)))$ auf $u^{-1}(p_1)$ für irgend einen Punkt p_1 konstant, sagen wir gleich r. Wir wählen $p \in X$. Zu zeigen ist, daß $f|u^{-1}(p)$ konstant ist. Aus Lemma 2 ergibt sich dann Lemma 14. Wir verbinden p_1 mit p durch einen Weg σ. Diesen Weg heben wir zu einem Weg σ^1 in $\tilde{X}$ hoch, der als Anfangspunkt $q_1 \in u^{-1}(p_1)$ besitzt. Der Endpunkt von σ liegt in $u^{-1}(p)$. Sei $f = {}^*u(g)$. Nun heben wir $*u(g) \circ \sigma$, der in $*u(g)(p_1) \in S^1$ beginnt und in $*u(g)(p) \in S^1$ endet, zu einem Weg σ'' in R mit Anfangspunkt $f(q^1)$ hoch. Der eindeutigen Wegehochhebung in R wegen, ist $\sigma'' = f \circ \sigma^1$. Der Endpunkt von σ'' ist r. Da $q_1 \in u^{-1}(p_1)$ beliebig ausgewählt war und es nur eine Hochhebung von $u^*(f) \circ \sigma$ mit Endpunkt r gibt,

ist $f|u^{-1}(p)$ konstant.

Aus Lemma 14 ergibt sich, daß C_p^o gerade aus allen Funktionen in C_p besteht, die auf $u^{-1}(p)$ konstant sind.

Der Homomorphismus

$$b_p = b \circ {}^*u^{-1} \circ \varkappa_{\tilde{X}} : C_p \longrightarrow \Pi^1(X)$$

(man beachte, daß ${}^*u^{-1}$ nur auf dem Bild von ${}^*u$ definiert ist) ist surjektiv und sein Kern besteht gerade aus C_p^o. Dieser faktorisiert zu einem Isomorphismus

$$\bar{b}_p : C_p/C_p^o \longrightarrow \Pi^1(X).$$

Fixieren wir einen Punkt $q \in u^{-1}(p)$ und bezeichnen mit C_p^q und M_p alle Funktionen aus C_p respektive C_p^o die auf q verschwinden, dann faktorisiert $b \circ {}^*u^{-1} \circ \varkappa_{\tilde{X}}|C_p^q$ offenbar zu einem Isomorphismus

$$15) \qquad \bar{b}_p : C_p^q \,/\, M_p \longrightarrow \Pi^1(X).$$

Damit kann $\Pi^1(X)$ als Quotient von Gruppen von Funktionen aus $C(\tilde{X})$ aufgefaßt werden. Um die Limesstruktur auf $\Pi^1(X)$ in Abhängigkeit der Faser $u^{-1}(p)$ beschreiben zu können (Lemma 15) interpretieren wir diese Gruppe nochmals um, und zwar als Gruppe von $\mathbb{Z}$-wertigen Funktionen von $u^{-1}(p)$.

Bekanntlich läßt sich $u^{-1}(p)$ mit der Fundamentalgruppe $\Pi_1(X,p)$ von X mit Aufpunkt p identifizieren, und zwar so, daß ein festgewählter Punkt, es sei q, zum Einselelement $e \in \Pi_1(X,p)$ wird ([Sp] chapter 2). Statt $u^{-1}(p)$ schreiben wir nun F_p. Diese Faser trägt also eine Gruppenstruktur mit $q \in F_p$ als Einselelement. Jede Abbildung $t \in G_p$ induziert dann einen Homomorphismus [Sp]

$$t_* : \Pi_1(X,p) \longrightarrow \Pi_1(S^1,1) ,$$

der die Homotopieklasse $[\sigma]$ eines jeden in p geschlossenen Weges σ nach $[f \circ \sigma]$ überführt. Identifizieren wir $\Pi_1(S^1,1)$ mit $\mathbb{Z}$ derart, daß das Einselelement in der Fundamentalgruppe mit $0 \in \mathbb{Z}$ zusammenfällt, erhalten wir eine Abbildung

$$t_* : F_p \longrightarrow \mathbb{Z},$$

die q nach Null abbildet. Bezeichnet f_t^q die eindeutig bestimmte Hochhebung von t, die in q verschwindet, so gilt, wie man leicht nachprüft,

$$t_* = f_t^q \mid F_p .$$

Folglich haben wir eine Abbildung

$$d_p : C_p^q \longrightarrow \mathrm{Hom}(F_p,\mathbb{Z}),$$

die jedem $f \in C_p^q$ die Einschränkung auf F_p zuordnet. Der Kern von d_p ist gerade M_p. Mithin faktorisiert d_p zu einem Monomorphismus

$$\bar{d}_p : C_p^q \mid M_p \longrightarrow \mathrm{Hom}(F_p,\mathbb{Z}).$$

Also haben wir einen natürlichen Monomorphismus

$$\bar{d}_p \circ \bar{b}_p^{-1} : \Pi^1(X) \longrightarrow \mathrm{Hom}(F_p,Z).$$

Wir bezeichnen das Bild von $\bar{d}_p$ mit $\Pi^1(X)_p$. Alle Homomorphismen aus $\Pi^1(X)_p$ lassen sich mithinein zu stetigen reellwertigen Funktionen aus $C_p^q \subset C(\tilde{X})$.

Nun versehen wir $\Pi^1(X)$ mit verschiedenen, für die natürliche technische Handhabung von $\Pi^1(X)$ angepaßten Limitierungen und zeigen, daß sie alle homöomorph sind. Die Gruppe $C_p^q \subset C(X)$ fassen wir als Teilraum von $C_c(\tilde{X})$ auf. Damit ist C_p^q eine Limesgruppe. Die Gruppe $\mathrm{Hom}(F_p, Z)$ und $\Pi^1(X)_p$ versehen wir je mit der Topologie der punktweisen Konvergenz und erhalten die topologischen Gruppen $\mathrm{Hom}_s(F_p, \mathbb{Z})$ und $\Pi_s^1(X)_p$. Die Gruppe $\Pi^1(X)$ versehen wir mit der von

$$b_p : C_p^q \longrightarrow \Pi^1(X)$$

induzierten Quotientenlimitierung. Diese Limitierung heiße $\Pi_{c_1}^1(X)$. Offenbar ist $id : \Pi_{c_1}^1(X) \longrightarrow \Pi_c^1(X)$ stetig.

Unser nächstes Ziel muß es sein, den stetigen Homomorphismus $\bar{d}_p \circ \bar{b}_p^{-1}$ genauer zu studieren.

Zunächst beschreiben wir eine bequeme Nullumgebungsbasis in $\mathrm{Hom}_s(F_p, \mathbb{Z})$. Zu irgend endlich viele Elemente $e_1, \ldots, e_n$ gibt es eine Nullumgebung $K \subset \mathrm{Hom}_s(F_p Z)$ mit $h(e_1) = \ldots = h(e_n) = o$ für alle $h \in K$. Also annulliert jeder Homomorphismus $h \in K$ die von $e_1, \ldots, e_n$ in F_p erzeugte Gruppe $H_{e_1, \ldots e_n}$. Offenbar ist

$$\{h \in \mathrm{Hom}_s(F_p, \mathbb{Z}) \mid h(H_{e_1, \ldots, e_n}) = o\}$$

eine Nullumgebung in $\mathrm{Hom}_s(F_p, \mathbb{Z})$. Das System

16) $\{\{h \in \mathrm{Hom}_s(F_p, \mathbb{Z}) \mid h(H) = o\} \mid H \subset F_p \text{ endlich erzeugt}\}$

ist eine Nullumgebungsbasis $\mathcal{B}$ der topologischen Gruppe $\mathrm{Hom}_s(F_p, \mathbb{Z})$. Diese Basis geschnitten mit $\Pi^1(X)_p$ ergibt eine Nullumgebungsbasis von $\Pi_s^1(X)_p$.

Lemma 15

Der Isomorphismus

$$\bar{d}_p \circ \bar{b}_p^{-1} : \Pi^1_{c_1}(X) \longrightarrow \Pi^1_s(X)_p$$

ist ein Homöomorphismus. Folglich ist

$$\bar{d}_p \circ \bar{b}_p^{-1} : \Pi^1_c(X) \longrightarrow \Pi^1_s(X)_p$$

ein bistetiger Isomorphismus.

Beweis: Wir wissen bereits, daß der Isomorphismus $\bar{d}_p \circ \bar{b}_p^{-1}$ stetig ist. Zum Nachweis der Stetigkeit der Umkehrabbildung konstruieren wir erst einen gegen $o \in C^q_p$ konvergenten Filter Φ, der durch $b_p|C^q_p$ auf die Nullumgebungsfilter $\mathfrak{U}(o)$ in $\Pi^1_s(X)_p$ abgebildet wird. Sei $H \subset F_p$ eine endlich erzeugte Gruppe. N_H bezeichne den kleinsten Normalteiler in F_p der H enthält. Wir sagen, daß N_H endlich erzeugt sei. Für jedes $h \subset \mathrm{Hom}(F_p,\mathbb{Z})$ gilt offenbar $h(N_H)=o$ genau dann, wenn $h(H)=o$. Die Gruppe $H^\perp$ umfasse alle Homomorphismen aus $\mathrm{Hom}_s(F_p,\mathbb{Z})$ die H annullieren. Der Raum X besitzt eine Überlagerung X_{N_H}, (die von $\tilde{X}$ überlagert wird) deren Decktransformationsgruppe gerade $F_p|N_H$ ist. Die Projektion von X_{N_H} auf X heiße u_{N_H}. Die Projektion von $\tilde{X}$ auf X_{N_H} sei mit v_{N_H} bezeichnet. Dann gilt $u_{N_H} \circ v_{N_H} = u$. Die Faser über $v_{N_H}(q)$ besteht gerade aus N_H. Nun sei $h \in (N_H)^\perp \cap \Pi^1(X)_p$. Jede Funktion $f \in C^q_p$ mit $f|F_p=h$ ist auf den Fasern über X_{N_H} konstant. Zu p und irgend welchen Punkten $p_i \in X$, wo $i=1,..,n$, wählen wir neben $q \in u^{-1}(p)$ je einen Punkt $p_i' \in u^{-1}(p_i)$. Weiter wählen wir um p sowie um p_i je eine abgeschlossene Umgebung (X ist regulär!) U_p resp. U_{p_i}, die bezüglich u von Blättern V^q_p resp. V_{p_i} überlagert werden, die auch Blätter bezüglich v_N sind und die q bzw. p_i' enthalten. Dabei sei N ein endlich erzeugter Normalteiler, für den alle $f \in C^q_p$ mit $f|F_p \in N^\perp$ auf

$U_p \cup \bigcup_{i=1}^{n} U_{p_i}$ durchfaktorisieren.

Mit
$$T^{N^\perp}_{p,p'_1,\dots,p'_n} \subset C^q_p$$

sei die Menge aller Funktionen gemeint, deren Restriktion auf jeder der Mengen $V^q_p, V_{p_1}, \dots, V_{p_n}$ verschwinden und deren Restriktionen auf F_p der Menge $N^\perp$ angehören.

Solche Funktionen gibt es: Sei $f \in C^q_p$ mit $f|F_p \in N^\perp$. In $C(X)$ gibt es, der Normalität von X wegen, eine Funktion g für die

$$u^*(g)\,|\,(V^q_p \cup \bigcup_{i=1}^{i=n} V_{p_i}) = -f\,|\,(V^q_p \cup \bigcup_{i=1}^{i=n} V_{p_i})$$

Also gehört $f + u^*(q)$ der obigen Menge $T^{N^\perp}_{p,p',\dots,p_n}$ an. Das System der Mengen dieser Form bildet eine Filterbasis $\mathcal{B}$: Seien $p_{1_i},\dots,p_{n_i}$, wo $i=1,\dots,m$ Punkte aus X. Weiter seien $N^\perp_1,\dots,N^\perp_m \in \mathcal{B}$. Offenbar gibt es einen endlich erzeugten Normalteiler $N \subset F_p$ mit $N^\perp \subset \bigcap_{i=1}^{m} N^\perp_i$ Wir bilden nun analog wie oben die Überlagerung X_N von X. Wiederum sind die Funktionen aus C^q_p, deren Restriktionen auf F_p den Normalteiler N annullieren, auf den Fasern über X_N konstant. N kann so groß gewählt werden, daß alle $f \in C^q_p$ mit $f \in N^\perp$ auf $U_p \cup \bigcup_{j,i} U_{p_{j_i}}$ durchfaktorisieren. Offenbar gilt

$$\bigcap_{i=1}^{m} T^{N_i^\perp}_{p,p_{1_i},\dots,p_{n_i}} \supset T^{N^\perp}_{p,p_{i_1},\dots,p_{n_m}}$$

Also ist $\mathcal{B}$ eine Filterbasis auf C^p_q. Der von dieser Basis in C^p_q erzeugte Filter heiße Φ. Er konvergiert in C^p_q nach 0: Denn seien $p_1 \in X$ und $q^1 \in U^{-1}(p_1)$ und σ ein Weg, der p_1 mit p verbindet. Wir heben diesen Weg zu einem Weg σ^1 in X mit Anfangspunkt q^1. Der Endpunkt q_E von σ^1 liegt in F_p. Der von q_E erzeugte Normalteiler in F_p heiße N. Es ist

$$T^{N^\perp}_{p,p_1} \in \mathcal{B}$$

Außerdem annullieren alle Funktionen in T^N_{p,p_1} das Blatt über p_1, das q^1 enthält. Folglich konvergiert Φ gegen $o \in C^p_q$. Nach Konstruktion konvergiert $b_p(\Phi)$ und es ist $b_p(\Phi) = \mathfrak{U}(o)$. Also ist

$$\bar{d}_p \circ \bar{b}_p^{-1} : \Pi^1_{c_1}(X) \longrightarrow \Pi^1_s(X)_p$$

bistetig. Daraus folgert man mühelos den Rest des Lemmas. Damit ist Lemma 15 bewiesen.

Wir bilden die Komposition aus

$$b : C(X,S^1) \longrightarrow \Pi^1_c(X)$$

und

$$\bar{d}_p \circ \bar{b}_p : \Pi^1_c(X) \longrightarrow \Pi^1_s(X)_p \subset \mathrm{Hom}_s(F_p,\mathbb{Z})$$

und erhalten

$$b^1 : C(X,S^1) \longrightarrow \Pi^1_s(X)_p \subset \mathrm{Hom}_s(F_p,\mathbb{Z}) .$$

Wir können Lemma 15 in folgendem festhalten:

<u>Corollar 16</u>

<u>Beide Abbildungen</u>

$$b^1 : C_c(X,S^1) \longrightarrow \Pi^1_s(X)_p \subset \mathrm{Hom}_s(F_p,\mathbb{Z})$$

<u>und</u>

$$b : C_c(X,S^1) \longrightarrow \Pi^1_c(X)$$

<u>sind Quotientenabbildungen.</u>

Im nächsten Paragraph soll die P_c-Reflexivität von $\Pi_s^1(X)_p$ (falls vollständig) nachgewiesen werden.

7) P_c - Reflexivität vollständiger Untergruppen von $C_s(Z,\mathbb{Z})$

Die limitierte Bruschlinski'sche Gruppe $\Pi_c^1(X)$ kann nach Lemma 15 als topologische Untergruppe von $Hom_s(F,\mathbb{Z})$ aufgefaßt werden. Um nun die P_c-Reflexivität dieser topologischen Gruppe nachzuweisen, studieren wir kurz die mit der Topologie der punktweisen Konvergenz versehene Gruppe $C_s(Z,\mathbb{Z})$ aller stetigen $\mathbb{Z}$-wertigen Funktionen eines topologischen Raumes Z. Die Gruppenoperationen sind punktweise definiert.

<u>Lemma 18</u> <u>Sei</u> Z <u>ein diskreter topologischer Raum</u>, $G \subset C_s(Z,\mathbb{Z})$ <u>eine abgeschlossene Untergruppe und</u> $h : G \longrightarrow S^1$ <u>ein stetiger Homomorphismus. Dann kann</u> h <u>zu einem stetigen Homomorphismus</u> $h^1 : C_s(Z,\mathbb{Z}) \longrightarrow S^1$ <u>derart erweitert werden, daß für ein</u> $g \in C(Z,\mathbb{Z})$ <u>das nicht</u> G <u>angehört</u> $h^1(g) \neq 1$ <u>gilt</u>.

<u>Beweis:</u> Sei $E \subset Z$ eine endliche Menge und $I(Z)$ das $\mathbb{Z}$-Ideal aller auf E verschwindenden Funktionen des Ringes $C(Z,\mathbb{Z})$. Die Operationen seien punktweise definiert. Das System

$$\{ I(E) \mid E \text{ endliche Teilmenge von } Z \}$$

ist eine Nullumgebungsbasis der topologischen Gruppe $C_s(Z,\mathbb{Z})$. Da letztere separiert ist und $G \subset C_s(Z,\mathbb{Z})$ abgeschlossen, gibt es eine endliche Menge $E^1 \subset Z$ mit $g+I(E^1) \cap G=\emptyset$. Zu jeder gleichmäßig überlagerten Umgebung V von $1 \in S^1$ gibt es eine endliche Menge $E \subset Z$ mit $h(I(E) \cap G) \subset V$ und $I(E) \subset I(E^1)$. Wenn nun V klein genug ist und $h(I(E) \cap G) \neq \{1\}$ angenommen wird, gibt es ein $g_1 \in I(E) \cap G$ und $n \in \mathbb{Z}$ mit $(h(g_1))^n \notin V$. Dem widerspricht aber $n \cdot g_1 \in I(E)$. Also faktorisiert h zu

$$\bar{h} : G/I(E) \cap G \longrightarrow S^1.$$

Der stetige Restriktionshomomorphismus $r_E : C_s(Z,\mathbb{Z}) \longrightarrow C_s(E,\mathbb{Z})$ induziert einen Isomorphismus

$$\bar{r}_E : C(Z,\mathbb{Z})/I(E) \longrightarrow C(E,\mathbb{Z}).$$

Da nun $C_s(E,\mathbb{Z})$ diskret ist, ist die von $r_E|G$ auf $G/I(E) \cap G$ induzierte Quotientenlimitierung diskret. Zu h gibt es einen Charakter h auf $\bar{r}_E(G/I(E) \cap G)$ derart, daß $\bar{h} = h \circ r_E$. Da $C_s(E,\mathbb{Z})$ diskret ist, läßt sich bekanntlich h zu $h_1 : C_s(E,Z) \longrightarrow S^1$ derart ausdehnen, daß $h_1(\bar{r}_E(q)) \neq 1$. Der Homomorphismus $h_1 \circ \bar{r}_E$ ist eine stetige Ausdehnung von h, die g annulliert.

Für einen diskreten Raum Z, ist $C_s(Z,\mathbb{Z})$ das $\mathbb{Z}$-fache kartesische Produkt von $\mathbb{Z}$. Für dieses gilt:

<u>Lemma 19</u> <u>Für jeden diskreten topologischen Raum</u> Z <u>ist</u> $C_s(Z,\mathbb{Z})$ P_c-<u>reflexiv</u>.

<u>Beweis:</u> Die Beweisführung ist eine reine Routinesache und soll hier deshalb nur skizziert werden. Sei $\mathcal{E}$ das System der endlichen Mengen von Z. Es ist $C_s(Z,\mathbb{Z})$ der projektive Limes $\varprojlim_{\mathcal{E}} C_s(E,\mathbb{Z})$ der diskreten Gruppen $C_s(E,\mathbb{Z})$. Daraus ergibt sich leicht, daß $\Gamma_c C_s(Z,\mathbb{Z})$ der induktive Limes $\varprojlim_{\mathcal{E}} \Gamma_c C_s(E,\mathbb{Z})$ der kompakten Gruppe $\Gamma_c C_s(E,\mathbb{Z})$ ist. Aus der P_c-Reflexivität von $C_s(E,\mathbb{Z})$ ergibt sich dann die P_c-Peflexivität von $C_s(Z,\mathbb{Z})$.

<u>Proposition 20</u> <u>Jede vollständige Untergruppe</u> $G \subset C_s(Z,\mathbb{Z})$ <u>ist für jeden topologischen Raum</u> Z P_c <u>-reflexiv</u>.

<u>Beweis</u>: Wir nehmen erst an, daß Z diskret ist. Die Inklusion $v : G \longrightarrow C_s(Z,\mathbb{Z})$ induziert die stetige Surjektion

$$^*v : \Gamma_c C_s(Z,\mathbb{Z}) \longrightarrow \Gamma_c G.$$

(Lemma 18), von der man unter Verwendung der Darstellung von $\Gamma_c C_s(Z,\mathbb{Z})$ als $\varinjlim_{\mathcal{E}} \Gamma_c C_s(E,\mathbb{Z})$ mühelos nachweist, daß sie eine Quotientenabbildung ist. Folglich ist der Monomorphismus

$$^{**}v : \Gamma_c \Gamma_c G \longrightarrow \Gamma_c \Gamma_c C_s(Z,\mathbb{Z})$$

ein Homöomorphismus auf einen Teilraum. Also ist $^{***}v$ der P_c -Reflexivität von $C_s(Z,\mathbb{Z})$ wegen surjektiv (Lemma 18). Aus der Kommutativität von

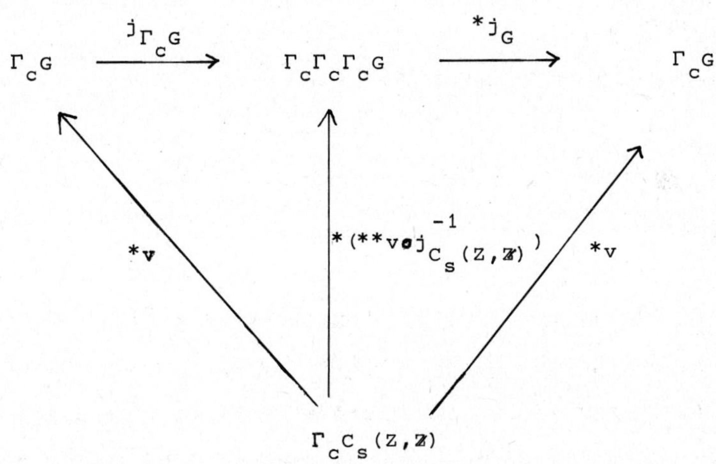

ergibt sich $^*j_G \circ j_{\Gamma_c G} = id_{\Gamma_c G}$, also insbesondere, daß $j_{\Gamma_c G}$ ein Homöomorphismus auf sein Bild ist. Weil $^{***}v$ eine Surjektion ist, muß $^*(^{**}v \circ j^{-1}_{C_s(Z,\mathbb{Z})})$ surjektiv sein. Aus der Surjektivität von

$*(**v \circ j^{-1}_{C_s(Z,\mathbb{Z})})$ folgert man, daß $j_{\Gamma_c G}$ ein bistetiger Isomorphismus ist. Nun verwendet man wiederum Lemma 18 und zeigt leicht die P_c-Reflexivität von G. Ist nun Z irgend ein topologischer Raum, so ist G eine abgeschlossene topologische Untergruppe von $C_s(Z_D,\mathbb{Z})$. Dabei bedeutet Z_D den zu Z assoziierten diskreten topologischen Raum. Daraus ergibt sich unmittelbar die Aussage in der obigen Proposition.

Damit folgern wir aus Proposition 20 und Lemma 15 endlich:

<u>Satz 21</u> $Hom_s(F_p,\mathbb{Z})$ <u>ist</u> P_c-<u>reflexiv</u>. <u>Die topologische Gruppe</u> $\Pi^1_c(X)$ <u>ist dann</u> (<u>und nur dann</u>) P_c-<u>reflexiv, wenn sie vollständig ist</u>.

Die folgenden beiden Sätze geben Bedingungen für die Kompaktheit von $\Gamma_c\Pi^1_c(X)$ an.

<u>Satz 22</u> $Hom_s(F_p,\mathbb{Z})$ <u>ist diskret, wenn</u> F_p <u>endlich erzeugt ist</u>. <u>Wenn</u> F_p <u>endlich erzeugt ist, dann sind</u> $\Gamma_c Hom_s(F_p,\mathbb{Z})$ <u>und</u> $\Gamma_c\Pi^1_c(X)$ <u>kompakt</u>.

<u>Beweis:</u> Mit Hilfe der in (16) dargestellten Umgebungsbasis folgert man sofort den ersten Teil der Aussage des obigen Satzes. Der zweite Teil ist eine Konsequenz des ersten zusammen mit der Tatsache, daß die Limitierung der stetigen Konvergenz auf $\Gamma_c Hom_s(F_p,\mathbb{Z})$ und $\Gamma_c\Pi^1_c(X)$ die Topologie der kompakten Konvergenz ist. Für die letztere sind aber diese Aussagen richtig [Po].

<u>Satz 23</u> <u>Für jeden kompakten topologischen Raum</u> X (der unsere generellen Voraussetzungen nicht zu erfüllen braucht) <u>ist</u> $\Pi^1_c(X)$ <u>diskret und damit</u> $\Gamma_c\Pi^1_c(X)$ <u>kompakt</u>.

<u>Beweis:</u> Sei Φ irgend ein in $C_c(X,S^1)$ gegen das Einselelement konvergenter Filter. Zu jedem Punkt $p \in X$ gibt es zu einer vorgegebenen gleichmäßig überlagerten Umgebung V von $1 \in S^1$ eine Umgebung $U(p)$ und ein $F \in \Phi$ mit

$$F(U(p)) \subset V.$$

Endlich viele Mengen aus $\{U(p) \mid p \in X\}$ überdecken X. Also gibt es ein $F_1 \in \Phi$ mit

$$F_1(X) \subset V.$$

Daraus aber folgt $F_1 \subset \varkappa_X C_c(X)$. Mithin hat Φ eine Basis in $\varkappa_X C_c(X)$. Daraus schließt man, daß $C_c(X,S^1) \mid \varkappa_X C(X)$ diskret ist.

8) Zur P_c - Reflexivität von $C_c(X,S^1)$

Sei $\Pi_c^1(X)$ vollständig. Die Folge

$$0 \longrightarrow \varkappa_X C_c(X) \xrightarrow{i} C_c(X,S^1) \xrightarrow{b} \Pi_c^1(X) \longrightarrow 0$$

ist exakt. Lemma 3 besagt, daß i ein Homomorphismus auf das Bild ist, während b nach Konstruktion eine Quotientenabbildung ist.

Nun dualisieren wir diese Folge und erhalten:

$$17) \quad 0 \longrightarrow \Gamma_c \Pi_c^1(X) \xrightarrow{*b} \Gamma_c C_c(X,S^1) \xrightarrow{*i} \Gamma_c \varkappa_X C_c(X) \longrightarrow 0$$

Zum Nachweis der Exaktheit dieser Folge verwendet man einerseits, daß b eine Quotientenabbildung ist und andererseits, daß nach Korollar 13, der Homomorphismus $*i$ eine stetige Surjektion ist. Durch nochmaliges Dualisieren erhalten wir dann:

$$0 \longrightarrow \Gamma_c\Gamma_c\varkappa_X C_c(X) \xrightarrow{**i} \Gamma_c\Gamma_c C_c(X,S^1) \xrightarrow{**b} \Gamma_c\Gamma_c\Pi^1_c(X) \longrightarrow 0 .$$

Diese Folge ist, wie man unter Verwendung von Korollar 13, der Kommutativität von

$$\begin{array}{ccc}
\Gamma_c\Gamma_c C_c(X,S^1) & \xrightarrow{**b} & \Gamma_c\Gamma_c\Pi^1_c(X) \\
\Big\uparrow {\scriptstyle j_{C_c(X,S^1)}} & & \Big\uparrow {\scriptstyle j_{\Pi^1_c(X)}} \\
C_c(X,S^1) & \xrightarrow{b} & \Pi^1_c(X)
\end{array}$$

und der P_c - Reflexivität von $\Pi^1_c(X)$ (Satz 21) schließt, exakt. Also ist

$$\begin{array}{ccccccccc}
0 & \longrightarrow & \Gamma_c\Gamma_c\varkappa_X C_c(X) & \xrightarrow{**i} & \Gamma_c\Gamma_c C_c(X,S^1) & \xrightarrow{**b} & \Gamma_c\Gamma_c\Pi^1_c(X) & \longrightarrow & 0 \\
 & & \Big\uparrow {\scriptstyle j_{\varkappa_X C_c(X)}} & & \Big\uparrow {\scriptstyle j_{C_c(X,S^1)}} & & \Big\uparrow {\scriptstyle j_{\Pi^1_c(X)}} & & \\
0 & \longrightarrow & \varkappa_X C_c(X) & \xrightarrow{i} & C_c(X,S^1) & \xrightarrow{b} & \Pi^1_c(X) & \longrightarrow & 0
\end{array}$$

ein kommutatives Diagramm exakter Folgen. Daraus folgt unter Benützung von Satz 9 und des Fünferlemmas, daß $j_{C_c(X,S^1)}$ ein Isomorphismus ist. Die Abbildung

$$j_X : X \longrightarrow \Gamma_c C_c(X,S^1),$$

definiert durch $j_X(p)(t) = t(p)$ für alle $p \in X$ und alle $t \in C_c(X,S^1)$, ist stetig und induziert einen stetigen Homomorphismus

$${}^*j_X : \Gamma_c\Gamma_c C_c(X,S^1) \longrightarrow C_c(X,S^1),$$

der jeden Charakter $\gamma \in \Gamma_c\Gamma_c C_c(X,S^1)$ nach $\gamma \circ j_X$ abbildet. Weil nun $^*j_X \circ j_{C_c(X,S^1)} = id$ ist, haben wir endlich

<u>Satz 24 Wenn</u> $\Pi_c^1(X)$ <u>vollständig ist</u> (<u>also etwa wenn</u> X <u>kompakt ist oder</u> $H_1(X,\mathbb{Z})$ <u>endlich erzeugt ist oder</u> $\Pi^1(X) = Hom(\Pi_1(X),\mathbb{Z})$, <u>was z.B. dann gilt</u>, <u>wenn</u> X <u>ein CW-Komplex ist</u>, s.§9), <u>dann ist</u> $C_c(X,S^1)$ P_c-<u>reflexiv</u>.

9) Zur Charaktergruppe von $C_c(X,S^1)$

Die Folge (7) besagt, daß $\Gamma_c C_c(X,S^1)$ eine (uns unbekannte) Erweiterung von $\Gamma_c\Pi_c^1(X)$ und $\Gamma_c\varkappa_X C_c(X)$ ist. Die Gruppe $\Gamma_c\varkappa_X C_c(X)$ ist durch Lemma 8 bestimmt. Wir werden deshalb $\Gamma_c\Pi_c^1(X)$ unter einschränkenden Voraussetzungen noch etwas genauer untersuchen. Dazu studieren wir erst $\Gamma_c Hom_s(F_p,\mathbb{Z})$. Die Gruppe F_p modulo der Kommutatorengruppe ist isomorph der ersten singulären Homologiegruppe $H_1(X,\mathbb{Z})$ mit Koeffizienten in $\mathbb{Z}$. Offenbar sind damit $Hom_s(F_p,\mathbb{Z})$ und $Hom_s(H_1(X,\mathbb{Z}),\mathbb{Z})$ bistetig isomorph.

Wir beginnen mit der Vorbereitung einer spezifischen Beschreibung von $\Gamma_c Hom_s(H_1(X,\mathbb{Z}),\mathbb{Z})$. Dabei setzen wir voraus, daß $H_1(X,\mathbb{Z})$ <u>endlich erzeugt sei</u>. Demnach gibt es also einen Homomorphismus ρ von einer freien, endlich erzeugten, kommutativen Gruppe H auf $H_1(X,\mathbb{Z})$. Die Basis von H heiße M.

Wenden wir uns nun $\Gamma_c\ Hom_s(H,\mathbb{Z})$ zu. Es ist $\mathbb{Z} \cong \Gamma(S^1,S^1)$. Mithin ist, da die Topologie der punktweisen Konvergenz auf $Hom(H,\mathbb{Z})$ mit der Limitierung der stetigen Konvergenz identisch ist, $\Gamma_c\ Hom_s(H,\mathbb{Z})$ bistetig isomorph $\Gamma_c\ B_c(H \times S^1,S^1)$. (Siehe [Bi,Ke]). Unter $B_c(H \times S^1,S^1)$ verstehen wir die Gruppe $B(H \times S^1,S^1)$ der stetigen S^1-wertigen Bihomomorphismen von $H \times S^1$ versehen mit der Limitierung der stetigen Konvergenz. Dabei trägt $H \times S^1$ die Produkttopologie. Das Tensorprodukt $H \otimes S^1$ kann in natürlicher Weise mit $\bigoplus_{\alpha \in M} S^1_\alpha$ identifiziert werden. Für $\alpha \in M$ bezeichnet S^1_α die Gruppe $\mathbb{Z} \cdot \alpha \otimes S^1$. Diese ist natürlich isomorph S^1. Die Gruppe $\bigoplus_{\alpha \in M} S^1_\alpha$ versehen wir nun mit der gröbsten Topologie für die die Projektion auf S^1_α für jedes $\alpha \in M$ stetig ist. Weil M endlich ist, wird $\bigoplus S^1_\alpha$ und damit $H \otimes S^1$ eine kompakte topologische Gruppe. Wir überlassen es dem Leser zu verifizieren, daß der natürliche Bihomomorphismus ξ von $H \times S^1$ nach $H \otimes S^1$ einen Isomorphismus

$$*\xi : \Gamma(H \otimes S^1) \longrightarrow B(H \times S^1, S^1).$$

definiert durch $*\xi(\gamma) = \xi \circ \gamma$ für jedes $\gamma \in \Gamma_c(H \otimes S^1)$, induziert. Dieser Homomorphismus ist, wie man aus allgemeinen Resultaten aus [Bi,Ke] ersehen kann, bistetig, falls beide Gruppen mit der Limitierung der stetigen Konvergenz versehen sind. Mithin ist

$$**\xi : \Gamma_c B_c(H \times S^1, S^1) \longrightarrow \Gamma_c \Gamma_c(H \otimes S^1)$$

ebenfalls ein bistetiger Isomorphismus. Nach der klassischen Pontrjaginschen Dualitätstheorie ist $H \otimes S^1$ bistetig isomorph zu $\Gamma_c \Gamma_c(H \otimes S^1)$ und damit auch zu $\Gamma_c B_c(H \times S^1, S^1)$.
Der Homomorphismus

$$\rho : H \longrightarrow H_1(X, \mathbb{Z})$$

induziert eine surjektive Abbildung

$$\rho \otimes id : H \otimes S^1 \longrightarrow H_1(X, Z) \otimes S^1.$$

Wir versehen $H_1(X, \mathbb{Z}) \otimes S^1$ mit der Quotiententopologie definiert durch $\rho \otimes id$. Dadurch wird $H_1(X, \mathbb{Z}) \otimes S^1$ zu einer kompakten topologischen Gruppe. Es ist

$$*(\rho \otimes id) : \Gamma_c(H_1(X, \mathbb{Z}) \otimes S^1) \longrightarrow \Gamma_c(H \otimes S^1)$$

ein Monomorphismus zwischen diskreten Gruppen. Folglich ist, wiederum nach der klassischen Pontrjaginschen Dualitätstheorie

$$^{**}(\rho \times \mathrm{id}) : \Gamma_c\Gamma_c(H \otimes S^1) \longrightarrow \Gamma_c\Gamma_c(H_1(X,\mathbb{Z}) \otimes S^1)$$

eine stetige Surjektion, ja sogar eine Quotientenabbildung (da $H_1(X,\mathbb{Z}) \otimes S^1$ kompakt ist). Daraus folgt, daß

$$j_{H_1(X,Z) \otimes S^1} : H_1(X,\mathbb{Z}) \otimes S^1 \longrightarrow \Gamma_c\Gamma_c(H_1(X,\mathbb{Z}) \otimes S^1)$$

ein bistetiger Isomorphismus ist.

Offenbar sind $\mathrm{Hom}_s(H_1(X,\mathbb{Z}),\mathbb{Z})$, $B_c(H_1(X,\mathbb{Z}) \times S^1,S^1)$ und $\Gamma_c(H_1(X,\mathbb{Z}) \otimes S^1)$ bistetig isomorph. Folglich sind auch $H_1(X,\mathbb{Z}) \otimes S^1$ und $\Gamma_c(\mathrm{Hom}_s(H_1(X,\mathbb{Z}),\mathbb{Z})$ bistetig isomorph. Zusammengefaßt haben wir mithin:

Lemma 24 <u>Wenn $H_1(X,\mathbb{Z})$ endlich erzeugt ist, dann ist $H_1(X,\mathbb{Z}) \otimes S^1$ kompakt und damit P_c-reflexiv. Überdies sind $\Gamma_c \mathrm{Hom}_s(H_1(X,\mathbb{Z}),\mathbb{Z})$ und $H_1(X,\mathbb{Z}) \otimes S^1$ bistetig isomorph.</u>

Wenn X ein CW-Komplex ist, so ist nach ([Sp] p.427) die Gruppe $\Pi^1(X)$ isomorph zu $\mathrm{Hom}(\Pi_1(X,p),\mathbb{Z})$ und es gilt $H^1(X,\mathbb{Z}) = \mathrm{Hom}(H_1(X,\mathbb{Z})\mathbb{Z}) =$
$= \mathrm{Hom}(\Pi_1(X,p),\mathbb{Z})$. Damit ergibt sich aus Lemma 24:

Satz 25 <u>Wenn der Raum X zusätzlich ein CW-Komplex ist und die Gruppe $H_1(X,\mathbb{Z})$ endlich erzeugt ist, dann sind einerseits $\Pi^1_c(X)$, $\Gamma_c(H_1(X,\mathbb{Z}) \otimes S^1)$ und $\Gamma_c \mathrm{Hom}_s(H_1(X,\mathbb{Z}),\mathbb{Z})$ und andererseits $\Gamma_c\Pi^1_c(X)$ und $H_1(X,\mathbb{Z}) \otimes S^1$ bistetig isomorph.</u>

LITERATURVERZEICHNIS

[Bi] E.Binz: "Continuous convergence on C(X)". Lecture Notes in Mathematics 1975, Springer-Verlag Berlin, Heidelberg, New York.

[Bi,Ke] E.Binz und H.H.Keller: "Funktionenräume in der Kategorie der Limesräume". Ann.Acad.Sci.Fenn.Ser. A I.383 (1966) 1-21.

[Bu] H.P.Butzmann: "Dualitäten in $C_c(X)$". Dissertation, Universität Mannheim (1971).

———:"Über die c-Reflexivität von $C_c(X)$". Comment. Math.Helv. 47 (1972), 92-101.

[Hu] S.T.Hu: "Homotopy Theory". Academic Press, New York and London (1959).

[Ku] K.Kutzler: "Eine Bemerkung über endlichdimendionale, separierte, limitierte Vektorräume". Arch.Math.XX, Fasc.2 (1969), 165-168.

[Po] L.S.Pontrjagin: "Topological Groups". Gordon and Breach, Science Publishers Inc. New York, London, Paris (1966).

[Sp] E.H.Spanier: "Algebraic Topology". Mc Graw-Hill Book Company, New York (1966).

SOME CARTESIAN CLOSED TOPOLOGICAL CATEGORIES OF CONVERGENCE SPACES

by Gérard BOURDAUD

1. CATEGORICAL FOREWORDS.

Let $\underline{C}$ be a cartesian closed topological category (8) ("catégorie complète fermée" in the sense of (1))(*). We denote $|X|$ the underlying set of an object X of $\underline{C}$ and Hom the internal hom-functor of $\underline{C}$.

Définition 1.1: A closed topological subcategory (CTSC) of $\underline{C}$ is a full subcategory $\underline{C}'$ of $\underline{C}$ which satisfies:

(CTSC I) If X and Y are objects of $\underline{C}'$, then Hom(X,Y) is an object of $\underline{C}'$.

(CTSC II) If X is the initial structure defined by a family of mappings $f_i: |X| \longrightarrow |X_i|$ and objects X_i of $\underline{C}'$, then X is an object of $\underline{C}'$.

The intersection of any family of CTSC of $\underline{C}$ is also a CTSC:

Definition 1.2: If $\underline{D}$ is any subcategory of $\underline{C}$, let $\hat{\underline{D}}$ be the intersection of all the CTSC of $\underline{C}$ which include $\underline{D}$. $\hat{\underline{D}}$ is called the CTSC of $\underline{C}$ spanned by $\underline{D}$.

(*) Despite it is now a classical terminology, we think that "topological category" is far from being satisfactory. It is rather ambiguous when we speak about categories of "quasi"-topologies which are...topological. Moreover, the term "topological category" was already used by EHRESMANN (7) to denote an internal category of Top .

L being a fixed object of $\underline{C}$ and X any object of $\underline{C}$ we have a canonical morphism $a_X : X \longrightarrow \mathrm{Hom}(\mathrm{Hom}(X,L),L)$, which maps x to : $f \longrightarrow f(x)$; let us denote 1X the initial structure defined on $|X|$ by this morphism. X is usually finer than 1X; let $\underline{C}(L)$ be the full subcategory of $\underline{C}$ whose objects X satisfy to X = 1X .

<u>Proposition 1.3</u>: $\underline{C}(L)$ is a CTSC of $\underline{C}$ which contains L; moreover, if X is any object of $\underline{C}$:

(i) Hom(X,Y) is an object of $\underline{C}(L)$ whenever Y is ,

(ii) 1X is a reflexion of X in $\underline{C}(L)$,

(iii) for all $Y \in \underline{C}(L)$, Hom(1X,Y) = Hom(X,Y) .

Proof: 1) Let $Y \in \underline{C}$ be the initial structure defined by mappings $f_i : |Y| \longrightarrow |Y_i|$ and objects Y_i of $\underline{C}(L)$. We have a commutative diagram:

$$\begin{array}{ccc} Y_i & \xrightarrow{\;a_i\;} & \mathrm{Hom}(\mathrm{Hom}(Y_i,L),L) \\ {\scriptstyle f_i}\uparrow & & \uparrow{\scriptstyle \tilde{f}_i} \\ Y & \xrightarrow{\;a\;} & \mathrm{Hom}(\mathrm{Hom}(Y,L),L) \end{array}$$

where $\tilde{f}_i$ is the morphism $\mathrm{Hom}(\mathrm{Hom}(f_i,L),L)$. If Z is any object of $\underline{C}$ and f a mapping : $|Z| \longrightarrow |Y|$ such that a o f is a morphism from Z to Hom(Hom(Y,L),L) , then

$$a_i \circ f_i \circ f = \tilde{f}_i \circ a \circ f$$

is a morphism from Z to $\mathrm{Hom}(\mathrm{Hom}(Y_i,L),L)$. But $Y_i \in \underline{C}(L)$ implies that $f_i \circ f$ is a morphism from Z to Y_i . Finally, f is a morphism: $Z \longrightarrow Y$. So $\underline{C}(L)$ satisfies to CTSC II .

2) Let $Y \in \underline{C}(L)$ and $X \in \underline{C}$, U = Hom(X,Y) . We take $Z \in \underline{C}$ and f any mapping from $|Z|$ to $|U|$, such that $a_U \circ f$ is a morphism from Z to Hom(Hom(U,L),L) . In order to prove that f is a morphism: $Z \longrightarrow U$,

it suffices to have a morphism:

$$\tilde{f} : X \times Z \longrightarrow Y$$
$$(x, z) \longrightarrow f_z(x) .$$

If $(x,g) \in |X| \times |\mathrm{Hom}(Y,L)|$, we have a morphism:

$$[x,g] : U \longrightarrow L$$
$$h \longrightarrow (g \circ h)(x) , \quad \text{hence :}$$
$$u : \mathrm{Hom}(\mathrm{Hom}(U,L),L) \longrightarrow \mathrm{Hom}(X \times \mathrm{Hom}(Y,L),L)$$
$$k \longrightarrow \left[(x,g) \longrightarrow k([x,g])\right]$$

If we compose u with $a_U \circ f$, we obtain a morphism:

$$Z \longrightarrow \mathrm{Hom}(X \times \mathrm{Hom}(Y,L),L)$$
$$z \longrightarrow \left[(x,g) \longrightarrow g(f_z(x))\right]$$

thus by adjunction: $X \times Z \longrightarrow \mathrm{Hom}(\mathrm{Hom}(Y,L),L)$

$$(x, z) \longrightarrow \left[g \longrightarrow g(f_z(x))\right]$$

which is precisely $a_Y \circ \tilde{f}$. Y being an object of $\underline{C}(L)$, $\tilde{f}$ is a morphism from $X \times Z$ to Y , q.e.d.

3) We prove now that $L \in \underline{C}(L)$. Let f be a mapping: $|Z| \longrightarrow |L|$ such that $a_L \circ f$ is a morphism : $Z \longrightarrow \mathrm{Hom}(\mathrm{Hom}(L,L),L)$; we have morphisms

$$Ev_1 : \mathrm{Hom}(\mathrm{Hom}(L,L),L) \longrightarrow L$$
$$g \longrightarrow g(id_L) ,$$

$$Ev_1 \circ a \circ f : Z \longrightarrow L$$
$$z \longrightarrow Ev_1(a(f(z))) = f(z) \qquad \text{q.e.d.}$$

4) From (1),(2),(3) we deduce that $1X \in \underline{C}(L)$; on the other hand, X is clearly finer than 1X . Let f be a morphism from X to $Y \in \underline{C}(L)$. If we compose $a_X : 1X \longrightarrow \mathrm{Hom}(\mathrm{Hom}(X,L),L)$ and

$$\mathrm{Hom}(\mathrm{Hom}(f,L),L) : \mathrm{Hom}(\mathrm{Hom}(X,L),L) \longrightarrow \mathrm{Hom}(\mathrm{Hom}(Y,L),L) ,$$

we obtain $a_Y \circ f : 1X \longrightarrow \mathrm{Hom}(\mathrm{Hom}(Y,L),L)$. Y being an object of $\underline{C}(L)$, it follows that f is a morphism from 1X to Y .

5) By adjunction, Hom(1X,Y) and Hom(X,Y) have the same underlying set; moreover, Hom(1X,Y) is finer than Hom(X,Y) . To prove the reverse, we have only to see that the evaluation map is a morphism:

$$\mathrm{Hom}(X,Y) \times 1X \longrightarrow Y \ ;$$

Ev is a morphism : $\mathrm{Hom}(X,Y) \times X \longrightarrow Y$, and so , functorially :

$$1\left[\mathrm{Hom}(X,Y) \times X\right] \longrightarrow 1Y \ .$$

$\underline{C}(L)$ being a CTSC , 1 is compatible with finite products , which implies : $1\left[\mathrm{Hom}(X,Y) \times X\right] = 1\mathrm{Hom}(X,Y) \times 1X$

$$= \mathrm{Hom}(X,Y) \times 1X \ ,$$

Hom(X,Y) being an object of $\underline{C}(L)$ by (2); for the same reason Y = 1Y and we obtain the required morphism .

2.CATEGORIES OF CONVERGENCE SPACES.

Let <u>QTop</u> be the cartesian closed topological category of <u>convergence spaces</u> and continuous mappings (N.B. Convergence spaces are usually called "espaces quasi-topologiques" by French authors). In <u>QTop</u> the structure of Hom(X,Y) is called the <u>continuous convergence</u>. We denote:

- Ω the topological space on $|\Omega| = \{0,1\}$ whose open sets are : $\emptyset$, $\{1\}$, $|\Omega|$.

- Λ the pretopological space on $|\Lambda| = \{0,1,2\}$ whose neighborhood filters are : $\underline{V}(0) = \{|\Lambda|\}$, $\underline{V}(1) = \{|\Lambda|\}$, $\underline{V}(2) = \{|\Lambda|, \{1,2\}\}$.

- R the usual space of real numbers .

Let <u>PTop</u> , <u>STop</u> , <u>Top</u> , <u>UTop</u> denote the full subcategories of <u>QTop</u> whose objects are , respectively :

- pseudo-topological (or Choquet) spaces,

- pretopological spaces (called "espaces semi-topologiques" in (4),(10),(3)),

- topological spaces,

- uniformizable topological spaces .

PTop is a CTSC of QTop , but not the three others ; the four subcategories are reflexive subcategories of QTop . If $X \in$ QTop , we denote sX , tX , ωX the reflexions of X in STop , Top , UTop resp . ωX (resp sX , tX) is the initial structure on |X| defined by all the continuous mappings from X to R (resp. Λ, Ω).

If F is a filter on |X|, we write: $F \xrightarrow[X]{} x$ when F converges to x in the space X . We define $\mathrm{Conv}_X F = \{x \in |X| / F \xrightarrow[X]{} x\}$.

Definition 2.1: Let K be any functor from QTop to Top , compatible with the forgetful functors . We say that $X \in$ QTop is K-closed-domained if, for each filter F on |X| , $\mathrm{Conv}_X F$ is KX-closed . A t-closed-domained space is simply called closed-domained .

Proposition 2.2: The subcategory of QTop whose objects are K-closed-domained spaces is a CTSC .

Proof: Let Z = Hom(X,Y) , where Y is a K-closed-domained space ; for $x \in |X|$, $ev_x : Z \longrightarrow Y$ is a continuous mapping , so is :

$$ev_x : KZ \longrightarrow KY \ .$$

By the definition of continuous convergence

$$\mathrm{Conv}_Z F = \bigcap_{\substack{x \in |X| \\ G \to x}} ev_x^{-1}(\mathrm{Conv}_Y \ Ev(F \ \ G)) \ .$$

$\mathrm{Conv}_Y Ev(F \times G)$ being KY-closed , each $ev_x^{-1}(\mathrm{Conv}_Y Ev(F \times G))$ is KZ-closed .

Let Y be the initial convergence space defined by mappings $f_i : |Y| \longrightarrow |Y_i|$ and K-closed-domained spaces Y_i .

$$\mathrm{Conv}_Y F = \bigcap_{i\in I} f_i^{-1}(\mathrm{Conv}_{Y_i} f_i F) .$$

f_i being continuous from KY to KY_i, $f_i^{-1}(\mathrm{Conv}_{Y_i} f_i F)$ is a KY-closed set, so is $\mathrm{Conv}_Y F$.

Remark: The subcategory of closed-domained (resp. ω-closed-domained) spaces is a CTSC which includes Top (resp. UTop) .

Definition 2.3: Let K be any functor from QTop to STop, compatible with the forgetful functors . We say that $X \in$ QTop is a K-regular space if , for each filter F on $|X|$ and $x \in |X|$, $\mathrm{Cl}_{KX} F \underset{X}{\rightarrow} x$ whenever $F \underset{X}{\rightarrow} x$. A s-regular space is simply called regular .

Here $\mathrm{Cl}_{KX} F$ denotes the closure of F in the pretopological space KX (Recall that STop is isomorphic with the category of closure spaces . The present notion of K-regularity is somewhat different from the K-regularity of COCHRAN and TRAIL (6). Besides s, t, ω, we define two functors from QTop to STop in the following manner :

Definition 2.4: Let X be a convergence space . We define the topological space $X^{\bullet}$ and pretopological space X^{*} by their closure operators :

- $\mathrm{Cl}_{X^*} A = \{ y \in |X| \,/\, \exists x \in A \cap \mathrm{Cl}_{sX} y\}$,
- $\mathrm{Cl}_{X^{\bullet}} A = \{ y \in |X| \,/\, \exists x \in A \cap \mathrm{Cl}_{tX} y\}$

for any $A \subset |X|$. The correspondances $X \rightarrow X^{*}$, $X \rightarrow X^{\bullet}$ define two functors from QTop to STop , called, respectively, the star-functor and the point-functor .

For details on X^* and $X^\bullet$, see (3),(4) (where "étoile-stable" stands for "star-regular").

<u>Proposition 2.5</u>: The subcategory of <u>QTop</u> whose objects are K-regular spaces is a CTSC .

Proof: Let $Z = Hom(X,Y)$, where Y is K-regular . First, we prove that, for all $A \subset |Z|$, $B \subset |Y|$,

(I) $Ev(Cl_{KZ}A \times B) \subset Cl_{KY}Ev(A \times B)$.

By the continuity of $ev_x : KZ \longrightarrow KY$, we have , for all $x \in |X|$,

$$ev_x(Cl_{KZ}A) \subset Cl_{KY}ev_x(A) .$$

Hence, for $g \in Cl_{KZ}A$, $x \in B$, $g(x) = ev_x(g) \in Cl_{KY}ev_x(A) \subset Cl_{KY}Ev(A \times B)$.

From (I) we deduce, for all filters F on $|Z|$, G on $|X|$:

(II) $Ev(\ Cl_{KZ}F \times G) \supset Cl_{KZ}Ev(F \times G)$, therefore :

$F \xrightarrow[Z]{} f \implies G \xrightarrow[X]{} x \;\; Ev(F \times G) \xrightarrow[Y]{} f(x)$

$\implies G \xrightarrow[X]{} x \;\; Cl_{KY}Ev(F \times G) \xrightarrow[Y]{} f(x)$ (since Y is K-regular)

$\implies G \xrightarrow[X]{} x \;\; Ev(Cl_{KZ}F \times G) \xrightarrow[Y]{} f(x)$ (from (II))

$\implies Cl_{KZ}F \xrightarrow[Z]{} f$ q.e.d.

Let Y the initial space defined by a family of mappings $f_i : |Y| \longrightarrow |Y_i|$ and K-regular spaces Y_i. By the continuity of $f_i : KY \longrightarrow KY_i$, we have, for all $A \subset |Y|$, $f_i(Cl_{KY}A) \subset Cl_{KY_i}(f_iA)$, so, for a filter F on $|Y|$,

$$f_i(Cl_{KY}F) \supset Cl_{KY_i}(f_iF) \quad \text{then :}$$

$F \xrightarrow[Y]{} x \implies \forall i \;\; f_iF \xrightarrow[Y_i]{} f_i(x)$

$\implies \forall i \;\; Cl_{KY_i}(f_iF) \xrightarrow[Y_i]{} f_i(x)$ (Y_i being K-regular)

$\implies \forall i \;\; f_i(Cl_{KY}F) \xrightarrow[Y_i]{} f_i(x) \implies Cl_{KY}F \xrightarrow[Y]{} x$ q.e.d.

3.CLOSED TOPOLOGICAL CATEGORIES SPANNED BY Top AND STop.

If C is any CTSC of QTop , two types of characterization can be expected for C :

- a categorical one , e.g. C is the CTSC spanned by a well-known subcategory of QTop ,

- an internal one : the objects of C are exactly those which satisfy to certain convergence properties.

By prop.1.3 , QTop(Λ) and QTop(Ω) are CTSC . The following theorems give the two expected characterizations for these CTSC.

Lemma 3.1: For all X,Y objects of QTop , there exists a family of topological spaces X_i and mappings g_i: Hom(X,Y) $\longrightarrow$ Hom(X_i,Y) , such that Hom(X,Y) is the initial space defined by this family .

For the proof see (10) (proof of prop.1.8) .

Theorem 3.2: QTop(Λ) is the CTSC spanned by STop . On the other hand, it is exactly the category of pseudo-topological spaces.

Proof: We remarked that each X $\in$ STop is the initial space defined by all the continuous mappings from X to Λ ; it follows that $\widehat{\text{STop}} \subset$ QTop(Λ) . Lemma 3.1 implies that, for any Y $\in$ QTop , Hom(Y,Λ) is the initial space defined by Hom(X_i,Λ) , where X_i (and $\bar{\Lambda}$!) are pretopological spaces; hence QTop(Λ) $\subset \widehat{\text{STop}}$.

PTop being a CTSC which includes STop , we have QTop(Λ) $\subset$ PTop ; for the reverse inclusion, see (4) (Theorem II.4.1).

Theorem 3.3: (ANTOINE (1), MACHADO (10), BOURDAUD (4)) QTop(Ω) is the CTSC spanned by Top . The objects of QTop(Ω) are called Antoine spaces . A space X is an Antoine space iff X is pseudo-topological, star-regular and closed-domained .

Proof: Again by lemma 3.1, we have QTop(Ω) = $\widehat{\text{Top}}$; topological spaces are pseudo-topological, star-regular, closed-domained : the same is true for objects of $\widehat{\text{Top}}$, by prop. 2.2, 2.5 . For the reverse inclusion, see (4) (Theorem I.4.4).

4. THE CATEGORY OF c-SPACES.

We give now a double characterization of QTop(R). If we restrict ourselves to Hausdorff spaces, an object of QTop(R) is a c-embedded space of BINZ or a c-space (9). Here we call a c-space any object of QTop(R) (even non Hausdorff ones). First, we need an analoguous to lemma 3.1 :

Lemma 4.1: If X is any convergence space and Y a T_1 convergence space, there exists a family of uniformizable topological spaces X_i and mappings g_i: Hom(X,Y) $\longrightarrow$ Hom(X_i,Y) such that Hom(X,Y) is the initial space defined by this family.

Proof: By lemma 3.1 and transitivity of initial structures, we can suppose that X is topological. Let x be a fixed point of X. The sets $(V \times V) \cup \Delta$, where V is a X-neighborhood of x and Δ the diagonal of $|X| \times |X|$, form an entourage bases on $|X|$. Let X(x) denote the uniformizable space induced by this uniform structure.

$X(x)$ has the following neighborhood filters:

- $\underline{V}_{X(x)}(y) = \underline{V}_X(x)$ if $\underline{V}_X(y) \supset \underline{V}_X(x)$,
- $\underline{V}_{X(x)}(y) = \dot{y}$ if $\underline{V}_X(y) \not\supset \underline{V}_X(x)$.

For a mapping $f : |X| \longrightarrow |Y|$, we have the following lemma:

<u>Lemma</u>: f is continuous from X to Y iff, for all $x \in |X|$, f is continuous from $X(x)$ to Y .

If f is continous from $X(x)$ to Y , we have :

$f(\underline{V}_X(x)) = f(\underline{V}_{X(x)}(x)) \xrightarrow[Y]{} f(x)$; thus f is continuous from X to Y . Let f be continuous from X to Y and $x \in |X|$. If $\underline{V}_X(y) \supset \underline{V}_X(x)$, we have $f(y) = f(x)$, Y being a T_1 space , thus :

$$f(\underline{V}_{X(x)}(y)) = f(\underline{V}_X(x)) \xrightarrow[Y]{} f(x) = f(y) .$$

If $\underline{V}_X(y) \not\supset \underline{V}_X(x)$, then $f(\underline{V}_{X(x)}(y)) = f(\dot{y}) = \dot{\widehat{f(y)}} \xrightarrow[Y]{} f(y)$.

By the previous lemma $|Hom(X,Y)| \subset |Hom(X(x),Y)|$ for all $x \in |X|$.

Let us consider the family of canonical injections :

$$i_x : |Hom(X,Y)| \longrightarrow |Hom(X(x),Y)| .$$

1) For each $x \in |X|$, i_x <u>is continuous</u> from $Hom(X,Y)$ to $Hom(X(x),Y)$.

Let $F \xrightarrow[Hom(X,Y)]{} f$. If $\underline{V}_X(y) \supset \underline{V}_X(x)$, we have :

$$Ev(F \times \underline{V}_{X(x)}(y)) = Ev(F \times \underline{V}_X(x)) \xrightarrow[Y]{} f(x) = f(y) ;$$

if $\underline{V}_X(y) \not\supset \underline{V}_X(x)$, we have :

$$Ev(F \times \underline{V}_{X(x)}(y)) = Ev(F \times \dot{y}) \xrightarrow[Y]{} f(y) .$$

Thus $F \xrightarrow[Hom(X(x),Y)]{} f$.

2) $Hom(X,Y)$ is the <u>initial structure defined by the</u> i_x .

Let Z be any convergence space and f a mapping $: |Z| \longrightarrow |Hom(X,Y)|$, with $i_x \circ f$ continuous from Z to $Hom(X(x),Y)$, for all $x \in |X|$. We have the following morphisms :

$$Z \longrightarrow Hom(X(x),Y) : z \longmapsto f_z ,$$

$$X(x) \longrightarrow Hom(Z,Y) : y \longmapsto [z \longmapsto f_z(y)] .$$

Y being T_1, so is $Hom(Z,Y)$; by the lemma, we have continuous mappings : $X \longrightarrow Hom(Z,Y) : y \longmapsto [z \longmapsto f_z(y)]$, thus

$$Z \longrightarrow Hom(X,Y) : z \longmapsto f_z \quad \text{q.e.d.}$$

Theorem 4.2: $\underline{QTop}(R)$ is the CTSC spanned by $\underline{UTop}$.

A uniformizable topological space X is the initial space defined by all the continuous functions from X to R , therefore $X \in \underline{QTop}(R)$, thus $\widehat{\underline{UTop}} \subset \underline{QTop}(R)$. Conversely, by lemma 4.1, for any $Y \in \underline{QTop}$, $Hom(Y,R) \in \widehat{\underline{UTop}}$, which implies that any c-space is an object of $\widehat{\underline{UTop}}$.

We shall give now an internal characterization of c-spaces. Such a characterisation was proposed by SCHRODER ((12) prop.3.5), but it involves a notion of "strong solid" which is far from being simple.

We introduce first a few notations and lemmas . If $X \in \underline{QTop}$ and $A \subset |X|$, CX will denote the space $Hom(X,R)$, cX the reflexion of X in $\underline{QTop}(R)$ (see prop.1.3) , $A°$ the set of functions $f \in |CX|$ such that $f(A) \subset [-1,1]$.

Lemma 4.3: If F is a filter on $|X|$ and H a filter on $|CX|$, $Ev(H \times F) \xrightarrow[R]{} 0$ iff , for each $r > 0$, there exists $A \in F$ such that $rA° \in H$.

Proof: $Ev(H \times F) \xrightarrow[R]{} 0$ iff :

$\forall\ r > 0\ \exists A \in F\ \exists\ B \in H\ :\ Ev(B \times A) \subset [-r,r]$, but :

$\exists\ B \in H\ :\ Ev(B \times A) \subset [-r,r] \Longleftrightarrow \exists\ B \in H\ :\ B \subset rA^{\circ} \Longleftrightarrow rA^{\circ} \in H$.

<u>Lemma 4.4</u>: Let F be a filter on $|X|$ and $x \in |X|$. $F \xrightarrow[cX]{} x$ iff $F \xrightarrow[\omega X]{} x$ and (P) for all filters H on $|CX|$ such that $H \xrightarrow[CX]{} 0$, we have $Ev(H \times F) \xrightarrow[R]{} 0$.

Proof: Let a be the canonical morphism from X to CCX . We have : $F \xrightarrow[cX]{} x \Longleftrightarrow aF \xrightarrow[CCX]{} a(x)$

$\Longleftrightarrow \forall H \xrightarrow[CX]{} f \quad Ev(aF \times H) \xrightarrow[R]{} f(x)$;

but $Ev(aF \times H) = Ev(H \times F)$, thus :

$F \xrightarrow[cX]{} x \Longleftrightarrow \forall\ H \xrightarrow[CX]{} f \quad Ev(H \times F) \xrightarrow[R]{} f(x)$.

If we take $H = \dot{f}$, we obtain , if $F \xrightarrow[cX]{} x$, $\forall f \in |CX|\quad fF \xrightarrow[R]{} f(x)$, which exactly means that $F \xrightarrow[\omega X]{} x$. On the other hand, if H is any filter which converges to 0 in CX, we have $Ev(H \times F) \xrightarrow[R]{} 0(x) = 0$.

Conversely, let (F,x) satisfy to $F \xrightarrow[\omega X]{} x$ and (P) . If $H \xrightarrow[CX]{} f$, then $H - \dot{f} \xrightarrow[CX]{} 0$, thus $Ev((H - \dot{f}) \times F) \xrightarrow[R]{} 0$; on the other hand $fF \xrightarrow[R]{} f(x)$. We have :

$Ev(H \times F) \supset Ev((H - \dot{f}) \times F) + fF$, which implies :

$Ev(H \times F) \xrightarrow[R]{} 0 + f(x) = f(x)$ q.e.d.

<u>Lemma 4.5</u>: Let X be a ω-closed-domained space , F a filter on $|X|$ and $x \in X$. If $F \xrightarrow[X]{} x$ and $F \xrightarrow[\omega X]{} x$, then $Conv_X F = \emptyset$.

Proof: Let z be a point of $\mathrm{Conv}_X F$; this set being an ω-closed set which do not contain x , there exists $f \in |CX|$ such that $f(x) = 0$ and $f(z) = 1$. $F \xrightarrow[\omega X]{} x \iff fF \xrightarrow[R]{} 0$ and $F \xrightarrow[X]{} z \iff fF \xrightarrow[R]{} 1$, which is a contradiction .

Theorem 4.6: A convergence space X is a c-space iff X is pseudo-topological, ω-regular and ω-closed-domained .

Proof: R has the three properties of the theorem: it is the same for any c-space, by prop.2.2, 2.5 .

Conversely, let X be a space which satisfies to the conditions of the theorem. In order to prove that X = cX , it suffices to see that, for each ultrafilter F on $|X|$, $F \underset{X}{\not\to} x$ implies $F \underset{cX}{\not\to} x$.

If $F \underset{\omega X}{\not\to} x$, then $F \underset{cX}{\not\to} x$, cX being finer than ωX . Let us suppose now that $F \xrightarrow[\omega X]{} x$; then, by lemma 4.5 , $\mathrm{Conv}_X F = \emptyset$. For any convergent filter G of X, $\mathrm{Cl}_{\omega X} G$ is a convergent filter of X, therefore $F \not\supset \mathrm{Cl}_{\omega X} G$ and there exists $A_G \in G$ such that $\mathrm{Cl}_{\omega X} A_G \notin F$. Let H be the filter on $|CX|$ spanned by the sets rA°_G , where $r > 0$ and G is a convergent filter of X . It is clear that $H \xrightarrow[CX]{} 0$.

Suppose if possible that $F \xrightarrow[cX]{} x$, then, by lemmas 4.3,4.4 , there exists an $A \in F$ such that $A^\circ \in H$, that is to say :

$$A^\circ \supset \bigcap_{i=1}^{n} rA^\circ_{G_i}$$, with $r > 0$ and $G_1, \ldots, G_n$ convergent filters of X .

Let $K = \bigcup_{i=1}^{n} \mathrm{Cl}_{\omega X} A_{G_i}$. Suppose that $z \in A - K$; K being an ω-closed set, it would exist $f \in |CX|$ such that $f(z) = 2$ and f vanishes on K : thus $f/r \in A^\circ_{G_i}$ for $i=1,\ldots,n$, $f \in A^\circ$ and $f(z) \in [-1,1]$, which is a contradiction.

Therefore $K \supset A$, which implies that $K \in F$. On the other hand, F being an ultrafilter, we have $X - Cl_{\omega X} A_G \in F$, thus: $X - K = \bigcap_{i=1}^{n} (X - Cl_{\omega X} A_{G_i}) \in F$. This is a contradiction.!

<u>Corollary 4.7</u>: Let X be any Hausdorff convergence space; then X is a c-space iff X is an ω-regular pseudo-topological space.

This result is due to KENT ((9) Theorem 2.4) and SCHRODER ((12) Theorem 3.6); they remarked that any ω-regular Hausdorff space is ω-Hausdorff, and so ω-closed-domained. In the pretopological case, it was already established by BUTZMANN and MÜLLER (5).

<u>Remarks</u>: SCHRODER ((12) Ex.4.3.) gave an example of a compact Hausdorff ω-regular space which is not pseudo-topological. BUTZMANN and MÜLLER (5) found a regular ω-Hausdorff topological space which is not ω-regular. There remains the following problem: find an ω-regular topological space which is not ω-closed-domained (Such a space must not be Hausdorff !).

5.CATEGORICAL REMARKS.

1. Are the conditions of prop.1.3 characteristic for categories of $\underline{C}(L)$-type ? More precisely, we ask the following: let $\underline{C}'$ be a reflexive CTSC of $\underline{C}$ which satisfies to (i) and (iii); does exist $L \in \underline{C}'$ such that $\underline{C}' = \underline{C}(L)$? The answer is not clear: indeed, we ignore if QTop itself is equal to QTop(L), for some convergence space L .

2. Let Ω_g be the two-element space with the indiscrete topology. In QTop , Ω_g appears to be a classifying object for regular monomorphisms (or kernels). It is clear that Ω_g is an object of the CTSC of QTop whose objects are, respectively:

- pseudo-topological spaces,
- Antoine spaces,
- c-spaces,
- K-regular spaces, for any K: QTop $\longrightarrow$ STop ,
- K-closed-domained spaces, for K: QTop $\longrightarrow$ Top with $K(\Omega_g) = \Omega_g$.

Thus, Qtop and each of these categories are quasi-topoi, in PENON's sense (11). This fact seems to us a sufficient reason to involve non-Hausdorff spaces in our considerations.

ooooooooooo

Bibliography:

(1) ANTOINE P., Etude élémentaire des catégories d'ensembles structurés, Bull. Soc. math. Belge, 18, n°2-4, 1966.

(2) ANTOINE P., Notion de compacité et quasi-topologie, Cah. Top. Géom. diff., 14, n°3, 1973.

(3) BOURDAUD G., Structures d'Antoine associées aux semi-topologies et aux topologies, Comptes rendus Acad. Sc. Paris, 279, série A, 1974, pp 591-594.

(4) BOURDAUD G., Espaces d'Antoine et semi-espaces d'Antoine, Cah. Top. Géom. diff., 15 (to appear).

(5) BUTZMANN H.P. and MÜLLER B., Topological c-embedded spaces, Man. Fak. Math. Univ. Mannheim, nr.31 (1972).

(6) COCHRAN A.C. and TRAIL R.B., Regularity and complete regularity for convergence spaces, Lecture notes 375.

(7) EHRESMANN C., Catégories topologiques, Indag. math., 28, n°1, 1966.

(8) HERRLICH H., Cartesian closed topological categories, Math. Coll. Univ. Capetown, 9, 1974.

(9) KENT D., Continuous convergence in C(X), Pacific jour. math.,52, n°2, 1974.

(10) MACHADO A., Espaces d'Antoine et pseudo-topologies, Cah. Top. Géom. diff., 14, n°3, 1973.

(11) PENON J., Quasi-topos,Cah. Top. Géom. diff., 14, n°2, 1973.

(12) SCHRODER M., Solid convergence spaces, Bull. Aust. Math. Soc., 8, 1973.

ooooooooooooo

Je remercie particulièrement Armando MACHADO et Philippe ANTOINE, dont les idées sont à l'origine des développements présentés ici.

TOPOLOGICAL FUNCTORS AND STRUCTURE FUNCTORS

G.C.L. Brümmer

0. Introduction In this paper we study the construction and some properties of functors F — we shall call them structure functors — which make the triangle

(1)

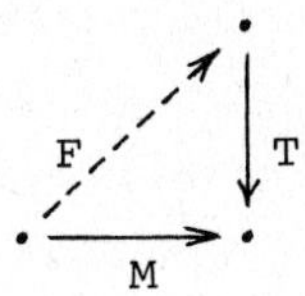

commute, where T and M are given faithful functors.

Structure functors arise naturally in the theory of the top categories of O. Wyler [44], [45] — (also discussed by, among others, H.-G. Ertel [10], R.-E. Hoffmann [17], S.H. Kamnitzer [26], T. Marny [28], [29], W. Tholen [38], M.B. Wischnewsky [41], [42], [43]) — and in other situations of greater or lesser generality treated by M. Hušek [22], [23], [24] as well as, for example, by P. Antoine [1], A.A. Blanchard [3], the author [5], [6], [7], C.R.A. Gilmour [12], J.W. Gray [13], M.N. Halpin [14], H. Herrlich [15], [16], R.-E. Hoffmann [20], A. Pultr [31], J.E. Roberts [33], J. Rosický [34], S. Salbany [35], [36], [37], and M.B. Wischnewsky [39], [40].

Research aided by grants to the author and to the Topology Research Group from the South African Council for Scientific and Industrial Research and from the University of Cape Town.

This paper is a sequel to the joint paper [8]. The characterization of topological functors in [8] accounts for the intimate relationship between topological functors and structure functors.

In the first section we reconsider the diagram

(2) 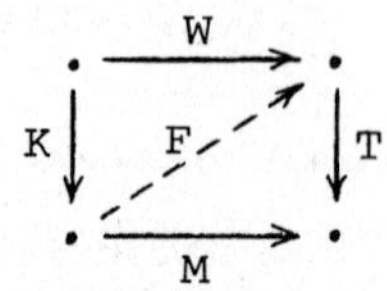

from [8] and drop the requirement that K be full. Then F as constructed in [8] ceases to be a diagonal in the strict sense, but is still characterized by an extremal property. (The construction goes back to M. Hušek's paper [24], was rediscovered in [5] and developed in [6].)

In the second and third sections we study the effect of K and W on the properties of F, in particular the property that F is right inverse to a given amnestic functor L. We find an appropriate context for the latter problem to be the case where L is followed by a functor M such that ML is topological and L sends ML-initial sources to M-initial sources. The concept of an M-spanning subcategory of the domain of M (a kind of density condition) enters into the study of right inverses of L, and therefore we have a fourth section leading to a characterization of M-spanning subcategories, overlapping somewhat with work of T. Marny [28],[29].

The results on preservation of initiality in section 2, though special cases of results of O. Wyler [44], are proved for the sake of exposition.

For basic properties of topological functors not discussed here and for more bibliography the reader is referred to H. Herrlich [15],

[16, appendix A], R.-E. Hoffmann[17], [19], [20], and further to the cited works on top categories. In particular [15], [28], [19] and [7] deal with relatively topological functors. (As our topological functors are those that are transportable and absolutely topological in the sense of Herrlich [15], our considerations are independent of any factorization structures in the base category.)

Many of the results come from my thesis [6], written under the supervision of Keith A. Hardie, to whom I wish to express my deep appreciation. The present exposition is however based on the new point of view offered by the diagonal diagram (2) above.

1. <u>Filling in triangles of faithful functors</u> Throughout this section we consider a faithful functor $T: \mathcal{A} \to \mathcal{C}$. For definitions we refer to [8], adding the observation that for objects A, B of $\mathcal{A}$ it is consistent with our definitions to write $A \leq_T B$ if there is a morphism $g: A \to B$ with $Tg = 1$.

Given a commutative square of faithful functors

$$\begin{array}{ccc} \mathcal{D} & \xrightarrow{W} & \mathcal{A} \\ {\scriptstyle K}\downarrow & & \downarrow{\scriptstyle T} \\ \mathcal{X} & \xrightarrow[M]{} & \mathcal{C} \end{array}$$

without assuming K full, we reconsider Hušek's [24] construction of a functor $F: \mathcal{X} \to \mathcal{A}$, as given in [8], as follows. For $X \in \mathrm{ob}\mathcal{X}$, supposing that the T-initiality problem

$$(MX,\quad Mf: MX \to MKD,\quad WD \mid D \in \mathrm{ob}\mathcal{D},\quad f \in \mathcal{X}(X, KD))$$

has a solution, we denote the solution by

$$(FX,\quad (Mf)': FX \to WD \mid D \in \mathrm{ob}\mathcal{D},\quad f \in \mathcal{X}(X, KD));$$

and for $g \in \mathcal{X}(X', X)$ we define Fg as follows:

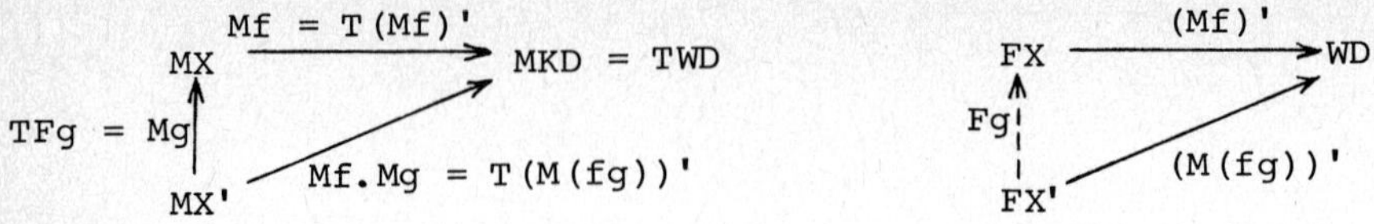

Since the left hand triangle commutes, there exists unique $Fg: FX' \to FX$ such that $TFg = Mg$.

This defines a functor $F: \mathfrak{X} \to \mathcal{A}$ provided each of the relevant T-initiality problems has a solution and provided, for each X, a choice is made among isomorphic candidates for FX. For this functor we shall use the notation

$$F = \langle T, M, K : W \rangle.$$

If T is amnestic, then F is uniquely determined by T, M, K, W.

1.1 <u>Proposition</u> Given a commutative square

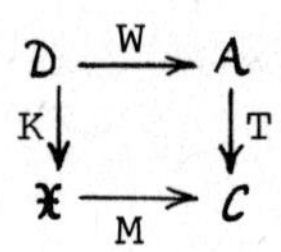

of faithful functors. If $F = \langle T, M, K : W \rangle$ exists, then F is a T-coarsest functor $\mathfrak{X} \to \mathcal{C}$ satisfying the conditions $TF = M$ and $FK \leq_T W$. If in addition K is full and T amnestic, then $FK = W$.

<u>Proof</u> Clearly $TF = M$. For fixed D_o in $\text{ob}\,\mathcal{D}$, the source $(FKD_o, (Mf)': FKD_o \to WD)$ is the solution to the T-initiality problem $(MKD_o, Mf: MKD_o \to MKD, WD)$ (all $D \in \text{ob}\,\mathcal{D}$, all $f: KD_o \to KD$). Taking in particular $f = 1_{D_o}$ we define $n_{D_o} = (M1_{D_o})'$. As $Tn_{D_o} = 1$ for arbitrary D_o, we have $FK \leq_T W$. If $G: \mathfrak{X} \to \mathcal{A}$ is any functor with $TG = M$ and $GK \leq_T W$, then we have $m: GK \to W$ with $Tm = (1)$ and considering the diagrams

$$X \xrightarrow{\ f\ } KD \qquad \begin{array}{ccc} MX & \xrightarrow{\ Mf\ } & MKD \\ 1 \Uparrow & & \Uparrow 1 \\ TGX & \xrightarrow[\ TGf\ ]{} & TKD \end{array} \qquad \begin{array}{ccc} FX & \xrightarrow{\ (Mf)'\ } & WD \\ p_X \uparrow & & \uparrow m_D \\ GX & \xrightarrow[\ Gf\ ]{} & GKD \end{array}$$

we obtain $p_X: GX \to FX$ with $Tp_X = 1$; thus $G \leq_T F$.

It was shown in [8] that if K is full and T amnestic, then $FK = W$.

1.2 <u>Proposition</u> Given faithful $T: \mathcal{A} \to \mathcal{C}$ and $M: \mathfrak{X} \to \mathcal{C}$. Then every functor F such that $TF = M$, is of the form $F = \langle T, M, K : W \rangle$ for suitable K and W (e.g. $K = I_{\mathfrak{X}}$ and $W = F$).

1.3 <u>Proposition</u> Given a commutative square of faithful functors

$$\begin{array}{ccc} \mathcal{D} & \xrightarrow{\ W\ } & \mathcal{A} \\ K \downarrow & & \downarrow T \\ \mathfrak{X} & \xrightarrow[\ M\ ]{} & \mathcal{C} \end{array}$$

let $\mathcal{D}_d$ be the discrete category with the same objects as $\mathcal{D}$, $K_d = K|\mathcal{D}_d$, $W_d = W|\mathcal{D}_d$. Then $F = \langle T, M, K : W \rangle$ if and only if $F = \langle T, M, K_d : W_d \rangle$.

2. <u>Going up a given triangle</u> In this section we consider two given amnestic functors $L: \mathcal{A} \to \mathfrak{X}$, $M: \mathfrak{X} \to \mathcal{C}$. Then ML is amnestic. We are interested in the functors $F: \mathfrak{X} \to \mathcal{A}$ for which $MLF = M$;

$$\begin{array}{ccc} & & \mathcal{A} \\ & {}^{F}\nearrow\!\swarrow_{L} & \downarrow ML \\ \mathfrak{X} & \xrightarrow[\ M\ ]{} & \mathcal{C} \end{array}$$

these are just the functors of the form $F = \langle ML, M, K : W \rangle$.

We shall say that L <u>preserves initiality for</u> M if L sends each ML-initial source to an M-initial source. (We shall briefly discuss the relation of this concept to the taut lifts of O. Wyler [41] at the end of this section.)

2.1 Proposition Let $F = \langle ML, M, K{:}W \rangle$.

(i) If $I_{\mathcal{X}} \leq_M LF$, then $K \leq_M LW$.

(ii) If $K \leq_M LW$ and L preserves initiality for M, then $I_{\mathcal{X}} \leq_M LF$.

Proof (i) From $I_{\mathcal{X}} \leq_M LF$ follows $K \leq_M LFK$ and from $FK \leq_{ML} W$ (1.1) follows $LFK \leq_M LW$; hence $K \leq_M LW$.

(ii) We have $t{:}K \to LW$ with $Mt = (1)$. The ML-initiality problem $(MX,\ Mf{:}MX \to MKD,\ WD \mid D \in \mathrm{ob}\,\mathcal{D},\ f{:}X \to KD)$ has the solution $(FX,\ (Mf)'{:}FX \to WD \mid \ldots)$. Hence the M-initiality problem $(MX,\ Mf,\ LWD \mid \ldots)$ has the solution $(LFX,\ L(Mf)' \mid \ldots)$.

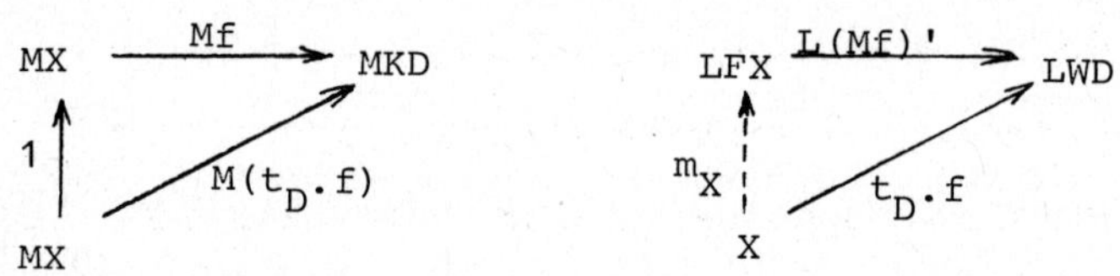

The definition of "solution" now yields $m_X{:}X \to LFX$ with $Mm_X = 1_{MX}$. Thus $I_{\mathcal{X}} \leq_M LF$.

We introduce a simplified form for an important special case of our notation, as follows. Any subcategory $\mathcal{U}$ of $\mathcal{A}$ determines an inclusion functor $W{:}\mathcal{U} \to \mathcal{A}$ and a functor $K = LW{:}\mathcal{U} \to \mathcal{X}$. In this case we abbreviate the notation $F = \langle ML, M, K{:}W \rangle$ to

$$F = [L, M{:}\mathcal{U}].$$

Use of the latter notation will always presuppose that $\mathcal{U}$ is a subcategory of $\mathcal{A}$. By 1.3, $[L, M{:}\mathcal{U}]$ is unchanged if we make $\mathcal{U}$ discrete, so we may regard $\mathcal{U}$ simply as a class of objects of $\mathcal{A}$.

2.2 Lemma If $F = [L, M{:}\mathcal{U}]$, then $FLA \leq_{ML} A$ for each $A \in \mathcal{U}$.

<u>Proof</u> As above, we have $F = \langle ML, M, K : W\rangle$ and by 1.1, $FK \leq_{ML} W$, i.e. $FLW \leq_{ML} W$.

2.3 <u>Theorem</u> Consider a functor $F : \mathfrak{X} \to \mathbf{A}$.

(i) If $I_{\mathfrak{X}} \leq_M LF$ and $FLF = F$, then there exists a class $\mathcal{U}$ of objects of $\mathbf{A}$ such that

$$F = [L, M : \mathcal{U}].$$

There is even a largest such class, namely

$$\hat{\mathcal{U}} = \{A \in \text{ob}\,\mathbf{A} \mid FLA \leq_{ML} A\}.$$

(ii) If $F = [L, M : \mathcal{U}]$ for some class $\mathcal{U} \subset \text{ob}\,\mathbf{A}$ and if L preserves initiality for M, then $I_{\mathfrak{X}} \leq_M LF$ and $FLF = F$.

<u>Proof</u> (i) With $\hat{\mathcal{U}}$ as above, for each $A \in \hat{\mathcal{U}}$ we have $t_A : FLA \to A$ with $MLt_A = 1$. We show that the ML-initiality problem $(MX, Mf : MX \to MLA, A \mid A \in \mathcal{U}, f : X \to LA)$ has the solution $(FX, t_A.Ff : FX \to A \mid \ldots)$. Consider a source $(Y, h_f : Y \to A \mid \ldots)$ and $u : MLY \to MX$ such that the left-hand triangle commutes:

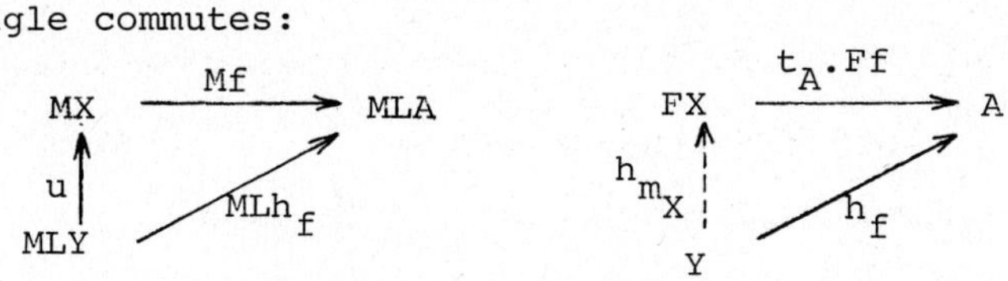

As $I_{\mathfrak{X}} \leq_M LF$, we have $m_X : X \to LFX$ with $Mm_X = 1$; and as $FL(FX) = FX$ we have $FX \in \hat{\mathcal{U}}$, so that m_X is one of the indexing f, giving us a special case of the left-hand triangle which shows that $u = MLh_{m_X}$. Hence $F = [L, M : \hat{\mathcal{U}}]$. If $\mathcal{U}$ is any class of objects of $\mathbf{A}$ such that $F = [L, M : \mathcal{U}]$, then by 2.2 $\mathcal{U} \subset \hat{\mathcal{U}}$.

(ii) Let $F = [L, M : \mathcal{U}]$ and let L preserve initiality for M. Considering the inclusion $W : \mathcal{U} \to \mathbf{A}$ and $K = LW$, we have from 2.1 (ii) that $I_{\mathfrak{X}} \leq_M LF$. So there is $m: I_{\mathfrak{X}} \to LF$ with $Mm = (1)$. By 2.2 we have $t_A: FLA \to A$ with $MLt_A = 1$. Now consider that FX and FLFX are given, respectively, by the ML-initiality problems

$$(MX,\ Mf: MX \to MLA,\ A \mid A \in \mathfrak{U},\ f: X \to LA)$$

and

$$(MLFX,\ Mg: MLFX \to MLA,\ A \mid A \in \mathfrak{U},\ g: LFX \to A).$$

In these problems we have $MLFX = MX$ and

$$M(LFX \xrightarrow{LFf} LFLA \xrightarrow{Lt_A} LA) = Mf$$

$$M(X \xrightarrow{m_X} LFX \xrightarrow{g} LA) = Mg$$

so that the two problems are identical. As ML is amnestic, it follows that $FX = FLFX$, whence $F = FLF$.

2.4 <u>Examples</u> <u>Unif</u>, <u>Prox</u>, <u>Creg</u> will denote the categories of uniform spaces, proximity spaces, completely regular topological spaces, respectively, without imposition of the T_o-separation axiom.

(i) If L:<u>Unif</u> $\to$ <u>Creg</u>, M:<u>Creg</u> $\to$ <u>Ens</u> are the usual forgetful functors, L preserves initiality for M. M and ML are topological.

(ii) If L:<u>Unif</u> $\to$ <u>Prox</u>, M:<u>Prox</u> $\to$ <u>Ens</u> are the usual forgetful functors, L does not preserve initiality for M. Still M and ML are topological.

(iii) In example (i), if F:<u>Creg</u> $\to$ <u>Unif</u> is the functor giving each space the discrete uniformity, then $MLF = M$ but not $I \leq_M LF$.

(iv) Let $\mathfrak{X}$ be the category of metric spaces and non-expansive mappings, $\mathcal{C}$ the category of metrizable topological spaces and continuous mappings, $L = I_{\mathcal{X}}$, $M: \mathcal{X} \to \mathcal{C}$ the functor assigning the metric topology, $F: \mathcal{X} \to \mathcal{X}$ given by $F(x,d) = F(X, \frac{1}{2}d)$. L and M are amnestic and L preserves initiality for M, and $I_{\mathcal{X}} \leq_M LF$ but $FLF \neq F$.

The assumption of preservation of initiality in Theorem 2.2 was the best possible in the following sense:

2.5 Proposition Let ML be topological. If for each class $\mathfrak{U}$ of objects of $\mathcal{A}$, putting $F = [L,M:\mathfrak{U}]$ we have $I_{\mathcal{X}} \leq_M LF$ and $FLF = F$, then L preserves initiality for M.

Proof Given an ML-initial source $(A, F_i: A \to A_i)_I$. The functor $F = [L,M:\{A_i\}_I]$ exists, $FLF = F$ and $I_{\mathcal{X}} \leq_M LF$, so we have $m: I_{\mathcal{X}} \to LF$ with $Mm = (1)$. By 2.2 we have $n_i: FLA_i \to A_i$ with $MLn_i = 1$. To see that the source $(LA, Lf_i: LA \to LA_i)_I$ is M-initial, consider any source $(X, g_i: X \to LA_i)_I$ and any $u: MX \to MLA$ such that the left-hand triangle commutes:

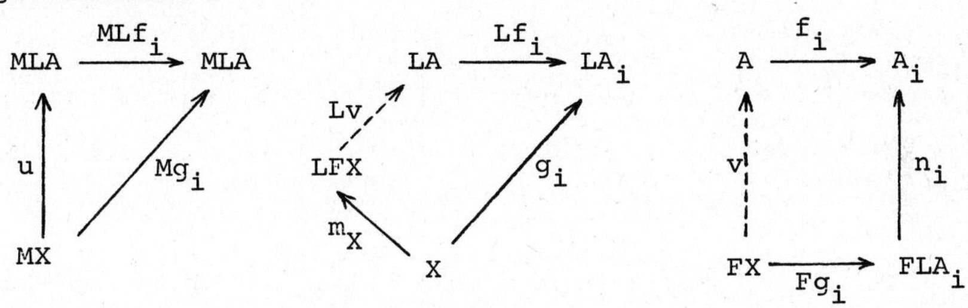

As $ML(n_i Fg_i) = Mg_i$, there exists $v: FX \to A$ such that $MLv = u$. Thus $M(Lv.m_X) = u$, and the source $(LA, Lf_i)_I$ is M-initial.

2.6 Theorem

(1) If there exists a functor $F: \mathcal{X} \to \mathcal{A}$ such that $I_{\mathcal{X}} \leq_M LF$ and $FL \leq_{ML} I_{\mathcal{A}}$, then L preserves initiality for M.

(2) If ML is topological and L preserves initiality for M, then there exists a functor $F: \mathfrak{X} \to \mathbf{A}$ such that $I_{\mathfrak{X}} \leq_M LF$ and $FL \leq_{ML} I_{\mathbf{A}}$.

(3) In both (1) and (2), $F = [L,M:\mathbf{A}]$.

Proof (3) From $I_{\mathfrak{X}} \leq_M LF$ follows $F \leq_{ML} FLF$ and from $FL \leq_{ML} I_{\mathbf{A}}$ follows $FLF \leq_{ML} F$, whence $FLF = F$. Thus by 2.3 $F = [L,M:\hat{\mathfrak{U}}]$ where $\hat{\mathfrak{U}} = \{A \in \mathrm{ob}\,\mathbf{A} \mid FLA \leq_{ML} A\} = \mathrm{ob}\,\mathbf{A}$.

(1) Given any ML-initial source $(A, f_i : A \to A_i)_I$ we may use $F = [L,M:\mathrm{ob}\,\mathbf{A}]$ and carry through the remainder of the proof of 2.5 to show that the source $(LA, Lf_i : LA \to LA_i)_I$ is M-initial.

(2) Let $F = [L,M:\mathbf{A}]$. By 2.3 $I_{\mathfrak{X}} \leq_M LF$, and by 1.1, since $F = \langle ML,M,L:I_{\mathbf{A}}\rangle$, we have $FL \leq_{ML} I_{\mathbf{A}}$.

2.7 Corollary Let $L: \mathbf{A} \to \mathfrak{X}$ be an amnestic functor. The following two statements are equivalent:

(1) $F = [L,I_{\mathfrak{X}}:\mathbf{A}]$;

(2) $I_{\mathfrak{X}} = LF$ and $FL \leq_L I_{\mathbf{A}}$.

Moreover they imply:

(3) F is the L-finest right inverse of L;

(4) F is left adjoint right inverse to L;

(5) F is fully faithful;

(6) L preserves initiality for any faithful functor $M: \mathfrak{X} \to \mathbf{C}$.

The conditions (4) and (5) form part of R.-E. Hoffmann's characterization (under side conditions) of topologicity [17, pp. 84-85], [20].

The condition (3) above does not imply (4):

2.8 Example The forgetful functor L:Unif → Prox has a unique right inverse but no left adjoint [24], [6].

A category $\mathfrak{X}$ is said to have trivial centre if there is just one natural transformation of the identity functor $I_{\mathfrak{X}}$ to itself. This occurs very often: a sufficient condition is that $\mathfrak{X}$ have a generator with just one endomorphism (remark due to K. A. Hardie, cf. [6].) More on this concept is to be found in R.-E. Hoffmann [18].

2.9 Proposition Let $L: \mathfrak{A} \to \mathfrak{X}$ be amnestic and let $\mathfrak{X}$ have trivial centre. If L has a left adjoint right inverse F, then $F = [L, I_{\mathfrak{X}} : \mathfrak{A}]$. Thus L has at most one left adjoint right inverse.

Proof The unit $r: I_{\mathfrak{X}} \to LF$ and counit $s: FL \to I_{\mathfrak{A}}$ of the adjunction satisfy $Ls_A . r_{LA} = 1_{LA}$ for $A \in \text{ob}\,\mathfrak{A}$. As $LF = I_{\mathfrak{X}}$, r is trivial, hence $Ls_A = 1_{LA}$, i.e. $FL \leq_L I_{\mathfrak{A}}$, so that by 2.7 $F = [L, I_{\mathfrak{X}} : \mathfrak{A}]$.

When $L: \mathfrak{A} \to \mathfrak{X}$ is topological, then $F = [L, I_{\mathfrak{X}} : \mathfrak{A}]$ exists. (The converse is not true: consider the forgetful functor from T_0-spaces, or T_1-spaces, ... to Ens.) By Antoine's theorem ([1], [33], [5], [17], cf. [8] for an external proof), L is cotopological. Thus the dual of $[L, I_{\mathfrak{X}} : \mathfrak{A}]$ - for which we have no notation: it may be described by the dual of Proposition 1.1 or by coinitiality construction - exists, so that by the dual of 2.7 we have:

2.10 Proposition If $L: \mathfrak{A} \to \mathfrak{X}$ is topological, then L preserves initiality and coinitiality for any faithful functor $M: \mathfrak{X} \to \mathfrak{C}$.

In this proposition we may drop the assumption that L is amnestic.

An important theorem of O. Wyler may in our terminology be phrased as follows.

2.11 <u>Theorem on "taut lifts" of O. Wyler</u> [44, 45]

Let $P:\mathcal{A} \to \mathcal{B}$ and $Q:\mathcal{X} \to \mathcal{C}$ be amnestic topological functors and Φ and R functors such that $Q\Phi = RP$.

Then the following two conditions are equivalent:

(1) Φ sends P-initial sources to Q-initial sources and S is left adjoint to R;

(2) Φ has a left adjoint Ψ such that $P\Psi = SQ$.

Moreover, under either condition it is possible to arrange the relation between the adjunctions as follows:

$$Q\theta = \eta Q \quad \text{and} \quad P\delta = \varepsilon P$$

with η (respectively ε) as unit (respectively counit) for $S \dashv R$ and θ, δ those for $\Psi \dashv \Phi$.

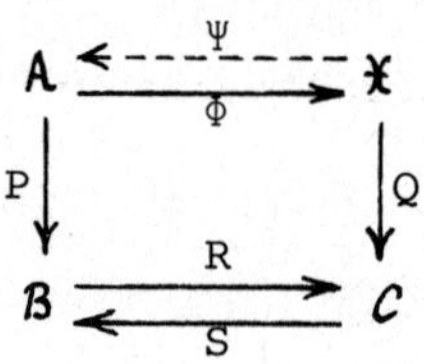

For other elaborations on the theorem, see [45], [29], [38]. For example, the implication from (1) to (2) does not need topologicity of Q.

The basic example [44] has P, Q, Φ , R the usual forgetful functors between $\mathcal{A}$ = {topological groups}, $\mathcal{B}$ = {groups}, $\mathcal{X}$ = <u>Top</u>, $\mathcal{C}$ = <u>Ens</u>.

Wyler [44,45] gives applications to topological algebra; for generalizations see [10], [38], [41], [42], [43].

Clearly our theorem 2.6 is essentially the case $R = I_{\mathcal{C}}$ of Wyler's theorem.

In proving the implication (1) ⇒ (2) in Wyler's theorem, we can construct the functor $\Psi: \mathcal{X} \to \mathcal{A}$ as follows. We have the counit $\varepsilon: SR \to I_{\mathcal{B}}$. For any object X of $\mathcal{X}$, we form the P-initiality problem

$$(SQX, \quad \varepsilon_{PA}.SQf, \quad A \mid A \in \operatorname{ob}\mathcal{A}, \quad f \in \mathcal{X}(X, \Phi A)).$$

The domain of its solution will be ΨX.

This exceeds the framework of our construction in section 1. In fact, considering the mentioned example of the topological groups, one readily sees that it is impossible in general to express Ψ in our notation as a functor of the form $\Psi = \langle P, M, K:W \rangle$.

3. <u>Right inverses of an amnestic functor</u> Let $L: \mathcal{A} \to \mathcal{X}$ be amnestic. From the following two special cases of the diagonal construction

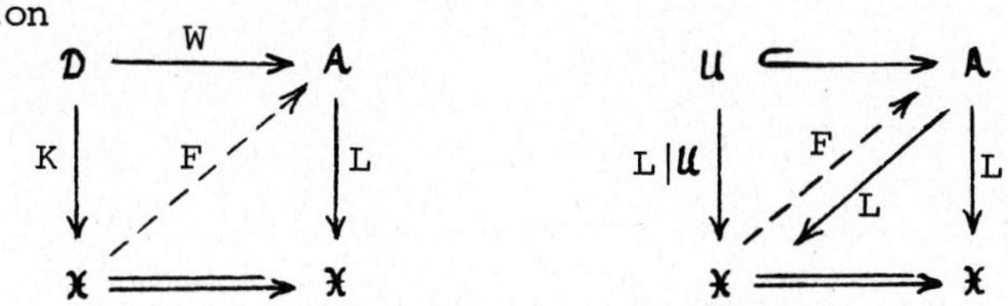

together with 1.2 one immediately sees that:

3.1 <u>Proposition</u> For $F: \mathcal{X} \to \mathcal{A}$ the following are equivalent:

(1) $LF = I_{\mathcal{X}}$;

(2) There exist functors K and W such that $LW = K$ and $F = \langle L, I_{\mathcal{X}}, K:W \rangle$;

(3) There exists a subcategory $\mathcal{U}$ of $\mathcal{A}$ such that $F = [L, I_{\mathcal{X}} : \mathcal{U}]$.

However, unless L is topological, these characterizations seem to be of little use in calculating those right inverses that may exist. Also, when $\mathcal{X}$ = Ens, the right inverses are so obvious as to have little interest, and when $\mathcal{X}$ is not Ens but some category of topological type, then it may be quite hard to verify that $L: \mathbf{A} \to \mathcal{X}$ is topological; a nice example is the forgetful functor from contiguity spaces to separated Lodato proximity spaces, shown by Bentley and Herrlich [2, theorem 7.3] to be topological. The straightforward but interesting situation often arises that L is not topological but there exists a functor $M: \mathcal{X} \to \mathcal{C}$ such that ML is topological and L preserves initiality for M, as is the case in the following examples (taking the usual forgetful functors):

3.2 (i) $\mathbf{A}$ = Unif, $\mathcal{X}$ = Creg, $\mathcal{C}$ = Ens;

(ii) $\mathbf{A}$ = Qun (quasi-uniform spaces [5]), $\mathcal{X}$ = Top, $\mathcal{C}$ = Ens;

(iii) $\mathbf{A}$ = Qun, $\mathcal{X}$ = Pcrg (pairwise completely regular bitopological spaces [27], [35], [5]), $\mathcal{C}$ = Ens;

(iv) $\mathbf{A}$ = Prox, $\mathcal{X}$ = Creg, $\mathcal{C}$ = Ens [21];

(v) $\mathbf{A}$ = {Lodato proximity spaces}, $\mathcal{X}$ = {R_o-spaces}, $\mathcal{C}$ = Ens [21];

(vi) $\mathbf{A}$ = Zero (zero-set spaces), $\mathcal{X}$ = Creg, $\mathcal{C}$ = Ens [12].

Herrlich [16, pp. 62, 80, 112-113] has argued convincingly against a simplistic view about certain topological categories being richer in structure than others; as "forgetful" functors now go in all directions, there seems to be a need for tools to handle them.

For the rest of this section we again consider two amnestic functors $L: \mathbf{A} \to \mathcal{X}$, $M: \mathcal{X} \to \mathcal{C}$.

We shall say that a subcategory $\mathcal{V}$ of $\mathfrak{X}$ is M-<u>spanning</u> if $I_{\mathfrak{X}} = [I_{\mathfrak{X}}, M:\mathcal{V}]$. Again only the objects of $\mathcal{V}$ are involved.

3.3 <u>Proposition</u> A subcategory $\mathcal{V}$ of $\mathfrak{X}$ is M-spanning if and only if for each object X of $\mathfrak{X}$ the source

$$(X,\ f:X \to V \mid V \in \mathrm{ob}\mathcal{V},\ \ f \in \mathfrak{X}(X,V))$$

is M-initial.

Two other characterizations (under side assumptions) follow later. For relations to J.R. Isbell's deeper concept of adequacy, see [32].

3.4 <u>Examples</u> In the following instances $\mathcal{V}$ is M-spanning:

(1) $M: \mathfrak{X} \to \mathcal{C}$ arbitrary, $\mathcal{V} = \mathfrak{X}$;

(2) M:<u>Top</u> → <u>Ens</u> the forgetful functor, $\mathcal{V}$ contains as object a Sierpiński space (two points with three open sets) or any space having a Sierpiński subspace;

(3) Write down the obvious proposition of the form "If $\mathcal{W}$ is M-spanning, so is every $\mathcal{V}$ such that ..." Hint: Compose M-initial sources [9, p.876];

(4) M:<u>Creg</u> → <u>Ens</u> forgetful, $\mathbb{R} \in \mathrm{ob}\mathcal{V}$;

(5) M:<u>Pcrg</u> → <u>Ens</u> forgetful, $\mathcal{V}$ has as object the real line with the two topologies

$$\{(-\infty,x) \mid -\infty \leq x \leq \infty\}, \qquad \{(x,\infty) \mid -\infty \leq x \leq \infty\}.$$

We recall that the notation $[L,M:\mathcal{U}]$ presupposes that $\mathcal{U}$ is a subcategory of $\mathcal{A}$.

3.5 <u>Theorem</u> Suppose $F = [L,M:\mathcal{U}]$. Then:

(1) If $LF = I_{\mathfrak{X}}$, then $L\mathcal{U}$ is M-spanning.

(2) If $L\mathcal{U}$ is M-spanning and L preserves initiality for M, then $LF = I_{\mathfrak{X}}$.

<u>Proof</u> (1) We show that the source $(X,\ f:X \to LA \mid A \in \mathcal{U},\ f \in \mathcal{X}(X,LA))$ is M-initial. Consider a source $(Y,\ g_f:Y \to LA \mid \ldots)$ and $u:MY \to MX$ such that the first triangle commutes:

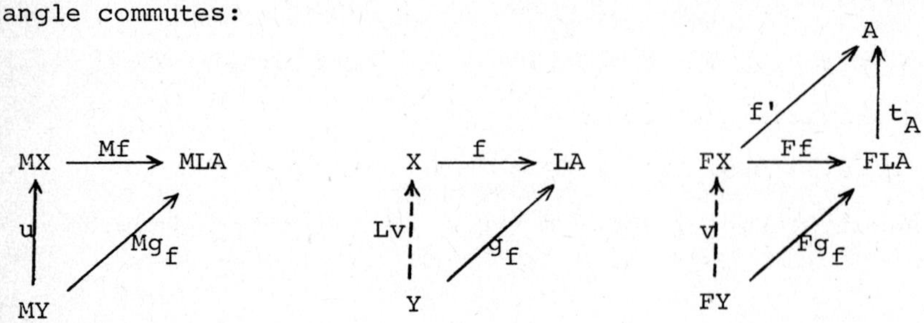

By 2.2 we have $t_A:FLA \to A$ with $MLt_A = 1$. Let $f' = t_A.Ff$. Then $MLf' = Mf$, and $ML(t_A.Fg_f) = Mg_f$. Hence there exists $v:FY \to FX$ with $MLv = u$. As $LF = I$, we have $Lv:Y \to X$ with $M(Lv) = u$, as required.

(2) $F = [L,M:\mathcal{U}]$ means that for each $X \in \text{ob}\,\mathcal{X}$ the source $(FX,\ (Mf)' \mid A \in \mathcal{U},\ f:X \to LA)$ solves the M-initiality problem $(MX,\ Mf \mid \ldots)$. As L preserves initiality for M, the source $(LFX,\ L(Mf)' \mid \ldots)$ is then M-initial; this means that $LF = [I_{\mathcal{X}},\ M:L\mathcal{U}]$, and by definition of spanning, $LF = I_{\mathcal{X}}$.

3.6 <u>Corollary</u> Every right inverse F of L is of the form $F = [L,M:\mathcal{U}]$ with $L\mathcal{U}$ M-spanning.

3.7 <u>Theorem</u> Let $\mathcal{V}$ be an M-spanning subcategory of $\mathcal{X}$, let L preserve initiality for M, and let $F:\mathcal{X} \to \mathcal{A}$ be a functor such that $I_{\mathcal{X}} \leq_M LF$. Then F is a right inverse of L if and only if $LFV = V$ for each V in $\mathcal{V}$.

<u>Proof</u> Let $I_{\mathcal{X}} \leq_M LF$ and $LFV = V$ for each V in $\mathcal{V}$. By 1.2 and 2.1, $F = \langle ML,M,K:W \rangle$ with $K \leq_M LW$. FX is given by solving the ML-initiality problem

$(MX,\ Mf{:}MX \to MKD,\ WD \mid D \in \mathrm{ob}\mathcal{D},\ f \in \mathcal{X}(X,KD))$

and as L preserves initiality for M, LFX is given by solving the M-initiality problem

(1) $(MX,\ Mf{:}MX \to MKD,\ LWD \mid D \in \mathrm{ob}\mathcal{D},\ f \in \mathcal{X}(X,KD))$.

In particular for $V \in \mathcal{V}$, since $LFV = V$, we then get V by solving the M-initiality problem

(2) $(MV,\ Mh{:}MV \to MKD,\ LWD \mid D \in \mathrm{ob}\mathcal{D},\ h \in \mathcal{X}(V,KD))$.

As $\mathcal{V}$ is M-sufficient, we get X by solving the M-initiality problem

(3) $(MX,\ Mg{:}MX \to MV,\ V \mid V \in \mathcal{V},\ g \in \mathcal{X}(X,V))$.

By the well known compositive property of initial sources [9, p.876] it follows from (2) and (3) that X is given by solving the M-initiality problem

(4) $(MX,\ Mh.Mg,\ LWD \mid D \in \mathrm{ob}\mathcal{D},\ V \in \mathcal{V},\ X \xrightarrow{g} V \xrightarrow{h} KD)$.

As $K \leq_M LW$, we may augment the family of all hg in (4) to the family of all $f{:}X \to KD$ (since M sends both to the same family). But then the problem (4) becomes the problem (1) and still has the solution X. Hence $X = LFX$.

It is important to observe that in the above theorem the condition $I_{\mathcal{X}} \leq_M LF$ cannot be weakened to $MLF = M$ [5, example 3.2]. Thus M-spanning subcategory is not an analogue of dense subspace.

3.8 <u>Proposition</u> Suppose $F = [L,M{:}\mathcal{U}]$. Then $LF = I_{\mathcal{X}}$ if and only if $F = [L,I_{\mathcal{X}}{:}\mathcal{U}]$.

<u>Proof</u> Routine calculation.

3.9 <u>Proposition</u> Suppose ML is topological, L preserves initiality for M, and L is M-spanning. Then $F = [L,M{:}\mathcal{A}]$ exists, satisfies $I_{\mathcal{X}} = LF$ and $FL \leq_L I_{\mathcal{A}}$, and is thus left adjoint right inverse to L.

<u>Proof</u> By 3.5, $LF = I$. Thus by 3.8, $F = [L, I_{\mathcal{X}} : \mathcal{A}]$ and 2.7 can be applied.

We observe that if F and G are right inverses of L, then

$$F \leq_L G \text{ if and only if } F \leq_{ML} G$$

so that we may write simply $F \leq G$. The class of all right inverses of L will henceforth be equipped with this ordering. Suitable set theory will safeguard operations in this class.

3.10 <u>Theorem</u> Suppose ML is topological, L preserves initiality for M, and $L\mathcal{A}$ is M-spanning. Then, in the class of all right inverses of L, every non-void subclass has an infimum.

<u>Proof</u> Express each F_i in the subclass in the form $F_i = [L, M : \mathcal{U}_i]$ with largest $\mathcal{U}_i$ (2.3), let $\mathcal{U}$ be the union of the $\mathcal{U}_i$, and let $F = [L, M : \mathcal{U}]$. By 3.5 each $L\mathcal{U}_i$ is M-spanning, hence so is $L\mathcal{U}$, and $LF = I$. Clearly $F \leq F_i$. If $G \leq F_i$ for each i, let $G = [L, M : \mathcal{V}]$ with largest $\mathcal{V}$. It follows by 2.3 that $\mathcal{U} \subset \mathcal{V}$. Hence $G \leq F$.

3.11 <u>Corollary</u> Any one of the following sets of conditions is sufficient for the class of all right inverses of L to form a large-complete lattice:

(1) ML is topological, L preserves initiality for M, $L\mathcal{A}$ is M-spanning, and L has a coarsest right inverse;

(2) ML is topological, L preserves initiality and co-initiality for M, and $L\mathcal{A}$ is M-spanning and M-cospanning;

(3) L is topological.

Call an object X_o of $\mathcal{X}$ an L,M-<u>pivot</u> if $\{X_o\}$ is M-spanning and there exists a unique object A_o of $\mathcal{A}$ with $LA_o = X_o$.

3.12 Proposition Let L preserve initiality for M and let $\mathfrak{X}$ have an L,M-pivot $X_o = LA_o$. If the functor $[L,M:\{A_o\}]$ exists, then it is the coarsest right inverse of L.

3.13 Examples of L,M-pivots Let $\mathcal{C}$ = Ens, let L and M be the usual forgetful functors, and let

(i) A = Unif, $\mathfrak{X}$ = Creg, $X_o = [0,1]$;

(ii) A = Qun, $\mathfrak{X}$ = Pcrg, X_o the set $[0,1]$

equipped with the subspace bitopology from the example in 3.4(5)[35];

(iii) A = Qun, $\mathfrak{X}$ = Top, X_o the Sierpiński space (cf. 3.4(2)).

3.14 Applications In Example 3.13(i), thinking of embedding in a uniform product of uniform intervals, one readily sees from 3.12 that the separated-completion reflector γ in Unif acts on the coarsest right inverse $\mathcal{C}^*$ of L:Unif $\to$ Creg to produce the Stone-Čech compactification, uniformized; in fact one has the well-known natural isomorphism $\gamma\mathcal{C}^* \simeq \mathcal{C}^*\beta$. There is an exactly analogous situation in Example 3.13(ii), discovered by Salbany [35] and further discussed by the author [6], in which the bitopological analogue of compactness is double compactness, i.e. the infimum (with respect to $\leq_M$, which means finer!) of the two topologies is compact in the usual sense, and the quasi-uniform analogue of completeness is double completeness, i.e. the infimum of the quasi-uniformity and its conjugate is a complete uniformity. For 3.13(iii), see [5]. Returning to Unif $\xrightarrow{L}$ Creg $\xrightarrow{M}$ Ens, and taking completions of other right inverses than the coarsest, one has reflections to generalized compactness properties. For analogous completions, see Narici, Beckenstein and Bachman [30].

3.15 Example If $X_o = LA_o$ is an $L, I_{\mathfrak{X}}$-pivot and $F = [L, I_{\mathfrak{X}}:\{A_o\}]$ exists, then not only is F the coarsest right inverse of L, but also F gives each object of $\mathfrak{X}$ the L-coarsest $\mathbf{A}$-structure. Hence, for instance, the forgetful functor Unif → Creg is not topological: not every completely regular space admits a coarsest uniformity.

We shall call $F:\mathfrak{X} \to \mathbf{A}$ a simple right inverse of L if there exists an object A of $\mathbf{A}$ such that $F = [L, I_{\mathfrak{X}}:\{A\}]$. Functors of this form are always right inverses of L (3.1), and in this form the definition of "simple" has the advantage of being intrinsic: there is no reference to an extraneous M.

3.16 Proposition (1) If F is a right inverse of L and $F = [L,M:\{A\}]$ where $A \in \mathrm{ob}\mathbf{A}$, then $F = [L, I_{\mathfrak{X}}:\{A\}]$, so that F is simple.

(2) Let $\mathbf{A}$ have small products and let ML preserve products. If $\mathfrak{U}$ is a set of objects of $\mathbf{A}$ such that the functor $F = [L,M:\mathfrak{U}]$ is a right inverse of L, then F is simple.

Proof (1) 3.8. (2) It is routine to check that if A is the product of the objects in the set $\mathfrak{U}$, then $F = [L,M:\{A\}]$. Then (1) applies.

The author does not know a suitable converse to the above proposition.

3.17 Examples (1) Consider the usual forgetful functors L:Qun → Top, M:Top → Ens. For each infinite cardinal number m let $\mathfrak{U}_m$ be the class of all quasi-uniform spaces of cardinality not exceeding m, and let $F_m = [L,M:\mathfrak{U}_m]$. As the class $\mathfrak{U}_m$ can be skeletized to a set and $L\mathfrak{U}_m$ is M-spanning, by 3.16 F_m is a simple right inverse of L. It was shown in [5] that if if $m < n$, then F_m is strictly coarser than F_n. Thus L has a proper class of right inverses, forming a large-complete lattice by 3.11. If G is any simple right inverse of L, say $G = [L, I_{\mathfrak{X}}:\{A\}]$, we may take m large enough that $A \in \mathcal{U}_m$, and as by 3.8 $F_m = [L, I_{\mathfrak{X}}:\mathcal{U}_m]$, it is clear that $F_m \leq G$. Hence the finest right inverse of L is not simple. Halpin [14] proved that there exists another non-simple right inverse of L: the finest which gives transitive quasi-uniformities (in the sense of [11]) to topological spaces.

(2) Similarly it can be shown that the forgetful functor Unif → Creg has a proper class of right inverses, of which the finest is non-simple.

4. Bireflectors as structure functors If $\mathcal{B}$ is a reflective subcategory of $\mathfrak{X}$ with reflection functor $R:\mathfrak{X} \to \mathcal{B}$, then we call the composite $\mathfrak{X} \xrightarrow{R} \mathcal{B} \hookrightarrow \mathfrak{X}$ a reflector in $\mathfrak{X}$. It is a bireflector if the universal morphisms $X \to RX$ are bimorphisms.

Bireflectors given by initiality are of widespread occurrence (cf., e.g., the talks [4] and [25] in the present conference). We apply our formalism to one aspect, and refer the reader to [16, Appendix A] for other.

Throughout this section $M:\mathfrak{X} \to \mathcal{C}$ will be an amnestic functor. M is called transportable [17], [16, p.110] if for each $Y \in \text{ob}\,\mathfrak{X}$, each $C \in \text{ob}\,\mathcal{C}$ and each isomorphism $g:C \to MY$, there exists $X \in \text{ob}\,\mathfrak{X}$ and an isomorphism $f:X \to Y$ such that $Mf = g$. Transportability is self-dual. Each topological functor is transportable. (The weaker definition of topologicity in [17] and [15] does not imply transportability.)

We consider the situation

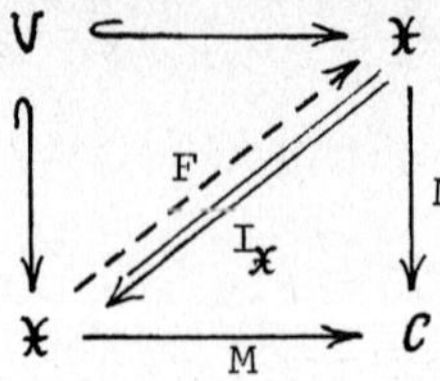

As $I_{\mathfrak{X}}$ preserves initiality for M, we have the following special case of Theorem 2.3:

4.1 <u>Proposition</u> For a functor $F:\mathfrak{X} \to \mathfrak{X}$ the following are equivalent:

(1) $FF = F$ and $I_{\mathfrak{X}} \leq_M F$;

(2) There exists a subcategory $\mathcal{V}$ of $\mathfrak{X}$ such that $F = [I_{\mathfrak{X}}, M{:}\mathcal{V}]$.

Moreover the largest subcategory $\hat{\mathcal{V}}$ for which condition (2) holds has object class $\{X \in \text{ob}\mathfrak{X} \mid FX = X\}$ and this class is isomorphism-closed.

<u>Proof</u> In 2.3, $\hat{\mathcal{V}}$ is isomorphism-closed by its maximality; also $\text{ob}\hat{\mathcal{V}} = \{X \mid FX \leq_M X\}$, which by $I_{\mathfrak{X}} \leq_M F$ reduces to $\{X \mid FX = X\}$.

4.2 <u>Proposition</u> If $FF = F$ and $I_{\mathfrak{X}} \leq_M F$, then F is a bireflector in $\mathfrak{X}$.

<u>Proof</u> We have $r_X{:}X \to FX$ with $Mr_X = 1$, so that r_X is a bimorphism and a reflection to the $\hat{\mathcal{V}}$ of 4.1.

The commonest examples of bireflectors $[I_{\mathfrak{X}}, M{:}\mathcal{V}]$ are

(i) <u>Top</u> onto <u>Creg</u>, with $\mathcal{V} = \{\mathbb{R}\}$ and $M{:}$<u>Top</u> $\to$ <u>Ens</u> forgetful;

(ii) <u>Unif</u> onto {Totally bounded uniform spaces}, with $\mathcal{V}$ consisting of the usual uniform space on $[0,1]$ and $M{:}$<u>Unif</u> $\to$ <u>Ens</u> forgetful.

4.3 <u>Proposition</u> Let M be transportable and send bimorphisms to isomorphisms. If F is a bireflector in $\mathcal{X}$ onto the subcategory $\mathcal{B}$, then the functor $[I_{\mathcal{X}}, M:\mathcal{B}]$ exists and is naturally isomorphic to F.

<u>Proof</u> We have the universal morphism $r_X: X \to FX$ such that Mr_X is iso. By the transportability there exists GX and an isomorphism $n_X: GX \to FX$ with $Mn_X = Mr_X$. For $f: X' \to X$ we let $Gf = n_X^{-1}.Ff.n_{X'}$. This defines a functor $G: \mathcal{X} \to \mathcal{X}$ and a natural isomorphism $n: G \to F$. Letting $s_X = n_X^{-1} r_X: X \to GX$ we have $Ms_X = 1$, so that $I_{\mathcal{X}} \leq_M G$. Hence $G \leq_M GG$. The reflection property of $s_{GX}: GX \to GGX$ gives $k_X: GGX \to GX$ with $k_X s_{GX} = 1$, whence $Mk_X = 1$. Hence $GG = G$. Letting $\mathrm{ob}\,\hat{\mathcal{V}} = \{X \mid GX = X\}$ we have by 4.1 $G = [I_{\mathcal{X}}, M:\hat{\mathcal{V}}]$, and as $\hat{\mathcal{V}}$ is iso-closed, $\mathrm{ob}\,\hat{\mathcal{V}} = \{X \mid FX \simeq X\} = \mathrm{ob}\,\mathcal{B}$, so that $G = [I_{\mathcal{X}}, M:\mathcal{B}]$.

4.4 <u>Proposition</u> Let M be transportable and send bimorphisms to isomorphisms, and let $F = [I_{\mathcal{X}}, M:\mathcal{V}]$. Then the bireflective hull of $\mathcal{V}$ has object class $\{X \in \mathrm{ob}\,\mathcal{X} \mid FX = X\}$.

<u>Proof</u> Let $\hat{\mathcal{V}}$ be the full subcategory with objects $X = FX$. Then $\hat{\mathcal{V}}$ is iso-closed and bireflective. Suppose $\mathcal{B}$ is any iso-closed bireflective subcategory between $\mathcal{V}$ and $\mathcal{X}$. By 4.3 the bireflector for $\mathcal{B}$ is $G = [I_{\mathcal{X}}, M:\mathcal{B}]$. If $X \in \mathrm{ob}\,\hat{\mathcal{V}}$, then $X = FX = [I_{\mathcal{X}}, M:\mathcal{V}]X$, which means that the source $(X, f:X \to V \mid V \in \mathcal{V},\ f \in \mathcal{X}(X,V))$ is M-initial. As $\mathrm{ob}\,\mathcal{V} \subset \mathrm{ob}\,\mathcal{B}$, also the source $(X, f:X \to B \mid B \in \mathrm{ob}\,\mathcal{B},\ f \in \mathcal{X}(X,B))$ is M-initial, which means that $X = GX$, so that $X \in \mathrm{ob}\,\mathcal{B}$.

4.5 <u>Theorem</u> Let $M: \mathcal{X} \to \mathcal{C}$ be topological and send bimorphisms to isomorphisms. Then a subcategory $\mathcal{V}$ of $\mathcal{X}$ is M-spanning if and only if the bireflective hull of $\mathcal{V}$ is $\mathcal{X}$.

Proof If V is M-spanning, then $I_{\mathfrak{X}} = [I_{\mathfrak{X}}, M:V]$ whence by 4.4 the bireflective hull of V is $\mathfrak{X}$. Conversely, as M is topological, $F = [I_{\mathfrak{X}}, M:V]$ exists. Now the bireflective hull of V is $\mathfrak{X}$, so that by 4.4 $\mathrm{ob}\,\mathfrak{X} = \{X \mid FX = X\}$, i.e. $F = I_{\mathfrak{X}}$ and V is M-spanning.

We observe that all proofs in this section go through if we replace bireflections by M-isoreflections, i.e. reflections whose universal morphisms are sent by M to isomorphisms. Then the assumption that M sends bimorphisms to isomorphisms becomes superfluous, and we have, for example:

4.6 Corollary Let $M: \mathfrak{X} \to \mathcal{C}$ be topological. Then a subcategory V of $\mathfrak{X}$ is M-spanning if and only if the M-isoreflective hull of V is $\mathfrak{X}$.

References

[1] Antoine, P. Etude élémentaire des catégories d'ensembles structurées. Bull. Soc. Math. Belg. 18(1966), 142-164 and 387-414.

[2] Bentley, H.L. and H. Herrlich. Extensions of topological spaces. Preprint.

[3] Blanchard, A.A. Structure species and constructive functors. Canad. J. Math. 26(1974), 1217-1227.

[4] Bourdaud, G. Some cartesian closed topological categories of convergence spaces. Proc. Conf. Mannheim 1975 on Categorical Topology

[5] Brümmer, G.C.L. Initial quasi-uniformities. Nederl. Akad. Wetensch., Proc. Ser. A 72 = Indag. Math. 31(1969), 403-409.

[6] Brümmer, G.C.L. A categorial study of initiality in uniform topology. Thesis, Univ. Cape Town 1971.

[7] Brümmer, G.C.L. Struktuurfunktore en faktorisering. Proc. S.Afr. Math. Soc. 4(1974), 81-83.

[8] Brümmer, G.C.L. and R.-E. Hoffmann. An external characterization of topological functors. Proc. Conf. Mannheim 1975 on Categorical Topology

[9] Čech, E. Topological spaces. Revised edition, Z. Frolík and M. Katětov, eds. Prague, London, New York 1966.

[10] Ertel, H.-G. Topologische Algebrenkategorien. Arch. Math. (Basel) 25(1974), 266-275.

[11] Fletcher, P. and W.F. Lindgren. Quasi-uniformities with a transitive base. Pacific J. Math. 43(1972), 619-631.

[12] Gilmour, C.R.A. Special morphisms for zero-set spaces. Bull. Austral. Math. Soc., to appear.

[13] Gray, J.W. Fibred and cofibred categories. Proc. Conf. on Categorical Algebra, La Jolla 1965, pp. 21-83. Springer-Verlag, Berlin 1966.

[14] Halpin, M.N. Transitive quasi-uniform spaces. M.Sc. thesis, Univ. Cape Town 1974.

[15] Herrlich, H. Topological functors. General Topology and Appl. 4(1974), 125-142.

[16] Herrlich, H. Topological structures. Math. Centre Tracts (Amsterdam) 52(1974), 59-122.

[17] Hoffmann, R.-E. Die kategorielle Auffassung der Initial- und Finaltopologie. Thesis, Univ. Bochum 1972.

[18] Hoffmann, R.-E. On the centre of a category. Math. Nachrichten, to appear.

[19] Hoffmann, R.-E. (E,M)-universally topological functors. Habilitationsschrift, Univ. Düsseldorf 1974.

[20] Hoffmann, R.-E. Topological functors admitting generalized Cauchy-completions. Proc. Conf. Mannheim 1975 on Categorical Topology

[21] Hunsaker, W.N. and P.L. Sharma. Proximity spaces and topological functors. Proc. Amer. Math. Soc. 45(1974), 419-425.

[22] Hušek, M. Generalized proximity and uniform spaces I. Comment. Math. Univ. Carolinae 5(1964), 247-266.

[23] Hušek, M. Categorial methods in topology. Proc. Symposium Prague 1966 on General Topology, pp. 190-194. New York, London, Prague 1967.

[24] Hušek, M. Construction of special functors and its applications. Comment. Math. Univ. Carolinae 8(1967), 555-566.

[25] Hušek, M. Lattices of reflective and coreflective subcategories of continuous structures. Proc. Conf. Mannheim 1975 on Categorical Topology

[26] Kamnitzer, S.H. Protoreflections, relational algebras and topology. Thesis, Univ. Cape Town 1974.

[27] Lane, E.P. Bitopological spaces and quasi-uniform spaces. Proc. London Math. Soc. (3) 17(1967), 241-256.

[28] Marny, T. Rechts-Bikategoriestrukturen in topologischen Kategorien. Thesis, Freie Univ. Berlin 1973.

[29] Marny, T. Top-Kategorien. Lecture Notes, Freie Univ. Berlin 1973.

[30] Narici, L., E. Beckenstein and G. Bachman. Some recent developments on repletions and Stone-Čech compactifications of 0-dimensional spaces. TOPO 72, pp. 310-321. Lecture Notes in Math. 378, Springer-Verlag, Berlin 1974.

[31] Pultr, A. On full embeddings of concrete categories with respect to forgetful functors. Comment. Math. Univ. Carolinae 9(1968), 281-305.

[32] Reynolds, G.D. Adequacy in topology and uniform spaces. TOPO 72, pp. 385-398. Lecture Notes in Math. 378, Springer-Verlag, Berlin 1974.

[33] Roberts, J.E. A characterization of topological functors. J. Algebra 8(1968), 181-193.

[34] Rosický, J. Full embeddings with a given restriction. Comment. Math. Univ. Carolinae 14(1973), 519-540.

[35] Salbany, S. Bitopological spaces, compactifications and completions. Thesis, Univ. Cape Town 1970. Math. Monographs Univ. Cape Town No. 1, 1974.

[36] Salbany, S. Quasi-uniformities and quasi-pseudometrics. Math. Colloq. Univ. Cape Town 6(1970-71), 88-102.

[37] Salbany, S. Lifting functors defined on separated subcategories. Math. Colloq. Univ. Cape Town 7(1971-72), 33-37.

[38] Tholen, W. Relative Bildzerlegungen und algebraische Kategorien. Thesis, Univ. Münster 1974.

[39] Wischnewsky, M.(B.) Partielle Algebren in Initialkategorien. Math. Zeitschr. 127(1972), 83-91.

[40] Wischnewsky, M.(B.) Initialkategorien. Thesis, Univ. München 1972.

[41] Wischnewsky, M.B. Generalized universal algebra in initialstructure categories. Algebra-Berichte No. 10. Verlag Uni-Druck, München 1973.

[42] Wischnewsky, M.B. Aspects of universal algebra in initialstructure categories. Cahiers Topologie Géom. Diff. 14(1974), 1-27.

[43] Wischnewsky, M.B. Coalgebras in reflective and coreflective subcategories. Algebra Universalis 4(1974), 328-335.

[44] Wyler, O. On the categories of general topology and topological algebra. Arch. Math. (Basel) 22(1971), 7-17.

[45] Wyler, O. Top categories and categorical topology. General Topology and Appl. 1(1971), 17-28.

Addendum: The following deals with top categories, in particular with O. Wyler's taut lifting theorem:

Harder, A. Topologische Kategorien. Diplomarbeit, Univ. Münster 1974.

Topology Research Group
University of Cape Town
Rondebosch 7700
South Africa

AN EXTERNAL CHARACTERIZATION OF TOPOLOGICAL FUNCTORS

G.C.L. Brümmer and R.-E. Hoffmann

0. Introduction

Our concept of topological functor coincides with the transportable absolutely topological functors in the sense of Horst Herrlich [5]. All topological functors are faithful [6], [5]. Our main result is that a faithful functor T is topological if and only if whenever K is a fully faithful functor and M and W are faithful functors such that the outer square in the diagram

(1)

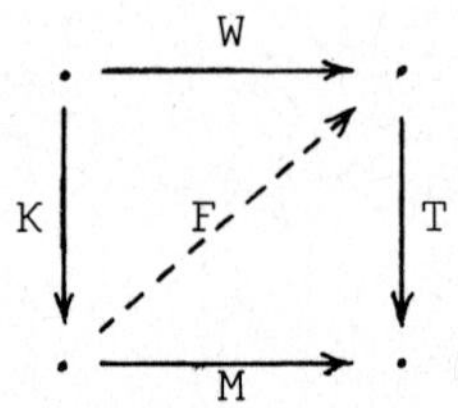

commutes, there exists a functor F making the diagram commute. As one implication is only strictly true up to natural isomorphism, we restrict ourselves to those faithful functors which lift isomorphisms uniquely.

The first author acknowledges financial aid from the South African Council for Scientific and Industrial Research and from the University of Cape Town to the Topology Research Group. Both authors acknowledge the hospitality of the Mathematisches Forschungsinstitut Oberwolfach.

We have to emphasize that our topological functors are not required to satisfy a fibre-smallness condition, but are assumed to solve initiality problems over arbitrary - including properly large - index classes. Only when dealing with classes of diagonals of the diagram (1) do we pass to a higher universe; the main theorem stays within a single universe.

Amongst the applications we give an external characterization of Oswald Wyler's [10] taut lifts, and the following result: A full subcategory $\mathcal{S}$ of a topological category $\mathbf{A}$ is topological if and only if there is a retraction of $\mathbf{A}$ onto $\mathcal{S}$ which respects underlying structure.

The diagonal F in the diagram (1) is in general not unique. In Proposition 1.2 we shall construct the T - coarsest diagonal. The construction was used in [7] (without emphasizing the diagonal situation) by Miroslav Hušek, who has kindly pointed out that the said proposition can be deduced from [7, Theorem 7, Corollary]. For expository reasons we again give a full proof.

Further applications of the Hušek construction, now explicitly from the diagonal point of view, are given in [3].

1. <u>Factorizing faithful functors</u>

We shall mostly work in a single universe $\mathfrak{U}$.

1.1 <u>Definitions</u> Let $T : \mathbf{A} \to \mathcal{C}$ be a functor.

(1) A <u>source</u> in $\mathcal{C}$ is a pair $(C, (f_i)_I)$ where $f_i : C \to C_i$

are $\mathcal{C}$-morphisms indexed over a $\mathcal{U}$-class I. The class I may be a proper class or a set, and it may be void. The usual notation for this source will be $(C, f_i : C \to C_i)_I$.

(2) A T-initiality problem consists of an indexed family $(A_i)_I$ of objects in $\mathbf{A}$ together with a source $(C, f_i : C \to TA_i)_I$ in $\mathcal{C}$. We shall denote this problem by $(C, f_i : C \to TA_i, A_i)_I$.

(3) A solution to the T-initiality problem in (2) is any source $(A, f_i' : A \to A_i)_I$ in $\mathbf{A}$ with the following properties:

(i) $TA = C$ and $Tf_i' = f_i$ for each i in I;

(ii) given any source $(B, g_i : B \to A_i)_I$ in $\mathbf{A}$ and any morphism $u : TB \to TA$ such that $f_i u = Tg_i$ for each i, there exists unique $v : B \to A$ such that $f_i' v = g_i$ for each i and $Tv = u$.

(4) The functor $T : \mathbf{A} \to \mathcal{C}$ is topological if each T-initiality problem has a solution (the index classes of the problems may be proper $\mathcal{U}$-classes and may be void).

(5) As it is known from [6] and [5] that each topological functor (even under a weaker definition) is faithful (provided the domain category has small hom-sets, which we assume), we shall henceforth consider only faithful functors. This induces a simplification in (3)(ii) above: the uniqueness of v and the equations $f_i' v = g_i$ follow from the other conditions.

(6) With Hušek [7] we shall call the functor $T : \mathbf{A} \to \mathcal{C}$ amnestic if it is faithful and, whenever f is an isomorphism with $Tf = 1$, then $f = 1$. Equivalently: T is faithful and each T-initiality problem has at most one solution.

(7) If $F_1, F_2 : \mathfrak{X} \to \mathbf{A}$ are functors, we write $F_1 \leq_T F_2$ and say

F_1 is T - <u>finer</u> than F_2, or F_2 is T - <u>coarser</u> than F_1, if there exists a natural transformation $n : F_1 \to F_2$ such that Tn is the identity natural transformation $TF_1 = TF_2$. Observe that if $F_1 \leq_T F_2$ and $F_2 \leq_T F_1$ and T is amnestic, then $F_1 = F_2$.

1.2 <u>Proposition</u> Let $T : \mathcal{A} \to \mathcal{C}$ be topological and amnestic. Let $M : \mathcal{X} \to \mathcal{C}$, $K : \mathcal{D} \to \mathcal{X}$ and $W : \mathcal{D} \to \mathcal{A}$ be faithful functors, with K full, such that $TW = MK$. Then there exists a functor $F : \mathcal{X} \to \mathcal{A}$ such that $TF = M$ and $FK = W$.

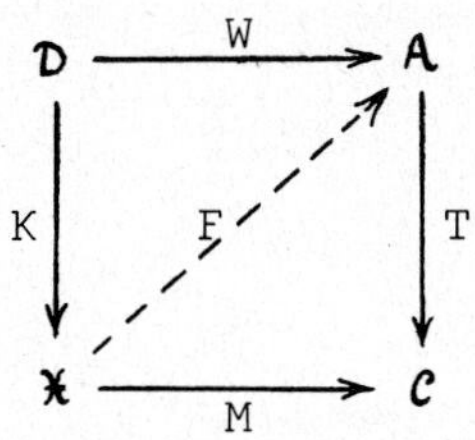

<u>Proof</u> For X in $\mathrm{ob}\,\mathcal{X}$ we define FX as follows: The T - initiality problem $(MX, Mf : MX \to MKD, WD)$ in which D ranges through $\mathrm{ob}\,\mathcal{D}$ and f ranges through the hom-set $\mathcal{X}(X, KD)$ has a unique solution which we denote by $(FX, (Mf)' : FX \to WD)$ with D and f ranging as said. For any morphism $g : X' \to X$ in $\mathcal{X}$ we define Fg as follows:

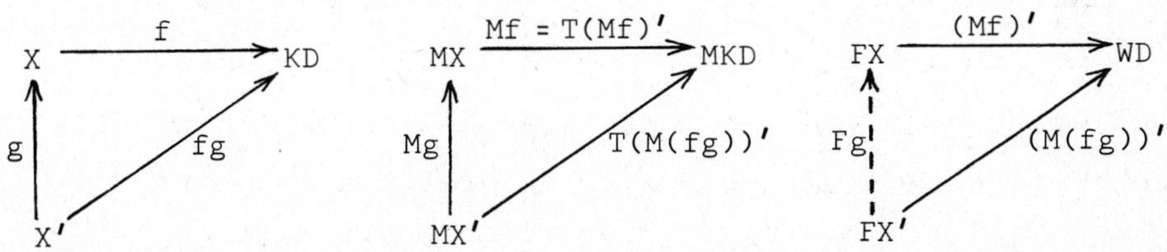

Indeed, since the source $(FX', (M(fg))' : FX' \to WD)$ with D and f ranging as above is such that $T(Mf)'.Mg = T(M(fg))'$ for each f, by 1.1(3)(ii) there exists unique $Fg : FX' \to FX$ with $TFg = Mg$. Functoriality of F is readily verified, and one has $TF = M$. To prove $FK = W$, consider a fixed object D_0 of $\mathcal{D}$, and all $f : KD_0 \to KD$, $D \in \text{ob}\,\mathcal{D}$. For each such f , since K is full, there exists unique $\bar{f} : D_0 \to D$ with $K\bar{f} = f$. The source $(WD_0, W\bar{f} : WD_0 \to WD)$ (indexed over f as before) is then the solution of the T-initiality problem $(MKD_0, Mf : MKD_0 \to MKD, WD)$; indeed, $TW\bar{f} = MK\bar{f} = Mf$ and if we are given any source $(A, g_f : A \to WD)$ and any $u : TA \to MKD_0$ such that $TW\bar{f}.u = Tg_f$ for each f , then $g_{1_{D_0}} : A \to WD_0$ is such that $Tg_{1_{D_0}} = u$. But by definition the source $(FKD_0, (Mf)' : MKD_0 \to WD)$ is the solution to the above problem; hence $WD_0 = FKD_0$. This with $TW = TFK$ gives $W = FK$, since T is faithful.

By a <u>diagonal</u> of the equation $TW = MK$ (or of the commutative square drawn as above) we shall mean any functor F such that $TF = M$ and $FK = W$.

1.3 Under the assumptions of Proposition 1.2, the functor $F : \mathfrak{X} \to \mathbf{A}$ constructed in the above proof is the T-coarsest diagonal of the given square.

<u>Proof</u> Given any diagonal $G : \mathfrak{X} \to \mathbf{A}$. For any $f : X \to KD$ the left hand triangle

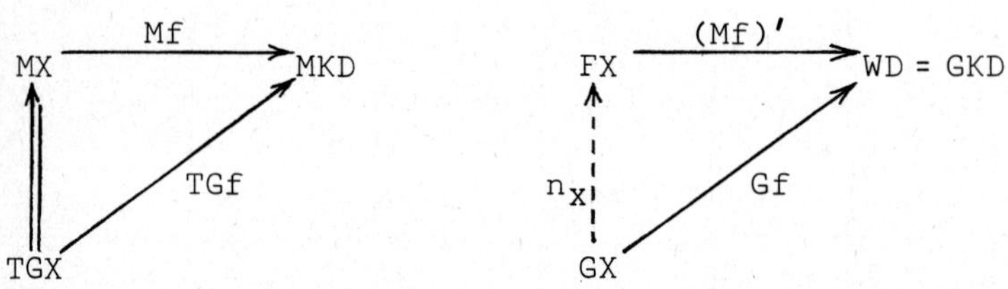

commutes. Hence there exists $n_X : GX \to FX$ with $Tn_X = 1$. Thus $G \leq_T F$.

1.4 We give some illustrations of the occurrence of the T - coarsest diagonal.

(1) Non-uniqueness: With T : Top → Ens the usual forgetful functor and ∅ the empty category, the diagram

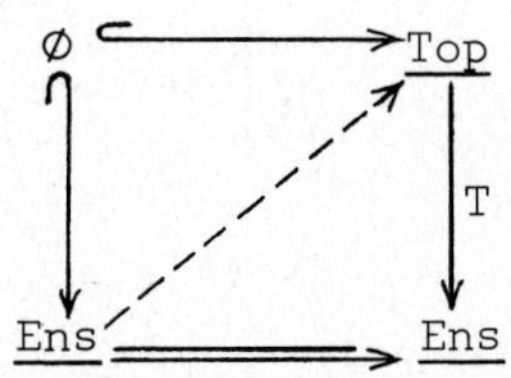

has exactly two diagonals Ens → Top [2, p.56].

(2) Inducing metric topology: Let $\mathcal{R}$ be the category having one object, the real line $\mathbb{R}$ with the usual metric, and morphisms the non-expansive mappings. Let Mne be the category of metric spaces and non-expansive mappings, and M : Mne → Ens and T : Top → Ens the forgetful functors. Let K : $\mathcal{R}$ → Mne be the inclusion functor and W : $\mathcal{R}$ → Top the functor which gives $\mathbb{R}$ the usual topology.

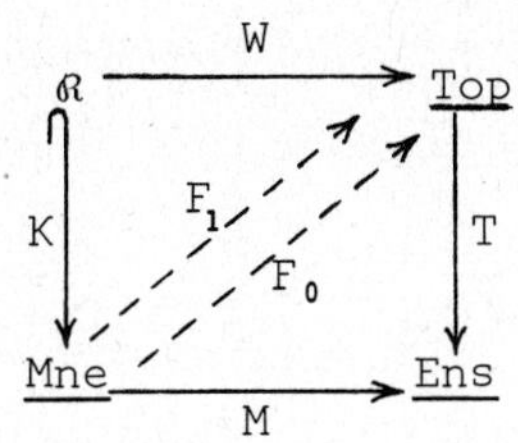

Then the T - coarsest diagonal F_1 : Mne → Top gives each metric space the metric topology, and the T - finest diagonal F_0 is distinct from F_1 .

Proof For a metric space (X, d) , consideration of the mappings d(a, -) : X → $\mathbb{R}$ shows that the metric topology on X gives the solution to the T - initiality problem posed by all non-expansive mappings $(X, d) \to K\mathbb{R}$ and $W\mathbb{R}$; thus $F_1(X, d)$ has the metric topology. On the other hand, F_0 is given by a T-coinitiality construction (dual to that of F_1), and one readily sees that F_0 gives each totally disconnected metric space the discrete topology.

(3) <u>Completing upward</u>

Each commutative triangle of faithful functors

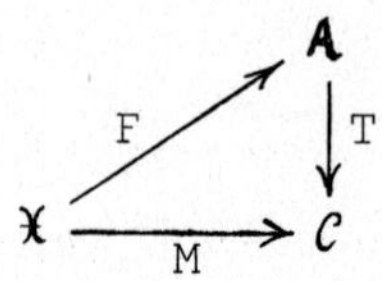

can be completed to a square of faithful functors,

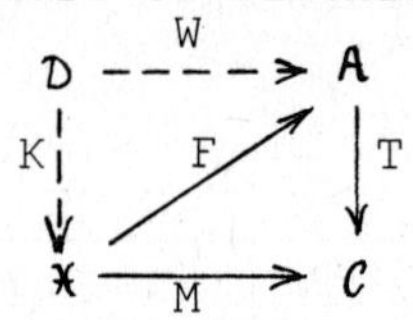

with K full, in which F is the T-coarsest diagonal (e.g. let $\mathcal{D} = \mathcal{X}$, $K = I_{\mathcal{X}}$, $W = F$).

(4) <u>Completing downward</u>

Each commutative triangle of faithful functors

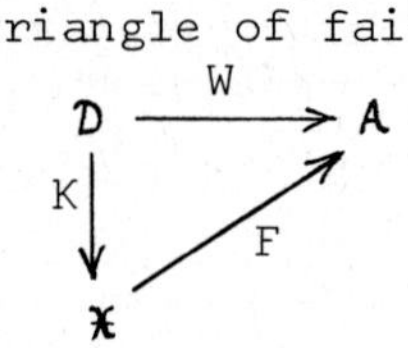

can be completed to a square of faithful functors,

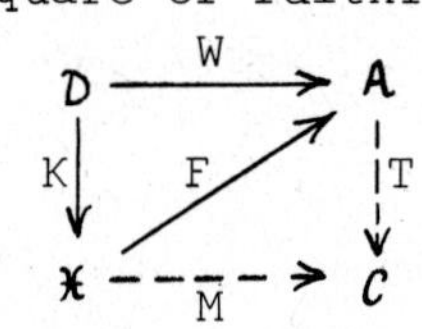

with T topological, in which F is the T-coarsest diagonal (e.g. let $\mathcal{C} = \mathcal{A}$, $T = I_{\mathcal{A}}$, $M = F$).

1.5 <u>Theorem</u> If $T : \mathcal{A} \to \mathcal{C}$ is an amnestic functor, then the following conditions are equivalent:

(1) T is topological;

(2) For arbitrary amnestic functors M, K, W with K full, such that the outer square in

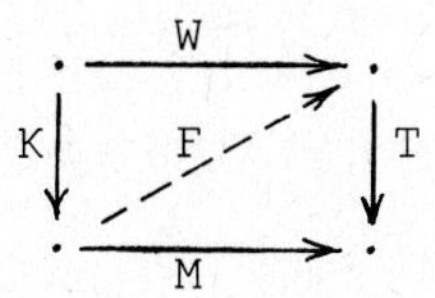

commutes, there exists a diagonal F of the above diagram;

(3) Same as (2) but with K a full embedding.

Moreover, if T is merely assumed faithful instead of amnestic, then still (2) ⇒ (3) ⇒ (1).

<u>Proof</u> By 1.2, (1) ⇒ (2). Trivially (2) ⇒ (3). To see that (3) ⇒ (1), assuming T faithful, consider a T-initiality problem $(C, f_i : C \to TA_i, A_i)_I$. We construct a category $\mathcal{X}$ as follows: We arrange that the classes $\mathrm{ob}\,\mathcal{A}$, $\{C\}$ and I are mutually disjoint and define

$$\mathrm{ob}\,\mathcal{X} = \mathrm{ob}\,\mathcal{A} \cup \{C\} \cup I .$$

The hom-sets in $\mathcal{X}$ are defined as follows (with arbitrary $A \in \mathrm{ob}\,\mathcal{A}$, $i \in I$):

$$\mathcal{X}(A, C) = \{A \xrightarrow{u} C \mid u \in \mathcal{C}(TA, C) \wedge (\forall\, j \in I)\ [\exists\, g \in \mathcal{A}(A, A_j) \quad Tg = f_j u]\} ;$$

(The notation $A \xrightarrow{u} C$ is a convenient way of writing a 3-tuple (A, u, C) and serves to make hom-sets disjoint. Observe that in the definition of $\mathcal{X}(A, C)$, as T is faithful, g is uniquely determined by A, u and j, so that we may write $g = g_{A,u,j}$) ;

$\mathfrak{X}(C, i) = \{C \xrightarrow{f_i} i\}$;

$\mathfrak{X}(A, i) = \{A \xrightarrow{g} i \mid g \in \mathbf{A}(A, A_i) \wedge [\exists\ u \in \mathfrak{X}(A, C) \quad Tg = f_i u]\}$;

$\mathfrak{X}(A, A) = \{1_A\}$; $\mathfrak{X}(C, C) = \{1_C\}$; $\mathfrak{X}(i, i) = \{1_i\}$;

$\mathfrak{X}(i, C) = \emptyset$; $\mathfrak{X}(C, A) = \emptyset$; $\mathfrak{X}(i, A) = \emptyset$.

As for composition in $\mathfrak{X}$, we posit that 1_A, 1_C and 1_i shall act as left and right identities at the corresponding objects; then only one case of composition remains to define:

$$A \xrightarrow{u} C \xrightarrow{f_i} i = A \xrightarrow{g_{A,u,i}} i .$$

Herewith $\mathfrak{X}$ is a category. We now define a functor $M : \mathfrak{X} \to \mathcal{C}$ as follows:

$MA = TA$; $MC = C$; $Mi = TA_i$;

$M1_A = 1_{TA}$; $M1_C = 1_C$; $M1_i = 1_{TA_i}$;

$M(A \xrightarrow{u} C) = u$; $M(C \xrightarrow{f_i} i) = f_i$; $M(A \xrightarrow{g} i) = Tg$.

Clearly M is a faithful functor, and as the isomorphisms in $\mathfrak{X}$ are identities, M is amnestic.

We define $\mathcal{D}$ to be the full subcategory of $\mathfrak{X}$ with $\mathrm{ob}\,\mathcal{D} = \mathrm{ob}\,\mathbf{A} \cup I$, and we let $K : \mathcal{D} \to \mathfrak{X}$ be the inclusion functor. Thus K is full and amnestic.

We define a functor $W : \mathcal{D} \to \mathbf{A}$ as follows:

$WA = A$; $Wi = A_i$; $W1_A = 1_A$;

$W1_i = 1_{A_i}$; $W(A \xrightarrow{g} i) = g$.

Then W is a faithful functor, and as the isomorphisms in $\mathcal{D}$ are identities, W is amnestic. Clearly $TW = MK$, so that by (3) there exists a functor $F : \mathfrak{X} \to \mathbf{A}$ with $TF = M$ and $FK = W$. We claim that the source $(FC, F(C \xrightarrow{f_i} i))_I$ in $\mathbf{A}$ is the solution to the given T-initiality problem $(C, f_i : C \to TA_i, A_i)_I$. Firstly, the codomain of $F(C \xrightarrow{f_i} i)$ is $Fi = FKi = Wi = A_i$, and $TFC = MC = C$, and $TF(C \xrightarrow{f_i} i) = M(C \xrightarrow{f_i} i) = f_i$. Secondly, given any source $(A, g_i : A \to A_i)_I$ in $\mathbf{A}$ and any $u : TA \to C$ with $Tg_i = f_i u$ for all i , then $A \xrightarrow{u} C \in \mathfrak{X}(A, C)$ so that $F(A \xrightarrow{u} C) \in \mathbf{A}(FA, FC) = \mathbf{A}(A, FC)$ (since $FA = FKA = WA = A$) , and $TF(A \xrightarrow{u} C) = M(A \xrightarrow{u} C) = u$. Q.E.D.

It is intended to present a proof of the following variant of the main result elsewhere:

1.6 <u>Theorem</u> Let T be a faithful functor. If the following condition is fulfilled, then T is topological:

For arbitrary faithful functors M and K , with K a full embedding, such that the outer square in

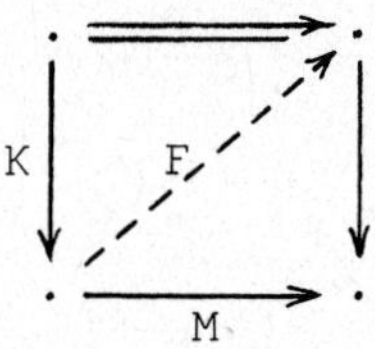

commutes, there exists a diagonal F of the above diagram.

For the sake of the following result we assume that our given universe $\mathfrak{U}$ is a set in a universe $\mathfrak{U}^+$.

1.7 <u>Proposition</u> Let the outer square in

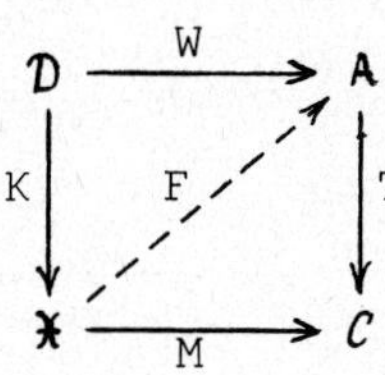

commute and let T be topological and amnestic, K fully faithful, W and M faithful. Under the ordering $\leq_T$, the class of all diagonals F forms a complete lattice in $\mathfrak{U}^+$.

Proof For any $C \in \mathrm{ob}\,\mathcal{C}$ the partial order $\leq_T$ on the T-fibre $\{A \in \mathrm{ob}\,\mathcal{A} \mid TA = C\}$ is such that T-infima of subclasses again belong to the fibre; indeed, taking the T-initiality problem $(C, 1_C : C \to TA_j, A_j)_J$ and calling its solution $(A, h_j : A \to A_j)_J$, we have $A = \inf_T\{A_j \mid j \in J\}$. Therefore, given diagonals F_i $(i \in I)$ we may define $F = \inf_T\{F_i \mid i \in I\}$ by $FX = \inf_T\{F_i X \mid i \in I\}$ and the corresponding action on morphisms. By 1.3 the class of diagonals also has a T-coarsest member; hence the class is a $\leq_T$-lattice in $\mathfrak{U}^+$.

2. Elementary consequences of the main theorem

2.1 Self-duality (Antoine [1]; cf. [9], [2], [6], [5])

If $T : \mathcal{A} \to \mathcal{C}$ is topological and amnestic, so is $T^{op} : \mathcal{A}^{op} \to \mathcal{C}^{op}$.

Proof The equivalent conditions in Theorem 1.5 are clearly self-dual.

The results in the following proposition are standard consequences of the diagonalization property in a generalized factorization system [8] (cf. also [4]).

2.2 Proposition For amnestic functors we have:

(1) The composite of two topological functors is

topological.

(2) The full topological functors are precisely the isomorphisms of categories.

(3) If ST and T are topological, so is S .

(4) If ST is topological and T a retraction, then S is topological.

(5) If ST is topological and S an embedding, then T is topological.

(6) In a pullback

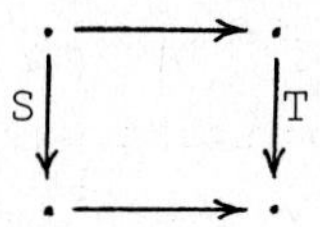

if T is topological, so is S .

(7) If $T_i : \mathcal{A}_i \to \mathcal{C}_i$ are topological, so is $\Pi T_i : \Pi \mathcal{A}_i \to \Pi \mathcal{C}_i$.

2.3 <u>Theorem</u> Let $T : \mathcal{A} \to \mathcal{C}$ be an amnestic topological functor, $\mathcal{S}$ a full subcategory of $\mathcal{A}$, and $J : \mathcal{S} \to \mathcal{A}$ the inclusion functor. The following conditions are equivalent:

(1) $TJ : \mathcal{S} \to \mathcal{A}$ is topological;

(2) There exists a functor $E : \mathcal{A} \to \mathcal{S}$ such that $EJ = I_{\mathcal{S}}$ AND $TJE = T$.

<u>Proof</u> (1) $\Rightarrow$ (2):

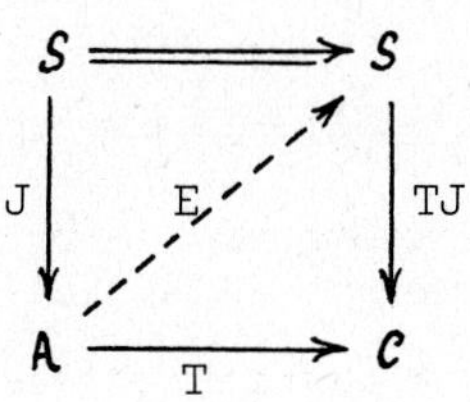

(2) ⇒ (1): Consider arbitrary faithful functors W, M, K with K full such that TJ.W = MK .

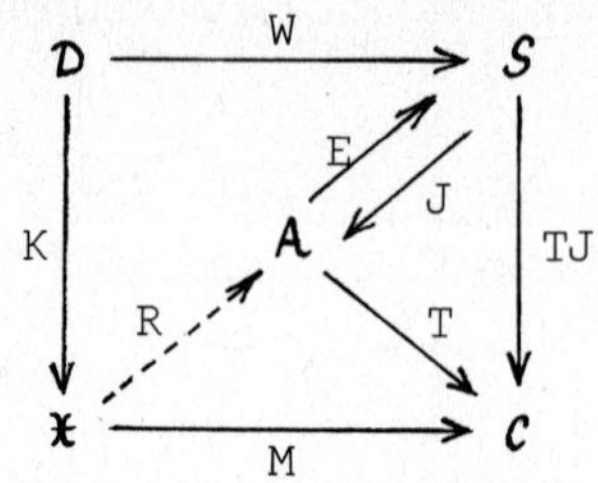

As T.JW = MK and T is topological, there exists $R : \mathcal{X} \to \mathcal{A}$ with TR = M and RK = JW . Now TJ.ER = TJE.R = TR = M , and ER.K = E.RK = E.JW = W , so that ER is a diagonal of the outer square, and TJ is topological.

3. Taut lifts

We give an external characterization of those functors which carry initial sources from one topological functor to another. These are essentially O. Wyler's taut lifts [10].

Given a functor $T : \mathcal{A} \to \mathcal{C}$, a source in $\mathcal{A}$ is called T - initial if it is a solution to a T - initiality problem. For simplicity of statement, we agree that all functors in the following theorem are amnestic.

3.1 Theorem Let $T : \mathcal{A} \to \mathcal{C}$ and $U : \mathcal{B} \to \mathcal{E}$ be topological, and V and L functors such that UV = LT . The following conditions are equivalent:

(1) V sends T - initial sources to U - initial sources;

(2) For each commutative outer square

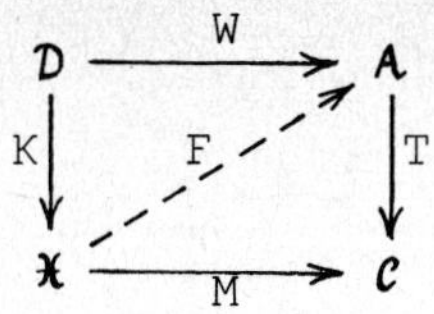

and for each diagonal $G : \mathcal{X} \to \mathcal{B}$ in

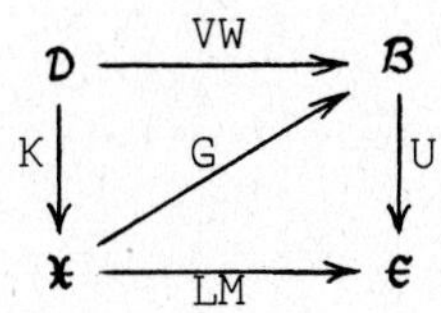

there exists a diagonal $F : \mathcal{X} \to \mathcal{A}$ in the former square such that $G \leq_U VF$.

<u>Proof</u>

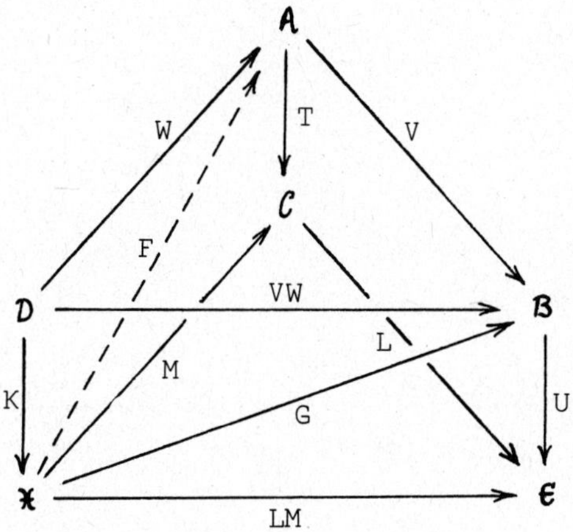

(1) ⇒ (2): Given the diagonal G between K and U , we let F be the T - coarsest diagonal between K and T . Condition (1) implies that the map $A \mapsto VA$ of any T - fibre to the corresponding U - fibre preserves infima. Hence VF is the U - coarsest diagonal between K and U , and thus $G \leq_U VF$.

(2) ⇒ (1): Given a T - initial source $(A_0, h_i : A_0 \to A_i)_I$ in $\mathbf{A}$, we write $TA_0 = C$, $Th_i = f_i$, and we construct $\mathcal{X}$, $\mathcal{D}$, K, M, W from the T - initiality problem $(C, f_i : C \to TA_i, A_i)_I$ as in the proof of 1.5.

We construct a functor $G : \mathfrak{X} \to \mathcal{B}$ as follows: let $GA = VA$ $(A \in \mathrm{ob}\,\mathfrak{A})$; let $Gi = VA_i$ $(i \in I)$; let the source $(GC, G(C \xrightarrow{f_i} i))_I$ be the solution to the U-initiality problem $(LC, Lf_i : LC \to UVA_i, VA_i)_I$ and accordingly let $G(A \xrightarrow{u} C)$ be given as follows: by the definition of $\mathfrak{X}(A, C)$ there exist $g_i \in \mathfrak{A}(A, A_i)$ with $f_i u = Tg_i$; then

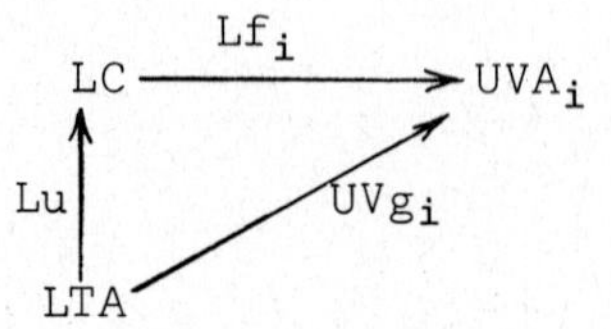

commutes and there exists unique $v : VA \to GC$ such that $Uv = Lu$; we let $G(A \xrightarrow{u} C) = v$. Finally for $A \xrightarrow{g} i$ in $\mathfrak{X}$, Gg is given by the appropriate composite, and immediately $G : \mathfrak{X} \to \mathcal{B}$ is an amnestic functor with $GK = VW$ and $UG = LM$. By (2) there now exists a diagonal F between K and T with $G \leq_U VF$. From the proof of 1.5 we know that $(FC, F(C \xrightarrow{f_i} i))_I$ is the solution to the T-initiality problem $(C, f_i, A_i)_I$; hence $FC = A_0$, $F(C \xrightarrow{f_i} i) = h_i$. We have $GC \leq_U VFC = GA_0$. Since $Th_i = f_i$, $A_0 \xrightarrow{1_{TA_0}} C$ occurs in $\mathfrak{X}(A_0, C)$ as a non-identity u , say, and $Gu : GA_0 \to GC$ is such that $UGu = LMu = L1_{TA_0} = 1$; hence $GA_0 \leq_U GC$ and thus $GC = GA_0$. Hence Vh_i and $G(C \xrightarrow{f_i} i)$ have the same domain and the same codomain; the faithful functor U sends both to the same morphism; hence $Vh_i = G(C \xrightarrow{f_i} i)$. Thus V has sent a given T-initial source to a U-initial source.

The authors acknowledge the benefit of conversations with Keith Hardie, Horst Herrlich and George Strecker during the preparation of this paper.

References

[1] Antoine, P. Etude élémentaire des catégories d'ensembles structurés. Bull. Soc. Math. Belg. 18(1966), 142-164.

[2] Brümmer, G.C.L. A categorial study of initiality in uniform topology. Thesis, Univ. Cape Town, 1971.

[3] Brümmer, G.C.L. Topological functors and structure functors. Proc. Conf. Mannheim 1975 on Categorical Topology.

[4] Freyd, P. and G.M. Kelly. Categories of continuous functors, I. J. Pure Appl. Algebra 2(1972), 169-191.

[5] Herrlich, H. Topological functors. General Topology and Appl. 4(1974), 125-142.

[6] Hoffmann, R.-E. Die kategorielle Auffassung der Initial- und Finaltopologie. Thesis, Univ. Bochum, 1972.

[7] Hušek, M. Construction of special functors and its applications. Comment. Math. Univ. Carolinae 8(1967), 555-566.

[8] Ringel, C.M. Diagonalisierungspaare, I, II. Math. Zeitschr. 117(1970), 249-266 and 122(1971), 10-32.

[9] Roberts, J.E. A characterization of initial functors. J. Algebra 8(1968), 181-193.

[10] Wyler, O. On the categories of general topology and topological algebra. Arch. Math. (Basel) 22(1971), 7-17.

G.C.L. Brümmer
Topology Research Group
University of Cape Town
Rondebosch 7700
South Africa

R.-E. Hoffmann
4 Düsseldorf
Mathematisches Institut der
Universität Düsseldorf
German Federal Republic

HOMOTOPY AND KAN EXTENSIONS

by
Allan Calder* and Jerrold Siegel

Let $\underline{T}' \subseteq \underline{T}$ be a pair of topological categories (i.e. full subcagegories of $\underline{\text{Top}}$, topological spaces and continuous maps). Let F be an arbitrary functor on T' having a right Kan extension F^K to $\underline{T}$. We will be concerned with questions of the following sort: Suppose F is a homtopy functor, what relation on the maps of $\underline{T}$ must F^K respect?

The prototype of the sort of result we have in mind is the classical theorem that the Čech extension (with respect to finite covers) of cohomology on polyhedra to normal spaces is a uniform homotopy invariant but not a homotopy invariant [3]. It is our intention to provide a general framework for understanding results of this sort.

Our method is as follows, we first consider general relations R on the maps of $\underline{T}'$. The Kan extension of R is defined to be the weakest relation R^K on the maps of $\underline{T}$ such that the Kan extension of every functor that respects R, respects R^K. A constructive definition of R^K is then given and some general observations are made. We then proceed to specific computations. We show, for example, that the Kan extension of homotopy on $\underline{P}_f$ to normal spaces, $\underline{N}$, is uniform homotopy over $\underline{P}_f$ (i.e. maps $f,g : X \to X'$ are uniformly homotopic over $\underline{P}_f$ iff πf and πg are uniformly homotopic for all $\pi : X' \to Q$, any map π into any Q in $\underline{P}_f$.)

We next show that the Čech extension (finite covers) [2] of a functor from $\underline{P}_f$ to $\underline{N}$ is naturally isomorphic to the Kan extension.

---

*Based on a talk given by the first named author at the conference.

Hence, in addition to having a different view of the Čech extension, we have a mild generalization of the classical result mentioned above.

Finally, we address ourselves to the question of when a particular homotopy functor has a Kan extension that is also a homotopy functor. The main result we obtain in this direction is that $F = [\ ,Y]$ (homotopy classes), then F^K is never a homotopy functor when Y is a finite complex with non-zero homology.

Notation Again, we will be concerned with Top, the category of topological spaces and continuous maps. $\underline{T}$, $\underline{T}'$, $\underline{T}''$, etc. will always denote some given full subcategory of Top. $\underline{P}_f$ ($\underline{P}_c$) will denote the category of finite(locally finite) polyhedra and continuous maps.

We will assume that functors below defined on $\underline{T}'$ ($F : \underline{T}' \to \underline{A}$) admit right Kan extensions [4] $F^K : \underline{T} \to \underline{A}$, a some suitable category usually Ens (sets) or Ens^{op}.

Definition 1 Given a category $\underline{T}$ a natural relation (on the maps of $\underline{T}$) is a functor $R : \underline{T} \to \tilde{\underline{T}}$, where $\tilde{\underline{T}}$ is a category with $Ob(\tilde{\underline{T}}) = Ob(\underline{T})$, R is the identity on objects and $R : Mor_T(X,Y) \to Mor_{\tilde{T}}(X,Y)$ is epi as a set map.

If $R(f) = R(g)$ we write $f \underset{R}{\sim} g$.

Finally, given a pair of categories $\underline{T}' \subseteq \underline{T}$ and a natural relation R on $\underline{T}$. R will be called the extension of $R|\underline{T}'$.

Examples (a) Homotopy induces a natural relation H on any category $\underline{T}$. We write $f \sim g$ suppressing the subsymbol.

Uniform Homotopy

(b) If $\underline{T}$ is a subcategory of the category $\underline{N}$ of normal spaces, we say f is uniformly homotopic to g as maps $X \to X'$ ($f \underset{\beta}{\sim} g$) if $\beta(f) \sim \beta(g)$ as maps $\beta X \to \beta X'$ where β is the Stone-Cech compactification. This particular form of the definition of uniform homotopy is convenient for our purposes and is easily seen to be equivalent to the Eilenberg-Steenrod definition [3]. We denote the natural relation by βH.

(c) <u>Homotopy over</u> $\underline{P}_f$ ($\underline{P}_c$). Two maps $g_1, g_2 : X \to X'$ are said to be <u>homotopic over</u> $\underline{P}_f$ if for any map $\pi : X' \to Q$ with Q in $\underline{P}_f$ we have $\pi g_1 \sim \pi g_2$. We write $g_1 \underset{f}{\sim} g_2$ ($g_1 \underset{c}{\sim} g_2$).

Similarly, for normal spaces we have uniform homotopy over $\underline{P}_f$ ($\underline{P}_c$) written $g_1 \underset{\beta f}{\sim} g_2$ ($g_1 \underset{\beta c}{\sim} g_2$). The induced relations are denoted by H_f, H_c, βH_f and βH_c respectively.

These last examples are non-trivial since if X is a single point and $X' = \{(x, \sin(\frac{1}{x}) \mid 0 < x \leq 1\} \cup \{(0,y) \mid 0 \leq |y| \leq 1\}$ then the maps $g_0(pt) = (1, \sin(1))$, $g_1(pt) = (0,0)$ are (uniformly) homotopic over $\underline{P}_f$ ($\underline{P}_c$) but not (uniformly) homotopic. Note however, $H_f | \underline{P}_f = H$.

Given any relation $R : \underline{T}' \to \underline{\tilde{T}}'$ there is a natural category theoretic extension to $\underline{T}$. It is this extension that will be of most concern to us in what follows.

<u>Definition 2</u> Let $\underline{T}' \subseteq \underline{\tilde{T}}$ and $R : \underline{T}' \to \underline{\tilde{T}}'$ be as above. <u>The Kan extension of R</u> (written R^K) is defined by setting $f \underset{R^K}{\sim} g$ iff for every functor F of the form $F = \tilde{F}R$ ($\tilde{F} : \underline{\tilde{T}} \to A$) such that F admits a Kan extension $F^K : \underline{T} \to A$ we have $F^K(f) = F^K(g)$.

The following relationships hold among the various notions mentioned above.

<u>Theorem 3</u> (a) For the pair $\underline{P}_f \subseteq \underline{\text{Comp}}_2$ (compact hausdorff) the Kan extension of homotopy is homotopy over P_f.

(b) For the pair $P_f \subseteq \underline{N}$ (normal) the Kan extension of homotopy is uniform homotopy over P_f.

Before giving a proof of Theorem 3 it is necessary to establish some elementary facts about natural relations.

<u>Definition 4</u> Given natural relations $R_1 : \underline{T} \rightarrow \underline{\tilde{T}}_1$ and $R_2 : \underline{T} \rightarrow \underline{\tilde{T}}_2$ we say $R_1 \leq R_2$ if R_1 factors through R_2 (i.e. $R_1 = QR_2$ for $Q : \tilde{T}_2 \rightarrow \tilde{T}_1$)

Note $Id : \underline{T} \stackrel{\bullet}{\rightarrow} \underline{T}$ is a natural relation and for any $R \leq Id$. We will make use of this trivial observation. We also use the following:

<u>Theorem 5</u> Let $\underline{T}' \subseteq \underline{T}$ and let $R : \underline{T}' \rightarrow \underline{\tilde{T}}'$ be a natural relation. Let $R_1 : T \rightarrow \tilde{T}_1$ be a natural relation such that $R^K \leq R_1$. Let $F : \underline{T}' \rightarrow A$ factor through R hence $R_1|\underline{T}'$. We denote by $\tilde{F}$ and $\tilde{F}_1$ the unique factorization ($F = \tilde{F}R = \tilde{F}_1 R_1|\underline{T}'$).

Then $F^K = \tilde{F}^K R = \tilde{F}_1{}^K R_1$, where = means natural equivalence of functors and $\tilde{F}^K$ and $\tilde{F}_1{}^K$ are computed over the appropriate categories.

Proof. Since we may choose $R_1 = R^K$ it suffices to check $F^K = \tilde{F}_1{}^K R_1$. This is immediate from the definition of K an extension [4 p. 232] since cones $\mu : c \stackrel{\bullet}{\rightarrow} F$ factor uniquely through the cones $\mu_1 : c \stackrel{\bullet}{\rightarrow} \tilde{F}_1$.

It is an obvious corollary of Theorem 5 that F^K factors through R_1. Again, we will make use of this observation.

We now give a constructive definition of R^K. We will base our computations on this definition.

Definition 6 Let $\underline{T}' \subseteq \underline{T}$ and $R : \underline{T}' \to \tilde{\underline{T}}'$ be as above. We define a natural relation $\hat{R}$ on $\underline{T}$ as follows. Given $f,g : X \to X'$ we say $f \underset{\hat{R}}{\sim} g$ iff for every map $\pi : X' \to Q$ with Q in $\underline{T}'$ there exists

$Q_1, Q_2, \ldots, Q_n$ in $\underline{T}'$ and maps

$\pi_i : X \to Q_i, \quad \phi_i : Q_i \to Q \quad .\ni.$

(1) $\phi_1\pi_1 = \pi f, \ \phi_n\pi_n = \pi g$

(2) for each $i \leq 1 < n$ either

(a) $\phi_i\pi_i = \phi_{i+1}\pi_{i+1}$ or

(b) $\pi_i = \pi_{i+1}$ and $\phi_i \underset{R}{\sim} \phi_{i+1}$.

Theorem 7 $\hat{R} = R^K$

Proof. To show $R^K \leq \hat{R}$ we must show that for every functor $F : \underline{T}' \to \underline{A}$ if F factors through R then F^K factors through $\hat{R}$. That is if $f \underset{\hat{R}}{\sim} g$ then $F^K(f) = F^K(g)$ which by the universal characterization of F^K is the same as showing $F^K(\pi f) = F^K(\pi g)$ for all Q in $\underline{T}'$ and all $\pi : X' \to Q$. This follows at once from the definition of $\hat{R}$ since either

(a) $F^K(\phi_i\pi_i) = F^K(\phi_{i+1}\pi_{i+1})$ since $\phi_i\pi_i = \phi_{i+1}\pi_{i+1}$ or

(b) $F^K(\phi_i\pi_i) = F(\phi_i)F^K(\pi_i) = F(\phi_{i+1})F^K(\pi_{i+1}) = F^K(\phi_{i+1}\pi_{i+1})$ since $\pi_i = \pi_{i+1}$ and $\phi_i \underset{R}{\sim} \phi_{i+1}$.

To show $\hat{R} \leq R^K$ we show $f \underset{\hat{R}}{\not\sim} g \Rightarrow f \underset{R^K}{\not\sim} g$. In particular, we exhibit a functor $F : \underline{T}' \to \underline{Ens}^{op}$ (op of sets) $.\ni.$ F factors through R yet $F^K(f) \neq F^K(g)$.

If $f \underset{\hat{R}}{\not\sim} g$, then there exists $\bar{Q}$ in $\underline{T}'$ and $\bar{\pi} : X' \to \bar{Q}$ such that there does not admit a factorization of the type in Definition 6. Let $\hat{R} : \underline{T} \to \hat{\underline{T}}$. On the category $\hat{\underline{T}}$ consider the $\underline{Ens}^{op}$ valued functor $\hat{F}_{\bar{Q}} = [\_,\bar{Q}]$. One verifies that $\bar{\pi}f \underset{\hat{R}}{\not\sim} \bar{\pi}g \Rightarrow [\bar{\pi}f] \neq [\bar{\pi}g]$ in $[X,Q]$ which, in turn, implies $\hat{F}_{\bar{Q}}\hat{R}(\bar{\pi}f) \neq \hat{F}_{\bar{Q}}\hat{R}(\bar{\pi}g) \Rightarrow \hat{F}_{\bar{Q}}\hat{R}(f) \neq \hat{F}_{\bar{Q}}\hat{R}(g)$.

Let $F_{\bar{Q}} = \hat{F}_{\bar{Q}} \hat{R}|\underline{T}'$. Since $\hat{R}|\underline{T}' = R$ we have trivially that $F_{\bar{Q}}$ factors through R.

Consider $F_{\bar{Q}}^K$. From the universal mapping properties of the K an extension, there is a natural transformation $\eta : \hat{F}_{\bar{Q}}\hat{R} \to F_{\bar{Q}}^K$ which is an equivalence on $\underline{T}'$, [4 p. 234].

One has $F_{\bar{Q}}^K(\pi f)\eta(X) = \eta(\bar{Q})\hat{F}_{\bar{Q}}R(\pi f) \neq \eta(\bar{Q})\hat{F}_{\bar{Q}}R(\pi g) = F_Q^K(\pi g)\eta(X)$ (since $\eta(\bar{Q})$ is an equivalence).

Hence, $F_{\bar{Q}}^K(\pi f) \neq F_{\bar{Q}}^K(\pi g)$ which again in turn implies $F_{\bar{Q}}^K(f) \neq F_{\bar{Q}}^K(g)$.

Given Theorem 7, it is now possible to complete the proof of Theorem 3.

Proof of 3a. Let H denote the natural relation of homotopy restricted to the category $\underline{P}_f$. For the category Comp_2 we must show that homotopy over P_f (H_f) is the same as $H^K = \hat{H}$.

$H_f \leq \hat{H}$: Trivial since Definition 6 gives a prescription for piecing together a homotopy $\pi f \sim \pi g$ for any $X' \overset{\pi}{\to} Q$.

$\hat{H} \leq H_f$: This follows from the following stronger result which we prove in an appendix.

Lemma 8 Let X be in $\underline{\text{Comp}}_2$ and Q' in $\underline{P}_f$ suppose we are given $h_1 \sim h_2 : X \to Q'$ then we can find Q' in $\underline{P}_f$ and maps $\pi : X \to Q$, $\phi_1, \phi_2 : Q \to Q'$ with

(1) $\phi_1\pi = h_1$, $\phi_2\pi = h_2$

(2) $\phi_1 \sim \phi_2$

Proof of 3b. Remembering that now $\hat{H}$ is the Kan extension over $\underline{N}$ we wish to show that $\hat{H} = \beta H_f$.

$\beta H_f \leq \hat{H}$: Let $h_1 \underset{\hat{H}}{\sim} h_2 : X \to X'$. Let $\pi : X' \to Q$. We must produce a homotopy $\beta(\pi h_1) \sim \beta(\pi h_2)$. Let $(\{Q_i\},\{\pi_i\},\{\phi_i\})$ be a factorization of πh_1

πh_2 (Definition 6). Noting that any map $\pi_i : X \to Q_i$ extends uniquely to a map $\beta(\pi_i) : \beta X \to Q_i$, we have $(\{Q_i\},\{\beta(\pi_i)\},\{\phi_i\})$ is a factorization of $\beta(\pi h_1)$ and $\beta(\pi h_2)$, thus $\beta(\pi h_1) \underset{\hat{H}}{\sim} \beta(\pi h_2)$, hence $\beta(\pi h_2) \sim \beta(\pi h_2)$.

$\hat{H} \leq \beta H_f$: Since βX is compact, we may use Lemma 8 to factor the homotopy for βX then restrict the factorization to X.

Question: Consider the pair $\underline{P}_c \subseteq \underline{Top}$. Let H now be homotopy on $\underline{P}_c$. Does $H^K = H_c$? As before, $H_c \leq \hat{H} = H^K$. However, for the reverse conclusion, we do not know whether the corresponding version of Lemma 8 holds. One possible approach is to use Milnors C.W. path bundle over a simplicial complex [5]. However, it is not known whether the total space is a simplicial complex as would be required by a proof using this approach.

Theorem 3 has several interesting consequences. First, let $\widetilde{\underline{Comp}_2}$, $\tilde{P}_f$ be the respective homotopy categories (i.e. the quotients of $Comp_2$, P_f by H). Let $F = \tilde{F}H$ on P_f. We then have

Theorem 9 $F^K = \tilde{F}^K H$, the homotopy Kan extension of $\tilde{F}$ to $\widetilde{\underline{Comp}_2}$ is naturally isomorphic to the factorization of the continuous Kan extension to $\underline{Comp}_2$ through the homotopy category.

Proof. Denoting $H|P_f$ by $\bar{H}$, Theorem 3 gives us $\bar{H}^K = H_f$ but trivially $H_f \leq H$. Now apply Theorem 5 to the relation $\bar{H}^K \leq H$.

Before preceeding to the next result, we need the following simple observation about Kan extensions from $\underline{P}_f$ to $\underline{N}$.

Theorem 10 On $\underline{N}$, $F^K(\beta X) = F^K(X)$.

Proof. For X in $\underline{N}$, Q in $\underline{P}_f$ and $\pi\colon X \to Q$, we have the unique factorization

$$\begin{array}{ccc} & \beta X & \\ {\scriptstyle i}\nearrow & & \searrow{\scriptstyle \beta(\pi)} \\ X & \xrightarrow{\ \pi\ } & Q \end{array}$$

Hence, i induces an isomorphism of comma categories $(\beta X \downarrow P_f) \cong (X \downarrow P_f)$ hence an isomorphism $F^K(\beta X) = F^K(X)$ [4].

We now briefly consider the Čech extension $\check{F}$ of the functor F (finite covers) from $\underline{P}_f$ to $\underline{N}$.

<u>Theorem 11</u> On $\underline{N}$, $\check{F} = F^K$.

Proof. The proof amounts to listing four natural equivalences:

$F^K(X) = F^K(\beta X)$	by Theorem 10
$F^K(\beta X) = \tilde{F}^K H(\beta X)$	by Theorem 9
$\tilde{F}^K H(\beta X) = \check{F}(\beta X)$	following Dold [2]
$\check{F}(\beta X) = \check{F}(X)$	[3]

The proof of this last equivalence amounts to a selection of a cofinal family of covers of X that extend in a suitable way to a cofinal family of covers of βX.

In light of Theorem 11, the <u>Question</u> raised earlier takes on added meaning for by [2], we also know that as extensions over $\underline{P}_c$ $\tilde{F}^K H = \check{F}$ (numerable covers). Hence, $\tilde{F}^K H = \tilde{F}^K$ would imply $F^K = \check{F}$ as extensions over $\underline{P}_c$ as well as over $\underline{P}_f$ suggesting a new point of view about the nature of Čech extensions.

Finally, we may also use Theorems 9 and 10 to obtain a geometric description of F^K if F is representable. For this, let $F\_ = [\_,Y]$, homotopy classes of maps into Y.

Theorem 12 For the pair $\underline{P}_f \subseteq \underline{N}$, consider the functor $F\_ = [\_,Y]$ on $\underline{P}_f$.

Then $F\_^K = [\beta\_,Y]$, homotopy classes of the Stone-Čech compactification of a space into Y.

Proof. Again $F^K(X) = F^K(\beta X) = \tilde{F}^K(\beta X)$. But, again, following Dold [2], $\tilde{F}^K(X) = [X,Y]$. Dold actually gives details for $\tilde{\underline{P}}_c \subseteq \widetilde{\underline{Top}}$ but as he observes ([2] 3.15) the results hold for other pairs, in particular $\tilde{\underline{P}}_f \subseteq \widetilde{\underline{Comp}}_2$.

As a final observation, we use Theorem 12 to study to what extent Theorem 3 is "best possible" for particular representable functors. More precisely, if F is a homotopy functor on $\underline{P}_f$, is F^K ever one on $\underline{N}$?

This question is studied in [1] for subcategories of $\underline{N}$. There we show that if Y is of finite type with $\pi_1(Y)$ finite, then $F\_^K = [\beta\_,Y]$ is, in fact, a homotopy functor on $\underline{N}_f$, the category of finite dimensional normal spaces. However, for $\underline{N}$ itself the situation seems quite different as the following negative result indicates.

Theorem 13 Let $F\_ = [\_,Y]$, where Y is in $\underline{P}_f$ and $H_n(Y) \neq 0$, some $n > 0$, (homology). Then $F\_^K = [\beta\_,Y]$ is not a homotopy functor.

Proof. Consider the standard based path space fibration $\Omega Y \to PY \xrightarrow{p} Y$. If $[\beta\_,Y]$ were a homotopy functor $\beta(p) : \beta PY \to Y$ would be homotopic to the constant map. Hence, lifting the reverse homotopy with initial map constant we could find a map $r : \beta PY \to PY$ such that $pr = \beta(p)$. Restricting to the fibre, we have a map

$s : \beta\Omega Y \to \Omega Y$ such that $si : \Omega Y \to \Omega Y$ ($i : \Omega Y \to \beta\Omega Y$) is an isomorphism in homotopy, (hence homology) and $si(\Omega Y)$ lies in a compact set of ΩY ($\beta\Omega Y$ compact). Thus the homology of Y must be 0 except in a finite number of dimensions and $\pi_1(Y)$ must be finite (see remarks above). But a simple spectral sequence argument on the fibration $\Omega Y \to PY \to Y$ if local coefficients are not present, or if necessary on $(\Omega Y)_0 = (\Omega\bar{Y}) \to P\bar{Y} \to \bar{Y}$, $\bar{Y}$ the simply connected covering complex of Y, shows that ΩY is never finite dimensional under our assumptions on Y.

APPENDIX

Proof of Lemma 8. Let $(U, \{V_\beta\}, \lambda)$ be an equilocally convex structure for Q' [6]. That is, $\Delta \subseteq U \subseteq Q' \times Q'$, where Δ is the diagonal and U is an open neighborhood of Δ in $Q' \times Q'$. $\{V_\beta\}$ is an open cover of Q' with $V_\beta \times V_\beta \subseteq U$ for all β. $\lambda : U \times I \to Q'$ is a homotopy of p_1 to $p_2 : U \to Q'$ where $P_1(a,b) = a$, $P_2(a,b) = b$, and $\lambda(V_\beta \times V_\beta \times I) = V_\beta$.

Since Q' is a finite simplicial complex we may subdivide it so finely that for every vertex b_i of the subdivision (henceforth denoted by Q') we have $\overline{\mathrm{Star}(b_i)} \subseteq V_\beta$ for some V_β.

Let $\hat{Q} = \bigcup_{\text{all } b_i} (\overline{\mathrm{Star}(b_i)} \times \overline{\mathrm{Star}(b_i)})$. Q is easily seen to be a subcomplex of $Q' \times Q'$ (after subdivision). $\hat{Q}$ has another property which is essential for our purposes. Let δ be a lebesgue number for the open cover $\{\mathrm{Star}(b_i)\}$. Let $|h_0(x) - h_1(x)| < \delta$ for all $x \in X$, then we are able to factor h_0, h_1 as required. In particular $\pi = (h_0, h_1) : X \to \hat{Q} \subseteq Q' \times Q'$ and $\phi_0 = p_1|\hat{Q}$, $\phi_1 = p_2|\hat{Q}$. Trivially $\phi_0\pi = p_1(h_0, h_1) = h_0$ and $\phi_1\pi = h_1$. Also, $\lambda|\hat{Q} \times I \to Q'$ is the desired homotopy.

In general, since X is compact, if H is the homotopy $h_0 \sim h_1$ we may choose $0 = t_0 < t_1 < \ldots < t_n = 1$ with $|H_{t_i}(x) - H_{t_{i+1}}(x)| < \delta$ for all $x \in X$.

Let $\bar{Q} = \prod_n \hat{Q}$, the product of n copies of Q with projections ρ_i, $1 \leq i \leq n$ onto the factors. Define $Q \subseteq \bar{Q}$ by $Q = \{\prod_{j=1}^{n} (s_{j_1}, s_{j_2}) \mid s_{j_2} = s_{(j+1)_1} \quad 1 \leq j \leq n-1\}$. Q may be seen to be a simplicial complex by considering it as a subcomplex of

$\prod_{n+1} Q'$ under the map $((r_0,r_1),(r_1,r_2),\ldots(r_{n-1},r_n)) \to (r_0,r_1\ldots r_n)$.

Define $\pi(x) = \prod_n (H_{t_i}(x), H_{t_{i+1}}(x))$, $0 \le i \le n-1$ and define

$\phi_0 = p_1\rho_1$ and $\phi_1 = p_2\rho_n$. Again trivially, $\phi_0\pi = h_0$ and $\phi_1\pi = h_1$. Finally, to check $\phi_0 \sim \phi_1$ we use the facts that $p_2\rho_i = p_1\rho_{i+1}$ and $p_1\rho_i \sim p_2\rho_i$ since, as above, $p_1 \sim p_2$.

BIBLIOGRAPHY

[1] A. Calder and J. Siegel, Homotopy and Uniform homotopy (to appear).

[2] A. Dold, Lectures on Algebraic Topology, Springer- Verlag (1972).

[3] S. Eilenberg and N. Steenrod, Foundations of Algebraic Topology, Princeton University Press (1952).

[4] S. MacLane; Categories for the Working Mathematician, Springer-Verlag (1971).

[5] J. Milnor, Constructions of universal bundles I, Ann. of Math. (63) 1956, pp. 272-284.

[6] ________ , On spaces having the homotopy type of a C.W. complex, Trans. A.M.S. (90) 1959, pp. 272-280.

Tensor products of functors on categories of Banach spaces

J. Cigler

1. Sketch of the situation:

In their fundamental paper [11] B.S. MITYAGIN and A.S. SHVARTS have laid the foundations for a theory of functors on categories of Banach spaces. The situation may be roughly described as follows: The family Ban of all Banach spaces becomes a category by choosing as morphisms all linear contractions, i.e. all bounded linear mappings $\varphi: X \to Y$ satisfying $\|\varphi\| \leq 1$. The set of all morphisms from X into Y may therefore be identified with the unit ball of the Banach space H(X,Y) of all bounded linear maps from X into Y. By a (covariant) functor F: Ban $\to$ Ban we mean a functor in the algebraic sense with the additional property that the mapping $f \to F(f)$ is a linear contraction from H(X,Y) into H(F(X), F(Y)) for all X,Y. The simplest examples are the functors Σ_A and H_A defined by $\Sigma_A(X) = A \hat{\otimes} X$ (i.e. the projective tensor product) and $H_A(X) = H(A,X)$.

By a natural transformation $\varphi: F_1 \to F_2$ we understand a natural transformation $\varphi = (\varphi_X)_{X \in \mathrm{Ban}}$ in the algebraic sense satisfying $\|\varphi_X\| \leq 1$ for all $X \in$ Ban. Thus the natural transformations from F_1 to F_2 form the unit ball of the Banach space Nat (F_1, F_2) of all natural transformations in the algebraic sense satisfying

$$\|\varphi\| = \sup_X \|\varphi_X\| < \infty .$$

Denote now by $\mathrm{Ban}^{\mathrm{Ban}}$ the category whose objects are all functors from Ban into Ban and whose morphisms are all natural transformations.

It is easy to verify that for each $A \in$ Ban and each functor F the equation

(1.1) $$\mathrm{Nat}\ (\Sigma_A, F) = H(A, F(I))$$

holds, where the (isometric) isomorphism is given by $\varphi \leftrightarrow \varphi_I$. (Here I denotes the one-dimensional Banach space). As a special case we get

(1.2) $$\mathrm{Nat}\ (\Sigma_A, \Sigma_B) = H(A,B)$$

for all $A,B \in$ Ban. This may be interpreted intuitively in the following way: The mapping $A \to \Sigma_A$ from Ban into Ban ^{Ban} is an "isometric embedding", or functors are generalized Banach spaces.

For the functor H_A we get the equation

(1.3) $$\mathrm{Nat}\ (H_A, F) = F(A)$$

given by $\varphi \leftrightarrow \varphi_A\ (1_A)$ (Yoneda lemma).
As a special case we get

(1.4) $$\mathrm{Nat}\ (H_A, H_B) = H(B,A)$$

for all $A,B \in$ Ban, which may analogously be interpreted to say, that Ban ^{op} is isometrically contained in Ban ^{Ban}.

It is now tempting to ask if it is possible to extend the natural mapping from Ban onto Ban ^{op} to a (contravariant) mapping from Ban ^{Ban} into itself. In other words:
Does there exist a contravariant functor D: Ban $^{\mathrm{Ban}} \to$ Ban ^{Ban} satisfying

1) $D\,\Sigma_A = H_A$ for all $A \in$ Ban
2) $\mathrm{Nat}\ (D\,F_1, D\,F_2) = \mathrm{Nat}\ (F_2, F_1)$ for all functors F_1, F_2.

If such a D would exist it would be uniquely determined by the equation

$$DF(A) = \mathrm{Nat}\ (H_A, DF) = \mathrm{Nat}\ (D\,\Sigma_A, DF) = \mathrm{Nat}\ (F, \Sigma_A).$$

Though it turns out that this functor DF does not satisfy 2) for all pairs of functors, it nevertheless proved to be of utmost importance for the theory. It is called the dual functor to F.

MITYAGIN and SHVARTS have begun to compute DF for some concretely given functors. These computations were rather long and cumbersome. There was missing some kind of formalism which would be able to reduce lengthly calculations to simple formulas. The purpose of this talk is to show that the concept of tensorproduct for functors provides us with such a formalism.

2. Functors as generalized Banach modules:

I want to indicate my main ideas by means of a simple analogy, which I find more illuminating than the corresponding abstract theory which would be required by contemporary mathematical standards. Let me state this analogy in the following form ([1], [2]):

"Functors are generalized Banach modules". This has of course been observed several times before, but nobody seems to have used this analogy in order to carry over Banach space theory to functors on categories of Banach spaces by using Banach modules as a sort of catalyst.

First some definitions: Let A be a Banach algebra. A Banach space V is called a left A-module if there is a bilinear operation $A \times V \to V$, written $(a,v) \to a\,v$, such that $b(a\,v) = (b\,a)v$ and $\|a\,v\|_V \leq \|a\|_A \, \|v\|_V$ for $a,b \in A$ and $v \in V$. A Banach space W is called right A-module if $w\,a$ is defined with similar properties.

A Banach space Z will be called A-B - bimodule if it is a left A-module and a right B-module and if furthermore these module operations commute:

$$(a\,z)b = a(z\,b).$$

In order to get a satisfying theory we have to assume that the Banach algebra A has approximate (left) identities. By this we mean a net (u_ι) of elements $u_\iota \in A$ satisfying $\|u_\iota\| \leq 1$ and $\lim u_\iota a = a$ for all $a \in A$.

The following theorem is well known:

Factorization theorem (Hewitt-Cohen): Let A be a Banach algebra with left approximate identity (u_ι) and let V be a left A-module. Then the following assertions are equivalent for an element $v \in V$:

1) There exist $a \in A$, $w \in V$ such that $v = a\,w$
2) $\lim \|u_\iota v - v\| = 0$.

The set of all such elements forms an A-submodule V_e of V which is called the essential part of V.

Let us now denote by $H_A(V_1,V_2)$ resp. $H^A(W_1,W_2)$ the Banach space of all left (resp. right) A-module-homomorphisms from V_1 into V_2 (resp. from W_1 into W_2). Of course $\varphi \in H_A(V_1,V_2)$ if and only if $\varphi \in H\ (V_1,V_2)$ and $\varphi(a\,v) = a\,\varphi(v)$ for all $a \in A$ and $v \in V_1$.

In the analogy mentioned above between Banach modules and functors on categories of Banach spaces the following notions correspond with each other:

Banach algebra A	full subcategory $\underline{K}$ of Ban
left A-module V	covariant functor $F\colon \underline{K} \to$ Ban
right A-module W	contravariant functor $G\colon \underline{K} \to$ Ban
$H_A(V_1,V_2)$	Nat (F_1,F_2)
$H^A(W_1,W_2)$	Nat (G_1,G_2).

To see this analogy let $F: \underline{K} \to$ Ban, $X, Y \in K$, $v \in F(X)$, and $a: X \to Y$. Set $a\,v = F(a)\,v$. Then $1_X v = F(1_X) v = v$, $\|a\,v\| = \|F(a)v\| \leq \|a\|\,\|v\|$.

If $b: Y \to Z$ is a morphism in $\underline{K}$ then

$$b(a\,v) = F(b)\ (F(a)v) = F(b\,a)v = (b\,a)v.$$

This shows in what sense a functor may be considered as a generalized Banach module.

Let now $\varphi: F_1 \to F_2$ be a natural transformation. Then

$$\varphi_Y\ (F_1(a)v) = F_2(a)\ \varphi_X(v)$$

or without indices $\varphi(a\,v) = a\,\varphi(v)$ which may serve as justification for interpreting natural transformations as generalized module-homomorphisms.

Once one has recognized this analogy it is easy to give further notions which correspond with each other.

An important example are Banach algebras with a left approximate identity and full subcategories of $\underline{A}$, where A denotes the full subcategory of Ban consisting of all Banach spaces satisfying the metric approximation property of Grothendieck. For in this case there is an approximate identity in the algebra $K(X,X)$ of all compact operators on X.

In this paper we want to generalize the following assertions for Banach modules (which may be found in M. RIEFFEL [13]) to functors on categories of Banach spaces:

a) For each right A-module W and each left A-module V there is a Banach space $W \hat{\otimes}_A V$, the tensor product of W and V, and a A-bilinear mapping $\omega: W \times V \to W \hat{\otimes}_A V$ such that the following condition holds:
For every Banach space Z and each A-bilinear mapping φ: $W \times V \to Z$ there is a uniquely determined continuous linear mapping $T_\varphi: W \hat{\otimes}_A V \to Z$ such that $\|T_\varphi\| = \|\varphi\|$ and such that the

diagram

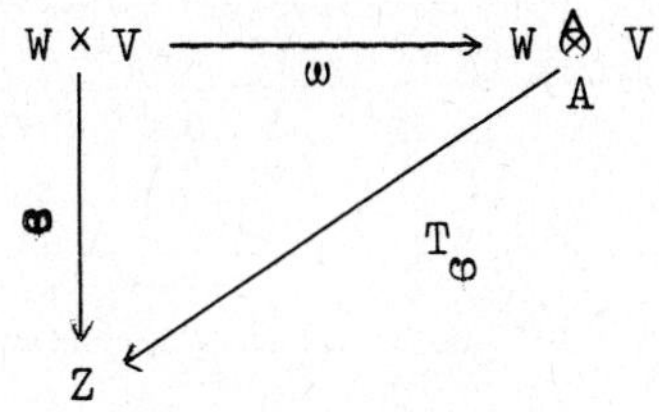

commutes. The pair $(W \hat{\otimes}_A V, \omega)$ is uniquely determined up to an isomorphism in Ban.

This tensor product is given by the formula $W \hat{\otimes}_A V = (W \hat{\otimes} V)/_N$, where N is the closed subspace of $W \hat{\otimes} V$ spanned by the elements of the form $wa \otimes v - w \otimes av$.

b) Let W be a right A-module, Z an A - B-bimodule and V a left B-module. Then

$$W \hat{\otimes}_A (Z \hat{\otimes}_B V) = (W \hat{\otimes}_A Z) \hat{\otimes}_B V$$

c) Let V be a left A-module, Z an A - B-bimodule and X a left B-module. Then the socalled exponential law holds:

$$H_A (Z \hat{\otimes}_B X, V) = H_B (X, H_A (Z,V)).$$

This isometry is natural in all variables.

d) If A has a unit element which acts as the identity on the left module V and the right module W then

$$A \hat{\otimes}_A V = V \text{ and } W \hat{\otimes}_A A = W.$$

e) If A has an approximate left identity, then

$$A \hat{\otimes}_A V = V_e.$$

3. Tensor products of functors:

Let $\underline{K}$ be a full subcategory of Ban containing I. Let $G: \underline{K} \to$ Ban be a contravariant and $F: \underline{K} \to$ Ban a covariant functor. Then $(G(Y) \times F(X))_{Y,X}$ is a contra-covariant bifunctor on $\underline{K} \times \underline{K}$. By a $\underline{K}$-bilinear mapping α from $G \times F$ into a Banach space Z we mean a family $(\alpha_X)_{X \in \underline{K}}$ of bilinear mappings

$$\alpha_X: G(X) \times F(X) \to Z \text{ with } \|\alpha\| = \sup_X \|\alpha_X\| < \infty$$

such that

$$\alpha_X \, (G(\varphi)g_Y, f_X) = \alpha_Y \, (g_Y, F(\varphi)f_X)$$

(or symbolically $\alpha(g\varphi, f) = \alpha(g, \varphi f)$) for all $g_Y \in G(Y)$, $f_X \in F(X)$, and $\varphi: X \to Y$.

If we introduce the bifunctor $(G(Y) \hat{\otimes} F(X))_{Y,X}$ then to a $\underline{K}$-bilinear map α corresponds a dinatural transformation $\overline{\alpha}: G(\cdot\cdot) \hat{\otimes} F(\cdot) \to Z$.

<u>Definition</u>: Let $F, G, \underline{K}$ be as above. By a tensor product $G \hat{\otimes}_{\underline{K}} F$ we mean a Banach space together with a dinatural transformation

$$\omega: G(\cdot\cdot) \hat{\otimes} F(\cdot) \to G \hat{\otimes}_{\underline{K}} F,$$

such that for each dinatural transformation $\psi: G(\cdot\cdot) \hat{\otimes} F(\cdot) \to Z$ into some Banach space Z, there exists a unique continuous linear mapping $T_\psi: G \hat{\otimes}_{\underline{K}} F \to Z$ such that the diagram

$$\begin{array}{ccc} G(\cdot\cdot) \hat{\otimes} F(\cdot) & \xrightarrow{\;\;\omega\;\;} & G \hat{\otimes}_{\underline{K}} F \\ \downarrow \psi & \swarrow T_\psi & \\ Z & & \end{array}$$

commutes and $\|T_\psi\| = \|\psi\| = \sup_X \|\psi_X\|_{X \in \underline{K}}$.

If such a tensor product $G \hat{\otimes}_{\underline{K}} F$ exists, it is uniquely determined, since for $Z = I$ the above condition means that $(G \hat{\otimes}_{\underline{K}} F)'$ coincides with the set $B(G,F)$ of all dinatural transformations from $G(\cdot\cdot) \hat{\otimes} F(\cdot)$ into I. It is easy to see that

$$B(G,F) = \text{Nat}(F,G') = \text{Nat}(G,F').$$

In order to show the existence of $G \hat{\otimes}_{\underline{K}} F$, define

$$\omega : G(\cdot\cdot) \hat{\otimes} F(\cdot) \to B(G,F)'$$

i.e. a family $\omega_X : G(X) \hat{\otimes} F(X) \to B(G,F)'$

by $\omega_X(g_X \otimes f_X)(\alpha) = \alpha_X(g_X \otimes f_X)$ for $\alpha \in B(G,F)$.

Then clearly ω is dinatural and

$$\|\omega_X(\Sigma\, g_X^k \otimes f_X^k)\| \leq \Sigma\, \|g_X^k\|\, \|f_X^k\|$$

which implies $\|\omega\| \leq 1$.

Let now $G \hat{\otimes}_{\underline{K}} F$ be the closure in $B(G,F)'$ of all finite linear combinations of elements of the form $\omega_X(g_X \otimes f_X)$. Then $G \hat{\otimes}_{\underline{K}} F$ is a Banach space. We assert that it has all properties required from a tensor product.

Let thus $\psi : G(\cdot\cdot) \hat{\otimes} F(\cdot) \to Z$ be a dinatural transformation.

Then we must have

$$T_\psi\Big(\sum_X \sum_k \omega_X(g_X^k \otimes f_X^k)\Big) = \sum_X \sum_k \psi_X(g_X^k \otimes f_X^k)$$

and

$$\|T_\psi\Big(\sum_X \sum_k \omega_X(g_X^k \otimes f_X^k)\Big)\| =$$

$$= \sup_{\|z'\| \leq 1} \Big|z' \circ \sum_X \sum_k \psi_X(g_X^k \otimes f_X^k)\Big| =$$

$$= \|\psi\| \sup_{\|z'\| \leq 1} \Big|\sum_X \sum_k \frac{z' \circ \psi_X}{\|\psi\|}(g_X^k \otimes f_X^k)\Big| \leq$$

$$\leq \|\psi\| \sup_{\|\alpha\| \leq 1} \Big|\alpha\Big(\sum_X \sum_k g_X^k \otimes f_X^k\Big)\Big| =$$

$$= \|\psi\|\, \Big\|\sum_X \sum_k \omega_X(g_X^k \otimes f_X^k)\Big\|$$

because $z' \circ \psi$ belongs to $B(G,F)$.

This reasoning implies also that T_ψ is well defined and linear.

All general theorems on tensor products can be reduced to the above definition.

It is easy to show (and has been shown in [1]) that

$$G \underset{\underline{K}}{\hat{\otimes}} F = \Big(\sum_{X \in \underline{K}} G(X) \hat{\otimes} F(X) \Big) / _N$$

where Σ denotes the coproduct in Ban and N is the closed subspace of this coproduct spanned by all elements of the form

$$\sum_k G(\varphi)\, g_Y^k \otimes f_X^k - \sum_k g_Y^k \otimes F(\varphi)\, f_X^k.$$

<u>Proposition</u>: Let $\beta : G_1 \to G_2$ and $\alpha : F_1 \to F_2$ be natural transformations. Then

$$\beta \otimes \alpha : G_1 \underset{\underline{K}}{\hat{\otimes}} F_1 \to G_2 \underset{\underline{K}}{\hat{\otimes}} F_2$$

is a continuous linear map satisfying $\|\beta \otimes \alpha\| \leq \|\beta\|\ \|\alpha\|$.

<u>Proof</u>: We define $\beta \otimes \alpha$ by

$$(\beta \otimes \alpha)(g_X \otimes f_X) = \beta(g_X) \otimes \alpha(f_X).$$

Then $\beta \otimes \alpha$ defines a dinatural transformation from the bifunctor

$$(G_1(Y) \hat{\otimes} F_1(X))_{Y,X}$$

into the Banach space $G_2 \underset{\underline{K}}{\hat{\otimes}} F_2$ satisfying $\|\beta \otimes \alpha\| \leq \|\beta\|\ \|\alpha\|$.

Therefore by the universal property above it defines also a continuous linear map

$$\alpha \otimes \beta : G_1 \underset{\underline{K}}{\hat{\otimes}} F_1 \to G_2 \underset{\underline{K}}{\hat{\otimes}} F_2$$

with the same norm.

We state now the theorems which we shall use in the sequel and whose proofs are easy consequences of the above reasonings, but rather lengthy and shall therefore be omitted. (They may be found in [3]).

Theorem 1: Let $\underline{K}$ and $\underline{L}$ be full subcategories of Ban,

$M: \underline{K} \times \underline{L} \to$ Ban a contra-covariant bifunctor,

$G: \quad \underline{L} \to$ Ban contravariant, and

$F: \quad \underline{K} \to$ Ban covariant.

Then the equation

$$G \hat{\otimes}_{\underline{L}} (M \hat{\otimes}_{\underline{K}} F) = (G \hat{\otimes}_{\underline{L}} M) \hat{\otimes}_{\underline{K}} F$$

holds.

Theorem 2: Let $\underline{K}$ and $\underline{L}$ be full subcategories of Ban,

$M: \underline{L} \times \underline{K} \to$ Ban a contra-covariant bifunctor, and

$F_1: \quad \underline{L} \to$ Ban and

$F_2: \quad \underline{K} \to$ Ban be covariant functors

Then the exponential law holds:

$$\operatorname{Nat}_{\underline{K}} (M \hat{\otimes}_{\underline{L}} F_1, F_2) = \operatorname{Nat}_{\underline{L}} (F_1, \operatorname{Nat}_{\underline{K}} (M, F_2)).$$

There is also a further generalization of Theorem 2 which turns out to be very useful.

Let $M: \underline{L} \times \underline{K} \to$ Ban be replaced by a functor with values in Ban_A^B the category of all A - B-bimodules and let $F_1: \underline{L} \to \text{Ban}_B$ and $F_2: \underline{K} \to \text{Ban}_A$ be functors into the categories of all left B-modules and left A-modules respectively.

Denote further by $[M \hat{\otimes}_{\underline{L}} F_1]_B$ the tensorproduct which is formed from the bifunctor $G(X) \hat{\otimes}_B F(Y)$ instead of $G(X) \hat{\otimes} F(Y)$; and let $\operatorname{Nat}_A(M, F_2)$ denote the space of all natural transformations which are at the same time A-module-homomorphisms. The we have

Theorem 3:

$$\operatorname{Nat}_{\underline{K}} {}_A ([M \hat{\otimes}_{\underline{L}} F_1]_B, F_2) = \operatorname{Nat}_{\underline{L}} {}_B (F_1, \operatorname{Nat}_{\underline{K}} {}_A (M, F_2)).$$

This equation is again natural in all variables.

4. Computation of tensor products:

4.1. Let $H: \underline{K} \times \underline{K} \to$ Ban be the restriction of the contra-covariant bifunctor $H(X,Y)$ to $\underline{K} \times \underline{K}$.

Let $G: \underline{K} \to$ Ban be a contravariant and

$F: \underline{K} \to$ Ban a covariant functor. Then the equations

$H \hat{\otimes}_{\underline{K}} F = F$ and $G \hat{\otimes}_{\underline{K}} H = G$ hold.

Proof: Let us prove the first assertion. The second one is derived in the same way.

It suffices to show that for each $A \in \underline{K}$ we have

$$H(\cdot,A) \hat{\otimes}_{\underline{K}} F(\cdot) = F(A).$$

By 3. we know that $H(\cdot,A) \hat{\otimes}_{\underline{K}} F(\cdot)$ is the closure in

$$\{B(H(\cdot,A),F(\cdot)')\}' = \{\operatorname{Nat}_{\underline{K}}(H(\cdot,A),F(\cdot)')\}' = F(A)''$$

(by the contravariant form of 1.3) of all finite linear combinations of elements of the form $\omega_X(\varphi_X \otimes f_X)$ for $\varphi_X \in H(A,X)$ and $f_X \in F(X)$. Now it is easy to see that this is just the functional on $F(A)'$ defined by the element $F(\varphi_X)f_X \in F(A)$. Since each $f_A \in F(A)$ can be written in this form, e.g. as $f_A = F(1_A)f_A$, it follows that $H(\cdot,A) \hat{\otimes} F(\cdot) = F(A)$, as asserted.

Remark: These equations reduce to the trivial relations $I \hat{\otimes} X = X = X \hat{\otimes} I$ for $K = \{I\}$ and are the analogs of 2d).

The analogs for the equations $H(I,X) = X$ are the Yoneda lemmas $\underset{\underline{K}}{\mathrm{Nat}}(H,F) = F$ and $\underset{\underline{K}}{\mathrm{Nat}}(H,G) = G$ (compare 1.3).

4.2. There are some very easy but useful consequences of 4.1. Let $\underline{K}$ and $\underline{L}$ be full subcategories of Ban and let $F_1 : \underline{K} \to \underline{L}$ and $F_2 : \underline{L} \to$ Ban be two covariant functors. Then the product F_2F_1 can be written either as tensorproduct or as space of natural transformations:

$$F_2F_1(\cdot) = H(\cdot\cdot, F_1(\cdot)) \underset{\underline{L}}{\hat{\otimes}} F_2(\cdot\cdot)$$

and

$$F_2F_1(\cdot) = \underset{\underline{L}}{\mathrm{Nat}}(H(F_1(\cdot),\cdot\cdot),F_2(\cdot\cdot)).$$

This simple observation implies e.g. that left Kan extensions can be written in the form of a tensor product. This is of course a well-known fact, but the proof becomes particularly simple:

$$\begin{aligned}\underset{\underline{K}}{\mathrm{Nat}}(F_1,F_2\,S) &= \underset{\underline{K}}{\mathrm{Nat}}(F_1(\cdot),\ \underset{\mathrm{Ban}}{\mathrm{Nat}}(H(S(\cdot),\cdot\cdot),F_2(\cdot\cdot))) = \\ &= \underset{\mathrm{Ban}}{\mathrm{Nat}}(H(S(\cdot),\cdot\cdot) \underset{\underline{K}}{\hat{\otimes}} F_1(\cdot),F_2(\cdot\cdot))\end{aligned}$$

and therefore

$$\mathrm{Lan}_S\, F_1(\cdot\cdot) = H(S(\cdot),\cdot\cdot) \underset{\underline{K}}{\hat{\otimes}} F_1(\cdot).$$

4.3. Another consequence is the equation

$$\underset{\mathrm{Ban}}{\mathrm{Nat}}(F_1,F_2\,\Sigma_A) = \underset{\mathrm{Ban}}{\mathrm{Nat}}(F_1H_A,F_2)$$

for all functors F_1,F_2: Ban $\to$ Ban.

This follows from the computation:

$$\underset{\bullet}{\mathrm{Nat}}(F_1(\cdot),\ F_2(A \hat{\otimes} \cdot)) =$$

$$= \underset{\bullet}{\mathrm{Nat}}(F_1(\cdot),\ \mathrm{Nat}(H(A \hat{\otimes} \cdot,\cdot\cdot),\ F_2(\cdot\cdot))) =$$

$= \underset{\cdot\cdot}{\mathrm{Nat}}\ (H(A \hat{\otimes} \cdot, \cdot\cdot) \underset{\cdot}{\hat{\otimes}} F_1(\cdot), F_2(\cdot\cdot)) =$

$= \underset{\cdot\cdot}{\mathrm{Nat}}\ (H(\cdot, H(A, \cdot\cdot)) \underset{\cdot}{\hat{\otimes}} F_1(\cdot), F_2(\cdot\cdot)) =$

$= \underset{\cdot\cdot}{\mathrm{Nat}}\ (F_1(H(A, \cdot\cdot)), F_2(\cdot\cdot)) =$

$= \mathrm{Nat}\ (F_1\ H_A, F_2).$

The essential point here is of course that Σ_A is left adjoint to H_A.

Further theorems of this type have been obtained by G. RACHER [12].

4.4. Let G be contravariant, F covariant and let Σ^A be the contravariant functor defined by $\Sigma^A(X) = A \hat{\otimes} X'$. Then the equations

$$G \underset{\mathrm{Ban}}{\hat{\otimes}} \Sigma_A = G(I) \hat{\otimes} A$$

and

$$\Sigma^A \underset{\mathrm{Ban}}{\hat{\otimes}} F = A \hat{\otimes} F(I)$$

hold. (P. MICHOR [9]).

<u>Proof</u>: $G \underset{\mathrm{Ban}}{\hat{\otimes}} \Sigma_A = G(\cdot) \underset{\mathrm{Ban}}{\hat{\otimes}} (\cdot \hat{\otimes} A) =$

$= G(\cdot) \underset{\mathrm{Ban}}{\hat{\otimes}} (H(I,\cdot) \underset{\{I\}}{\hat{\otimes}} A) =$

$= (G(\cdot) \underset{\mathrm{Ban}}{\hat{\otimes}} H(I,\cdot)) \underset{\{I\}}{\hat{\otimes}} A = G(I) \hat{\otimes} A.$

The second equation follows in the same way.

4.5. For the next result we need the notion "functor of type Σ".

Let F: Ban → Ban be covariant and $A \in$ Ban.

The equation (1.1)

$$\mathrm{Nat}(\Sigma_A, F) = H(A, F(I))$$

allows the following interpretation. The functor $A \to \Sigma_A$ from Ban into $\mathrm{Ban}^{\mathrm{Ban}}$ is the left adjoint of the forgetful functor

$F \to F(I)$. The counit $\varepsilon^F : \Sigma_{F(I)} \to F$ of this adjointness relation is given by

$\varepsilon^F_X (\Sigma f_i \otimes x_i) = \Sigma F(x_i) f_i$ for $f_i \in F(I)$ and $x_i \in X = H(I,X)$.

The closure $F_e(X)$ of all elements of the form $\Sigma F(x_i)f_i$ in $F(X)$ defines a functor F_e, the essential part of F. If $F = F_e$, then F is called essential or of type Σ.

It is a well-known fact (cf. e.g. V.L. LEVIN [7]), that for every X the restriction of ε^F_X to the algebraic tensor product $F(I) \otimes X$ is injective and that $F(X)$ induces on this tensor product a reasonable norm α in the sense of A. GROTHENDIECK [5].

For contravariant functors G: Ban $\to$ Ban we have the equation $\mathrm{Nat}\,(\Sigma^A, G) = H(A, G(I))$. The counit ε^G is given by $\varepsilon^G_X (\Sigma g_i \otimes x_i') = \Sigma G(x_i')g_i$ and the essential part G_e is again defined as the closure in G of the image by ε^G.

In terms of tensor products of functors F is of type Σ if and only if $H(\cdot,X) \underset{\{I\}}{\hat{\otimes}} F(\cdot) \to H(\cdot,X) \underset{\mathrm{Ban}}{\hat{\otimes}} F(\cdot)$ is epi. An analogous relation holds for contravariant functors.

By dualizing one gets the condition that F is essential if and only if the mapping $F(X)' \to H(X, F(I)')$ defined by

$$f_X' \to (x \to F(x)' f_X')$$

is injective, etc.

If F_1 is essential and F_2 arbitrary, then

$\mathrm{Nat}\,(F_1, F_2) = \mathrm{Nat}\,(F_1, (F_2)_e)$. Furthermore

$\mathrm{Nat}\,(F_1, F_2) \to H\,(F_1(I), F_2(I))$

is injective.

Proposition: (P. MICHOR [9]): Let G be contravariant, F covariant and one of type Σ. Then

$$G \underset{Ban}{\hat{\otimes}} F = G(I) \overline{\underset{\alpha}{\otimes}} F(I)$$

where $\overline{\underset{\alpha}{\otimes}}$ denotes the completion of the algebraic tensorproduct with respect to a reasonable tensor norm α.

Proof: Suppose without loss of generality that F is essential. Since

$$(G \underset{Ban}{\hat{\otimes}} F)' = \underset{Ban}{Nat}(F,G') \to H(F(I), G(I)') = (G(I) \hat{\otimes} F(I))'$$

is injective, the mapping

$$G(I) \hat{\otimes} F(I) \to G \underset{Ban}{\hat{\otimes}} F$$

has dense image.

Consider the natural transformations $G(I) \hat{\otimes} \cdot' \to G(\cdot)$ and $G(\cdot) \to H(\cdot, G(I))$ where the first coincides with ε^G and the second one with $g_X \to (x \to G(x)g_X)$. Both have norm ≤ 1 and induce therefore linear contractions

$$G(I) \hat{\otimes} F(I) = \Sigma^{G(I)} \hat{\otimes} F \to G \hat{\otimes} F \to H(\cdot, G(I)) \hat{\otimes} F = F(G(I)).$$

Since $F(G(I)) \to F(I) \hat{\hat{\otimes}} G(I)$ has dense image (F is of type Σ) and these mappings act as the identity on $G(I) \otimes F(I)$, the proof is finished.

4.6. Let $X \in \underline{A}$. Then there is a net (u_ι) of finite-dimensional operators on X such that $\|u_\iota\| \leq 1$ and such that $u_\iota x \to x$ for all $x \in X$.

Lemma 1: Let F be of type Σ and $X \in \underline{A}$. Then for $f \in F_e(X)$ the equation $\lim_\iota F(u_\iota) f = f$ holds.

Proof: It is obviously sufficient to consider an f of the form

$$f = \Sigma\, F(x_k) f_k.$$

Then

$$\|F(u_\iota)f - f\| = \|\Sigma(F(u_\iota x_k)f_k - F(x_k)f_k)\| \leq \Sigma\, \|u_\iota x_k - x_k\|\, \|f_k\| \to 0.$$

For contravariant functors a similar lemma holds. Let $X' \in \underline{A}$. For every given finite set $\{x_1', \ldots, x_n'\} \subseteq X'$ and each $\varepsilon > 0$ there is a finite-dimensional $u\colon X' \to X'$ such that $\|u\| \leq 1$ and $\|u\, x_i' - x_i'\| < \varepsilon$. Let $u = \Sigma\, x_i'' \otimes \overline{x_i}'$, where (x_i'') and $(\overline{x_i}')$ are linearly independent. There are $x_{i\alpha} \in X$ such that $\lim x_{i\alpha} = x_i''$ in the topology $\sigma(X'', X')$ and such that $\|\Sigma\, x_{i\alpha} \otimes \overline{x_i}'\| \leq \|u\| \leq 1$. (cf. D. DEAN [4]). Therefore there exists α such that $v'_\alpha = \Sigma\, x_{i\alpha} \otimes \overline{x_i}$ is the dual operator to some $v_\alpha \in H(X,X)$ and such that $\|v'_\alpha\, x_i' - x_i'\| < \varepsilon$ and $\|v_\alpha\| \leq 1$.

In other words: If $X' \in \underline{A}$, there exists a net (v_ι) of finite-dimensional operators on X such that $\|v_\iota\| \leq 1$ and such that $v_\iota'\, x' \to x'$ for all $x' \in X'$.

Lemma 2 (V. LOSERT): Let $X' \in \underline{A}$ and G a contravariant functor of type Σ. Let (v_ι) be a net with the property stated above. Then $\lim G(v_\iota) g = g$ for all $g \in G(X)$.

Proof: Without loss of generality $g = \Sigma\, G(x_k') g_k$.

Then

$$\|G(v_\iota)g - g\| = \|\Sigma(G(v_\iota)G(x_k')g_k - G(x_k')g_k)\| = \|\Sigma(G(v_\iota' x_k')g_k - G(x_k')g_k)\| \leq$$

$$\leq \Sigma \|v_\iota' x_k' - x_k'\|\, \|g_k\| \to 0.$$

We can now prove the following

Proposition: If $X \in \underline{A}$ then $(\cdot' \,\hat{\otimes}\, X) \underset{\mathrm{Ban}}{\hat{\otimes}} F(\cdot) = F_e(X)$; and if $X' \in \underline{A}$ then $G(\cdot) \underset{\mathrm{Ban}}{\hat{\otimes}} (X' \,\hat{\otimes}\, \cdot) = G_e(X)$ for each contravariant functor G and each covariant functor F on Ban.

Proof: It suffices to prove the first assertion. The second one is similar.

Note first that by 4.5. the functor $(\cdot' \hat{\otimes} X) \underset{\text{Ban}}{\hat{\otimes}} F(\cdot)$ is essential since $.' \hat{\otimes} X$ is essential.

Since the inclusion $F_e \to F$ is a natural transformation and induces therefore a linear contraction

$(\cdot' \hat{\otimes} X) \underset{\text{Ban}}{\hat{\otimes}} F_e(.)$ into $(\cdot' \hat{\otimes} X) \underset{\text{Ban}}{\hat{\otimes}} F(\cdot)$ and since there is a linear contraction of the second space into $F_e(X) = H(\cdot,X) \underset{\text{Ban}}{\hat{\otimes}} F_e(\cdot)$,

it suffices to show that the canonical map

$$(\cdot' \hat{\otimes} X) \underset{\text{Ban}}{\hat{\otimes}} F_e(\cdot) \to H(\cdot,X) \underset{\text{Ban}}{\hat{\otimes}} F_e(\cdot)$$

is an isometric isomorphism.

It suffices to show that the adjoint map

$$\text{Nat}\,(H(\cdot,X),F_e(\cdot)') \to \text{Nat}\,(\cdot' \hat{\otimes} X,\ F_e(\cdot)')$$

is an isometric isomorphism onto. We know that it is injective.

We show first that it is isometric. Let $\alpha \in \text{Nat}\,(H(\cdot,X),F_e(\cdot)')$.

Then $\|\alpha\| = \|\alpha_X(1_X)\|$. It suffices therefore to show that

$$\sup_\iota \|\alpha_X(u_\iota)\| = \|\alpha_X(1_X)\|.$$

But

$$\sup_\iota \|\alpha_X(u_\iota)\| = \sup_\iota \sup_{\substack{f \in F_e(X) \\ \|f\| \le 1}} |\langle f,\alpha_X(u_\iota)\rangle| =$$

$$= \sup_\iota \sup_{\|f\| \le 1} |\langle f, F(u_\iota)'\,\alpha_X(1_X)\rangle| =$$

$$= \sup_f \sup_\iota |\langle F(u_\iota)\,f,\ \alpha_X(1_X)\rangle| =$$

$$= \sup_f |\langle f,\ \alpha_X(1_X)\rangle| = \|\alpha_X(1_X)\|.$$

Now it remains to show that each $\beta \in \text{Nat}\,(\cdot' \hat{\otimes} X,\ F_e(\cdot)')$ is the image of some α.

The net $\beta_X(u_\iota) \in F_e(X)'$ is bounded. We may therefore assume from the beginning that it is w^*-convergent. Let

$$f' = \lim \beta_X(u_\iota) \in F_e(X)'.$$

Then to f' corresponds $\alpha \in \mathrm{Nat}\ (H(\cdot,X),\ F_e(\cdot)')$.

We then have for $Y \in \mathrm{Ban}$, $f_Y \in F_e(Y), \xi_Y \in Y' \hat{\otimes} X$:

$$\langle f_Y, \beta_Y(\xi_Y)\rangle = \lim_\iota \langle f_Y, \beta_Y(u_\iota \xi_Y)\rangle =$$

$$= \lim_\iota \langle f_Y, F(\xi_Y)'\ \beta_X(u_\iota)\rangle = \lim_\iota \langle F(\xi_Y)f_Y, \beta_X(u_\iota)\rangle =$$

$$= \langle F(\xi_Y)f_Y, f'\rangle = \langle f_Y, \alpha_Y(\xi_Y)\rangle; \text{ qed.}$$

Remark: This proposition may be regarded as analogon of 2)e).

4.7. If $X \notin \underline{A}$ then in general

$$LF(X) = (\cdot' \hat{\otimes} X) \underset{\mathrm{Ban}}{\hat{\otimes}} F(\cdot)$$

does not coincide with $F_e(X)$.

The functor LF has been introduced in another way by C. HERZ and J. WICK-PELLETIER [6] and has been called by them "computable part" of F.

It has been further studied by P. MICHOR [10].

It would be interesting to study the following generalization $L^{\underline{K}}F$ of LF in more detail: Let $\underline{K}$ be a full subcategory of Ban and let

$$H^{\underline{K}}(A,X) = H(\cdot,X) \underset{\underline{K}}{\hat{\otimes}} H(A,.).$$

Then $H^{\underline{K}}$ is a contra-covariant functor on Ban. For $\underline{K} = \{I\}$ we have $H^{\{I\}}(A,X) = A \hat{\otimes} X$, for $\underline{K}$ = Ban it coincides with $H(A,X)$ and for $\underline{K}$ = Fin, the full subcategory of all finite-dimensional spaces we have $H^{\underline{K}}(A,X) = A' \hat{\otimes} X$.

Thus LF(X) could be generalized to

$$L^{\underline{K}} F(X) = H^{\underline{K}}(\cdot,X) \hat{\otimes}_{Ban} F(\cdot) .$$

5. Computation of some dual functors:

Here I want to indicate how tensor products may be used to compute the dual functor for some concrete functors.
I choose as example functors defined by sequence spaces. The same method applies also to functors defined by measurable functions but the details are more intricate (and will be given in another paper).

Let me introduce first of all the notion of a sequence space:
A linear subspace $n \subseteq l^{\infty}$ will be called a sequence space if

1) n is an l^{∞}-module (as defined in 2.)
2) $e_i \in n$ and $\|e_i\|_n = 1$ for $e_i = (0,\dots,1,0,\dots)$
3) $\|x\|_n = \sup_k \|u_k x\|_n$, $u_k = (1,1,\dots,1,0,0,\dots)$.

Proposition 1: For every sequence space n we have

$$l^1 \subseteq n \subseteq l^{\infty} \quad \text{and} \quad \|x\|_{\infty} \le \|x\|_n \le \|x\|_1 .$$

Proof: Trivial.

A sequence space n is called minimal if $n = c_o \hat{\otimes}_{l^{\infty}} n$, where c_o is the subalgebra of l^{∞} consisting of all null-sequences. An equivalent condition is that the finite sequences are dense in n.

With the sequence space n we associate the functor $n(\cdot)$ on Ban defined by

$n(X) = \{x = (x_k): x_k \in X, (\|x_k\|) \in n\}$ with norm $\|(x_k)\|_{n(X)} = \|(\|x_k\|)\|_n$.
We want to compute the dual functor $(D\, n(\cdot))(X)$.
This has already been done by MITYAGIN - SHVARTS with a complicated proof. The use of tensor products reduces it almost to a triviality.

<u>Proposition 2</u>: $n(X) = l^{\infty}(X) \hat{\otimes}_{l^{\infty}} n$

<u>Proof</u>: let $\omega : l^{\infty}(X) \times n \to n(X)$ be the l^{∞}-bilinear map $(f,\varphi) \to f\varphi$

We have to show that for each Banach space Z and each l^{∞}-bilinear map $\alpha : l^{\infty}(X) \times n \to Z$ there exists a unique continuous map $T_{\alpha} : n(X) \to Z$ such that the diagram

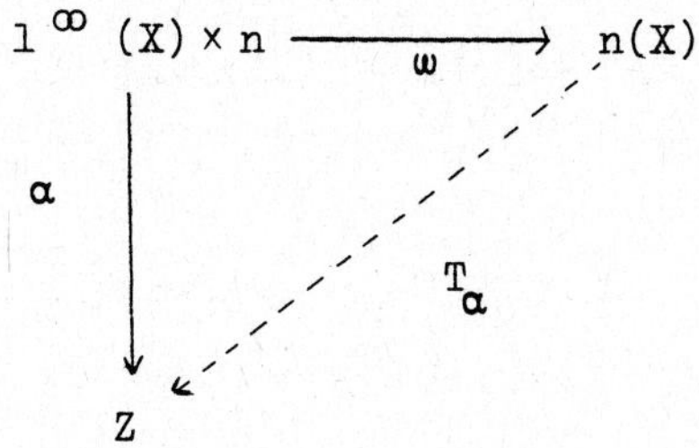

commutes and $\|T_{\alpha}\| = \|\alpha\|$.

Now $\omega(f_1,\varphi_1) = \omega(f_2,\varphi_2)$ implies $\alpha(f_1,\varphi_1) = \alpha(f_2,\varphi_2)$ because this means $f_1\,\varphi_1 = f_2\,\varphi_2$ and therefore

$$\alpha(f_1,\varphi_1) = \alpha(f_1 \frac{\varphi_1}{|\varphi_1|+|\varphi_2|}, |\varphi_1|+|\varphi_2|) = \alpha(f_2 \frac{\varphi_2}{|\varphi_1|+|\varphi_2|}, |\varphi_1|+|\varphi_2|) = \alpha(f_2,\varphi_2).$$

Every element $w \in n(X)$ is the image of an element of $l^{\infty}(X) \times n$, e.g. of $(\frac{w}{\|w\|}, \|w\|)$.

Now define T_{α} by $T_{\alpha}(f\,\varphi) = \alpha(f,\varphi)$. Then T_{α} is well defined and linear because

$$T_{\alpha}\,(\Sigma\, f_i\,\varphi_i) = T_{\alpha}\,(\frac{\Sigma\, f_i\,\varphi_i}{\Sigma\,|\varphi_k|}, \Sigma\,|\varphi_k|) =$$

$$= \alpha(\frac{\Sigma\, f_i\,\varphi_i}{\Sigma\,|\varphi_k|}, \Sigma\,|\varphi_k|) = \sum_i \alpha(\frac{f_i\,\varphi_i}{\Sigma|\varphi_k|}, \Sigma\,|\varphi_k|) = \sum_i T_{\alpha}\,(f_i\,\varphi_i)$$

Now $T_{\alpha}\,(\omega(f,\varphi)) = \alpha(f,\varphi) \Rightarrow \|\alpha\| \leq \|T_{\alpha}\|$.

On the other hand

$\|T_\alpha(w)\| = \|T_\alpha\ (\frac{w}{\|w\|} \cdot \|w\|)\| = \|\alpha\ (\frac{w}{\|w\|}, \|w\|)\| \leq \|\alpha\|\ \|\ \|w\|\ \|_n$

$\Rightarrow \|T_\alpha\| \leq \|\alpha\|$.

Corollary: For n minimal we have $n(X) = c_o(X) \underset{l^\infty}{\hat{\otimes}} n$.

Proof: $n = c_o \underset{l^\infty}{\hat{\otimes}} n$

$$\Rightarrow l^\infty(X) \underset{l^\infty}{\hat{\otimes}} n = l^\infty(X) \underset{l^\infty}{\hat{\otimes}} (c_o \underset{l^\infty}{\hat{\otimes}} n) = (l^\infty(X) \underset{l^\infty}{\hat{\otimes}} c_o) \underset{l^\infty}{\hat{\otimes}} n =$$

$$= c_o(X) \underset{l^\infty}{\hat{\otimes}} n.$$

Proposition 3: Let $n^x = H_{l^\infty}(n, l^1)$. Then n^x consists of all

$\mu = (\mu_k) \in l^\infty$ such that $\Sigma\ |\lambda_k u_k| < \infty$ for all $(\lambda_k) \in n$ with

$$\|\mu\| = \sup_{\|\lambda\|_n \leq 1} \|\lambda\ \mu\|_1.$$

Furthermore $n^x(A) = H_{l^\infty}(n, l^1(A))$ for $A \in$ Ban.

Proof: Let $\varphi \in H_{l^\infty}(n, l^1(A))$.

$\Rightarrow \varphi(\lambda) e_k = \varphi(\lambda\ e_k) = \varphi(\lambda_k\ e_k) = \lambda_k\ e_k\ \varphi(e_k)$

Now $e_k\ \varphi(e_k) = (0, \ldots, a_k, 0, \ldots) = a_k\ e_k$, $a_k \in A$.

$\Rightarrow \varphi(\lambda) = (\lambda_1 a_1, \lambda_2 a_2, \ldots)$

$$\|\varphi\| = \sup_{\|\lambda\|_n \leq 1} \Sigma\ |\lambda_k|\ \|a_k\| = \|(\|a_k\|)\|_{n^x}$$

$\Rightarrow (a_k) \in n^x(A)$.

Proposition 4: For n minimal we have $n^x(A') = n(A)'$.

Proof: $n^x(A\ ) = H_{l^\infty}(n, l^1(A')) =$

$= H_{l^\infty}(n, c_o(A)') = H_{l^\infty}(n, H(c_o(A), I)) = H(c_o(A) \underset{l^\infty}{\hat{\otimes}} n, I) = n(A)'$.

<u>Theorem</u>: $D(n(\cdot))(A) = n^x(A)$.

<u>Proof</u>: First observe that $D\, l^\infty = l^1$ because $l^\infty(X) = H_{l^1}(X)$, and

$$D\, H_{l^1}(X) = \Sigma_{l^1}(X) = l^1 \hat{\otimes} X = l^1(X).$$

Then (by Theorem 3)

$$D(n(\cdot))(A) = \mathrm{Nat}\,(n(\cdot),\ A \hat{\otimes} \cdot) =$$

$$= \mathrm{Nat}\,(l^\infty(\cdot) \underset{l^\infty}{\hat{\otimes}} n,\ A \hat{\otimes} \cdot) =$$

$$= H_{l^\infty}(n,\ \mathrm{Nat}\,(l^\infty(\cdot),\ A \hat{\otimes} \cdot)) =$$

$$= H_{l^\infty}(n,\ D\, l^\infty(\cdot)\ (A)) =$$

$$= H_{l^\infty}(n,\ l^1(A)) = n^x(A).$$

<u>Remark</u>: The dual functor satisfies the relation

$$\mathrm{Nat}\,(F_1,\ D\, F_2) = \mathrm{Nat}\,(F_2,\ D\, F_1).$$

It is given by $\varphi \leftrightarrow \bar{\varpi}$ such that

$$(\varphi_X(f_X))_Y\ (f_Y) = (\overline{\varpi_Y}\ (f_Y))_X\ (f_X).$$

Therefore also $\mathrm{Nat}\,(F,\ D^2F) = \mathrm{Nat}\,(DF,\ DF)$.

To the identity on the right side corresponds a natural transformation $j: F \to D^2F$ on the left. The functor F is called reflexive if j_X is an isometry of $F(X)$ onto $(D^2F)(X)$ for all $X \in \mathrm{Ban}$.

The above methods can be used to show that Dn is always reflexive. MITYAGIN and SHVARTS have conjectured that every dual functor is reflexive. It has recently been proved by V.LOSERT [8] that their conjecture is not true: Consider the functor $A \overset{\wedge}{\otimes} \cdot$. Then the dual dunctor coincides with the integral operators $I(A,\cdot)$. LOSERT has shown that there are Banach spaces A such that $I(A,\cdot)$ is not reflexive.

B I B L I O G R A P H Y

[1] J. CIGLER, Funktoren auf Kategorien von Banachräumen, Monatshefte für Mathematik 78 (1974), 15 - 24

[2] J. CIGLER, Duality for functors on Banach spaces, preprint 1973

[3] J. CIGLER, Funktoren auf Kategorien von Banachräumen, Lecture Notes Univ. Wien 1974

[4] D.W. DEAN, The equation $L(E,X^{**}) = L(E,X)^{**}$ and the principle of local reflexivity, PAMS 40 (1973), 146 - 148

[5] A. GROTHENDIECK, Résumé de la théorie metrique des produits tensoriels topologiques, Bol.Soc.Matem. Sao Paulo 7 (1952), 1 - 79

[6] C. HERZ - J. WICK PELLETIER, Dual functors and integral operators in the category of Banach spaces, preprint 1974

[7] V.L. LEVIN, Tensor products and functors in categories of Banach spaces defined by KB-lineals, Transl.Moscow Math.Soc. 20 (1969), 41 - 77

[8] V. LOSERT, Dissertation Univ. Wien 1975

[9] P. MICHOR, Zum Tensorprodukt von Funktoren auf Kategorien von Banachräumen, Monatshefte für Mathematik 78 (1974), 117 - 130

[10] P. MICHOR, Tensor products, operator ideals and functors on categories of Banach spaces, Univ. of Warwick 1975

[11] B.S. MITYAGIN - A.S. SHVARTS, Functors on categories of Banach spaces, Russian math. Surveys 19 (1964), 65 - 127

[12] G. RACHER, Dissertation Univ. Wien 1974

[13] M.A. RIEFFEL, Induced Banach representations of Banach algebras and locally compact groups, J. Functional Analysis 1 (1967), 443 - 491.

DUALITY OF COMPACTOLOGICAL AND LOCALLY COMPACT GROUPS

J. B. Cooper and P. Michor

We develop a duality for compactological groups, based on a concrete realization of the dual category of the category of compactological spaces in terms of a mixed topology on $\ell^{\infty}(S)$. The dual of a compactological group appears as a Hopf-algebra with mixed topology.

We are able to treat the following notions in terms of the dual: characterization of locally compact groups, Bohr compactification, almost periodic functions, Pontryagin duality, and some connections with representation theory.

§1. Preliminaries and Notation

1.1. Compactological spaces: The model for a compactological space is a topological Hausdorff space together with the collection of all its compact subsets, disregarding its original topology. So a compactological space S is a set S together with a collection $\mathcal{K}(S)$ of subsets of S, each $K \in \mathcal{K}(S)$ bearing a compact (Hausdorff) topology τ_K such that

(1) $\mathcal{K}(S)$ is closed under formation of finite unions and taking closed subsets.

(2) for each $K \subset L$; $K, L \in \mathcal{K}(S)$ the inclusion $K \to L$ is continuous.

The category CPTOL of compactological spaces has as morphisms maps $f: S \to T$ such that for each $K \in \mathcal{K}(S)$ there is $L \in \mathcal{K}(T)$ with $f(K) \subset L$ and $f|K:K \to L$ is $\tau_K - \tau_L$ - continuous.

By $C^\infty(S)$ we mean the vector space of all bounded complex valued functions on S whose restrictions to each $K \in \mathcal{K}(S)$ are τ_K-continuous. A compactological space is said to be regular, if $C^\infty(S)$ separates points on S. We denote the full subcategory of regular compactological spaces by RCPTOL. We note that compactological spaces may be regarded as formal inductive limits of systems of compact spaces. For more information see BUCHWALTER.

<u>1.2. The category</u> MIXC*: Objects are triples $(E, \|.\|, \tau)$ where E is a commutative involutive algebra with unit over C, $\|.\|$ is a norm on E and τ is a locally convex topology on E such that:

(1) $B_{\|.\|} = \{x \in E: \| x \| \leq 1\}$ is bounded and complete for τ

(2) τ may be defined by a family of seminorms P on E such that $p(xy) = p(x)p(y)$, $p(1) = 1$ and $p(x^*x) = p(x)^2$ holds for all $x, y \in E$ and $p \in P$ and $\| x \| = \sup \{p(x), p \in P\}$ for all $x \in E$.

Morphisms are multiplicative linear maps, respecting involution and unit, contractive for the norm and continuous for the locally convex topology.

MIXC* may be regarded as the category of formal projective limits of systems of commutative C^*-algebras with unit. For more information see COOPER 1975.

<u>1.3. The category</u> MIXTOP: Objects are triples $(E, \|.\|, \tau)$ where E is a vector space (over C), $\|.\|$ is a norm on E and τ is a locally convex topology on E such that $B_{\|.\|} = \{x \in E, \|x\| \leq 1\}$ is τ-bounded. Morphisms are linear maps, contractive for $\|.\|$ and continuous for $\tilde{\iota}$. $(E, \|.\|, \tau)$ is said to be complete, if $B_{\|.\|}$ is τ-complete. Then $(E, \|.\|)$ is a Banach space. The complete objects in MIXTOP are exactly the formal projective limits of systems of Banach spaces.

1.4. The functor C^{∞}:

Let S be a compactological space. Let $C^{\infty}(S)$ be the space of all bounded complex valued functions on S whose restrictions to all $K \in \mathcal{K}(S)$ are τ_K-continuous. Consider the following structures on $C^{\infty}(S)$: $\| \ \|$ - the supremum norm

τ - the topology of uniform convergence on members of $\mathcal{K}(S)$.

Then $(C^{\infty}(S), \|.\|, \tau)$ is an object of MIXC*. If φ: S → T is a CPTOL-morphism, then $C^{\infty}(\varphi)$: $C^{\infty}(T) \to C^{\infty}(S)$, given by $x \to x \circ \varphi$, is a MIXC*-morphism.

We have constructed a contravariant functor C^{∞}: CPTOL → MIXC*

1.5. The functor M_{γ}:

Let $(E, \|.\|, \tau)$ be an object of MIXC*.

Denote by $M_{\gamma}(E)$ the set of all MIXC*-morphisms E → C. We equip it with the following compactology: members of $\mathcal{K}(M_{\gamma}(E))$ are the weak*-closed subsets of $M_{\gamma}(E)$, whose restriction to $B_{\|.\|}$ is τ-equicontinuous, and they bear the restriction of the weak*-topology. If φ: E → F is a MIXC*-morphism, then $M_{\gamma}(\varphi)$: $M_{\gamma}(F) \to M_{\gamma}(E)$, given by $f \to f \circ \varphi$, is a CPTOL-morphism. We have constructed a contravariant functor MIXC* → RCPTOL.

<u>1.6. Proposition</u>:(COOPER 1975): The categories $MIXC^*$ and RCPTROL are quasi-dual to each other under the functors C^∞ and M_γ.

The maps $\delta: s \to (x \to x(s))$ gives a natural isomorphism $S \to M_\gamma C^\infty(S)$ for each $S \in$ RCPTOL. We call it the Dirac transformation.

The map $\hat{}: x \to (f \to f(x))$ give a natural isomorphism $E \to C^\infty M_\gamma(E)$ for each $E \in MIXC^*$. We call it the Gelfand-Naimark transformation. These two maps produce the quasi-duality.

1.7. <u>The tensor product in</u> $MIXC^*$ <u>and</u> MIXTOP:

The category RCPTOL has products (the obvious ones), so $MIXC^*$ as the (quasi)-dual category has coproducts. The γ-tensorproduct, which we will now describe, is an explicit construction of the coproduct in $MIXC^*$.

Yet $(E, \|\cdot\|_E, \tau_E)$ and $(F, \|\cdot\|_F, \tau_F)$ be two objects of $MIXC^*$. We consider the following structures on $E \otimes F$, the vector-space tensor product of E and F: $\|\cdot\|^{\hat{}}$-the inductive tensor product of the norms $\|\cdot\|_E$, $\|\cdot\|_F$ (i.e. that induced by the operator norm via the embedding $E \otimes F \to L(E',F)$.

$\tau = \tau_E \hat{\otimes} \tau_F$ -the inductive tensor product of the locally convex topologies τ_E, τ_F.

Let B denote the closure of $\{u \in E \otimes F, \|u\|^{\hat{}} \leq 1\}$ in the completion of $(E \otimes F, \tau_E \hat{\otimes} \tau_F)$ and let $E \hat{\otimes}_\gamma F$ denote the subspace $\bigcup_{n>0} nB$ of this completion, and let $\|\cdot\|^{\hat{}}$ be the $(E \hat{\otimes}_\gamma F, \|\cdot\|^{\hat{}}, \tau_E \hat{\otimes} \tau_F)$

is again an object of MIXC*. The same construction works for MIXTOP. A result to be found in COOPER 1975 asserts that $E \hat{\otimes}_\gamma F \cong C^\infty(M_\gamma(E) \times M_\gamma(F))$ in MIXC*, so the γ-tensor product is the coproduct in MIXC*.

<u>1.8.</u> <u>The strict topology:</u>

If $(E, \|.\|, \tau)$ is an element of MIXC*, let $\gamma = \gamma[\|.\|, \tau]$ be the finest locally convex topology on E which agrees with τ on $B_{\|.\|}$; γ is a complete topology.

We note that $\gamma[\|.\|^{\wedge}, \tau_E \hat{\otimes} \tau_F] = \gamma[\|.\|_E, \tau_E] \hat{\otimes} \gamma[\|.\|_F, \tau_F]$ on $E \hat{\otimes}_\gamma F$.

For further details see COOPER 1975.

The same definition holds of course for objects of MIXTOP.

<u>1.9.</u> It is well known, that locally compact topological spaces are k-spaces, i.e. their topology is uniquely determined by their natural compactology. In this spirit we can regard the category of locally compact topological spaces (we call it LOCCOMP) as a full subcategory of RCPTOL and we will speak of <u>locally compact compactological spaces</u>.

<u>1.10.</u> Let $(E, \|.\|, \tau)$ be an object of MIXC* and let P be a defining family of C^*-seminorms on E (i.e. a family P satisfying 1.2. (2)). If $p \in P$ we denote by I_p the ideal $\{x \in E: p(x) = 0\}$ and by A_p its annihilator in E, i.e. $A_p = \{y \in E: y\, I_p = 0\}$.

E is said to be <u>perfect</u> (APOSTOL 1971) if the sum $\sum_{p \in P} A_p$ is γ-dense (1.8) in E.

If S is a regular compactological space and if $K \in \mathcal{K}(S)$ and p_K the associated C^*- seminorm, then $I_{p_K} = \{x \in C^\infty(S) : x|K = 0\}$ and $A_{p_K} = \{y \in C^\infty(S) : y(s) = 0 \text{ for } s \notin K\}$ So $\sum_{K \in \mathcal{K}(s)} A_{p_K}$ is the subspace $C_c(S)$ of functions in $C^\infty(S)$ with compact support (i.e. $x \in C_c(S)$ iff there is $K \in \mathcal{K}(S)$ with $x(s) = 0$ for $s \notin K$).

1.11. Proposition (COOPER 1975) Let S be a regular compactological space. S is locally compact if and only if $C^\infty(S)$ is perfect.

§2. Compactological groups and duality

2.1. Definition. A compactological group is a group in the category CPTOL of compactological spaces, i.e. it is a quadruple (S,m,e,i) where S is a compactological space and in: $S \times S \to S$, $e: I \to S$, $i: S \to S$ are CPTOL-morphisms so that the following diagrams commute:

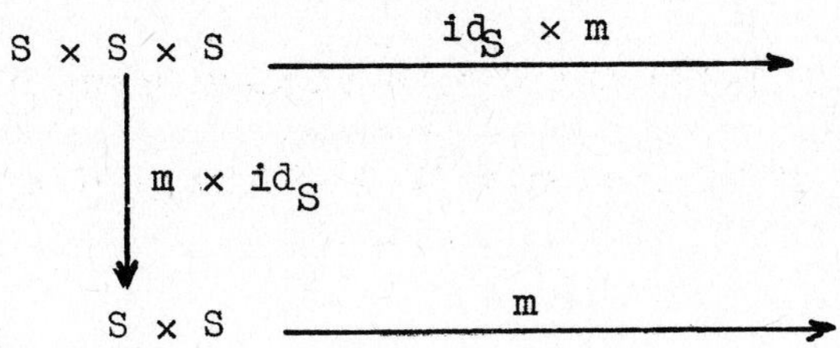

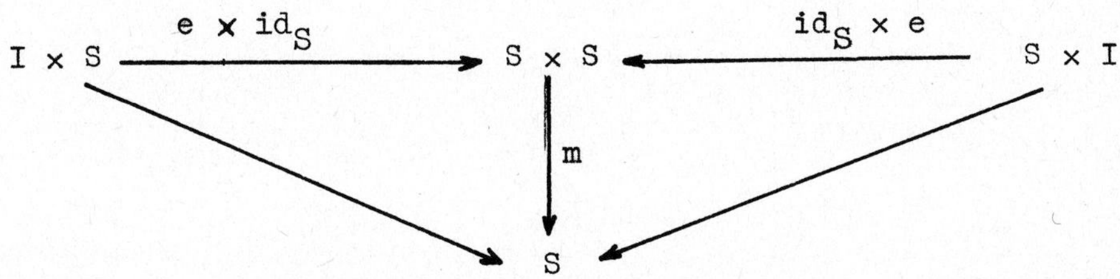

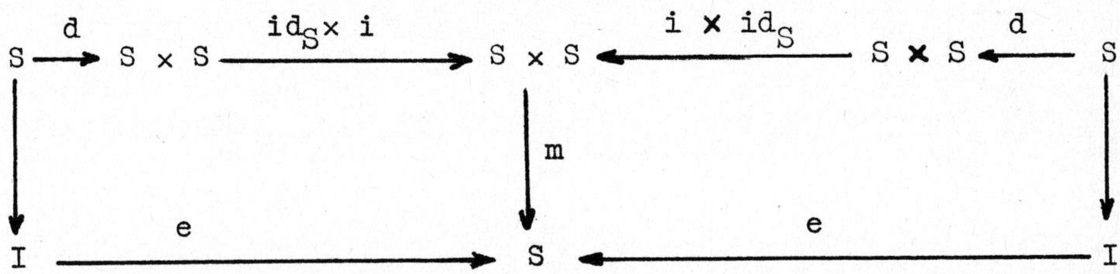

(I is the final object of CPTOL, i.e. the one-point set, d is the diagonal map). The compactological groups form a category which we denote by GCPTOL (the morphisms are those CPTOL-morphisms which respect the maps (m,e,i). A compactological group is said to be <u>regular</u> if its underlying compactological space is regular, i.e. if $C^{\infty}(S)$ separates S. GRCPTOL denotes the full subcategory of regular compactological groups.

__2.2.__ __Problem__: Do there exist compactological groups which are not regular?

__2.3.__ __Definition__: $CMIXC^{*}$ denotes the category of cogroups in $MIXC^{*}$. Thus an object of $CMIXC^{*}$ is a quadruple (E,c,η,a) where E is an object of $MIXC^{*}$, and $c: E \to E \hat{\otimes}_{\gamma} E$, $\eta\ E \to C$, $a\ E \to E$ are $MIXC^{*}$-morphisms so that the following diagrams commute:

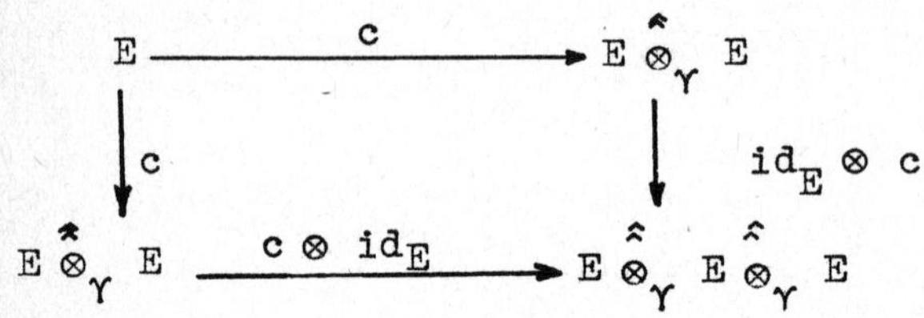

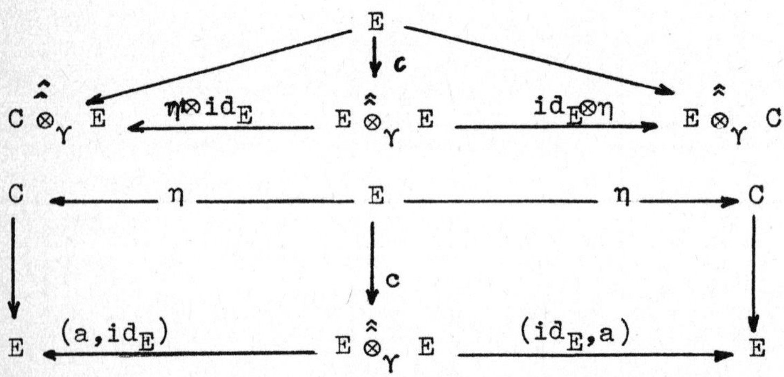

((a,id_E)) denotes the canonical morphism from $E \hat{\otimes}_{\gamma} E$, the coproduct in $MIXC^{*}$ (cf 1.6) into E, defined by the maps a and id_E; C is the initial object of $MIXC^{*}$).

2.4. If (S,m,e,i) is a compactological group, then clearly $(C^\infty(S),C^\infty(m),C^\infty(e),C^\infty(i))$ is a cogroup in $MIXC^*$. Clearly C^∞ lifts to a functor GCPTOL → $CMIXC^*$.

If on the other hand (E,c,η,a) is a cogroup in $MIXC^*$, then again $(M_\gamma(E),M_\gamma(c),M_\gamma(a))$ is a regular compactological group. M_γ lifts to a functor $CMIXC^*$ → GRCPTOL.

Proposition: The functors C^∞ and M_γ induce a duality between $CMIXC^*$ and GRCPTOL.

2.5. A compactological group (S,m,e,i) is locally compact if and only if its dual $C^\infty(S)$ is perfect. (cf 1.11).

§3. The Bohr Compactification

3.1. Let (E,c,η,a) be a cogroup in $MIXC^*$. If we consider the C^*-algebra $(E,\|.\|)$, then $M_\gamma(E,\|.\|)$ is a compact topological space, in fact, the Stone-Čech compactification of $M_\gamma(E)$. It is, however, in general not a topological group, since $(E,\|.\|)$ is not a cogroup in the category C^* of commutative C^*-algebras with unit, since $E \hat{\hat{\otimes}} E \neq E \hat{\otimes}_\gamma E$, where $E \hat{\otimes} E$, is the C^*-algebra tensor product or the inductive tensor product of Banach spaces.

Let $\tilde{c} = (id_E \otimes a) \circ c$.

3.2. Lemma: There is a largest C^*-subalgebra $\tilde{\tilde{E}}$ of E with the property that $\tilde{c}(\tilde{\tilde{E}}) \subset \tilde{\tilde{E}} \hat{\hat{\otimes}} \tilde{\tilde{E}}$. $\tilde{\tilde{E}}$, with the induced norm and cogroup structure is a cogroup in C^*. The assignment $E \to \tilde{\tilde{E}}$ is functorial.

Proof: For each ordinal α, we define a subalgebra E_α of E inductively by

$$E_o := E$$

$$E_\alpha := \tilde{c}^{-1}(E_\beta \hat{\hat{\otimes}} E_\beta) \quad (\alpha = \beta + 1)$$

$$E_\alpha := \cap \{E_\beta, \beta < \alpha\} \quad (\alpha \text{ is a limit ordinal}).$$

Then the family $\{E_\alpha\}$ is eventually stationary and we denote its limit by $\tilde{\tilde{E}}$. Then $\tilde{\tilde{E}}$ is C^*-subalgebra of E with the desired properties.

3.3. The functor $E \to \tilde{\tilde{E}}$ is a "forgetful functor" from $CMIXC^*$ into CC^*. We denote it by U. We can now define a functor $B := M_\gamma \circ U \circ C^\infty$ from GCPTOL into GCOMP, the category of compact groups. If S is a compactolbgical group, we call $B(S)$ the Bohr-compactification of S. There is a natural morphism $j_S: S \to B(S)$, $j_S = M_\gamma(\lambda)$, where $\lambda: \widetilde{C^\infty(S)} \to C^\infty(S)$ is the embedding. j_S has dense image, since λ is injective.

3.4. Proposition: B(S) has the following universal property: every GCPTOL-morphism from S into a compact group factorises over j_S.

Proof: If $\phi: S \to T$ is a GCPTOL-morphism, where T is a compact group, then $C^\infty(\phi): C(T) \to C^\infty(S)$ is a CMIXC*-morphism. Since $UC(T) = C(T)$ and U is a functor, acting on morphisms by restricting them, we conclude that $\widetilde{C^\infty(\phi)} = C^\infty(\phi)$ maps C(T) into $\widetilde{C^\infty(S)}$, i.e. factors over $\lambda: \widetilde{C^\infty(S)} \to C^\infty(S)$.

So $\phi = \tilde{\phi} \circ j_S$, $\tilde{\phi} = M_\gamma\, \widetilde{C^\infty(\phi)}$.

3.5. If S is a compactological group and $x \in C^\infty(S)$, then we define $(L_a x)(s) = x(as)$, $(R_a x)(s) = x(sa)$ for a, s $\in$ S. $x \in C^\infty(S)$ is said to be (left) almost periodic if $\{L_a x, a \in S\}$ is relatively norm-compact in $C^\infty(S)$. We denote by AP(S) the set of left almost periodic functions on S.

With induced norm AP(S) is C^*-subalgebra of $C^\infty(S)$.

3.6. Lemma: $\widetilde{C^\infty(S)} \subset AP(S)$.

Proof: If $x \in \widetilde{C^\infty(S)}$, there is an $\tilde{x} \in C(B(S))$ so that $x = \tilde{x} \circ j_S$ by definition of B(S). Then $L_a x = (L_{j_S(a)}\, \tilde{x}) \circ j_S$ and the result follows from the fact that $\tilde{x}$ is almost periodic on B(S).

3.7. Lemma: Let S be a set, M a finite dimensional subalgebra of BF(S), the C^*-algebra of bounded, complex valued functions on S. Then there is a finite subset S_1 of S so that $x \to x|S_1$ is a C^*-isomorphism from M onto $C(S_1)$.

Proof: S_1 is the subset of S on which not all $x \in M$ vanish. The result follows then from the uniqueness of C^*-norms.

3.8. Lemma: $\tilde{c}\,(AP(S)) \subset AP(S) \hat{\otimes} AP(S)$.

Proof: Let $x \in AP(S)$. If $\varepsilon > 0$ there is a finite subset M of AP(S) so that

$\{L_a x: a \in S\} \subset M + \varepsilon B_{\|\ \|}$. By 3.7 we can choose

$\{s_1,\ldots,s_n\} \subset S$ and a basis $(x_1,\ldots,x_n\}$ of the linear span of M in $C^\infty(S)$ so that $x_i(s_j) = \delta_{ij}$ and

$\| x \| = \sup \{ |x(s_i)|: i = 1,\ldots,u\}$ holds for all x in the linear span of M.

Now if $a \in S$ there is $\{\lambda_1,\ldots,\lambda_n\} \subset C$ so that $|x(as) - \sum_{i=1}^{n} \lambda_i x_i(s)| \leq \varepsilon$ for each $s \in S$. Hence $|x(as_i) - \lambda_i| \leq \varepsilon$ for $i = 1,\ldots,n$.

Then

$$|x(as^{-1}) - \sum_{i=1}^{n} x(as_i)\, x_i(s^{-1})| \leq$$

$$|x(as^{-1}) - \sum_{i=1}^{n} \lambda_i x_i(s^{-1})| + \|\sum_{i=1}^{n} (\lambda_i - x(as_i))x_i\| \leq 2\varepsilon$$

Thus $\| \tilde{c}(x) - \sum_{i=1}^{n} R_{s_i} x \otimes x_i \| \leq 2\varepsilon$,

so $\tilde{c}(x)$ is in the norm closure of $AP(S) \otimes AP(S)$ in $C^\infty(S \times S)$, i.e. in $AP(S) \hat{\otimes} AP(S)$.

3.9. Proposition: $\widetilde{C^\infty}(S) = AP(S)$.

Proof: 3.6. and 3.8.

§4. The Algebra $M_t(S)$ and representations

4.1. Let (E,c,η,a) be a cogroup in $MIXC^*$. Equip E with the strict topology $\gamma\,[\|\cdot\|,\tau]$ (1.8) and let E_γ be its dual.

Define a multiplication on E_γ' in the following way: if $f,g \in E_\gamma'$, then $f * g$ be given by $x \to f \otimes g(c(x))$.

Proposition: If E is a cogroup in $MIXC^*$, then E_γ' is Banach algebra with identity. It is commutative if E is commutative.

4.2. Let S be a compactological space.

A premeasure on S is a member of the projective limit of the system $\{\tilde{f}_{K_1 K_2}: M(K_1) \to M(K_2): K_2 \subset K_1, K_1, K_2 \in \mathcal{K}(S)\}$ where $M(K)$ denotes the space of all Radon -measures on K.

If $\mu = \{\mu_K\}_{K\in\mathcal{K}}$ is a premeasure on S and $|\mu_K|^*$ denotes the outer measure on K defined by $|\mu_K|$ then we define, for a set $C \subset S$ $|\mu|^*(C) = \sup\{|\mu_K|^*(C \cap K): K \in \mathcal{K}(S)\}$. A premeasure μ on S is said to be tight if for each $\varepsilon > 0$ there is a $K \in \mathcal{K}$ so that $|\mu|^*(S\backslash K) < \varepsilon$. Equivalent is the existence of an increasing sequence K_n in $\mathcal{K}(S)$ with $|\mu|^*(S\backslash K_n) \to 0$.

We denote by $M_t(S)$ the space of all tight (pre) measures on S.

If $x \in C^\infty(S)$ and $\mu \in M_t(S)$, then the limit $\lim\limits_{n\to\infty} \int x|K_n \, d\mu_{K_n}$ exists and is independent of the particular choice of the sequence K_n.

We write $\int x \, d\mu$ for this limit.

A premeasure $\mu = \{\mu_K\}$ is said to have compact support if there is a $K \in \mathcal{K}(S)$ so that $|\mu|^*(S\backslash K) = 0$. The space $M_o(S)$ of premeasures with compact support is identifiable with $\bigcup_{K\in\mathcal{K}} M(K)$.

Proposition: (COOPER) If S is a regular compactology, then the dual of $[C^\infty(S), \tau_{K(S)}]$ is naturally isomorphic to $M_o(S)$ under the bilinear form $(x,\mu) \to \int x \, d\mu$; and the dual of $[C^\infty(S), \gamma[\|.\|, \tau]]$ is naturally isomorphic to $M_t(S)$ under the bilinear form

$$(x,\mu) \to \int x \, d\mu.$$

So $C^\infty(S)_\gamma' = M_t(S)$.

4.3. If S is a compactological group, then we can give an explicit description of the multiplication * (3.1) in $M_t(S)$: If $x \in C^\infty(S)$, $\mu, \nu \in M_t(S)$, then we have

$$\begin{aligned}\int x \, d(\mu * \nu) &= (\mu \otimes \nu)(c(x)) \\ &= \int x(s.t) \, d(\mu \otimes \nu)(s.t) \\ &= \iint x(s.t) \, d\mu(s) d\nu(t),\end{aligned}$$

i.e. we have the ordinary convolution.

4.4. If S is a regular compactological space and E a complete object of MIXTOP, then define $C^\infty(S;E)$ as the space of all $\|.\|$-bounded maps $f: S \to E$, equipped with the pointwise linear structure and the following mixed structure:

$\|.\|$ - the norm $\|f\| = \sup \{\|f(s)\|_E, s \in S\}$.

τ - the topology of uniform τ_E-convergence on members of $\mathcal{K}(S)$.

Proposition: If S is a regular compactological space, than the embedding $\delta : S \to M_t(S)$ has the following universal property: for every complete object E of MIXTOP and every $f \in \mathcal{C}^\infty(S;E)$ there is a unique $T \in L(M_t(S);E)$ which extends f via δ.

This gives rise to natural equivalences

$\mathcal{C}^\infty(S;E) \cong \mathcal{C}^\infty(S) \hat{\otimes} \gamma\, E \cong L(M_t(S);E)$.

4.5. If S is a compactological group and E a complete object in MIXTOP, then $\mathcal{C}^\infty(S,E)$ has a natural map c_E: $\mathcal{C}^\infty(S;E) \to \mathcal{C}^\infty(S\times S;E)$, given by $c_E : x \mapsto ((s,t) \mapsto x(st))$;

i.e. $c_E = c \otimes id_E : \mathcal{C}^\infty(S) \hat{\hat{\otimes}}_\gamma E \to \mathcal{C}^\infty(S) \hat{\otimes}_\gamma \mathcal{C}^\infty(S) \hat{\hat{\otimes}}_\gamma E$.

Let $(E, \|.\|, \tau)$ be a complete algebra with unit e in MIXTOP, i.e. there is an associative multiplication $m: E \times E \to E$ so that $(E, \|.\|)$ is a Banach algebra and $m|_{B_{\|.\|} \times B_{\|.\|}}$ is $\tilde{\tau} \times \tilde{\tau}$ -continuous.

Then $\mathcal{C}^\infty(S;E)$ has a natural "multiplication".

m_E: $\mathcal{C}^\infty(S;E) \times \mathcal{C}^\infty(S;E) \to \mathcal{C}^\infty(S \times S;E)$, given by

m_E: $(x,y) \mapsto ((s,t) \mapsto m(x(s),y(t)))$.

We say that an element $x \in \mathcal{C}^\infty(S;E)$ is primitive, if $c_E(x) = m_E(x,x)$.

4.6. Proposition: Under the identification $\mathcal{C}^\infty(S;E) \cong L(M_t(S),E)$ the primitive elements correspond to the Banach algebra morphisms. $x \in \mathcal{C}^\infty(S;E)$ induces a unit preserving operator if and only if $x(e) = e_E$.

Proof: Suppose that $x \in \mathcal{C}^\infty(S;E)$ is primitive. The image T_x of x in $L(M_t(S);E)$ is defined by $\mu \mapsto \int_S x\, d\mu$.

Then $T_x\,(\mu * \nu) = \int_S x\, d(\mu * \nu) =$

$$\int_{S\times S} c_E(x)\, d\,(\mu \otimes \nu) = \int_{S\times S} x \otimes x\, d(\mu \otimes \nu)$$

$$= \int_S x\, d\mu \int_S y d\nu = T_x(\mu)\, T_x(\nu).$$

On the other hand, if $T \in L(M_t(S);E)$, then $T = T_x$ where $x = T \circ \delta$.

Then $c_E(x)(s,t) = x(st) = T(\delta_{st})$

$$= T(\delta_s * \delta_t) = m(T(\delta_s),T(\delta_t))$$

$$= m(x(s),x(t)) = m_E(x,x)(s,t).$$

4.7. Corollary: Let S be a regular compactological group, X a Banach space. Then there is a one-one correspondence between

(i) the set of strongly continuous representations of S in X.

(ii) the unit preserving Banach algebra morphisms in $L(M_t(S);E)$

(iii) the primitive elements x of $\mathcal{C}^\infty(S;E)$ with $x(e) = e_E$.

(E denotes the object $(L(X,X), \|.\|, \tau_s)$ of MIXTOP - τ_s is the strong operator topology).

§5. Pontryagin duality

5.1. Let (E,c,η,a) be an object of $CMIXC^*$. $x \in E$ is called *strongly primitive* if

(i) $c(x) = x \otimes x$

(ii) $\eta(x) = 1$

(iii) $a(x) = x^{-1}$ (in E).

We denote by $P(E)$ the set of strongly primitive elements of E.

5.2. *Proposition.* $P(E)$, with the topology and the multiplicative structure induced from $(E,\gamma(\|\cdot\|_E,\tau_E))$, is a topological group. It is contained in $\{x \in E: \|x\| = 1\}$.

Proof: $P(E)$ is closed under multiplication:

Let $x, y \in P(E)$, then

$c(xy) = c(x)c(y) = (x \otimes x)(y \otimes y) = xy \otimes xy.$

$\eta(xy) = \eta(x)\,\eta(y) = 1.1 = 1$

$a(xy) = a(x)a(y) = x^{-1}\,y^{-1} = (xy)^{-1}$

(E is commutative: the C^*-algebra part).

The constant function 1 is a unit for $P(E)$.

If $x \in P(E)$, then $a(x) = x^{-1}$ is an inverse for x. Since multiplication is $\gamma\,[\,\|\cdot\|_E, \tau_E)$ continuous, $(P(E),\gamma[\,\|\cdot\|_E,\tau_E\,])$ is a topological group. Since $\|x\|\,\|x\| = \|x \otimes x\| = \|c(x)\| \leq \|x\|$ and $1 = \|1\| = \|y(x)\| \leq \|x\|$ we conclude that $\|x\| = 1$ for all $x \in P(E)$.

<u>5.3.</u> <u>Definition</u>: Let S be a commutative compactological group. A <u>character</u> on S is a GCPTOL-morphism from S into the circle group T. The set $\hat{S}$ of all characters form a group. $\hat{S}$, with the topology of uniform convergence on members of $\mathcal{K}$ (S), is a topological group, and it is complete in this uniformity.

<u>5.4.</u> <u>Proposition</u>: Let S be a commutative, compactological group. The $\hat{S} = P(C^\infty(S))$ (a topological group).

<u>Proof</u>: If $x \in P(C^\infty(S))$, then $\|x\| = 1 = \|x^{-1}\|$, so x takes its values in T.

$c(x)(s,t) = x(st)$ and

$(x \otimes x)(s,t) = x(s)x(t)$ show that the strong primitivity of x is equivalent to its being a character.

Since $\gamma[\|.\|, \tau] = \tau$ on $B_{\|.\|}$ we see that even the topologies coincide.

<u>5.5.</u> <u>Corollary</u>: Let S be a commutative, regular compactological group. The $\hat{S}$ separates S if and only if the vector space generated by $P(C^\infty(S))$ is γ-dense in $C^\infty(S)$.

REFERENCES

C. APOSTOL — B^*-algebras and their representations, Jour. Lond. Math. Soc. (2) 3 (1971) 30-38.

H. BUCHWALTER — Topologies, bornologies et compactologies, Lyon

J. B. COOPER — The mixed topology and applications.

H. HEYER — Dualität lokalkompakter Gruppen, Springer Lecture Notes Nr. 150 (Berlin, 1970).

E. HEWITT, K.A. ROSS — The Tannaka-Krein duality theorems, Jahresber. D.M.V. 71 (1969) 61-83.

Abstract harmonic analysis II (Berlin, 1970).

K. H. HOFMANN — The duality of compact semigroups and C^*-bigebras, Springer Lecture Notes Nr. 129 (Berlin 1970).

P. MICHOR — Duality in Groups, unpublished note - 1972.

J.W. NEGREPONTIS — (J.W. Pelletier) Duality in analysis from the point of view of triples, Jour. of Algebra 19 (1971) 228-253.

N. NOBLE — k-groups and duality, Trans. Amer. Math. Soc. 151 (1970) 551-561.

D. W. ROEDER — Category theory applied to Pantryagin duality, Pac. J. Math. 52 (1974) 519-527.

S. SANKARAN, S.A. SELESNICK — Some remakrs on C^*-bigebras and duality, Semigroups Forum 3 (1971) 108-129.

S. A. SELESNICK — Watts cohomology for a class of Banach algebras and the duality of compact abelian groups, Math. Z. 130 (1973) 313-323.

M. TAKESAKI — Duality and von Neumann algebras (in "Lectures on OPerator Algebras" - Springer Lecture Notes Nr. 247 - Berlin, 1972).

PRODUCTS AND SUMS IN THE CATEGORY OF FRAMES

by

C. H. Dowker, Birkbeck College, University of London

and

Dona Strauss, University of Hull

Spaces and frames. The open sets of a topological space X form a complete lattice tX in which the join is the union, while the meet is the interior of the intersection:

$$\bigvee_\alpha G_\alpha = \bigcup G_\alpha, \qquad \bigwedge G_\alpha = \operatorname{int} \bigcap G_\alpha .$$

Many of the properties of a topological space X, e.g. the properties of being compact, regular, normal, etc., are actually properties of the topology tX. Consequently much of what claims to be topology is actually lattice theory.

A frame L is a complete lattice with infinite distributivity

$$a \wedge \bigvee b_\alpha = \bigvee a \wedge b_\alpha .$$

It follows that

$$a \vee (b \wedge c) = (a \vee b) \wedge (a \vee c).$$

A frame map $f : L \longrightarrow M$ is a function commuting with finite meets and with arbitrary joins. Thus the topology tX of a space X, that is its lattice of open sets, is a frame. Each continuous function $f : X \longrightarrow Y$ induces a frame map $tf : tY \longrightarrow tX$, where $tf(G) = f^{-1}G$. Thus there is determined a contravariant functor

$$t : \underline{\text{Top}} \longrightarrow \underline{\text{Fr}}$$

from the category of topological spaces to the category of frames. The functor t has a right adjoint

$$s : \underline{\text{Fr}} \longrightarrow \underline{\text{Top}},$$

([1], Theorem 8), where sL is a space whose points are the prime elements of L or, equivalently, the points are the maps of L to the two element frame $\{0, 1\}$.

<u>The dual of the category of frames.</u> We realize the dual category $\underline{\mathbf{Fr}}^{op}$ as a concrete category in the following way.

Given a frame map $f : L \longrightarrow M$, let $g : M \dashrightarrow L$ be the function for which

$$g(u) = \bigvee\{x \in L : f(x) \leq u\}. \qquad (A)$$

Since f commutes with joins

$$fg(u) = \bigvee\{f(x) : f(x) \leq u\} \leq u.$$

Hence $g(u) = \max\{x : f(x) \leq u\}$. If $f(x) \leq u$ then $x \leq g(u)$. Also, if $x \leq g(u)$, $f(x) \leq fg(u) \leq u$. Thus

$$f(x) \leq u \quad \text{iff} \quad x \leq g(u).$$

Since $f(a) = \bigwedge\{u \in M : f(a) \leq u\}$,

$$f(a) = \bigwedge\{u \in M : a \leq g(u)\}. \qquad (B)$$

In lattice theory the relative semicomplement $b*c$ in L is defined by

$$b*c = \max\{a \in L : a \wedge b \leq c\}.$$

A complete lattice L has semicomplements $b*c$ for all $b, c \in L$ iff L is a frame. The semicomplement has the property that

$$a \wedge b \leq c \quad \text{iff} \quad a \leq b*c.$$

The function g determined by (A) has the following properties :

(i) $g(\bigwedge_\alpha u_\alpha) = \bigwedge_\alpha g(u_\alpha)$,

(ii) if $u \neq 1$, $g(u) \neq 1$,

(iii) $a * g(u) \leq g(f(a) * u)$,

where f is given by (B) in terms of g.

We give the proof only of (iii). Let $b = a * g(u)$. Then $a \wedge b \leq g(u)$. Hence

$$f(a) \wedge f(b) = f(a \wedge b) \leq u,$$

that is $f(b) \leq f(a) * u$. Therefore $b \leq g(f(a) * u)$, that is

$$a * g(u) \leq g(f(a) * u).$$

We define a function $g : M \dashrightarrow L$, where M and L are frames, to be an <u>antimap</u> if it satisfies conditions (i), (ii), (iii) above. We use broken arrows for antimaps.

Given an antimap $g : M \dashrightarrow L$, the function $f : L \longrightarrow M$ determined by (B) is a frame map, that is,

(1) $f(\bigvee a_\alpha) = \bigvee f(a_\alpha)$,

(2) $f(1) = 1$,

(3) $f(a \wedge b) = f(a) \wedge f(b)$.

We give the proof only for (3). Let $v = f(a \wedge b)$. Then $a \wedge b \leq g(v)$. Hence

$$b \leq a * g(v) \leq g(f(a) * v),$$

$$f(b) \leq f(a) * v,$$

$$f(a) \wedge f(b) \leq v = f(a \wedge b).$$

Since the reverse inequality is trivial, $f(a \wedge b) = f(a) \wedge f(b)$.

The correspondence between maps $f : L \longrightarrow M$ and antimaps $g : M \dashrightarrow L$ gives an isomorphism of the dual category $\underline{Fr}^{op}$ to the category of frames and antimaps. Identifying under this isomorphism, f^{op} is realized as the function $g : M \dashrightarrow L$ given by (A).

It can easily be verified that f is injective iff f^{op} is surjective, while f is surjective iff f^{op} is injective.

For each frame map $f : L \longrightarrow M$, the image $f(L)$ is a frame, and f factors uniquely into the surjective map $f_1 : L \longrightarrow f(L)$ followed by the inclusion $j : f(L) \longrightarrow M$. Then $g = f^{op} = f_1^{op} j^{op}$. The injective antimap $f_1^{op} : f(L) \dashrightarrow L$ factors uniquely into an isomorphism $h : f(L) \longrightarrow gf(L)$ followed by the inclusion antimap $i : gf(L) \dashrightarrow L$. The inclusion maps (antimaps) are taken as subobjects in $\underline{Fr}$ (respectively $\underline{Fr}^{op}$). The duals of inclusions may be taken as quotients, that is representative surjections. Thus each map f or antimap g can be factored uniquely

$$f = j \circ h^{op} \circ i^{op}, \qquad g = i \circ h \circ j^{op},$$

into a quotient followed by an isomorphism followed by the inclusion of a subobject.

<u>Products</u>. Let (L_α) be a family of frames. In <u>Ens</u> there is a product set $\prod L_\alpha$ and projections $\pi_\beta : \prod L_\alpha \to L_\beta$. Let $a \leq b$ in the product set if $a_\alpha \leq b_\alpha$, i.e. $\pi_\alpha a \leq \pi_\alpha b$, for all α. With this order $\prod L_\alpha$ is a frame and the projections π_α are frame maps. Let K be a frame and let $f_\alpha : K \to L_\alpha$ be frame maps. The unique function $f : K \to \prod L_\alpha$, for which $\pi_\alpha f = f_\alpha$, is a frame map. Hence the frame $\prod L_\alpha$ with the projections π_α is the product in the category of frames.

If (X_α) is a family of topological spaces and $i_\beta : X_\beta \to \sum X_\alpha$ are the canonical injections into the topological sum, the frame maps $ti_\beta : t\sum X_\alpha \to tX_\beta$ induce an isomorphism $j : t\sum X_\alpha \to \prod tX_\alpha$.

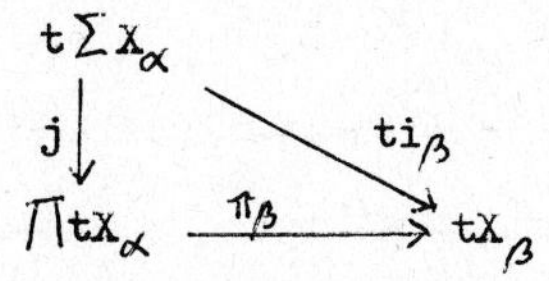

Although, as appears from the above brief description, products of frames are trivial, they form the basis of the following construction.

<u>Construction</u>. Let B be the subset of $\prod L_\alpha$ consisting of all b such that either $b_\alpha = 1$ for all but a finite number of α, or $b_\alpha = 0$ for all α. Then B is a subframe of $\prod L_\alpha$. If $i : B \to \prod L_\alpha$ is the inclusion, the composites $\pi_\alpha i : B \to L_\alpha$ are surjective. A subset G of B is called <u>decreasing</u> if whenever $a \in G$ and $b \in B$ with $b \leq a$, then $b \in G$. Clearly every intersection of decreasing sets is decreasing and every union of decreasing sets is decreasing. Thus the set T of all decreasing sets of B is a topology for B. The function $\varphi : T \to B$ for which $\varphi G = \bigvee G$ is a surjective frame map. Hence $\pi_\alpha i \varphi : T \to L_\alpha$ is surjective. Let $\gamma_\alpha : L_\alpha \dashrightarrow T$ be the corresponding injective antimap ; $\gamma_\alpha = (\pi_\alpha i \varphi)^{op}$.

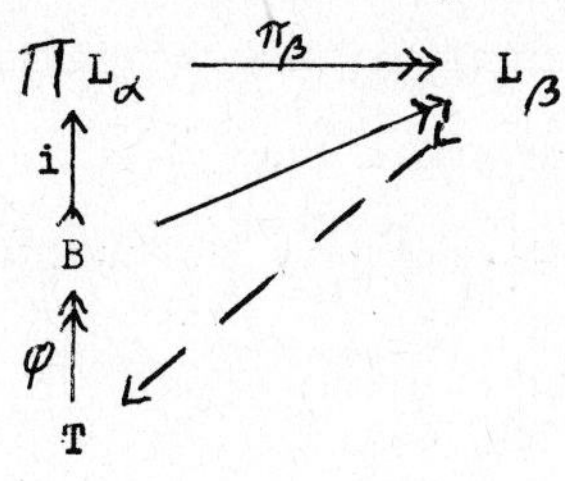

Then the functions γ_α have the following universal property. If $f_\alpha : L_\alpha \longrightarrow M$ are frame maps, there exists a unique frame map $g : T \longrightarrow M$ with commutativity $f\gamma_\alpha = f_\alpha$. Thus T has the universal property for a coproduct except that the morphisms γ_α are in the wrong category.

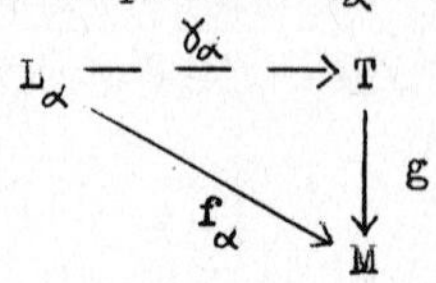

Coproducts. That coproducts exist in the category of frames was announced in [5] and was proved in S. Papert's unpublished dissertation [6]. There is still no published proof, but a detailed proof is contained in our paper [3], to appear.

Here it is sufficient to state that the frame T constructed above has a greatest quotient $\theta : T \twoheadrightarrow L$ such that the composites $h_\alpha = \theta\gamma_\alpha$ are frame maps. It can be shown that L with the maps $h_\alpha : L_\alpha \longrightarrow L$ is the coproduct or sum $\sum L_\alpha$ of the frames L_α.

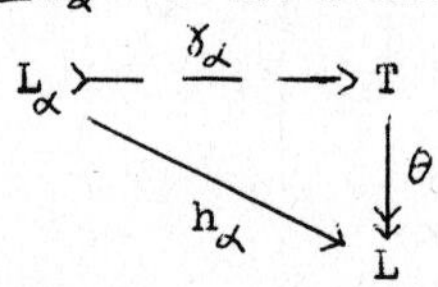

If (X_α) is a family of spaces and $\pi_\beta : \prod X_\alpha \longrightarrow X_\beta$ are the projections of the product space, there is induced a surjective frame map $u : \sum tX_\alpha \longrightarrow t\prod X_\alpha$ which in general is not an isomorphism.

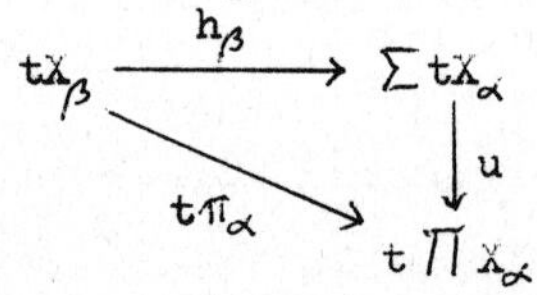

The map u is however an isomorphism if each X_α is a compact Hausdorff space. It is an isomorphism also if the set of indices α is countable and each X_α is a locally compact Hausdorff space.

Elements of sums. The injective antimap $\varphi^{op} : B \dashrightarrow T$ maps each $b \in B$ to the set

$$\varphi^{op}(b) = \{c \in b : c \leq b\}.$$

Such sets form a base for the topology T. Thus if $G \in T$,

$$G = \bigvee_{b \in G} \varphi^{op}(b).$$

For $u \in L$ let $H = \theta^{op}u = \bigvee_{\theta G \leq u} G$. Since θ is surjective, $\theta H = u$. Then

$$H = \bigvee_{\hat{b} \leq u} \varphi^{op}(b),$$

where

$$\hat{b} = \theta\varphi^{op}b = \bigwedge_{\alpha} h_{\alpha}b_{\alpha},$$

this meet being finite since for each b, either all but a finite number of b_{α} are 1 or all b_{α} are 0. Then

$$u = \bigvee_{\hat{b} \leq u} \bigwedge_{\alpha} h_{\alpha}b_{\alpha}.$$

Any frame maps $f_{\alpha} : L_{\alpha} \longrightarrow K$ induce a frame map $f : L \longrightarrow K$ where

$$f(u) = \bigvee_{\hat{b} \leq u} \bigwedge_{\alpha} f_{\alpha}b_{\alpha}.$$

Properties preserved by sums. A frame L is called compact if each family (a_{α}) of elements of L such that $\bigvee a_{\alpha} = 1$ has a finite subfamily (a_{α_i}) such that $\bigvee a_{\alpha_i} = 1$. A frame L is called regular if for each $u \in L$, there is a family (a_{α}) of elements of L such that $\bigvee a_{\alpha} = u$ and for each α, $a_{\alpha}* \vee u = 1$, where

$$a_{\alpha}* = a_{\alpha}*0 = \max\{x \in L : a_{\alpha} \wedge x = 0\}.$$

Similarly the definitions of other properties of the topologies of spaces are extended to define the corresponding properties of frames.

Each sum of compact frames is compact (S. Papert, [7]). Each sum of regular frames is regular.

Each sum of paracompact regular frames is paracompact (Isbell, [4]). Each sum of Lindelöf regular frames is Lindelöf. Each sum of fully normal frames is fully normal. (By A. H. Stone's theorem, fully normal is the same as paracompact and normal, which is weaker than paracompact and regular.

The properties of being paracompact regular, Lindelöf regular or fully normal are not preserved in topological products. Thus the points of a topological space are not only irrevelant to such properties as normal, Lindelöf, paracompact, etc., which are properties of the topology of the space. The presence of the points leads us to work in the wrong category, and thus get less satisfactory theorems.

<u>Application to the transfer of algebraic structures.</u> Let $\underline{LKT_2}$ be the category of locally compact Hausdorff spaces and continuous functions. Since, for countable products $\prod X_i$, $u : \sum tX_i \longrightarrow t\prod X_i$ is an isomorphism,

$$Op\ t : \underline{LKT_2} \longrightarrow \underline{Fr}^{op}$$

preserves countable products, up to equivalence.

$$\begin{array}{ccc} [\sum L_\alpha, K] & \xrightarrow{\quad H_K(h_\beta) \quad} & \\ v \downarrow & \searrow & \\ \prod [L_\alpha, K] & \xrightarrow[\pi_\alpha]{} & [L_\beta, K] \end{array}$$

Since, for each frame K, the map $v : H_K(\sum L_\alpha) \longrightarrow \prod H_K(L_\alpha)$ is an isomorphism, the functor

$$H_K\ Op : \underline{Fr}^{op} \longrightarrow \underline{Ens}$$

preserves products, up to equivalence. Hence

$$H_K\ t : \underline{LKT_2} \longrightarrow \underline{Ens}$$

preserves countable products, and in particular finite products. Therefore $H_K\ t$ transfers algebraic structures ([8], Theorem 11.3.4).

Let $C(K) = [t\,\mathbb{R}, K]$, that is $C(K)$ is the set of frame maps of the topology of the reals to the frame K. In particular, if $K = tX$ for some space X, t maps $[X, \mathbb{R}]$ bijectively to $[t\,\mathbb{R}, tX]$ ([2], Theorem 1), so $C(tX)$ may be regarded as the set of continuous real functions on X. $C(K)$ is the generalization from topologies to frames.

Since $H_K\ t$ transfers algebraic structures, the algebraic structure of $\mathbb{R}$ is taken by $H_K\ t$ to $[t\,\mathbb{R}, K] = C(K)$. Thus $C(K)$ is a ring and a lattice.

There is an injective morphism from the ring $\mathbb{R}$ to the ring $C(K)$, taking real numbers to 'constants'; see below. With multiplication by constants, $C(K)$ is an algebra and a vector lattice.

<u>Binary operations in $C(K)$.</u> We describe more explicitly the binary operations transferred by $H_K\ t$ from $\mathbb{R}$ to $C(K)$.

Let $p : \mathbb{R}^2 \longrightarrow \mathbb{R}$ be a continuous function. Typical cases: $p(x, y)$ may be $x + y$ or xy or $x \vee y$ or $x \wedge y$. Then tp is a frame map

$$tp : t\,\mathbb{R} \longrightarrow t\,\mathbb{R}^2$$

or, if we identify $t\,\mathbb{R}^2$ with $t\,\mathbb{R} + t\,\mathbb{R}$ under the isomorphism

$u : t\mathbb{R} + t\mathbb{R} \longrightarrow t\mathbb{R}^2$,

$$tp : t\mathbb{R} \longrightarrow t\mathbb{R} + t\mathbb{R}.$$

Applying H_K we have

$$H_K(tp) : [t\mathbb{R} + t\mathbb{R}, K] \longrightarrow [t\mathbb{R}, K]$$

or, if we identify $[t\mathbb{R} + t\mathbb{R}, K]$ with $[t\mathbb{R}, K] \times [t\mathbb{R}, K]$,

$$H_K(tp) : [t\mathbb{R}, K] \times [t\mathbb{R}, K] \longrightarrow [t\mathbb{R}, K].$$

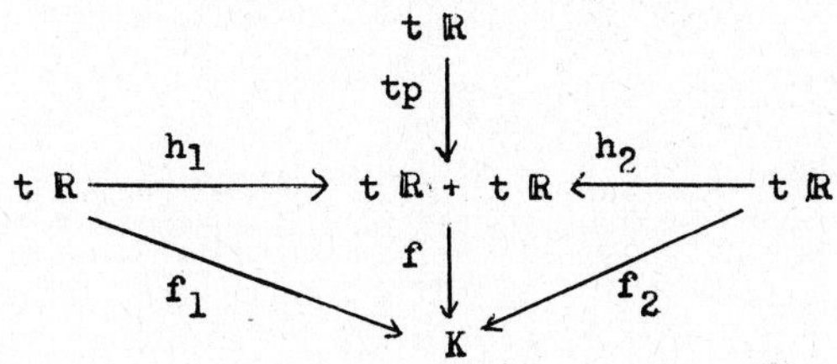

For $f_1, f_2 \in [t\mathbb{R}, K]$, let $f : t\mathbb{R} + t\mathbb{R} \longrightarrow K$ be the map for which $fh_i = f_i$. Then

$$H_K(tp)(f_1, f_2) = H_K(tp)(f) = f \circ tp.$$

When there is no danger of misunderstanding we may write p for $H_K(tp)$. Thus for any maps $f_1, f_2 : t\mathbb{R} \longrightarrow K$, we have

$$p(f_1, f_2) = f \circ tp : t\mathbb{R} \longrightarrow K.$$

If $p(x, y) = x + y$, we write $f_1 + f_2$ instead of $p(f_1, f_2)$, etc. Thus we have

$$f_1 + f_2 : t\mathbb{R} \longrightarrow K,$$

$$f_1 f_2 : t\mathbb{R} \longrightarrow K,$$

$$f_1 \vee f_2 : t\mathbb{R} \longrightarrow K,$$

$$f_1 \wedge f_2 : t\mathbb{R} \longrightarrow K.$$

The lattice operations give an order in $C(K)$. Thus $f_1 \leq f_2$ if $f_2 = f_1 \vee f_2$.

<u>Formulas for binary operations.</u> We write

$$r^- = \{x \in \mathbb{R} : x < r\}, \qquad r^+ = \{x \in \mathbb{R} : x > r\}$$

for the members of the usual subbase for $\mathbb{R}$. Then, for example,

$$t(+)(r^-) = \{(x, y) : x + y < r\},$$

$$t(\cdot)(r^-) = \{(x, y) : xy < r\},$$

$$t(\vee)(r^-) = \{(x, y) : x < r \text{ and } y < r\},$$

$$t(\wedge)(r^-) = \{(x, y) : x < r \text{ or } y < r\}.$$

For each $p : \mathbb{R}^2 \longrightarrow \mathbb{R}$,

$$t(p)(U) = \bigvee_{(V,\, W)^\wedge \leq t(p)(U)} h_1 V \wedge h_2 W$$

and hence

$$p(f_1, f_2)(U) = f(t(p)(U)) = \bigvee_{V \times W \subset t(p)(U)} f_1 V \wedge f_2 W.$$

Applying this formula to particular binary operations p gives, for example, the following :

(i) $(f_1 + f_2)(r^-) = \bigvee_{x + y \leq r} f_1(x^-) \wedge f_2(y^-)$,

$(f_1 + f_2)(r^+) = \bigvee_{x + y \geq r} f_1(x^+) \wedge f_2(y^+)$,

(ii) $(f_1 \vee f_2)(r^-) = f_1(r^-) \wedge f_2(r^-)$,

$(f_1 \vee f_2)(r^+) = f_1(r^+) \vee f_2(r^+)$,

(iii) $(f_1 \wedge f_2)(r^-) = f_1(r^-) \vee f_2(r^-)$,

$(f_1 \wedge f_2)(r^+)) = f_1(r^+) \wedge f_2(r^+)$,

(iv) $f_1 \leq f_2$ iff $f_2(r^-) \leq f_1(r^-)$ for all $r \in \mathbb{R}$,

$f_1 \leq f_2$ iff $f_1(r^+) \leq f_2(r^+)$ for all $r \in \mathbb{R}$.

<u>Constants in $C(K)$.</u> For $a \in R$, let $k_a : t\mathbb{R} \longrightarrow K$ be the function for which $k_a(G) = 1$ if $a \in G$, $k_a(G) = 0$ if $a \notin G$. Then k_a is a frame map. One easily verifies that

$$k_a = k_b \text{ iff } a = b,$$

$$k_{a+b} = k_a + k_b,$$

$$k_{ab} = k_a k_b,$$

$$k_{a \wedge b} = k_a \wedge k_b,$$

$$k_{a \vee b} = k_a \vee k_b.$$

Thus k is an embedding of R in $C(K)$. Identifying a with k_a, we may write $a : t\mathbb{R} \longrightarrow K$. Then for $f : t\mathbb{R} \longrightarrow K$,

$f \leq a$ iff $f(r^-) = 1$ for all $r > a$,

$f \leq a$ iff $f(r^+) = 0$ for all $r > a$,

$f \geq a$ iff $f(r^-) = 0$ for all $r < a$,

$f \geq a$ iff $f(r^+) = 1$ for all $r < a$.

One can verify that if $a > 0$,

$$af(r^-) = f((\tfrac{r}{a})^-), \qquad af(r^+) = f((\tfrac{r}{a})^+),$$

while if $a < 0$,

$$af(r^-) = f((\tfrac{r}{a})^+), \qquad af(r^+) = f((\tfrac{r}{a})^-).$$

The bounded elements of $C(K)$. We shall say that $f \in C(K)$ is bounded if there exist $a, b \in \mathbb{R}$ such that $a \leq f \leq b$. We write $C^*(K)$ for the set of bounded elements of $C(K)$. It is easy to see that $C^*(K)$ is a subalgebra and a vector sublattice of $C(K)$.

We introduce a norm in $C^*(K)$ as follows :

$$\| f \| = \inf \{a : -a \leq f \leq a\}.$$

With this norm $C^*(K)$ is a normed algebra. We shall show that it is a Banach algebra.

$C^*(K)$ is complete. Indeed, let (f_n) be a Cauchy sequence in $C^*(K)$. For each $r \in \mathbb{R}$, let

$$\gamma_r = \bigvee_m \bigwedge_{n>m} f_n(r^-), \qquad \eta_r = \bigvee_m \bigwedge_{n>m} f_n(r^+).$$

We shall verify that $\gamma_r \wedge \eta_r = 0$, while if $r < s$, $\gamma_s \vee \eta_r = 1$. First

$$\gamma_r \wedge \eta_r = \bigvee_k \bigvee_m \bigwedge_{n>k,p>m} (f_n(r^-) \wedge f_p(r^+))$$
$$\leq \bigvee_k \bigvee_m \bigwedge_{n>\max(k,\, m)} (f_n(r^-) \wedge f_n(r^+)) = 0.$$

If $r < s$, choose u, v with $r < u < v < s$, and choose ε so that

$$0 < \varepsilon < \min (u - r, s - v).$$

Let m be such that if $n > m$ then

$$-\varepsilon < f_n - f_m < \varepsilon .$$

Then

$$f_m(u^+) = (f_n + (f_m - f_n))(u^+)$$
$$= \bigvee_{x+y \geq u} (f_n(x^+) \wedge (f_m - f_n)(y^+)).$$

If $x + y \geq u$ and $x < r$, then $y > \varepsilon$ and $(f_m - f_n)(y^+) = 0$. Hence

$$f_m(u^+)) \leq \bigvee_{x \geq r} f_n(x^+) = f_n(r^+).$$

Similarly it may be shown that, when $n > m$,

$$f_m(v^-) \leq f_n(s^-).$$

Hence

$$\gamma_s \vee \eta_r \geq \bigwedge_{n>m} f_n(s^-) \vee \bigwedge_{n>m} f_n(r^+)$$

$$\geq f_m(v^-) \vee f_m(u^+) = 1.$$

Using these properties of γ_r and η_r, one can show, as in the proof of [2], Theorem 3, that there exists a frame map

$$f : t\mathbb{R} \longrightarrow K$$

such that for all $r \in R$,

$$f(r^-) = \bigvee_{s<r} \gamma_s, \qquad f(r^+) = \bigvee_{s>r} \eta_s.$$

The frame map with this property is clearly unique. It is to be shown that the sequence (f_n) converges to f.

Let $\varepsilon > 0$. Choose n_o so that if $m, n > n_o$,

$$-\varepsilon < f_m - f_n < \varepsilon .$$

For $r \in \mathbb{R}$ and $n > n_o$,

$$(f - f_n)(r^-) = \bigvee_{x+y\leq r} (f(x^-) \wedge (-f_n)(y^-))$$

$$= \bigvee_{x+y\leq r} ((\bigvee_m \bigwedge_{p>m} f_p(x-)) \wedge (-f_n)(y^-)).$$

If $x + y \leq r$ and $p > n_o$,

$$f_p(x^-) \wedge (-f_n)(y^-) \leq (f_p - f_n)(r^-) \leq -\varepsilon(r^-).$$

Hence

$$\bigvee_m \bigwedge_{p>m} (f_p(x^-)) \wedge (-f_n)(y^-) \leq -\varepsilon(r^-),$$

and hence

$$f - f_n \geq -\varepsilon.$$

Similarly, using $(f - f_n)(r^+)$, one can show that

$$f - f_n \leq \varepsilon .$$

The proof of the completeness of $C^*(K)$ outlined here uses special properties of frames. We do not know a category theory argument for the transfer of the Banach space structure from $\mathbb{R}$ to $C^*(K)$.

REFERENCES

1 C. H. Dowker and Dona Papert, Quotient frames and subspaces, Proc. London Math. Soc. 16 (1966), 275-296.

2 C. H. Dowker and Dona Papert, On Urysohn's lemma, General topology and its relations to modern analysis and algebra II, Prague, 1966, 111-114.

3 C. H. Dowker and Dona Strauss, Sums in the category of frames, to appear.

4 J. R. Isbell, Atomless parts of spaces, Math. Scandinav. 31 (1972), 5-32.

5 Dona Papert and S. Papert, Sur les treillis des ouverts et les paratopologies, Séminaire C. Ehresmann 1957/58, exp. no. 1 (Paris, 1959).

6 S. Papert, The lattices of logic and topology, Cambridge University Ph. D. dissertation, 1959.

7 S. Papert, An abstract theory of topological spaces, Proc. Cambridge Phil. Soc. 60 (1964), 197-203.

8 H. Schubert, Categories, Springer-Verlag, Berlin Heidelberg New York, 1972.

CATEGORICAL METHODS IN DIMENSION THEORY

Roy Dyckhoff

St. Andrews University, Fife, Scotland.

The classical conception of dimension theory has been substantially augmented by cohomological methods, and within the last fifteen years by the theory of sheaves also. Modern interest in sheaf theory is based on the categorical properties of sheaves, in that they form an elementary topos, and on the logical properties, in that the topos provides a language in which certain sheaves can be described more simply than is classically possible; for example, the sheaf of germs of (continuous) real-valued maps is internally the real number object, hence classical but intuitionistically valid theorems about real numbers can therefore be exploited. The main purpose of this paper is to ask the question "in what sense is dimension of a space an internal property of the topos of sheaves over the space", and to give evidence that the question is not uninteresting.

In investigation of this subject, various approaches closely related to other work reported at this conference have been fruitful, and we therefore begin with a discussion of two subjects familiar to categorical topologists: factorisation theory and projective resolutions. We report on our work linking the two; the former is convenient in our study of dimension-changing maps, and the latter has a homological interpretation via sheaf theory.

Section three is a survey of basic dimension theory with sheaves of coefficients; no novelty is claimed for this, but it is useful background to the rest as well as an interesting way of fixing notation. We give here the relationship between projective resolutions and dimension. Section four is a study of dimension-changing closed maps; here the novelty is not so much in the theorems of dimension theory but rather in the relation of categorical and logical concepts to the theory and the ensuing simplifications. Specifically, we illustrate some internal category theory in an analysis of the monadic resolution of a proper map, by associating to a map a directed sheaf in essentially the same way as to a compact Hausdorff space of dimension zero one may associate a directed system of finite discrete spaces; we then represent the monadic resolution of the map as an internal direct limit of Čech resolutions of locally finite closed covers, with an application to dimension-raising maps.

No bibliography for such a diffuse subject can be exhaustive, hence we have chosen to refer mainly to survey articles rather than to original sources, with a few exceptions to be up to date. The reference [77] refers to the [1977] publication of the last-mentioned author, or the first in that year, and [77a] to the next in that year. Certain publication dates are necessarily conjectural.

§1. Factorisation theory.

To avoid confusion with Benabou [67] and to avoid saying "bicategory in the sense of Isbell [63] or Kennison [68] without epi- or mono- assumptions", we use the term Factorisation category for the following basic concept: a triple $(\mathcal{C}, \mathcal{P}, \mathcal{Q})$ where $\mathcal{P}$ and $\mathcal{Q}$ are classes of morphisms, in the category $\mathcal{C}$, each closed under composition and containing all isomorphisms, such that

i) Any morphism of $\mathcal{C}$ has an essentially unique $(\mathcal{Q}, \mathcal{P})$-factorisation

$$f = pq, \qquad p \in \mathcal{P}, \qquad q \in \mathcal{Q}.$$

ii) When $pf = pg$ and $fq = gq$ for $p \in \mathcal{P}$, $q \in \mathcal{Q}$, then $f = g$.

As in Herrlich [68] 7.2.3, i) implies that $\mathcal{C}$ satisfies the $(\mathcal{Q}, \mathcal{P})$-diagonal condition, and ii) implies uniqueness of such a diagonal. Since $\mathcal{P}$ determines $\mathcal{Q}$, we use $\mathcal{P}$-factorisation for brevity. Some would call $\mathcal{C}$ a $(\mathcal{Q}, \mathcal{P})$-category. For general theorems about such categories consult e.g. Kennison [68], Herrlich [68], [72], Ringel [70], [71], Strecker [74], [76], and Dyckhoff [72], [76].

Apart from onto-factorisations and the like, one of the first significant examples of this type is Eilenberg's light-factorisation in compact metric spaces [34], due also to Whyburn simultaneously (see [63] for applications to topological analysis). The essential content of this was extracted by Michael's [64] light-factorisation in T_1-spaces; in the same vein are Collins' dissonant-factorisation [71], [71a] in $\mathcal{J}$ = top. spaces, Strecker's superlight-factorisation in $\mathcal{J}$ [74], Henriksen and Isbell's perfect-factorisation [58] in

Tychonoff spaces, Herrlich's $\mathcal{K}$-perfect factorisation in Hausdorff spaces [71a] (where $\mathcal{K}$ = compact Hausdorff spaces), and the rather more complex factorisation theorem in T_0-spaces with skeletal maps due to Blaszczyk [74], where $\mathcal{Q}$ = maps g with $g^{-1}\overline{G} = \overline{g^{-1}G}$ for open G, and skeletal maps are the same as Herrlich's demi-open maps (Harris [71]).

Our aim here is to introduce a few more of these theorems, some being of particular interest in dimension theory, and others being included to complete various unfilled pictures. Michael's theorem decomposes the domain of a map into the space of components of fibres of the map; a similar theorem for T_1-spaces is valid (Dyckhoff [74a]) for the T_1-reflection of the space of quasi-components of fibres with a strange topology, where Fox's spreads (Michael [64]) (= decomposing, or separating, maps in Russia) play the role of $\mathcal{P}$. The T_1-assumption for these results is more or less vital: it ensures that the induced map from the middle space to the range is light (resp. a spread). A map $f: X \longrightarrow Y$ is <u>proper</u> iff perfect and separated (i.e. $f \times i_Z$ and $X \longrightarrow X \times_f X$ are all closed); the following result is easy and a good substitute for both these theorems in the non T_1-case:

<u>Proposition 1.1</u>. $(\mathcal{TP}, \mathcal{L}, \mathcal{M})$ is a factorisation category, where $\mathcal{TP}$ denotes the category of spaces and proper maps, $\mathcal{L}$ (resp. $\mathcal{M}$) the light (resp. monotone) maps in $\mathcal{TP}$. □

Many generalisations of perfect maps (cf Strecker [76]) are used, some being based on Herrlich's $\mathcal{K}$-perfect factorisation. Nevertheless, proper maps, including perfect maps of Hausdorff spaces, behave rather well:

<u>Proposition 1.2</u>. (Dyckhoff [72], [76], cf also Herrlich [72]).

$(\mathcal{T}, \mathcal{P}, \mathcal{A})$ is a factorisation category, where $\mathcal{T}$ denotes the category of top. spaces, $\mathcal{P}$ the proper maps. □

The maps $\mathcal{A}$ herein are those satisfying a unique diagonal condition wrt $\mathcal{P}$, which for our purposes is more useful than any intrinsic characterisation; we call $\mathcal{A}$ the <u>improper</u> maps in [76], the term <u>antiperfect</u> being used in [72] for improper maps of Hausdorff spaces. For a map $f: X \longrightarrow Y$ of T_2-spaces, the proper-factorisation has as middle space the T_2-reflection of the space of ultrafilters on X with convergent image in Y (Dyckhoff [74a]); extension outside T_3-spaces is unlikely by Wyler [71].

The property, or class, $\mathcal{R}$ of mappings we call <u>hereditary</u> iff whenever $f: X \longrightarrow Y$ has $\mathcal{R}$ and $U \subseteq Y$ is open, then $f^{-1}U \longrightarrow U$ has $\mathcal{R}$. Almost all the above mentioned classes $\mathcal{Q}$ in the factorisation categories are hereditary, as well as (dense maps), (quotient maps). The main exception is (dense, $\mathcal{K}$-extendable maps) in Hausdorff spaces, by a counterexample based on a Tychonoff-like corkscrew due to Herrlich.

<u>Proposition 1.3.</u> Improper maps form a hereditary class.

<u>Proof.</u> (Dyckhoff [76], [76a]). □

Perhaps this gives some justification to our preference for the proper-factorisation over the $\mathcal{K}$-perfect factorisation; note that the dense $\mathcal{K}$-extendable maps of Tychonoff spaces form a hereditary class, by the same proof (Dyckhoff [74a]).

In the study of dimension-changing properties of proper mappings, it is useful to recall by (1.2) that such a map has a monotone factor and a light factor; in the classical (Nagami [70]) terminology, the former "lowers" dimension, the latter "raises" it. Our three propositions imply

<u>Proposition 1.4.</u> $(\mathcal{J}, \mathcal{PL}, \mathcal{H})$ is a factorisation category, where $\mathcal{PL}$ is the class of proper light maps, and $\mathcal{H}$ is a hereditary class (namely, the dense, hereditarily 2-extendable maps). □

In our later application of the logic of sheaves to the analysis of a mapping, that is, the systematic quantification over all open sets, this hereditary property may appear more natural and desirable.

For convenience we here list the factorisation categories referred to:

<u>Proposition 1.5.</u> The following are factorisation categories:

i) Top. spaces, 1-1 maps, quotient maps.

ii) Top. spaces, embeddings, onto maps.

iii) Top. spaces, closed embeddings, dense maps.

iv) Top. spaces, dissonant maps, concordant quotients (= 2-extendable quotients).

v) Top. spaces, superlight maps, submonotone quotients.

vi) Top. spaces, proper maps, improper maps.

vii) T_1-spaces, light maps, monotone quotients.

viii) T_1-spaces, spreads, hereditarily concordant quasiquotients.

ix) T_2-spaces, perfect maps, antiperfect maps.

x) T_2-spaces, $\mathcal{K}$-perfect maps, dense, $\mathcal{K}$-extendable maps.

xi) Tychonoff spaces, perfect maps, dense C^*-embeddings.

xii) Top. spaces and proper maps, light maps, monotone maps.

xiii) T_0-spaces and skeletal maps, irreducible r.o. minimal maps, e.d. preserving maps. □

Categorical methods are available for proving most of the above results, but it is generally best to proceed directly, Parts vi), ix), and x) are the odd cases, and our next section is designed to show precisely why the categorical defining property of, e.g. improper maps, is more useful than any intrinsic characterisation.

For other types of factorisation theorem connected with dimension, see Kulpa [70] and Lisitsa [73].

§2. Projective resolutions.

Gleason's classic [58] on projective topological spaces pointed the way to a profusion of generalisations of his construction of projective covers in $\mathcal{K}$(= compact Hausdorff spaces) to wider topological categories, the most significant being due to Banaschewski [68] whose work simultaneously covers topological algebras, and to Ponomarev [64], whose theory of absolutes (regarded in [65] by Aleksandroff as 'one of the most outstanding achievements of set-theoretic topology in the last decade") drops naturally out of the spectral methods used to study perfectly n-dimensional spaces. Hager [71] and Mioduszewski and Rudolf [69] can be used as bases for further reading. For all these authors, a projective cover of a space X means a proper irreducible onto map RX ⟶ X, where RX is extremally disconnected and thus projective wrt proper onto maps; such a space RX may be constructed when X is Hausdorff as a space of convergent maximal open filters on X, e.g. as in Banaschewski [68];

note also Gonshor's construction [73] by non-standard analysis.

Henriksen and Jerison [65] noted the non-functoriality of this notion, and Herrlich [71] asked essentially both for this to be rectified and for the construction to be done for non-Hausdorff spaces. Shukla and Srivastava [75] solve the former problem with the idea of "stable coreflectivity" of the extremally disconnected spaces in the Hausdorff spaces, there are solutions by Mioduszewski and Rudolf [69] of the former by restriction to skeletal mappings, and by Blaszczyk [74], [74a] of the latter in the category of T_0-spaces. Nevertheless, in Dyckhoff [72] and [76] we have proposed the following alternative solution to these problems:

Proposition 2.1. There is a comonad (P,p,q) on $\mathcal{T}$ such that

i) each counit map p_X: PX —>> X is proper, onto, and

ii) PX is always projective in $\mathcal{T}\mathcal{P}$, and thus extremally disconnected.

Proof. Apply (1.2) to the discrete modification DX —>> X, taking its proper factor p_X: PX —>> X. By diagram chasing, P is a functor, p a natural transformation, and PX projective. The comultiplication P —> P^2 comes from an induced factorisation theorem in the category of comonads on $\mathcal{T}$: cf Dyckhoff [76] or Diers [73] for details. When X is regular, we may think of PX as the space of convergent ultrafilters on X with DX embedded therein as the principal ultrafilters (Dyckhoff [74a]). A point of PX, *qua* ultrafilter on X, is also an ultrafilter on DX and thus on a subset of PX; on close analysis, this is the comultiplication PX —> PPX at X. □

Rainwater's methods in [59] apply to our construction PX —> X in the Hausdorff case, reducing it to the irreducible map RX —>> X considered above; RX is a minimal (by Zorn) closed subspace of PX mapping onto X.

Corollary 2.2. Every Hausdorff space has a projective cover. □

As noted above, this corollary has been proved for T_0-spaces by Blaszczyk [74a], yet we do not know whether the T_0-axiom is necessary. At the price of irreducibility, our construction is both functorial and valid for all spaces; both constructions evidently have their value.

The *semi-simplicial category* Δ consists of the finite ordinals and order preserving maps; these maps factorise into face maps and degeneracies (MacLane [71]). A functor $\Delta^{op} \longrightarrow \mathcal{T}$ is a semi-simplicial diagram in $\mathcal{T}$; we can represent it as a sequence of spaces and maps

$$\cdots X_n \cdots\cdots\cdots \overset{\cdots\cdots}{\underset{\cdots}{\rightrightarrows}} X_3 \;\overset{\longrightarrow}{\underset{\longleftarrow}{\rightrightarrows}}\; X_2 \;\underset{\longleftarrow}{\rightrightarrows}\; X_1 \longrightarrow X_0$$

with relations between the face maps $\longrightarrow$ and degeneracies $\longleftarrow$. Such diagrams, or their duals, arise naturally in constructions of (co)homology theories (MacLane [71], Barr, Beck [69], Duskin [75], for example).

<u>Theorem 2.3</u>. (Dyckhoff [76]). For every space X, there is a semi-simplicial diagram in $\mathcal{TP}$:

$$\cdots\cdots\cdots P^2X \;\underset{\longleftarrow}{\rightrightarrows}\; PX \longrightarrow X$$

natural in X, all face maps being proper onto, each P^iX being projective.

<u>Proof</u>. Iterate the comonad of (2.1) in the standard way (MacLane [71]). □

We refer to this diagram as the <u>projective resolution of X</u>; in sheaf theory, (Godement [64]), "resolution" denotes exactness of a sequence of sheaves, and our use of the word seems presumptuous. We shall see in the next section how this diagram converts a sheaf on X into an exact sequence of sheaves in the most natural way, and regard this as justification for our usage. Indeed, we must do some sheaf theory before we can relate the projective resolution of X to dimension.

§3. Sheaves and dimension.

This section aims to outline very briefly the fundamentals of sheaf theory and cohomological dimension, both to fix our notation and as a basis for further reference. For details of the sheaf theory, we refer to Godement [64] and to Bredon [67], and for the application to dimension, to Kuz'minov [68]. Hurewicz, Wallman [48], Nagami [70], and Pears [75] are excellent references for topological dimension theory, and Kodama's appendix to Nagami's book is a good exposition of sheafless cohomological dimension theory; see also Bokshtein [66]. The application of sheaves to dimension is essentially due to Sklyarenko [62], [64]. We refer to Freyd [72], Kock and Wraith [71], Wraith [75], and Fourman [76] for the theory of topoi.

A sheaf on X is a local homeomorphism $p: S \longrightarrow X$; for $B \subseteq X$, the sections of p over B form a set S[B], with a restriction map $S[B] \longrightarrow S[C]$ for $B \supseteq C$. A presheaf on X is a functor to Sets from the category $\mathcal{O}(X)^{op}$ of open sets of X with a morphism $U \longrightarrow V$ iff $U \supseteq V$. The canonical presheaf of the sheaf S is the presheaf $\tilde{S}: U \longmapsto S[U]$. Certain presheaves may be thought of as sheaves by

Proposition 3.1. $\tilde{\ }: Sh(X) \longrightarrow Psh(X)$, $S \longmapsto \tilde{S}$, has an exact left adjoint. □

Proposition 3.2. Let $f: X \longrightarrow Y$ be a map and $f_*: Sh(X) \longrightarrow Sh(Y)$ the (direct image) functor with f_*S the sheaf on Y associated by (3.1) to the presheaf $U \longmapsto S[f^{-1}U]$. Then f_* has an exact left adjoint f^*.

Proof. (see Grothendieck [72] IV.4.1). For a sheaf $T \longrightarrow Y$, $f^*T \longrightarrow X$ is just the pullback by f. □

Proposition 3.3. (Grothendieck [72]). Sh(X) and Psh(X) are elementary topoi.

Proof. Let $p: S \longrightarrow X$ be a sheaf on X; then p_*p^* is right adjoint to $- \times S$, so Sh(X) is cartesian closed. As subobject classifier, take $X \longrightarrow \Omega$, where Ω is the sheaf on X associated by (3.1) to the presheaf $U \longmapsto \mathcal{O}(U)$. Similarly for presheaves. □

In any category with pullbacks, (abelian) groups, rings, modules, partially ordered sets, categories, etc. are all definable diagrammatically; in any topos, they are also models for the appropriate axioms written in the internal language of the topos (Mulvey [74]). The above propositions permit the application of topos theory in the study of a topological space - they assert the existence of morphisms between the topoi Sh(X), Psh(X), Sh(Y). Now it is clear what an abelian group object in Sh(X) must be, we call it an abelian sheaf, denoting the category of abelian sheaves on X by Ash(X), (and Apsh(X) for abelian presheaves); it is not only an abelian category but also an abelian category object in Sh(X) (if we ignore the foundational difficulty). Injectives are definable from the external (or internal) hom-functors Hom (or $\mathcal{H}$om) in Ash(X) - fortunately, if Hom(-,I) is exact then so is $\mathcal{H}$om(-,I); in either sense, we have

Proposition 3.4. (Grothendieck). There are enough injective abelian sheaves on X.

Proof. (see Godement [64]). True for singleton X (MacLane [63]), hence true for discrete X. Now use (3.2) and preservation of injectives by any functor with an exact left adjoint. □

The <u>Grothendieck cohomology</u> of X is the sequence H^* of right derived functors of the global section functor H^0: Ash(X) $\longrightarrow$ Ash(pt) = Ab. grps., $A \longmapsto A[X]$. Interestingly, this factors through Apsh(X). Let $U \subseteq X$ be open; there is an exact functor $-_U$: Ash(X) $\longrightarrow$ Ash(X) which <u>concentrates</u> a sheaf on U; formally, if i: $X\backslash U \longrightarrow X$ is the embedding, then $A_U = \ker(A \longrightarrow i_* i^* A)$. We say an abelian sheaf A is <u>soft</u> iff restriction $A[X] \longrightarrow A[F]$ is onto for all closed $F \subseteq X$. An exact sequence of abelian sheaves $0 \longrightarrow A \longrightarrow A_0 \longrightarrow A_1 \longrightarrow \ldots.$ is said to be a <u>resolution of A</u>.

<u>Theorem 3.5</u>. (Zarelua [69]). Let A be an abelian sheaf on the paracompact space X, and $n \geq 0$. The following are equivalent, and if they hold we say <u>dim(X,A) $\leq$ n</u>:

i) $H^i(A_U) = 0 \quad \forall\, i > n, \quad \forall$ U open in X,

ii) For some, or any, resolution $0 \longrightarrow A \longrightarrow A_0 \longrightarrow \ldots \longrightarrow A_n \longrightarrow 0$ of A, with A_i soft for all $i < n$, then A_n is soft. □

In particular, A is soft iff $H^i(A_U) = 0$ for all $i > 0$, for all open $U \subseteq X$, and then dim(X,A) = 0. The theorem is fundamental both in the reduction of problems in higher dimensions to problems in dimension zero (analogous to the role of decomposition theorems in the dimension of metric spaces (Nagami [70])), and in its conversion of the algebraic problem of computation of cohomology groups to the topological problem of extending sections over closed sets. For example, it thus suffices to prove the next two theorems for dimension zero, by gradual extension of sections via Zorn:

<u>Proposition 3.6</u>. Let A be an abelian sheaf on paracompact X, $F \subseteq X$, and (F_α) a locally countable closed cover of X. Then

i) dim(F,A) $\leq$ dim(X,A) provided F is closed in X or X is totally normal,

ii) dim(X,A) $\leq$ sup. dim(F_α,A).

<u>Proof</u>. See Kuz'minov [68]. We make little distinction between a sheaf on X and its restriction to a subspace F or F_α of X. □

Since every open cover of a paracompact space has a locally finite closed refinement, there is an obvious corollary.

<u>Theorem 3.7</u>. (Kuz'minov and Liseikin [71]). Let (A_λ) be a direct system of

abelian sheaves on paracompact X. Then $\dim(X, \varinjlim A_\lambda) \leq \sup.\dim(X,A_\lambda)$. □

For compact Hausdorff spaces and hereditarily compact spaces this evidently follows from Godement [64]; in fact, for such spaces cohomology H^* commutes with direct limits of sheaves, a result typical of many in algebraic geometry (Grothendieck [72] VI). According to Bredon [67], H^* does not commute with direct limits of sheaves even on a space as nice as $\mathbb{R}$; it is essentially the characterisation (3.5) of dimension in terms of soft sheaves which permits a result such as 3.7.

The relationship between cohomological dimension and covering dimension is due to Aleksandroff and Dowker:

Theorem 3.8. Let X be a paracompact space of finite covering dimension. Then $\dim X = \dim(X, \mathbb{Z})$.

Proof. $\geq$ is easy (Godement [64]), once the Čech cohomology theory has been shown identical with Grothendieck's, since it is based on coverings. For both parts, see Kodama 36.15 (Nagami [70]) or Kuz'minov [68] II §7. Bokshtein [56] has a bypass to the Hopf extension theorem. □

Clearly there are decent sheaves A for which $\dim(X,A) \neq \dim X < \infty$: for example, the sheaf $\mathcal{R}$ of Dedekind real numbers on a paracompact space is soft (Huber [61]), and a converse result holds (Zakharov [74]), that for locally compact X, softness of this sheaf implies paracompactness, if we reinterpret softness as zero-dimensionality. See also Mulvey [76], for a proof that X is paracompact iff $\mathcal{R}$ is a projective $\mathcal{R}$-module. The whole point of using sheaves as coefficient groups is not just that lots of different dimension functions thereby arise, but, for example, that a map $f: X \longrightarrow Y$ can turn the trivial sheaf $\mathbb{Z}$ on X into the non-trivial sheaf $f_*\mathbb{Z}$ on Y. It is vital for us to relate $\dim(X,A)$ to $\dim(X,\mathbb{Z})$ and thus to dimX; we state the relationship in the next result, proved for example in Kuz'minov [68], the essentials being a lemma of Grothendieck and (3.7); the idea is to use (3.7) and Zorn to find a maximal subsheaf A' of A for which $\dim(X,A') \leq \dim(X,\mathbb{Z})$ and show $A' = A$. Of course, the theorem is unnecessary for finite dimensional X if we use the proof in (3.8) that $\dim(X,A) \leq \dim X$ and the equality in (3.8).

<u>Theorem 3.9</u>. Let A be an abelian sheaf on paracompact X. Then $\dim(X,A) \leq \dim(X,\mathbb{Z})$. □

For the representation of cohomology with coefficients in a group as the homotopy classes of maps into Eilenberg-MacLane spaces, see Huber [61], Bartik [68] or Goto [67].

Fundamental to the above discussion is the existence, for an abelian sheaf on X, of a soft resolution $0 \longrightarrow A \longrightarrow A_0 \longrightarrow A_1 \longrightarrow \ldots$, finite or infinite. For paracompact X, there is a variety of constructions of such resolutions: injective, canonical flabby, semi-simplicial flabby (Godement [64]), all of which are soft since on paracompact X injective and flabby imply soft, or by the easy

<u>Lemma 3.10</u>. Let f: X ⟶ Y be continuous, Y paracompact. Then f_*A is soft when A is soft. □

Let us reconsider the "projective resolution" of a space X defined in §2: an abelian sheaf A on X can be pulled back to a sheaf p_i^*A on P^iX, and the direct image $p_{i*}p_i^*A$ formed on X; the face maps can be transferred too, and we have a (co)semi-simplicial diagram in Ash(X)

$$0 \longrightarrow A \longrightarrow p_{1*}p_1^*A \overset{\longrightarrow}{\underset{\longleftarrow}{\longrightarrow}} \cdots\cdots \quad \overset{\vdots}{\underset{\longleftarrow}{\Longrightarrow}} p_{n*}p_n^*A \overset{\Longrightarrow}{\underset{\longleftarrow}{\vdots}} \cdots\cdots \quad .$$

Let us remove the degeneracies ⟵ and simplify the face maps ⟶ by taking alternating sums:

<u>Theorem 3.11</u>. The sequence $0 \longrightarrow A \longrightarrow \ldots\ldots \longrightarrow p_{n*}p_n^*A \longrightarrow \ldots\ldots$ is a resolution of A; all the sheaves $p_{n*}p_n^*A$ are soft when X is paracompact, and thus the cohomology $H^*(A)$ is the global cohomology of the semi-simplicial complex (2.3) with coefficients A.

<u>Proof</u>. Dyckhoff [76]: by (3.10) and softness of all sheaves on paracompact extremally disconnected spaces. □

Our [76] discusses also the interesting analogy with cohomology of groups, which is definable both from projectives and injectives; so is the cohomology of sheaves by (3.11), bearing in mind that since Ash(X) lacks projectives we construct them in the category of spaces and proper maps, over X, instead.

<u>Corollary 3.12</u>. Let X be finite dimensional paracompact.

Then dim $X \leq n$ iff $\ker(p_{n+1*}p_{n+1}{}^{*}\mathbb{Z} \longrightarrow p_{n+2*}p_{n+2}{}^{*}\mathbb{Z})$ is soft. □

The theorem and its corollary give some point to Gleason's suggestion [58] that his projective spaces have homological significance; but we do not claim, for example, that they provide any method for the calculation of cohomology or the estimation of dimension: for that the Čech theory and Huber's representation theorem are more appropriate. There may be a link between our theory restricted to compact Hausdorff spaces and the injective C*-algebras (cf Gleason [58]).

Our next section will discuss finite-to-one closed mappings and certain resolutions associated thereto; meanwhile, there is the following amusing and easy observation concerning projective covers and a "dimension" function: let <u>edimX ≤ n</u> iff every locally finite closed cover of X has a locally finite closed refinement of order ≤ n+1.

<u>Proposition 3.13.</u> (Dyckhoff [74a]) edimX ≤ n iff the projective cover RX ⟶>> X has order ≤ n+1; but edim$\mathbb{R}$ is infinite. □

<u>Problem 3.14.</u> What significance have the degeneracies in the projective resolution of a space?

<u>§4. Dimension-changing closed maps.</u>

Proper maps preserve many topological properties, particularly those concerning coverings, but not of course dimension: the classical theorems of Hurewicz on the relationship between the dimensions of two spaces related by a proper map are well known, and of importance both in the theory of product spaces and in attempts (e.g. Nagami [62]) to base dimension theory on the minimum order ΔX of a proper map onto a given space X with a domain of dimension zero. See Nagami [70] for the classical theory, modernised, Filippov [72] for recent contributions based on reduction to the metric case, and Pears and Mack [74] for a study of ΔX among other dimension functions. The most precise forms of these theorems were originally obtained by the theory of sheaves, in Sklyarenko [62] and Zarelua [69]; they are very fine illustrations of the power of cohomological methods. Our aim in this section is to make their proofs more accessible to the category theorist, and in particular to illustrate the advantage of using the concept of a topos.

Let F: $\mathcal{A} \longrightarrow \mathcal{B}$ denote the restriction to abelian objects of a geometric morphism of topoi, where $\mathcal{A}$ has enough injectives; let $A \in \mathcal{A}$, $n \geq 0$. We say F is n-exact at A iff the right derived functors R^pF vanish at A for all $p > n$.

Lemma 4.1. Let F: $\mathcal{A} \longrightarrow \mathcal{B}$, G: $\mathcal{B} \longrightarrow \mathcal{C}$ be two such morphisms, where F is n-exact at A and G is m-exact everywhere. Then GF is (m+n)-exact at A.

Proof. See Grothendieck's SGA4 V 0.3 [72a]. Apply F to an injective resolution of A and apply G to the image of that from the nth kernel onwards. □

Theorem 4.2. (Hurewicz, Sklyarenko). Let f: X ⟶ Y be closed, X paracompact, A an abelian sheaf on X. Then $\dim(X,A) \leq \sup(\dim(f^{-1}y,A): y \in Y) + \dim(Y,\mathbb{Z})$.

Proof. Apply to $F = f_*$, $G = t_*$, where t is the terminal map Y ⟶ point. □

We have deliberately removed the Leray spectral sequence implicit in (4.1) to stress the simplicity of the argument. By (3.8) and (3.9) and elementary arguments, all dimension functions here can be replaced by "dim". Then, the theorem is true (Pasynkov [65]) for X normal but Y paracompact, by consideration of the $\mathcal{K}$-perfect factor of f, which has the same dimension as f for X normal. Filippov [72] shows that normality of Y cannot replace paracompactness.

Corollary 4.3. Let X be compact, Y paracompact. Then $\dim(X \times Y) \leq \dim X + \dim Y$. □

Morita's work [73] suggests that 4.3 is true for all Tychonoff spaces; in any case, proofs of 4.3 for more general spaces generally involve reduction to easy special cases such as compact × paracompact.

For an improper map f: X ⟶ Y (of paracompacta), dim X = dim Y, since the map βf between the Čech extensions is an isomorphism. Let f: X ⟶ Y be a proper map, f = gh its light-factorisation by (1.1). By (4.2), dim Z ≤ dim Y: one says that g: Z ⟶ Y raises dimension.

Problem 4.4. What conditions on a proper map guarantee dimY ≤ dimX, and is there a factorisation category involving the maps which raise (resp. lower) dimension?

The fundamental theorem on dimension-raising maps is

Theorem 4.5. (Hurewicz, Zarelua). Let f: X ⟶⟶ Y be closed, onto, where X is paracompact, A an abelian sheaf on Y. Let order(f) = $\sup(|f^{-1}y| : y \in Y) = k + 1$.

Then

$$\dim(Y,A) \leq k + \dim(X,f^*A). \square$$

Corollary 4.6. Let $f: X \twoheadrightarrow Y$ be closed onto, where X is Tychonoff, Y normal. Then $\dim Y \leq k + \dim X$, where order (f) = k+1.

Proof. Suppose f has order $\leq$ k+1; by lemma 4.1 of Pears and Mack [74], so does the Cech extension βf, to which we apply (4.5). $\square$

We sketch the proof of (4.5) in categorical terms to suggest that there is a natural way to look at the problem. Let $f: X \longrightarrow Y$ be any map; there is a monad (f_*f^*,u,m) on Sh(Y), or on Ash(Y), induced by the change of base adjunction (3.2). When f is onto, $u: I \longrightarrow f_*f^*$ is a monomorphism and when f is a montone quotient, u is an isomorphism. Indeed if we form the semi-simplicial functor from the monad in the usual way (see Barr, Beck [69], or MacLane [71]), and in Ash(Y) convert by alternate summation to a complex, we have

Lemma 4.7. When f is onto, then the sequence

$$0 \longrightarrow I \longrightarrow f_*f^* \longrightarrow \cdots \longrightarrow (f_*f^*)^n \longrightarrow \cdots$$

is exact; when f is a monotone quotient, the arrows of the sequence are alternately isomorphisms and zero.

Proof. To prove exactness of the sequence of the stalks over $y \in Y$ of images of a sheaf A, pick any $x \in f^{-1}y$ and use evaluation at x to contract the sequence: cf Godement [64] Appx. The last part depends on monotone quotients forming a hereditary class (§1) and lightness of structure maps of sheaves. $\square$

Thus for an onto map $f: X \twoheadrightarrow Y$ there is a monadic resolution functor M_f on Ash(Y), taking the sheaf A to the resolution

$$0 \longrightarrow A \longrightarrow f_*f^*A \longrightarrow \cdots \longrightarrow (f_*f^*)^nA \longrightarrow \cdots .$$

Godement's semi-simplicial flabby resolution ([64], Appx.) is an example, and the original inspiration for monads, being based on the map $DX \longrightarrow X$. For two other examples, consider the Čech resolution (Godement [64]) determined by an open, or locally finite closed cover - bearing in mind that a cover $\mathfrak{M}$ of X induces an onto map $\oplus\mathfrak{M} \twoheadrightarrow X$, open or proper respectively. We shall be concerned with such resolutions determined by proper maps; meanwhile, we note that the discrete modification map $DX \longrightarrow X$ and its proper factor $PX \longrightarrow X$

induce distinct resolution functors on Ash(X), both of which for X paracompact have the same global cohomology, by proof similar to (3.11).

Problem 4.8. If gf and g are onto and f is improper, does f induce an equivalence between the cohomology sheaves of the resolutions for gf and g?

There are two other important descriptions of the monadic resolution of a proper map; the first gives us a geometrical picture, the second is an aid to calculation.

Theorem 4.9. (from Grothendieck's SGA4 p. 141 [72a]). Let $f: X \twoheadrightarrow Y$ be a proper map. Then the following two resolution functors are isomorphic:

i) $0 \longrightarrow I \longrightarrow f_*f^* \longrightarrow \dots \longrightarrow (f_*f^*)^n \longrightarrow \dots\dots$,

ii) $0 \longrightarrow I \longrightarrow f_*f^* \longrightarrow \dots \longrightarrow f_{n*}f_n^{\ *} \longrightarrow \dots\dots$,

where $f_n: X \times_f X \times_f \dots\dots \times_f X$ (n factors) $\longrightarrow Y$ and the maps in ii) are alternating sums of face maps derived from the 'resolution of Y":

$$\dots\dots \longrightarrow X \times_f X \times_f X \begin{smallmatrix}\longleftarrow\\ \longrightarrow\\ \longrightarrow\\ \longrightarrow\\ \longleftarrow\end{smallmatrix} X \times_f X \begin{smallmatrix}\longrightarrow\\ \longrightarrow\\ \longleftarrow\end{smallmatrix} X \longrightarrow Y. \ \Box$$

Corollary 4.10. The resolution (4.9.i) has a simplicial structure.

Proof. ii) has, from the symmetric group acting on each n-fold product $X \times_f \dots \times_f X$. □

Corollary 4.11. The monadic resolution of a proper map f has an "alternating subresolution" $\tilde{M}_f$.

Proof. For a sheaf A on Y, $(f_*f^*)^n A$ has stalk at $y \in Y$ the abelian group $A_y^{(f^{-1}y)^n}$, which has a subgroup consisting of all the alternating elements; over y, the sequence of such subgroups is exact by standard homological methods (Godement [64]); since any such (alternating) element has a local representation over a nbd of y, which can be made alternating when the nbd is taken small enough, the "alternating subgroups" form a subsheaf of $(f_*f^*)^n A$. □

For example, the alternating Čech resolution determined by a locally finite closed cover; a similar argument works for open covers. Macdonald [68] has a useful description of these.

Corollary 4.12. Let $f: X \twoheadrightarrow Y$ be a proper map of order $\leq k+1$. Then the alternating resolution by f of any abelian sheaf A on Y vanishes after the term

$(f_*f^*)^{k+1}A$. □

We now come to the representation of the monadic resolution of a proper map in calculable terms; the following procedure applied to any map gives a result dependent only on the proper light factor. Let $f: X \longrightarrow Y$ be a map, which, by (4.7) we shall assume to be onto. A <u>partition of f over open $U \subseteq Y$</u> is a locally finite closed cover $\mathcal{M}$ of U and a factorisation $f^{-1}U \rightarrowtail \oplus\mathcal{M} \twoheadrightarrow U$, the first factor being dense. Since we consider many partitions, the index λ will denote the partition $f^{-1}U_\lambda \rightarrowtail \oplus\mathcal{M}_\lambda \longrightarrow U_\lambda$. We say of two partitions λ, μ over U that $\lambda \leq \mu$ iff there is a factorisation

$$\begin{array}{ccc} f^{-1}U & \longrightarrow & \oplus\mathcal{M}_\mu \\ \downarrow & \swarrow & \downarrow \\ \oplus\mathcal{M}_\lambda & \twoheadrightarrow & U \end{array}$$

Note that $\oplus\mathcal{M}_\mu \twoheadrightarrow \oplus\mathcal{M}_\lambda$ is onto and is the unique map, if any, making the diagram commute; hence $\leq$ is a partial order on <u>$\Lambda_f(U)$</u>, the set of partitions of f over U. Let (U_α) be an open cover of open $U \subseteq Y$, and let $\lambda_\alpha \in \Lambda_f(U_\alpha)$ be given for each α in such a way that λ_α and λ_β always agree over $U_\alpha \cap U_\beta$. Then the partitions patch together, forming uniquely a partition λ of f over U. Hence Λ_f is a sheaf of partially ordered sets on Y.

In the internal language of the topos of sheaves over X, due to Benabou, but see Mulvey [74] for a convenient account, having the open sets of X as truth values in the intended interpretation, and therefore intuitionistic, we can say that Λ_f is a partially ordered object of Sh(Y). To say that Λ_f is <u>directed</u> would be to say that given λ, μ in $\Lambda_f(U)$, there is a cover (U_α) of U and ν_α in $\Lambda_f(U_\alpha)$ with $\nu_\alpha \geq$ both λ and μ on U_α: formally,

$$\forall \lambda \, \forall \mu \, \exists \nu : (\lambda \leq \nu) \wedge (\mu \leq \nu).$$

By using pullbacks it is easy to see that this is true even with $U_\alpha = U$ and all ν_α equal: see Dyckhoff [74] for the details in different terminology. Thus Λ_f is "directed", and thus a "filtered category": see Gray [75] for internal category theory.

In the language of set theory, a <u>natural resolution on $\mathcal{A}$</u> is a functor from an abelian category $\mathcal{A}$ to the category of exact sequences in $\mathcal{A}$, taking A to the sequence $0 \longrightarrow A \longrightarrow A_0 \longrightarrow A_1 \longrightarrow \ldots\ \ldots$, and $f: A \longrightarrow A'$ to a morphism of exact sequences over f. Now we consider the topos Sh(Y); let $\mathcal{R}$ denote the internal category, for which $\mathcal{R}(U)$ denotes the set of all natural resolutions on Ash(U): this is a presheaf, and actually a sheaf, and thus an object of Sh(Y); it is also a category object, or internal category. We have described above an (internal) functor $M: \Lambda_f \longrightarrow \mathcal{R}$, taking, over open U, the partition λ to the monadic resolution M_λ determined by $\oplus \mathcal{M}_\lambda \longrightarrow\!\!\!\!> U_\lambda = U$.

The simplest example of a topos is that of finite sets; this is neither complete nor cocomplete, <u>except internally</u>, - it has finite limits and colimits. More generally, every topos is internally complete and cocomplete, and externally complete and cocomplete iff it is a Sets-topos (i.e. there is a morphism to Sets, i.e. ordinary sets can be pictured inside the topos, by pulling back along the morphism). Thus, in topos theory, the fundamental notion of, e.g. cocompleteness, is internal; it is accidental that Sh(Y) is also externally cocomplete. Moreover, colimits over "filtered categories" commute with finite limits, in particular, direct limits inside Sh(Y) commute with finite limits (cf Johnstone [74]).

Once these rather complex ideas are absorbed, the following result is trivial, both parts of (1.4) being borne in mind:

<u>Theorem 4.13</u>. (Dyckhoff [74], [74a]). Let $f: X \longrightarrow\!\!\!\!> Y$ be an onto map. Then the diagram $M: \Lambda_f \longrightarrow \mathcal{R}$ has a colimit, the monadic resolution of the proper light factor of f. □

<u>Corollary 4.14</u>. Let $f: X \longrightarrow\!\!\!\!> Y$ be proper, light, onto. Then its monadic resolution is representable as an <u>internal</u> direct limit of Čech resolutions determined by partitions of f. □

Our [74] does this as a representation theorem for f as a partial inverse limit of simple maps, i.e. maps looking like finite closed covers. We have replaced "finite" by "locally finite" to ensure that our Λ_f is actually a sheaf. Partial inverse limits are to inverse limits as partial products (Pasynkov [65]) are to ordinary products.

Now recall 3.7: simply, that over a paracompact space, a direct limit of soft sheaves is soft. This result was first proved by Zarelua [69] with a mild restriction on the bonding maps, but in a more general form, which can be stated as a theorem about internal direct limits. We conjecture that there is a formula in the language of sheaves over X, a paracompact space, asserting of an abelian sheaf over X that it is soft. Martin Hyland (private communication) has given an intuitionistically valid proof of the lemma (15.10 in Bredon [67]) of Grothendieck referred to as the key ingredient in Theorem 3.9. We note also that Deligne, in Grothendieck's SGA4 V Appx. [72a], studies internal filtered colimits and extends to topoi a theorem of Lazard, that every flat module is a direct limit of free modules of finite type. Mulvey's work [74] on internal descriptions of rings shows that such formulae should be intuitionistically provable in the topos language; this seems feasible for the Deligne-Lazard theorem and both ambitious and fascinating for the dimension theory. Thus the problem is this: give an internal proof of

Theorem 4.15. (Zarelua [69]) An internal direct limit of soft sheaves, over a paracompact space, is soft, provided the system is regular (bonding maps are monos, and the cover (U_α) of U in the condition for directness is just (U). □

A subsidiary problem is to remove the regularity.

Corollary 4.14 has an alternating version: this and 4.15 are the main ingredients of the proof of (4.5). To estimate dim(Y,A), resolve A by the alternating monadic resolution, calculate the dimension of Y with coefficients in the terms of the resolution (which are soon zero), and use long exact cohomology sequences: see our [74] for details, where we regret that it was not made clear that all resolutions considered are alternating.

Thus the monadic resolution of a proper map is a natural tool for studying the dimension-raising properties, both for geometrical reasons (4.6), and because internally it is like a Čech resolution. The resolution, and its alternating form, determine spectral sequences; in the case of a finite sheeted regular covering map between locally contractible paracompacta, we obtain the Cartan spectral sequence of the action of a finite group on a space; see Skordev [70], [71]. One wonders about profinite groups in this context: see Grothendieck [72a] VIII. For the map induced by a locally finite closed cover, the Leray spectral sequence is obtained (to be distinguished from the Leray s.s. of a proper map): thus the Zarelua spectral sequence is a generalisation of one of Leray's, as Grothendieck implicitly suggested in [72a] VIII 8.1.

We conclude by repeating the question: in what sense is the dimension of X expressible in the language of sheaves on X? We can say, for example, that X is of dimension zero iff the sheaf $\mathbb{Z}$ of integers is soft; is that so expressible, and what about higher dimensions? There is indeed the characterisation (3.5) in terms of exact sequences; is there anything more explicit? What, for example, is the sheaf analogue of Katetov dimension (see Gillman and Jerison [60] or Pears [75]) - the analytic dimension defined in terms of generators of certain subrings of C(X)? Some attempt has been made on this problem by Fourman [75], but the results are not yet quite well enough related to dimension even on standard spaces.

BIBLIOGRAPHY

Aleksandroff, P.S.: On some basic directions in general topology, Russian Math. Surveys 19 (1964), 6.1-39.

_____ : corrections to above, ibid 20 (1965) 1.177-178.

Banaschewski, B.: Projective covers in categories of topological spaces and topological algebras, Proc. Kanpur Top. Conference (1968), Academia, Prague, 63-91.

Barr, M., Beck, J.: Homology and standard constructions, Lecture Notes in Mathematics 80, Springer-Verlag (1969), 245-335.

Bartik, V.: Aleksandrov-Čech cohomology and mappings into Eilenberg-MacLane polyhedra, Math. USSR Sbornik 5 (1968), 221-228.

Benabou, J.: Introduction to bicategories, Lecture Notes in Mathematics 47, Springer-Verlag (1967), 1-77.

Blaszczyk, A.: A factorisation theorem and its application to extremally disconnected resolutions. Colloq. Math. 28 (1974), 33-40.

_____ : Extremally disconnected resolutions of T_0-spaces, ibid 32 (1974a), 57-68.

Bokshtein, M.F.: A new proof of the fundamental theorem of homological dimension theory, Moskov. Gos. Univ. Uch. Zap. 181 Mat. 8 (1956) 13-44.

_____ : The homological theory of dimension, Russian Math. Surveys 21 (1966), 7-12.

Bredon, G.E.: Sheaf theory, McGraw-Hill, New York (1967).

Collins, P.J.: Concordant mappings and the concordant-dissonant factorisation of an arbitrary continuous function, Proc. Amer. Math. Soc. 27 (1971), 587-591.

_____ : Connection properties in topological spaces, Mathematika Balkancia 1, (1971a) 44-51.

Diers, Y.: Complétion monadique, C.R. Acad. Sci. Paris 276 (1973) A1397-A1400.

Duskin, J.: Simplicial methods and the interpretation of "triple" cohomology, Mem. Amer. Math. Soc. 163 (1975).

Dyckhoff, R.: Factorisation theorems and projective spaces in topology, Math. Zeitschrift 127 (1972), 256-264.

_____ : Perfect light maps as inverse limits, Quart. J. Math. Oxford (2), 25 (1974), 441-449.

_____ : Topics in general topology: bicategories, projective covers, perfect mappings and resolutions of sheaves, thesis, Oxford (1974a).

_____ : Projective resolutions of topological spaces, J. Pure and Applied Algebra 7 (1976), 115-119.

_____ : Categorical Cuts, to appear (1976a).

Eilenberg, S.: Sur les transformations continues d'espaces metriques compacts, Fund. Math. 22 (1934), 292-296.

Filippov, V.: On the dimension of closed mappings, Sov. Math. Dokl. 13 (1972), 895-900.

Fourman, M.P.: Comparaison des réelles d'un topos: structures lisses sur un topos élémentaire, Amiens 1975, to appear in Cah. Top. Geom. Diff.

__________ : The logic of topoi, Handbook of mathematical logic (ed. Barwise), North Holland (1976).

Freyd, P.: Aspects of topoi, Bull. Austral. Math. Soc. 7 (1972), 1-76.

Gillman, L., Jerison, M.: Rings of continuous functions, Van Nostrand, Princeton (1960).

Gleason, A.: Projective topological spaces, Illinois J. Math. 2 (1958), 482-9.

Godement, R.: Théorie des faisceaux, Hermann, Paris (1964).

Gonshor, H.: Projective covers as subquotients of enlargements, Israel J. of Math. 14 (1973), 257-261.

Goto, T.: Homotopical cohomology groups of paracompact spaces, Sci. Rep. Tokyo Kyoiku Daigaku, Sect. A, 9 (1967), 21-27.

Gray, J.W.: Formal Category Theory II, Springer Lecture Notes, to appear (1975?).

Grothendieck, A., et al.: Théorie des topos et cohomologie etale des schemas, (SGA4), Lecture notes in mathematics 269, Springer-Verlag (1972).

____________________ : ibid. 270 (1972a).

Hager, A.W.: The projective resolution of a compact space, Proc. Amer. Math. Soc. 28 (1971), 262-266.

Harris, D.: Katetov extension as a functor, Math. Ann. 193 (1971), 171-175.

Henriksen, M., Isbell, J.R.: Some properties of compactifications, Duke Math. J. 25 (1958), 83-105.

Henriksen, M., Jerison, M.: Minimal projective extensions of compact spaces, Duke Math. J. 32, (1965), 291-295.

Herrlich, H.: Topologische Reflexionen und Coreflexionen, Lecture notes in Mathematics 78, Springer-Verlag, Heidelberg (1968).

__________ : Categorical topology, General topology and its applications 1, (1971) 1-15.

__________ : A generalisation of perfect maps, Proc. Third Prague Top. Symposium 1971, Academia Prague (1973) 187-192.

__________ : Perfect subcategories and factorisations, Proc. Hungarian Top. Conference 1972 = Topics in topology, (ed. A. Czaszar), Coll. Math. Soc. Janos Bolyai 8, North-Holland, Amsterdam (1974).

Huber, P.J.: Homotopical cohomology and Čech cohomology, Math. Ann. 144 (1961), 73-76.

Hurewicz and Wallman, Dimension theory, Princeton (1948).

Isbell, J.R.: Subobjects, adequacy, completeness, and categories of algebras, Rozprawy Mat. 38, (1963) 1-32.

Johnstone, P.: Aspects of internal category theory, thesis, Cambridge (1974).

Kennison, J.F.: Full reflective subcategories and generalised covering spaces, Illinois J. Math. 12, (1968) 353-365.

Kock, A., Wraith, G.C.: Elementary topoi, Aarhus lecture notes 30 (1971).

Kulpa, W.: Factorisation and inverse expansion theorems for uniformities, Colloq. Math. 21, (1970) 217-227.

Kuz'minov, V.I.: Homological dimension theory, Russ. Math. Surveys 23 (1968), 5.1-45.

__________, Liseikin, V.D.: The softness of an inductive limit of soft sheaves, Siberian Math. J. 12 (1971) 820-821.

Lisitsa, Y.: Extension of continuous maps and the factorisation theorem, Siberian Math. J. 14 (1973), 90-96.

Macdonald, I.G.: Algebraic geometry, Benjamin, New York (1968).

MacLane, S.: Homology, Springer-Verlag, Heidelberg (1963).

__________: Categories for the working mathematician, Springer-Verlag, Heidelberg (1971).

Michael, E.: Cuts, Acta Math. 111 (1964), 1-36.

Mioduszewski, J., Rudolf, L.: H-closed and extremally disconnected Hausdorff spaces, Dissertationes Mathematicae 66, (1969) 1-55.

Morita, K.: On the dimension of the product of Tychonoff spaces, General topology and its applications 3 (1973), 125-134.

Mulvey, C.: Intuitionistic algebra and representations of rings, Mem. Amer. Math. Soc. 148 (1974), 3-57.

__________: Compact ringed spaces, (1976) (preprint).

Nagami, K.: Mappings defined on 0-dimensional spaces and dimension theory, J. Math. Soc. Japan 14 (1962) 101-117.

__________: Dimension theory, Academic Press, New York (1970).

Pasynkov, B.A.: Partial topological products, Trans. Moscow Math. Soc. 13, (1965) 153-272.

__________: On a formula of Hurewicz, Vestnik Mosk. Gos. Univ. ser. 1. Math. Mech. 20 (1965), (4) 3-5.

Pears, A.R.: Dimension theory of general spaces, Cambridge (1975).

__________, Mack, J.: Closed covers, dimension, and quasi-order spaces, Proc. London Math. Soc. (3) 29 (1974), 289-316.

Ponomarev, V.: Projective spectra and continuous mappings of paracompacta, Amer. Math. Soc. Translat. Ser. 2 39 (1964).

Rainwater, J.: A note on projective resolutions, Proc. Amer. Math. Soc. 10 (1959), 734-735.

Ringel, C.M.: Diagonalisierungspaare I, Math. Zeitschrift 117 (1970), 249-266.

__________ : ibid. II, ibid. 122 (1971), 10-32.

Scott, D.S., Fourman, M.P.: Sheaves and logic, preprint (Oxford, 1975).

Shukla, W., Srivastava, A.: Local reflectivity + stable reflectivity = reflectivity, Gen. top. and its applications 5 (1975), 61-68.

Sklyarenko, E.G.: A theorem on maps which lower dimension, Bull. Acad. Polon. des Sciences, ser. Sci., Math., etc. 10 (1962), 429-432.

__________ : Some applications of the theory of sheaves in general topology, Russ. Math. Surveys 19 (1964), 6.41-62.

Skordev, G.S.: On resolutions of continuous mappings, Math. USSR Sbornik 11 (1970), 491-506.

__________ : Resolutions corresponding to closed mappings, ibid. 15 (1971), 227-240.

Strecker, G.: Component properties and factorisations, Mathematical Centre Tracts 52 Amsterdam (1974), 123-140.

__________ : Perfect morphisms, these proceedings (1976?).

Whyburn, G.T.: Analytic topology, Amer. Math. Soc. Colloquium Publ. 28 (1963).

Wraith, G.C., et al: Model Theory and Topoi, Lecture Notes in Mathematics 445, Springer-Verlag (1975).

Wyler, O.: A characterisation of regularity in topology, Proc. Amer. Math. Soc. 29 (1971), 588-590.

Zakharov, V.K.: Isomorphism of the homology groups of a locally compact space and groups of module extensions, Siberian Math. J. 15 (1974), 670-673.

Zarelua, A.V.: Finite-to-one mappings of topological spaces and cohomology manifolds, Siberian Math. J. 10 (1969), 45-63.

Envelopes in the category of Kakutani-M-spaces

by

Jürgen Flachsmeyer

Introduction

Every compact Hausdorff space X is uniquely determined (up to a homeomorphism) by its system of all real valued continuous functions on X regarded as an algebra, resp. a vector lattice, resp. a lattice, resp. a Banach space etc. If these corresponding algebraic structures are abstractly characterizable as special structures[1)] then this gives a dual equivalence of the category COMP of all compact Hausdorff spaces to the category of such special algebraic structures. Here we are interested in the category KAKUMI of all Kakutani-M-spaces with unit. For the subcategory COMP_0 of COMP consisting of all compact Boolean spaces (zero-dimensional spaces) there is a dual equivalence to the category BOOLALG of all Boolean algebras. It may be asked for concepts and theorems concerning Boolean algebras which can be generalized for MI-spaces (the objects in KAKUMI).

Our paper is a contribution to such a program. In 1950 Sikorski [17] has studied dense subalgebras A of Boolean algebras B. From a categorical point of view dense embeddings in BOOLALG are envelopes. A subalgebra A of a Boolean algebra B is called dense iff under each non-zero element of B lies a non-zero-element of A. Sikorski has shown that dense embeddings preserve all existing extremas and every element of B is the supremum (resp. infimum) of ele-

---

1) For example: In the case of algebras these are the self-adjoint commutative Banach algebras with unit.
In the case of normed vector lattices these are the Kakutani-M-spaces with unit.

ments from X. Our theorem 1, (1),(2), generalizes this fact. With the help of theorem 2 we conclude that for every embedding of a MI-space into another exist maximal envelopes. The theorem 3 shows that the injective envelopes in KAKUMI are characterized in the same manner as the injective envelopes in BOOLALG.

1. Preliminaries

A vector lattice M over the real field R is called a Banach-lattice iff M is endowed with a complete norm $||\cdot||$ and this norm is monotone in the following sense:

$$|x| \leq |y| \Rightarrow ||x|| \leq ||y|| \quad \text{for all } x,y \in M .$$

A Banach-lattice M is called a Kakutani-M-space iff there holds the following M-condition:

$$x \geq 0,\ y \geq 0 \Rightarrow ||\, x \vee y \,|| = ||x|| \vee ||y|| \quad \text{for all } x,y \in M .$$

(whereby $\vee$ means the supremum in M resp. in R).
A Kakutani-M-space M is said to have a strong order unit u iff there is a greatest element u in the unit ball $\{x \mid ||x|| \leq 1\}$. For $M \neq \{0\}$ this order unit satisfies $||u|| = 1$ and it is unique. These M-spaces with unit - abbreviated as MI-spaces - are abstract descriptions of the concrete Banach-lattices $C(X,R)$ of all continuous realvalued functions on compact Hausdorff spaces X. Namely, by a wellknown representation theorem of Kakutani [12] (see also [15]) to every MI-space M corresponds a topological unique representation space X (compact T_2) such that M is linear lattice isomorphic isometric to $C(X,R)$ and the unit goes to the unit. Let us call this correspondence $M \longrightarrow X$ that sends each MI-space to the representation space the *Kakutani-functor*. By a MI-homomorphism is meant a linear continuous lattice homomorphism of one MI-space into another which preserves the unit. Then they must have norm 1. KAKUMI denotes the category of all MI-spaces as objects and the MI-homomorphism as

mosphisms. Via the Kakutani-functor the category KAKUMI is dual equivalent to the category COMP of all compact Hausdorff-spaces with continuous maps.

Now let us note the meaning of some categorical notions in KAKUMI. (For category theory we refer to [11], [14] and [15]).

Isomorphism	linear isometric lattice isomorphism preserving the unit
monomorphism	linear isometric lattice injection preserving the unit (embedding)
epimorphism	linear lattice surjection of norm 1 preserving the unit
injective object M	each MI-homomorphism $h : A \longrightarrow M$ of a MI-subspace A of a MI-space B can be extended to a MI-homomorphism of B into M.

The internal characterization of the injective MI-spaces is as follows. *A MI-space M is injective iff M is an order complete MI-spaces, e.g. every bound set in M has a supremum and an infimum.* This follows from the Gleason characterization [9] of the projective objects in COMP as the extremally disconnected spaces and the Stone-Nakano theorem [13], [19], which says that the MI-space $C(X,R)$ is order complete iff X is extremally disconnected.

Remark: We remember that the first theorem with respect to injectivity in some category is Sikorski's theorem on extensions of homomorphisms in the category of Boolean algebras [16]. This theorem stated that every complete Boolean algebra is injective.

2. Envelopes

The categorical concept of an *envelope* (Semadeni [15]) gives for KAKUMI the following interpretation: By an envelope of a MI-space A is meant an embedding $A \xrightarrow[\varepsilon]{} B$ such that for every homomorphism $B \xrightarrow[h]{} C$ for which $h \circ \varepsilon$ is an embedding h itself is an embedding. If we regard A as a subspace of B then the inclusion $A \hookrightarrow B$

is an envelope iff each extension of any MI-isomorphism $A \longrightarrow C$ to a MI-homomorphism of B into C is itself a MI-isomorphism.

To have internal characterizations of envelopes we prove the following:

Proposition 1 *Let be A a MI-subspace of the MI-space B. Then the following conditions are equivalent:*

(1) Every element of B is the supremum of elements of A.

(2) Every element of B is the infimum of elements of A.

(3) Every open internal $]0,x[$ in B, $x > 0$, contains elements of A.

Proof: In vector lattices it holds for subsets S :
$x = \sup S \Longleftrightarrow -x = \inf(-S)$. Thus (1) and (2) are equivalent.
(1) $\Rightarrow$ (3). If $x > 0$ and $x = \sup\{y \mid y \in A, 0 \leq y \leq x\}$, then there must be one $y_o \in A : 0 < y_o \leq x$, otherwise $x = 0$. For such y_o is $0 < \frac{1}{2} y_o < x$.
(3) $\Rightarrow$ (1) Let be $x \in B$ we may assume $x > 0$. The general case follows by translation with a suitable $\lambda \cdot 1$ (1 the unit in B).
$S := \{y \mid y \in A, 0 \leq y \leq x\}$. Let be z an upper bound of S. We will show that $z \leq x$ implies $z = x$. Assume, $z < x$. Then for $x-z > 0$ is an element $v \in A$: $0 < v < x-z$. Then $v+y \leq z$ for all $y \in S$ and we get $2v < x-y$ for all $y \in S$, e.g. $2v + S \subset S$. Then $2v + y \leq z$ for all $y \in S$ and we get $3v < x-y$ for all $y \in S$, e.g. $3v + S \subset S$. By induction it follows that (*) $kv+S \subset S$ for every natural number k. Now a vector lattice is archimedean but this gives a contradiction to (*).

Theorem 1 *Let be A a MI-subspace of the MI-space B. Then the following are equivalent:*

(1) $A \hookrightarrow B$ is an envelope

(2) A is order dense in B in the sense of proposition 1.

(3) Every norm closed non-zero lattice ideal I in B contains non-zero elements of A.

(4) If $B = C(X,\mathrm{R})$ and Y is a proper closed subset of X then A contains a non-zero function which vanishes on Y.

Proof: (1) $\Leftrightarrow$ (2) The equivalence of (1) and (2) follows by a dual argument and a theorem proved in [8]. $A \hookrightarrow B$ gives by the Kakutani-functor a surjection of the representation space of B onto the representation space of A. Now envelopes in KAKUMI means coenvelopes in COMP . Coenvelopes are described by irreducible surjections (see [15, p.446]). But the irreducibility of a surjection in COMP is equivalent to the order density of the corresponding embedding of the associated function lattices (see[8]).

(1) $\Rightarrow$ (3) The norm closed lattice ideals I with $1 \notin I$ are the cernels of MI-homomorphisms φ .

Now $I \cap A = \{0\} \Leftrightarrow \varphi|_A$ is a monomorphism.

Thus (1) $\Leftrightarrow$ (3)

(3) $\Leftrightarrow$ (4) The proper closed subsets Y of X are in one-one correspondence to the proper norm closed lattice ideals I of $C(X,\mathrm{R})$.

$$I = \{f \mid f \in C(X,\mathrm{R}),\ f \equiv 0 \text{ on } Y\}.$$

Remarks: 1. The notion of an envelope is firstly due to Eckmann and Schopf [6] for modules. The envelopes are there called *essential extensions* and they are defined through the corresponding property (3) of the preceding theorem.

2. By the equivalence of the categories KAKUMI and C^*- ALG of C^*- algebras with unit and their homomorphisms the equivalence of (1),(3) and (4) is only a translation of the corresponding properties formulated by Gonshor [10].

In the next statement we give a test for beeing an envelope by further reduction to norm dense subspaces.

Theorem 2. *Let be* A *a MI-subspace of the MI-space* B *with a given vector sublattice* C *of* B *containing* A. *The norm closure* $\overline{C}$ *of* C *in* B *is an envelope of* A *iff* A *is dense in* C *in the sense that for all* $x \in C$

$$\sup_C(]\leftarrow, x] \cap A) = x = \inf_C([x, \rightarrow[\cap A) .$$

Proof: We have only to prove that from the density of A in C follows the density with respect to $\overline{C}$. First we show that for every $x \in C$: $\sup_{\overline{C}} (] \leftarrow, x] \cap A) = x$ and then we show it for arbitrary $x \in \overline{C}$. We interprete the situation for $B = C(X,\mathbb{R})$.

1. Let be $f \in C$ and h an upper bound in $\overline{C}$ of $U(f) :=] \leftarrow, f] \cap A$, with $h \leq f$. It should be proved $h = f$. We have a sequence $h_n \in C$ with $h_n \longrightarrow h$ uniformly. For a suitable subsequence h_{n_ν} holds $h \leq h_{n_\nu} + \frac{1}{\nu}$. Now $h_{n_\nu} + \frac{1}{\nu} \longrightarrow h$ and $h_{n_\nu} + \frac{1}{\nu} \in C$. Observe $h_n \wedge f \rightarrow h \wedge f$. Therefore we may start with a sequence $h_n \in C$ such that $h \leq h_n \leq f$ and $h_n \longrightarrow h$. Every h_n is an upper bound for $U(f)$. This gives $f \leq h_n$ and therefore $h = f$.
2. Now consider $f \in \overline{C}$. Let be h an upper bound of $U(f)$. We have a sequence $f_n \in C$ with $f_n \longrightarrow f$ uniformly. It must be shown $f \leq h$. Assume, there is a point $x \in X$: $f(x) > h(x)$. Take an $\varepsilon > 0$ with (*) $f(x) - \varepsilon > h(x)$. Now for almost all n holds (**) $f - \varepsilon < f_n - \frac{\varepsilon}{2} \leq f$. The functions $f_n - \frac{\varepsilon}{2}$ belong to C, therefore from the first part follows $f_n - \frac{\varepsilon}{2} = \sup_{\overline{C}} U(f_n - \frac{\varepsilon}{2})$. Then from (**) we get $h \geq f_n - \frac{\varepsilon}{2}$ for every n. This contradicts (*).

From the preceding theorem we conclude the

Corollary *Let be* A *a MI-subspace of the MI-space* B *and let be* $A \subset A_\alpha \subset B$ *an increasing family of MI-subspaces of* B *all envelopes of* A. *Then the norm closure* $\overline{\cup A_\alpha}$ *is an envelope of* A.

Proof: $C := \cup A_\alpha$ is a vector sublattice of B. A_α being envelopes of A implies that for every $x \in C$ holds $\sup U(x) = x = \inf O(x)$,

where $U(x) =]\leftarrow, x] \cap A$ and $O(x) = [x, \rightarrow[\cap A$.
By theorem 2 $\overline{C}$ is an envelope of A.

Corollary *If A is a MI-subspace of the MI-space B, then a maximal envelope of A in B exists.*

Proof: The preceding corollary with the Zorn lemma gives this result.

Remark In the paper [10] of Gonshor this was shown in the category C^*ALG by another approach.

3. Injective envelopes

For every object in KAKUMI exists an injective envelope and this is unique up to an isomorphism. Therefore it is also named the injective hull. The existence and the unity follows from the corresponding facts in the category COMP. Every object in COMP has a projective coenvelope and this is unique (Gleason [9]). It can be constructed for $X \in$ COMP as the natural projection $p : pX \longrightarrow X$ of the Stone representation space pX of the Boolean algebra $R_o(X)$ of all regular open (closed) sets. The injective envelope of $M \in$ KAKUMI is then the Dedekind-MacNeille completion of M by cuts. Analog to the characterization of the injective hull of a Boolean algebra in BOOLALG holds the following:

Theorem 3. *Let be A a MI-subspace of the MI-space B. Then the following are equivalent:*

(1) $A \hookrightarrow B$ is an injective envelope

(2) a) B is order complete

b) The injection $A \hookrightarrow B$ is an order complete isomorphism, e.g. for every family $F \subset A$ for which A-sup F resp. A-inf F exist holds A-sup F=B - sup F resp. A-infF = B-infF.

c) There is no proper complete MI-subspace of B containing A.

Proof: From the category COMP follows by dualizing that every envelope $A \hookrightarrow B$ can be realized in the injective hull. But the Dedekind-MacNeille completion of arbitrary ordered sets preserves all existing extrema (see [1]).
Therefore every envelope $A \hookrightarrow B$ preserves all existing extrema. Thus from (1) follows 2)a) and b).

c) No proper complete MI-space C of B contains A, because the injective hull is the smallest injective extension. Now let be holds (2). We take a maximal envelope $A \subset C \subset B$ from A in B. This C fulfills a) and b). Then by c) it must be $C = B$.

Remark: For the category BOOLALG is known the notion of the m-completion given by Sikorski [18] (m-a given infinite cardinal). In our paper [8] we have shown that this notion can be generalized to the category KAKUMI . Thus for every object in KAKUMI is a (unique) m-injective envelope. This gives for every object in COMP a (unique) m-extremally disconnected irreducible preimage.

Our next theorem looks for the special MI-space $C(X,\mathrm{R})$ over a hyperstonian space X. This gives a generalization of a theorem of Gonshor ([10, Theorem 7]) if it will be translated in the category C^*-ALG.

Theorem 4 Let be A a MI-subspace of the MI-space $B = C(X,\mathrm{R})$ over a hyperstonian space X. Then the following are equivalent:

(1) The injection $A \hookrightarrow B$ is an injective envelope

(2) For every closed set $Y \subset X$ for which not all hyperdiffuse measures on X vanish exists a non-zero functions $f \in A$ which vanishes outside from Y.

Proof: A MI-space B is isomorphic to $C(X,\mathrm{R})$ over a hyperstonian X iff B is the second dual of a MI-space C (see for hyperstonian spaces Dixmier [5]). Then B is order complete. For every nonvoid set G in X exists a nontrivial hyperdiffuse measure μ on X with supp $\mu \subset G$. Therefore a closed set $Y \subset X$ for which not all hyperdiffuse measures vanish has a nonvoid interior (equivalently: contains a nonvoid clopen set).

Now under every indicator function of a nonvoid clopen set lies a positive non-zero function of the subspace A iff $A \hookrightarrow B$ is an envelope (theorem 1,(2). Thus we have (1) $\Leftrightarrow$ (2).

References

[1] B.Banaschewski: Hüllensysteme und Erweiterungen von Quasi-Ordnungen. Z. math. Logik Grundl. Math. 2, (1956), 35-46.

[2] B.Banaschewski: G.Bruns: Categorical characterization of the MacNeille completion. Archiv Math. 18, (1967), 369-377.

[3] G.Birkhoff: Lattice theory 3. ed.Amer.Math.Soc.Colloq. Publ.(1967).

[4] R.P.Dilworth: The normal completion of the lattice of continuous functions. Amer. Math. Soc. 68, (1950), 427-438.

[5] J.Dixmier: Sur certains espaces considérés par M.H.Stone. Summa Brasil. Math. 2, (1951), 151-182.

[6] B.Eckmann: A.Schopf: Über injektive Moduln. Arch. Math. 4, (1953), 75-78.

[7] C.Faith: Lectures on injective modules and quotient rings. Lecture Notes in Math., Springer, (1967).

[8] J.Flachsmeyer: Dedekind MacNeille extensions of Boolean algebras and of vector lattices of continuous functions and their structure spaces. General Topology and its Appl. (to appear).

[9] A.M.Gleason: Projective topological spaces. Ill.J.Math. 2, (1958), 482-489.

[10] H.Gonshor: Injective hulls of c algebras. Trans.Amer. Math. Soc. 131, (1968), 315-322.

[11] H.Herrlich: G.E.Strecker: Category theory. Allyn and Bacon Inc. Boston (1973).

References

[12] S.Kakutani: Concrete representation of abstract (M)-spaces. Ann. of Math. 42, (1941), 994-1024.

[13] H.Nakano: Über das System aller stetigen Funktionen auf einem topologischen Raum. Proc. Imp. Acad. Tokyo, 17, (1941), 308-310.

[14] Z.Semadeni: Projectivity, injectivity and duality. Dissertationes Math.35, (1963), 1-47.

[15] Z.Semadeni: Banach spaces of continuous functions I. PWN Warszawa (1971).

[16] R.Sikorski: A theorem on extension of homomorphisms. Ann.Soc.Pol.Math. 21, (1948), 332-335.

[17] R.Sikorski: Cartesian products of Boolean algebras. Fund. Math. 37, (1950), 125-136.

[18] R.Sikorski: Boolean algebras. Berlin-Göttingen-Heidelberg.

[19] M.H.Stone: Boundedness properties in function lattices. Canad. J. Math. 1, (1949), 176-186.

COMPACTLY GENERATED SPACES AND DUALITY

by Alfred Frölicher

1. Introduction

Duality theory within the classical topological frame-work (e.g. for topological vector spaces E) does not give very satisfactoy results. A main reason for this is the fact that there is no good function-space topology available (e.g. such that the canonical map of E into its bidual E** is always continuous). It therefore seams advantageous to use a cartesian closed category instead of the category of topological spaces. This has been done very successfully by E. Binz who used the cartesian closed category of limit-spaces and so obtained many interesting and useful results on duality [1] . General considerations with an arbitrary cartesian closed category have been made by D. Franke for the case of algebras [4] . We shall use the cartesian closed category $\underline{K}$ of compactly generated spaces.

Only in special cases (e.g. for vector spaces), the dual of an object X of a category $\underline{A}$ is an object of the same category. In general, one has contravariant functors

$$\mathcal{F} : \underline{A} \to \underline{B} \quad \text{and} \quad \mathcal{G} : \underline{B} \to \underline{A}$$

and $\mathcal{F}X$ is called the dual, $\mathcal{G}\mathcal{F}X$ the bidual of the object X.

We shall work with categories $\underline{A}$ and $\underline{B}$ whose objects are sets with an algebraic structure of some type and a compatible compactly generated topology. It is essential, that "compatible" means the continuity of the algebraic operations with respect to the categorical, i.e. the compactly generated product and not with respect to the product topology. Using the cartesian closedness of $\underline{K}$ we then get in each case morphisms $X \to \mathcal{G}\mathcal{F}X$ forming a natural transformation of the identity functor $1_{\underline{A}}$ of $\underline{A}$ into $\mathcal{G} \circ \mathcal{F}$.

The categories we are going to examine are examples of so-called enriched categories, consisting of a category $\underline{A}$ together with a faithful functor into a cartesian closed category. We omit a general categorical outline in this direction, and shall directly examine the following categories $\underline{A}$.

- the category of real (or complex) compactly generated vector spaces ;
- the category of compactly generated spaces (without additional algebraic structure) ;
- the category of compactly generated *-algebras.

Other categories shall be investigated later in the same way ; in particular the category of compactly generated abelian groups and that of compactly generated groups.

The particular problems which shall be discussed here can be summarized as follows. A duality $\underline{A} \overset{\mathcal{F}}{\rightarrow} \underline{B} \overset{\mathcal{G}}{\rightarrow} \underline{A}$ with a natural transformation $1_{\underline{A}} \rightarrow \mathcal{G} \circ \mathcal{F}$ being established one asks for further information on the morphisms $X \rightarrow \mathcal{G}\mathcal{F}X$. We shall give necessary and sufficiant conditions on X in order that $X \rightarrow \mathcal{G}\mathcal{F}X$ shall be

(a) a monomorphism (an injective map) ;

(b) an extreme monomorphism (X has the compactly generated topology induced by the injection into the bidual $\mathcal{G}\mathcal{F}X$) ;

(c) an isomorphism (bijective and bicontinuous).

In case (b), X is called imbeddable ; in case (c) it is called reflexive.

Once the imbeddable resp. reflexive objects are determined, one will ask for properties of the full subcategories of $\underline{A}$ formed by these. Some results and some problems of this sort shall be mentioned.

Similar duality problems have been examined from a slightly different point of view by H. Buchwalter [2] : he used in one of the involved categories $\underline{A}$, $\underline{B}$ a topological, in the other a compactological structure. In this way he obtained excellent results, and many of the methods he developed in his proofs have been crucial in order to obtain certain results presented in the following.

2. Generalities on compartly generated spaces.

Compactly generated spaces, also called K-spaces, were introduced by Kelley[9] and have been studied and used in many articles as for example in [7][10] [11].

If X is an object of the category $\underline{T_2}$ of Hausdorff spaces, we can form a new Hausdorff space kX by putting on the underlying set of X the topology induced by the inclusions of the compact subspaces of X. It is easily verified that X and kX have the same compact subspaces. Therefore kkX = kX ; and for f : X → Y continuous, f : kX → kY is also continuous. Hence one has a functor $k : \underline{T_2} \to \underline{T_2}$ satisfying $k^2 = k$. A compactly generated space is an object X of $\underline{T_2}$ for which kX = X ; the full subcategory of $\underline{T_2}$ formed by these will be called $\underline{K}$. The functor k yields a functor $k : \underline{T_2} \to \underline{K}$ which is a retraction and an ad - joint of the inclusion functor $i : \underline{K} \to \underline{T_2}$. It then follows that K is a complete and cocomplete category. As an adjoint, k commutes with limits ; in particular the categorical product $X \pi Y$ of two objects of $\underline{K}$ is given by $X \pi Y = k\ (X \times Y)$, where $X \times Y$ denotes the topological product.

Most T_2-spaces, in particular all sequential and hence all metrizable spaces, are in $\underline{K}$; a simple counter-example is the weak topology of ordinary Hilbert space.

We write $C_{co}(X,Y)$ for the space of continuous maps $X \to Y$ with the compact-open topology and $\mathcal{C}(X,Y) = k\ C_{co}(X,Y)$. Then we get a bifunctor

$$\mathcal{C} : \underline{K}^{op} \times \underline{K} \to \underline{K}$$

and for X,Y,Z in $\underline{K}$ one has the following universal property : a map $f : X \to \mathcal{C}(Y,Z)$ is continuous if and only if the map $\tilde{f} : X \pi Y \to Z$, defined by $\tilde{f}(x,y) = f(x)(y)$, is continuous. Hence for each Y, the functor $\mathcal{C}(Y,-)$ is adjoint to the functor $-\pi Y$, and this caracterizes $\mathcal{C}$ up to an isomorphism. The existence of a functor $\mathcal{C}$ with this property is expressed by saying that $\underline{K}$ is a certain closed category.

3. Duality for compactly generated vector spaces.

This duality was examined in [6]; more results and all proofs can be found there. We consider the category KV of compactly generated vector spaces ; the objects are vector spaces E with a compactly generated topology such that $E \pi E \xrightarrow{+} E$ and $\mathbb{R}\pi E \xrightarrow{\cdot} E$ are continuous, the morphisms are the continuous linear maps. $\mathbb{R}$ could be replaced by $\mathbb{C}$. The dual E* of E is again in KV : it is the space of continuous linear functions $E \to \mathbb{R}$ with the universal compactly generated function space topology. So we have for this case A = B = KV and $\mathcal{F}E = \mathcal{G}E = E^*$. Using the universal property of the function space topology one easily shows that the canonical map $e_E : E \to E^{**}$ is continuous. These morphisms form a natural transformation of the functor $1_{\underline{KV}}$ into the functor "bidual".

If E is in KV, the convex neighborhoods of zero in E form a basis for the filter of zero-neighborhoods of a locally convex topology on the underlying vector space ; the so obtained locally convex space is denoted by cE. For a continuous linear map $f : E_1 \to E_2$, also $f : cE_1 \to cE_2$ is continuous. By means of the theorem of Hahn-Banach one gets immediately :

(3.1) $\quad e_E : E \to E^{**}$ injective $\iff$ cE separated

We restrict now to objects E satisfying cE separated and we denote by KVs the full subcategory of KV formed by these. With LCV denoting the category of separated locally convex spaces we have functors

$$\underline{KVs} \underset{k}{\overset{c}{\rightleftarrows}} \underline{LCV}$$

and one easily shows the following : k is adjoint to c ; ckc = c ; kck = k. We can state now the main results :

(3.2) $e_E : E \to E^{**}$ a subspace $\iff$ kcE = E

(3.3) $e_E : E \to E^{**}$ a homeomorphism $\iff$ kcE = E and cE complete

We remark that by "a subspace" we mean that E has the compactly generated topology induced by the injection e_E ; this topology has the universal subspace property within compactly generated spaces and can be obtained by applying k to the subspace topology.

Since $E^* = k\,L_{co}(E\,;\mathbb{R})$ and since $kck = k$, any dual and in particular any bidual is kc-invariant. But this property goes over to subspaces. Hence the condition in (3.2) is necessary. In order to show that it is sufficient, one makes use of the theorem of bipolars. The proof of (3.3) uses Grothendiecks caracterization of the completeness of a locally convex space.

By (3.2), the category of imbeddable compactly generated vector spaces is the full subcategory of KVs formed by the objects invariant under kc. Using the above functors KVs $\xrightarrow{c}$ LCV $\xrightarrow{k}$ KVs and their properties one sees that this full subcategory of KVs is isomorphic to the full subcategory LCV° of LCV formed by the objects invariant under ck, and also that ck yields a retraction and adjoint to the inclusion functor LCV° $\rightarrow$ LCV. Therefore the completeness and cocompleteness of the category LCV yields the same properties for LCV°, and one has the first of the following results :

(3.4) a) The imbeddable compactly generated vector spaces form a complete and cocomplete category ;

b) If $E_1,\ldots,E_n$, F are imbeddable, the space $\mathcal{L}(E_1,\ldots,E_n\,;F)$ of multilinear maps with its universal compactly generated topology is also imbeddable ;

c) One has a bifunctor $\otimes$ satisfying

$$\mathcal{L}(E_1 \otimes E_2\,;E_3) \cong \mathcal{L}(E_1\,;\,\mathcal{L}(E_2\,;E_3))\ ;$$

d) Products " $\prod$" and coproducts " $\coprod$ " satisfy

$$\left(\coprod_{i\in I} E_i\right)^* \cong \prod_{i\in I} E_i^* \quad \text{and} \quad \left(\prod_{i\in I} E_i\right)^* \cong \coprod_{i\in I} E_i^*$$

e) If E,F are imbeddable, then also $\mathcal{C}(E,F)$.

From the caracterization (3.3) it follows that all Fréchet spaces and in particular all Banach spaces are reflexive compactly generated vector spaces. For a reflexive E, $\mathcal{C}(E,\mathbb{R})$ can be shown to be also reflexive. It is not known however, whether reflexivity carries over from E, F to $\mathcal{L}(E\,;F)$; if it would, then the same would follow easily for $\mathcal{C}(E\,;F)$. As to the question of completness and cocompleteness of the category of reflexive compactly generated vector spaces, there is no problem with products and coproducts : (3.4 d) shows that they exist and are the same as in the category of imbeddable spaces. But it is not known either, whether kernels (and hence equalizers) exist. To summarize : one has not obtained for the reflexive objects as good categorical properties as for the imbeddable ones.

4. Reflexive vector spaces and calculus.

Compactly generated vector spaces have been very successfully used by U. Seip for calculus [10] . The obtained theory is not only more general than classical calculus for Banach spaces, but gives much better results, in particular with respect to functions spaces. In fact, for admissible spaces $E, F, \mathcal{O} \subset E$ open and $o \leqslant k \leqslant \infty$, the function space $\mathcal{C}^k(\mathcal{O}, F)$ formed by the maps $\mathcal{O} \to F$ of class C^k and equiped with a natural compactly generated C^k-topology, is again admissible. It turned out that a convenient notion of "admissible" is the following : E = kcE and cE sequentially complete. Therefore the admissible spaces are all imbeddable, and in fact they are very close to the reflexive ones. The reason for imposing on the imbeddable spaces E the additional condition "cE sequentially complete" is the fact that by this condition one gets a full subcategory having the same excellent properties as the category of imbeddable ones (cf. (3.4)). The crucial difference is that for a sequentially complete locally convex space L, the space ckL is also sequentially complete, while L complete does not imply ckL complete. Otherwise one would have worked with the reflexive spaces as the admissible ones.

One used to say that for a differentiation theory within a given class of topological (or similar) vector spaces, two somehow artificial choices must be made : the remainder condition and the topology (or respective structure) on the spaces L(E ;F). From a new point of view, both become, as we shall indicate, very natural within Seip's calculus. Usually the central role in calculus is attributed to the operator $f \mapsto f'$, where $f : E \to F$ and $f' : E \to L(E;F)$. However one could as well work with the operator $f \mapsto Tf$, where $Tf : TE \to TF$ is defined by $TE = E \,\pi\, E$ and $Tf(x,h) = (f(x), f'(x)(h))$. The operator T has at least the big advantage of being functorial ; and instead of involving L(E;F) which later requires an additional structure (topology), it only brings in the products $E \pi E$, $F \pi F$ which have their natural categorically determined structure. It therefore seems more natural to define, for a differentiable f, "continuously differentiable" resp. "twice differentiable" by imposing "Tf continuous" resp. "Tf differentiable". In Seip's theory however, these conditions become equivalent to "f' continuous" resp. " f' differentiable", provided one uses on L(E ;F) the universal compactly generated topology. Furthermore, in analogy with results of H. Keller [8] one can show : in order that $f : \mathcal{O} \to F$

(where $\mathcal{O} \subset E$ open) is C^1 (i.e. continuously differentiable) it is sufficient (and of course also necessary) that there exists a continuous map $f' : \mathcal{O} \to \mathcal{L}(E;F)$ such that the "weak" Gâteau condition

$$\lim_{\lambda \to o} \frac{f(a + \lambda x) - f(a)}{\lambda} = f'(a)(x)$$

is satisfied for $a \in \mathcal{O}$, $h \in E$; the "weak" here shall indicate that the limit is taken with respect to the weak topology of F. The proof uses the modern form of the mean value theorem. The result shows, that for the definition of "continuously differentiable on an open set" everything becomes natural : the structure on L(E ; F)as already indicated, and the remainder condition also, since the above Gâteau condition is the weakest reasonable one.

If F is reflexive, continuous differentiability of a map can be caracterized by means of its behaviour with respect to the C^1-functions ; using the preceding result one can show :

(4,1) $f : \mathcal{O} \to F$ is C^1 if and only if for each $\varphi \in \mathcal{C}^1(F, R)$, $f^* \varphi \in \mathcal{C}^1(\mathcal{O}, R)$ and $f^* : \mathcal{C}^1(F, R) \to \mathcal{C}^1(\mathcal{O}, R)$ is continuous.

We recall that $\mathcal{C}^1(\mathcal{O}, R)$ has the natural compactly generated topology which takes care of the functions and their derivatives ; it is the coarsest making continuous the map $\mathcal{C}^1(\mathcal{O}, R) \ \pi \ T\mathcal{O} \to TF$ defined by $(f,x,h,) \mapsto Tf(x,h)$. An analogous caracterization of C^k-maps is obtained by induction. The interesting part of (4.1) is the sufficiency of the condition. The only elements φ in $\mathcal{C}^1(F, R)$ which are available for the proof are the elements of F*. The remark that in the Gâteau condition the weak topology can be used becomes crucial.

Since reflexivity would also have other great advantages in analysis, it seems very desirable to find a category of reflexive vector spaces which is big enough and has good categorical properties (possibly those of (3.4)). There might be a chance to get such a category if one starts with an other cartesian closed category of topological spaces instead of K (cf. [12]).

5. Duality for compactly generated spaces.

For details and proofs we refer to [5] . We take now for A the category K of compactly generated spaces without additional algebraic structure, and for B the category KA of compactly generated unitary real algebras (analogous results hold if one replaces $\mathbb{R}$ by $\mathbb{C}$). One has contravariant functors

$$\mathcal{F} : \underline{K} \to \underline{KA} \qquad \text{and} \qquad \mathcal{H} : \underline{KA} \to \underline{K}$$

where, for X in K and A in KA, $\mathcal{F}X$ is the algebra $\mathcal{C}(X,\mathbb{R})$ of continuous functions $X \to \mathbb{R}$ and $\mathcal{H}A$ is the set of continuous unitary algebra homomorphisms $A \to \mathbb{R}$; $\mathcal{F}X$ and $\mathcal{H}A$ are equipped with their universal compactly generated function space topologies. The canonical map $e_X : X \to \mathcal{H}\mathcal{F}X$, also called the Dirac transformation of X, is always continuous. In order to study further properties of e_X, some considerations on uniform and on completely regular spaces are useful.

The classical functor t : Unif $\to$ Top which associates to a uniform space the underlying topological space has an coadjoint u : Top $\to$ Unif ; for T in Top, uT is the underlying set of T with the finest uniforme structure such that each uniform neighborhood of the diagonal Δ_T is a neighborhood of Δ_T, and uf = f for a morphism f, since $f : T_1 \to T_2$ continuous implies $f : uT_1 \to uT_2$ uniformly continuous. A topological space T is called topologically complete if uT is complete. It is known (see e.g. Problem L(d) Chap. 6 in [9]) that all paracompact and in particular all metrizable spaces are topologically complete. One easily verifies utu = u and tut = t. The identity map T $\to$ tuT is always continuous, but the topology of tuT can be strictly coarser than that of T.

We denote by Ks the full subcategory of K whose objects X satisfy tuX separated. Putting tuX = rX and remarking that rX is completely regular one has functors

$$\underline{Ks} \underset{k}{\overset{r}{\rightleftarrows}} \underline{CR}$$

where CR denotes the category of separated completely regular spaces. For these functors one has : k is adjoint to r ; rkr = r ; krk = k.

The following results are formally completely analogous to those for compactly generated vector spaces (cf. (3.1) to (3.3)) ; the proofs however are quite different.

(5.1) $e_X : X \to \mathcal{K}\mathcal{F}X$ injective $\Leftrightarrow$ rX separated.

Supposing in the following this condition to be satisfied, one has furthermore :

(5.2) $e_X : X \to \mathcal{K}\mathcal{F}X$ a subspace $\Leftrightarrow$ krX = X ;

(5.3) $e_X : X \to \mathcal{K}\mathcal{F}X$ a homeomorphism $\Leftrightarrow$ krX = X and X topologically complete.

Of course, since utu = u, the condition "X topologically complete" is equivalent to the condition that the associated completely regular space rX is topologically complete.

For the categorical properties of the imbeddable objects X one shows, as in the analogous situation for vector spaces, that they form a category isomorphic to that of the completely regular spaces T satisfying rkT = T, and this last category is a reflective subcategory of the complete and cocomplete category CR. This yields the first of the following results :

(5.4) a) The category of imbeddable compactly generated spaces is complete and cocomplete.

b) For Y imbeddable, $\mathcal{C}(X,Y)$ is also imbeddable, and hence the category of imbeddable compactly generated spaces is cartesian closed by means of the restriction of the bifunctor $\mathcal{C}$.

(5.3) shows, that the conditions for being reflexive are not very restrictive ; e.g. all compactly generated paracompact spaces, in particular all metrizable spaces, are reflexive. It is not known however whether the category of reflexive spaces has as good categorical properties as that of the imbeddable ones.

6. Duality for compactly generated *-algebras.

Duality for this case is being studied by D. Favrot [3] . We give a summary of the results be obtained so far. The category A is now the category KA* of compactly generated *-algebras, whose objects are the unitary complex algebras with a compatible compactly generated topology and a continuous involution $x \mapsto x^*$. To such an algebra A one associates a locally multiplicatively convex *-algebra cA by putting on the underlying algebra the topology determined by all continuous semi-norms $p : A \to \mathbb{R}$ satisfying $p(x.y) \leqslant p(x).p(y)$ and $p(x.x^*) = p^2(x)$ (and hence $p(x^*) = p(x)$). Conversely, to a locally multiplicatively convex separated *-algebras B one associates, by refining its topology by means of the functor k, a compactly generated *-algebra kB. One has contravariant functors

$$\mathcal{H} : \underline{KA^*} \to \underline{K} \qquad \text{and} \qquad \mathcal{F} : \underline{K} \to \underline{KA^*}$$

where, for A in KA* and X in K, $\mathcal{H}A$ is the space of all continuous algebra-homomorphisms $h : A \to \mathbb{C}$ satisfying $h(a^*) = \overline{h(a)}$, and $\mathcal{F}X$ is the *-algebra of all continuous functions $X \to \mathbb{C}$, $\mathcal{H}A$ and $\mathcal{F}X$ being equipped with their universal compactly generated function space topologies.

For all A in KA* the canonical map $e_A : A \to \mathcal{F}\mathcal{H}A$, also called the Gelfand transformation of A, is continuous.

(6.1) $e_A : A \to \mathcal{F}\mathcal{H}A$ injective $\Longleftrightarrow$ A separated.

Supposing this condition satisfied, we have furthermore :

(6.2) $e_A : A \to \mathcal{F}\mathcal{H}A$ an extreme monomorphism $\Longleftrightarrow kcA = A$

(6.3) $e_A : A \to \mathcal{F}\mathcal{H}A$ an isomorphism $\Longleftrightarrow kcA = A$ and cA complete.

This result considerably improves the classical theorem of Gelfand-Naïmark concerning Banach*-algebras. The proof uses the theory of Gelfand and the methods used by Buchwalter [2] in his investigations of the Gelfand transformation by means of compactologies.

BIBLIOGRAPHY

[1] Binz E.- Continuous Convergence on C(X). - Lecture Notes in Mathematics 469 - Springer, Berlin-Heidelberg - New York 1975.

[2] Buchwalter H.- Topologie et compactologies - Publ. Dept. Math. Lyon - t. 6-2 - 1-74 (1969).

[3] Favrot D.- Thesis - University of Geneva (in preparation).

[4] Franke D.- Funktionenealgebren in kartesisch abgeschlossen Kategorien - Dissertation - Freie Universität - Berlin (1975).

[5] Frölicher A.- Sur la transformation de Dirac d'un espace à génération compacte - Publ. Dept. Math. Lyon t. 10-2, 79-100 (1973).

[6] Frölicher A.- Jarchow W.- Zur Dualitätstheorie kompakt erzeugter und lokalkonvexer Vektorraüme - Comm Math. Helv. Vol. 47 - 289-310 (1972).

[7] Gabriel P. and Zisman M.- Calculus of fractions and homotopy theory - Ergebn. der Math 35 - Springer, Berlin-Heidelberg - New York 1967.

[8] Keller HH.- Differential Calculus in Locally Convex Spaces - Lecture Nores in Mathematics 417 - Springer - Berlin-Heidelberg New York 1974.

[9] Kelley J.L.- General Topology - Van Nostrand - New York 1955.

[10] Seip U.- Kompakt erzeugte Vektorraüme und Analysis - Lecture Notes in Mathematics 273 - Springer - Berlin-Heidelberg-New York 1972.

[11] Steenrod N. - A convenient category of topological spaces - Mich. Math. Journ. 14 - 133-152 - (1967).

12 Wyler O.- Convenient categories for topology. General topology Vol. 3, 225-242 (1973).

Some Topological Theorems which Fail to be True

by

Horst Herrlich

Consider the following statements:

(1) Products of paracompact topological spaces are paracompact.

(2) Products of compact Hausdorff spaces with normal topological spaces are normal.

(3) Subspaces of paracompact (normal) topological spaces are paracompact (normal).

(4) $\dim (X \times Y) \leq \dim X + \dim Y$ for non-empty paracompact topological spaces X and Y.

(5) $\dim X = \dim Y$ for dense subspaces X of regular topological spaces Y.

(6) $\dim X \leq \dim Y$ for subspaces X of topological spaces Y.

(7) Continuous maps from dense subspaces of topological spaces into regular topological spaces have continuous extensions to the whole space.

(8) $X^{Y \times Z} \approx (X^Y)^Z$ for topological spaces X, Y and Z.

(9) Products of quotient maps between topological spaces are quotient maps.

Although we would like the above statements to be true, we know that none of them are--provided such operations as the formation of products, subspaces and function spaces are performed, as usual, in the category Top of topological spaces and continuous maps. However, there exist settings--more appropriate it would seem--in which the above statements are valid. The category Top can be decently embedded in larger, more convenient categories such that, when the mentioned operations are performed in the larger category, the above statements are not only true but, in fact, special cases of more general theorems.

Especially simple and convenient settings for theorems such as the above are the category S-Near of semi-nearness spaces and nearness preserving maps, introduced under the name Q-Near in [27] , resp. the isomorphic category of merotopic spaces, introduced by M. Katětov [33] in a very important but hardly noticed paper already ten years ago, and various full subcategories of S-Near. Especially, statements (1)-(7) are true in the bireflective full subcategory Near of S-Near, whose objects are all nearness spaces, and statements (8)-(9) are true in the bicoreflective full subcategory Grill of S-Near, whose objects are all grill-determined semi-nearness spaces.

Some of our results are new, others are just reinterpretations of known facts.

1. Structures Induced by Topologies

With any topological space $\underline{X} = (X, cl)$ there can be associated various structures, e.g.

(1) a nearness structure, consisting of all collections $\mathcal{A}$ of subsets of X with $\bigcap cl\mathcal{A} \neq \emptyset$

(2) a covering structure, consisting of all covers of X, which are refined by some open cover of $\underline{X}$

(3) a convergence structure, consisting of all convergent filters in $\underline{X}$.

In case, $\underline{X}$ is a topological R_o-space, i.e. if $x \in cl\{y\} \Leftrightarrow y \in cl\{x\}$ each of the above structures determines the topology of $\underline{X}$ and hence all the other structures. From now on, topological space means topological R_o-space, and Top denotes the category of topological (R_o-)spaces and continuous maps. (For a slightly more complicated setting covering the non-R_o-case see D. Harris [24] and K. Morita [41].)

There are natural constructions, which--when applied to topological spaces--automatically yield more general types of structures, e.g.

(1) if $\underline{X}$ is a topological space, and S is a subset of X, then the

nearness structure, consisting of all collections $\mathcal{A}$ of subsets of S with $\bigcap cl_{\underline{X}}\mathcal{A} \neq \emptyset$, is in general not topological

(2) if $\underline{X}_1$ and $\underline{X}_2$ are topological spaces, then the covering structure on $X_1 \times X_2$, consisting of all covers, which can be refined by some cover of the form $\{A_1 \times A_2 \mid A_i \in \mathcal{A}_i\}$, where the $\mathcal{A}_i$ are open covers of $\underline{X}_i$, is in general not topological

(3) if $\underline{X}$ and $\underline{Y}$ are topological spaces, $C(\underline{X},\underline{Y})$ is the set of all continuous maps from $\underline{X}$ into $\underline{Y}$, and $e: X \times C(\underline{X},\underline{Y}) \to Y$ is the evaluation map, defined by $e(x, f) = f(x)$, then the convergence structure on $C(\underline{X}, \underline{Y})$, consisting of those filters $\mathcal{F}$ which have the property that for every convergent filter $\mathcal{G}$ in $\underline{X}$ the filter generated by $e(\mathcal{G},\mathcal{F}) = \{e(G \times F) \mid G \in \mathcal{G}, F \in \mathcal{F}\}$ converges in $\underline{Y}$, is in general not topological.

In order to find supercategories of Top which are closed under the above constructions, we need to concentrate only on one of the three types of structures described above: nearness structures, covering structures and convergence structures resp., since--as has been shown in [27]--they are all equivalent, i.e. just different facets of the same type of structure.

2. (Semi-)Nearness Spaces

For any set X, denote by PX the set of all subsets of X. A semi-nearness structure on X is a collection ξ of subsets of PX, satisfying the following axioms:

(N1) If $\mathcal{A} \subset PX$, $\mathcal{A}$ corefines* $\mathcal{B}$, and $\mathcal{B} \in \xi$ then $\mathcal{A} \in \xi$

(N2) If $\mathcal{A} \subset PX$ and $\bigcap\mathcal{A} \neq \emptyset$ then $\mathcal{A} \in \xi$

(N3) If $\mathcal{A} \subset PX$, $\mathcal{B} \subset PX$, and $\{A \cup B \mid A \in \mathcal{A}, B \in \mathcal{B}\} \in \xi$ then $\mathcal{A} \in \xi$ or $\mathcal{B} \in \xi$

(N4) $\emptyset \in \xi$ and $\{\emptyset\} \notin \xi$.

A nearness structure on X is a semi-nearness structure ξ on X,

---

* $\mathcal{A}$ corefines $\mathcal{B}$ iff for each $A \in \mathcal{A}$ there exists $B \in \mathcal{B}$ with $B \subset A$.

satisfying the additional axiom:

(N5) If $\mathcal{A} \subset PX$ and $\{cl_\xi A \mid A \in \mathcal{A}\} \in \xi$ then $\mathcal{A} \in \xi$, where $cl_\xi A = \{x \in X \mid \{A,\{x\}\} \in \xi\}$.

A pair (X, ξ) is called a <u>(semi-)nearness space</u> provided ξ is a (semi-)nearness structure on X. A map $f: (X, \xi) \to (Y, \eta)$ between semi-nearness spaces is called <u>nearness preserving</u> provided $\mathcal{A} \in \xi$ implies $\{fA \mid A \in \mathcal{A}\} \in \eta$ The category of all semi-nearness spaces and nearness preserving maps is denoted by <u>S-Near</u>, its full subcategory, consisting of all nearness spaces, is denoted by <u>Near</u>.

The categories <u>S-Near</u> and <u>Near</u> are known to be well-behaved categories ([25], [27]). Here we need only the following facts:

(1) <u>Near</u> has products.

(2) If (X, ξ) is a nearness space and S is a subset of X, then $\xi_S = \{\mathcal{A} \subset PS \mid \mathcal{A} \in \xi\}$ is a nearness structure on S and (S, ξ_S) is called the <u>nearness-subspace</u> of (X, ξ) determined by S.

3. <u>Topological Spaces and Nearness Spaces</u>

If $\underline{X} = (X, cl)$ is a topological space, then $\xi = \{\mathcal{A} \subset PX \mid \cap\{clA \mid A \in \mathcal{A}\} \neq \emptyset\}$ is a nearness structure on X. As is easily seen this correspondence is functorial and, in fact, gives rise to a full embedding of <u>Top</u> into <u>Near</u> . A nearness space (X, ξ) belongs to the image of the above embedding iff it satisfies the following axiom:

(N6) If $\mathcal{A} \in \xi$ then $\cap\{cl_\xi A \mid A \in \mathcal{A}\} \neq \emptyset$.

Nearness spaces, satisfying condition (N6), will be called <u>topological nearness spaces</u>. We may identify each topological space with its associated topological nearness space, and from now on we will call such spaces topological spaces. Also we will identify <u>Top</u> with the full subcategory <u>T-Near</u> of <u>Near</u> whose objects are the topological (nearness) spaces. Vice versa, we may associate with any nearness space $\underline{X} = (X,\xi)$

a topological space $T\underline{X} = (X, cl) = (X, \xi_t)$ defined by $cl = cl_\xi$, resp., $\xi_t = \{\mathcal{A} \subset PX \mid \bigcap\{cl_\xi A \mid A \in \mathcal{A}\} \neq \emptyset$, which we may call the underlying topological space (= the topological coreflection) of $\underline{X}$. The category Top is bicoreflective in Near and the bicoreflection has just been described.

Consequently colimits in Top are formed in the same way as in Near, but limits are formed differently: a limit in Top is obtained by forming it first in Near and then passing over to its underlying topological space. Especially:

(1) If $(\underline{X}_i)_{i\in I}$ is a family of topological spaces, and $\underline{X}$ is their product in Near, then the underlying topological space $T\underline{X}$ of $\underline{X}$ is the product of the family $(\underline{X}_i)_{i\in I}$ in Top

(2) If $\underline{X}$ is a topological space, S is a subset of X, and $\underline{S}$ is the nearness subspace of $\underline{X}$ determined by S, then the underlying topological space $T\underline{S}$ of $\underline{S}$ is the topological subspace of $\underline{X}$ determined by S.

As is well known and as the introductory examples demonstrate, products and subspaces are ill behaved, when obtained in Top. The following results indicate that they are much better behaved when performed in Near. It seems that in passing from a nearness space to its underlying topological space too much valuable information gets lost.

4. Paracompact Spaces

We will use the terms paracompact and fully normal synonymously, i.e. a topological space $\underline{X}$ is called paracompact provided every open cover of $\underline{X}$ is star-refined by some open cover of $\underline{X}$. In order to define paracompactness for nearness spaces, we need a suitable equivalent for open covers in a nearness space. Let $\underline{X} = (X, \xi)$ be a nearness space, then $\mathcal{A} \subset PX$ will be called a $\underline{X}$-cover iff $\{X - A \mid A \in \mathcal{A}\} \notin \xi$. Every $\underline{X}$-cover is a cover of X. Moreover every $\underline{X}$-cover is refined by some open cover of X (with resp. to the underlying topology), and the

X-covers of a topological space are characterized by this condition. A nearness space X is called paracompact provided it satisfies the following condition:

(N7) Every X-cover is star-refined by some X-cover.

A topological space is paracompact in the nearness-sense iff it is paracompact in the topological sense.

For any paracompact nearness space X, the set of all X-covers forms a uniform structure on X in the sense of J.W. Tukey. This correspondence is easily seen to be functorial, and, in fact, gives rise to an isomorphism between the category of paracompact nearness spaces and nearness preserving maps and the category of uniform spaces and uniformly continuous maps. We may identify each paracompact nearness space with its associated uniform space. Then the paracompact topological spaces are precisely those nearness spaces which are simultaneously topological and uniform. No wonder that many topologists have found them so attractive. On the other hand they are badly behaved with respect to any of the standard constructions. No wonder, topological spaces are bicoreflective in Near and paracompact (=uniform) nearness spaces are bireflective in Near (as proved in [25]), and-- formation of the intersection of a bicoreflective subcategory with a bireflective subcategory usually ruins all the constructions. But, since paracompact nearness spaces form a bireflective subcategory of Near, any product of paracompact nearness spaces (especially any product of paracompact topological spaces), taken in Near, is again paracompact (but generally no longer topological), and any subspace of a paracompact nearness space (especially of a paracompact topological space), taken in Near, is again paracompact (but generally no longer topological). This proves that assertion (1) and the paracompact part of assertion (3) are true, if interpreted in Near.

The reason why the nearness product of two paracompact topological spaces is again paracompact, but the topological product generally fails to be so, may be easier understood if we observe that it is hardly possible to describe the open covers of a product in any decent way by means of the open covers of the factors, whereas there is a very simple description of the $(\underline{X} \times \underline{Y})$-covers by means of the $\underline{X}$-covers and the $\underline{Y}$-covers: $\mathcal{A} \subset P(X \times Y)$ is a $(\underline{X} \times \underline{Y})$-cover iff there exist a $\underline{X}$-cover $\mathcal{B}$ and a $\underline{Y}$-cover $\mathcal{C}$ such that $\{B \times C \mid B \in \mathcal{B}, C \in \mathcal{C}\}$ refines $\mathcal{A}$.

5. Extensions of Maps. Complete and Regular Spaces

Every uniformly continuous map from a dense subspace of a uniform space into a complete uniform space has a uniformly continuous extension to the whole space. This well-known theorem of A. Weil seems to have no direct topological counterpart. Translated into the nearness language we obtain a straightforward "topological" application: Every nearness preserving map from a dense nearness-subspace of a paracompact topological space into a paracompact topological space has a continuous extension to the whole space.

A. Weil's theorem has a natural generalization, whose topological application is essentially assertion (7). It can be found in [26], but was observed much earlier in a different context by K. Morita [41].

If $\underline{X} = (X, \xi)$ is a nearness space, then $\mathcal{A} \subset PX$ is called a $\underline{X}$-cluster provided $\mathcal{A} \neq \emptyset$ and $\mathcal{A}$ is a maximal element of ξ, ordered by inclusion. A nearness space $\underline{S}$ is called complete provided it satisfies the following condition:

(N8) Every $\underline{X}$-cluster has an adherence point.

If $\underline{X} = (X, \xi)$ is a nearness space, $A \subset X$ and $B \subset X$, then $A <_{\underline{X}} B$ iff $\{A, X - B\} \notin \xi$. A nearness space $\underline{X} = (X, \xi)$ is called regular provided it satisfies the following axiom:

(N9) Iff $\mathcal{A} \subset PX$ and $\{B \subset X \mid A <_{\underline{X}} B \text{ for some } A \in \mathcal{A}\} \in \xi$ then $\mathcal{A} \in \xi$.

A topological space is regular in the nearness sense iff it is regular in the topological sense. Every paracompact nearness space is regular. The following is the announced generalization of A. Weil's result: Every nearness preserving map from a dense nearness-subspace of a nearness space into a complete regular nearness space has a nearness preserving extension to the whole space. As topological application we obtain the following interpretation of assertion (7): Every nearness preserving map from a dense nearness-subspace of a topological space into a regular topological space has a continuous extension to the whole space.

6. Normal Spaces

A topological space $\underline{X}$ is called normal provided each finite open cover of $\underline{X}$ is star-refined by some finite open cover of $\underline{X}$. This can be most easily expressed in the realm of nearness space by introducing the concept of a contigual nearness space first. A nearness space (X,ξ) is called contigual provided it satisfies the following condition:

(N1o) If $\mathcal{A} \subset PX$, and every finite subset of $\mathcal{A}$ belongs to ξ, then $\mathcal{A}$ belongs to ξ.

A topological space is contigual iff it is compact. A uniform space is contigual iff it is totally bounded (=precompact). Contigual nearness spaces form a bireflective subcategory of Near. For any nearness space $\underline{X} = (X, \xi)$ its contigual reflection is defined by $C\underline{X}=(X,\xi_C)$ with $\mathcal{A} \in \xi_C$ iff each finite subset of $\mathcal{A}$ belongs to ξ. A subset $\mathcal{A}$ of PX is a $C\underline{X}$-cover iff it is refined by some finite $\underline{X}$-cover. Now, normality of a nearness space $\underline{X}$ could be defined by requiring the contigual reflection $C\underline{X}$ of $\underline{X}$ to be paracompact or, equivalently (see [25]), to be regular. Since this property does not imply regularity of $\underline{X}$, we just add it. Hence a nearness space $\underline{X}$ is called normal provided it satisfies the following condition:

(N11) The space $\underline{X}$ and its contigual reflection $C\underline{X}$ are regular.

Since the contigual reflector preserves paracompactness, every paracompact space is normal. By definition, every normal space is regular. A topological space is normal as a nearness space iff it is normal as a topological space.

In order to study products of normal spaces we have to investigate the behaviour of the contigual reflector C with respect to products. Unfortunately C does not preserve products. E.g. if $\underline{X}=(X,\xi)$ is a paracompact nearness space, which is not contigual, then $C(\underline{X} \times \underline{X}) \neq C\underline{X} \times C\underline{X}$. To see this, let $\mathcal{A}$ be a $\underline{X}$-cover, which cannot be refined by a finite $\underline{X}$-cover, and let $\mathcal{B}$ be a $\underline{X}$-cover, which star-refines $\mathcal{A}$. Then $\{(X \times X)\setminus \Delta X, \cup\{B \times B | B \in \mathcal{B}\}\}$ is a $C(\underline{X} \times \underline{X})$-cover, but not a $(C\underline{X} \times C\underline{X})$-cover. But we have the following result:

<u>Theorem</u>: Let $\underline{X}$ and $\underline{Y}$ be nearness spaces. If $\underline{X}$ is contigual, then

$$C(\underline{X} \times \underline{Y}) = \underline{X} \times C\underline{Y} .$$

<u>Proof</u>: Obviously, every $(\underline{X} \times C\underline{Y})$-cover is a $C(\underline{X} \times \underline{Y})$-cover. Since any nearness space $\underline{X}$ is uniquely determined by the set of all $\underline{X}$-covers, it remains to show the converse. If $\mathcal{A}$ is a $C(\underline{X} \times \underline{Y})$-cover, then $\mathcal{A}$ is refined by some finite $(\underline{X} \times \underline{Y})$-cover $\mathcal{B}$. Hence there exists a finite $\underline{X}$-cover $\mathcal{C}$ and a $\underline{Y}$-cover $\mathcal{D}$ such that $\{C \times D | C \in \mathcal{C}, D \in \mathcal{D}\}$ refines $\mathcal{B}$. For each $B \in \mathcal{B}$ and each $C \in \mathcal{C}$ define $E(B, C) = \{y \in Y | C \times \{y\} \subset B\}$. Then, for each $C \in \mathcal{C}$, $F_C = \{E(B, C) | B \in \mathcal{B}\}$ is refined by $\mathcal{D}$ and finite, hence a $C\underline{Y}$-cover. Consequently, by axiom (N3), $\mathcal{F} = \wedge\{\mathcal{F}_C | C \in \mathcal{C}\}$, defined by $\mathcal{F} = \{\cap \mathcal{G} | \mathcal{G} \in \prod_{C \in \mathcal{C}} F_C\}$, is a $C\underline{Y}$-cover. Since $\{C \times F | C \in \mathcal{C}, F \in \mathcal{F}\}$ refines $\mathcal{B}$, and hence $\mathcal{A}$, we conclude that $\mathcal{A}$ is a $(\underline{X} \times C\underline{Y})$-cover.

Let us call a nearness space <u>proximal</u> provided it is contigual and regular. Observing that regularity is preserved under products ([25]), we obtain as an immediate corollary of the above theorem that products of proximal nearness spaces with normal nearness spaces are normal. An application to topological nearness spaces yields assertion (2): Products of compact Hausdorff spaces with normal topological spaces are normal. Observing that nearness-subspaces of regular spaces are regular and that whenever $\underline{X}$ is a nearness-subspace of $\underline{Y}$, then $C\underline{X}$ is a nearness-subspace of $C\underline{Y}$, we obtain the missing half of assertion (3): nearness-subspaces of normal topological spaces are normal.

7. Dimension Theory for Nearness Spaces

Topological dimension theory suffers from the fact that no dimension function has yet been found that behaves decently for a reasonable class of topological spaces which is essentially bigger than the class of metrizable topological spaces. If we consider e.g. the Lebesgue covering dimension dim, which coincides with the large inductive dimension Ind for metrizable spaces, we observe deficiencies such as the following:

(1) As the Tychonoff plane shows, there exist zero dimensional compact Hausdorff spaces having subspaces which are not normal and hence of positive dimension, and as the example of Dowker [16] shows there exist zero dimensional spaces having normal subspaces with positive dimension.

(2) As an example of E. Michael [37] shows, there exist a zero dimensional metrizable space and a zero dimensional paracompact space whose product is not normal and hence of positive dimension.

(3) As a theorem of Noble [46] shows, for any zero dimensional noncompact space, there exists some power which is not normal and hence of positive dimension.

Because of the above deficiencies, several topologists have tried with limited success to modify the definition of the covering dimension slightly, e.g. by considering cozero-set covers (e.g. L.Gillman and M. Jerison [22]) or normal open covers (e.g. M. Katetov [32], Yu. M. Smirnov [53], K. Morita [43]). But, as the above examples demonstrate clearly, it is not so much the dimension function, which is at fault; it is the construction of subspaces and products in Top, in other words: it is the category Top itself. This, in fact, has been observed before, e.g. by Nagami [44]. The crucial misunderstanding seems to be that topologists have usually tried to find a solution inside Top, whereas it seems that a solution can only be found outside Top. In fact, there seems to be a rather decent dimension theory for nearness spaces which extends the dimension theory for proximity spaces, due to Yu. M. Smirnov [53], and the dimension theory for uniform spaces, due to J. R. Isbell [29-31], which, indeed, are both highly satisfactory.

Definition: A nearness space $\underline{X}$ has dimension at most n provided every $\underline{X}$-cover can be refined by a $\underline{X}$-cover of order at most n + 1. dim $\underline{X}$ is the smallest natural number n such that $\underline{X}$ has dimension at most n, provided such a number exists, otherwise dim $\underline{X} = \infty$.

Except for the empty space, the above nearness-dimension coincides

(1) for proximal nearness spaces (=proximity spaces) with the δ-dimension of Yu. M. Smirnov [53]

(2) for paracompact nearness spaces (=uniform spaces) with the large dimension of J. R. Isbell [29],

(3) for paracompact topological spaces with the Lebesgue covering dimension (see C.H. Dowker [15], K. Morita [40], J.R. Isbell [30] and B. A. Pasynkov [48]).

Before we present results, we like to mention some natural modifications of the dimension function. If $\underline{A}$ is a (co-)reflective subcategory of Near with (co)reflector A: Near $\to \underline{A}$ then the $\underline{A}$-dimension of a nearness space $\underline{X}$ may be defined by $\dim_A \underline{X} = \dim A\underline{X}$. Obviously, for any space $\underline{X}$ in $\underline{A}$ we have $\dim_A X = \dim X$. We mention especially the contigual dimension $\dim_C \underline{X}$ and the proximal dimension $\dim_P \underline{X}$.

For a nearness space $\underline{X}$ we have:

(1) if $\underline{X}$ is topological, then $\dim_C \underline{X}$ is the Lebesgue covering dimension of $\underline{X}$

(2) if $\underline{X}$ is topological, then $\dim_P \underline{X}$ is the modified covering dimension of $\underline{X}$, introduced by M. Katětov [32] and independently by Yu. M. Smirnov [53]

(3) if $\underline{X}$ is paracompact (=uniform), then $\dim_C \underline{X} = \dim_P \underline{X}$ is the uniform dimension $\delta d\underline{X}$ of $\underline{X}$, introduced by Yu. M. Smirnov [53].

Moreover, $\dim_C \underline{X} \leq \dim \underline{X}$, for every nearness space $\underline{X}$. This follows immediately from the fact that for every $\underline{X}$-cover $\mathcal{A}$ of order at most n, which refines a finite $\underline{X}$-cover $\mathcal{B}$, there exists a finite $\underline{X}$-cover $\mathcal{C}$ of order at most n, such that $\mathcal{A}$ refines $\mathcal{C}$ and $\mathcal{C}$ refines $\mathcal{B}$. For paracompact topological spaces $\underline{X}$ we have $\dim_C \underline{X} = \dim \underline{X}$. For paracompact nearness spaces $\underline{X}$ with $\dim \underline{X} < \infty$ we have $\dim \underline{X} = \dim_C \underline{X}$, but there exist paracompact nearness spaces $\underline{X}$ with $\dim_C \underline{X} = 0$ and $\dim \underline{X} = \infty$ (J. R. Isbell [30]). P. Alexandroff's long line is an example of a normal topological space $\underline{X}$ with $\dim_C \underline{X} = 0$ and $\dim \underline{X} = \infty$.

__Proposition:__ If $\underline{X}$ is a nearness-subspace of $\underline{Y}$, then $\dim \underline{X} \leq \dim \underline{Y}$.

Moreover, if $\underline{X}$ is a nearness-subspace of $\underline{Y}$, then $C\underline{X}$ is a nearness-subspace of $C\underline{Y}$, which implies $\dim_C \underline{X} \leq \dim_C \underline{Y}$. The topological application of this result is our assertion (6). In the following, we will restrict our investigations to the dimension function dim.

There is a partial converse to the above proposition, asserting that for sufficiently big nearness-subspaces $\underline{X}$ of $\underline{Y}$ we have $\dim \underline{X} = \dim \underline{Y}$. For this we need some preparation. If (Y, η) is a nearness space, and $A \subset X \subset Y$, let $OP_X A = \mathrm{int}_{\underline{Y}}(A \cup (Y \setminus X))$ denote the largest open subset B of $\underline{Y}$ with $B \cap X = \mathrm{int}_{\underline{X}} A$.

__Definition:__ Let $\underline{X} = (X, \xi)$ and $\underline{Y} = (Y, \eta)$ be nearness spaces, and let X be a subset of Y. Then the injection $\underline{X} \hookrightarrow \underline{Y}$ is called a __strict extension__ provided the following equivalent conditions are satisfied.

(1) $\mathcal{A} \subset PY$ belongs to η iff $\{B \subset X | A \subset cl_\eta B$ for some $A \in \mathcal{A}\}$

belongs to ξ

(2) $\mathcal{A} \subset P\underline{Y}$ is a $\underline{Y}$-cover iff $\{B \subset X \mid OP_X B \subset A$ for some $A \in \mathcal{A}\}$ is a $\underline{X}$-cover.

Every strict extension is a dense nearness-embedding and the converse is true for regular nearness spaces (H.L. Bentley and H. Herrlich [6]). If $\underline{X} \hookrightarrow \underline{Y}$ is a strict extension and $\mathcal{A}$ is a $\underline{X}$-open $\underline{X}$-cover then $\mathcal{B} = \{OP_X A \mid A \in \mathcal{A}\}$ is a $\underline{Y}$-cover, and $\mathcal{A}$ and $\mathcal{B}$ have the same order. Therefore:

Proposition (1) If $\underline{X} \to \underline{Y}$ is a strict extension, then $\dim \underline{X} = \dim \underline{Y}$

(2) If $\underline{X}$ is a dense nearness-subspace of a regular nearness space $\underline{Y}$, then $\dim \underline{X} = \dim \underline{Y}$.

For every nearness space $\underline{X}$ there exists a complete nearness space $\underline{X}^*$ and a strict extension $\underline{X} \hookrightarrow \underline{X}^*$, called the completion of $\underline{X}$ (see [25]). Therefore:

Proposition: If $\underline{X} \hookrightarrow \underline{X}^*$ is the completion of $\underline{X}$, then $\dim \underline{X} = \dim \underline{X}^*$.

This proposition has a number of obvious corollaries:

(1) For a topological space $\underline{X}$, $(C\underline{X})^*$ is its Wallman--compactification $\omega\underline{X}$, hence $\dim \omega\underline{X} = \dim_C \underline{X}$.

(2) for a topological space $\underline{X}$, $(P\underline{X})^*$ is its Čech-Stone compactification $\beta\underline{X}$, hence $\dim \beta\underline{X} = \dim_P \underline{X}$.

(3) for a uniform space $\underline{X}$, $\underline{X}^*$ is its completion $\gamma\underline{X}$, hence $\dim \gamma\underline{X} = \dim \underline{X}$.

(4) for a uniform space $\underline{X}$, $(C\underline{X})^*$ is its Samuel-compactification $\beta\underline{X}$, hence $\dim \beta\underline{X} = \dim {}_C\underline{X}$.

(5) For a proximity space $\underline{X}$, $\underline{X}^*$ is its Smirnov compactification $u\underline{X}$, hence $\dim u\underline{X} = \dim \underline{X}$.

Since the class of paracompact nearness spaces is closed under the formation of products in Near the following proposition is just a restatement of a result stated as exercise in the book of J.R. Isbell ([31] , p. 94):

Proposition: If $\underline{X}$ and $\underline{Y}$ are paracompact nearness spaces, then $\dim(\underline{X} \times \underline{Y}) \leq \dim \underline{X} + \dim \underline{Y}$.

The following result is trivial:

Proposition: Products of arbitrary families of zero dimensional nearness spaces are zero dimensional.

Much more can be said and even more can be asked, but we leave the matter here.

8. Function Spaces

For any pair $(\underline{X}, \underline{Y})$ of topological spaces, the set $C(\underline{X},\underline{Y})$ of all continuous maps from $\underline{X}$ to $\underline{Y}$ can be supplied in several ways with a topological structure and thus made into a topological space, denoted by $\underline{Y}^{\underline{X}}$. Among these structures, the compact-open topology is rather decently behaved, provided $\underline{X}$ is a locally compact Hausdorff space. But for arbitrary topological spaces, none of the topological structures on function sets $C(\underline{X}, \underline{Y})$ is sufficiently well behaved, e.g. such that, for any triple $(\underline{X}, \underline{Y}, \underline{Z})$ of topological spaces, the spaces $\underline{X}^{\underline{Y}\times\underline{Z}}$ and $(\underline{X}^{\underline{Y}})^{\underline{Z}}$ are naturally isomorphic (see R.F.Arens [3]). In other words, the category Top fails to be cartesian closed. Because of this deficiency, several authors have constructed better behaved substitutes for Top, which are usually either subcategories or supercategories of Top. Among the cartesian closed subcategories of Top, the one which has been used most often especially in homotopy theory and topological algebra is the coreflective hull of all compact Hausdorff spaces in the category Haus of Hausdorff topological spaces, i.e. the full subcategory of Haus, whose objects are the quotients of locally compact Hausdorff spaces (see e.g. R. Brown [11], E. J. Dubuc and H. Porta [17] , W. F. LaMartin [36] , E. C. Nummela [47], N. E. Steenrod [55]) resp. the slightly larger category of compactly generated spaces which is the coreflective hull of all compact Hausdorff spaces in Top (R. M. Vogt [56] and O. Wyler [58]) . The main disadvantage of this category is that it is rather awkward to describe it (e.g. no convenient axiomatic description is known) and to prove even the basic theorems. Among the cartesian closed supercategories of Top, the category of quasi-topological spaces, introduced by E. Spanier [54], is unnecessarily big, e.g. the quasitopolgies on a fixed set, in general form a proper class. Also, no axiomatic description of quasi-topological spaces is known.

Fortunately there exist cartesian closed supercategories of Top, which not only can be described axiomatically in a very elegant manner, but also can be obtained from Top in a rather natural way. First, any topological space $\underline{X}$ is completely described by the set of all convergent filters in $\underline{X}$ (in other words: by its convergence structure), and a map f: $\underline{X} \to \underline{Y}$ between topological spaces is continuous iff it preserves convergence, i.e. if for any filter F, which converges in $\underline{X}$, the filter generated by fF converges in $\underline{Y}$.

Second, for any pair $(\underline{X}, \underline{Y})$ of topological spaces, the set $C(\underline{X}, \underline{Y})$ can be supplied in at least two rather natural ways with a "weakest convergence structure", such that the evaluation map
$e: \underline{X} \times C(\underline{X}, \underline{Y}) \to \underline{Y}$, defined by $e(x,f) = f(x)$ preserves convergence. Just call a filter $\mathcal{F}$ on $C(\underline{X}, \underline{Y})$

(a) convergent provided, for any convergent filter $\mathcal{G}$ in $\underline{X}$, the filter generated by $e(\mathcal{G} \times \mathcal{F}) = \{e[G \times F] \; G \in \mathcal{G}, F \in \mathcal{F}\}$ converges in $\underline{Y}$

(b) Convergent to a point $f \in C(\underline{X}, \underline{Y})$ provided, for any filter $\mathcal{F}$ converging to a point x in $\underline{X}$, the filter generated by by $e(\mathcal{G} \times \mathcal{F})$ converges to the point $f(x)$ in $\underline{Y}$.

The latter convergence structure (b) on C(X,Y) has been introduced in special settings and in the realm of sequences instead of filters already by K. Weierstrass [57] under the name "gleichmäßige Konvergenz in jedem Punkt", by P. Du Bois Reymond [10] and H. Hahn [23] under the name "stetige Konvergenz", and has been shown by C. Carathéodory [12] to be the proper kind of convergence in the theory of complex functions. In full generality and in the realm of Moore-Smith sequences it was introduced by O. Frink [2o] and analyzed by R.Arens and J. Dugundji [4]. Finally in the realm of filters it has been introduced by G. Choquet [14] under the name "pseudo-convergence uniforme locale", by H. Schaefer [52] as "stetige Konvergenz", by A. Bastiani [5] as "quasi-topologie de la convergence locale", by C.H. Cook and H. R. Fischer [13] as the structure of "continuous convergence", by H. Poppe [49] as "stetige Konvergenz", and by E. Binz and H.H. Keller [9] as "Limitierung der stetigen Konvergenz". In case, $\underline{X}$ is a locally compact Hausdorff space, it is topological, in fact a filter $\mathcal{F}$ converges to f with resp. to (b) iff $\mathcal{F}$ converges to f in the compact-open topology on $C(\underline{X}, \underline{Y})$, but in general the convergence structure (b) is not induced by any topology on $C(\underline{X}, \underline{Y})$. The former convergence structure (a) has been introduced by M. Katětov [33]. At first glance it may seem that the latter structure (b) is more informative than the former (a), but it is just the other way around: (a) cannot be recovered from (b), but (b) can be recovered from (a). A filter $\mathcal{F}$ converges to f with resp. to (b) iff the filter $\{F \in \mathcal{F} \mid f \in F\}$ converges with resp. to (a).

Hence we have found two natural constructions which, when applied to topological spaces, yield structures more general than topologies. Depending on the kind of convergence structure ((a) or (b)) and the axioms, we want to impose on these structures, we obtain several cartesian closed supercategories of Top:

(1) filter-merotopic spaces (M. Katětov [33])

(2) convergence spaces (D. Kent [34], L.D. Nel [45])

(3) limit spaces = espaces quasi-topologiques (H. J. Kowalsky [35], H.R. Fischer [18], A. Bastiani [5], C.H. Cook and H.R. Fischer [13], E. Binz and H.H. Keller [9], E. Binz [8], A. Frölicher and W. Bucher [21], A. Machado [38], and others)

(4) espaces pseudo-topologiques (= L^*-Räume) (G. Choquet [14], H. Poppe [48], A. Machado [38], L.D. Nel [45])

(5) espaces épitopologiques (P. Antoine [1], A. Machado [38]).

All of these categories are closed under the construction of function spaces (b), moreover the filter-merotopic spaces (1) are closed under construction (a). If we restrict our attention to spaces satisfying a weak separation axiom, which corresponds to the R_o-axiom in topology (if a filter $\mathcal{F}$ converges and $x \in \bigcap \mathcal{F}$, then $\mathcal{F}$ converges to x), then each of the above categories contains all subsequent ones as full subcategories (W. A. Robertson [51]), and (5) is the smallest cartesian closed topological subcategory of (2) containing Top (A. Machado [38]). Moreover (1), and hence all of the mentioned categories, can--by means of a very simple construction--be fully embedded into S-Near. Hence the nearness concept provides a suitable framework for the investigation of function spaces too.

Definitions: (1) A filter-merotopic structure on X is a set of filters on X (called convergent filters) such that the following axioms hold:

(F1) If a filter $\mathcal{F}$ converges, and a filter $\mathcal{G}$ is finer than $\mathcal{F}$, then $\mathcal{G}$ converges.

(F2) For every $x \in X$, the filter $\{A \subset X | x \in A\}$ converges.

(2) A filter-merotopic space is a pair $\underline{X} = (X, \gamma)$, where X is a set and γ is a filter-merotopic structure on X.

(3) A map $f: \underline{X} \to \underline{Y}$ between filter-merotopic spaces is called continuous provided, for any convergent filter $\mathcal{F}$ in $\underline{X}$, the filter generated by $f\mathcal{F}$ converges in $\underline{Y}$.

(4) The category of filter-merotopic spaces and continuous maps is denoted by Fil.

To obtain the embedding of Fil into S-Near, observe, that, for a topological space $\underline{X} = (X,\xi)$, a subset $\mathcal{A}$ of PX has an adherence point, iff the subset $\sec \mathcal{A} = \{B \subset X \mid A \cap B \neq \emptyset$ for each $A \in \mathcal{A}\}$ of PX converges. If γ is a filter-merotopic structure on X, then $\xi = \{\mathcal{A} \subset PX \mid \sec \mathcal{A}$ contains some $\mathcal{B} \in \gamma\}$ is a semi-nearness structure on X, and this correspondence gives rise to a full embedding of Fil into S-Near (in fact, it induces an isomorphism between the category of all merotopic spaces and S-Near (see H. Herrlich [26])). By the above observation this embedding leaves topological spaces fixed. The image under this embedding of Fil has been characterized by W.A. Robertson [5o]. A non-empty collection $\mathcal{A} \subset PX$ is called a grill on X provided $\sec \mathcal{A}$ is a filter, i.e. provided (1) $\emptyset \notin \mathcal{A}$ and (2) $A \cup B \in \mathcal{A}$ iff $A \in \mathcal{A}$ or $B \in \mathcal{A}$. Then a semi-nearness space $\underline{X} = (X, \xi)$ is the image of a filter-merotopic space iff $\underline{X}$ is grill-determined, i.e. iff $\underline{X}$ satisfies the following axiom:

(N12) For any non-empty $\mathcal{A} \in \xi$ there exists a grill $\mathcal{B} \in \xi$ on X with $\mathcal{A} \subset \mathcal{B}$

Therefore the bicoreflective, full subcategory Grill of S-Near, consisting of all grill-determined semi-nearness spaces, is a cartesian closed supercategory of Top, which not only contains the above mentioned categories (2)-(5) as nicely embedded subcategories, but also the category of all contigual nearness spaces and hence espec. the category of proximity spaces (=totally bounded uniform spaces), which is not contained in any of the other categories (2)-(5). That Grill is cartesian closed and has a number of other pleasant properties, e.g. that, in Grill, arbitrary products commute with quotients, and finite products commute with direct limits, so that especially our assertions (8) and (9) are true if interpreted in Grill, can be seen with little effort directly starting from scratch, as demonstrated by H. L. Bentley, H. Herrlich and W. A. Robertson [7]. There it is also shown that Grill, compared with Top, is not too big: every grill-determined semi-nearness space is a quotient (in S-Near) of a nearness-subspace of some topological space, or--the other way around--

a semi-nearness subspace of a quotient (in S-Near) of some topological space. If the quotients (in S-Near) of topological spaces are called convergence spaces-- and that is what they are-- and the subspaces (in S-Near resp. Near) of topological spaces are called subtopological, then a semi-nearness space is topological iff it is a subtopological convergence space. Again, much more can be said, but we leave the matter here.

References

[1] P. Antoine, Étude élémentaire des categories d'ensembles structurés, Bull. Soc. Math. Belg. 18 (1966), 142-164 and 387-414.

[2] P. Antoine, Notion de compacité et quasi-topologie. Cahiers de Topol. et Géom. Diff. 14 (1973).

[3] R.F. Arens, A topology for spaces of transformations, Ann. Math. 47 (1946), 48o-495.

[4] R. Arens and J. Dugundji, Topologies for function spaces, Pacific J. Math. 1 (1951), 5-31.

[5] A. Bastiani, Applications différentiables et variétés différentiables de dimension infinie, J. Analyse Math. 13 (1964), 1-114.

[6] H.L. Bentley and H. Herrlich, Extensions of topological spaces, Proc. Memphis Conf. Topol. 1975.

[7] H.L. Bentley, H. Herrlich and W.A. Robertson, Convenient categories for topologists, Comment. Math. Univ. Carolinae.

[8] E. Binz, Bemerkungen zu limitierten Funktionenalgebren, Math. Ann. 175 (1968), 169-184.

[9] E. Binz and H. H. Keller, Funktionenräume in der Kategorie der Limesräume, Ann. Acad. Sci. Fenn. Sec. AI 383 (1966), 1-21-

[1o] P. Du Bois-Reymond, Über den Convergenzgrad der variablen Reihen und den Stetigkeitsgrad der Funktionen zweier Argumente, J.reine angew. Math. 1oo (1887), 331-358.

[11] R. Brown, Function spaces and product topologies, Quart. J.Math. Oxford (2) 15 (1964), 238-25o.

[12] C. Carathéodory, Stetige Konvergenz und normale Familien von Funktionen, Math. Ann. 1o1 (1929), 515-533.

[13] C.H. Cook and H.R. Fischer, On equicontinuity and continuous convergence, Math. Ann. 159 (1965), 94-1o4.

[14] G. Choquet, Convergences, Ann. Inst. Fourier 23 (1947/48), 57-112.

[15] C.H. Dowker, Mapping theorems for non-compact spaces, Amer.J. Math. 69 (1947), 2oo-242.

[16] C.H. Dowker, Local dimension of normal spaces, Quart. J.Math. (2) 6 (1955), 1o1-12o.

[17] E.J. Dubuc and H. Porta, Convenient categories of topological algebras, and their duality theory, J. Pure Appl. Algebra 1 (1971), 281-316.

[18] H.R. Fischer, Limesräume, Math. Ann. 137 (1959), 269-3o3.

[19] D. Franke, Funktionenalgebren in cartesisch abgeschlossenen Kategorien, Thesis, Free Univ. Berlin 1975.

[2o] O. Fink, Topology in lattices, Trans. Amer. Math. Soc. 51 (1942), 569-582.

[21] A. Frölicher and W. Bucher, Calculus in vector spaces with norm, Lecture Notes Math. (1966).

[22] L. Gillman and M. Jerison, Rings of Continuous Functions, Van Nostrand 196o.

[23] H. Hahn, Theorie der reellen Funktionen, Springer Berlin 1921.

[24] D. Harris, Structures in Topology, Mem. Am. Math. Soc. 115 (1971).

[25] H. Herrlich, A concept of nearness, Gen. Topol. Appl. 5 (1974), 191-212.

[26] H. Herrlich, On the extendibility of continuous functuons, Gen. Topol. Appl. 5 (1974), 213-215.

[27] H. Herrlich, Topological structures, Math. Centre Tract 52 (1974), 59-122.

[28] H. Herrlich, Cartesian closed topological categories, Math. Colloq. Univ. Cape Town 9 (1974), 1-16.

[29] J. R. Isbell, Zero-dimensional spaces, Tohoku Math. J. 7 (1955), 1-8.

[3o] J.R. Isbell, On finite-dimensional uniform spaces, Pacific J.Math. 9 (1956), 1o7-121.

[31] J.R. Isbell, Uniform spaces, Amer. Math. Soc. Math.Surveys 12 (1964).

[32] M. Katĕtov, A theorem on the Lebesgue dimension, Časopis Pĕst. Mat. Fys. 75 (195o), 79-87.

[33] J. Katĕtov, On continuity structures and spaces of mappings, Comment. Math. Univ. Carolinae 6 (1965), 257-278.

[34] D. Kent, Convergence functions and their related topolgies, Fund. Math. 54 (1964), 125-133.

[35] J. J. Kowalsky, Limesräume und Komplettierung, Math. Nachr. 12 (1954), 3o1-34o.

[36] W.F. LaMartin, k-groups, Thesis, Tulane Univ. 1973.

[37] A. Machado, Quasi-variétés complexes, Cahiers de Top. et Géom. Diff. 11 (197o), 231-279.

[38] A. Machado, Espaces d'Antoine et pseudo-topologies, Cahiers de Top. et Géom. Diff. 14 (1973).

[39] E. Michael, The product of a normal space and a metric space need not be normal, Bull. Amer. Math. Soc. 69 (1963), 375-376.

[4o] K. Morita, On the dimension of normal spaces I, Japan J. Math. 2o (195o), 5-36.

[41] K. Morita, On the simple extension of a space with respect to a uniformity I-IV, Proc. Japan Acad. 27 (1951) 65-72, 13o-137, 166-171, resp. 632-636.

[42] K. Morita, On the product of paracompact spaces, Prod. Japan Acad. 39 (1963), 559-563.

[43] K. Morita, On the dimension of the product of Tychonoff spaces, Gen. Topol. Appl. 3 (1973), 125-133.

[44] K. Nagami, Dimension Theory, Academic Press, 197o.

[45] L. D. Nel, Initially structured categories and cartesian closeness, Canad. J. Math.

[46] N. Noble, Products with closed projections II, Trans. Amer. Math. Soc. 16o (1971), 169-183.

[47] E.C. Nummela, K-groups generated by K-spaces, preprint.

[48] B. A. Pasynkov, On the spectral decomposition of topological spaces, Mat. Sb. 66 (1o8) (1965), 35-79.

[49] H. Poppe, Stetige Konvergenz und der Satz von Ascoli und Arzelà, Math. Nachr. 3o (1965), 87-122.

[5o] R. Pupier, Méthodes fonctorielles en topologie générale, Thesis, Univ. Lyon 1971

[51] W.A. Robertson, Convergence as a nearness concept, Thesis, Carleton Univ. 1975.

[52] H. Schaefer, Stetige Konvergenz in allgemeinen topologischen Räumen, Archiv Math. 6 (1955), 423-427.

[53] Yu. M. Smirnov, On the dimension of proximity spaces, Math. Sb. 38 (1956), 283-3o2.

[54] E. Spanier, Quasi-topologies, Duke Math. J. 3o (1963), 1-14.

[55] N. E. Steenrod, A convenient category of topological spaces, Michigan Math. J. 14 (1967), 133-152.

[56] R. M. Vogt, Convenient categories of topological spaces for homology theory, Archiv Math. 22 (1971), 545-555.

[57] K. Weierstrass, Werke 2, (188o), 2o3.

[58] O. Wyler, Convenient categories for topology, Gen. Topol. Appl. 3 (1973), 225-242.

Topological Functors Admitting Generalized Cauchy-Completions

by

Rudolf-E. Hoffmann

Mathematisches Institut der
Universität Düsseldorf

In order to describe the ideas to be investigated in this paper , let us start with two different aspects (A,B) , which - at first sight - seem to be rather unrelated . Then (C) we shall briefly develop a program of research in categorical topology , into which the above aspects fit , thus exhibiting their inner relationship . In D we give a short conspectus of contents .

Some of the assertions of the following introduction are not repeated in the text (in particular some material from A) : they are immediate consequences of results stated explicitly .

A

We are interested in those functors which are obtained by restricting a topological functor to a full reflective subcategory of its domain. The result we have does not describe these functors completely, however it shows that this class of functors is a very large class including many functors which are of specific interest *) :

A functor V: $\underline{C} \to \underline{D}$ is a "reflective restriction" of a topological functor, provided (1.8)

(1) V is faithful

(2) V is right adjoint

(3) the class Epi$\underline{C}$ induces a "factorization of cones" in $\underline{C}$

Condition (3) can be replaced by $(3_1) \wedge (3_2)$, which implies (3)

(3_1) $\underline{C}$ is co-complete (i.e. $\underline{C}$ has small colimits)

(3_2) $\underline{C}$ is co-well-powered.

(1) and (2) above are necessary conditions. The same holds for (3_1), if $\underline{D}$ is co-complete . However this is not true for (3_2), even for $\underline{D}$ = Ens (being ... and co-well-powered): H.Herrlich [He2] gives an example of an epi-reflective subcategory of Top being not co-well-powered; a result of the same type ("an unpleasant theorem") was obtained by O. Wyler for a category of limit spaces [Wy4] .

As a consequence of the above result we obtain:

If V: $\underline{C} \to \underline{D}$ satisfies (1), (2), (3) or $(3_1) \wedge (3_2)$, and - in addition:

(4) $\underline{D}$ is complete,

then also $\underline{C}$ is complete (1.8).

The proof of this result sheds some light on the influence of the duality theorem for topological functors to the problem, under which conditions a co-complete and co-well-powered category is complete (this influence was for a special case already observed in our doctoral thesis [Ho1] 4.2.7). Let $\underline{C}$ satisfy $(3_1) \wedge (3_2)$: Putting $\underline{D}$ = M-th power of Ens, i.e. assuming the existence of a generating set M of objects in $\underline{C}$ (such that the canonically induced faithful functor $\underline{C} \to \mathrm{Ens}^M$ has a left adjoint, since $\underline{C}$ is co-complete), the above result becomes a well known criterion (1.11, usually formulated in the dual way) - cf. [Sb1] 16.4.8. **) .

Condition (3) in the above result is not necessary,

even not in order to make sure that the functor constructed from V in section 1 is topological. Proving an "iff"-statement for this functor to be topological, (3) has to be replaced by a condition saying something like "V admits a 'relative factorization of cones'", i.e. one has to generalize appropriately H. Herrlich's ideas [He3] on factorization of morphisms $B \to FA$ to the situation with cones. (Similar ideas for the case of morphisms in W. Tholen [Th]). However, instead of formulating explicitly what means "V-relative factorization of cones", we prove an implicit characterization (1.5) generalizing our result in [Ho3] on factorizations of cones.

A tripleable functor with co-domain Ens is a reflective restriction of a topological functor with codomain Ens by means of a construction of M. Barr [Brr], which is different from ours:

The forgetful functor from Comp = {compact T_2-spaces and continuous maps} → Ens is tripleable; Comp is co-well-powered. Barr's construction embeds Comp into Top, whereas our construction embeds Comp into the category of totally bounded (not necessarily separated) uniform spaces and uniformly continuous maps. (Barr's construction has recently been generalized to some base categories different from Ens by S.H. Kamnitzer [Ka] .)

B Considering our characterization of topological functors in [Ho1] that the following conditions are equivalent:

(a) T: $\underline{X} \to \underline{Y}$ is topological,

$\underline{Y}$ is co-complete and co-well-powered,

(b) $\underline{X}$ is co-complete and co-well-powered

T is faithful, preserves colimits, and has a fully faithful left adjoint,

it becomes clear, that we cannot expect to rediscover such things like T_i-spaces (i=0,1,2,3) or Cauchy-complete separated uniform spaces in the general setting of <u>all</u> topological functors. However one can try to restrict this class suitably to rediscover some of these things on this general level.

In [Ho5] we have singled out those topological functors having "(weakly) separated objects" (= T_o-objects). Here we are interested in the general idea of Cauchy-completions. The reconstruction process used in solving question (A) provides the adequate framework for Chauchy-completions; our aximatization emphasizes the aspect one was originally interested in in point set topolgy: for a separated uniform space X there is a u n i q u e Cauchy-complete separated space Y (up to...) admitting a dense uniform embedding. Furthermore the nice behaviour of uniformly continuous maps with respect to dense extensions carries over to the general case. So the universal property of the Cauchy-completion is - in a sense - only a byproduct.

Most of the work, of course, has to be done in order to verify the examples (cf. sections 3,4).

C The interplay between the ideas of (A) and (B) fits into a general program we had a presentiment of in the introduction of our Habilitationsschrift [Ho5], which follows the same program as the present paper.

1) In order to find the "reflection" of a property, say, of some topological spaces, in the general setting of topological functors, one should try to derive a <u>categorically formulated</u> result on the behaviour of these spaces with

respect to all spaces.

2) Then one should axiomatize those functors obtained by restricting topological functors to those classes to be described in step 1, i.e. g e n e r a l i z e the concept of topological functor.

3) Now one needs a procedure reconstructing topological functors from those functors described in step 2 .

4) In order to insure that the examples one had in mind when starting the investigation fit into the framework developped in step 3, one has to prove some results on the "internal structure" of those topological functors obtained by the reconstruction process in step 3.

5) Furthermore one is interested in the question how to modify a given topological functor into a topological functor obtained by the procedure of step 3. For "(generalized) Cauchy-completions" we do not have any answer to this, whereas in [Ho5] we obtained a rather good answer for (E,M)-universally topological functors with co-domain <u>Ens</u>.

We plan to elaborate a more detailed exposition of this program [Ho8] .

D In section 0 we briefly give the basic definition of the concept of topological functor. In section 1 we establish the fundamental construction (1.4) and characterize internally those topological functors obtained by this construction (1.12). As "applications" we obtain the characterization of "factorizations of cones" in [Ho3] (1.6) and the well-known theorem of completeness of co-complete, co-well-powered categories with a generating set (1.11). Furthermore we prove an extension theorem (1.14). In section 2 we investigate the relationship between those topological functors characterized in section 1 and (E,M)-universally topolgical functors introduced in [Ho5] . By the criterion thus obtained

(2.4) it is easy to do the verification for the examples to be discussed in section 3 (namely by means of results of [Ho5] and by some lemmata being "topological" in charactor - some of them known, others until now unknown).

The most significant of these examples should be mentioned here:

The category of Cauchy-complete separated uniform spaces "generates" (by the above mentioned procedure) the category Unif of uniform spaces and uniformly continuous maps (3.2).

The category of sober spaces "generates" the category Top (3.1).

The category $^q\underline{Ban}_{\mathbb{K}}$ of (quasi-)Banach-$\mathbb{K}$-spaces "generates" the category $^q\underline{n\text{-}Vec}_{\mathbb{K}}$ of quasi-normed $\mathbb{K}$-vector spaces ($\mathbb{K}$ denotes a subfield of the field of complex numbers) (3.5).

In section 4 we discuss some examples which do not fit into the above framework. In particular, we use an ad hoc-version of Cauchy-complete (separated) object for topological functors in order to verify (by an explicit characterization of these objects) that the topological functors in question do not "admit (generalized) Cauchy-completions". We briefly comment the problem to find for a given topological functor a "best approximation admitting (generalized) Cauchy-completions" (which is for from being solved).

---

*) Cf. the footnote to 1.13

**) Furthermore one obtains the co-completeness theorem proved by Herrlich-Strecker [HeS] and Pumplün-Tholen [P T] :

Let $V:\underline{C} \to \underline{D}$ be a faithful right adjoint functor , let $\underline{C}$ be complete , wellpowered and co-wellpowered (this implies : Epi$\underline{C}$ induces a factorization of cones in $\underline{C}$) . If $\underline{D}$ is co-complete , then so is $\underline{C}$ (being a reflective full subcategory of a "topological category over $\underline{D}$ ") - cf. 1.8 .

§ 0

0.1 A cone $(C,\lambda:C_\Sigma \to T)$ in a category $\underline{C}$ is said to be a <u>V-co-identifying</u> (= "<u>V-co-idt.</u>") <u>cone</u> with respect to a functor $V:\underline{C} \to \underline{D}$, iff whenever $(V*\lambda)\, u_\Sigma = V*\eta$ for some cone $(X,\eta:X_\Sigma \to T)$ and some morphism $u:VX \to VC$, then there is a morphism $h:X \to C$ in $\underline{C}$ being unique with respect to the following properties

(1) $Vh = u$

(2) $\eta = \lambda h_\Sigma$.

$\emptyset$-indexed cones $(C,\emptyset)$ in $\underline{C}$ are interpreted as objects of $\underline{C}$: C is said to be a <u>V-co-discrete</u> <u>object</u> , iff $(C,\emptyset)$ is V-co-idt. *) .

A V-co-identifying cone indexed by a graph with exactly one vertex and without arrows is identified with a $\underline{C}$-morphism, which is said to be a <u>V-co-identifying</u> <u>morphism</u>.

A diagram $T:\Sigma \to \underline{C}$ and a cone $(D,\mu:D_\Sigma \to VT)$ form a "V-datum" $(T;D,\mu)$. $(C, \lambda:C_\Sigma \to T;h)$ is called a V-co-idt. lift of $(T;D,\mu)$, iff (C,λ) is V-co-idt. and $h:VC \to D$ is an isomorphism with $\mu h_\Sigma = V*\lambda$. A V-co-idt. lift is "unique up to an isomorphism" (analogous to limits).

0.2 A functor $V:\underline{C} \to \underline{D}$ is said to be a <u>topological</u> <u>functor</u>, provided that

(1) every V-datum $(T;D,\mu:D_\Sigma \to T)$ with Σ $\underline{U}$-small and discrete (i.e. without arrows) has a V-co-idt. lift;

(2) V satisfies a "smallness condition" (for functors): whenever $M \subseteq \mathrm{Ob}\underline{C}$ consists of non-isomorphic objects, which are mapped by V into objects isomorphic to some $Y \in \mathrm{Ob}\underline{D}$, then M is $\underline{U}$-small .

(1) imitates "existence of the 'initial' topology" in <u>Top</u> - cf. Bourbaki

Topological functors abound in point set topology and other branches of mathematics concerned with "elementary structures": E.g. the usual forgetful functors <u>Top</u> $\to$ <u>Ens</u>,

---

***) i.e. iff for every $X \in \mathrm{Ob}\underline{C}$ the mapping $[X,C] \to [VX,VC]$ induced by V is bijective .**

Unif → Ens, g-Alg → Ens, Preord → Ens are topological. Of course, there are also topological functors with a base category different from Ens : E.g. TopGr (topological groups and continuous homomorphisms) → Gr (groups and homomorphisms), etc. . However, monadic functors are not topological except for the identity([Ho1] 6.8.1).

A lot of examples are to be found in [Ho1, Ho5, Ro, Wy1] etc., cf. also sections 3,4.

0.3 Topological functors are automatically faithful [Ho1]. If $V:\underline{C} \to \underline{D}$ is topological, then $V^{op}:\underline{C}^{op} \to \underline{D}^{op}$ is too (duality theorem; [An , Ro]), i.e. V admits identifying lifts.

0.4 A topological functor $V:\underline{C} \to \underline{D}$ has a fully faithful right adjoint assigning to $D \in Ob\underline{D}$ a V-co-discrete object "over" it; because of the duality (0.3) V has also a fully faithful left adjoint.

Let T be a diagram in $\underline{C}$, and let (D,μ) be a limit of VT in $\underline{D}$: If $(C,\lambda;h)$ is a V-co-idt. lift of $(T;D,\mu)$, then (C,λ) is a limit of T. An analogous statement holds for colimits.

0.5 Let $F:\underline{A} \to \underline{B}$ be a topological functor. A full subcategory $\underline{X}$ of $\underline{A}$ is called an F-co-identifying subcategory (= a top subcategory [Wy1]), provided that whenever $(C,\lambda:C_{\Sigma} \to T)$ with Σ $\underline{U}$-small and discrete and T has values only in $\underline{X}$, then $C \in Ob\underline{X}$ (hence $\underline{X}$ is closed under isomorphisms).

Intersections of F-co-idt. subcategories are F-co-idt.; the F-co-idt. hull of $K \subseteq Ob\underline{A}$ consists of those objects A of $\underline{A}$ admitting an F-co-idt. cone $(A,\psi:A_{\Sigma} \to T)$ with $Te \in K$ for every $e \in \Sigma$ - where Σ is $\underline{U}$-small and discrete : this is the smallest F-co-idt. subcategory of $\underline{A}$ containing K . If $\underline{B}$ = Ens, then F-co-idt. = bireflective and F-idt. = bi-coreflective.

Of course, the restriction of F to $\underline{X}$ is a topological functor into $\underline{B}$.

A detailed investigation of the above concepts can be found in our doctoral thesis [Ho1] .

§ 1

1.0 We are interested in those functors which are obtained by restricting a topological functor to a full reflective subcategory of its domain. We cannot solve the problem of describing these functors in full generality*). However, there is a procedure reconstructing topological functors from a fairly wide class of functors of the above mentioned type. (A special case of this construction can be found in [Kn1].) So we have changed the standpoint of our investigation:

We look for conditions on a functor V being necessary and sufficient, such that the functor U obtained by this c a n o n i c a l reconstruction is topological.

From the well known properties of topological functors the following is immediate

1.1 Lemma: If V is the restriction of a topological functor to a full reflective subcategory, then

1. V is faithful

2. V has a left adjcint.

1.2 Now we sketch the construction to be investigated:

Let V: $\underline{C} \to \underline{D}$ be faithful

(1) Objects of $\underline{A}$ are pairs (f: D $\to$ VC,C), such that for every pair g, h: C $\to$ X in $\underline{C}$ V(g)f = V(h)f implies g = h. These pairs (f,C) are called "V-epimorphisms".

*) cf. 1.13

(2) Morphisms $(f,C) \to (f'\colon D' \to VC', C')$ in $\underline{A}$ are pairs $(a\colon D \to D',\ b\colon C \to C')$ with $f'a = V(b)f$. Composition in $\underline{A}$ is defined componentwise.

(Of course, one has to make the hom-sets disjoint from each other.)

(3) $U\colon \underline{A} \to \underline{D}$ maps $(f\colon D \to VC, C)$ to D and (a,b) to a.

(4) The embedding $F\colon \underline{C} \to \underline{A}$ is given by $C \longmapsto (id_{VC}, C)$, $(b\colon C \to C') \longmapsto (Vb, b)$, since V is faithful; F is full.

(5) The universal morphism of the adjunction $\underline{C} \to \underline{A}$ is given by the commutative square

$$\begin{array}{ccc} D & \xrightarrow{\ f\ } & VC \\ {\scriptstyle f}\downarrow & & \| \\ VC & = & VC \end{array}$$

1.3 Lemma: Let $V\colon \underline{C} \to \underline{D}$ be a faithful functor, let $L\colon \underline{D} \to \underline{C}$ be a left adjoint of V, and let $\eta\colon id_{\underline{C}} \to VL$ be a unit of the adjunction:

$(f\colon D \to VC, C)$ is a V-epimorphism, i.e. an object of $\underline{A}$ (as in 1.2), iff the morphism $h\colon LD \to C$ satisfying $V(h)\eta_D = f$ is an epimorphism.

It turns out that the question, whether U (in 1.2) is topological or not, is strongly related to the problem, whether Epi$\underline{C}$, the class of epimorphisms in $\underline{C}$, induces a factorization of cones in $\underline{C}$ (cf [Ho3], see below). This suggests the following modification (generalization) of the above construction - replacing Epi$\underline{C}$ by some class J.

1.4 Let $V\colon \underline{C} \to \underline{D}$ be a functor, let $L\colon \underline{D} \to \underline{C}$ be a left adjoint of V, and let $\eta\colon id_{\underline{D}} \to VL$ and $\varepsilon\colon LV \to id_{\underline{C}}$ be a unit and, resp., co-unit of the adjunction satisfying

$$(V * \varepsilon)(\eta * V) = id_V$$

$$(\varepsilon * L)(L * \eta) = id_L\ .$$

Let J be a class of epimorphisms in $\underline{C}$ with

(i) $\text{Iso}\underline{C} \subseteq J \subseteq \text{Epi}\underline{C}$,

(ii) J is compositive,

(iii) $\varepsilon_C \in J$ for every $C \in \text{Ob}\underline{C}$

Since J consists of epimorphisms, (iii) implies that V is faithful (cf. [Sb1] ↑6.5.3)

$$(f: D \to VC, C) \text{ with } f \in \text{Mor}\underline{D},\ C \in \text{Ob}\ \underline{C}$$

is an object of $\underline{A}_J$, iff the morphism $h: LD \to C$ in $\underline{C}$ with $V(h)\eta_D = f$ belongs to J.

A morphism $(f: D \to VC, C) \to (f': D' \to VC', C')$ in $\underline{A}_J$ is a pair (a,b) with $a: D \to D'$ in $\underline{D}$, $b: C \to C'$ in $\underline{C}$ satisfying $f'a = V(b)f$. Composition is defined componentwise. (Hom-sets have to be made disjoint to each other.)

The full embedding $F_J: \underline{C} \to \underline{A}_J$ is given by $C \longmapsto (id_{VC}, C)$ (since $\varepsilon_C \in J$ and $V(\varepsilon_C)\eta_{VC} = id_{VC}$, (id_{VC}, C) is admissible) and by $b \longmapsto (Vb, b)$.

The functor $U_J: \underline{A}_J \to \underline{D}$ is defined by $(f: D \to VC, C) \longmapsto D$, $(a,b) \longmapsto a$.

The universal morphism of the adjunction $\underline{C} \longrightarrow \underline{A}_J$ is given by the commutative square

$$\begin{array}{ccc} D & \xrightarrow{\quad f \quad} & VC \\ {\scriptstyle f}\downarrow & & \| \\ VC & = = = & VC \end{array}$$

1.5 <u>Theorem:</u>

Let $U: \underline{C} \to \underline{D}$, L, J, $\underline{A}_J$, U_J, F_J as above. Then the following conditions (a), (b), (c) are equivalent:

(a) (i) U_J is topological;

(ii) "$pq \in J$ with $q: L(D) \to C$ in J and $D \in \text{Ob}\ \underline{D}$" implies $p \in J$.

(b) (i) For every morphism $f: D \to X$ in $\underline{D}$ and every morphism $k: LD \to Y$ in J there is a pushout in $\underline{C}$

$$\begin{array}{ccc} LD & \xrightarrow{Lf} & LX \\ k\downarrow & & \downarrow l \\ Y & \xrightarrow[g]{} & Q \end{array} \qquad \text{with } l \in J.$$

(ii) For $D \in Ob\underline{D}$ and $k_i\colon LD \to X_i$ in J $(i \in I; I \in \underline{U}; I \neq \emptyset)$ there is a multiple pushout $l_i\colon X_i \to Q$ in $\underline{C}$ with $l_i \in J$.

(iii) The J-quotients of every LD form a $\underline{U}$-set.

(c) Let P_J denote those cones $(A, \{m_i\colon A \to A_i\}_{i \in I})$ in $\underline{C}$ with $I \in \underline{U}$, which do not factor over a non-isomorphic J-morphism $x\colon A \to X$:

(i) Every cone $(LD, \{f_i\colon LD \to X_i\}_I)$ with $I \in \underline{U}$ factors $f_i = p_i k$ with $k\colon LD \to Q$ in J and $(Q, \{p_i\}_I)$ in P_J.

(ii) For every commutative diagram in $\underline{C}$

$$\begin{array}{ccc} LD & \xrightarrow{k} & X \\ Lg\downarrow & & \\ LD' & & \downarrow f_i \\ j\downarrow & & \\ Q & \xrightarrow[p_i]{} & Y_i \end{array}$$

with $g\colon D \to D'$ in $\underline{D}$, $k \in J$, $j \in J$, $(Q, \{p_i\colon Q \to Y_i\}_{i \in I}) \in P_J$ there is a morphism $h\colon X \to Q$ with $hk = jL(g)$.

(iii) The J-quotients of every LD form a $\underline{U}$-set.

(c) (ii) is some kind of "diagonal condition".

<u>Proof:</u> The fibre of U_J at the object $D \in Ob\underline{D}$ consists of pairs $(f\colon D \to VX, X)$, such that the induced morphisms $LD \to X$ in $\underline{C}$ belong to J. This establishes a bijective correspondence between a skeleton of the U_J-fibre of D and the J-quotients of LD in $\underline{C}$.

(a) ⟹ (b):

Let $((Vk)\eta_D, Y) \in \mathrm{Ob}\,\underline{A}_J$ and let $f: D \to X$ be a morphism in $\underline{D}$ with domain $D = U_J((Vk)\eta_D, Y)$. Since U_J lifts isomorphisms and U_J is supposed to be topological, there is a U_J-identifying morphism in $\underline{A}_J$ with domain $((Vk)\eta_D, Y)$, which is taken by U_J into f: this induces a commutative square

$$\begin{array}{ccc} LD & \xrightarrow{Lf} & LX \\ {\scriptstyle k}\downarrow & & \downarrow{\scriptstyle l} \\ Y & \xrightarrow[g]{} & Q \end{array}$$

in $\underline{C}$ with $l \in J$.

It is easily shown by the universal property of U_J-identifying morphisms that this square is a pushout. Instead of giving an explicit proof, we draw a picture (lying in $\underline{D}$)

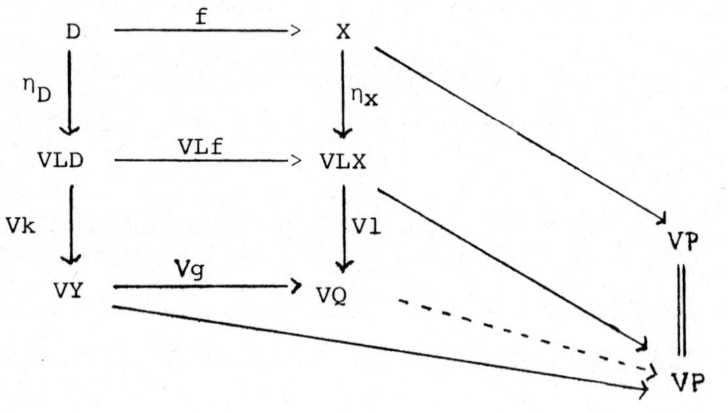

(id_{VP}, P) has to be considered as an object of $\underline{A}_J$.

Similarly, a family $k_i: LD \to X_i$ of J-quotients ($i \in I$; $I \in \underline{U}$; $I \neq \emptyset$) is interpreted as a family of objects $(V(k_i)\eta_D, X_i)$ of $\underline{A}_J$ in the U_J-fibre of D.

Since U_J lifts isomorphisms and U_J is topological, there is a U_J-identifying cone $\{\phi_i\}_{i \in I}$ in $\underline{A}_J$ with $((Vk_i)\eta_D, X_i)$ = domain of ϕ_i and $U_J(\phi_i) = id_D$.

This cone obviously induces a cone $\{l_i: X_i \to Q\}_{i \in I}$ in $\underline{C}$ with $l_i k_i$ independent from $i \in I$ and $l_i k_i \in J$, since if represents an object of $\underline{A}_J$. Because of (a) (ii) we have $l_i \in J$. It is shown by the universal property of the U_J-

identifying cone in $\underline{A}_J$ that $\{l_i: X_i \to Q\}_{i \in I}$ is a multiple pushout of $\{k_i: LD \to X_i\}_{i \in I}$.

(b) $\Longrightarrow$ (c):

(i) Let $(LD, \{f_i: LD \to X_i\}_{i \in I})$ with $I \in \underline{U}$ be a cone in $\underline{C}$.

Let $\{k_n: LD \to Q_n\}_{n \in N}$ denote the family of all J-quotients of LD such that every f_i factors over k_n.

Then f_i factors over the co-intersection $k: LD \to Q$ of the k_n with $k = h_n k_n$, i.e. $f_i = p_i k$ for some $p_i: Q \to X_i$.

Now let $p_i = m_i e$ with $e \in J$, then $ek = k_n \in J$ (w.l.o.g.) for some $n \in N$. Consequently $h_n ek = h_n k_n = k$, hence

$h_n e = id_Q$, i.e. e is an epic coretraction, i.e. an isomorphism

(ii) Now let $g: D \to D'$ in $\underline{D}$, and let $j: LD' \to Q$ in J, $k \in J$, and let

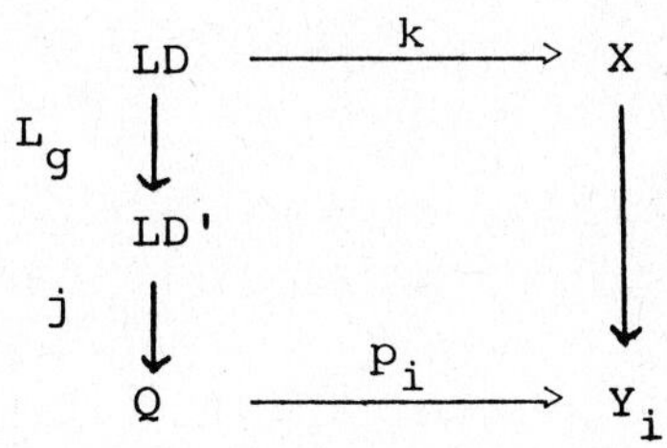

commute

with $D \in \mathrm{Ob}\,\underline{D}$, $i \in I$ $(I \in \underline{U})$, $(LD', \{p_i\}_I) \in \underline{P}_J$.

Applying (successively) (b) (i) and (b) (ii), we find pushouts for (Lg,k) and (j,k') and induced morphisms

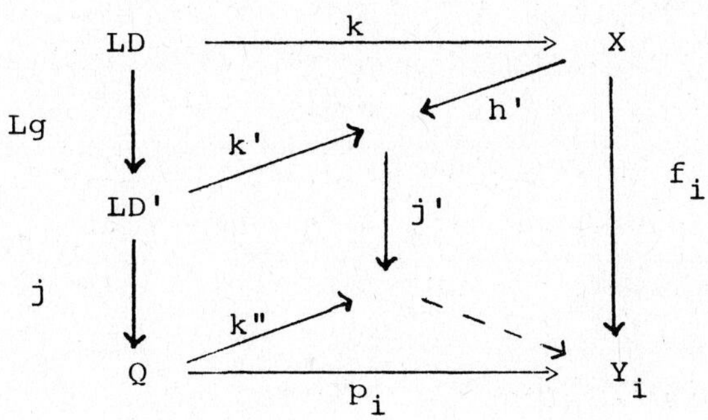

with k', k" $\in$ J.

Since $(Q, \{p_i\}_I) \in \underline{P}_J$, k" is an isomorphism;

hence $h := k''^{-1} j'h'$ satifies $hk = jL(g)$.

(c) $\Longrightarrow$ (a):

(i) We verify the existence of U_J-co-identifying lifts: Instead of giving a detailed comment, we draw a picture. Let $D_n \in \text{Ob}\,\underline{D}$, $k_n: LD_n \to X_n$ in J, $g_n: D \to D_n$ in $\underline{D}$, $f: E \to D$ in $\underline{D}$, $k: LE \to X$ in J, and let $f_n: X \to X_n$ with $V(f_n)V(k)\eta_E = V(k_n)\eta_{D_n} g_n f$ for every $n \in N$.

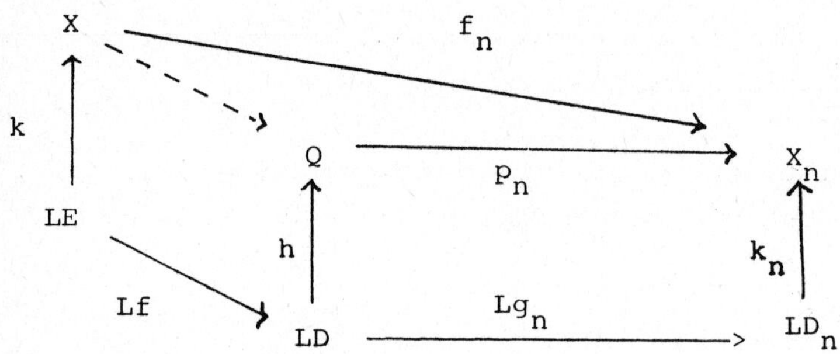

Let $k_n Lg_n = p_n h$ with $h \in J$, $(Q, \{p_n\}_N) \in P_J$. By (c) (ii) we can fill in the diagonal - proving the universal property of $\{(g_n, p_n): (D, h: LD \to Q) \to (D_n, k_n: LD_n \to X_n)\}_{n \in N}$.

(The same proof applies to the case $N = \emptyset$; only the notation has to be changed slightly.)

(ii) Let $D \in \text{Ob}\,\underline{D}$, $q: L(D) \to C$ in J and $pq \in J$, then $p = sj$ with $j \in J$ and $s \in P_J$. Because of $pq \in J$ and $jq \in J$, we can fill in the diagonal h in

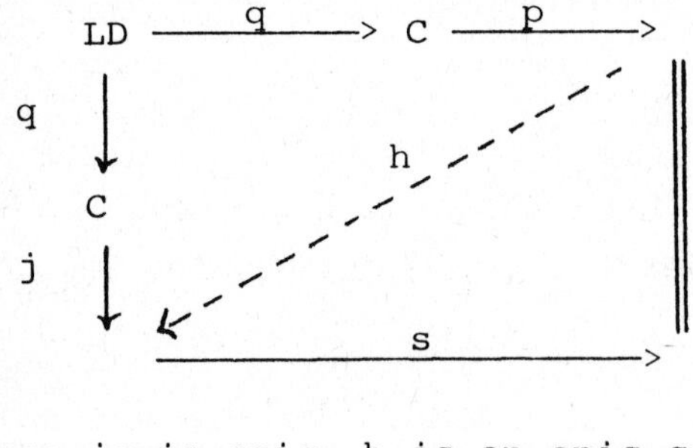

Since jq is epic, h is an epic coretraction, i.e. an isomorphism, hence so is s.
This completes the proof.

1.6 Remark:

We want to apply 1.5 to $\underline{C} = \underline{D}$, $V = id_{\underline{D}}$:

(1) 1.5 (a) (i) states that the functor "domain" from the category of J-morphisms to $\underline{C}$ is topological.

(2) 1.5 (c) gives a suitable formulation of the concept of "factorization of cones" (J, P_J) ([Ho3,4]) slightly modifying the definitions given in [He2,Mar,Ca2; furthermore cf. Ti].

(3) 1.5 (b) becomes the characterization of factorizations (E,M) of cones obtained in [Ho3]:

Let E be a compositive class of epimorphisms in $\underline{C}$ with Iso $\underline{C} \subseteq E$. E induces a factorization of cones, iff

(i) every "span" $\begin{array}{ccc} A & \longrightarrow & C \\ {\scriptstyle e}\downarrow & & \\ B & & \end{array}$ with $e \in E$

has a pushout $\begin{array}{ccc} A & \longrightarrow & C \\ {\scriptstyle e}\downarrow & & \downarrow{\scriptstyle f} \\ B & \longrightarrow & X \end{array}$ with $f \in E$,

(ii) for any collection $\{e_i: A \to A_i\}_{i \in I}$ of E-morphisms with $I \in \underline{U}$, $I \neq \emptyset$ there is a multiple pushout $\{g_i: A_i \to B\}_{i \in I}$ with $g_i \in E$,

(iii) $\underline{C}$ is E-co-well-powered.

In the following we return to the case $J = Epi\underline{C}$ (but we shall apply 1.6). We refer to U in 1.1 as the canonical extension of V.

1.7 Lemma:

Let C, D, V, L, η, U as in 1.1, 1.2, 1.3: If U is topological, then C is co-well-powered.

Proof:

Let C ∈ Ob C and let e: C → X be epic in C, then there is a morphism j: LVC → X with $V(j)\eta_{VC} = V(e)$. Since V is faithful and e is epic, j is epic. Thus an injection from the quotients of C to the quotients of LVC is defined . Now 1.5(b)(iii) applies .

1.8 Corollary :

Let V:C → D be a faithful right adjoint functor . If the class EpiC of epimorphisms in C induces a factorization of cones in C , then the above constructed "c a n o n i c a l e x t e n s i o n" U is topological , and V is the restriction of U to a full reflective subcategory .

If D is complete (resp. co-complete , resp. well-powered) , then so is C .

Proof : By 1.6(3) and 1.5(b) U is topological . If D is complete , resp. co-complete , resp. well-powered , then so is A (since it is the domain of the topological functor U) . Since C is a full reflective subcategory of A , the same holds for C .

That EpiC induces a factorization of cones in C is guaranteed by either of the following conditions (a) and (b):

(a) C is co-complete and co-well-powered .

(b) C is complete , well-powered , and co-wellpowered .

Cf. [Ho3] , (a) is immediate from 1.6(3) .

By virtue of 1.7 now we have

1.9 Theorem :

Let V:C → D be a faithful right adjoint functor , and let U denote its canonical extension :

(a) Let D be co-complete : U is topological , iff C is co-complete and co-well-powered .

(b) Let $\underline{D}$ be complete and well-powered : U is topological , iff $\underline{C}$ is complete , well-powered , and co-well-powered .

1.1o Remark :

From 1.7 it is immediate that the "canonical extension" does not solve the problem of reconstructing a topological functor from a reflective restriction of some topological functor in full generality (cf. 1.13) :

Let $\underline{X}$ be not co-well-powered (e.g. the well-ordered class of ordinals , which is - as a category - co-complete), then $id_{\underline{X}}$ is topological , in particular it is a reflective restriction of a topological functor .

The result 1.8 above shows that the interrelationship between "completeness" and "co-completeness" of categories is somehow related to the"duality theorem" (o.3) . In particular , the techniques developped here imply the following

1.11 Corollary (well known, [Sb1] 16.4.8):

If $\underline{C}$ is co-complete, co-well-powered, and if there is a generating $\underline{U}$-set $\underline{S}$ in $\underline{C}$, then $\underline{C}$ is complete and well-powered.

Proof: Let $\underline{S}$ denote simultaneously the set and the associated discrete category. The collection

$\{Hom(C,-): \underline{C} \to \underline{Ens}\}_{C \in \underline{S}}$ of functors induces a functor $G: \underline{C} \to Ens^{\underline{S}}$ (being the $\underline{S}$-th power of $\underline{Ens}$), which is faithful and admits a left adjoint $L: \underline{Ens}^{\underline{S}} \to \underline{C}$, L takes $(M^C)_{C \in \underline{S}}$ to $\coprod_{C \in \underline{S}} (\coprod_{M^C} C)$. By 1.9 there is a topological functor $U: \underline{A} \to \underline{Ens}^{\underline{S}}$ and a full reflective embedding $F: \underline{C} \to \underline{A}$ with G = UF. Since $\underline{Ens}^{\underline{S}}$ is complete and wellpowered, the domain $\underline{A}$ of U is too, and so is the full reflective subcategory $\underline{C}$ of $\underline{A}$.

Finally we want to give an intrinsic characterization of those topological functors U obtained by the above procedure (with J = Epi$\underline{C}$).

1.12 Theorem:

A topological functor $T: \underline{X} \to \underline{D}$ is "induced" by a functor $V: \underline{C} \to \underline{D}$ in the sense of 1.2, i.e. there is an equivalence $S: \underline{X} \to \underline{A}$ with $US = T$, iff there is a full isomorphism - closed reflective subcategory $\underline{C}$ of $\underline{X}$, such that (1) and (2) are satisfied:

(1) For every $X \in Ob\underline{X}$ the universal morphisms η_X of the adjunction $\underline{C} \to \underline{X}$ are T-co-identifying.

(2) If $f: X \to C$ in $\underline{X}$ with $C \in Ob\underline{C}$ is a T-co-identifying morphism, such that $gf = hf$ implies $g = h$ provided that the co-domain of g, h is in $\underline{C}$ (i.e. f is 'epic with respect' to $\underline{C}$), then the morphism p with $p\eta_X = f$ is an isomorphism.

Proof of 1.12

(a) Let $T = US$. W.l.o.g. we can assume $S = id_{\underline{A}}$.

(1) The universal morphism η_X is given by

$$\begin{array}{ccc} D & \xrightarrow{\ h\ } & VB \\ {\scriptstyle f_i}\downarrow & & \downarrow\| \\ VB & \xrightarrow{\ =\ } & VB \end{array}$$

provided that $X = (h: D \to VB, B)$. Consequently η_X is U_J-co-identifying by the explicit description of U_J-co-identifying lifts in the proof of 1.5 (c) $\Longrightarrow$ (a).

(2) That $f: X \to C$ in $\underline{X}$ with $C \in Ob\,\underline{C}$ is a T-co-identifying morphism means that in

$$\begin{array}{ccc} D & \xrightarrow{\ g\ } & VC \\ {\scriptstyle h}\downarrow & & \| \\ VB & \xrightarrow[\ Vp\ ]{} & VC \end{array}$$

with $X = (h: D \to VB, B)$ and with $f = (g, p)$ p is in P_J, i.e. does not factor over a non-isomorphic epimorphism.

The condition in 1.12(2) concerning f says that p is epic in $\underline{C}$, hence p is an isomorphism and we have $p\eta_X = f$.

(b) Now let $T: \underline{X} \to \underline{D}$ be a topological functor with $\underline{C} \to \underline{X}$ satisfying the conditions in 1.12. Let $V := T|\underline{C}$, then V satisfies all of the conditions required in 1.5. $S: \underline{X} \to \underline{A}$ takes $X \in \mathrm{Ob}\, \underline{X}$ to $(T\eta_X, B)$, where B denotes the codomain of η_X; S takes $f: X \to X'$ to (Tf, g) with $g: B \to B'$ satisfying $g\eta_X = \eta_{X'}f$. W.l.o.g. $\eta_C = id_C$ for $C \in \mathrm{Ob}\, \underline{C}$, hence $US = T$. Since T is faithful, S is too.

Let

$$\begin{array}{ccc} TX & \xrightarrow{\;h\;} & TX' \\ {\scriptstyle T\eta_X}\downarrow & & \downarrow{\scriptstyle T\eta_{X'}} \\ TB & \xrightarrow[\;Tg\;]{} & TB' \end{array}$$

commute;

Since $\eta_{X'}$ is T-co-identifying, there is a morphism $f: X \to X'$ with $Tf = k$, hence $S(f) = (h,g)$, i.e. S is full.

Let $(h: D \to VB, B)$ be an object of $\underline{A}$. There is a T-co-identifying morphism $u: X \to B$ and an isomorphism $i: D \to TX$ with $h = T(u)i$. Becauce of condition 1.12 (2) there is an isomorphism p with $p\eta_X = u$. By definition of S (i, p^{-1}) becomes an isomorphism in $\underline{A}$ from $(h: D \to VB, B)$ to $S(X)$.

1.13 Problems:

(a) Let $T: \underline{X} \to \underline{Y}$ be a topological functor, and let $\underline{K}$ be a full reflective subcategory of $\underline{X}$. We do not know whether there is a suitable J in $\underline{K}$ satisfying the conditions in 1.5 for $V := T|\underline{K}$.

(b) There seems to be a "natural" idea of "semi-topological" functor . Let $T: \underline{X} \to \underline{Y}$ be topological, and let $\underline{K}$ be a full reflective subcategory of $\underline{X}$, then the functor $V := T|\underline{K}: \underline{K} \to \underline{Y}$ satisfies (1) and (2):

(1) The induced functors $V/S: \underline{K}/S \to \underline{Y}/TS$ are right adjoint for every diagram S in $\underline{K}$ with a $\underline{U}$-small discrete domain.

(2) V satisfies the "smallness condition".

We do not know, whether functors V satisfying (1) and (2) are always obtained by a "reflective restriction" of some topological functor - cf . [Ho2] *) .

A uniformly continuous map from a dense subspace of a uniform space into a Cauchy-complete separated uniform space is known to have a unique uniformly continuous extension. This result can be generalized (with respect to 3.2) slightly in the framework of those topological functors described in 1.12 (furthermore cf. H.Herrlich[He9]).

An object K of a category $\underline{K}$ is said to be "injective with respect to a class Q of morphisms in $\underline{K}$", iff Hom(q,K) is surjective for every $q \in Q$, i.e. every span

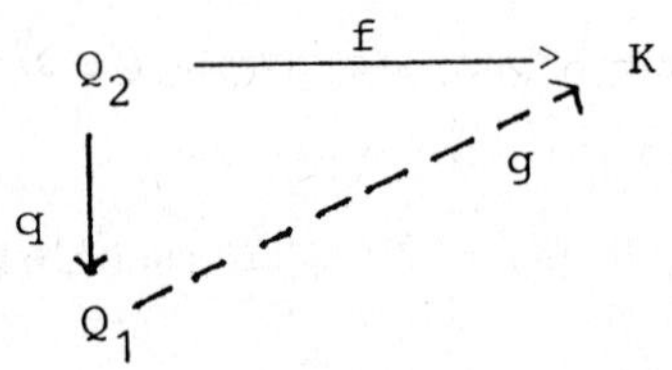

is made commutative by some $g: Q_1 \to K$: "f is extendible over q"

1.14 Theorem ("Extendibility"):

Let $U: \underline{A} \to \underline{D}$ be a topological functor and let $\underline{C}$ be a full reflective subcategory of $\underline{A}$ satisfying the conditions in 1.12. Let η denote the unit of the adjunction $\underline{C} \to \underline{A}$.
If $q: X \to Y$ is a U-co-idt. morphism being 'epic with respect to $\underline{C}$' in $\underline{A}$, then every morphism $f: X \to C$ with $C \in Ob\underline{C}$ is extendible over q, i.e. $gq = f$ for some $g: Y \to C$ (which is uniquely determined):
$C \in Ob\underline{C}$ is injective with respect to q (which is not supposed to be monic!).

*) We are going to give a solution of these problems in [Ho9] , which , however , is somehow "formal" in character , since it "leaves" the universe $\underline{U}$ (!) .

Proof:

$\eta_Y q$ is a U-co-idt. morphism being ' epic with respect to $\underline{C}$ the co-domain of which is in $\underline{C}$, hence by 1.12 (2) there is an isomorphism p with $p\eta_X = \eta_Y q$. Since there is a morphism h with $h\eta_X = f$, we put $g := hp^{-1}\eta_Y$, hence we have $gq = hp^{-1}\eta_Y q =$ $= h\eta_X = f$. Since q is ' $\underline{C}$-epic ', g is unique.

1.15 Proposition :

Let $U:\underline{A} \to \underline{D}$ be a topological functor , and let $\underline{C}$ be a full reflective subcategory of $\underline{A}$ with unit η , such that all of the conditions in 1.12 are satisfied . Let Q denote the class of those U-co-idt. morphisms in $\underline{A}$ which are 'epic with respect to $\underline{C}$ ' .

(1) Let $g:X \to Y$ in Mor$\underline{A}$: If every $f:X \to C$ in Mor$\underline{A}$ with $C \in$ Ob$\underline{C}$ is uniquely extendible over g , then $g \in Q$.

(2) If $A \in$ Ob$\underline{A}$ is Q-injective , then A is (isomorphic to) an object of $\underline{C}$.

Proof :

(1) Since the "extensions" are unique , g is $\underline{C}$-epic . Since $\eta_X : X \to C$ is extendible over g , there is a morphism $h:Y \to C$ with $hg = \eta_X$. Because of the universal property of η_X , resp. $\eta_Y : Y \to C'$ there are morphisms $a:C \to C'$ and $b:C' \to C$ with $a\eta_X = \eta_Y g$ and $b\eta_Y = h$. $ba\eta_X = b\eta_Y g = hg = \eta_X$ implies $ba = id_C$. Since $\eta_Y g$ is $\underline{C}$-epic , $ab\eta_Y g =$ $= ahg = a\eta_X = \eta_Y g$ implies $ab = id_{C'}$. Consequently g is U-co-idt. , since it is the first factor in $\eta_Y g = a\eta_X$: a (being an isomorphism) and η_X are U-co-idt. , hence $\eta_Y g$ is , and - since U is faithful - so is g .

(2) Since $\eta_A \in Q$, there is a morphism $h:C \to A$ rendering commutative

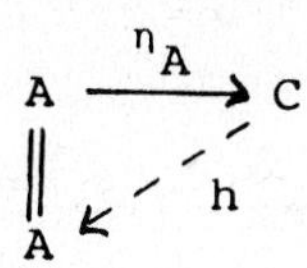

We have $\eta_A h \eta_A = \eta_A$; this implies $\eta_A h = id_C$; hence η_A is an isomorphism .

§ 2

In [He2] H. Herrlich has introduced the concept of (E,M)-topological functor T: $\underline{X} \to \underline{Y}$, where (E,M) denotes a factorization of cones in $\underline{Y}$; here we use our modification of the concept of "factorization" proposed in [Ho3], cf. 1.6; hence we have to modify the concept of (E,M)-topological functor accordingly (more detailed comments on this problem can be found in [Ho2]).

T: $\underline{X} \to \underline{Y}$ is (E,M)-topological, iff

1. every T-datum $(S;D,\mu)$ with $(D,\mu) \in M$ (i.e. the domain of S is $\underline{U}$-small and didcrete, and (D,μ) does not factor over a non-isomorphic E-morphism) has a T-co-idt. (≡T-co-identifying) lift,
2. T satisfies the "smallness condition" for functors .

In [He2] 9.1 it is shown that (E,M)-topological functors with co-domain $\underline{Y}$ are exactly those functors which are obtained (up to an equivalence) by restricting topological functors with co-domain $\underline{Y}$ to a full , isomorphism-closed , reflective subcategory of their domain , such that the universal morphisms of the adjunctions are mapped (by the topological functor) into E . In our "Habilitationsschrift"[Ho5] we have shown that those topological functors obtained by the construction [He2] 9.1 play an important role in categorical topology :
These data are the adequate framework to formulate a separation axiom appropriate to many topological structures used in mathematics, namely T_0: A.S. Davis [D] realized that the usual separation axioms (epi-reflective subcategories of Top being not bi-reflective) are "intersections" of T_0 and a "wider" property, namely a bi-reflective subcategory ("Davis' correspondence" - the categorial interpretation was given first in [Mar]; cf. [Ho5] § 2).
The universal property of the above mentioned construction (proved in [Ho5] § 1) has suggested to call these topological functors (E,M)-universally topological functors. In this section we want to investigate whether the functors U described in 1.5 are (E,M)-universally topological. This will give us sufficient information on the construction in section 1 to study individual examples (in section 3).

2.1

In the following we need the fundamental criterion for (E,M)-universality (here it may be considered as a definition):

A topological functor R: $\underline{K} \to \underline{L}$ is (E,M)-universally topological - where (E,M) denotes a factorization of cones in $\underline{L}$, iff there is a full, isomorphism - closed, reflective subcategory $\underline{B}$ of $\underline{K}$ with universal morphisms ρ_K ($K \in \text{Ob}\ \underline{K}$) satisfying:

1. ρ_K is R-co-identifying.

2. $R(\rho_K) \in E$.

3. If $f: K \to B$ in $\underline{K}$ with $Rf \in E$, $B \in \text{Ob}\ \underline{B}$ is R-co-identifying then there is an isomorphism k with $k\rho_K = f$.

By these condition $\underline{B}$ is uniquely described:

$K \in \text{Ob}\ \underline{K}$ belongs to $\underline{B}$, iff every R-co-identifying morphism f with domain K and with $Rf \in E$ is an isomorphism. The objects of $\underline{B}$ are called "(U;E,M)-separated" : i.e. if $(A, \{f_i: A \to A_i\}_{i \in I})$ in $\underline{K}$ with $A \in \text{Ob}\ \underline{B}$ is R-co-idt., then $(RA, \{Rf_i\}_I)$ belongs to M.

2.2 Examples:

(a) Let $\underline{L}$ = Ens and E = {surjective maps}, hence M = {joint-injective families of maps, i.e. point-separating families with $\underline{U}$-small index sets} :

Now e.g. the usual forgetful functors U from Top (topological spaces and continuous maps), Unif (uniform spaces and uniformly continuous maps), Preord (preordered sets and isotone maps), etc. to Ens are (E,M)-universally topological.

(U;E,M)-separated means in Top "T_0", in Unif "separated" (hence the name "~-separated"), in Preord "antisymmetric".

(b) Let τ denote a "type" of universal algebras, then the forgetful functor from topological τ-algebras (and continuous τ-homomorphisms) to τ-algebras is universally topological with respect to the factorization of cones induced by E = {surjective τ-homomorphisms}.

Much more examples can be found in [Ho5], furthermore cf section 3. In particular, the situation $\underline{L} = \underline{Ens}$ is investigated in [Ho5] § 3: a basic "approximation theorem" is shown.

2.3 Theorem:

Let V: $\underline{C} \to \underline{D}$ be a faithful right adjoint functor, let Epi$\underline{C}$ induce a factorization of cones in $\underline{C}$, and let U: $\underline{A} \to \underline{D}$ be the functor obtained from V by the procedure described in 1.4; by 1.8 U is topological; let γ denote the unit of the adjunction $\underline{C} \to \underline{A}$.

Let (E,M) denote a factorization of cones:

(a) U is (E,M)-universally topological; the (U;E,M)- separated objects of $\underline{A}$ are those (f: D $\to$ VC,C) with f $\in$ M, i.e. those A $\in$ Ob$\underline{A}$ with $U\gamma_A \in$ M.

(b) The full subcategory $\underline{B}$ of (U;E,M)-separated objects of $\underline{A}$ is co-well-powered, provided that $\underline{D}$ is M_1-well-powered (where M_1 denotes the 1-indexed cones, i.e. morphisms, in M). The embedding $\underline{C} \to \underline{B}$ preserves epimorphisms, provided that M_1 consists of monomorphisms.

By the way, that M consists of mono-cones means exactly that $\underline{D}$ has co-equalizers and that every co-equalizer belongs to E; however, this condition is stronger than $M_1 \subseteq$ Mono $\underline{D}$ (e.g. let $\underline{D}$ be a group $\neq$ {O}) - cf. [Ho3].

Proof:

(a) Let A be an arbitrary object of $\underline{A}$: γ_A: A $\to$ C factors γ_A = me with U(e) $\in$ E, U(m) $\in$ M, m being U-co-identifying. Let ρ_A := e and let B denote the co-domain of e. Since m is U-co-identifying and since m is 'epic with respect to $\underline{C}$' according to 1.12 (2) (because γ_A is too), there is an isomorphism p with $p\gamma_B$ = m, hence $U(\gamma_B) \in$ M, i.e. B $\in$ Ob$\underline{B}$. Now let B' $\in$ Ob$\underline{B}$, let u: A $\to$ B' be a morphism in $\underline{A}$, and let $\gamma_{B'}$: B' $\to$ C', then there is a morphism h: C $\to$ C' with $h\gamma_A = \gamma_{B'}u$. Because of the diagonal condition now we find d: B $\to$ B' rendering

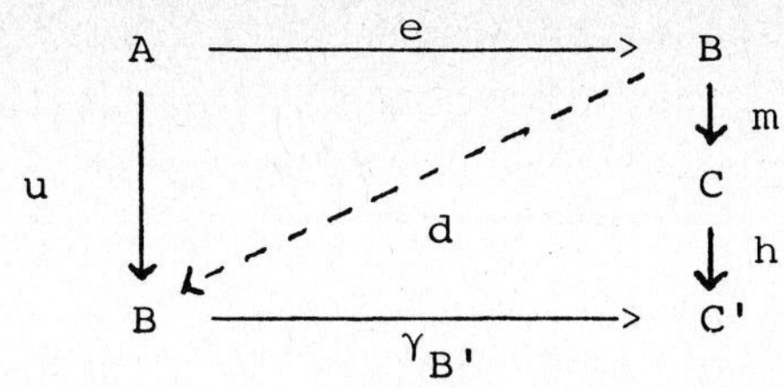

commutative

(since $\gamma_{B'}$ is U-co-identifying), i.e. we have $d\rho_A = u$. Consequently ρ_A is the universal morphism of the adjunction $\underline{B} \to \underline{A}$. It is easy to see, that 2.1 (3) is satisfied: Let f: A → X be U-co-idt. in $\underline{A}$ with $X \in Ob\underline{B}$, $U(f) \in E$; now $\gamma_X f$ satisfies the conditions of 1.12 (2), hence there is an isomorphism p with $\gamma_A = p\gamma_X f$. Since $U(p\gamma_X) \in M$ and $U(f) \in E$, there is an isomorphism q with $qf = \rho_A$.

(b) Let $B \in Ob\underline{B}$ and let g: B → X be an epimorphism in $\underline{B}$. There is a morphism h with $h\gamma_B = \gamma_X g$. Since γ_C is epic with respect to $\underline{C}$, $\gamma_X g$ is too, hence h is an epimorphism in $\underline{C}$. Since $\underline{D}$ is M_1-well-powered, and since γ_X is a U-co-idt. morphism with $U(\gamma_X) \in M$, those $\underline{B}$-quotients g' of B satisfying $h\gamma_B = \gamma_X \cdot g'$ form a $\underline{U}$-set. Consequently the class of $\underline{B}$-quotients of B is obtained as a union of a $\underline{U}$-small number of $\underline{U}$-small sets, since $\underline{C}$ is co-well-powered.

Now let g: X → Y be an epimorphism in $\underline{C}$, let (f: D → VC, C) be an object of $\underline{B}$, i.e. $f \in M$, and let (a_1, b_1), (a_2, b_2) be morphisms $(1_{VY}, Y) \to (f, C)$ with $(a_1, b_1) \circ (Vg, g) = (a_2, b_2) \circ (Vg, g)$. Then we have $fa_1 = V(b_1)$, $fa_2 = V(b_2)$, and $b_1 g = b_2 g$; consequently $b_1 = b_2$ (g is epic in $\underline{C}$) and $f a_1 = f a_2$. Since f is monic, we have $a_1 = a_2$, i.e. (Vg, g) is an epimorphisms in $\underline{B}$.

Now we want to reformulate the inner description of U in 1.12 in the framework of (E, M)-universally topological functors. This criterion will enable us to do the verification for the examples in section 3 by means of lemmata being "natural" topological statements.

2.4 Theorem

Let (E,M) be a factorization of cones in D, such that M consists of mono - cones.

U: A → D is reconstructed as in 1.2 from some faithful right adjoint functor V: C → D, such that Epi C induces a factorization of cones in C, iff

1. U is (E,M)-universally topological

2. Let B denote the full subcategory of (U;E,M)-separated objects of A. B contains a full reflective subcategory C with:

   (a) The universal morphisms λ_B: B → C for the adjunction C → B are U-co-identifying B-epimorphisms (and mapped by U into M)

   (b) If f: Y → C in B is a U-co-identifying B-epimorphism (with Uf ∈ M) and if C ∈ Ob C, then there is an isomorphism j with $j\lambda_B = f$.

   (c) Epi C the class of epimorphisms in C, induces a factorization of cones in C.

   (...) above contains consequences of the other assumptions - because of 2.1.

Proof:

By 2.3 " ... ⟹ 1., 2." is shown: Remember that $\lambda_B = \gamma_B$ and that the embedding C → B preserves C-epimorphisms (!).

In order to show the other implication notice that $\gamma_A = \lambda_B \rho_A$, where ρ_A: A → B, γ_A: A → C denotes the universal morphism of the adjunktion B → A and, resp., C → A. Consequently γ_A is U-co-identifying. Let f: X → C in A be U-co-identifying and epic with respect to C, let C ∈ Ob C, then f = me , such that m: Y → C is U-co-idt., U(m) ∈ M, U(e) ∈ E. Since B is stable under M_U-cones (i.e. U-co-idt. cones which are taken by U into M-cones, cf. [Ho5]), Y belongs to B (because of C ∈ Ob B). Since f is U-co-idt., e is too, hence there is an isomorphism j: Y' → Y with $e = j\rho_X$. mj is U-co-idt. with co-domain C ∈ Ob C and domain Y' in $\underline{B}$. Since M consists of mono-cones, f is not only epic with respect to $\underline{C}$, but also epic with respect to $\underline{B}$, hence so is mj. Consequently we have an isomorphism h with $h\lambda_{Y'} = mj$, hence $h\gamma_X = h\lambda_{Y'}\rho_X = mj\rho_X = me = f$. Now 1.12 applies.

The above criterion splits up the question whether a given "structure" is obtained by the reconstruction process described in 1.2 into several steps:

1) Verify that U: $\underline{A} \to \underline{D}$ is topological, i.e. construct U-co-idt. lifts (or U-idt. lifts) and check "smallness condition"

2) Find the (U;E,M)-separated objects and look, whether they fulfill the conditions in 2.1

3) Describe the objects of $\underline{C}$ and check the conditions in 2.4 (2).

It will be evident from the examples in section 3 that this splitting gives the natural approach to practical verifications.

The theoretical relevance of the reconstruction problem in 1.12, 2.4 is in part based on the following observation:

2.5 Corollary:

Let (E,M) denote a factorization of cones in $\underline{D}$ satisfying

1) M_1 consists of monomorphisms

2) $\underline{D}$ is M_1-well-powered.

(M_1 denotes the class of morphisms, i.e. 1-indexed cones in M.) If a topological functor U: $\underline{A} \to \underline{D}$ is reconstructed from V: $\underline{C} \to \underline{D}$ as in 2.4, then $\underline{C}$ is uniquely determined up to an equivalence by U: $\underline{A} \to \underline{D}$:

A $\in$ Ob$\underline{A}$ is isomorphic to an object of $\underline{C}$, iff A satisfies

1. A is (U; E,M)-separated, i.e. A $\in$ Ob$\underline{B}$ (where $\underline{B}$ is defined as in 2.4)

2. Whenever f: A → B with B ∈ Ob$\underline{B}$ is U-co-idt. (hence Uf ∈ M) and epic in $\underline{B}$ (!), then f is an isomorphism.

The examples to be discussed in section 3 suggest to call the objects of $\underline{C}$ "U-complete" : $\underline{C}$ does not depend on the special choice of (E,M). U is said to "admit generalized C-completions" (instead of "Cauchy - completions").

Proof:

(a) Let C ∈ Ob$\underline{C}$. If f: C → B with B ∈ Ob$\underline{B}$ is U-co-identifying and epic in $\underline{B}$, then the same holds for $\gamma_B f$, hence there is an isomorphism j with $j\gamma_C = \gamma_B f$. Since γ_C is an isomorphism, $\gamma_B f$ is too, hence f is an epic coretraction = isomorphism.

(b) Now let A ∈ Ob$\underline{A}$ satisfy 1) and 2). γ_A: A → C is U-co-identifying and epic with respect to $\underline{C}$, hence epic in $\underline{B}$ (since $\underline{C}$ is mono-reflective in $\underline{B}$!). Consequently γ_A is an isomorphism.

2.6 Remark:

Extendibility in the category Unif is sometimes (cf. [**He9**]) formulated with respect to not necessarily separated Cauchy-complete uniform spaces:

A uniformly continuous map from a dense subspace of X into a Cauchy-complete uniform space extends to X.

In order to recapture this in the framework of section 1, let R: $\underline{A}$ → Ens (!) be a topological functor admitting C-completions. Let E = class of epimorphisms in Ens, and let M be the (corresponding) class of joint-injective cones (of discrete type).

Let ρ denote the unit as in 2.2. Let us call A ∈ Ob$\underline{A}$ quasi-C-c o m p l e t e, iff the co-domain of ρ_A is C-complete. Now we have the following extension theorem:

Let f: A → B be a morphism in $\underline{A}$, let B be quasi-C-complete. Let q: A → A* be a R-co-idt. morphism with Rq being injective, and let q be epic with respect to the class of C-complete objects in $\underline{A}$ (or, equivalently, epic with respect to the lass of (R;E,M)-separated objects in $\underline{A}$), shortly, let

$q: A \to A^*$ be a "d e n s e e m b e d d i n g":

Then $f: A \to B$ admits an "extension" $g: A^* \to B$ with $gq = f$.

Proof:

By 1.14 there is an "extension" $h: A^* \to C$ rendering commutative

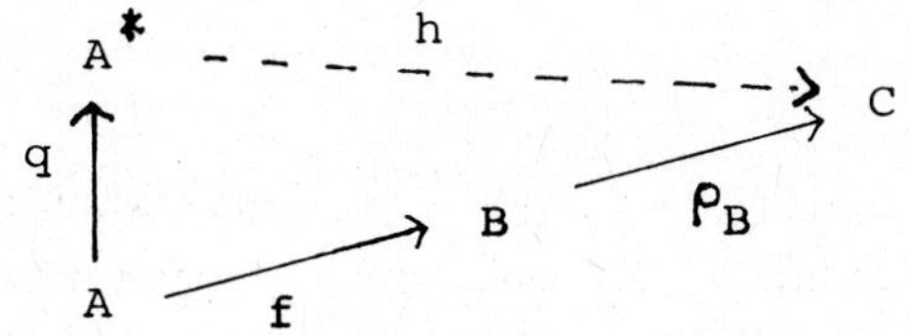

since C is C-complete.

We define a map $u: RA^* \to RB$ by $u(Rq(x)) = Rf(x)$ (since Rq is injective) and $u(y) \in (R\rho_B)^{-1}[\{Rh(y)\}]$ (arbitrary choice; since $R\rho_B$ is surjective, this set is non-empty) for every $y \in RA^*$ with $y \neq Rq(x)$ for every $x \in RA$.

Obviously $uRq = Rf$ and $R(\rho_B)u = Rh$, hence there is a morphism $g: A^* \to B$ with $Rg = u$. Since R is faithful, we have $gq = f$.

There is an analogue of the extendibility theorem 1.14 for (E,M)-universally topological functors $R: \underline{K} \to \underline{L}$:

Let $K \in Ob\underline{K}$ be (R;E,M)-separated, let $d: Q_2 \to Q_1$ be R-co-idt. with $Rd \in E$, then every morphism $f: Q_2 \to K$ is extendible over d, i.e. there is a morphism $g: Q_1 \to K$ with $gd = f$. (The proof is analogous to 1.14.)

This suggests the following modification of the criterion 2.1.

2.7 Theorem

Let $R: \underline{K} \to \underline{L}$ be a topological functor, let (E,M) denote a factorization of cones in $\underline{L}$, and let Σ denote the class of all those R-co-idt. morphisms which are taken by R into E:

R is (E,M)-universally topological, iff for every object $K \in Ob\underline{K}$ there is a Σ-injective object A in $\underline{K}$ and a Σ-morphism $K \to A$. An object $B \in Ob\underline{K}$ is Σ-injective, iff it is (R;E,M)-separated.

Proof:

By our remark above, (for some (E,M)-universally topological functor R) **every (R;E,M)-separated object is** Σ-injective. Now let $B \in \mathrm{Ob}\underline{B}$ be Σ-injective. Since $\rho_B \in \Sigma$ (ρ as in 2.1), there is a morphism g with $g\rho_B = id_B$, hence ρ_B is an epic co-retraction, i.e. an isomorphism. Consequently B is (R;E,M)-separated.

Now let $R: \underline{K} \to \underline{L}$ be topological. Let $\mathrm{Ob}\underline{B} = \{\Sigma\text{-injective objects}\}$, let $\underline{B}$ denote the corresponding full subcategory of $\underline{K}$. For $K \in \mathrm{Ob}\underline{K}$ let ρ_K denote a Σ-morphism with Σ-injective co-domain. For a span

$$\begin{array}{ccc} B & & B^* \\ {\scriptstyle \rho_K}\uparrow & \nearrow {\scriptstyle f} & \\ K & & \end{array}$$

with $B^* \in \mathrm{Ob}\underline{B}$, we have a unique morphism $g: B \to B^*$ with $g\rho_K = f$, since B^* is Σ-injective and since ρ_K is epic. If f is in Σ, then there is also a morphism $h: B^* \to B$ with $hf = \rho_K$, since B is Σ-injective. The right cancellation property of ρ_K and f now implies $hg = id_B$ and $gh = id_{B^*}$, i.e. g is an isomorphism. This completes the proof.

There is an analogous theorem for topological functors $U: \underline{A} \to \underline{D}$ admitting generalized Cauchy-completions: This is less satisfactory than 2.7, since it involves $\underline{C}$ on both sides of the equivalence (in order to define what is "$\underline{C}$-epic").

§ 3

3.0 In this section we want to give significant examples of the construction explained in section 1 and 2. The criterion 2.4 helps us to do the verification:

(E,M)-universality of the topological functors considered below has been shown in [Ho5] - of course, characterizing those objects "separated" with respect to (E,M); some of the conditions in 2.1 and 2.4 are then guaranteed by well known results from point set topology [Pr,Sb2]. On the other hand 2.1 and 2.4 have suggested new results in point set topology, in particular on sober spaces.

We think that drawing pictures is more helpful for the reader to understand the situation than giving sophisticated explications. The pictures below are to be interpreted as

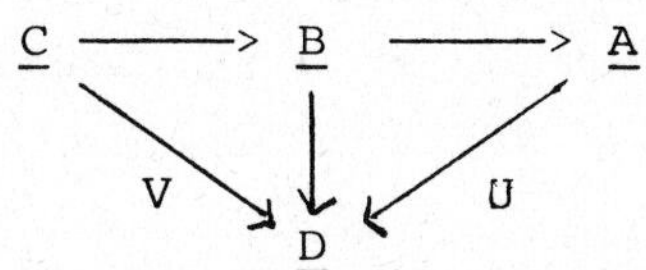

with C, B, A and the forgetful functors V,U, etc. as in 2.3 and 2.4. In Ens we refer to the factorization (E,M) with E={epimorphisms} = {surjective maps}, hence M ={joint injective cones (of discrete type), i.e."point-separating families of maps"}.

3.1

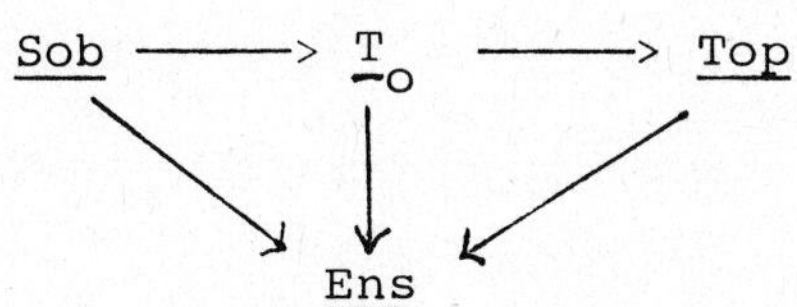

Top denotes the category of topological spaces and continuous maps, T_o its full subcategory of T_o-spaces, Sob its full subcategory of 'sober spaces' (cf. [Ar,Bln]). A topological space X is "sober", iff every irreducible, closed, non-empty subset A of X has a unique "generic" point p, i.e. a point p with cl{p} = A. Every topological space Y admits a universal "sobrification" ${}^sY$: points of ${}^sY$ are irreducible, closed, non-empty subsets of Y; for an open set O in Y let ${}^sO := \{ A \in {}^sY \mid A \cap O \neq \emptyset \}$ be open in ${}^sY$, $\{{}^sO \mid O$ open in Y$\}$ is the topology (on) ${}^sY$; the universal morphism $i_Y: Y \to {}^sY$ is given by $p \longmapsto cl\{p\}$ (cl is to be interpreted in Y), this morphism is an embedding and an epimorphism in T_o provided that Y is T_o. In [Ho6] we have shown that every embedding $X \to Y$ being an epimorphism in T_o with X being sober is necessarily a homeomorphism. Thus by 2.5 the U-complete objects are exactly the sober spaces - provided Top → Ens "admits generalized C-completions".

For the convenience of the reader we sketch the way the statement 'U: Top → Ens admits generalized C-completions' has to be verified. We do so, in order to make clear, what this statement means explicitly in an individual situation.

3.1.1 A space X is (U;E,M) - separated, iff $x,y \in X$ with cl{x} = cl{y} implies x = y, i.e. iff X is T_o:

(a1) Let Y be a space : for $a,b \in Y$ let $a \approx b$, iff cl{a} = cl{b}. The quotient space $Y_o := Y/_{\approx}$ is T_o, the canonical projection $p: Y \to Y_o$ is open and closed, and p induces the initial topology on Y, i.e. p is U-co-identifying and $U(p) \in E$.

(a2) Let $g: Y \to Z$ be a continuous map into a T_o-space Z, then there is a unique map $h: Y_o \to Z$ with hp = g, h is continuous.

(a3) Let $g: Y \to Z$ be a continuous surjection into a T_o-space Z inducing the initial topology on Y, then the morphism $h: Y_o \to Z$ with hp = g according to (a2) is a homeomorphism.

Consequently U: Top → Ens is (E,M)-universally topological with E = {surjective maps} by 2.1.

(b 1) A continuous embedding $A \to B$ into a T_o-space B is epic in T_o, iff A is b-dense in B, i.e B is the b-closure of A in B:

$x \in cl(A \cap cl\{x\})$ for every $x \in B$. $\underline{T}_o$ is co-well-powered and co-complete. Cf. [Bar1, Sk, Ho6] .

(b 2) Every T_o-space X admits a 'sobrification' ${}^sX$, with a canonical embedding $i_X: X \to {}^sX$ having the universal property; i_X is a $\underline{T}_o$-epic embedding (cf. [Ar, N1, Ho6]) .
The last statement to be verified - corresponding to 2.4.2 (b) - seems to be a new result on the universal sobrification of T_o-spaces :

3.1.2 Proposition :

Let X be a b-dense subspace of the sober space Y, and let f denote the embedding $X \to Y$, then the (unique) continuous morphism $g: {}^sX \to Y$ with $g \circ i_X = f$ is a homeomorphism. Roughly speaking: Y is the universal sobrification of any b-dense subspace.

Proof:

The induced morphism $g: {}^sX \to Y$ takes $A \in {}^sX$ to the generic point of $cl\,A := cl\,f[A] \subseteq Y$. cl is always interpreted in Y, the embedding $f: X \to Y$ is sometimes neglected in the notation.

(a) Let $A' \neq A$, $A' \in {}^sX$, then $cl\,A' \neq cl\,A$, since $A = X \cap B$, $A' = X \cap B'$ for some closed sets B, B' in Y. Consequently g is injective.

(b) Now let $y \in Y$, then $A := X \cap cl\{y\}$ is closed in X and irreducible: Let O, O' be open in Y with $X \cap cl\{y\} \cap O \neq \emptyset$ and $X \cap cl\{y\} \cap O' \neq \emptyset$, then $y \in O$ and $y \in O'$, hence $X \cap cl\{y\} \cap O \cap O' \neq \emptyset$, since X is b-dense in Y; in particular $X \cap cl\{y\} \neq \emptyset$. In order to make sure that $clA = cl\{y\}$, i.e. $g(A) = y$, it is sufficient to have $y \in clA$, i.e. $y \in cl(X \cap cl\{y\})$; this is immediate from ' X b-dense in Y '. Consequently g is surjective.

(c) Let U be open in X, i.e. $U = X \cap O$ with O open in Y. We want to show that $g[{}^sU] = O$, i.e. g is open: If $y \in g[{}^sU]$, then there is an $A \in {}^sX$ with $A \cap U \neq \emptyset$ and $g(A) = y$, i.e. $clA = cl\{y\}$. Since $cl\{y\} \cap O \cap X = clA \cap U \neq \emptyset$, we have $cl\{y\} \cap O \neq \emptyset$, i.e.

$y \in O$. If $z \in O$, then $cl\{z\} \cap O \cap X \neq \emptyset$, hence $cl\{z\} \cap X$ is an element of ${}^{s}U$, $cl\{z\} \cap X$ is taken by g to z - according to (b), hence $z \in g[{}^{s}U]$.

3.1.3 Remark

(a) We need not verify that (U;E,M)-separatedness means T_o, i.e. a space X satisfies T_o, iff a (family of) continuous map(s) with domain X inducing the initial topology on X is (joint) injective: this is a consequence of (a1), (a2), (a3). Similary the characterization of sober spaces in [Ho6] "a T_o-space X is sober, iff every b-dense embedding of X into a T_o-space is a homeomorphism" is a consequence of (b1), (b2) and 3.1.2.

(b) Checking individual examples, one often realizes a "natural" separation axiom which turns out to be (U;E,M)-separatedness. For Ens-valued topological functors U represented by a terminal object t (i.e. $U \cong Hom(t,-)$) we have given a criterion more easily to be verified([Ho5] §3). On the other hand generally it seems to be difficult to find the U-complete objects.

(c) The theorem on the existence of the sobrification can be combined with 3.1.2 to a statement being more "topological in character": For a T_o-space X there is up to a homeomorphism a unique sober space Y admitting a b-dense embedding $i_X: X \to Y$. i_X is a unit of the adjunction $\underline{Sob} \to \underline{T_o}$.

Thus the splitting up of the concept of "admitting generalized C-completions" in section 2 becomes evident to be just the adequate formulation of the categorical framework of the above mentioned facts from point set topology.

In the following we shall omit details, but our above remarks will carry over to the examples below (with obvious modifications).

3.2

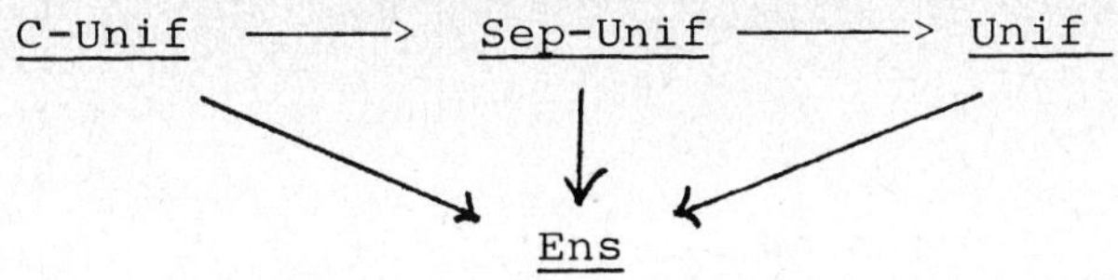

Unif denotes the category of uniform spaces and uniformly continuous maps, Sep-Unif = (its full subcategory of) separated uniform spaces, C-Unif = Cauchy-complete separated uniform spaces. 'Epic' in Sep-Unif means 'dense' [Pr]; for a separated uniform space X there is up to a uniform homeomorphism a unique Cauchy-complete separated uniform space Y admitting a dense uniformly continuous embedding $u_X: X \to Y$; $u_X: X \to Y$ is the universal morphism of the adjunction C-Unif $\to$ Sep-Unif.

3.3

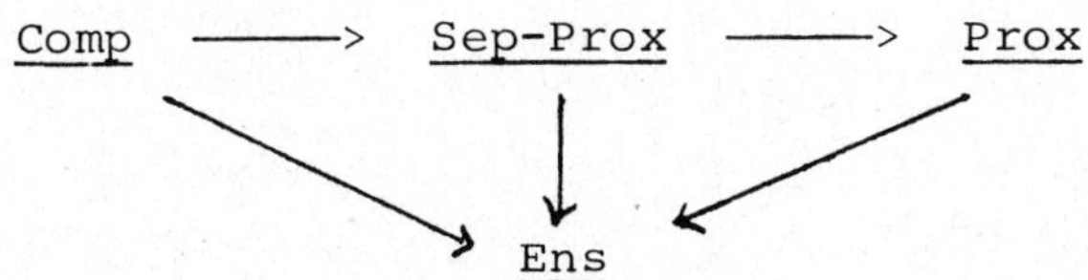

Prox denotes the category of totally bounded uniform spaces, i.e. (uniformizable) "proximity spaces" and uniformly continuous maps, Sep-Prox = separated totally bounded uniform spaces, Comp = compact Hausdorff spaces (admitting a unique compatible uniformity, continuous maps become uniformly continuous).
Since there is a full embedding Comp $\to$ C-Unif, the verification is given by 3.2 realizing that V-epimorphisms are dense subsets of compact spaces (one can verify this in the category CReg of completely regular T_2-spaces and continuous maps, since Comp is mono-reflective in CReg : 'CReg-epi' means 'dense').

3.4

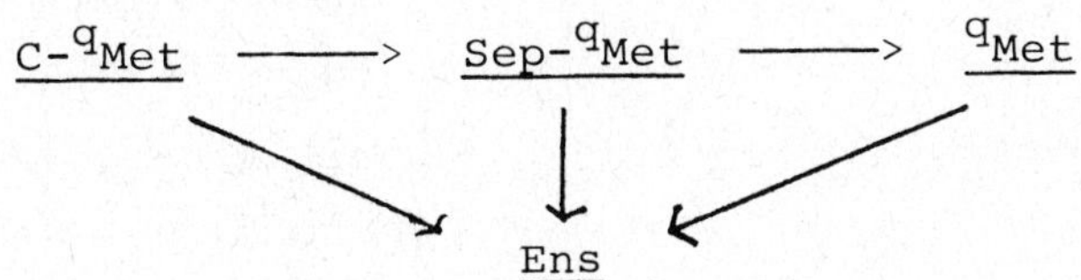

Let X be a set, a mapping $d: X \times X \to [0,\infty]$ is said to be a quasi-metric (= q-metric), iff

(1) $d(x,y) = d(y,x)$ ("symmetric")

(2) $d(x,x) = 0$

(3) $d(x,y) \leq d(x,z) + d(z,y)$ ("triangle inequality")

for any elements $x,y,z \in X$; (X,d) is said to be a q-metric (= quasimetric) space; a mapping $f\colon (X,d) \to (X',d)$ is said to be non-expansive, iff $d(x,y) \geq d'(fx, fy)$ for any $x,y \in X$.

$\underline{{}^q\text{Met}}$ denotes the category of q-metric spaces and non-expansive maps, $\underline{\text{Sep-}{}^q\text{Met}}$ = separated q-metric spaces ($d(x,y) = 0$ implies $x=y$), $\underline{\text{C-}{}^q\text{Met}}$ = Cauchy-complete separated q-metric spaces : it should be clear what "Cauchy-sequence" in (X,d) means. 'Epic' in $\underline{\text{Sep-}{}^q\text{Met}}$ means 'dense' (with respect to the induced topology) - cf 3.4.1 below. For any separated q-metric space (X,d) there is a unique Cauchy - complete separated q-metric space (Y,d') admitting a dense embedding of (X,d) into (Y,d'): the space of (equivalence classes of) Cauchy - sequences in (X,d) where d' is defined in the usual way. The canonical embedding $(X,d) \to (Y,d')$ taking x to the constant sequence (x_n) with $x_n := x$ is the universal morphism of the adjunction $\underline{\text{C-}{}^q\text{Met}} \to \underline{\text{Sep-}{}^q\text{Met}}$.

3.4.1 <u>Lemma:</u>

A morphism $f\colon (M,d) \to (N,t)$ in $\underline{\text{Sep-}{}^q\text{Met}}$ is an epimorphism, iff $f[M]$ is dense in (N,t) with respect to the usual topology on N induced by t.

<u>Proof:</u>

Let f be not dense, w.l.o.g. f is an inclusion of M into N, the closure cl is interpreted in the topology on N induced by t.

From the sum $N \amalg N := \{(i,n) \mid n \in N,\ i=1,2\}$ we obtain a quotient set identifying "corresponding" points of $\mathrm{cl}M \amalg \mathrm{cl}M$: $(i,n) \approx (i',n')$ iff $n=n'$ and $i=i'$,

or $n=n'$ and $n \in \mathrm{cl}M$.

On the quotient set $Q = N \amalg N/_{\approx}$ we define a separated q-metric by

$$h\,[(i,n),\ (i,n')] = t(n,n')$$

and, if $i \neq i'$

$$h[(i,n),\ (i',n')] = \inf_{x \in clM} t(n,x) + t(x,n').$$

For $n \in clM$ (or $n' \in clM$) we have $h[(i,n),\ (i',n')] = t(n,n')$ because of $t(n,x) + t(x,n') \geq t(n,n')$, hence the definition of h is compatible with $\approx$, i.e. h becomes a map $Q \times Q \to [0,\infty]$.

For $n \in N - clM$ and $i \neq i'$ we have

$$h[(i,n),\ (i',n')] = \inf_{x \in clM} t(n,x) + t(x,n')$$

$\geq t(n,clM) > 0$, hence (Q,h) is separated.

Let $j_1, j_2\colon N \to N \amalg N$ denote the canonical injections, $p\colon N \amalg N \to Q$ the canonical projection, then $pj_1, pj_2\colon (N,t) \to (Q,h)$ are non-expansive, $pj_1 \neq pj_2$ and $pj_1 f = pj_2 f$.

The reserve assertion is immediate from the faithful functor $\underline{\text{Sep-}^{q}\text{Met}} \to \underline{T}_2 :=$ {T_2-spaces and continuous maps} reflecting epimorphisms, since 'epic' in $\underline{T}_2$ means dense.

3.5

$$\begin{array}{ccccc} \underline{^{q}\text{Ban}_{\mathbb{K}}} & \longrightarrow & \underline{\text{Sep-}^{q}\text{n-Vec}_{\mathbb{K}}} & \longrightarrow & \underline{^{q}\text{n-Vec}_{\mathbb{K}}} \\ & \searrow & \downarrow & \swarrow & \\ & & \underline{\text{Vec}_{\mathbb{K}}} & & \end{array}$$

Let $\mathbb{K}$ denote a subfieldt of $\mathbb{C}$, e.g. $\mathbb{K} = \mathbb{R}$. The factorization (E,M) of cones in $\underline{\text{Vec}_{\mathbb{K}}}$ = {vector spaces over $\mathbb{K}$ and $\mathbb{K}$-linear maps} we refer to is induced by E = {surjective $\mathbb{K}$-linear maps}, hence M = {joint injective $\underline{U}$-small-indexed cones of $\mathbb{K}$-linear maps, i.e. cones $(A,\{f_i\colon A \to A_i\}_{i \in I})$ in $\text{Vec}_{\mathbb{K}}$ with $I \in \underline{U}$ and $\bigcap_{i \in I} \text{kernel } f_i = \{0\}$ }.

Let X be a $\mathbb{K}$-linear space. A map $\|?\|\colon X \to [0,\infty]$ is called a quasi-norm (=q-norm) on X, provided that

$$\begin{aligned} \|0\| &= 0 \\ \|\lambda x\| &\leq |\lambda| \cdot \|x\| \\ \|x+y\| &\leq \|x\| + \|y\| . \end{aligned}$$

$(X, \|\,.\,\|)$ is said to be a quasi-normed (=q-normed) $\mathbb{K}$-space.

$\underline{{}^q n\text{-}Vec_{\mathbb{K}}}$ denotes the category of quasi-normed $\mathbb{K}$-spaces and $\|\,.\,\|$-decreasing $\mathbb{K}$-linear maps, $\underline{Sep\text{-}{}^q n\text{-}Vec_{\mathbb{K}}}$ = separated quasi-normed $\mathbb{K}$-spaces ("separated" means: $\|x\| = 0$ implies $x = 0$), $\underline{{}^q Ban_{\mathbb{K}}}$ = quasi-Banach-$\mathbb{K}$-spaces, i.e. those separated (!) quasi-normed $\mathbb{K}$-spaces, such that the (canonically) induced quasi-metric is Cauchy-complete.

If $(X, \|\,.\,\|)$ is a separated quasi-normed $\mathbb{K}$-space, then the canonical Cauchy-completion of the induced quasi-metric space (cf 3.4) becomes in a canonical way a $\mathbb{K}$-space, and, moreover, a quasi-Banach-$\mathbb{K}$-space.

The verification is similar to 3.4. We still have to show 3.5.1.

3.5.1 Lemma:

Let $f: (X, \|\,.\,\|) \to (Y, \|\,.\,\|)$ be a morphism in $\underline{Sep\text{-}{}^q n\text{-}Vec_{\mathbb{K}}}$, i.e. $\|f(x)\|_Y \leq \|x\|_X$.

f is an epimorphism in $\underline{Sep\text{-}{}^q n\text{-}Vec_{\mathbb{K}}}$, iff f is dense, i.e. $f[X]$ is dense in $(Y, \|\,.\,\|)$ with respect to the topology on Y canonically induced by $\|\,.\,\|_Y$.

Proof:

Let us assume that f is not dense, and let w.l.o.g f be an embedding, i.e. $X \subsetneq Y$.

$A := \mathrm{cl}X$ is a $\mathbb{K}$-linear subspace of Y (cl is to be interpreted in the topology induced on Y by $\|\,.\,\|$). The factor $\mathbb{K}$-space $Y/_A$ becomes a quasi-normed $\mathbb{K}$-space by putting

$$\|[y]\| := \inf_{a \in A} \|y+a\|$$

where $[y] := \{y+a \mid a \in A\} \in Y/A$.

If $\|[y]\| = 0$, then there is a sequence $(a_n)_{n \in \mathbb{N}}$ in A with $\|y+a_n\|$ tending to 0, hence a_n converges to $-y$, i.e. $-y$ and $y \in A$, since A is closed (and a $\mathbb{K}$-linear subspace), i.e. $[y] = [o]$, i.e. $({}^Y/A, \|.\|)$ is separated.

Of course, both the canonical projection $p: (Y,\|.\|) \to ({}^Y/A,\|.\|)$ and the null homomorphism $n: (Y,\|.\|) \to ({}^Y/A,\|.\|)$ are $\|.\|$-decreasing, hence f is not an epimorphism in <u>Sep-^{q}n-Vec$_{\mathbb{K}}$</u> because of $n\cdot f = p\cdot f$, $n \neq p$.

The reverse assertion is immediate from 3.4.1, since there is a faithful functor <u>Sep-^{q}n-Vec$_{\mathbb{K}}$</u> - <u>Sep-^{q}Met</u> (reflecting epimorphisms).

3.5.2 <u>Remark:</u>

In the appendix (§5) of [Ho5] we have shown that the forgetful functor U: <u>^{q}n - Vec$_{\mathbb{K}}$</u> $\to$ <u>Vec$_{\mathbb{K}}$</u> is not only (E,M)-universally topological with (U;E,M)-separated = separated, but U is also co-universally topological with respect to the following co-factorization (E',M') of co-cones in <u>Vec$_{\mathbb{K}}$</u>:

E' = {<u>U</u>-small-idexed co-cones in <u>Vec$_{\mathbb{K}}$</u>, such that the union of the images "generate" the co-domain},

M' = {injective $\mathbb{K}$-homomorphisms}.

The (U;E',M')-co-separated objects of <u>^{q}n-Vec$_{\mathbb{K}}$</u> are exactly those $(X, \|.\|)$ with $\|x\| \neq \infty$ for every $x \in X$.

3.5.3 <u>Remark</u>:

The above examples 3.2 - 3.5 exhibit the common features of Cauchy-completions in the framework of uniform spaces, q-metric spaces, and q-normed $\mathbb{K}$-linear spaces.

(What about Cauchy-completions for nearness spaces in the sense of H. Herrlich [He8]?)

The program, to find a common setting of the idea of completion, goes back to an early paper of G.Birkhoff [Bi]. Some of the examples above suggest that the results in section 2 are somehow related to Rattray's [Ra], but there seems to be no straightforward implication.

3.6

The functor $id_{\underline{Ens}}$: $\underline{Ens} \to \underline{Ens}$ is, of course, (E,M)-universally topological: the separated objects are the sets of cardinality at most 1; moreover $id_{\underline{Ens}}$ admits generalized C-completions: the C-complete objects are the sets of cardinality 1.

More generally, $id_{\underline{X}}$ admits generalized C-completions, provided $\underline{X} = \coprod_{I \in I} \underline{X}_i$ and every $\underline{X}_i$ has a terminal object C_i: Let $\underline{C}$ be the (discrete) full subcategory of $\underline{X}$ consisting of the C_i's and let $V: \underline{C} \to \underline{X}$ denote the embedding.

3.7 Let $\underline{L}$ be a complete lattice (which is canonically interpreted as a category). Applying the construction of section 1 to the constant functor $\underline{L} \to \underline{Ens}$ which takes everything to a one-element-set and, resp., to its identity, one obtains the following category $\underline{L\text{-}Ens}$:

Objects of $\underline{L\text{-}Ens}$ are pairs (M,i) with $M \in Ob\underline{Ens}$, $i \in \underline{L}$, $\underline{L\text{-}Ens}$-morphisms f: (M,i) $\to$ (M',i') are maps f: M $\to$ M' with $i \leq i'$ (in $\underline{L}$). The forgetful functor U: $\underline{L\text{-}Ens} \to \underline{Ens}$ takes (M,i) to M and f: (M,i) $\to$ (M',i') to f. U is a topological functor admitting generalized C-completions:

(M,i) is C-complete, iff cardM = 1; (M,i) is separated, iff cardM $\leq$ 1.

Furthermore, U: $\underline{L\text{-}Ens} \to \underline{Ens}$ is also co-universally topological with respect to the co-factorization (E',M') of co-cones in $\underline{Ens}$ with M' = {injective maps} and E' = {joint-surjective families of maps} = = {$\underline{U}$-small-indexed epi-co-cones in $\underline{Ens}$} : (M,i) is co-separated, iff M = $\emptyset$.

The above topological functors L-Ens → Ens (for complete lattice L) are characterized by the fact that they are induced by constant (contravariant) topological theories in the sense of O.Wyler [Wy1,2], i.e. constant functors t: Ens^{op} → C-Ord (= complete lattices and inf-preserving maps) taking everything to the lattice L and, resp., its identity.

We plan to prove elsewhere that every topological functor with co-domain Ens, which lifts isomorphisms uniquely, being both universal and co-universal (with respect to the above co-factorization in Ens), is necessarily of the above described type for some lattice L (which is- up to an isomorphism - uniquely determined).

§ 4

In this section we want to give "unpleasant" examples for our theory of "generalized Cauchy-completions". We shall use the same kind of general notation and pictures as in section 3. The functors to Ens are to be understood as the usual forgetful functors, the embeddings as the obvious embeddings.

There are two different types of these examples.

A The first kind of examples we mean fits too smoothly into the concept of "topological functor admitting generalized Cauchy-completions": Every separated object is already Cauchy-complete.

This turns out to be equivalent to the condition that epimorphisms in the category of (U;E,M)-separated objects are mapped by U into E.

Thus C-completion does not give any "improvement" of these separated "structures".

B The second kind of examples are (E,M)-universally topological functors which do not admit generalized C-completions. However, using the description of C-complete objects given in section 2 as an ad hoc-definition, one arrives at some kind of objects which in these particular situations are of interest in themselves.

The examples given in the following are (E,M)-universally topological functors with codomain Ens and E = Epi(Ens). We freely use the concrete description of (U;E,M)-separated objects either given explicitly in [Ho5] by direct verification (in particular, by means of the approximation theorem [Ho5] § 3 - thus computation becomes trivial) or to be immediately followed from the generalized Davis' correspondence theorem [Ho5] § 2.

A

4.1 Theorem:

Let U : A → D be a topological functor "admitting generalized Cauchy-completions", and let (E,M) be a factorization of cones in D satisfying

$M_1 \subseteq$ Mono D

Then the following conditions (a) and (b) are equivalent:

(a) Every (U;E,M)-separated object is U-complete

(b) In the full subcategory B of A consisting of the (U;E,M)-separated objects of A every B-epimorphism is taken by U into E.

Proof:

(b) ⇒ (a): Let η denote the unit of the adjunction C → A. For B∈Ob B $T\eta_B \in E$ by wirtue of (b) (because η_B is epic in B!). Conrequently η_B is T-co-idt. morphism with $T\eta_B \in E \cap M_1$ = Iso D, hence η_B is a isomorphism.

(b) ⇒ (a):

Let f : X → Y be an epimorphism in B with Uf∉E. Let Uf = me with $m \in M_1$, e∈E, then there is a U-co-idt. morphism g : B → Y in A and a morphism h : X → B with gh = f and $Ug \in M_1$, Uh∈E, in particular Ug∉E (otherwise Uf = UgUh∈E).

In particular we have B∈Ob B, because B is "closed" in A under all those (not necessarily U-co-idt.!) cones which are taken by U into M-cones. Now we see that B is (U;E,M)-separated, but not U-complete, since g : B → Y is epic in B (in particular, with respect to C) and U-co-idt.

4.2

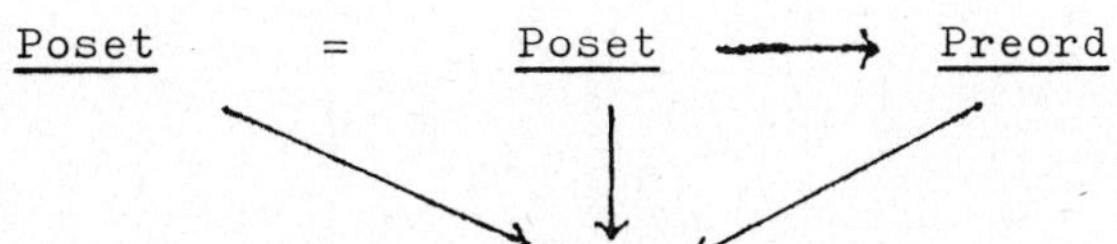

Preord denotes the category of pre-ordered (= quási-ordered) sets and isotone maps. The objects of Poset are the partially ordered sets (pre-order with antisymmetry).

Every preordered set gives rise to a topology on it, called the associated Alexandroff-discrete space; thus Preord becomes a full bi-coreflective subcategory of Top, namely the (bi)-coreflective hull of the Sierpinski

space (this is shown in[Ho7]); "antisymmetry" now means "T_o".
In consequence, (E,M)-universality of U : Preord → Ens is also a consequence of[Ho5]3.14(once (E,M)-universality of Top → Ens is verified).

4.2.1

Lemma:

Every Poset-epimorphism is surjective.

This result is due to [B B] (I saw that after having worked out the proof) .

Thus our concept of "C-completion" is not anyhow related to some "completion" of posets (which - usually - means a left adjoint from Poset into a subcategory of some complete lattices which is n o t a full subcategory of Poset).

4.3

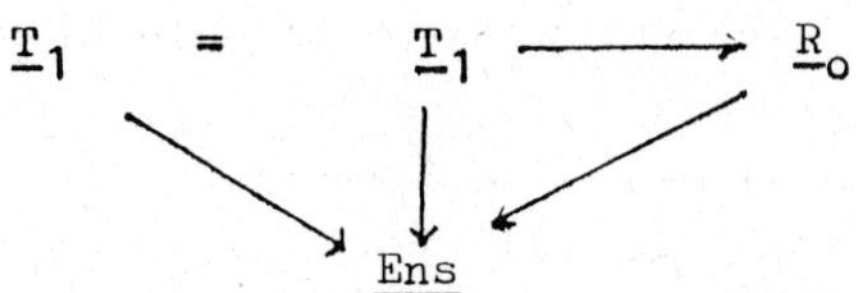

$\underline{R}_o$ means the category of R_o-spaces and continuous maps. A topological space is said to be R_o (Davis [D]), iff the Sierpinski space (2 points and 3 open sets) is not embeddable into it , i.e. $cl\{x\} \neq cl\{y\}$ implies $cl\{x\} \cap cl\{y\} = \emptyset$ (cf [Ho5]); $T_1 = R_o \wedge T_o$.

"Epi" in $\underline{T}_1$ means "surjective" ([Ko , Bu]) .

4.4

If $U : \underline{A} \to \underline{D}$ is (E,M)-universally topological, if $E = \text{Epi}\ \underline{D}$, ($M_1 \subseteq \text{Mono}\ \underline{D}$), and if - in addition - $T = U|\underline{B} : \underline{B} \to \underline{D}$ is (absolutely) topological (in particular, T preserves epimorphisms), then we arrive at

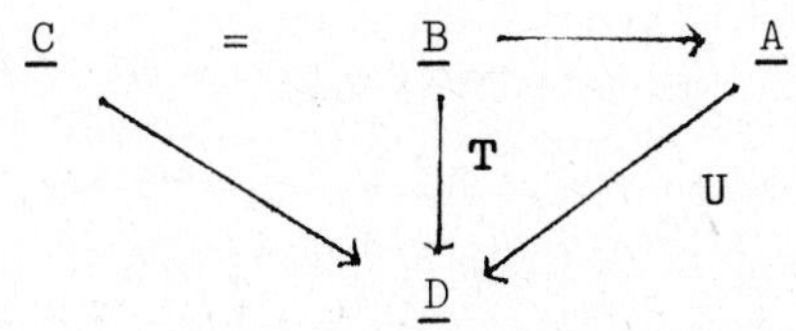

I.e. in particular, U admits generalized C-completions.

Proof:

$(f : D \to TB, B)$ is a T-epimorphisms, iff f is an epimorphism in $\underline{D}$. Thus the (E,M)-universal extension of T coincides with the canonical extension of T. ($M_1 \subseteq \text{Mono}\ \underline{D}$ guarantees that $\underline{C}$ is uniquely determined by $U : \underline{A} \to \underline{D}$).

Now we come to the second kind (B) of "unpleasant" examples.

4.5

Let $\underline{T}_{3a}$ denote the category of T_{3a}-spaces and continuous maps; the forgetful functor $U : \underline{T}_{3a} \to \underline{\text{Ens}}$ is (E,M)-universally topological with $E = \text{Epi}(\underline{\text{Ens}})$; the (U;E,M)-separated objects are exactly the completely regular spaces ($\underline{\text{C Reg}}$).

Using the description of U-complete objects as an ad hoc-definition we arrive at:

A T_{3a}-space X is "complete", iff X is T_o ("separated") and every dense embedding of X into any separated T_{3a}-space is an isomorphism (recall that "epi" in $\underline{\text{CReg}}$ means "dense"), i.e. iff X is compact (:= compact $\wedge\ T_2$).

However, applying the canonical extension to $\underline{\text{Comp}} \to \underline{\text{Ens}}$, we arrive at the category $\underline{\text{Prox}}$ (cf. 3.3), which is different from $\underline{T}_{3a}$:

Let $\mathbb{N}$ denote the discrete space of natural numbers, and let $\mathbb{N}^*$ be its Alexandrov-compactification, then $\mathbb{N}^*$ is not the Stone-Cech-compactification $\beta\mathbb{N}$ of the dense subspace $\mathbb{N}$ (in contrast to 2.4.2 b), since card $\beta\mathbb{N}$ = exp exp card $\mathbb{N}$.

In consequence $\underline{T}_{3a} \to \underline{\text{Ens}}$ does not admit generalized C-completions.

4.6

Let $\underline{R}_1$ denote the category of R_1-spaces and continuous maps, i.e. those spaces whose universal T_o-quotient is T_2 (i.e."Hausdorffsch") - cf [D]. The forgetful functor $U : \underline{R}_1 \to \underline{Ens}$ is (E,M)-universally topological with $E = \mathrm{Epi}(\underline{Ens})$: the (U;E,M)-separated objects are exactly the T_2-spaces (= "separated spaces").

R_1-complete = T_2-closed since "epi" in $\underline{T}_2$ means dense.
The T_2-closed spaces do not form a full reflective subcategory of $\underline{T}_2$ - cf.[He1]. However for every T_2-space there is a distinguished T_2-closed extension, the so-called Katětov-extension (cf. [Ha]).

4.7

The examples above (4.5 , 4.6) easily generalize to the following situation :

Let $\underline{X}$ be a full epi-reflective subcategory of $\underline{Top}$, let $\underline{X}$ consist of T_o-spaces (i.e. $\underline{X}$ is not bi-reflective in $\underline{Top}$) , and let $\underline{Y}$ denote the bi-reflective hull of $\underline{X}$ in $\underline{Top}$. The objects of $\underline{X}$ are the "T_o-separated" objects of $\underline{Y}$ - according to [Mar] or [Ho5] §2 :

If "epi" in $\underline{X}$ means "dense" , then

"$\underline{Y}$-complete" = "$\underline{X}$-closed" .

(Sometimes it may be convenient to write "$\underline{X}$-complete" instead of "$\underline{Y}$-complete".)

(a) This is true e.g. for $\underline{X}$ = regular T_2-spaces .

(b) Recently J.Schröder [Sd] has pointed out that for $\underline{X}$ = Urysohn spaces (= T_{2a}-spaces) "epi" is weaker than "dense" . We wonder , whether the $\underline{R}_{1a}$-complete spaces coincide with the Urysohn-closed spaces , or , if they do not , whether they behave more nicely than the Urysohn-closed spaces do . Cf. [He7] .
($\underline{R}_{1a}$:= bi-reflective hull of $\underline{T}_{2a}$ in $\underline{Top}$)

4.8

Finally , let us briefly comment the problem of finding an "approximating" topological functor admitting generalized C-completions for those topological functors described in 4.5 sqq. .

(a) For $\underline{T}_{3a} \longrightarrow \underline{Ens}$ (4.5) the answer seems to be "natural" :

$\underline{Prox} \longrightarrow \underline{Ens}$ (3.3) , since $\underline{Comp}$ is reflective in $\underline{CReg}$, and since the canonical extension of $\underline{Comp} \longrightarrow \underline{Ens}$ yields $\underline{Prox} \longrightarrow \underline{Ens}$.

(b) In case of $\underline{T}_2$ (4.6) the question is open :

What about the category of "T_2-closed extensions of R_1-spaces *) " , is it topological over $\underline{Ens}$, etc. ? This would be analogous to the situation with proximity spaces in (a) , which can be viewed at as (not necessarily injective) "compactifications" of T_{3a}-spaces .
Cf. [Blz,Ha,He7,Po]

What about a general theory of "approximation" with respect to topological functors admitting generalized Cauchy-completions ?

---

*) Such an "extension" f of an R_1-space X is to be understood as a continuous map $f:X \longrightarrow Y$ furnishing X with the initial topology , $f[X]$ dense in Y , Y being T_2-closed . The morphisms of this category are , of course , the extendible continuous maps . The forgetful functor is supposed to be the functor "domain" .

Some remarks on the bibliography :

I. Papers concerned with (concepts strongly related to) the concept of t o p o l o g i c a l f u n c t o r or - at least - working with some variation of this concept :

An 1-2 , Brr (implicitly) , Be , Bor , Bod , Brü , B H ,
Ca 1 , Er 1-2 , Fa , He 2 , He 1o , Ho 1 , Ho 2 , Ho 3 ,
Ho 4 , Ho 5 , Ho 7 , Ho 8 , Ho 9 , HuSh , Hu 1-5 , Ka ,
Kn 1 , Mn1 , Mn 2 , Mn 3 , Mar , M P , Mü , N 2 , O , Ro ,
Sh 1 , Wi 1-5 , Wy 1-2 , Wy 3 ;
furthermore cf. Bob 1 , Bob 2 , Ce , Pu , Ta .

We have tried to make this list as complete as possible (the list is based on Brü , Ho 1 , Wy 1 , and , in particular , on Ho 5) .

II. Papers concerned with f a c t o r i z a t i o n

(a) of m o r p h i s m s :

Bar 2 , He 3 , He 5 , He 6 , Kel , Kn 2 , Ri , So , Th ;

(b) of c o n e s :

Ca 2 , Fa (implicitly) , He 1 , Ho 2 , Ho 3 , Ho 4 , Ho 5 ,
Mar , Ti (implicitly) .

III. Without any claim for completeness (!) we mention the following papers (including some material) on f i b r a t i o n s :

Eh , Ga 1-2 , Go , La , Le , (So) .

(Here , topological functors are , of course , excluded .)

IV. Papers on " s o b e r " s p a c e s (= " p c - s p a c e s " , i.e. "point closure spaces") :

Ar , Bar 2 (implicitly) , Bln , Ho 6 , Ho 1o , N 1 , N W , Sk (implicitly) .

V. Papers on e p i m o r p h i s m s in some "familiar" categories :

Bar 1 , Bu , (Ko) , (Pr) , Sd , Sk .

Bibliography :

Ad Adnadžević , D.: Some properties of A-spaces . Soviet Math. Dokl. 14,492-496 (1973)

An 1 Antoine ,Ph.: Extension minimale de la catégorie des espaces topologiques . Compt.Rend.Acad.Sci.Paris , sér.A , 262, 1389-92 (1966)

2 - - : Etude élémentaire des catégories d'ensembles structurés . Bull.Soc.Math.Belgique 18,142-164 and 387-414 (1966)

Ar Artin , M. , A.Grothendieck , J.Verdier : Theorie des topos et cohomologie étale des schémas . Lect.Notes in Math. 269 , Berlin - Heidelberg - New York : Springer 1972 (revised version of SGA 4 1963/64)

B B Banaschewski, B. and G.Bruns : Categorical Characterization of the Mac Neille completion . Arch.d.Math.18,369-377 (1967)

Bar 1 Baron , S.: Note on epi in T_o . Canad.Math.Bull. 11, 5o3-5o4 (1968)

2 -- : Reflectors as compositions of epi-reflectors . Trans.AMS. 136,499-5o8 (1969)

Brr Barr , M.: Relational algebras . Reports of the Midwest Category Seminar IV . Lect. Notes in Math. 137,39-55 . Berlin - Heidelberg - New York : Springer 197o

Be Bentley , H.L.: T-categories and some representation theorems . Portugaliae Mathematica 32,2o1-222 (1973)

Bi Birkhoff , G.: The meaning of completeness . Ann.of Math. 38,57-6o (1937)

Bln Blanksma ,T. : Lattice characterizations and compactifications . Doctoral dissertation . Rijksuniversiteit te Utrecht 1968

Blz Błaszczyk , A.: Remarks on a generalization of completeness in the sense of Čech . Colloqu.Math. 29,219-222 (1974)

Bor Borceux , F.: Structure initiales et finales . Rapport 22 (1972) . Séminaire de Math. Pure , Université Catholique de Louvain (Heverlé / Belgique)

Bob Bourbaki , N.: Eléments de mathématique :

1 Livre III : Topologie générale , Chap.IX . Paris : Hermann 1948

2 Livre I : Théorie des ensembles . Paris : Hermann 1957

Bod Bourdaud , G.: Foncteurs à structures initiales . Esquisses mathématiques (Paris) 8 (197o)

Brü Brümmer , G.C.L. : A categorical study of initialtity in uniform topology . Thesis Cape Town 1971

B H - - and R.-E.Hoffmann : External characterization of topological functors (this Proceedings)

Bu Burgess , W. : The meaning of mono and epi in some familiar categories . Canad.Math.Bull. 8,759-769 (1965)

Ca 1 Căsănescu , V.E. : Familles initiales et finales . Revue Roumaine Math. pur. appl. 17,829-836 (1972)

2 - - : Catégories à images et coimages . Revue Roumaine Math. pur. appl. 18,1171-1193 (1973)

Ce Čech , E. : Topological spaces . Revised edition by Z.Frolík and M.Katětov . London - New - York - Sidney : Interscience 1966

Cs Császár , A.: Foundations of general topology . Oxford - London - New York - Paris : Pergamon Press 1963

D Davis , A.S.: Indexed systems of neighborhoods for general topological spaces . Amer.Math.Monthly 68,886-893 (1961)

Eh Ehresmann , Ch.: Catégories et structures . Paris : Dunod 1965 (and bibliography)

Er 1 Ertel , H.G. : Algebrenkategorien mit Stetigkeit in gewissen Variablenfamilien . Dissertation Düsseldorf 1972

2 - - : Topologische Algebrenkategorien . Arch.d.Math. 25,266-275 (1974)

Fa Faehling , P.: Kategorien mit ausgezeichneten Morphismenklassen . Dissertation FU Berlin 1973

Fr Freyd , P.: Abelian categories . Evanston - London : Harper and Row 1964

Ga 1 Gray , J.W.: Fibred and cofibred categories . Proc. Conf.La Jolla 1965 on Categorical Algebra , pp.21 - 83. Berlin - Heidelberg - New York : Springer 1966

2 The categorical comprehension scheme . Category Theory , Homology Theory and Their Applications III . Lect.Notes in Math. 99,242-312 . Berlin - Heidelberg - New York : Springer 1969

Go Grothendieck , A.: Catégories fibrées et descente . Séminaire de géométrie algébrique de l'I.H.E.S. (Paris) 1961

Ha Harris , D.: Katětov extension as a functor . Math.Ann. 193,171-175 (1971)

He 1 Herrlich , H.: Topologische Reflexionen un Coreflexionen . Lect.Notes in Math. 78 , Berlin - Heidelberg - New York : Springer 1968

2 Topological functors . Gen.Top.Appl.4,125-142 (1974)

3 Factorizations of morphisms B → FA .Math.Z.114,18o-186 (197o)

4 Epireflective subcategories of TOP need not be cowellpowered . Prelimnary report

5 A generalization of perfect maps . Proc.Third Prague Symp.Gen.Top.(1971) 187-191 , Prague 1972

6 Perfect subcategories and factorizations . Colloquia Math.Soc.J.Bolyai , 8.Topics in Topology , Keszthely (1972) , 387-4o3 (1974)

7 Regular-closed , Urysohn-closed and completely Hausdorff-closed spaces . Proc.Amer.Math.Soc. 26, 695-698 (197o)

8 A concept of nearness . Gen.Top.Appl. 4,191-212 (1974)

9 On the extendibility of continuous functions . Gen.Top.Appl. 4,213-215

1o Cartesian closed topological categories . Math.Colloq. Univ.Cape Town 4 (1974)

HeS - - and G.Strecker : Category Theory . Boston : Allyn and Bacon 1973

Ho 1 Hoffmann , R.-E.: Die kategorielle Auffassung der Initial- und Finaltopologie . Dissertation Bochum 1972

2 Semi-identifying lifts and a generalization of the duality theorem for topological functors . Math. Nachrichten (to appear)

3 Factorization of cones

4 Topological functors and factorizations . Arch.d.Math. 26,1-7 (1975)

5 (E,M)-universally topological functors . Habilitationsschrift Universität Düsseldorf 1974

6 Charakterisierung nüchterner Räume . Manuscripta Math. 15,185-191 (1975)

7 A characterization of a class of categories of topological spaces

8 Eine Problemstellung in der Kategoriellen Topologie (in preparation)

9 Completion of faithful functors (in preparation)

1o Sobrification of partially ordered sets (in preparation)

HuSh Hunsaker , W.N. , and P.L.Sharma : Universally initial functors in topology . Manuscript Southern Illinois University , Carbondale / Illinois . (1974)

Hu 1 Hušek,M.: S-categories . Comm.Math.Univ.Carol. 5,37-46 (1964)

2 Generalized proximity and uniform spaces I . Comm.Math. Univ.Carol. 5,247-266 (1964)

3 Categorial methods in topology . Proc.Symp.Prague 1966 on General Topology , pp.190 -194 . New York - London - Prague 1966

4 Construction of special functors and its applications . Comm.Math.Univ.Carol. 8,555-566 (1967)

5 Categorial connections between generalized proximity spaces and compactifications . Proc.Symp.Berlin 1967 on Extension Theory of Topol. Structures , pp.127-132 . Berlin : VEB-Verl.d.Wiss. 1969

Ka Kamnitzer , S.H.: Protoreflections , relational algebras and topology . Thesis Cape Town 1974

Kel Kelly , M.G.: Monomorphisms , epimorphisms , and pullbacks . J.Aust.Math.Soc. 9,124-142 (1969)

Kn 1 Kennison , J.F.: Reflective functors in general topology and elsewhere . Trans.Amer.Math.Soc.118,303-315 (1965)

2 - - : Full reflective subcategories and generalized covering spaces . Illinois J.Math. 12,353-365 (1968)

Ko Kowalsky , H.-J.: Kategorien topologischer Räume . Math.Z. 77,249-272 (1961)

La Lavendhomme , R.: Cohomologie des catégories . Université Catholique de Louvain (Heverlé / Belgique)

Le Leroux , P.: Structure et sémantique abstraites : extension à des catégories de morphismes d'une paire de foncteurs adjoints . Lect.Notes in Math. 195 , pp.154-198 . Berlin - Heidelberg - New York : Springer 1966

ML Mac Lane , S.: Categories for the working mathematician . Berlin - Heidelberg - New York : Springer 1971

Mn 1 Manes , E.G.: Compact Hausdorff objects . Gen.Top.Appl. 4,341-36o (1974)

2 - - : A pullback theorem for triples in a lattice fibering with applications to algebra and analysis . Algebra Univ. 2,7-17 (1972)

3 --- : Algebraic Theories . Springer Graduate Texts in Math. (to appear)

Mar Marny , Th.: Rechts-Bikategoriestrukturen in topologischen Kategorien . Dissertation FU Berlin 1973

M P Menu , J. and A.Pultr : On categories which are determined by poset- and set-valued functors . Comment.Math.Univ.Carol. 15,665-678

Mü Müller , H.: Informal Manuscript . Universität Bielefeld (1974)

N 1 Nel , L.D.: Lattices of lower semi-continuous functions and assocated topological spaces . Pac.J.Math.4o,667-673 (1972)

2 - - : Initially structured categories and cartesian closedness Manuscript (1975)

N W - - and R.G.Wilson : Epireflections in the category of T_0-spaces . Fund.Math. 75,69-74 (1972)

O Osius , G.: Eine axiomatische Strukturtheorie . Dissertation FU Berlin 1969

Po Porter , J.R.: Lattices of H-closed extensions . Bull. de l'Acad.Polon.des Sci. 22,831-837 (1974)

Pr Preuß , G.: Allgemeine Topologie . Berlin - Heidelberg - New York : Springer 1972

Pu Pupier , R.: Methodes fonctorielles en topologie générale. Université de Lyon I . 1971

P T Pumplün , D. and W.Tholen : Covollständigkeit vollständiger Kategorien . Manuscripta Math. 11,127-14o (1974)

Ra Rattray , B.: Torsion Theories in Non-additive Categories . Manuscripta Math. 12,285-3o5 (1974)

Ri Ringel , C.M.: Diagonalisierungspaare I. Math.Z. 117, 249-266 (197o) . II. Math.Z. 122,1o-32 (1971)

Ro Roberts , J.E.: A characterization of initial functors . J.of Algebra 8,181-193 (1968)

Sd Schröder , J.: Epi und extremer Mono in $\underline{T}_{2a}$. Arch.d.Math. 25,561-565 (1974)

Sb 1 Schubert , H.: Categories . Berlin - Heidelberg - New York : Springer 1972

2 - - : Topologie . Stuttgart : Teubner 1964

Sh 1 Shukla , W.: On top categories . Thesis Indian Institute of Technology Kanpur 1971

2 - - and A.K.Srivastava : Local reflectivity + stable reflectivity = reflectivity .Gen.Top.Appl. 5,61-68 (1975)

Sk Skula , L.: On a reflective subcategory of the category of all topological spaces . Trans.Amer.Math.Soc. 142,37-41 (1969)

So Sonner , J.: Canonical categories . Proc.Conf. La Jolla 1965 on Categorical Algebra , pp.272-294 . Berlin - Heidelberg - New York : Springer 1966

Sr Srivastava , A.K.: Non-genuine adjunctions . Thesis Indian Institute of Technology New Delhi 1974

Ta Taylor , J.C.: Weak families of maps . Canad.Math.Bull. 8,771-781 (1965)

Th Tholen , W.: Relative Bildzerlegungen und algebraische Kategorien . Dissertation Münster 1974

Ti Tiller , J.: Unions in E-M categories and coreflective subcategories . Comm.Math.Univ.Carol.15,173-187 (1974)

Wi 1 Wischnewsky , M.B.: Initialkategorien . Dissertation München 1972

2 Partielle Algebren in Initialkategorien . Math.Z. 127,83-91 (1972)

3 Algebra-Berichte 1o and 16 (1973)

4 Coalgebras in reflective and coreflective subcategories Algebra Universalis 4,328-335 (1974)

5 On the boundedness of topological categories . Manuscripta Math. 12,2o5-215 (1974)

Wy 1 Wyler , O.: On the categories of general topology and topological algebra . Arch.d.Math.12,7-17 (1971)

2 - - : Top categories and categorical topology . Gen.Top.Appl. 1,17-28 (1971)

3 - - : Quotient maps . Gen.Top.Appl. 3,149-16o (1973)

4 - - : An unpleasant theorem on limit spaces (unpublished)

CATEGORY THEORETICAL METHODS IN TOPOLOGICAL ALGEBRA

Karl Heinrich Hofmann

INTRODUCTION

My contributions to this conference will consist of some comments on the way category theoretical thinking has had an impact on the mathematics of topological algebra. Traditionally, topological algebra is understood to concern the study of algebraic structures enriched by the presence of topology and the continuity of algebraic operations; this includes the theory of topological groups, semigroups, rings, fields. But I also include areas of what is generally called functional analysis. In this way Banach algebra, operator algebras, representation theory, harmonic analysis all become subsumed under the concept of topological algebra. Right in the beginning, the name of "topological algebra" is, therefore recognized itself as a very general label. As far as category theory is concerned, many will maintain that it permeates everything mathematical. However, what matters here is the analysis of functors, their preservation properties, the theme of limits and that of monoidal structures with its variations, and how all of this is to be applied and utilized in solving problems inside topological algebra.

I am, therefore, not speaking as a professional category theoretician, since I cannot claim to be one, but as one whose experience lies in topological algebra. In allusion to the title of MacLane's book [23], I might say that I will concentrate on category theory for the working topological algebraist and analyst.

It appears to me that there is a certain difference between pure category theory and applied category theory (that is, the technique of applying functorial methods to "concrete" problems arising in various branches of mathematics) which in character resembles that which has traditionally been felt as the difference between pure and applied mathematics. The pure categorist will seek the highest level of generality and abstraction which will cover the largest possible area he can conceive of (and then some). I visualize the problem of the working mathematician who seeks out functorial methods to help him solve his problems somewhat differently. His problem is to find the "right pitch of generality", a term which I hope will be accepted as an undefined concept. The degree of generality will in any case be "right" if it adequately covers the problems at hand, hopefully allows to attack related and even more general ones; but the concrete root of the functorial concepts used should still be visible and serve as the guideline for the abstract functorial setting which is used. I consider it my task to explain by illustration what is meant.

Evidently, my selection cannot be encyclopedic even for the case of topological algebra. It is clearly determined by some topics I have been interested in myself together with my colleagues and students, and I disclaim any attempt towards completeness.

ACKNOWLEDGEMENTS

To Ernst Binz and Horst Herrlich go my thanks for having invited me to a conference which consolidates categorical topology as a new discipline: on one hand it has emerged as a new field of applications of functor theory and on the other it provides new impulses to research in general topology, a branch of mathematics which, not so long ago, was considered as rather stable (see Dieudonné [9]). I further express my thanks to the National Science Foundation for having supported through the years some of the research on which I will report, and to Deborah Casey and Meredith Mickel for typing the manuscript of this article.

TABLE OF CONTENTS

CHAPTER I. LIMITS IN TOPOLOGICAL ALGEBRA

Through the work of von Neumann and Weyl it was known in the twenties that every compact group could be "approximated" by compact linear and therefore by compact Lie groups. In the early original proofs of the duality theorem for locally compact abelian groups by Pontryagin, van Kampen and Weil it was used, at least implicitly that every locally compact abelian group could be, in the same vein, be approximated by certain elementary abelian groups build up from vector groups, finite dimensional torus groups, and finitely generated discrete groups. As far as the general structure theory of (let us say, connected) locally compact groups is concerned, half a century of development culminated in the fifties in the results of Gleason, Montgomery, Yamabe, which showed that every such group is a projective limit of Lie groups.

Much of the philosophy of structural investigations of locally compact groups and transformation groups is based on the idea to utilize the wealth of information available for Lie groups (J. Dieudonné [9, p. 77]: Les groupes de Lie sont devenus le centre des mathématiques; on ne peut rien faire de sérieux sans eux), and to lift this information to the general case by passing to the limit; even before the final solution to Hilbert's Fifth Problem through Gleason, Montgomery and Yamabe this philosophy was well established through Iwasawa's key work on locally compact groups [21]. This technique has become absolutely indispensible for almost anything that has to do with locally compact groups, be it structure theory, representation theory, harmonic analysis, probability theory, or topological structure theory. Similar situations as those indicated for topological groups arise, e.g., in the theory topological semigroups, for which S. Eilenberg showed as early as 1937 [10] that every compact topological semigroup can be approximated by metric ones equipped with subinvariant metrics. More generally, they arise in universal algebra where in many varieties of compact universal algebras, all zero dimensional algebras are profinite, such as is exemplified by rings, semigroups and lattices [25]. It is noteworthy question concerning compact topological algebras to determine why for certain varieties all compact zero dimensional are profinite while this fails for others. Banaschewski gave me an example of a variety defined by one unary operation in 1970 in which not every compact zero dimensional algebra was profinite; at this conference Linton showed me this behavior in the variety of Jonnson-Tarski algebras (see Linton's paper in these Proceedings); the first systematic studies in the general area are recent and are due to Choe [3,4]. Related results are in [1].

While the formation of projective limits is a traditional technique, its fully functorial aspects were not fully utilized in topology and topological algebra until the late sixties. A typical application of the functorial scheme of projective systems in topology is that of shape theory for which I refer to the contribution of Mardešić in these Proceedings. Some applications in topological algebra which originated just a little earlier I will describe later. It should be emphasized, however, that the principle of these considerations was clearly formulated as early as 1945 in Eilenberg's and MacLane's lead article on category theory [11] (whose existence seems to be much better recalled than its content).

One might ask at this point what might be so special about projective limits in topological algebra when category theory provides in fact a general theory for all limits. The simple answer is that in topological algebra numerous important functors arise naturally, often as left adjoints which do not at all preserve arbitrary limits but which do preserve projective limits. A prime example is the Čech cohomology functor on compact spaces or algebras [14,19]; in analysis the functor which associates with a topological group its enveloping von Neumann algebra preserves strict projective limits of locally compact groups (we will speak of *strict projective* limits if all occurring maps are proper and surjective) [13]; the mapping cylinder functor for compact spaces and semigroups preserves projective limits, projective limits of inverse systems with surjective maps [20]; Pommer's functor is another example [26]. In fact, for the compact situation, Stralka and I gave a general lemma which illustrates how the preservation of (strict) projective limits may arise [20, p. 225].

In this chapter we analyse the functorial methods which are necessary to deal with limit constructions in topological algebra and describe some of the applications.

LIMITS

In order to fix notation, let us consider categories X and A; for the most part we imagine X to be small although formally in the definitions this plays not much of a role, but does so when it comes to the existence of limits for functors defined on X with values in A. Let A^X be the category of functors $f: X \longrightarrow A$ with natural transformations $\alpha: f \longrightarrow g$ between such functors as morphisms.

1.1 DEFINITION. i) The functor $X \longrightarrow A$ whose value is constantly equal to $A \in \text{ob } A$ on objects and constantly equal to $1_A \in \text{morph } A$ on morphism is called *the constant functor* on X with values A and it is written $A_X: X \longrightarrow A$.

ii) A functor $f: X \longrightarrow A$ is said to have a *limit* if there is an object $\lim f \in \text{ob } A$ and a natural transformation $\lambda^f: (\lim f)_X \longrightarrow f$ such that for every natural transformation $\alpha: A_X \longrightarrow f$ there is a unique morphism $\alpha': A \longrightarrow \lim f$ in A such that $\lambda^f \alpha'_X = \alpha$. We call λ^f the *limit morphism*, and $\lim f$ the *limit object*. □

We note that the assignment $\alpha \longmapsto \alpha': A^X(A_X, f) \longrightarrow A(A, \lim f)$ is a natural bijection with inverse $\phi \longmapsto \lambda^f \phi_X$. Thus we obtain immediately

1.2 REMARK. *If every functor* $f: X \longrightarrow A$ *has a limit, then the functor* $A \longmapsto A_X: A \longrightarrow A^X$ *has a (right) adjoint* $\lim: A^X \longrightarrow A$. □

In particular, the limit construction is functorial. A typical domain category X which may occur in topological algebra (and elsewhere) is a partially ordered set. We recall:

1.3 CONVENTION. Every *poset* (= partially order set) $(X, \leq)$ is in particular a category with the elements of X as objects and an arrow $x \longrightarrow y$ between x and y if and only if $y \leq x$. A function $f: X \longrightarrow Y$ between posets is a functor iff it preserves the partial order. If $f: X \longrightarrow Y$ is such a functor, then $\lim f$ exists iff $\sup f(X)$ exists and $\lim f = \sup f(X)$. □

A pair $(A, |\ |)$ consisting of a category A and a faithful functor $|\ |: A \longrightarrow \underline{\text{Set}}$ is called a *concrete* or *set based* category. Virtually all categories of interest in topological algebra are set based. A morphism is a set based category is called *injective* [resp. *surjective*, *bijective*] if $|\ |$ is an injective [sur-, bijective] function.

1.4 DEFINITION. A *projective system* is a functor $f\colon X \longrightarrow A$ where X is an upwards directed poset. If f is a projective system, then $\lim f$ is a *projective limit*. If $(A,|\ |)$ is a set based category, then f is called a *strict projective system* if all morphisms λ^f_x, $x \in X$ are surjective, and, accordingly, $\lim f$ is called a *strict projective limit*. □

Typically, one might have $X = \mathbb{N}$, $X = \{1,2,3,\ldots\}$ with its natural order. If, by way of example, A is the category Comp G of compact groups, then $A^X = \text{Comp G}^{\mathbb{N}}$ is the category, whose objects are the inverse sequences

$$G_1 \leftarrow G_2 \leftarrow \ldots$$

of compact groups; morphisms are sequences of morphisms $\alpha_n\colon G_n \longrightarrow H_n$ such that the infinite ladder with these as rungs commutes. The projective limit of each of these sequences exists, and by 1.2 we know that $\lim$ is in fact a functor, which of course one verifies immediately by direct inspection.

TRANSFORMATION OF DOMAIN

In practice however, it occurs frequently that we are given a *transformation of index categories*, i.e., a functor $f\colon X \longrightarrow Y$ and a diagram, i.e., a functor $F\colon Y \longrightarrow A$, and we would like to compare the limits of $Ff\colon X \longrightarrow A$ and of $F\colon Y \longrightarrow A$, if they exist. We first make the following simple observation:

1.5 LEMMA. *The assignment* $F \longmapsto Ff\colon \text{ob}\, A^Y \longrightarrow \text{ob}\, A^X$ *and* $\alpha \longmapsto \alpha f\colon \text{morph}\, A^Y \longrightarrow \text{morph}\, A^X$ *(where* $(\alpha f)_x = \alpha_{f(x)}$*) is a functor* $A^Y \longrightarrow A^X$ *which we will denote by* A^f. *In particular, we have a natural map of sets*

$$A^f_{F,G}\colon A^Y(F,G) \longrightarrow A^X(Ff,Gf).\ \square$$

If $A \in \text{ob}\, A$, then, in particular, we note that $A^f(A_Y) = A_Y f = A_X$. If both $F\colon Y \longrightarrow A$ and $Ff\colon X \longrightarrow A$ have a limit, then by the universal property of the limit of F there is a unique $\mathcal{A}$-morphism ${}_F f\colon \lim F \longrightarrow \lim Ff$ such that $\lambda^F f = \lambda^{Ff}({}_F f)_{\mathcal{X}}$:

$$\begin{array}{ccc} \lim F & \xrightarrow{\quad {}_F f \quad} & \lim Ff \\ & {\scriptstyle \lambda^F f}\searrow \quad \swarrow{\scriptstyle \lambda^{Ff}} & \\ & Ff & \end{array}$$

(Note that in the top row of this diagram we left off the designation $(-)_{\mathcal{X}}$ indicating that we have in fact a diagram in $\mathcal{A}^{\mathcal{X}}$; we will continue to do this if no confusion is likely to arise.)

One observes directly the following Lemma

1.6 LEMMA. *If* $f\colon \mathcal{X} \longrightarrow \mathcal{Y}$ *and* $F\colon \mathcal{Y} \longrightarrow \mathcal{A}$ *are functors such that* F *and* Ff *have limits, then there is a unique morphism* ${}_F f\colon \lim F \longrightarrow \lim Ff$ *characterized by any of the two properties:*

(a) $\lambda^F f_{\mathcal{X}} = \lambda^{Ff}({}_F f)_{\mathcal{X}}$

(b) *The diagram*

$$\begin{array}{ccc} \mathcal{A}(A, \lim F) & \xrightarrow{\quad \cong \quad} & \mathcal{A}^{\mathcal{Y}}(A_{\mathcal{Y}}, F) \\ {\scriptstyle \mathcal{A}(A,{}_F f)}\downarrow & & \downarrow{\scriptstyle \mathcal{A}^f} \\ \mathcal{A}(A, \lim Ff) & \xrightarrow{\quad \cong \quad} & \mathcal{A}^{\mathcal{X}}(A_{\mathcal{X}}, Ff) \end{array}$$

commutes for all $A \in \operatorname{ob} \mathcal{A}$.

If $g\colon \mathcal{Y} \longrightarrow \mathcal{Z}$ *and* $G\colon \mathcal{Z} \longrightarrow \mathcal{A}$ *are functors such that* G, Gg, Ggf *have limits, then* ${}_G(gf) = {}_{Gg}f\,{}_G g$. □

Let us briefly pause to illustrate the significance for the concrete applications:

If, e.g., $G_1 \longleftarrow G_2 \longleftarrow \ldots \longleftarrow G$ is a limit diagram of compact groups, it may become necessary to consider a subsequence $G_{n_1} \longleftarrow G_{n_2} \longleftarrow \ldots \longleftarrow H$ with its limit, and to compare the limits. If the original sequence is formalized by the functor $Ff\colon \mathbb{N} \longrightarrow$ Comp G, then the subsequence is described by the functor $Ff\colon \mathbb{N} \longrightarrow$ <u>Comp G</u> with the indexing functor $f\colon \mathbb{N} \longrightarrow \mathbb{N}$, $f(k) = n_k$. The lemma says that there is a natural map $G \longrightarrow H$ with certain properties. Evidently, one raises the question when this natural map is an isomorphism, as it should be in this case.

In order to answer this question we return to Lemma 1.6. If

${}_F f\colon \lim F \longrightarrow \lim Ff$ is an isomorphism, then clearly

$$\mathcal{A}(A,{}_F f)\colon \mathcal{A}(A,\lim F) \longrightarrow \mathcal{A}(A,\lim Ff)$$

is an isomorphism for all A. By the dual of the Yoneda Lemma (which one finds in any text on category theory) this condition is also sufficient. By Lemma 1.6, the function $\mathcal{A}(A,{}_F f)$ is an isomorphism iff $\mathcal{A}^f(A_{\mathcal{Y}},F)$ is an isomorphism. On the other hand, if we assume that $\mathcal{A}^f$ is an isomorphism, then, in view of the limit property, from the diagram in 1.6 (b) we conclude that $\lim F$ exists iff $\lim Ff$ exists. We summarize:

1.7 <u>LEMMA</u>. *Consider the following statements*

(1) $\mathcal{A}^f\colon \mathcal{A}^{\mathcal{Y}}(A_{\mathcal{Y}},F) \longrightarrow \mathcal{A}^{\mathcal{X}}(A_{\mathcal{X}},Ff)$ *is bijective for all* $A \in \text{ob}\,\mathcal{A}$.

(2) $\lim F$ *exists iff* $\lim Ff$ *exists and if so, then* ${}_F f\colon \lim F \longrightarrow \lim Ff$ *is an isomorphism.*

Then (1) *implies* (2), *and if one of* $\lim F$ *or* $\lim Ff$ *exists then* (1) *and* (2) *are equivalent.* □

It is now clear that one needs a handy sufficient condition for condition (1) in 1.7 to be satisfied.

<u>COFINALITY</u>

1.8 <u>DEFINITION</u>. A functor $f\colon \mathcal{X} \longrightarrow \mathcal{Y}$ is called *cofinal* if the following two conditions hold:

(i) For each $y \in \text{ob}\,\mathcal{Y}$ there is a morphism $f(x) \longrightarrow y$.

(ii) For each pair of morphisms $\phi_j\colon f(x_j) \longrightarrow y$ in $\mathcal{Y}$, $j = 1,2$ there is a pair of morphisms $\psi_j\colon x \longrightarrow x_j$ in $\mathcal{X}$, $j = 1,2$ such that $\phi_1 f(\psi_1) = \phi_2 f(\psi_2)$.

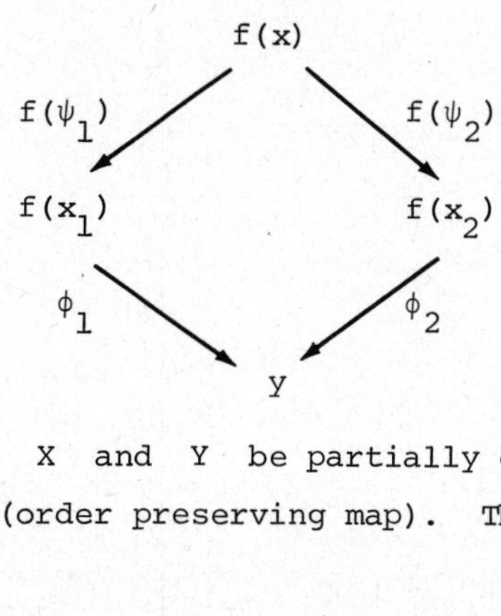

1.9 <u>EXAMPLE</u>. i) Let X and Y be partially ordered sets and $f\colon X \longrightarrow Y$ a functor (order preserving map). Then f is cofinal iff for each $y \in Y$ there is an x with $y \leq f(x)$, and if X is (upwards) directed.

ii) If $\mathcal{X}$ has pull-backs and f has an adjoint g then f is cofinal. (For each y there is the counit $\varepsilon_y\colon f(g(y)) \longrightarrow y$ of the adjunction; if $\phi_j\colon f(x_j) \longrightarrow y$, $j = 1,2$ are given, then the adjunction yields maps $\phi_j'\colon x_j \longrightarrow g(y)$; let

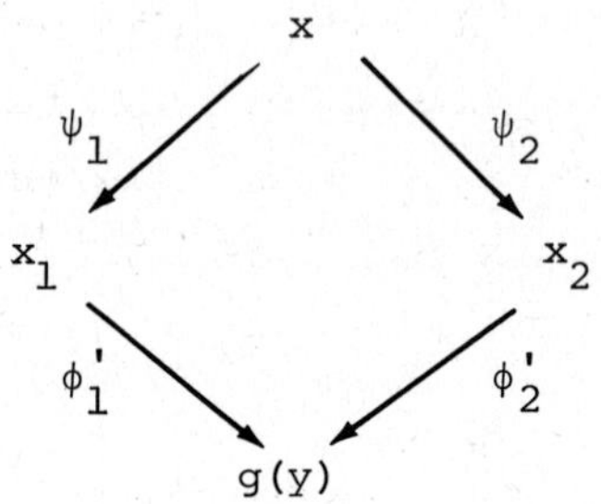

be the pull-back and verify 1.8.(ii).) □

The crucial fact on cofinality is the following:

1.10 PROPOSITION. *If* $f\colon \mathcal{X} \longrightarrow \mathcal{Y}$ *is cofinal, and* $F\colon \mathcal{Y} \longrightarrow \mathcal{A}$ *is a functor, then* F *has a limit iff* Ff *has a limit and if these exist, then*

$$F^f\colon \lim F \longrightarrow \lim Ff$$

is an isomorphism.

Proof. By Lemma 1.7 it suffices to show that $\mathcal{A}^f\colon \mathcal{A}^{\mathcal{Y}}(A_{\mathcal{Y}},F) \longrightarrow \mathcal{A}^{\mathcal{X}}(A_{\mathcal{X}},Ff)$ is an isomorphism. Condition 1.8.(i) is readily applied to show that $\mathcal{A}^f$ is injective (if $\alpha f = \beta f$, then for each $y \in \mathrm{ob}\,\mathcal{Y}$ there is a morphism $\pi\colon fx \longrightarrow y$, so that $\alpha_y = \alpha_{fx} = (\alpha f)_x = (\beta f)_x = \beta_{fx} = \beta_y$); on the other hand, condition 1.8.(ii) suffices to show the surjectivity of A^f (if $\gamma\colon A_{\mathcal{X}} \longrightarrow Ff$ is given, use 1.8.(ii) to observe that for each $y \in \mathrm{ob}\,\mathcal{Y}$ and each morphism $\phi\colon fx \longrightarrow y$ the composition $(F\phi)\gamma_x$ is independent of ϕ and defines an $\alpha\colon A_{\mathcal{Y}} \longrightarrow F$ with $A^f\alpha = \gamma$.) □

Any example of the type discussed after Lemma 1.6 can be completely answered by Proposition 1.10. We will now discuss other examples more systematically. By contrast with our first typical example of limits occurring in topological algebra, the projective limits, we encounter in here a potentially very large diagram whose limit we wish to calculate. The tool to achieve this is the cone category.

CONE CATEGORIES

Suppose that we are given a functor $J: \mathcal{A}_o \longrightarrow \mathcal{A}$ (which most frequently is the inclusion functor of a full subcategory). We begin by taking a fixed object A in $\mathcal{A}$ and by defining the *cone category* (A,J) over J with vertex A: Its objects are pairs $(\phi,x) \in \text{morph}\,\mathcal{A} \times \text{ob}\,\mathcal{A}_o$ where $\phi: A \longrightarrow Jx$, its morphisms $(\phi,x) \longrightarrow (\psi,y)$ are $\mathcal{A}_o$-maps $m: x \longrightarrow y$ such that $(Jm)\phi = \psi$

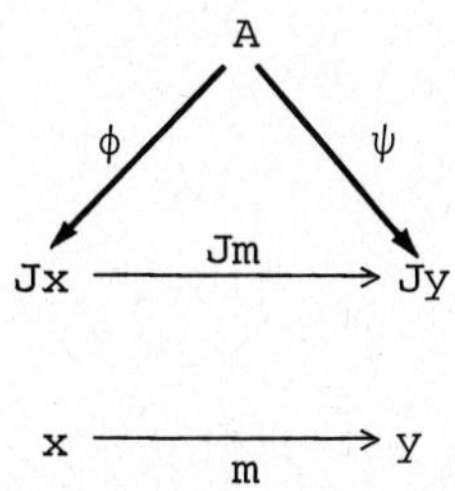

The cone category is a special case of the more general *comma category* invented by Lawvere. The assignment which associates with an object $(\phi,x) \in \text{ob}\,(A,J)$ the element $x \in \text{ob}\,\mathcal{A}_o$ and with a morphism $m: (\phi,x) \longrightarrow (\psi,y)$ in (A,J) the morphism $m: x \longrightarrow y$ in $\mathcal{A}_o$ is a functor $P_A: (A,J) \longrightarrow \mathcal{A}_o$. Let us now suppose that $\pi: A \longrightarrow A'$ is a morphism in $\mathcal{A}$. Then there is a functor $(\pi,J): (A',J) \longrightarrow (A,J)$ given by $(\pi,J)(\phi,x) = (\phi\pi,x)$ and $(\pi,J)(m) = m$.

In order to understand how the assignment $A \longmapsto (P_A: (A,J) \longrightarrow \mathcal{A}_o)$ is functorial we should expand our concept of functor categories. In the first part of this chapter we considered functor categories $\mathcal{A}^{\mathcal{X}}$; for many purposes this is not sufficient; we have to allow the variation of the domain category X. Typically, the diagrams P_A have the variable domain (A,J). A similar phenomenon appears if we associate with a compact group G the system of all closed normal subgroups N such that G/N is a Lie group. For all of these purposes we need the category of "all diagrams", barring set-theoretical difficulties.

FUNCTOR CATEGORIES

The objects of the category which we are to describe will be functors $F: \mathcal{X} \longrightarrow \mathcal{A}$ with a fixed category $\mathcal{A}$. In order to avoid set theoretical qualms we assume that $\mathcal{X}$ is small. There are two ways of transforming functors:

i) If $f\colon \mathcal{Y} \longrightarrow \mathcal{X}$ is a functor and $F\colon \mathcal{X} \longrightarrow \mathcal{A}$ an object, we have an object $Ff\colon \mathcal{Y} \longrightarrow \mathcal{A}$ and we consider

$$\begin{array}{ccc} Y & \xrightarrow{\quad Ff \quad} & A \\ {\scriptstyle f}\Big\downarrow & & \Big\Downarrow \\ X & \xrightarrow[\quad F \quad]{} & A \end{array}$$

as a transformation from F to Ff (!) which we denote with a vertical double arrow

$$\begin{array}{c} Ff \\ {\scriptstyle f}\Big\Downarrow \\ F \end{array}$$

ii) If $G_1, G_2\colon Y \longrightarrow A$ are two objects, then we may have natural transformation $G_1 \xrightarrow{\alpha} G_2$.

A pair

$$\begin{array}{ccc} (\alpha, f) \; = \; Ff & \xrightarrow{\quad \alpha \quad} & G \\ {\scriptstyle f}\Big\Downarrow & & \\ F & & \end{array}$$

of such transformation we will declare to be a morphism $F \longrightarrow G$ from F to G. We have to explain, how these morphisms compose. For this purpose let $(\alpha,f)\colon F \longrightarrow G$ and $(\beta,g)\colon G \longrightarrow H$ be two morphisms. The following scheme explains the composition law $(\beta,g)(\alpha,f) = (\beta(\alpha g), fg)\colon F \longrightarrow H$

$$\begin{array}{ccccc} Ffg & \xrightarrow{\quad \alpha g \quad} & Gg & \xrightarrow{\quad \beta \quad} & H \\ {\scriptstyle g}\Big\Downarrow & & {\scriptstyle g}\Big\Downarrow & & \\ Ff & \xrightarrow{\quad \alpha \quad} & G & & \\ {\scriptstyle f}\Big\Downarrow & & & & \\ F & & & & \end{array}$$

It is routine to check that we have indeed defined a category. We record:

1.11 <u>DEFINITION</u>. If A is a category, then $A^{\underline{\mathrm{cat}}}$ denotes the category whose objects are functors $F\colon X \longrightarrow A$ with small domains and whose morphisms are pairs $(\alpha, f)\colon F \longrightarrow G$ where $f\colon \mathrm{dom}\, G \longrightarrow \mathrm{dom}\, F$ is a functor and $\alpha\colon Ff \longrightarrow G$ a natural transformation; the law of composition is defined by $(\beta, g)(\alpha, f) = (\beta(\alpha g), fg)$. The category is called the *general functor category over* A. ▯

We note that there are variations of this theme obtained by reversing various arrows in the appropriate spots. For the purpose of topological algebra, the version which we introduced is the most convenient one. One should note the similarity of the law of composition in $A^{\underline{\mathrm{cat}}}$ with that of a semidirect product, taking into account the contravariance in the second argument. This law of composition is also familiar in sheaf and bundle theory.

We observe that the simple functor categories of 1.1 are subcategories of $A^{\underline{\mathrm{cat}}}$. For the singleton category $A = \mathbb{1}$ we retrieve the category $\underline{\mathrm{cat}} \cong \mathbb{1}^{\underline{\mathrm{cat}}}$ of (small) categories and functors.

We observed right in the beginning that the formation of limits is functorial on the simplest instance of a functor category. Now we have to see to what extent the limit functor (1.2) is functorial on the general functor category (1.11) or a subcategory thereof.

<u>LIMITS REVISITED</u>

Let C be a full subcategory of $A^{\underline{\mathrm{cat}}}$ such that for all morphisms

$$(\alpha, f) = \begin{array}{ccc} Ff & \xrightarrow{\ \alpha\ } & G \\ {\scriptstyle f}\Big\Downarrow & & \\ F & & \end{array} \qquad \text{in morph } C$$

the limits $\lim F$, $\lim Ff$ and $\lim G$ exist. As an example we might take for C the subcategory $A^{\underline{\mathrm{dir}}}$ of all projective systems (where <u>dir</u> is the category of up-directed sets and order preserving maps (see 1.3)) or where C might be the category of all strict projective systems (1.4).

Suppose that

$$(\alpha, f) = \begin{array}{ccc} Ff & \longrightarrow & G \\ {\scriptstyle f}\Big\Downarrow & & \\ F & & \end{array}$$

is a morphism $F \longrightarrow G$ in C. Then we define a morphism $\lim(\alpha,f)\colon \lim F \longrightarrow \lim G$ by

$$\begin{array}{ccc} \lim Ff & \xrightarrow{\lim \alpha} & \lim G \\ {\scriptstyle F^f}\uparrow & \nearrow{\scriptstyle \lim(\alpha,f)} & \\ \lim F & & \end{array}$$

(see 1.2, 1.6). A commutative "diagram"

$$\begin{array}{ccc} Ffg & \xrightarrow{\alpha g} & Gg \\ {\scriptstyle g}\Downarrow & & \Downarrow{\scriptstyle g} \\ Ff & \xrightarrow[\alpha]{} & G \end{array}$$

of functor transformations gives rise to a commutative diagram

$$\begin{array}{ccc} \lim Ffg & \xrightarrow{\lim \alpha g} & \lim Gg \\ {\scriptstyle Ff^g}\uparrow & & \uparrow{\scriptstyle G^g} \\ \lim Ff & \xrightarrow{\lim \alpha} & \lim G \end{array}$$

This remark allows us to verify readily that

$$\lim(\beta(\alpha g), fg) = \lim(\beta, g)\lim(\alpha, f).$$

Thus we have:

1.12 PROPOSITION. *Let* C *be a full subcategory of* $A^{\underline{cat}}$ *such that for each* $(f,\alpha)\colon F \longrightarrow G$ *in* C *the limits* $\lim F$, $\lim Ff$, $\lim G$ *exist. Then the prescription* $\lim(\alpha,f) = (\lim \alpha)_F{}^f$ *defines a functor* $\lim\colon C \longrightarrow A$. □

1.13 PROPOSITION. *Let* $J\colon A_o \longrightarrow A$ *be a functor (with* A_o *small). Let* $\underline{cone}$ *be the full subcategory of* $\underline{cat}$ *generated by all cone categories* (A,J), $A \in \text{ob } A$. *Then there is a functor* $\Delta\colon A \longrightarrow A_o^{\underline{cone}}$ *given by* $\Delta(A) = P_A\colon (A,J) \longrightarrow A_o$, $\Delta(\pi\colon A \longrightarrow A') = (1_{P_{A'}}, (\pi, J))$:

$$\begin{array}{ccc} (A',J) & \xrightarrow{\ 1\ } & (A',J) \\ {\scriptstyle (\pi,J)}\Big\| & & \\ (A,J) & & \end{array} \quad . \ \square$$

1.14 <u>LEMMA</u>. *If* $\phi\colon A \longrightarrow B$ *is a functor, then there is a functor* $\phi^{\underline{cat}}\colon A^{\underline{cat}} \longrightarrow B^{\underline{cat}}$ *given by* $\phi^{\underline{cat}}(F) = \phi F$, $\phi^{\underline{cat}}(\alpha,f) = (\phi\alpha,f) =$

$$\begin{array}{ccc} \phi F f & \xrightarrow{\ \phi\alpha\ } & \phi G \\ {\scriptstyle f}\Big\| & & \\ \phi F & & \end{array}$$

If $\mathcal{D}$ *is a full subcategory of* <u>cat</u>, *then* $\phi^{\underline{cat}}$ *restricts and corestricts to a functor* $\phi^{\mathcal{D}}$. $\square$

Suppose that under the conditions of 1.13 the category A satisfies the hypothesis that all functors $f\colon (A,J) \longrightarrow A$ have a limit for all $A \in \mathrm{ob}\, A$, then the composition

$$A_o^{\underline{cone}} \xrightarrow{\ J^{\underline{cone}}\ } A^{\underline{cone}} \xrightarrow{\ \lim\ } A$$

is a well-defined functor. If $A \in \mathrm{ob}\, A$, then we have a natural transformation $\alpha^A\colon A_{(A,J)} \longrightarrow JP_A$ given by $\alpha^A_{(\phi,x)} =$ $\phi\colon A \longrightarrow JP_A(\phi,x) = Jx$. Hence there is a unique morphism $\rho_A\colon A \longrightarrow \lim JP_A = (\lim J^{\underline{cone}}\,\Delta)(A)$ such that $(\lambda^{JP}A)\rho_A = \alpha^A$, by the universal property of the limit (1.1). The following is then directly verified:

1.16 <u>LEMMA</u>. $\rho\colon Id_A \longrightarrow \lim J^{\underline{cone}}\,\Delta(A)$ *is a natural transformation. If* $A \in A_o$ *then* $\lim J^{\underline{cone}}\,\Delta(A) = \lim JP_{JA} = JA$ *and* $\rho_{JA} = 1_{JA}$, *i.e.,* $\rho J = 1\colon J \longrightarrow J$. $\square$

If J is an inclusion functor, then ρ_A is an isomorphism iff the cone (A,J) is a limit cone, i.e., iff $A \cong \lim JP_A$. This means that every object A in A can be canonically approximated by elements in A_o. We therefore are led to the following definition:

1.16 <u>DEFINITION</u>. A functor $J\colon A_o \longrightarrow A$ is called *canonically dense* (and if J is an inclusion functor then A_o is called *a canonically dense subcategory*) if $\rho: \mathrm{Id}_A \longrightarrow \lim J^{\underline{\text{cone}}} \Delta$ is a natural isomorphism. Here $\underline{\text{cone}}$ is the full subcategory of $\underline{\text{cat}}$ generated by all cones (A,J), and A is assumed to have limits for all functors $f: (A,J) \longrightarrow A$. □

Note that we do not require here that A_o be small.

We illustrate this functorial set-up in terms of a basic concept in topological group theory which we touched upon in the introduction, the concept of approximating groups via projective limits.

PRO-P-GROUPS

We will speak of a "*property*" of topological groups while having in mind properties such as being abelian, finite, a compact Lie group, a Lie group and so on. In our present framework it is most convenient to describe a "property" by singling out a full subcategory P of the category Top G of topological groups. The statement $G \in \mathrm{ob}\, P$ can then be formulated as saying "G has property P"; we will also say that G is a P-group.

1.17 DEFINITION. Let P be a full subcategory of Top G. Assume the following hypotheses:

(1) All singleton groups are in P.

(2) If $G \in \mathrm{ob}\, P$ and $H \cong G$, then $H \in \mathrm{ob}\, P$.

Then P is called a *property of topological groups*, and a group $G \in \mathrm{ob}\, P$ is called a *P-group*.

For each $G \in$ Top G set $N_P(G) = \{N: N$ is a normal subgroup of G such that G/N is a P-group$\}$. Note that $G \in N_P(G)$ because of (1) above. Evidently G is a P-group iff $1 \in N_P(G)$. We say that G is a *pro-P-group* if the following conditions are satisfied:

(i) $N_P(G)$ is a filter basis.

(ii) If $\psi: G \longrightarrow K$ is a Top G morphism and K is a P-group, then $G/\ker \psi$ is also a P-group (i.e., $\ker \psi \in N_P(G)$).

(iii) The family $q_N: G \longrightarrow G/N$, $N \in N_P(G)$ is a limit natural transformation (for the projective system $N \longrightarrow G/N: N_P(G)^{op} \longrightarrow$ Top G).

The essential condition (iii) may be rephrased somewhat briefly in the form $G = \lim G/N$, $N \in N_P(G)$; by (i) this limit is a strict projective limit (1.4).

The full subcategory in Top G of all pro-P-groups is called P_{pro}. □

1.18 REMARK. If A is any variety of algebraic structures and Top A denotes the category of all topological A-algebras and continuous A-morphisms, one can equally well consider a property P in Top A and define $N_P(A)$ = {R: R is a congruence on A such that A/R is a P-algebra}; this allows to define *pro-P-algebras* through conditions (i), (ii), (iii) of 1.17 (with "P-group" replaced by "P-algebra"). □

Note. The category P_{pro} of pro-P-objects is not to be mixed up with the category of projective systems in a category T which in Mardešić' contribution in these proceedings is denoted Pro T; in our own notation this category would be denoted T^{dir} where dir is the category of all up-directed sets (see 1.3, 1.4).

The following is a list of properties in topological group and algebra theory and their associated pro-P-objects

P	P_{pro}
compact groups	compact groups
compact Lie groups	compact groups
almost connected[1] finite dimensional Lie groups	contains all almost connected[1] compact groups
elementary abelian groups[2]	contains all locally compact abelian groups
finite groups (rings, lattices, semigroups)	compact 0-dimensional groups (rings, lattices, semigroups)
finite semilattices	complete algebraic semilattices[3]
finite algebras of a given type	pro-finite algebras of a given type[4]
compact matrix semigroups	compact pro-matrix semigroups (Peter-Weyl semigroups)

[1] A topological group G is almost connected if G/G_o is compact, where G_o is the component of the identity.

[2] A topological abelian group G will be called elementary here if it is of the form $(\mathbb{R}/\mathbb{Z})^m \times \mathbb{R}^n \times D$ with natural numbers m, n and a discrete abelian group D.

[3] See [18].

[4] Recall the question we raised in the introduction: Under which conditions is a compact 0-dimensional algebra pro-finite?

We now show that $\mathcal{P}$ is canonically dense in $\mathcal{P}_{\underline{pro}}$. Let $J: \mathcal{P} \longrightarrow \mathcal{P}_{\underline{pro}}$ be the inclusion functor and $P_G: (G,J) \longrightarrow \mathcal{P}$ the projection functor which we introduced in the context of cone categories. We now define a functor $f = f_g: N_{\mathcal{P}}(G)^{op} \longrightarrow (G,J)$ as follows:

For objects: $f(N) = (q_N, G/N)$, where $q_N: G \longrightarrow G/N$ is the quotient map.

For morphisms:

$f(M \subseteq N) =$ 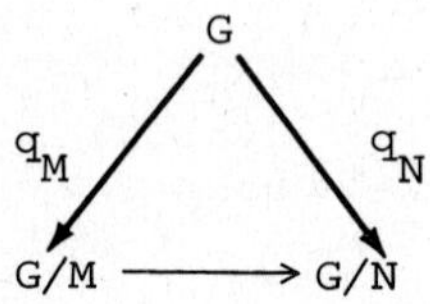

We claim that f is cofinal (1.8). Firstly, let (ϕ,H), $\phi: G \longrightarrow H$ be an arbitrary object of the cone category (G,J). We factorize

(F) 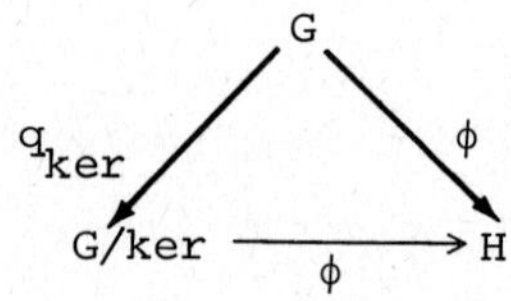

By 1.17 (ii) we know that $\ker \phi \in N_{\mathcal{P}}(G)$; thus (F) constitutes a morphism $f(\ker \phi) \longrightarrow (\phi,H)$. Secondly, suppose that $f(N_j) \longrightarrow (\phi,H)$, $j = 1, 2$ are two morphisms in (G,J). This means that we have a diagram

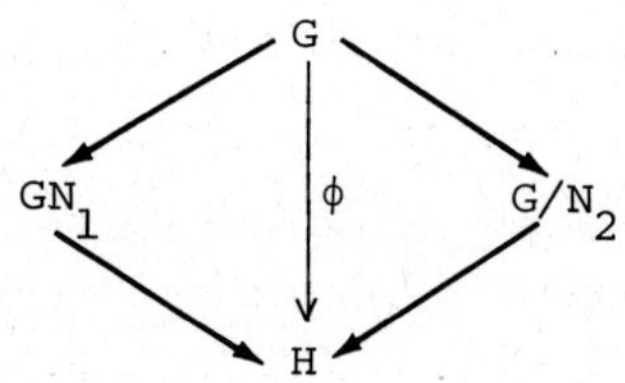

It was assumed that $N_{\mathcal{P}}(G)$ is a filterbasis; hence there is an $N \in N_{\mathcal{P}}(G)$ with $N \subseteq N_j$, $j = 1, 2$. Then there is a complementation of the diagram

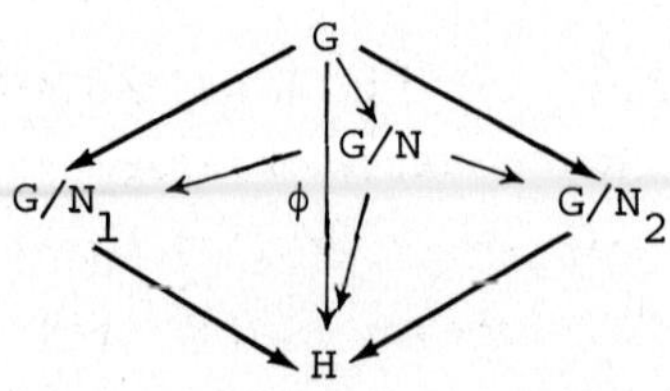

which gives the desired diagram

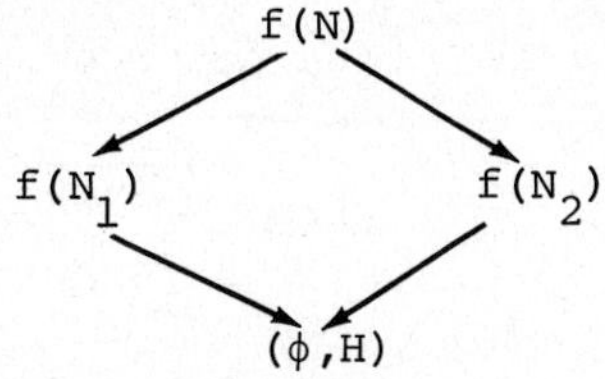

By 1.17 (iii) we have $G = \lim JD_G$ with the projective system $D_G = P_G f\colon \mathcal{N}_{\mathcal{P}}(G) \longrightarrow$ <u>Top G</u>. It then follows from 1.10 that $\lim JP_G$ exists and is (naturally isomorphic to) G. More precisely, the diagram

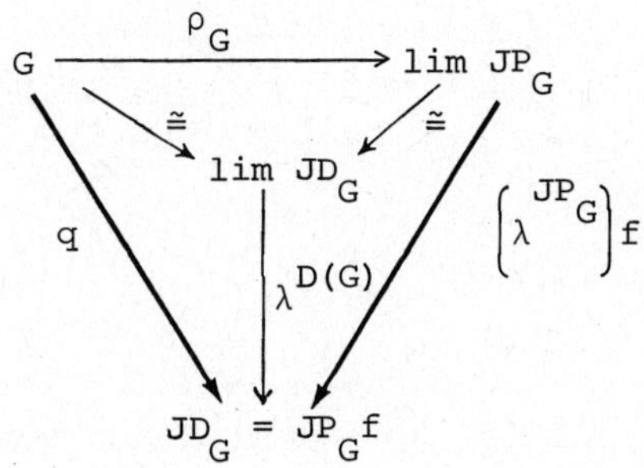

shows that ρ_G is an isomorphism which we had to prove.

This example motivates the following definition:

1.18 <u>DEFINITION</u>. Let $\mathcal{D}$ be a class of functors $X \longrightarrow A_o$ with some small X. A functor $J\colon A_o \longrightarrow A$ is called *$\mathcal{D}$-dense* (and if J is an inclusion functor then A_o is called *$\mathcal{D}$-dense* in A) if for each object A in A there is a functor $D_A\colon X \longrightarrow A_o$ in $\mathcal{D}$ and a cofinal functor $f\colon X \longrightarrow (A,J)$ such that $D_A = P_A f$ and that $\lim JD_A \cong A$, under an isomorphism induced by some natural transformation $\alpha\colon A_X \longrightarrow JD_A$. If D is the class of all [strictly] projective systems then J is [*strictly*] *pro-dense*. □

1.19 <u>PROPOSITION</u>. *Any $\mathcal{D}$-dense functor is canonically dense.*

Proof. Adjust the diagram preceding 1.18 to the general case. □

The significance of the concept of pro-density by comparison with the simpler idea of canonical density is rooted firstly in the prevalence of projective limits in topological algebra, secondly and primarily, in the occurrence of functors which are not limit preserving but do preserve projective limits, thirdly, that in the version of

1.18 the question of the smallness of A_o does not present any worry. We illustrate this in the following proposition, which summarizes our work on the example of $\mathcal{P}$-topological groups:

1.20 <u>PROPOSITION</u>. *Let* $\mathcal{P}$ *be a property of topological groups. Then* $\mathcal{P}$ *is pro-dense in* $\mathcal{P}_{\underline{pro}}$. *In fact,* $\mathcal{P}$ *is* $\mathcal{D}$*-dense in* $\mathcal{P}_{\underline{pro}}$ *for the class of all projective systems* $\mathcal{D} = \{G/N: N \in \mathcal{N}_{\mathcal{P}}(G)\}$, $G \in \mathcal{P}_{\underline{pro}}$. $\square$

With no effort at all this Proposition may be formulated for any class of topological algebras in place of topological groups by replacing normal subgroups with congruences. Recall that its scope is illustrated by the table following 1.18.

We point out that in the definition 1.18 of pro-density we do not require that there is a functorial assignment $A \longmapsto D_A: \mathcal{A} \longrightarrow \mathcal{A}_o^{\underline{dir}}$. It is, however, true (and not hard to prove) that in the case of a property $\mathcal{P}$ of topological groups, the assignment $G \longmapsto D_G$ which we used in the discussion preceding 1.18 extends to a functor $\mathcal{P}_{\underline{pro}} \longrightarrow \mathcal{P}^{\underline{dir}}$ in such a fashion that the composition

$$\mathcal{P}_{\underline{pro}} \longrightarrow \mathcal{P}^{\underline{dir}} \xrightarrow{J^{\underline{dir}}} \mathcal{P}_{\underline{pro}}^{\underline{dir}} \xrightarrow{\lim} \mathcal{P}_{\underline{pro}}$$

(with the inclusion $J: \mathcal{P} \longrightarrow \mathcal{P}_{\underline{pro}}$) is naturally isomorphic to the identity. It was this fact which was used for cohomology calculations in [14] and [19]. We will see however, that only the information built into definition 1.18 is needed for these applications I have in mind.

If $\mathcal{A}$ is a category given by the poset $\{x,y,z\}$ with $x, y \leq z$ and no other non-equality relations, then the (discrete) full subcategory $\mathcal{A}_o$ containing x, y is canonically dense, but not pro-dense in $\mathcal{A}$:

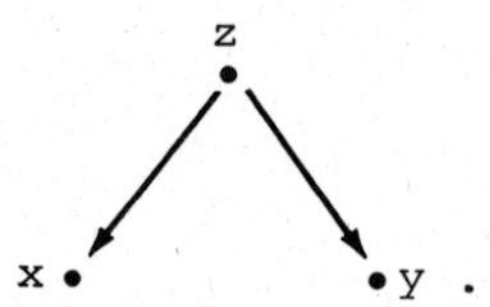

Before we conclude this section, we mention in passing a purely topological situation which together with its dual fits into the present framework: Let J: <u>Comp</u> $\longrightarrow$ <u>Top</u> be the inclusion of the category of compact Hausdorff spaces into the category of topological spaces. The natural map $\rho_X: X \longrightarrow \lim JP_X$ is the Stone-Čech compactification; it is an isomorphism (i.e., a homeomorphism) exactly if X is completely regular. Dually, if we consider the co-cone category (J,X) of all pairs (Y,JY⟶X) (and the appropriate morphisms), then the natural map colim $JP_X \longrightarrow X$ is a homeomorphism iff X is a compactly generated space (i.e., a k-space). Thus Comp is canonically co-dense in k. If X is a weakly separated compactly generated space (i.e., a t_2 k-space), then the upwards directed family of compact Hausdorff subspaces of X and their respective inclusion maps provide a direct (= co-projective) system which is cofinal in (J,X) and whose colimit is isomorphic to X. Thus Comp is pro-co-dense in kt_2. I pose the question whether or not the t_2 k-spaces are *precisely* those k-spaces which are colimits of a direct system of compact Hausdorff spaces.

CONTINUITY OF FUNCTORS

Just as in topology, it is the preservation of limits under suitable functions which makes the limit concept particularly fruitful. The appropriate functions in our present context are functors.

In 1.6 we considered situations

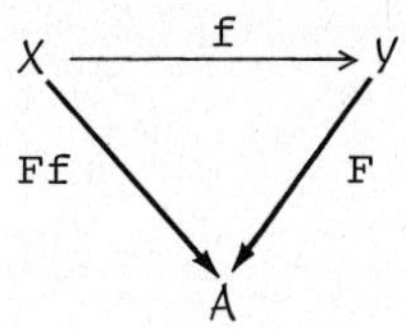

and compared the limits lim F and lim Ff, if they existed. Now we consider the situations

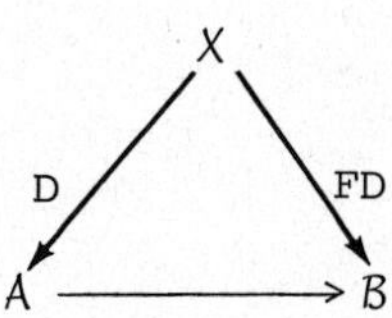

and compare F(lim D) and lim FD, if the limits exist.

1.21 LEMMA. *Let* $D: X \longrightarrow A$ *and* $F: A \longrightarrow B$ *be functors such that* lim D *and* lim DF *exist. Then there exists a unique morphism*

$$F_D: F(\lim D) \longrightarrow \lim FD$$

such that $\lambda^{FD}(F_D)_X = F(\lambda^D)$

$$\begin{array}{ccc} F(\lim D) & \longrightarrow & \lim FD \\ & {\scriptstyle F(\lambda^D)}\searrow \quad \swarrow{\scriptstyle \lambda^{FD}} & \\ & FD & \end{array} \quad .$$

Also, the following diagram commutes:

$$\begin{array}{ccc} \mathcal{A}(A, \lim D) & \xrightarrow{\;F\;} & \mathcal{B}(FA, F\lim D) \\ & & \downarrow {\scriptstyle \mathcal{B}(FA, F_D)} \\ \downarrow \cong & & \mathcal{B}(FA, \lim FD) \\ & & \downarrow \cong \\ \mathcal{A}^X(A_X, D) & \xrightarrow[F^X]{} & \mathcal{B}^X((FA)_X, FD) \end{array}$$

1.22 <u>DEFINITION</u>. Let $\mathcal{D}$ be a class of functors $X \to \mathcal{A}$ with small X. We say that a functor $F: \mathcal{A} \to \mathcal{B}$ is *$\mathcal{D}$-continuous*, if the following two conditions are satisfied:

(i) For each $D \in \mathcal{D}$ the existence of $\lim D$ implies that of $\lim FD$.

(ii) The morphism $F_D: F(\lim D) \to \lim FD$ is an isomorphism.

We say that a $\mathcal{D}$-continuous F is *continuous* if $\mathcal{D}$ is the class of all functors (with small domain) and that F is *pro-continuous* if $\mathcal{D}$ is the class of all up-directed sets (check 1.3!) If $\mathcal{A}$ is a concrete category (relative to a suitable grounding), then F is *strictly pro-continuous* if $\mathcal{D}$ is the class of a strict projective systems (see 1.4). []

Following 1.35 we will have a list of important functors which are pro-continuous (or strictly pro-continuous) without being continuous.

The following Lemma is often useful to determine continuity properties of functors:

1.23 <u>LEMMA</u>. *Let* $f: X \to Y$ *be cofinal and let* $D: Y \to \mathcal{A}$ *be a functor with a limit. If* $F: \mathcal{A} \to \mathcal{B}$ *is a functor such that* $F_{Df}: F\lim Df \to \lim FDf$ *is an isomorphism. Then* $F_D: F\lim D \to \lim FD$ *is an isomorphism.*

Proof. We operate in the following commuting diagram

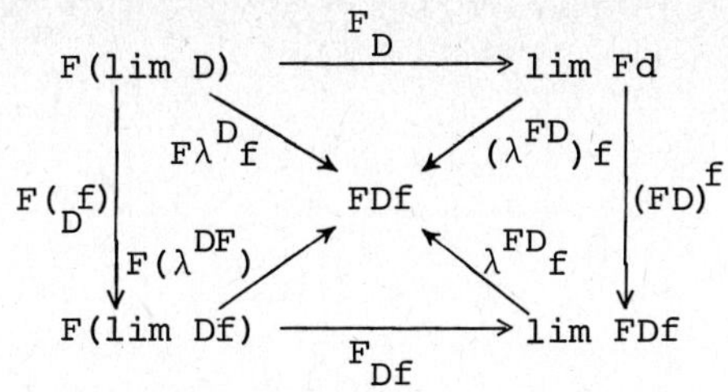

By the cofinality of f we know that ${}_Df$ and ${}_{FD}f$ (hence also $F({}_Df)$) are isomorphisms. By hypothesis F_{DR} is an isomorphism. It follows that F_D is an isomorphism. ▯

As a corollary, we formulate the following proposition:

1.24 <u>PROPOSITION</u>. *Let* $F: \mathcal{A} \longrightarrow \mathcal{B}$ *be a* $\mathcal{D}$*-continuous functor. Then* F *preserves the limits of all functors* $D: \mathcal{Y} \longrightarrow \mathcal{A}$ *which allow a cofinal functor* $f: \mathcal{X} \to \mathcal{Y}$ with $Df \in \mathcal{D}$: ▯

For example, if F is pro-continuous, then $\lim FD \cong F \lim D$ for all D allowing a cofinal $f: X \longrightarrow \mathrm{dom}\, D$ with $X \in$ <u>dir</u>.

A core question is now to what extent a continuous (procontinuous) functor is determined on a dense (pro-dense) subcategory. This is a *uniqueness* question. Subsequently one has to answer the question whether a functor defined on a dense subcategory can be continuously extended; this is an *existence* problem.

1.25 <u>PROPOSITION</u>. [14] *Let* $J: \mathcal{A}_o \longrightarrow \mathcal{A}$ *be dense (*$\mathcal{D}$*-dense). If* $F,G: \mathcal{A} \longrightarrow \mathcal{B}$ *are two continuous (*$\mathcal{D}$*-continuous) functors,* $\alpha: FJ \longrightarrow GJ$ *a natural transformation, then there is a unique natural transformation* $\alpha': F \longrightarrow G$ *such that* $\alpha = \alpha' J$. *In other words,* $F \mapsto FJ: \mathcal{B}^{\mathcal{A}} \longrightarrow \mathcal{B}^{\mathcal{A}_o}$ *induces a bijection*

$$\mathcal{B}^{\mathcal{A}}_{\mathrm{cont}}(F,G) \longrightarrow \mathcal{B}^{\mathcal{A}_o}(FJ,GJ)$$

Proof. We consider the following diagram involving an object $A \in \mathrm{ob}\,\mathcal{A}$.

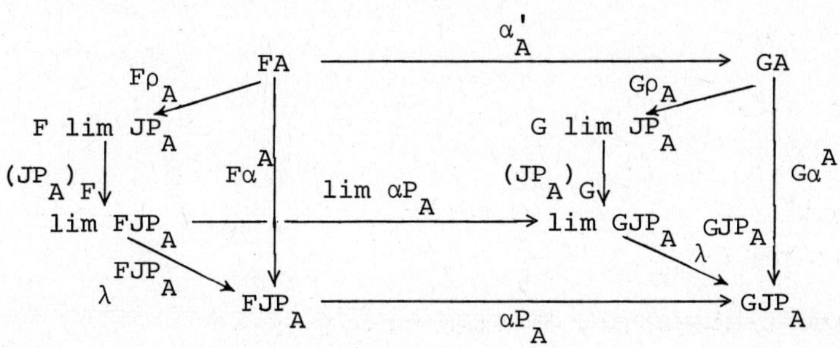

By the density hypothesis on J, the map $\rho_A: A \longrightarrow \lim JP_A$ is an isomorphism (1.16 and 1.18), whence $F\rho$, $G\rho$ are natural isomorphisms. By the continuity assumptions and 1.25 $(JP_A)_F$ and $(JP_A)_G$ are isomorphisms (1.24). The diagram therefore defines a unique α' whose naturality is readily checked. Since $\lim JP_{JA} = JA$ for $A \in A_o$ and $\rho_{JA} = 1_{JA}$, we deduce $\alpha' J = \alpha$. For every natural transformation $\alpha'': F \longrightarrow G$ with $\alpha'' J = \alpha$, we have a commutative diagram

$$\begin{array}{ccc} FA & \xrightarrow{\alpha'_A} & GA \\ \downarrow & & \downarrow \\ FJP_A & \xrightarrow[\alpha'' JP_A = \alpha P_A]{} & GJP_A \end{array}$$

in which the vertical arrows are limit natural transformations. It follows that $\alpha'' = \alpha'$. ▯

1.26 <u>COROLLARY</u>. *Let* $J: A_o \longrightarrow A$ *be dense (D-dense). If* $F,G: A \longrightarrow B$ *are two continuous (D-continuous) functors such that* $FJ \cong GJ$, *then* $F \cong G$. ▯

Once again, let us draw attention to the fact that many important functors are [strictly] pro-continuous, but not continuous; therefore, 1.25 and 1.26 are a first indication of the usefulness of pro-density. A couple of applications of this result in topological algebra are typical. The functor H on the category of Hausdorff spaces into the category of graded R-algebras over a commutative ring R given by Čech cohomology and cup product is pro-continuous. In view of 1.20, we have the following result:

1.27 <u>LEMMA</u>. *If* P *is a property of topological groups satisfying, then the Čech cohomology* H *of pro-P-groups is uniquely determined (up to natural isomorphy) on the category of* P-*groups*. ▯

For example, the Čech cohomology algebra functor of compact groups is uniquely determined on the category of compact Lie groups.

If B: <u>Comp G</u> $\longrightarrow k$ denotes the classifying space functor of Milgram for topological groups into the category of compactly generated spaces, then B is pro-continuous. We let $h = HB$ and call h the *algebraic cohomology functor*; then we have

1.28 LEMMA. *The algebraic cohomology functor* h *on compact groups is uniquely determined (up to isomorphy) on the category of Lie groups.* □

Another important example arises in duality of groups, if the duality is implemented by a hom-functor.

The essential feature here is that for a given category $\mathcal{A}$ of topological groups, say, there is a contravariant functor $S: \mathcal{A} \longrightarrow \mathcal{B}^{op}$ into a category $\mathcal{B}$ with a grounding functor $V: \mathcal{B} \longrightarrow \underline{Set}$ such that $VS(G) = \mathcal{A}(G,K)$ with a distinguished object K of $\mathcal{A}$ and that secondly there is a functor $T: \mathcal{B}^{op} \longrightarrow \mathcal{A}$ such that with a suitable grounding functor $U: \mathcal{A} \longrightarrow \underline{Set}$ and a distinguished object L of $\mathcal{B}$ one has $UT(H) = \mathcal{B}(H,L)$. In many instances there is an isomorphism $U(K) \cong V(L)$ so that K and L represent "the same object viewed in different categories A and B." The functor S is assumed to be left adjoint to T. The prime example is the case A = B = category of abelian topological groups with $SG = \mathrm{Hom}(G,\mathbb{R}/\mathbb{Z})$, where the hom-set is given pointwise addition and inversion and the topology of uniform convergence on compact sets. In this instance, we take T = S. (See [16], Chapter 0 for details.)

As a left adjoint, S will preserve all colimits; there is no category theoretical reason why it should preserve any limits. Nevertheless, it happens in important situations that it will preserve certain projective limits. Let us discuss this situation more in some detail:

1.29 LEMMA. *If* G *is a pro-*$\mathcal{P}$*-group and* $\mathcal{N}_{\mathcal{P}}(G)$ *is the associated filter basis of closed normal subgroup, then for every neighborhood* U *of* 1 *in* G *there is an* $N \in \mathcal{N}_{\mathcal{P}}(G)$ *such that* $N \subseteq U$.

Proof. We consider the morphism $g \mapsto (gN)_{N\in\mathcal{N}}: G \longrightarrow \prod_{\mathcal{N}} G/N$. By hypothesis, it is an isomorphism onto its image which is the set G' of all $(g_N)_{N\in\mathcal{N}} \in \prod_{\mathcal{N}} G/N$ with $\pi_{NM}(g_M) = g_N$, where $\pi_{NM}(gM) = gN$ for $M \subseteq N$. Thus the image U' of U is a neighborhood of the identity; hence there is a finite set $\mathcal{F} \subseteq \mathcal{N}$ such that $G' \cap \prod_{\mathcal{N}\setminus\mathcal{F}} G/M$ is contained in U' by the definition of the product topology. Since $\mathcal{N}$ is a filter basis, there is an $N \in \mathcal{N}$ which is contained in $\cap\,\mathcal{F}$. If $g \in N$, then $gM = 1_M \in G/M$ for $M \in \mathcal{F}$, whence $(gM)_{M\in\mathcal{N}} \in G' \cap \prod_{\mathcal{N}\setminus\mathcal{F}} G/M \subseteq U'$, and thus $g \in U$. □

1.30 LEMMA. *Let* G *be a pro-P-group and* K *a group without small subgroups. Then for any morphism* $f\colon G \to K$ *there is an* $N \in \mathcal{N}_P(G)$ *such that there is a factorization*

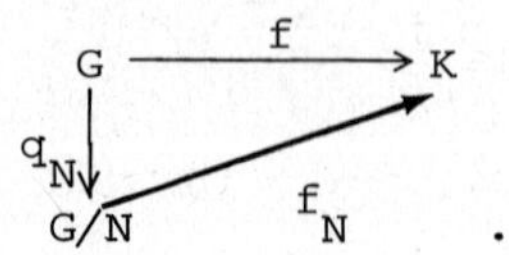

Proof. The group K has a neighborhood V of the identity in which $\{0\}$ is the only subgroup. Let $U = f^{-1}(V)$. By Lemma 1.29 there is an $N \in \mathcal{N}_P(G)$ with $N \subseteq U$. By the definition of U we then have $f(U) = \{0\}$. This implies the factorization as asserted. □

Remark. Recall that a Lie group has no small subgroups. In fact, a locally compact group has no small subgroups if and only if it is a Lie group. In particular, $K = \mathbb{R}/\mathbb{Z}$ is a group without small subgroups.

1.31 DEFINITION. We say that a function $f\colon X \to Y$ between topological spaces *allows the lifting of compact sets* iff f maps the set of compact subsets of X *surjectively* onto the set of compact subsets of Y. □

Note that any proper map certainly allows the lifting of compact sets.

1.32 LEMMA. *Let* $q_N^* = \mathrm{Hom}(q_N, K)\colon \mathrm{Hom}(G/N,K) \to \mathrm{Hom}(G,K)$, *where the hom-sets are equipped with the topology of uniform convergence on compact sets. Suppose that* K *has no small subgroups. Then* $\mathrm{Hom}(G,K) = \cup\{\mathrm{im}\, q_N^*\colon N \subseteq \mathcal{N}\}$. *If all* $q\colon G \to G/N$ *allow the lifting of compact sets, then,* $\mathrm{Hom}(G,K)$ *has the colimit topology, i.e., a set* W *is open in* $\mathrm{Hom}(G,K)$ *iff* $q_N^{*-1}(W \cap \mathrm{im}\, q_N^*)$ *is open for all* N.

Proof. The fact that $\mathrm{Hom}(G,K)$ is the union of the $\mathrm{im}\, q_N^*$ is an immediate consequence of Lemma 1.30. The topology of $\mathrm{Hom}(G,K)$ is generated by the sets $W(C,U)$ where C is compact in G and U open in K. We note that $q_N^{*-1}(W(C,U)) = W(q_N(C),U)$. If all q_N allow the lifting of compact sets, then every compact set C_N of G/N is one of the form $q_N(C)$ with a compact set $C \subseteq q_N^{-1}(C_N)$. The assertion then follows. □

Let us apply this to the duality of abelian groups.

Let $\mathcal{A}$ = <u>Top Ab</u> [resp., <u>k Ab</u>] the category of topological abelian groups [resp., abelian k-groups (i.e., group objects in the category of compactly generated spaces and continuous maps)] and let $^\wedge : \mathcal{A} \longrightarrow \mathcal{A}^{op}$ be the functor given by $\mathrm{Hom}(G, \mathbb{R}/\mathbb{Z})$ with the compact open topology [resp. $k(\mathrm{Hom}(G, \mathbb{R}/\mathbb{Z}))$]. Let $\eta_G : G \longrightarrow G^{\wedge\wedge}$ be the natural transformation given by $\eta_G(g)(\phi) = \phi(g)$. Let $\mathcal{A}_d$ be the full subcategory of all $G \in \mathrm{ob}\,\mathcal{A}$ for which η_G is an isomorphism. Then $^\wedge$ is left adjoint to $^\wedge : \mathcal{A}^{op} \longrightarrow \mathcal{A}$ and the restriction and corestriction of $^\wedge$ to $\mathcal{A}_d$ induces a duality of $\mathcal{A}_d$ with itself.

1.33 <u>PROPOSITION</u>. *If $\mathcal{P}$ is a property of $\mathcal{A}$-objects and if all $\mathcal{P}$-objects are in $\mathcal{A}_d$, then any pro-P-object G is in $\mathcal{A}_d$, provided all $q_N : G \longrightarrow G/N$, $N \in \mathcal{N}_{\mathcal{P}}(G)$ allow the lifting of compact sets* (1.31).

Proof. Suppose that G is a pro-P-group such that all $N \in \mathcal{N}_{\mathcal{P}}(G)$ are compact. By 1.32, the functor $^\wedge : \mathcal{A} \longrightarrow \mathcal{A}^{op}$ preserves projective limits of the type $\lim G/N$, $N \in \mathcal{N}_{\mathcal{P}}(G)$. Since $^\wedge : \mathcal{A}^{op} \longrightarrow \mathcal{A}$ is an adjoint, it preserves all limits. Hence

(i) $(\lim G/N)^{\wedge\wedge} \longrightarrow (\mathrm{colim}\ (G/N)^\wedge)^\wedge \longrightarrow \lim\ (G/N)^{\wedge\wedge}$

is an isomorphism. But by hypothesis on $\mathcal{P}$ all morphisms

(ii) $\eta_{G/N} : G/N \longrightarrow (G/N)^{\wedge\wedge}$

are isomorphisms. Since both the identity functor and the functor $^{\wedge\wedge}$ preserve projective limits of the type $\lim G/N$ (by (i)) we may apply 1.26 to the inclusion functor J of the full category of $\mathcal{P}$-objects in $\mathcal{A}$ into the category of all pro-$\mathcal{P}$-objects G in $\mathcal{A}$ for which all q_N allow the lifting of compact sets and conclude that

(iii) $\eta_{\lim G/N} : \lim G/N \longrightarrow (\lim G/N)^{\wedge\wedge}$

is an isomorphism. But by the definition of pro-$\mathcal{P}$-objects we know that $G \longrightarrow \lim G/N$ is an isomorphism. Thus by naturality, from the diagram

$$\begin{array}{ccc} G & \longrightarrow & \lim G/N \\ {\scriptstyle \eta_G}\downarrow & & \downarrow{\scriptstyle \eta_{\lim G/N}} \\ G^{\wedge\wedge} & \longrightarrow & (\lim G/N)^{\wedge\wedge} \end{array}$$

we conclude that η_G is an isomorphism. □

For locally compact abelian groups in particular one applies this with the property P defined by

$$G \in \mathrm{ob}\, P \Longleftrightarrow G \text{ is a Lie group of the form } \mathbb{R}^m \times (\mathbb{R}/\mathbb{Z})^n \times D \text{ with a discrete group } D.$$

A proof of Pontryagin duality is then obtained by proving the following two steps:

(a) Show that every locally compact abelian group is a pro-P-group with all $N \in N_P(G)$ compact.

(b) All P-groups G have duality (i.e., η_G is an isomorphism for $G \in \mathrm{ob}\, P$).

This program was carried through by Roeder [28]. We will return to this matter at the end of the second section.

There is, once more, a version of 1.29-1.32 which is applicable to universal topological algebra rather than to groups. The filter of normal subgroups N is once again replaced by a filter of congruences N the neighborhood of the identity U used in 1.30 has to be replaced by a neighborhood of the diagonal in $G \times G$. The object K in 1.31 is replaced by any compact universal algebra in the class under consideration which possesses a neighborhood of the diagonal in $K \times K$ without any congruence other than the diagonal itself. With these modifications in mind, we give the following example:

1.34 <u>PROPOSITION</u>. *Let* A = <u>Top Sl</u> [*resp.* <u>kSl</u>] *be the category of topological semilattices (idempotent commutative monoids)* [*resp.* k-*semilattices*] *and let* $\wedge\colon A \longrightarrow A^{op}$ *be the functor given by* Hom (G,2) *with the compact open topology* [*resp.* k(Hom(G,$\mathbb{R}/\mathbb{Z}$))], *where* $2 = \{0,1\}$ *is the two element semilattice. Let* $\eta_G\colon G \longrightarrow G^{\wedge\wedge}$ *be the natural transformation given by* $\eta_G(g)(\phi) = \phi(g)$. *Let* A_d *be the full subcategory of all* $G \in \mathrm{ob}\, A$ *for which* η_G *is an isomorphism. Then* $^\wedge$ *is left adjoint to* $^\wedge\colon A^{op} \longrightarrow A$ *and the restriction and corestriction of* $^\wedge$ *to* A_d *induces a duality with itself.*

If P *is a property of* A*-objects and if all* P*-objects are in* A_d, *then any pro-*P*-object* G *is in* A_d, *provided all quotient maps* $q_N\colon G \longrightarrow G/N$ *allow the lifting of compact sets* (1.31). ▯

This has been applied to establish the duality between the category of discrete and that of compact zero dimensional semilattices by Hofmann, Mislove and Stralka [18]. The method extends, however, to all those locally compact topological semilattices which are prodiscrete, among others. For further results on dualities in universal algebra see e.g., Davey [8].

Let us mention a final example of a pro-continuous functor on locally compact groups which was introduced recently by Greene [13]. Let $A = \underline{\text{Loc C G}}$ be the category of locally compact groups and proper homomorphisms, W^* the category of von Neumann algebras and normal $*$-morphisms.

1.35 PROPOSITION. *The functor* $W: \underline{\text{Loc C G}} \longrightarrow W^*$ *which associates with a group* G *its enveloping* W^**-algebra (the double dual of its* C^**-enveloping algebra, which in turn is the* C^**-enveloping algebra of* $L^1(G)$*) preserves strict projective limits with proper limit maps. It is , therefore, determined (up to isomorphism) by its action on Lie groups.* □

Note that once again, W is a left adjoint and would not normally preserve any limits.

Let us record a list of pro-continuous functors [respectively, strictly pro-continuous functors, designated by [s]] which are not continuous:

domain category	codomain category	functor
spaces	(graded modules [rings])^{op}	Čech cohomology
compact spaces	compact spaces	mapping cylinder [s][1]
compact monoids	t_2 k-spaces	universal and classifying constructions, E, B
compact groups	compact spaces	space of closed (topologically) subnormal subgroups[2]
locally compact groups and proper morphisms	von Neumann algebras	$W(G) = C^*(L^1(G))^{**}$ [s]
<u>Top Ab</u> [kAb]	<u>Top Ab</u> [kAb]	$\text{Hom}(-,\mathbb{R}/\mathbb{Z})$ $[\text{kHom}(-,\mathbb{R}/\mathbb{Z})]$[3]
compact semigroups	compact semigroups	$\text{Hom}(-,\{0,1\})$

1 See Hofmann and Stralka [20].

2 See Pommer, [26].

3 $\mathcal{D}$-continuous in the sense of 1.32.

KAN EXTENSIONS

The second basic question which arises in the context is the continuous extension of functors. The answer to this question is provided by the formalism originally established by Kan.

One considers, precisely, as in the discussion of density (which is in fact a special case of what follows) a functor $J: A_o \longrightarrow A$ which frequently is the inclusion of a full subcategory. Suppose that we are further given a functor $F: A_o \longrightarrow B$. The question is whether we have an extension of F to a functor $F^*: A \longrightarrow B$ with $F^*J \cong F$. In general, this is a bit too much to ask. One introduces therefore the following universal concept:

1.36 DEFINITION. Suppose that there is a natural transformation $\varepsilon^F: F^*J \longrightarrow F$ such that the function $\phi \longmapsto \varepsilon^F(\phi J): B^A(G,F^*) \longrightarrow B^{A_o}(GF,F)$ is a bijection. (Equivalently, for every functor $G: A \longrightarrow B$ and every natural transformation $\psi: GJ \longrightarrow F$ there is a unique natural transformation $\psi': G \longrightarrow F^*$ such that $\psi = \varepsilon^F(\psi' J)$.) Then F^* is called a (*right*) *Kan extension of* F (over J). □

Notice that, apart from set theoretic considerations, $F \longmapsto F^*$ is a right adjoint to $G \longmapsto GJ$. The usual adjoint formalism shows that a Kan extension, if it exists, is unique (up to natural isomorphism).

One has the following existence theorem:

1.37 PROPOSITION. *Let* $J: A_o \longrightarrow A$ *and* $F: A \longrightarrow B$ *be functors. Suppose that*

(lim) $\quad \lim\, [(A,J) \xrightarrow{P_A} A_o \xrightarrow{F} B]$ *exists for each* $A \in \mathrm{ob}\, A$.

Then there exists a right Kan extension $F^*: A \longrightarrow B$ *such that* $F^*A = \lim FP_{(A,J)}$ *and that* $\varepsilon^F: F^*J \longrightarrow F$ *is an isomorphism.* □

1.38 SUPPLEMENT. *The following hypotheses are sufficient for* (lim) *to be satisfied:*

(i) A_o *is equivalent to a small category and* B *is complete.*

(ii) J *is* D*-dense* (1.18) *and* B *is* D*-complete.* □

We notice that a functor $J: A_o \longrightarrow A$ is canonically dense only if the identity functor $1_A: A \longrightarrow A$ is the Kan extension of $J: A_o \longrightarrow A$. If $\lim JP_{(A,J)}$ exists for all A, then this condition is also sufficient.

The uniqueness of the Kan extension yields the following uniqueness theorem immediately:

1.39 COROLLARY. *Let* $J: A_o \longrightarrow A$ *and* $G: A \longrightarrow B$ *be functors satisfying at least one of the following conditions:*

(i) A_o *is equivalent to a small category* J *is dense, and* G *is continuous.*

(ii) J *is* $\mathcal{D}$*-dense and* G *is* $\mathcal{D}$*-continuous.*

Then G *is the Kan extension of* GJ. □

As an example, we note that the pro-continuous Čech cohomology functor on compact groups is the Kan extension of the (singular) cohomology functor on compact Lie groups. Note that on compact manifolds both Čech and singular cohomology agree. The functor W of 1.35 is the Kan extension of its restriction to the subcategory of Lie groups.

As an example of an application to the existence theorem, we note the following theorem on the existence of Lie algebras for arbitrary pro-Lie groups:

1.40 PROPOSITION. *Let* Lie *be the category of Lie groups (finite or infinite dimensional) and* Lie pro *the category of pro-Lie-groups. Let* Lie Alg *be the category of locally convex topological Lie algebras over the reals. Then the Lie algebra functor* L: Lie ⟶ Lie Alg *is the restriction of a unique functor* $\bar{L}$: Lie_{pro} ⟶ Lie Alg *such that* $\bar{L}$ *is the Kan extension of* L. □

For locally compact groups the Lie algebra functor was directly constructed by Lashof [22]. Some recent information on the category of generalized Lie groups was given by Chen and Yoh [2] who developed a theory I had outlined in my Tulane Lecture Notes on Compact Groups.

CHAPTER 2. MONOIDAL CATEGORIES AND FUNCTORS IN TOPOLOGICAL ALGEBRA

It is natural that categories with some additional element of structure should play a particular role in concrete applications. In the first section we have seen this exemplified for categories equipped with "projective limits". In the present chapter we discuss categories $\mathcal{A}$ with a multiplication i.e. an associative binary functor $\otimes : \mathcal{A} \times \mathcal{A} \to \mathcal{A}$; for the most part we will assume also that the functor is commutative and has an identity, in which case we will speak of a monoidal category. Monoidal categories open the door to such applications in topological algebra as duality theories between groups and operator algebras.

Unfortunately for the exposition, there are some delicate points in the foundation of the theory of monoidal categories which have to be explained, although in no application I know of do these fine points cause serious difficulties. The problem arises if one wants to explain what associativity, commutativity, identity element means for functors. The relevant concept of equality of functors is that of natural isomorphy. In order to have an idea what this means for the definition of associativity or commutativity, one need only consider the cartesian products of sets $(A \times B) \times C$ and $A \times (B \times C)$ which may be naturally "identified" but which are not equal; similar things can be said on the tensor product of vector spaces $(A \otimes B) \otimes C$ and $A \otimes (B \otimes C)$. The background theory which takes care of this problem in the context of multiplications in a category is the theory of *coherence* which is due to MacLane. We describe a recent presentation of coherence which illustrated more clearly, that coherence is not a category theoretical question at all, but one which belongs to the proper domain of universal algebra and combinatorics. The theory we present is due to D. Wallace. [31]

COHERENCE

We first focus on the simplest case of an associative multiplication without commutativity of the presence of the identity; the basic ideas are most easily explained in this case, while all conceptual complications are already present.

Suppose that M is a category and $\otimes : M \times M \to M$ is a binary functor. We assume that $\otimes$ is *associative*, i.e. satisfies

(M-1) *There is a natural isomorphism*

$$\alpha_{A,B,C} : A \otimes (B \otimes C) \to (A \otimes B) \otimes C.$$

Each such morphism is called an *associativity map*. A reparenthesizing of a product of more than three factors can generally be obtained by a composite application of associativity maps. We consider the situation of four factors and postulate.

(M-2) The following diagram commutes for all $A,B,C,D \in$ ob M

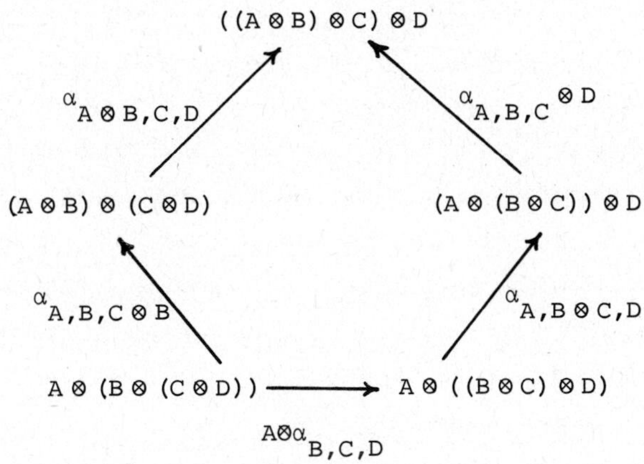

This diagram is called the *pentagon diagram*.

MacLane's first coherence theorem expresses the fact that *all* diagrams formed from associativity maps in arbitrarily many factors commute, provided (M-2) holds. The delicate point is to make precise what is meant by "all diagrams in M formed by associativity maps". Moreover, the particular nature of the category M evidently plays no role in this at all. The problem therefore is to give a presentation in the framework of universal algebra which does not involve any particular category.

We begin by considering the free binary algebra $\langle x \rangle$ in one variable. The elements $q \in \langle x \rangle$ are (non-associative) words of n letters $n = 1,2,3,\ldots$ on the alphabet $\{x\}$:

n = 1: x

n = 2: xx

n = 3: (xx)x , x(xx)

n = 4: ((xx)x)x, (x(xx))x, (xx)(xx),x((xx)x), x(x(xx)),...

Words are multiplied by juxtaposition. Another important operation is substitution: Let $p,q \in \langle x \rangle$; then $q_{(i)}(p)$ is the word obtained by substituting the word p in place of the i-th letter of the word q. Example: p = xx, q = x(xx), $q_{(1)}(p)$ = (xx)(xx), $q_{(2)}(p)$ = x((xx)x), $q_{(3)}(p)$ = x(x(xx)). The algebra $\langle x \rangle$ describes formally all possibilities to place parentheses in a word of n letters in a meaningful way. We now construct a directed graph: The vertices are the points of $\langle x \rangle$; the edges are the pairs $(q_{(i)}(q_1(q_2q_3)), q_{(i)}((q_1q_2)q_3)$ considered as arrows from the first component to the second. We write $q_{(i)}(\xi_{q_1,q_2,q_3})$ for such an edge, and we call the graph which we produced the graph over $\langle x \rangle$. We recall at this point that a groupoid is a category in which every morphism is an isomorphism. One shows that every directed graph generated a free groupoid whose objects are the vertices of the graph and whose morphisms are words in the edges and their formal inverses, whereby the words represent chains of arrows which can be concatenated by virtue of the target of the first arrow being the source of the next and so on. The free groupoid is indeed characterized by a universal property which makes the assignment of a free groupoid to a directed graph a left adjoint to the forgetful functor which considers a groupoid as a directed graph. Evidently, this description can be formalized (see [23], [31]).

Now let G be the free groupoid generated by the graph over $\langle x \rangle$. An object $q_{(i)}(q_1(q_2(q_3q_4)))$ of G has a particular type of automorphism arising from the pentagon diagram, namely

$$\pi = q_{(i)}(\xi_{(q_1q_2),q_3,q_4})q_{(i)}(\xi_{q_1,q_2,(q_3q_4)})q_{(i)}(q_1\xi_{q_2,q_3,q_4})^{-1} q_{(i)}(\xi_{q_1,(q_2q_3),q_4})^{-1} q_{(i)}(\xi_{q_1,q_2,q_3}q_4)^{-1} .$$

Every such automorphism is called a *pentagon cycle*. Evidently, in a groupoid, a *conjugate* of an isomorphism ψ is any isomorphism of the form $\phi\psi\phi^{-1}$ in the groupoid. Now we have Wallace's Theorem which is explains which the pentagon diagram plays such an important role:

2.1 LEMMA. (Wallace's First Coherence Thoerem). *Every automorphism in the groupoid* G *is a finite composition of conjugates of pentagon cycles.*

The proof is long and technical. □

We remark that the groupoid G is the disjoint union of the full subcategories G^n, n = 1,2,... spanned by the words in n letters as objects. These subcategories are called the homogeneous components of G; they are in fact the connected components of G. Now let X be an arbitrary set or class. Let $X^n = X \times \ldots \times X$ be the set [class] of n-letter (associative) words on the alphabet X. We form the category $G * X = (G^1 \times X^1) \cup (G^2 \times X^2) \cup \ldots$. We can visualize the objects of $G^n \times X^n$ as n-letter (non-associative) words on the alphabet X. The category $G * X$ is a free construction in a sense we will explain presently. First a formal definition:

2.2 DEFINITION. A pair $(M,\otimes)$ consisting of a category M and a binary functor $\otimes : M \times M \to M$ is called a *pre-multiplicative category* if (M-1) is satisfied. We then write also $(M,\otimes,\alpha)$ where specification of the data is required. It is called a *multiplicative category*, if, in addition (M-2) is satisfied. □

If $(M,\otimes)$ is a premultiplicative category and $q \in \langle x \rangle$, $A_1 \ldots A_n \in$ (ob M^n, then $w = q(A_1,\ldots,A_n)$ is defined inductively: If $q = q_1,q_2$ then $w = q_1(A_1,\ldots,A_n) \otimes q_2(A_{m+1},\ldots,A_n)$. Similarly for a morphism $\phi_j : A_j \to A_j'$ we define $q(\phi_1,\ldots,\phi_n)$.

2.3 LEMMA. *Let* $(M,\otimes)$ *be a pre-multiplicative category. Then there is a unique functor* $F : G *$ ob $M \to M$ *such that* $F(q,A_1A_2\ldots A_n) = q(A_1,A_2,\ldots,A_n)$, *and that* $F(q_{(i)}(\xi_{q_1,q_2,q_3}),$ $A_1\ldots A_n) = q_{(i)}(A_1,\ldots,A_{i-1}, \xi_{q_1}(A'),q(A''),q(A'''),A_m,\ldots,A_n)$ *where* A',A'',A''' *represent appropriate subsequences of the sequence* $A_1,\ldots,A_n$. □

The functor F is called the *functor associated with the pre-multiplicative category* $(M,\otimes,\alpha)$.

In our present discussion, MacLane's First Coherence Theorem takes the following form

2.3 PROPOSITION. *Let* $(M,\otimes,\alpha)$ *be a pre-multiplicative category. Then the following statements are equivalent:*

(1) $(M,\otimes)$ *is multiplicative.*

(2) *The associated functor* $F : G * \text{ob}\, M \to M$ *maps every pentagon cycle to an identity.*

(3) *The associated functor* $F : G * \text{ob}\, M \to M$ *maps every automorphism to an identity.*

Proof. (1) $\iff$ (2) is immediate from the definitions, and (3) $\Rightarrow$ (2) is trivial. But (2) $\Rightarrow$ (3) follows from Wallace's Lemma 2.1. □

We have now cleanly formalized the idea of "all possible diagrams formed by the associativity morphisms": They are represented by the diagrams in the groupoid $G * X$; to say that all those diagrams commute in M is precisely the statement (2).

We now sketch what has to be added if commutativity and identities enter the picture.

2.4 DEFINITION. A pre-multiplicative category $(M,\otimes,\alpha)$ is called *pre-commutative,* if

(M-3) *there is a natural involutive isomorphism* $\kappa_{A,B} : A \otimes B \to B \otimes A$ (involutive: $\kappa_{A,B}^{\;2} = 1$).

It is called a *commutative multiplicative category*, if in addition to (M-1,2,3) the following condition is satisfied:

(M-4) *For all* $A,B,C \in \text{ob}\, M$,

$$\begin{array}{ccc}
A \otimes (B \otimes C) & \xrightarrow{\;\alpha\;} & (A \otimes B) \otimes C \\
{\scriptstyle A \otimes \kappa}\downarrow & & \downarrow{\scriptstyle \kappa} \\
A \otimes (C \otimes B) & & C \otimes (A \otimes B) \\
{\scriptstyle \alpha}\downarrow & & \downarrow{\scriptstyle \alpha} \\
(A \otimes C) \otimes B & \xrightarrow[\kappa \otimes \alpha]{} & (C \otimes A) \otimes B
\end{array}$$

commutes.

The diagram in (M-4) is called a *hexagon diagram.*

We construct a groupoid G' whose objects are the elements of X and whose arrows are those of the groupoid G constructed for the proof of the first coherence theorem plus the ones generated freely by these and the arrows $q_{(i)}(\eta_{q_1,q_2}) : q_{(i)}(q_1 q_2) \to q_{(i)}(q_2 q_1)$ and their inverses.

Analogously to the introduction of the pentagon cycles we introduce a class of automorphisms of G called the hexagon cycles. Then we have

2.5 LEMMA. (Wallace's Second Coherence Theorem). *Every automorphism in the groupoid* G' *is a finite composition of conjugates of pentagon cycles and of hexagon cycles.* □

Once again, the proof of this Lemma is main burden of the coherence d for commutative multiplicative categories.

As before, for any set X or class X, we form the category $G' * X = G'^1 \times X^1) \cup (G'^2 \times X^2) \cup \ldots$ for each pre-commutative, pre-multiplicative category $(M,\otimes,\alpha,\kappa)$ we introduce the unique *associated functor* $F' : G' * \text{ob}\, M \to M$, which extends F and satisfies $F'(q_{(i)}(\eta_{q_1,q_2}, A_1 \ldots A_n) = q_{(i)}(A_1,\ldots,A_{i-1}, \kappa_{A',A''}, A_m,\ldots,A_n)$ where A', A'' represent appropriate subsequences of the sequence $A_1,\ldots,A_n$. Then we have the Second Coherence Theorem:

2.6 PROPOSITION. *Let* $(M,\otimes\ \alpha\ \kappa)$ *be a pre-commutative and pre-multiplicative category. Then the following statements are equivalent:*

(1) $(M,\otimes,\alpha,\kappa)$ *is commutative and multiplicative.*

(2) *The associated functor* F' *maps every automorphism to an identity.* □

The final build-up for a monoidal category is as follows.

Let M be a category with a bi-functor $M \otimes M \to M$. We say that there are *identity elements* if

(M-5) *there is an object* E *and there are natural isomorphisms*

$$\lambda_A : E \otimes A \longrightarrow A \quad \text{and} \quad \rho_A : A \otimes E \longrightarrow A.$$

Suppose that M is also pre-commutative and pre-multiplicative; then we consider the following condition:

(M-6) *For all objects* $A, B \in M$ *the following diagrams commute:*

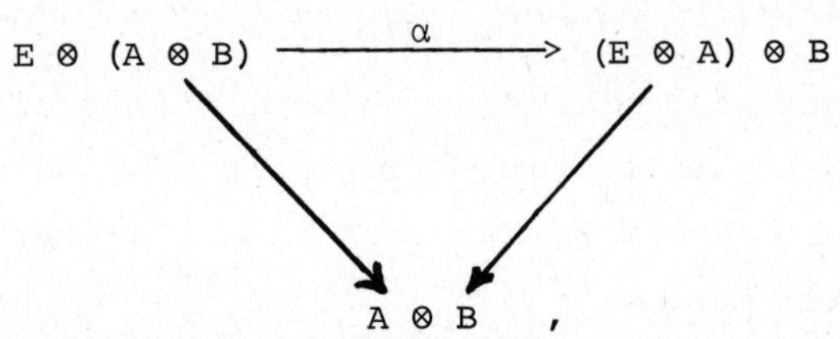

two similar diagrams with $A \otimes (E \otimes B)$, *resp.* $A \otimes (B \otimes E)$ *in place of* $E \otimes (A \otimes B)$, and the diagram

$$\begin{array}{ccc} E \otimes A & \xrightarrow{\kappa} & A \otimes E \\ & {}_{\lambda}\searrow \qquad \swarrow_{\rho} & \\ & A & \end{array}$$

2.7 <u>DEFINITION</u>. A category $\mathcal{M}$ together with a bi-functor satisfying (M-1,3,5) is a *premonoidal category* $(M,\otimes,\alpha,\kappa,\lambda,\rho)$. A *monoidal category* is a pre-monoidal category satisfying (M-1,...,6).

We hasten to remark that one might insist that a monoidal category should not be expected to be commutative, and indeed there are significant enough applications to warrant this more general concept of a monoidal category (compare [23]). However, since the type of monoidal category we encounter in the practive of topological algebra are always commutative, we chose this definition as a convenience of our definition.

In order to formulate the coherence theorem for monoidal categories, we consider the free groupoid $\mathcal{G}'$ constructed above. Let (X,e) be a set with a distinguished base point. We enlarge the groupoid $\mathcal{G}' * X$ by adding the following morphisms and their inverses plus all the arrows they generate freely together with the existing arrows:

$$q_{(i)}(\sigma) : (q_{(i)}(xx), x_1, \ldots x_{i-1} e x_i \ldots x_n) \to (q, x_1, \ldots x_{i-1} x_i \ldots x_n)$$

$$q_{(i)}(\tau) : (q_{(i)}(xx), x_1 \ldots x_i e x_{i+1} \ldots x_n) \to (q, x_1 \ldots x_i x_{i+1} \ldots x_n)$$

Denote the groupoid so obtained by $\mathcal{G}(X,e)$. In addition to the penta-

In addition to the pentagon and hexagon cycles in $\mathcal{G}(X,e)$ we introduce four types of additional automorphisms arising from (M-6) and we call these *triangle cycles*. Then we have

2.8 <u>LEMMA</u>. *Every automorphism in* $\mathcal{G}(X,e)$ *is a composition of conjugates of triangle, pentagon, and hexagon cycles.*

To any monoidal category we associate a unique functor $F'' : \mathcal{G}(\text{ob } \mathcal{M}, E) \to M$ which extends $\mathcal{G}'$ and maps $q_{(i)}(\sigma)$ to the morphism $q_{(i)}(\lambda_{A_i})$ and $q_{(i)}(\tau)$ to the morphism $q_{(i)}(\rho_{A_i})$.

The coherence theorem for monoidal categories then reads

2.9 PROPOSITION. (Coherence for monoidal categories). *Let* $(M,\otimes,\alpha,\kappa,\lambda,\rho)$ *be a pre-monoidal category. Then the following are equivalent*

(1) M *is monoidal.*

(2) F'' *maps all automorphisms to identities.* □

CARTESIAN CATEGORIES

As a first remark we note

2.10 PROPOSITION. *If* A *is a category with finite products, then there are natural isomorphisms* $(A \times B) \times C \to A \times (B \times C)$ *and* $A \times B \to B \times A$ *relative to which* A *is a commutative multiplicative category. If* A *has a terminal object* E *then projections* $E \times A \to A$, $A \times E \to A$ *make* A *into a monoidal category*

Dual statements hold in a category with coproducts [respectively, with coproducts and initial objects]. □

For the purposes of this exposition, we will call a monoidal category $(A,\times)$ arising from a category with products and terminal objects a *cartesian category* (which is not to be mixed up with the concept of a cartesian closed category, in which $A \mapsto A \times B$ has an adjoint).

We will give a list of monoidal categories which are of significance in topological algebra:

Category	multiplication	identity object	cartesian
spaces	$\times$ (product)	singleton	yes
pointed spaces	$\vee$ (coproduct)	singletons	co-cartesian
[pointed] k-spaces	same	same	same
R-modules	$\otimes_R$ (tensor product)	R	no
commutative R-algebra with identity	$\otimes_R$	R	co-cartesian
Banach spaces	$\hat{\otimes}$ (projective tensor product)	$\mathbb{R}$, $\mathbb{C}$	no
C*-algebras	$\otimes^*$ (C*-tensor product[1])	$\mathbb{C}$	no
W*-algebras	$\otimes_{W^*}$ (W*-tensor product[2])	$\mathbb{C}$	no

Considering the rather formidable formalities which are involved to even talk about coherence it may come as a relief to the applied category theoretician to notice that almost all pre-monoidal categories arising naturally are automatically monoidal. In other words, the automatic presence of coherence is the natural phenomenon in concrete situations.

MONOIDS

Monoidal categories are the abstract setting in which one may define monoids comonoids and groups. We discuss this in the following:

2.11 DEFINITION. i) Suppose that (M, x) is a monoidal category. Then a *monoid* in M is a pair of morphisms

$$A \otimes A \xrightarrow{m} A \xleftarrow{\mu} E$$

---

[1]) Several choices are possible and reasonable; the one with the expected universal properties is defined by $A \otimes^* B = C^*(A \hat{\otimes} B)$, where $C^*(-)$ is the C*-enveloping algebra.

[2]) Several choices are possible and reasonable; the one with the expected universal properties is due to Dauns [7]; the analysts generally prefer a spatial version.

such that m is associative, i.e. the following diagram commutes:

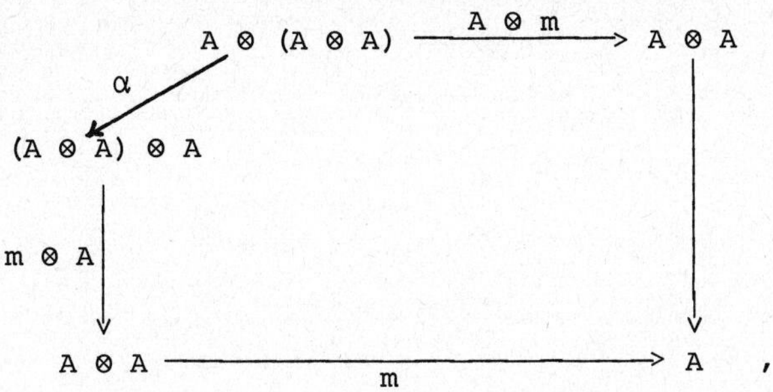

and that u is an identity, i.e. the following diagram commutes:

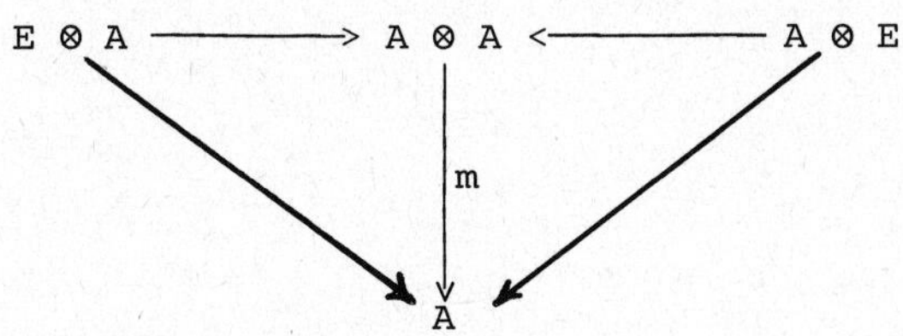

A monoid is *commutative* iff

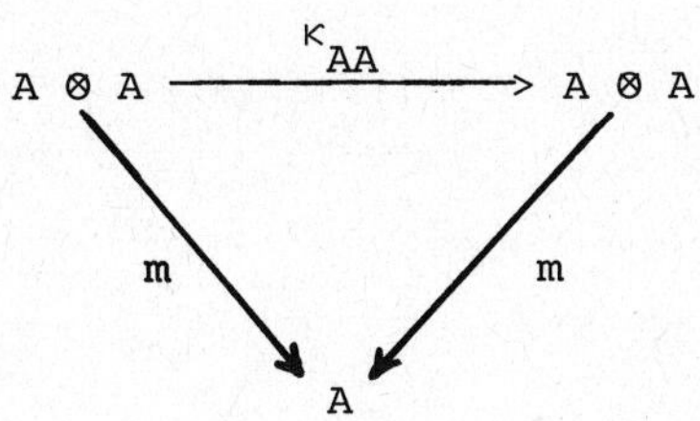

commutes.

A *morphism of monoids* is a morphism $f : A \to B$ in M such that

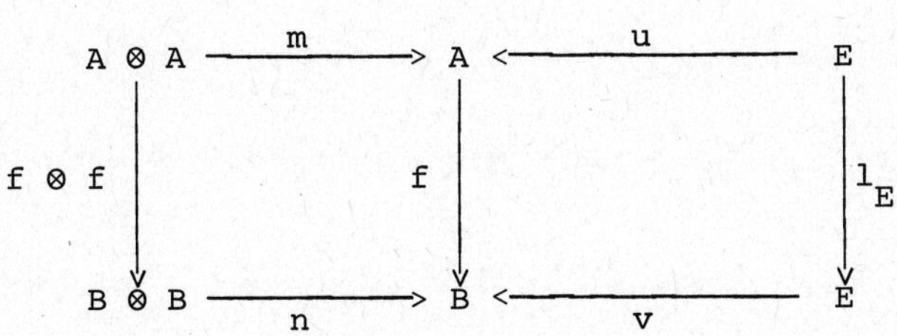

commutes. The monoids in together with the monoid morphisms form a category Mon M. The full subcategory of commutative monoids is called MonAb M. A monoid in M^{op} is called a *comonoid*. One defines Comon M = (Mon $M^{op})^{op}$.

2.12 <u>LEMMA</u>. *If* $(M, \times)$ *is a cartesian category then*

$$A \times A \xleftarrow{\text{diag}} A \xrightarrow{\text{const}} E$$

is a (co-)commutative comonoid, and every comonoid is of this form. □

The associativity and commutativity map in the category allow the definition of a unique isomorphism

$$\mu_{A,B,C,D} : (A \otimes B) \otimes (C \otimes D) \longrightarrow (A \otimes C) \otimes (B \otimes D)$$

called the *middle two exchange*. If m and n are multiplications on A and B, respectively, one shows that

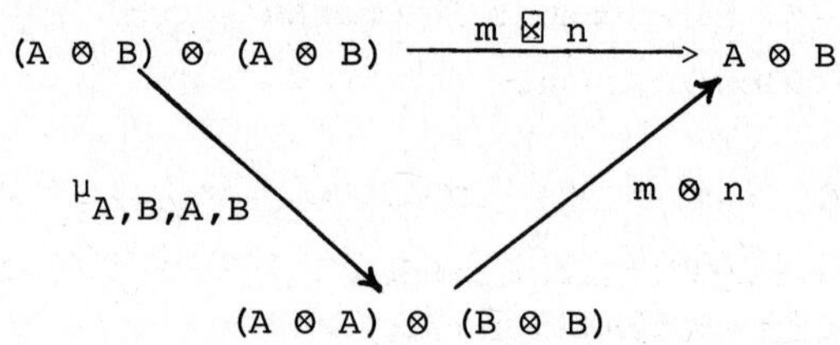

is an associative multiplication on $A \otimes B$ and that for identities $u : E \to A$, $v : E \to B$ the diagram

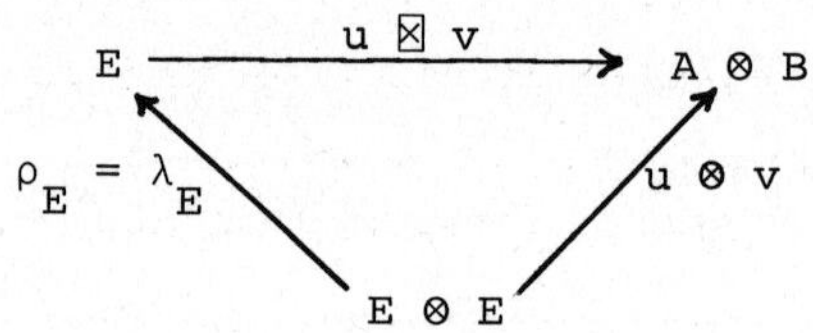

defines an identity. Thus we have obtained a monoid $(m,u) \boxtimes (n,v) = (m \boxtimes n, u \boxtimes v)$ on $A \otimes B$. Specifically, one can prove

2.13 <u>PROPOSITION</u>. $\mathrm{Mon}(M,\boxtimes)$ *is a monoidal category.* □

(Proof e.g. in [17]).

Let us illustrate what we obtained in a couple of examples. A monoid in the monoidal category of R-modules (over some commutative ring R) is given by R-module maps

$$A \otimes_R A \xrightarrow{m} A \xleftarrow{u} R$$

and thus is nothing but a ring with identity. In a similar vein, a monoid in the category of Banach algebras with the projective tensor product is a Banach algebra. Thus, if $\mathcal{B}$ denotes the monoidal category of Banach spaces with the projective tensor product, then Mon $\mathcal{B}$ is the category of unital Banach algebras (and identity preserving morphisms), and MonAb $\mathcal{B}$ is the category of unital commutative Banach algebras.

Proposition 2.13 seems to open up a cornucopia of monoidal categories arising from a given one by iteration of the formation of monoids. We discuss in the following that in reality this situation is harmless.

Form a monoid $A \otimes A \xrightarrow{m} A \xleftarrow{u} E$ in a monoidal category, m is a morphism of monoids iff the monoid is commutative. If $A \otimes A \xrightarrow{m} A \xleftarrow{u} E$ is a commutative monoid in M then $A \otimes A \xrightarrow{n} A \xleftarrow{v} E$ is a monoid in MonAB $\mathcal{M}$ iff $m = n$, $u = v$. In fact we have the following result

2.14 PROPOSITION. [17] *Let* $\mathcal{M}$ *be a monoidal category. Then the categories* Mon Mon $\mathcal{M}$, MonAb $\mathcal{M}$, Mon MonAb $\mathcal{M}$ *are all isomorphic (which is a bit stronger than equivalent).* □

For instance, a monoid in the category of rings with identity and a commutative ring are one and the same thing. In topological algebra one encounters a situation where the first link seems to be missing: Consider the category C* of C*-algebras with identity: Then Mon C* is isomorphic to the category C*Ab of commutative C*-algebras with identity and that of commutative monoids over itself, but there does not seem to be a monoidal category M such that C* ≅ Mon $\mathcal{M}$. A similar statement applies to the category W* of W*-algebras.

LIST of monoids

M	⊗	Mon	Mon Mon M
Cartesian	×	Usual monoids in M	Comm. monoids in M
Set		monoids	"
Top	×	top. monoids	"
k	×	k-monoids	"
Comp	×	comp. monoids	"
R-modules	$\otimes_R$	unital R-algebras	Comm. Unital R-algebras
Banach spaces	$\hat{\otimes}$ (proj.)	unital Banach algebras	Comm. unital Banach algebras
C*-algebras	$\otimes^*$	Comm. unital C*-algebras	
W*-algebras	$\otimes_{W^*}$	Comm. unital W*-algebras	

BIMONOIDS

For applications it is important to recognize that monoids and comonoids occur simultaneously. For this purpose let us formulate the Lemma

2.15 LEMMA. *Let*

$$A \otimes A \xrightarrow{\ m\ } A \xleftarrow{\ u\ } E$$

be a monoid in M *and*

$$A \otimes A \xleftarrow{\ c\ } A \xrightarrow{\ k\ } E$$

be a comonoid. Then the following statements are equivalent:

(1) *The following diagram commutes*

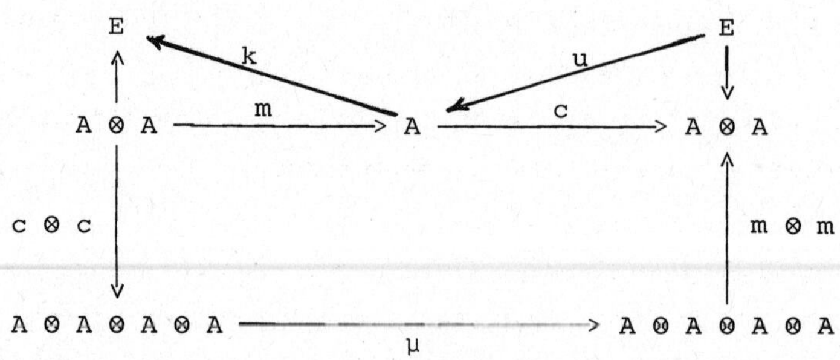

(2) c *and* k *are morphisms of monoids (i.e.* c,k ∈ Mon M)

(3) (c,k) ∈ ob Comon Mon M

(4) m *and* u *are morphisms of comonoids (i.e.* m,u ∈ Comon M)

(5) (m,u) ∈ ob Mon Comon M. □

This leads to the following definition

2.16 DEFINITION. A configuration

$$E \xleftarrow{k} A \xrightarrow{c} A \otimes A \xrightarrow{m} A \xleftarrow{u} E$$

in a monoidal category is called a *bimonoid* if it satisfies the condition in 2.15. We call A the underlying object. If (k',c',m',u') is a bimonoid with underlying object A', then f : A → A' is a *morphism of bimonoids* if it is simultaneously a morphism of monoids and comonoids. The class of bimonoids and their morphisms is a category Bim M. □

2.17 PROPOSITION. *The categories* Mon Comon M, Comon Mon M *and* Bim M *are isomorphic.* □

Let us observe for the purpose of applications, that in the case of cartesian categories we obtain nothing new:

2.18 PROPOSITION. *If* (M,×) *is a cartesian category, then* Bim M *and* Mon M *are isomorphic.*

Indeed, for any monoid $M \times M \xrightarrow{m} M \xleftarrow{u} E$ the data

$$E \xleftarrow{\text{const}} M \xrightarrow{\text{diag}} M \times M \xrightarrow{m} M \xleftarrow{u} E$$

constitute a bimonoid, and every monoid morphism is a bimonoid morphism. However, by 2.12 every bimonoid is of this form. □

Let us consider an example in a purely algebraic context:

If M is the category of commutative rings with identity then the tensor product is the coproduct. By 2.18 any comonoid is automatically a bimonoid in M, hence in particular a bimonoid in the category of abelian groups and the tensor product. Such bimonoids are also called *bigebras*.

It is a noteworthy phenomenon that in most natural concrete examples of bimonoids I know either the comultiplication or the multiplication is commutative; however, I do not know which conclusion along these lines would follow from the axioms. We will find it easy after the introduction of multiplication preserving functors to produce in the monoidal category of (finite dimensional) vector spaces with the tensor product, bimonoids A which are cocommutative but not commutative ((finite) monoid algebras), and bimonoids B which have the reverse property (their duals). Since the category of bimonoids is monoidal, we can form tensor products $A \boxtimes B$ and obtain *bimonoids which are neither commutative nor cocommutative.*

We note that each bimonoid has a canonical involutive endomorphism of bimonoids $p : A \to A$ given by $A \xrightarrow{k} E \xrightarrow{u} A$. (In the case of a cartesian category this is the constant endomorphism.)

GROUPS

Very frequently, a bimonoid is enriched by an additional element of structure which, in the case of the cartesian category Set, makes the difference between a monoid and a group.

2.19 DEFINITION. Let

$$E \xleftarrow{k} A \xrightarrow{c} A \otimes A \xrightarrow{m} A \xleftarrow{u} E$$

be a bimonoid. An *inversion* is a morphism $s : A \to A$ which satisfies the following conditions:

1) s is an endomorphism of the comonoid (c,k), i.e.

$$\begin{array}{ccccc}
E & \xleftarrow{k} & A & \xrightarrow{c} & A \otimes A \\
{\scriptstyle 1}\Vert & & \downarrow{\scriptstyle s} & & \downarrow{\scriptstyle s \otimes s} \\
E & \xleftarrow[k]{} & A & \xrightarrow[c]{} & A \otimes A
\end{array}$$

commutes.

2)

$$\begin{array}{ccc}
A \otimes A & \xrightarrow{s \otimes 1} & A \otimes A \\
{\scriptstyle c}\uparrow & & \downarrow{\scriptstyle m} \\
A & \xrightarrow[p]{} & A
\end{array}$$

commutes.

One proves the following facts:

2.20 LEMMA. [17] *Let* s *be an inversion on a bimonoid* A. *Then*

(i) s *is an involution (i.e.* $s^2 = 1$*).*

(ii)
$$\begin{array}{ccc} A\otimes A & \xrightarrow{1\otimes s} & A\otimes A \\ {\scriptstyle c}\uparrow & & \downarrow{\scriptstyle m} \\ A & \xrightarrow[p]{} & A \end{array}$$

commutes.

(iii) s *is uniquely determined (i.e. if* s,s' *are inversions then* s = s'.*)*

s *is an antimorphism of the monoid* (m,u), *i.e.*

$$\begin{array}{ccccc} & & A\otimes A & \xrightarrow{s\otimes s} & A\otimes A \\ & \nearrow^{\kappa} & & & \\ A\otimes A & & & & \downarrow{\scriptstyle m} \\ {\scriptstyle m}\downarrow & & & & \\ A & & \xrightarrow[s]{} & & A \end{array}$$
□

2.21 LEMMA. [17] *If* A *and* B *are bimonoids and* f : A → B *is a bimonoid morphism, and if* s *and* t *are inversions on* A *and* B *respectively, then the diagram*

$$\begin{array}{ccc} A & \xrightarrow{s} & A \\ {\scriptstyle f}\downarrow & & \downarrow{\scriptstyle f} \\ B & \xrightarrow[t]{} & B \end{array}$$

commutes, i.e. f *preserve inversion automatically.* □

So we are ready for the definition of a group:

2.22 DEFINITION. Let M be a monoidal category. Then a *group* in M is a bimonoid with an inversion. The full subcategory of groups in Bim M will be called Gr M. *Cogroups* are defined dually. □

PROBLEM. It now seems indeed plausible that the comultiplication in a group would have to be commutative as a consequence of the axioms, but I do not know whether this is the case.

If M is the cartesian category of sets, topological spaces, k-spaces, differentiable (analytic) manifolds, etc. then Gr M is the category of groups, topological groups, k-groups, Lie groups, etc.

A group object in the monoidal category $(R\text{-}\underline{\mathrm{Mod}}, \otimes_R)$ of R-modules over a commutative ring R with identity is called a *Hopf-algebra*; the inversion is then called an *antipode*. Other types of group objects we will see after the introduction of monoidal functors.

MONOIDAL FUNCTORS

We saw the significance of continuous or procontinuous functors in the first section; in a similar vein, the true importance of the concepts which were introduced so far is revealed in the consideration of functors between monoidal categories preserving the monoidal structure in one way or another.

2.23 DEFINITION. Let $(M,\otimes)$ and $(N,\circ)$ be monoidal categories (with associativity morphisms α^M, α^N etc.). A functor $F : M \to N$ is called *left monoidal* if there are natural transformations

$$\mu_{A,B} : F(A \otimes B) \to FA \circ FB \quad \text{and} \quad \xi : FE_M \to E_N$$

which are compatible with coherence in the sense that various diagrams involving $\alpha, \kappa, \lambda, \rho$ as well as μ and ξ commute for which the following is representative:

$$\begin{array}{ccccc} F(A_1 \otimes (A_2 \otimes A_3)) & \xrightarrow{\mu} & FA_1 \circ F(A_2 \circ A_3) & \xrightarrow{1 \circ \mu} & FA_1 \circ (FA_2 \circ FA_3) \\ \downarrow{\scriptstyle \alpha^M} & & & & \downarrow{\scriptstyle \alpha^N} \\ F((A_1 \otimes A_2) \otimes A_3) & \xrightarrow[\mu]{} & F(A_1 \otimes A_2) \circ FA_3 & \xrightarrow[\mu \circ 1]{} & (FA_1 \circ FA_2) \circ FA_3 . \end{array}$$

We call F *right monoidal*, if there are natural transformations $\nu_{A,B} : FA \circ FB \to F(A \otimes B)$, $\eta : F_N \to FE_M$ for which the corresponding diagrams commute. Finally, and most importantly, we call F *monoidal* if F is left monoidal and μ and ξ are isomorphisms. □

We observe that if natural transformation μ and ξ arise in concrete situations, then they will always be compatible with coherence. In particular we note

2.24 REMARK. *Every functor between cartesian categories preserving finite products is monoidal. In particular, every additive functor between abelian categories is monoidal.* □

LIST of monoidal functors

Domain Category	Codomain Category	Functor	Left Monoidal	Right Monoidal
(Set, x)	(R-Mod, $\otimes_R$)	F(=free functor)	+	+
(Set, ⨿)	(R-Mod, ⊕)	"	+	+
(R-Mod, ⊕)	(Grad R-Mod, $\otimes_R$)	Λ(= exterior alg.)	+	+
"	"	S(= symm. alg.)	+	+
(R-LieAlg, x)	(R-AssocAlg, $\otimes_R$)	U(= enveloping alg.)		
(Comp, x)	(C*, ⊗*)	X ⟼ C(X)	+	+
(C^*Ab_1, ⊗*)	(Comp, x)	Spec	+	+
(C*, ⊗*)	(W*, $\otimes_W$*)	( )**	+	+
(Comp G, x)	(C-Vect, ⊗)	R(= representative functions)	+	+
(Mon Top, x)	(Top, x)	E(= universal space Milgram)	+	+
([Top], x)	(Grp, x)	π_n	+	+
(Comp, x)	(Grad R-Mod,)	H(= Čech cohom.)	+	-
		over a field	+	+
(R-Mod, ⊗)	(Set, x)	U(= forgetful f.)	-	+
(R-Mod, ⊕)	(Grad R-Alg ⊗)	T(= tensor alg.)	+	-

CONSTRUCTION OF MONOIDAL FUNCTORS

There are several canonical ways how monoidal functors arise from other, often more simple ones. The first arises in adjoint situations.

2.25 PROPOSITION. *Let* M *and* N *be monoidal categories and let* $F : M \to N$ *be left adjoint to* $U : N \to M$. *Then the following are equivalent:*

(1) F *is left monoidal.*

(2) U *is right monoidal.*

The proof is natural but requires lengthy diagram chasing for the verification of all of the details. (See e.g. [17]). □

The last two lines in the list of monoidal functors exemplify the situation described in Proposition 2.25 and show that in general one may not expect much more, although in specific situations one of the two adjoints may in fact be monoidal (as in the first eight lines of the list).

The second way to produce monoidal functors is by lifting monoidal functors to monoid categories.

2.26 PROPOSITION. *Let* $U : N \to M$ *be a right monoidal functor between monoidal categories. Then there is a unique right monoidal functor* $\underline{Mon}\ U : \underline{Mon}\ N \to \underline{Mon}\ M$ *such that for the grounding functor* $|\,| : \underline{Mon}\ N \to N$ *one has* $|\underline{Mon}\ U| = U$. *In a similar way, and left monoidal functor* $F : M \to N$ *defines a left monoidal functor* $\underline{CoMon}\ F : \underline{CoMon}\ M \to \underline{CoMon}\ N$.

For the proof one takes a monoid $(m,u) = (B \otimes B \xrightarrow{m} B \xleftarrow{u} E_N)$ and defines $\underline{Mon}\ (m,u) = (UB \otimes UB \to U(B \otimes B) \to UB \leftarrow UE_M \leftarrow E_N)$. This assignment is functorial. The assertions of the proposition then have to be verified in detail by diagram chasing [17]. □

2.27 COROLLARY. *Any monoidal functor* $F : M \to N$ *induces unique monoidal functors*

$\underline{Bim}\ F : \underline{Bim}\ M \to \underline{Bim}\ N,$

$\underline{Gr}\ F : \underline{Gr}\ M \to \underline{Gr}\ N,$

$\underline{CoGr}\ F : \underline{CoGr}\ M \to \underline{CoGr}\ N.$

This set-up is exemplified by the free functor $F : \underline{Set} \to R\text{-}\underline{Mod}$ which is monoidal as functor $(\underline{Set}, \times) \to (R\text{-}\underline{Mod}, \otimes_R)$. The category $\underline{Mon}\ \underline{Set} = \underline{Bim}\ \underline{Set}$ is the category $\underline{Mon}$ of ordinary monoids, and the category $\underline{Bim}(R\text{-}Mod)$ is the category $R\text{-}\underline{Big}$ of R-bigebras. Thus F induces a functor $\underline{Bim}\ F : \underline{Mon} \to R\text{-}\underline{Big}$ which is nothing else than the monoid algebra functor. Similarly $\underline{Gr}\ \underline{Set}$ is just the category $\underline{Group}$ of groups and $\underline{Gr}\ R\text{-}\underline{Mod}$ is the category $R\text{-}\underline{Hopf}$ or Hopf algebras, and the functor $\underline{Gr}\ F : \underline{Group} \to R\text{-}\underline{Hopf}$ is the group algebra functor.

However, the applications demand stronger results. One is the following:

2.28 THEOREM. *Let* $(M,\otimes)$ *and* $(N,\otimes)$ *be monoidal categories and* $F : M \to N$ *a left adjoint of* $U : N \to M$. *If* F *is monoidal, then* $\underline{Mon}\ F : \underline{Mon}\ M \to \underline{Mon}\ N$ *is left adjoint to* $\underline{Mon}\ U : \underline{Mon}\ N \to \underline{Mon}\ M$.

Notice that U is right monoidal by 2.25, whence $\underline{Mon}\ U$ is well defined by 2.26. The proof of the theorem [17] is by verification through diagram chasing. □

A parallel theorem treats the situation that $\underline{CoMon}\ F$ is left adjoint to $\underline{CoMon}\ U$; however, here the situation is more complicated and the proofs are more difficult.

2.29 THEOREM. [17] *Let* $(M,\otimes)$ *and* $(N,\otimes)$ *be monoidal categories and* $F : M \to N$ *left adjoint to* $U : N \to M$. *Suppose that* F *is left monoidal and that the following hypotheses are satisfied: i)* N *has pull backs and intersections of countable towers (this is clearly satisfied if* N *is complete) ii)* $\otimes : M \times M \to M$ *preserves monics and intersections of countable towers. iii) the natural morphisms* $\nu : UA \otimes UB \to U(A \otimes B)$ *and* $\eta : E_M \to UE_N$ *are monic. Then* $\underline{\text{CoMon}}\ F : \underline{\text{CoMon}}\ M \to \underline{\text{CoMon}}\ N$ *has a right adjoint* $\underline{\text{Pr}} : \underline{\text{CoMon}}\ N \to \underline{\text{CoMon}}\ M$. *Specifically, there is a function* $P : \text{CoMon}\ N \to M$ *and a natural transformation* $\pi : P \to U$ *such that for any* $(c,k) = (A \otimes A \xleftarrow{c} A \xrightarrow{k} E) \in \text{ob}\ \underline{\text{CoMon}}\ N$ *we have*

$$\underline{\text{Pr}}(c,k) = (PA \otimes PA \xleftarrow{\bar{c}} PA \xrightarrow{\bar{k}} E)$$

and there is a commutative diagram

$$\begin{array}{ccccc} PA \otimes PA & \xleftarrow{\bar{c}} & PA & \xrightarrow{\bar{k}} & E \\ \downarrow & & & & \\ UA \otimes UA & & \pi \downarrow & & \downarrow \eta \\ \downarrow & & & & \\ U(A \otimes A) & \xleftarrow[Uc]{} & UA & \xrightarrow[Uk]{} & UE \end{array} . \qquad \square$$

The functors P and Pr are constructed [17] by a pull-back procedure, which is modelled after a construction I used in the duality theory of (compact monoids and) groups to derive functorially the Mostow-Hochschild Hopf algebra $R(G)$ of a compact group from the C*-Hopf algebra $C(G)$ (see [15]). The same process was used by Michor in his contribution to these Proceedings. $\square$

Let us illustrate the theorem by a few other examples which are purely algebraic and therefore are simpler.

a) Let $F : \underline{\text{Mon}} \to \underline{\text{Alg}}$ be the monoid algebra functor (which itself is the lifting of the free functor $F_o : \underline{\text{Set}} \to \underline{\text{Vect}}$) for a fixed ground field K. Then by 2.28, F is left adjoint to the underlying monoid functor $U : \underline{\text{Alg}} \to \underline{\text{Mon}}$. We induce a functor $F = \underline{\text{CoMon}}\ \underline{\text{Mon}} \to \underline{\text{CoMon}}\ \underline{\text{Alg}}$, $\bar{F} = \underline{\text{CoMon}}\ F$. But $\underline{\text{Mon}}$ is cartesian, whence $\underline{\text{CoMon}}\ \underline{\text{Mon}} = \underline{\text{Bim}} = \underline{\text{Mon}}$ by 2.17 and 2.18. Further, $\underline{\text{CoMon}}\ \underline{\text{Alg}} = \underline{\text{CoMon}}\ \underline{\text{Mon}}\ \underline{\text{Vect}} = \underline{\text{Bim}}\ \underline{\text{Vect}} = \underline{\text{Big}}$ is the category of bigebras over K. Thus $\overline{F}$: Mon $\to$ Big is the monoid bigebra functor. According to Theorem 2.29 it has an adjoint. For a bigebra $A \otimes A \xleftarrow{c} A \xrightarrow{k} K$ we define $PA = \{a \in A: c(a) = a \otimes a,\ a \neq 0\}$. Then PA is a submonoid of UA (the underlying multiplicative monoid of A) giving an inclusion $\pi_A : PA \to UA$ of monoids, and $Pr(c,k) = (PA \times PA \xleftarrow{diag} PA \xrightarrow{const} 1)$, is a comonoid in Mon, i.e. a bimonoid in Set (which, in this case, is the same as a monoid by 2.18). The Monoid PA is called the monoid of *monoidal elements* of the bigebra (c,k) (sometimes also the monoid of *primitive elements*, but this is in conflict with the standard notation in example b) which follows).

b) U : LieAlg $\to$ Alg be the universal enveloping algebra functor. Warning: This functor is traditionally called U, but is a left adjoint with right adjoint L : Alg $\to$ LieAlg which associates with an associative algebra A the Lie algebra defined on the vector space of A by [a,b] = ab - ba; therefore, the present U corresponds to F in Theorem 2.29, while L corresponds to U in Theorem 2.29! The functor U is monoidal as functor (LieAlg, $\times$) $\to$ (Alg, $\otimes$), and (LieAlg, $\times$) is cartesian, whence CoMon LieAlg = LieAlg. Recall CoMon Alg = Big. We induce a functor $\overline{U}$: LieAlg $\to$ Big, which by Theorem 2.29 has a left adjoint. For a bigebra $A \otimes A \xleftarrow{c} A \xrightarrow{k} K$ we define $PA = \{a \in A : c(a) = a \otimes 1 + 1 \otimes a\}$. Then PA is a Lie subalgebra of LA, giving an inclusion $\pi_A : PA \to LA$ of Lie algebras, and $Pr(c,k) = (PA \otimes PA \xleftarrow{\bar{c}} PA \xrightarrow{\bar{k}} K)$ is a bigebra for a suitable comultiplication $\bar{c}$ and augmentation $\bar{k}$ induced by the diagram

$$\begin{array}{ccccc}
PA \otimes PA & \xleftarrow{\bar{c}} & PA & \xrightarrow{\bar{k}} & K \\
\downarrow{\scriptstyle \pi_A \otimes \pi_A} & & & & \| \\
LA \otimes LA & & \downarrow{\scriptstyle \pi_A} & & \| \\
\downarrow{\scriptstyle \cong} & & & & \| \\
L(A \otimes A) & \xleftarrow[Lc]{} & LA & \xrightarrow[Lk]{} & K
\end{array} .$$

The Lie algebra PA is called the Lie algebra of *primitive elements* of the bigebra A.

DUALITY THEORIES IN TOPOLOGICAL ALGEBRA

The framework which we have described in 2.28, 2.29 is at the root of virtually all those duality theorems in topological algebra which establish a duality between some category of topological monoids or groups on one hand and some sort of topological bigebras, respectively Hopf algebras on the other; in turn, some of the classical duality theories such as Pontryagin duality for compact abelian groups, Tannaka or Hochschile-Mostow duality for compact groups may be deduced from the former.

As a typical example we note the duality for compact monoids (see [15]). There is a duality $C : \underline{Comp} \to \underline{C^*Ab}^{op}$ $Spec : \underline{C^*Ab}^{op} \to \underline{Comp}$ between compact spaces and commutative unital C*-algebras given by the Gelfand-Naimark formalism. Both functors C and Spec are monoidal between $(\underline{Comp}, \times)$ and $(\underline{C^*Ab}, \otimes^*)$. Then by 2.25-3.28 there are dualities

$$C : \underline{Mon}\ \underline{Comp} \to \underline{Mon}(\underline{C^*Ab}^{op}) = (\underline{CoMon}\ \underline{C^*Ab})^{op}$$

(i)

$$Spec : (\underline{CoMon}\ \underline{C^*Ab})^{op} \to \underline{Mon}\ \underline{Comp},$$

$$C : Gr\ \underline{Comp} \to Gr(\underline{C^*Ab}^{op}) = (\underline{CoGr}\ \underline{C^*Ab})^{op}$$

(ii)

$$Spec : (\underline{CoGr}\ \underline{C^*Ab})^{op} \to \underline{Gr}\ \underline{Comp},$$

where, in order to simplify notation, we also write C in place of $\underline{Mon}$ C etc. In [15] I called the category $\underline{CoMon}\ \underline{C^*}$ the category of C*-bigebras, and $\underline{CoMon}\ \underline{C^*Ab}$ the category of commutative C*-bigebras. The category $\underline{CoGr}\ \underline{C^*Ab}$ of cogroups in $\underline{C^*Ab}$ should be called the category of C*-*co-Hopf algebras*.
A more general variant of this theory has now been developed by Cooper and Michor (see [5], [6], and Michor's contribution in these Proceedings).

Another example, which needs to be fully developed from this viewpoint departs from the free functor $F : \underline{TopG} \to W^*$ which associates with a topological group the "W*-group-algebra"; this functor is defined as the left adjoint to the grounding functor $U : W^* \to \underline{TopG}$ which associates with a W*-algebra A the group UA of all unitary elements of A in the ultraweak topology. I believe that the functor F is monoidal $(\underline{TopG}, \times) \to (W^*, \otimes_{W^*})$ where $\otimes_{W^*}$ is the Dauns tensor product for W*-algebras [7]; in the absence of any proof on record let me formulate this as a conjecture. Given this conjecture we can carry out an analogue of the algebraic example a) above: Since (TopG, ×) is monoidal we have $\underline{CoMon}\ \underline{TopG} = \underline{TopG}$. One then obtains a functor $\overline{F} = \underline{CoMon}\ F : \underline{TopG} \to \underline{CoMon}\ W^*$. Theorem 2.29

should apply to show that $\overline{F}$ has an adjoint. One has to concretely identify PA in this situation. The objects in CoMon W* have been called W*-*Hopf algebras*. Some indications are in Dauns' paper; the full features of this program need to be worked out. Since the all W*-algebras carry a canonical predual along, the theory is particularly rich in this context since the *predual* A_* of an $A \in$ ob CoMon W* is a unital Banach algebra; there are extensive studies of the "duality" between A and A_* on the part of operator theoreticians notably by Takesaki [29], and Vainerman and Kac [30], Enock and Schwartz [12], but the functorial aspects have not been fully investigated.

EXTENSION THEOREMS FOR MONOIDAL FUNCTORS

We conclude our sampling of applications of monoidal functors in topological algebra by indicating a parallel to the continuous extension of functors which we discussed in Section 1.

For a monoidal category M, the category CoMonAb MonAb $\mathcal{M}$ is denoted BimAb $\mathcal{M}$. We remark that BimAb $\mathcal{M}$ shares certain features with abelian categories.

2.29 PROPOSITION. *If* $\mathcal{M}$ *is a monoidal category and* $\mathcal{B}$ = BimAb $\mathcal{M}$, *then for each pair* A, B *of objects in* $\mathcal{B}$ *the set* $\mathcal{B}(A,B)$ *is a commutative monoid w.r.t. to the addition defined by*

$$\begin{array}{ccc} A & \xrightarrow{f+g} & B \\ {\scriptstyle c}\downarrow & & \uparrow{\scriptstyle m} \\ A\otimes A & \xrightarrow[f\otimes g]{} & B\otimes B \end{array}$$

and to the identity given by $A \xrightarrow{k} E \xrightarrow{u} B$. *In other words,* $(A,B) \longmapsto \mathcal{B}(A,B): \mathcal{B}^{op} \times \mathcal{B} \longrightarrow$ MonAb Set *is a functor; i.e.,* $(\mathcal{B},\otimes)$ *is a semiadditive monoidal category.* □

With semiadditive categories, we have a matrix calculus (see Mitchell [24]). We formulate the following definition:

2.30 DEFINITION. If $\mathcal{B}$ is any semiadditive category, then the *matrix category* Matr $\mathcal{B}$ is the category, whose objects are n-tuples $(A_1,\ldots,A_n) \in (\text{ob } \mathcal{B})^n$, n = 1, 2, ... and whose morphisms $(A_1,\ldots,A_m) \longrightarrow (B_1,\ldots,B_n)$ are m by n matrices

$(f_{jk})_{j=1,\ldots,m,\ k=1,\ldots,n}$, $f_{jk}: A_j \longrightarrow B_k$ in $\mathcal{B}$ with matrix multiplication as composition. ▯

The matrix calculus in a semiadditive monoidal category $\mathcal{B}$ is then expressed by the following Lemma:

2.31 LEMMA. *If* G *is a full subcategory of* $\mathcal{B}$, *then there is a functor* $S: \underline{\text{Matr}}\ G \longrightarrow B$ *given by* $S(A_1, \ldots, A_n) = A_1 \otimes \ldots \otimes A_n$ *and by a suitable definition for morphisms (which is modelled after* Ab*).* ▯

DEFINITION. We say that $\mathcal{B}$ is *freely generated* by $G \subseteq \mathcal{B}$ if $S: \underline{\text{Matr}}\ G \longrightarrow \mathcal{B}$ is an equivalence. ▯

Let us tabulate a couple of examples.

List of Additive Categories Which Are Freely Generated

$\underline{M} = \text{Bim}_{ab}\ \underline{M}$	$\otimes$	ob $\underline{G}$
R-Mod_{fin}^{1} for R a principal ideal domain	+	{R/I: I ideal of R}
Ab_{fin}^{1}	+	{cyclic groups}
K-Vect_{fin}^{1}	+	{K}
Comp. conn. ab. Lie groups	×	$\{\mathbb{R}/\mathbb{Z}\}$
loc. comp. ab. Lie groups	×	$\{\mathbb{R}, \mathbb{R}/\mathbb{Z},$ discrete abelian groups}

We then have the following Kan extension theorem whose rudiments I developed in [14] and which in this form was given by Mostert and myself in [19]:

2.32 THEOREM. *Let* $(\mathcal{B},\otimes)$ *be a semiadditive monoidal category* $\mathcal{B} = \underline{\text{BimAb}}\ M$ *for a monoidal category* M *which is freely generated by* $G \subseteq \mathcal{B}$. *Let* $J: G \longrightarrow \mathcal{B}$ *be the inclusion functor. Then every functor* $F: G \longrightarrow \underline{C}$ *into a semiadditive monoidal category* $(C,\otimes)$ *has a unique extension* $F^*: \mathcal{B} \longrightarrow C$ *such that there is a natural isomorphism* $\varepsilon: F^*J \longrightarrow F$ *and that* F^* *is monoidal. The function*

---

1 index signals finitely generated objects

$\phi \mapsto \phi(\varepsilon J): C^B(H,F^*) \longrightarrow C^G(HJ,F)$ *is bijective (i.e.,* F^* *is the left Kan extension of* F *over* J (1.36). □

In particular, a monoidal functor from $\mathcal{B}$ into some category $\mathcal{C}$ = BimAb $\mathcal{N}$ is uniquely determined by its action on $\mathcal{G}$, and is in fact, the Kan extension of its restriction.

We may combine the extension and uniqueness theorems of Sections 1 and 2 and obtain the following Corollary:

2.33 PROPOSITION. *Let* $\mathcal{M}$ *and* $\mathcal{N}$ *be monoidal categories and let* $\mathcal{G} \subseteq \mathcal{B}$ *be full subcategories of* BimAb $\mathcal{M}$ *such that* $\mathcal{B}$ *is monoidal freely generated by* $\mathcal{G}$ *and prodense in* BimAb $\mathcal{B}$. *If* BimAb $\mathcal{N}$ *is pro-complete, then every functor* $F: \mathcal{G} \longrightarrow$ BimAb $\mathcal{N}$ *has a Kan extension* F^*: BimAb $\mathcal{M} \longrightarrow$ BimAb $\mathcal{N}$ *over the inclusion* $\mathcal{G} \longrightarrow$ BimAb $\mathcal{M}$, *and every procontinuous monoidal functor* G: BimAb $\mathcal{M} \longrightarrow$ BimAb $\mathcal{N}$ *is the Kan extension of its restriction to* $\mathcal{G}$. *The assertion also holds if* BimAb *is replaced by* GrAb. □

The proof is a simple application of 1.37-1.39 and 2.32.

Let us take for M e.g., the category CompAb_o of compact connected abelian groups. Then BimAb $\mathcal{M}$ = GrAb $\mathcal{M}$ = $\mathcal{M}$ since CompAb_o is cartesian and cocartesian. For N we take the opposite of the category of graded abelian groups. Relative to the tensor product of graded groups, $\mathcal{N}$ is monoidal when given the commutativity involution $\kappa_{A,B}: A \otimes B \longrightarrow B \otimes A$ defined by $\kappa^{p+q}_{A,B}(a^p \otimes b^q) = (-1)^{pq} b^q \otimes a^p$. With this convention, the commutative monoids N are precisely the anti-commutative graded algebras (which are characterized by $a^p b^q = (-1^{pq} b^q a^p)$. Typical examples are the exterior algebra $\wedge M$ generated by a module M in degree 1 and the symmetric algebra SM generated by a module M in degree 2. The category GrAb N is the category of commutative and co-commutative graded Hopf-algebras Hopf.

We consider the Čech cohomology functor $H: \mathcal{M} \longrightarrow \mathcal{N}$ and the induced functor GrAb $\mathcal{M}$ = $\text{CompAb}_o \longrightarrow$ GrAb N = Hopf which we will still denote with H. Similarly, we have the algebraic cohomology functor $h: \text{CompAb}_o \longrightarrow$ Hopf. Both H and h are monoidal and procontinuous [19]. By the table preceding 2.32, the category CompAb_o is freely generated by the full subcategory containing the single object $\mathbb{R}/\mathbb{Z}$, i.e., by End $\mathbb{R}/\mathbb{Z}$. By 2.33, this means that H and h are uniquely determined by their action on the single object $\mathbb{R}/\mathbb{Z}$ and the single morphism $1_{\mathbb{R}/\mathbb{Z}}$ which generates End $\mathbb{R}/\mathbb{Z}$ as an additive group. It is not too difficult, to show that the action of H on the circle group is the same as that of the monoidal procontinuous functor $G \mapsto \Lambda \hat{G}$ (with $\hat{G}$ in degree 1). It follows from 2.33 that $H(G,\mathbb{Z}) \cong \Lambda \hat{G}$ as Hopf algebras. Similarly one treats h and arrives at

2.34 EXAMPLE. *The Čech cohomology integral graded Hopf algebra* HG *of a compact connected abelian group is naturally isomorphic to* $\Lambda \hat{G}$ *the exterior algebra generated by* $\hat{G}$ *in degree 1. The algebraic cohomology Hopf algebra* $hG = HBG$ *of a compact connected abelian group is naturally isomorphic to* $S\hat{G}$ *the symmetric algebra generated by* $\hat{G}$ *in degree 2.* □

The details are given in Mostert's and my book on cohomology theories, where the much more complicated non-connected case is treated also by similar methods [19]. Example 2.34, however, illustrates the remarkable situation that relevant functors may be determined completely by their behavior on *one single object*. A similar situation occurs in one of the more recent proofs of Pontryagin duality of locally compact abelian groups [28]. Here we consider the cartesian and cocartesian category of locally compact abelian groups LCAb which then agrees with GrAb(LCAb). The functor $\wedge$: LCAb $\to$ LCAb is additive (i.e., monoidal). Let $\mathcal{P}$ be the monoidal subcategory of LCAb containing all $G \cong \mathbb{R}^m \times (\mathbb{R}/\mathbb{Z})^n \times D$ with D discrete. By the table following 2.31, $\mathcal{P}$ is freely generated by the full subcategory $\mathcal{G}$ containing $\mathbb{R}$, $\mathbb{R}/\mathbb{Z}$, and discrete abelian groups. In order to see that all groups in $\mathcal{P}$ have duality, it then suffices to know by 2.32, that $\eta_G: G \to \hat{\hat{G}}$ is an isomorphism for $G = \mathbb{R}$, $\mathbb{R}/\mathbb{Z}$ or discrete. The first two cases are straightforwardly verified, as is the case that G is cyclic. Hence η_G is an isomorphism for finitely generated G. The category of finitely generated abelian groups is co-pro-dense in the category of all abelian groups Ab. The functor $\wedge\wedge$: Ab $\to$ Ab preserves direct limits with injective maps (proof via 1.32); one concludes through the continuity arguments of Section 1 that η_G is an isomorphism for all discrete G. Thus by 1.33, the duality theorem for LCA is completely reduced to the proof of the statement that B is strictly prodense in LCA.

REFERENCES

1. Bulman-Fleming S., and H. Werner, *Equational compactness in quasi-primal varieties*, preprint 1975, 22 pp.

2. Chen S., and R. W. Yoh, *The category of generalized Lie groups*, Trans. Amer. Math. Soc. 199 (1974), 281-294.

3. Choe, T. H., *Zero-dimensional compact association distributive universal algebras*, Proc. Amer. Math. Soc. 42 (1974), 607-613.

4. Choe, T. H., *Injective and projective zero-dimensional compact universal algebras*, Alg. Univ. 1976.

5. Cooper, J. B., *Remarks on applications of category theory to functional analysis*, preprint 1974, 17 pp.

6. Cooper, J. B., and P. Michor, *Duality of compactological and locally compact groups*, preprint 1975, 19 pp.

7. Dauns, J., *Categorical* W*-*tensor product*, Trans. Amer. Math. Soc. 166 (1972), 439-440.

8. Davey, B. A., *Duality theory for quasi-varieties of universal algebras*, Dissertation, U. Manitoba 1974.

9. Dieudonné, J., *Orientation générale des mathématiques pures en 1973*, Gazette des Mathématiciens, Soc. Math. France, Octobre 1974, 73-79.

10. Eilenberg, S., *Sur les groupes compacts d'homéomorphies*, Fund. Math. 28 (1937), 75-80.

11. Eilenberg, S., and S. MacLane, *General theory of natural equivalences*, Trans. Amer. Math. Soc. 58 (1945), 231-294.

12. Enock, M. and F. M. Schwartz, *Une dualité dans les algèbres de von Neumann*, C. R. Acad. Sc. Paris 277 (1973), 683-685.

13. Greene, W. A., W* *preserves projective limits*, Preprint.

14. Hofmann, K. H., *Categories with convergence, exponential functors, and the cohomology of compact abelian groups*, Math. Z. 104 (1968), 106-140.

15. Hofmann, K. H., *The duality of compact semigroups and* C*-*bigebras*, Lecture Notes in Math. 129, Springer-Verlag, New York, 1970.

16. Hofmann, K. H. and K. Keimel, *A general character theory for partially ordered sets and lattices*, Memoir Amer. Math. Soc. 122, 1972, 121 pp.

17. Hofmann, K. H. and F. LaMartin, *Monoidal categories and monoidal functors*, Seminar Notes Tulane University 1971, 103 pp. (limited circulation).

18. Hofmann, K. H., M. Mislove, and A. Stralka, *The Pontryagin Duality of Compact 0-Dimensional Semilattices and its Applications*, Lecture Notes in Mathematics 396, 1974.

19. Hofmann, K. H., and P. S. Mostert, *Cohomology Theories for Compact Abelian Groups*, Dt. Verl. d. Wiss., Berlin and Springer-Verlag, Heidelberg, 1974.

20. Hofmann, K. H., and A. Stralka, *Mapping cylinders and compact monoids*, Math. Ann. 205 (1973), 219-239.

21. Iwasawa, K. *On some types of topological groups*, Am. Math. 50 (1949), 507-558.

22. Lashof, R. K., *Lie algebras of locally compact groups*, Pac. J. Math. 7 (1957), 1145-1162.

23. MacLane, S., *Categories for the working mathematician*, Springer-Verlag, New York, 1971.

24. Mitchell, B., *Theory of Categories*, Academic Press, New York, 1965.

25. Numakura, K., *Theorems on compact totally disconnected semigroups and lattices*, Proc. Amer. Math. Soc. 8 (1957), 623-626.

26. Pommer, H., *Projektive Limites kompakter Räume*, Topology 10 (1971), 5-8.

27. Roeder, D. W., *Functorial characterizations of Pontryagin duality*, Trans. Amer. Math. Soc. 154 (1971), 151-175.

28. Roeder, D. W., *Category theory applied to Pontryagin duality*, Pac. J. Math. 52 (1974), 519-527.

29. Takesaki, M., *Duality and von Neumann algebras, in Lectures on Operator Algebras*, Lect. Notes Math. 247, Springer-Verlag, New York 1972, 665-786.

30. Vainerman, L. I. and G. I. Kac, *Nonunimodular ring groups, and Hopf-von Neumann algebras*, Dold. Akad. Nauk SSSR 211 (1973), 1031-1034; Soviet Math. Doklady 14 (1973), 1144-1148.

31. Wallace, D., *Permutation groupoids*, Dissertation Tulane University, 1976.

LATTICES OF REFLECTIONS AND COREFLECTIONS IN CONTINUOUS STRUCTURES

by

Miroslav Hušek, Praha

This contribution reflects recent investigation of reflective and coreflective subcategories of Unif and Top due to Z.Frolík, A.W.Hager, the author and others. The first part is a general study of reflections and the remaining part deals with special applications in Top and Unif.

In our approach we shall not look at reflectivity from the standpoint of subcategories but from the standpoint of functors. The background for our procedure will be any theory where we may speak about category of functors between two given categories. (It is true that a more careful approach will avoid using such categories.) For terminology see [HS_2], [Č], [I_1]. We shall use the term topological category [He_3] instead of S-category defined in [$Hu_{1,2}$]; we suppose for the corresponding forgetful functor to have small preimages (i.e., $F^{-1}(A)$ are sets). But for initial or final objects from [He_3] we will use the term projectively or inductively generated objects, [Č], [$Hu_{1,2}$].

I. General part

If not stated otherwise, a subcategory means a full subcategory (thus reflections and coreflections are onto full subcategories).

Clearly, every reflector $F : \mathcal{A} \longrightarrow \mathcal{A}$ determines (up to isomorphism) the corresponding natural transformation $\eta : 1_{\mathcal{A}} \longrightarrow F$ and, in the sequel, we shall not distinguish between F and η in this case. Since all the natural transformations of the identity functor $1_{\mathcal{A}}$ are quasi-ordered by the reflexive and transitive relation $\varkappa < \varkappa'$ (i.e., $\varepsilon \cdot \varkappa = \varkappa'$ for a natural transformation ε), we get that all the reflectors are quasi-ordered. As one can easily show, this quasi-order on natural transformations becomes an order if the transformations are reflections or epitransformations. If $\varkappa : 1_A \longrightarrow G$, $\varkappa' : 1_A \longrightarrow G'$ are natural transformations such that, for a functor $U, U\varkappa = U\varkappa' = 1$, then the relation $\varkappa < \varkappa'$ is a special case of an order $G < G'$ from [Hu$_3$].

The main aim of this paper is to construct reflectors and coreflectors by means of the above order-structure. To avoid difficulties we shall restrict our consideration to epireflectors. Usually, there are no difficulties with coreflectors because in most continuous structures they are lower modifications (i.e., if $\eta : F \longrightarrow 1_A$ is the coreflection, then F is concrete and all η_X are identity mappings of the underlying sets).

Suppose now that A is co-well-powered, has cointersections and a terminal object T; denote by $R_e(A)$ (or R_e) all the epireflectors in A and by $T_e(A)$ (or T_e) all the epitransformations of 1_A (thus $R_e \subset T_e$ and T_e is ordered by $<$).

<u>Proposition 1.</u> T_e is complete with respect to $<$ and R_e is meet-stable in T_e.

<u>Proof:</u> The identity transformation $1 : 1_A \longrightarrow 1_A$ is the first transformation in T_e and the constant transformation $1_A \longrightarrow \{T\}$ is the last one. If $\mathcal{S} \subset T_e$, then for each object X of A, there is a representative set $\{X \xrightarrow{\eta_X} FX\}$ of all $\eta_X, \eta \in \mathcal{S}$. Put φ_X, GX to be a cointersection of this set. The obvious extension to morphisms defines a functor G, and $\varphi : 1_A \longrightarrow G$ is sup $\mathcal{S}$ in T_e. If $\eta \in T_e$, $\eta : 1_A \longrightarrow F$, we shall denote $\bar{\eta} = \sup \{\eta_\xi | \xi$ is an ordinal$\}$, where $\eta_o = \eta$, $\eta_{\xi+1} = F_{\eta_\xi} \circ \eta$ and for limit ξ, η_ξ is the cointersection of $\{\eta_\zeta | \zeta < \xi\}$. Then $\bar{\eta}$ is an epireflection smaller than any epireflection bigger than η, which entails the second assertion of our Proposition.

The epireflective modification of η will be denoted by $\bar{\eta}$ or $\bar{F}$

as in the preceding proof, and similarly $\underline{\eta}$, $\underline{F}$ for monoreflective modifications. Instead of epireflectors we may consider other reflections as bireflections or upper modifications, with a necessary change of conditions put on $\mathcal{A}$ in Proposition 1, e.g. for upper modifications we need almost that $\mathcal{A}$ is a topological category (in this case sup is just a supremum of structures on the same underlying sets).

Various constructions of reflections and coreflections are known: epireflective or coreflective hull of a subcategory, F-coarse or F-fine structures for a functor F, $\mathcal{A}$-fine structures for a subcategory $\mathcal{A}$, and just recently found reflections F_+ and coreflections F_- for a functor F. We shall try to bring these constructions together, to show that they have a common basis. First we shall look at the above constructions in more details.

(A) The epireflective hull of a subcategory $\mathcal{A}$ of a category $\mathcal{B}$ is the smallest epireflective subcategory in $\mathcal{B}$ containing $\mathcal{A}$(equivalently: the corresponding epireflection is the biggest epireflection in $\mathcal{B}$ being identity on $\mathcal{A}$). It may be proved more: the corresponding epireflection is the biggest epiadjunction being identity on $\mathcal{A}$. The details of this last assertion and of other results about "biggest" and "smallest" adjunctions and about factorizations of adjunctions (decompositions into coreflections and reflections) will appear in Comment. Math. Univ. Carolinae.

(B) The second above mentioned construction are F-fine structures: If $F : \mathcal{A} \longrightarrow \mathcal{B}$ is a concrete functor of concrete categories, we say that an object A in $\mathcal{A}$ is F-fine if $\mathcal{A}(A,X) = \mathcal{B}(FA,FX)$ for all objects X in $\mathcal{A}$. In many cases, F-fine objects form a coreflective subcategory of $\mathcal{A}$.

If each nonvoid fiber $F^{-1}[B]$ contains an F-fine object, then F has a concrete left adjoint H; HF is the coreflector onto F-fine structures and simultaneously the smallest coreflector G_1 with $FG_1=F$ and the biggest coreflector G_2 such that $G_2 = F'F$ for a functor F'.

It may happen that for a given F both G_1 , G_2 exist even if F has no left adjoint (then G_1 , G_2 are different). The functor G_2 gives rise the above F-fine coreflector and G_1 the functor F_- (see (D)).

Example 1 Put $\mathcal{A}$=Unif, $\mathcal{B}$=Prox, F the canonical functor $\mathcal{A} \longrightarrow \mathcal{B}$. Both functors G_1 , G_2 exist: $G_1 = 1_{\mathcal{A}}$ (see Theorem1), $G_2 = p_f$ (the proximally fine coreflector - [Po], [Hu_2]).

Example 2 Put $\mathcal{A}$=Unif and let $\mathcal{B}$ be the category with all uniform spaces as objects and with morphisms $\mathcal{B}(X,Y)$ = Top (X,Y) $\cup$ {f $\in$ Set (X,Y) | card f[X] $\leq$ 2}. Let F be the natural embedding of $\mathcal{A}$ into $\mathcal{B}$. Then, on X with card X > 2, G_1 is the topologically fine coreflector, and G_2 is the uniformly discrete coreflector. This example contradicts the assertion 4.1 (c), part (iii) in [Ha_4].

The F-fine and F-coarse uniform spaces were investigated mainly by Z.Frolík and A.W.Hager (see Literature). From categorical point of view they were studied by J.Vilímovský in [V_2].

(C) The most known and probably most important nontrivial example of $\mathcal{A}$-fine structures are metric-fine uniform spaces [Ha_3], i.e. those uniform spaces X for which Unif(X,M)=Unif (X,t_fM) for all metrizable spaces M (t_f is the topologically fine coreflection). These spaces form a coreflective subcategory of Unif with the coreflector which is <u>the biggest coreflector agreeing with</u> t_f <u>on metric spaces.</u> One can easily generalize this procedure to other classes of spaces and other coreflections - see [Ha_4] for Unif and [V_1] for general categories. (As just a construction, this method first appeared in [I_3], [Ke_2].)

(D) The last construction mentioned at the beginning was found by Z.Frolik several months ago. It seems to be important because of interesting connections with other known reflections. If F : $\mathcal{A} \longrightarrow \mathcal{B}$ is a concrete functor of concrete categories, then F_- is <u>the smallest coreflector in</u> $\mathcal{A}$ <u>such that</u> $F F_- = F$ <u>and</u> F_+ <u>is the biggest reflec-</u>

tor such that $F F_+ = F$. In [F_7] one can find several important examples, e.g. if $\mathcal{A}$=Unif, F=coz (i.e., FX = (X,coz X), coz X are all preimages of open sets in reals R by morphisms $X \longrightarrow R$), then F_- is the metric-fine coreflector from (C).

Now the promised general constructions:

Definition Suppose that $F : \mathcal{B} \longrightarrow \mathcal{K}$ is a functor, $\mathcal{B}$ is a subcategory of $\mathcal{A}$. Then we denote

$$F_-(\mathcal{R}_e) = \inf \{H \in \mathcal{R}_e(\mathcal{A}) \mid H_{\mathcal{B}} = F' F \text{ for an } F' , \; F' > 1 \text{ if } F > 1 \},$$

$$F_+(\mathcal{R}_e) = \overline{\sup} \{H \in \mathcal{R}_e(\mathcal{A}) \mid F = F' H_{\mathcal{B}} \text{ for an } F' , \; F' > 1 \text{ if } F > 1 \},$$

and similarly for monocoreflections

$$F_+(\mathcal{C}_m) = \sup \{H \in \mathcal{C}_m(\mathcal{A}) \mid H_{\mathcal{B}} = F' F \text{ for an } F' , \; F' < 1 \text{ if } F < 1 \},$$

$$F_-(\mathcal{C}_m) = \inf \{H \in \mathcal{C}_m(\mathcal{A}) \mid F = F' H_{\mathcal{B}} \text{ for an } F' , \; F' < 1 \text{ if } F < 1 \}.$$

In special cases one can give to the functors $F_\pm(\mathcal{R})$, $F_\pm(\mathcal{C})$ other equivalent forms. To make our exposition simpler we shall suppose now that all the categories and functors investigated are concrete and $\mathcal{A}$ is a topological category; we shall look at upper and lower modifications instead of epireflections and monocoreflections (briefly $F_\pm(\mathcal{R})$, $F_\pm(\mathcal{C})$).

Assume first that $F > 1$ (i.e., $FX > X$ for each X from $\mathcal{B}$, or equivalently, the identity mapping of the underlying set of X is $\mathcal{A}$-morphism $X \longrightarrow FX$). Then

$$F_-(\mathcal{R}) = \min \{H \in \mathcal{R} \mid H_{\mathcal{B}} = HF\} = \min \{H \in \mathcal{R} \mid H_{\mathcal{B}} > F\} .$$

If, moreover, $F^2 = F$, then

$$F_-(\mathcal{R}) = \min \{H \in \mathcal{R} | HX > \sup F^{-1}FX \text{ for each } X \text{ from } \mathcal{B}\} .$$

If F is surjective on objects and $\mathcal{B} = \mathcal{A}$, then

$$F_-(\mathcal{R}) = \min \{H \in \mathcal{R} | \mathcal{A}(X,HY) = \mathcal{K}(FX,FHY) \text{ for each } X,Y \}.$$

Dually for $F_+(\mathcal{C})$:

$$F_+(\mathcal{C}) = \max \{H \in \mathcal{C} | H_{\mathcal{B}} = HF\} = \max \{H \in \mathcal{C} | H_{\mathcal{B}} < F \} \text{ if } F < 1 ,$$

$$F_+(\mathcal{C}) = \max \{H \in \mathcal{C} | \mathcal{A}(HX,Y) = \mathcal{K}(FHX,FY) \text{ for each } X,Y\} \text{ if } \mathcal{A} = \mathcal{B},\ F[\mathrm{obj}\ \mathcal{A}] = \mathrm{obj}\ \mathcal{K} .$$

We see that if $\mathcal{A} = \mathcal{B}$, then $F_-(\mathcal{R})$, $F_+(\mathcal{C})$ generalize F-coarse and F-fine modifications from (B). If, moreover $\mathcal{K} = \mathcal{A}, F > 1$, then $F_-(\mathcal{R}) = \bar{F}$, $F_+(\mathcal{C}) = \underline{F}$ (for $\bar{F}, \underline{F}$ see the remark following Proposition 1).

If $\mathcal{B}$ is a subcategory of $\mathcal{A}$ and F a restriction of a modification in $\mathcal{A}$, then $F_-(\mathcal{R})$ describe the situations of (A), (C). Put $F = \tilde{F}_{\mathcal{B}}$, where $\tilde{F} \in \mathcal{R}$; then

$$F_-(\mathcal{R}) = \min \{H \in \mathcal{R} | H_{\mathcal{B}} = F \} ,$$

$$F_+(\mathcal{R}) = \max \{H \in \mathcal{R} | H_{\mathcal{B}} = F \}$$

and similarly for coreflections. Thus in this case, $F_-(\mathcal{R})$ is the smallest upper modification agreeing with $\tilde{F}$ on $\mathcal{B}$ and $F_+(\mathcal{R})$ is such biggest upper modification. Specially, if $\tilde{F} = 1_{\mathcal{A}}$ then $F_+(\mathcal{R})$ is the reflection corresponding to the bireflective hull of $\mathcal{B}$ in $\mathcal{A}$.

As for the remaining case $\mathcal{A} = \mathcal{B}$, $F_+(\mathcal{R})$, $F_-(\mathcal{C})$, we easily see that

$$F_+(\mathcal{R}) = \overline{\sup} \{H \in \mathcal{R} \mid FH = F\}$$

$$F_-(\mathcal{C}) = \underline{\inf} \{H \in \mathcal{C} \mid FH = F\} ;$$

no other expressions are known (see (D)).

We shall see in the next part that we can compute almost all the

$F_{\pm}(R)$, $F_{\pm}(C)$ provided F is a modification in Top or Unif.

At the end of the first part only several words to nonconcrete functors F on a topological category A. In that case $F_+(C) = \sup \emptyset$; if $F \in R_e$ and R is a nice subconglomerate of R_e (e.g. R are all bireflections in A), then $F_-(R) = \inf \emptyset$ if $F \notin R$, $F_-(R) = F$ if $F \in R$ and $F_+(R)$ is the R-reflective hull of F.

II. UNIF and TOP

In this part we shall try to show how to use the first general part in special categories. The procedures in Unif will be described in more details.

First we need a nice class of uniform spaces generating all uniform spaces. The class of all metrizable spaces is not convenient for our purposes because metrizable spaces may be very wild (this is true for any class epireflective hull of which is Unif). We shall use spaces mentioned in $[I_1]$, $[\check{C}]$ the coreflective hull of which is Unif. For a set P and a filter X on P denote by P_X the following uniform space: the underlying set is $(0,1) \times P$ and the uniformity has a base of covers $\{(i,p) \mid i \in (0,1),\ p \in P - x\} \cup \{((0,p),(1,p)) \mid p \in x\}$, $x \in X$ (i.e., $P_X = \inf \{P_x \mid x \in X\}$, where $P_x = \sum_{P-x} (0,1) + \sum_{x} \widetilde{(0,1)}$ and $(0,1)$ is uniformly discrete, $\widetilde{(0,1)}$ indiscrete two-point space).

Every uniform space (X,U), U the filter of uniform neighborhoods of diagonal, is a quotient of $(X \times X)_U$ along the map $f(0,(x,y)) = x$, $f(1,(x,y)) = y$, ($[I_1]$, $[\check{C}]$).

The usage of spaces P_X is very wide; we will show their applications but first their properties:

(1) P_X is uniformly 0-dimensional (base of decompositions) complete space; it is Hausdorff iff X is free, the topology of P_X is a sum of indiscrete spaces;

(2) Any strictly finer uniformity on P_X is complete uniformly

0-dimensional and induces another proximity than $P_{\mathcal{X}}$.

To have the property that ind $\{P_{\mathcal{X}}\}$ = Unif, it suffices to take only ultrafilters $\mathcal{X}$; as P. Simon noticed, these spaces are atoms in the ordered set of all uniformities on the set $(0,1) \times P$. There exist also atoms of another character; the problem of these atoms was investigated by J.Pelant and J.Reiterman in $[PR_{1,2}]$ and was shown to be nontrivial (in contrast to Top) - e.g., if $\mathcal{X}$ is an ultrafilter on N and the uniformity on N has a base of covers $(X) \cup \{(n) \mid n \in N - X\}$, $X \in \mathcal{X}$, then this uniform space is an atom iff $\mathcal{X}$ is selective.

Before we apply the spaces $P_{\mathcal{X}}$ we must know some facts about the order structure of $\mathcal{C}$ and $\mathcal{R}_e$ in Unif. There is the largest bireflector F in Unif smaller than indiscrete I : F = pz = zp, where p is the precompact modification and z the uniformly 0-dimensional modification, and the largest epireflector F different from I, const : F = h p z, where h stands for Hausdorff; p is the largest upper modification in Unif_H. Between both p and pz, there is a proper class of different bireflectors. The least bireflector 1_{Unif} is infimum of a decreasing class of bireflectors (e.g. cardinal reflectors or point-cardinal reflectors).

As for $\mathcal{C}$, there is the smallest coreflector F = k bigger than discrete D : $F(X,\mathcal{U})$, (here $\mathcal{U}$ are uniform neighborhoods of diagonal) has a base $\cap\mathcal{U}$. Again 1_{Unif} is supremum of an increasing class of coreflectors, moreover, there is no maximal coreflector in $\mathcal{C} - (1_{\mathrm{Unif}})$. Indeed, for such a maximal F there is an ultrafilter $\mathcal{X}$ on a P with $FP_{\mathcal{X}} = DP_{\mathcal{X}}$ and if $Q_{\mathcal{Y}}$ is another atom, then $P_{\mathcal{X}} \in \mathrm{ind}\,(Q_{\mathcal{Y}})$ implies $Q_{\mathcal{Y}} \in \mathrm{ind}\,(P_{\mathcal{X}})$; but there is no such $P_{\mathcal{X}}$ because $P_{\mathcal{X}} \in \mathrm{ind}\,(Q_{\mathcal{Y}})$ means exactly that $f\mathcal{Y} = \mathcal{X}$ for an $f : Q \longrightarrow P$.

(a) The spaces $P_{\mathcal{X}}$ may be used to give simple examples of proximities having no finest compatible uniformity: if $(X,\mathcal{U})$ is not proximally fine then the proximity of $(X \times X)_{\mathcal{U}}$ has such a property because $(X \times X)_{\mathcal{U}}$ is a minimal uniformity compatible with its proximity (use the property (2)). Even if $(X,\mathcal{U})$ is topologically fine, it may happen that $(X \times X)_{\mathcal{U}}$ is not proximally fine (first notice that if $\mathcal{X}$ is e.g. a free ultrafilter on $N \times N$ not containing diagonal and such that $pr_1\mathcal{X} = pr_2\mathcal{X}$, then $P_{\mathcal{X}}$ is not proximally fine; now take $N \times N$ with all its points isolated except $(0,0)$ with neighborhoods $(0,0) \cup X, X \in \mathcal{X}$; then $(N \times N \times N \times N)_{\mathcal{U}}$, $\mathcal{U}$ the topologically fine uniformity on this topology of $N \times N$, contains $(N \times N)_{\mathcal{X}}$ as a retract). It is proved in $[PR_2]$ that e.g. for selective ultrafilters on $N, N_{\mathcal{X}}$ are proximally fine. Of course, if $(X,\mathcal{U})$ is metrizable or has a linearly ordered base then $(X \times X)_{\mathcal{U}}$ has the same property and is thus proximally fine.

(b) The property (2) implies that there is no coreflector except identity preserving proximities (answer to a question by J.Vilímovský) Moreover, there is no coreflector except identity preserving uniformly 0-dimensional precompact Hausdorff modification.

Next, a coreflector F is not the identity iff $FP_{\mathcal{X}} = DP_{\mathcal{X}}$ for an atom $P_{\mathcal{X}}$. These two facts imply

<u>Theorem 1</u>. If $F \in \mathcal{R}_e$, then $F_+(\mathcal{R}_e) = F_-(\mathcal{R}_e) = F$ and $F_-(\mathcal{C})=1$ if $F \neq I$, const, $F_-(\mathcal{C}) = D$ if $F = I$ or $F = \text{const}$.

If $F \in \mathcal{C}$, then $F_+(\mathcal{C}) = F_-(\mathcal{C}) = F$ and $F_-(\mathcal{R}_e) = I$ if $F \neq 1$, $F_-(\mathcal{R}_e) = 1$ if $F = 1$.

The remaining cases $F_+(\mathcal{C})$ for $F \in \mathcal{R}_e$ and $F_+(\mathcal{R}_e)$ for $F \in \mathcal{C}$ can have various values. The only general assertions about them are the following:

if F is a bireflector, then $F_+(\mathcal{C}) = 1$ if $F = 1$, $F_+(\mathcal{C}) = D$ if F is not smaller than p, and $F_+(\mathcal{C}) > p_f$ (proximally fine) if F is smaller than p;

if F is a coreflector, then $F_+(\mathcal{R}_e) = I$ if $F = D$,

$F_+(\mathcal{R}_e) < p$ if $F \neq D$.

For instance, $k_+(\mathcal{R}_e) = p$ and thus the precompact modification can be defined by means of a trivial coreflection k.

For the category Unif_H, the results are almost the same (one must exclude I from the consideration).

(c) The spaces P_X may be used to a simple proof of the fact that the only subcategory of Unif which is both reflective and coreflective in Unif is the whole category (clearly, every P_X must belong to the subcategory). In fact, we can prove more:

Theorem 2. Let $\mathcal{A}$ be a reflective subcategory of Unif such that either $\mathcal{A}$ contains all uniformly discrete spaces or $\mathcal{A}$ is closed-hereditary. If $\mathcal{B}$ is both reflective and coreflective subcategory of $\mathcal{A}$, then $\mathcal{B} = \mathcal{A}$.

The proof of the last result is more complicated and uses modified spaces P_X constructed by means of inverse limits and the procedure contains special construction of inverse limits of a sink (not of a source). Instead of to be closed-hereditary it suffices to suppose for $\mathcal{A}$ that any its object has a base of uniform neighborhoods of diagonal composed of objects of $\mathcal{A}$ and, if $\mathcal{A} \subset \mathrm{Unif}_H$, that a discrete two-point space belongs to $\mathcal{A}$. For details see [Hu_5]. I do not know whether these last properties of $\mathcal{A}$ imply that $\mathcal{A}$ contains all uniformly discrete spaces or that $\mathcal{A}$ is closed-hereditary. Perhaps, it will be interesting to recall examples from [Hu_5] showing that there are coreflective or reflective subcategories of Unif containing nontrivial both reflective and coreflective subcategories:

Example 3. Let C be the de Groot's strongly rigid metrizable noncompact space and u a collection of uniformities on C that is meet-stable in the collection of all uniformities on C. The full subcategory $\mathcal{A}(u)$ of Unif composed of all products of uniformities from u form a reflective subcategory of Unif (the proof is similar to the Herrlich's proof that all the products of C in Top form a reflective subcategory of Top, [He_1]). The subcategory A(topologically fine uniformity on C) is both reflective and coreflective in A(all uniformities on C).

Example 4. Let A be the full subcategory of Unif composed of uniform spaces $(X,\mathcal{U})$, $\mathcal{U}$ uniform neighborhoods of diagonal, with the property that the intersection of equivalences from $\mathcal{U}$ belongs to $\mathcal{U}$. If B is the full subcategory of A generated by $(X,\mathcal{U})$ with $\cap\, \mathcal{U} \in \mathcal{U}$ then B is coreflective and bireflective in A. Also all uniformly discrete spaces are coreflective and epireflective in A.

Example 4 shows that a nice topological category may contain a nontrivial both coreflective and reflective subcategory (see also Example 6). It also implies that results similar to Theorem 2 but for coreflective A are not so general, e.g. if A is a coreflective hull in Unif of some complete uniformly 0-dimensional spaces, then the result of Theorem 2 holds for this A.

(d) We shall look at the similar problem as in (c) for nonfull subcategories. One can deduce from results in [V_2] that if a nonfull subcategory B of A is epireflective in A then there is a full coreflective subcategory C of B which is also a full epireflective subcategory of A (this also follows from factorizations of adjunctions - see (A) in Part I).

Theorem 3. Let Unif be both coreflective and epireflective object-full subcategory of a concrete category A. Then either $A \cong$ Unif or $A \cong$ Set .

Proof: Suppose that Unif is both coreflective and epireflective in A with F and G the corresponding bireflector or bicoreflector in Unif. If $F = I$ then $G = D$ and $A \cong$ Set. If $F \neq I$ then $G = 1_{Unif}$ because $FG = F$ (see Theorem 1) and, hence, $F = 1_{Unif}$.

The similar result for $Unif_H$ was earlier proved by J.Vilímovský with the only alternative $A \cong \mathrm{Unif}_H$ (in the proof always $F \neq I$).

(e) If F is defined on Unif, then each F-fine space is the finest member in Unif of all "F-isomorphic" structures on the same set (i.e., if X is F-fine then $X = \min F^{-1}FX$). If F is the canonical functor Unif $\longrightarrow$ Prox, then, conversely, $X = \min F^{-1}FX$ implies that X is proximally fine. That this converse result is not true in general shows the answer to the question by P.Pták for F to be the modification assigning to each X the uniformity with the base of all finite-dimensional convers of X . One can easily prove that all atoms P_X have properties $FP_X = P_X$, $P_X = \inf F^{-1}FP_X$. If any P_X is F-fine then also any its quotient is F-fine, thus any uniform space is F-fine - a contradiction.

(f) At the end quite different application of spaces P_X . For a long time it was an open problem whether product of proximally fine spaces is always proximally fine. We shall show here an example that such a product need not be proximally fine. For details of all results in this section se [Hu_6].

<u>Example 5.</u> Let $(X,\mathcal{U})$ be a uniform space, S the uniformly discrete space with the underlying set $X \times X - (1_X)$ and T be the set $X \times X - (1_X) \cup (t)$, $t \notin X \times X$, endowed with the fine uniformity of the topology having only one accumulation point t with the base of neighborhoods $\{U - (1_X) \cup (t) \mid U \in \mathcal{U}\}$.

Then $(X \times X)_{\mathcal{U}}$ is a retract of $S \times T$ and, consequently, $S \times T$ is not proximally fine provided $(X,\mathcal{U})$ is not proximally fine.

We shall only mention positive results. The proofs use factorizations of proximally continuous maps defined on subspaces of products of uniform spaces.

<u>Theorem 4.</u> Suppose that X,Y are proximally fine spaces. Then $X \times Y$ is proximally fine provided one of the following conditions hold:

(i) X,Y have linearly ordered bases;

(ii) X is precompact;

(iii) X has a linearly ordered base and intersection of α uniform neighborhoods of 1_Y is again a uniform neighborhood of 1_Y (here α is the least infinite cardinal such that any uniform cover of X has a subcover of cardinality less than α).

The part (ii) generalizes the result from $[I_2]$ (both X,Y are precompact) and for compact X it was earlier proved by another method in [Ku].

Theorem 5. A product of proximally fine spaces is proximally fine iff any finite subproduct has this property.

Thus any uniform space can be embedded into a proximally fine space (product of pseudometrizable spaces) and, consequently, injective spaces are proximally fine.

We shall look now at similar procedures in Top. The role of $P_{\mathcal{X}}$ will be played by $R_{\mathcal{X}}$, where R is a set, $\mathcal{X}$ a filter on R, all the points of R except one, say r_o , are isolated, neighborhoods of r_o are $(r_o) \cup X, X \in \mathcal{X}$. Again, if $\mathcal{X}$ are ultrafilters then $R_{\mathcal{X}}$ are atoms in Top (all the atoms in this case) and ind $\{R_{\mathcal{X}}\}$ = Top. The spaces $R_{\mathcal{X}}$ are paracompact spaces; they are Hausdorff iff $\mathcal{X}$ is free.

In $\mathcal{R}_b$(Top) there are the biggest bireflector z in $\mathcal{R}_e$ - (I) (z for zerodimensional) and the smallest bireflector s in $\mathcal{R}_b - (1_{Top})$ (s for symmetric). In $\mathcal{C}$(Top) there is the smallest coreflector F in $\mathcal{C}$ - (D) (FX has as a base all the intersections of open sets from X) and no biggest coreflector in $\mathcal{C} - (1_{Top})$. For more details see $[He_2]$.

By the same methods as in Unif we can prove:

Theorem 6. If F is a bireflector, then $F_+(\mathcal{R}_e) = F_-(\mathcal{R}_e) = F$, $F_+(\mathcal{C}) = D$ if $F \neq 1$ and $F_+(\mathcal{C}) = 1$ if $F = 1$, $F_-(\mathcal{C}) = 1$ if $F \neq I$ and $F_-(\mathcal{C}) = D$ if $F = I$.

If $F \in \mathcal{C}$, then $F_+(\mathcal{C}) = F_-(\mathcal{C}) = F$, $F_+(\mathcal{R}_e) = 1$ if $F \neq D$ and $F_+(\mathcal{R}_e) = I$ if $F = D$, $F_-(\mathcal{R}_e) = I$ if $F \neq 1$ and $F_-(\mathcal{R}_e) = 1$ if $F = 1$.

Similarly for Top_H, Top_{Unif}, etc.

We see that unlike Unif, in Top even $F_+(\mathcal{C})$ for $F \in \mathcal{R}_e$ and $F_+(\mathcal{R}_e)$ for $F \in \mathcal{C}$ have trivial values.

Theorem 7. Let $\mathcal{A}$ be a reflective subcateogry of Top containing a two-point space and such that any X in $\mathcal{A}$ is locally $\mathcal{A}$. Then any both reflective and coreflective subcategory $\mathcal{B}$ of $\mathcal{A}$ coincides with $\mathcal{A}$.

Theorem 8. Let Top be both reflective and coreflective object-full subcategory (in general, not full) of a concrete category $\mathcal{A}$. Then either $\mathcal{A} \cong$ Top or $\mathcal{A} \cong$ Set.

Theorem 7 improves the corresponding result from [Ka] ($\mathcal{A}$ a bireflective subcategory of Top), where an example of a full subcategory $\mathcal{A}$ of Top is given containing a nontrivial both reflective and coreflective subcategory - but $\mathcal{A}$ is neither reflective nor coreflective subcategory of Top. We shall show that such $\mathcal{A}$ may be found coreflective in Top or epireflective in Top_H (for the proofs and details see [Hu_5]) :

Example 6. Let $\mathcal{A}$ be the category of locally connected spaces and $\mathcal{B}$ its full subcategory composed of all spaces the collections of open and closed sets of which coincide. Then $\mathcal{A}$ and $\mathcal{B}$ are coreflective in Top and $\mathcal{B}$ is bireflective in $\mathcal{A}$. If $\mathcal{B}'$ are all discrete spaces then $\mathcal{B}'$ is coreflective in Top and epireflective in $\mathcal{A}$.

The condition on $\mathcal{A}$ in Theorem 7 is satisfied if $\mathcal{A} \subset Top_{Reg}$ and $\mathcal{A}$ is closed-hereditary. The next example shows that Top_H contains an epireflective subcategory with a nontrivial coreflective and epireflective subcategory.

Example 7. Let $\mathcal{A}$ be the epireflective hull in Top_H of

S_{ω_1+1} (S_{ω_1+1} is the space of countable ordinals T_{ω_1} together with ω_1 and neighborhoods of ω_1 are sets $(\omega_1) \cup [\eta,\omega_1] \cap T^o_{\omega_1}, \eta < \omega_1$, where $T^o_{\omega_1}$ are isolated numbers) and let $\mathcal{B}$ be the epireflective hull in Top_H of T_{ω_1}. Then $\mathcal{B}$ is both coreflective and epireflective in $\mathcal{A}$.

The corresponding Theorem 8 in Top_H is trivial because there is no nontrivial bireflection in Top preserving the Hausdorff property [HS_1].

We have seen in the previous part that it is important to be familiar with atoms, coatoms in concrete categories, whether the category is atomic, coatomic, because these facts then imply corresponding properties in $\mathcal{C}$, $\mathcal{R}_e$, etc. I would like to mention now one interesting question about the order structure of $\mathcal{R}_e(Top_H)$. There is the biggest epireflection const (onto singletons), the biggest epireflection β_o in $\mathcal{R}_e$ - (const) (onto compact 0-dimensional spaces), a maximal epireflection β_N in $\mathcal{R}_e$ - (const, β_o) (onto N-compact spaces) - see e.g.[H_e2]. There was a problem on existence of other maximal epireflections in $\mathcal{R}_e$ - (const, β_o). R.Blefko, [B], proved that the epireflection onto T_{ω_1} - compact spaces is not maximal. J.Pelant, [Pe], proved that any epireflection onto T_{ω_1} - compact spaces is smaller than a maximal epireflection in $\mathcal{R}_e$ - (const, β_o) and in [HP] it was proved in addition to it that any epireflection in Top_H onto a subcategory containing a uniformizable space which is not strongly countably compact (closure of a countable subset is not compact) is smaller than a maximal epireflection in $\mathcal{R}_e$ - (const, β_o). Moreover, these maximal epireflections were characterized as those onto X-compact spaces where $N \subset X \subsetneqq \beta N$, $\beta_X N = X$ (just take for X the given epireflection of N). There is at least $2^{2^{\omega_o}}$ of such maximal epireflections. For higher cardinals we were able to prove only one implication ($\Longrightarrow$) - one must add here a condition that X contains closures in βD, D discrete, of all its subsets A with card A < card D. The other implication was proved only in special cases because in our proof we need that Souslin numbers of certain subsets of X are card D, which is not true in general.

With T_{w_1}, there is connected also another problem: Is any 0-dimensional perfect image of a T_{w_1} - compact space again T_{w_1} - compact? For the motivations and connections see [Hu_4]. The authors of [RSJ] disproved my conjecture that any such image of T_{w_1} is again T_{w_1}; they characterized all perfect images of $[T_{w_1}]$ and one can prove from their result that all are T_{w_1} - compact. Perhaps it will be of help to notice that a 0-dimensional X is T_{w_1} - compact iff any maximal filter of clopen sets in X with linear w_1-intersection property is fixed ($\mathcal{X}$ has linear w_1-intersection property if any decreasing subcollection $\{X_\eta | \eta < w_1\}$ of $\mathcal{X}$ has a nonvoid intersection).

At the end we mention some facts about the order structure of the conglomerate $\mathcal{R}$ of all reflections in Top. The situation at the bottom is the same as in $\mathcal{R}_e$(Top), i.e., the epireflections onto symmetric spaces of T_o-spaces are minimal reflections in $\mathcal{R}-(1_{Top})$, moreover, any reflection different from the identity follows one of the two minimal reflections. Further, there is no reflective subcategory strictly between Top_{T_1} and Top_{T_o} or Top_{T_1} and Top_{T_o}. The top of $\mathcal{R}$ was described by H.Herrlich in a letter to the author (May 1974). The situation is quite different from the top in $\mathcal{R}_e$ (in $\mathcal{R}_e$ there is a counterpart to the bottom: {singletons} is strictly followed by {indiscrete spaces} and {0-dimensional T_o-spaces} and these two classes are strictly followed by {0-dimensional spaces}). There is a proper class of maximal reflections in $\mathcal{R}$ - (const) : any strongly rigid T_2-space gives rise to such a reflection (onto powers of X - see [He_1]). The corresponding categories are composed only of connected T_2-spaces. That there is a proper class of strongly rigid T_2-spaces was proved in [KR], [T]; perhaps it is worth to mention some of the Trnkova's results because not all of them are published:

There is a strongly rigid proper class of paracompact connected and locally compact T_2-spaces or of unions of compact and metrizable spaces; under the assumption (M) there is a strongly rigid proper class of metrizable spaces or of compact T_2-spaces. Now back to $\mathcal{R}$. There is only one maximal reflection in $\mathcal{R}$ - (const) onto a subcategory containing a nonconnected space: onto compact 0-dimensional T_o-spaces. There is only one maximal reflection in $\mathcal{R}$ - (const) onto a category containing a non-T_o-space: onto indiscrete spaces. H.Herrlich added a question whether there are other maximal reflections in $\mathcal{R}$ - (const). I can add only those onto powers of a strongly rigid T_1- space, because the proof in [He_1] works also for T_1-spaces. An example of a strongly rigid T_1-space which is not Hausdorff was communicated to me by V.Trnková (take four disjoint subcontinua of the Cook continuum, in any of them pick out two points a_i, b_i, $i=1,\dots,4$, and double a_1 into a_1', a_1''; now put together a_1' and a_2, a_1'' and a_3, b_1 and b_4, b_2 and b_3 and a_4). Almost all the preceding results follow from the following easy considerations ($\mathcal{A}$ is a reflective subcategory of Top with the reflection η):

(a) If there is an $X \in \mathcal{A}$, $x, y \in X$ such that the subspace (x,y) is indiscrete, then η is bireflection and $\mathcal{A}$ contains all indiscrete spaces;

(b) If there is an $X \in \mathcal{A}$, $x, y \in X$ such that (x,y) is connected T_o, then all η_Y are projectively generating surjections and $\mathcal{A} \supset \mathrm{Top}_{T_o}$;

(c) If there is an $X \in \mathcal{A}$, $x, y \in X$ such that (x,y) is discrete, then all η_Y, Y 0-dimensional, are embeddings and are dense if $(x,y) \in \mathcal{A}$; then $\mathcal{A}$ contains all compact 0-dimensional T_o-spaces.

LITERATURE

[B] Blefko R.: Some classes of E-compactness, Austr.Math.J. (1972), 492-500.

[Č] Čech E.: Topological spaces, Academia Prague 1966 (revised edition by M.Katětov, Z. Frolík)

[F_1] Frolík Z.: Basic refinements of uniform spaces. Topology Conf. Pittsburgh 1972, Lecture Notes in Math. 378 (1974, 140-158.

[F_2] Frolík Z.: Three uniformities associated with uniformly continuous functions, Symposia Math.

[F_3] Frolík Z.: Interplay of measurable and uniform spaces, Top. and its Appl. Budva 1972 (Beograd 1973), 98-101.

[F_4] Frolík Z.: Locally e-fine measurable spaces, Trans. Amer. Math. Soc. 196 (1974), 237-247.

[F_5] Frolík Z.: A note on metric-fine spaces, Proc. Amer. Math. Soc. 46 (1974), 111-119.

[F_6] Frolík Z.: Measure-fine uniform spaces, Seminar Abstract Analysis 1974 (preprint).

[F_7] Frolík Z.: Cozero refinements of uniform spaces, Seminar Uniform Spaces 1975 (preprint).

[Ha_1] Hager A.W.: Three classes of uniform spaces, Proc. 3rd Prague Top.Symp. 1971 (Academia Prague 1972), 159-164.

[Ha_2] Hager A.W.: Measurable uniform spaces, Fund. Math. 77 (1972), 51-73.

[Ha_3] Hager A.W.: Some nearly fine uniform spaces, Proc.London Math. Soc.28 (1974), 517-546.

[Ha_4] Hager A.W.: Vector lattices of uniformly continuous functions and some categorical methods in uniform spaces, Topology Conf. Pittsburgh 1972, Lecture Notes in Math. 378 (1974), 172-187.

LITERATURE

[I_2] Isbell J.R.: Spaces without large projective subspaces, Math. Scand. 17 (1965), 89-105.

[I_3] Isbell J.R.: Structure of categories, Bull. Amer. Math. Soc. 72 (1966), 619-655.

[Ka] Kannan V.: Reflexive cum coreflexive subcategories in topology, Math. Ann. 195 (1972), 168-174.

[KR] Kannan V.: Rajagopalan M.: Constructions and applications of rigid spaces I (preprint).

[Ke_1] Kennison J.F.: Reflective functors in general topology and elsewhere, Trans. Amer. Math. Soc. 118 (1965), 303-315.

[Ke_2] Kennison J.F.: A note on reflection maps, Ill. J. Math. 11 (1967), 404-409.

[Ků] Kůrková V.: Concerning products of proximally fine uniform spaces, Seminar Uniform Spaces 1974 (Prague 1975), 159-171.

[Pe] Pelant J.: Lattices of E-compact spaces, Comment. Math. Univ. Carolinae 14 (1973), 719-738.

[PR_1] Pelant J.: Reiterman J.: Atoms in uniformities, Seminar Uniform Spaces 1974 (Prague 1975), 73-81.

[PR_2] Pelant J.: Reiterman J.: Which atoms are proximally fine? Seminar Uniform Spaces 1975 (preprint).

[Po] Poljakov V.Z.: Regularity, products and spectra of proximity spaces, Doklady Akad. Nauk SSSR 154 (1964), 51-54

[RSJ] Rajagopalan M.: Soundararajan T.: Jakel D.: On perfect images of ordinals, Report of Memphis St. Univ. 74/16

[R_1] Rice M.D.: Metric-fine uniform spaces (to appear)

LITERATURE

[He_1] Herrlich H.: On the concept of reflections in general topology, Proc.Symp. Berlin 1967 (Berlin 1969), 105-114.

[He_2] Herrlich H.: Topologische Reflexionen und Coreflexionen, Lecture Notes in Math. 78 (1968)

[He_3] Herrlich H.: Topological structures, Math.Centre Tracts 52 (Amsterdam 1974), 59-122.

[HS_1] Herrlich H.: Strecker G.E.: H-closed spaces and reflective subcategories, Math. Ann. 177 (1968), 302-309.

[HS_2] Herrlich H.: Strecker G.E.: Category theory, Allyn and Bacon (Boston, 1973).

[Hu_1] Hušek M.: S-categories, Comment. Math. Univ. Carolinae 5 (1964), 37 - 46.

[Hu_2] Hušek M.: Categorical methods in topology, Proc. 2nd Prague Top. Symp. (Academia Prague 1967), 190-194.

[Hu_3] Hušek M.: Construction of special functors and its applications, Comment. Math. Univ. Carolinae 8 (1967), 555-566.

[Hu_4] Hušek M.: Perfect images of E-compact spaces, Bull. Acad. Polon. Sci. 20 (1972), 41-45.

[Hu_5] Hušek M.: Reflective and coreflective subcategories of Unif and Top, Seminar Uniform Spaces 1974 (Prague 1975), 113-126.

[Hu_6] Hušek M.: Factorizations of mappings on products of uniform spaces, Seminar Uniform Spaces 1974 (Prague 1975), 173-190.

[HP] Hušek M.: Pelant J.: Note about atom-categories of topological spaces, Comment. Math. Univ. Carolinae 15 (1974), 767-773.

[I_1] Isbell J.R.: Uniform spaces, Amer. Math. Soc. (Providence 1964).

LITERATURE

[R_2] Rice M.D.: Metric-fine, proximally fine, and locally fine uniform spaces (to appear).

[T] Trnková V.: Non-constant continuous mappings of metric or compact Hausdorff spaces, Comment. Math. Univ. Carolinae 13 (1972), 283-295.

[V_1] Vilímovský J.: Generation of coreflections in categories, Comment. Math. Univ. Carolinae 14 (1973), 305-323.

[V_2] Vilímovský J.: Categorical refinements and their relation to reflective subcategories, Seminar Uniform Spaces 1974 (Prague 1975), 83-111.

Pro-categories and shape theory

by

Sibe Mardešić [1)]

Shape theory is a modification of homotopy theory created with the scope of obtaining a theory more applicable to spaces with bad local properties. In systematic manner the theory was initiated by K.Borsuk in his talks delivered at the Symposium on infinite-dimensional topology, Baton Rouge, Louisiana, 1968 [4] and the Topology symposium, Herceg-Novi, 1968 [1]. His first technical paper was [2]. Borsuk wanted a coarser classification than the homotopy type which would make e.g. the "Polish circle" equivalent to the circle. Instead of maps between metric compacta X, Y embedded in the Hilbert cube Q , Borsuk considered fundamental sequences of maps $(f_1,f_2...):X \to Y$. These are sequences of maps $f_n:Q \to Q, n \in N$, such that every neighborhood V of Y admits a neighborhood U of X and an integer n_V with the property that $f_n|U \simeq f_m|U$ in V for $n,m \geq n_V$. Fundamental sequences are composed coordinatewise. Two fundamental sequences $(f_n),(g_n):X \to Y$ are considered homotopic provided every neighborhood V of Y admits a neighborhood U of X and an integer n_V such that $f_n|U \simeq g_n|U$ in V for $n \geq n_V$. The relation $\simeq$ is an equivalence relation.

---

1) Presented at the Conference on Categorical Topology, Mannheim, 21.-25. VII, 1975.

Every map $f : X \to Y$ admits an extension $\tilde{f} : Q \to Q$ and $(\tilde{f}, \tilde{f}, \ldots)$ is a fundamental sequence $X \to Y$ whose homotopy class is independent of the extension $\tilde{f}$. In particular, the identity $1 : X \to Y$ determines a class of fundamental sequences $(1) : X \to X$. Two compacta $X, Y \subseteq Q$ are said to have the same shape, $sh(X) = sh(Y)$, provided there are fundamental sequences $(f_n) : X \to Y$, $(g_n) : Y \to X$ such that $(g_n)(f_n) \simeq (1)$, $(f_n)(g_n) \simeq (1)$. If X and Y have the same homotopy type, $X \simeq Y$, then $sh(X) = sh(Y)$. Borsuk has also shown that for ANR's $sh(X) = sh(Y)$ implies $X \simeq Y$. The notion of shape does not depend on the embeddings of X and Y in Q.

In 197o the author and J.Segal have noticed that the main notions of Borsuk shape theory admit a rather elegant description using inverse systems of ANR's [14], [15]. The essential reason for this is that X and Y are intersections of closed neighborhoods in Q which can be chosen in such a manner that they are ANR's. This approach was also developed for compact Hausdorff spaces.

W. Holsztyński gave the first axiomatic description of the shape category of Hausdorff compacta [8]. This is a category having all Hausdorff compacta for objects, the morphisms are shape maps and correspond to Borsuk's homotopy classes of fundamental sequences. Two Hausdorff compacta have the same shape provided they are isomorphic objects in the shape category.

A shape theory for metric spaces patterned after Borsuk's approach was deviced by R.H. Fox [7].

In 1973 the author has described the shape category $\mathcal{S}$ for topological spaces [1o]. The same category has been described independently by G.Kozlowski (unpublished), J.H Le Van [9] and C. Weber [19].

The objects of $\mathcal{S}$ are all topological spaces. In order to define the morphisms of $\mathcal{S}$ called shape maps, one considers the category $\mathcal{W}$ whose objects are all spaces having the homotopy type of a CW-complex and the morphisms are homotopy classes of maps. One considers the functors $[X,.] : \mathcal{W} \to \underset{\sim}{\mathrm{Ens}}$, $[Y,.] : \mathcal{W} \longrightarrow \underset{\sim}{\mathrm{Ens}}$, where $[X,P]$, $P \in \mathrm{Ob}\ \mathcal{W}$, denotes the set of homotopy classes of maps $X \longrightarrow P$. A shape map $F: X \longrightarrow Y$ is a natural transformation $[Y,.] \longrightarrow [X,.]$. In other words, f assigns to every homotopy class $\eta \in [Y,P]$ a homotopy class $f(\eta) \in [X,P]$. If $\eta' \in [Y,P']$, $\mu \in [P',P]$, $\mu\eta' = \eta$, then $\mu f(\eta') = f(\eta)$.

Recently K. Morita [16] has noticed that the notion of a shape map in the above sense can be described using inverse systems in the category $\mathcal{W}$ essentially in the same way as in the ANR-system approach to shape of Hausdorff compacta of J.Segal and the author ([14] and [1o], Section 7). This can be described very conveniently using the notion of pro-category of a given category. We follow here [11] where these notions are described in sufficient generality for our purposes.

Let $\mathcal{K}$ be an arbitrary category. With $\mathcal{K}$ one associates a new category pro $(\mathcal{K})$ whose objects are all inverse systems $\underline{X} = (X_\lambda, p_{\lambda\lambda'}, \Lambda)$ in $\mathcal{K}$ over all directed sets $(\Lambda ; \leq)$. A map of systems $\underline{X} \to \underline{Y} = (Y_\mu, q_{\mu\mu'}, M)$ consists of a function $f : M \longrightarrow \Lambda$ and of a collection of morphisms $f_\mu : X_{f(\mu)} \longrightarrow Y_\mu$, $\mu \in M$, in $\mathcal{K}$ such that for $\mu \leq \mu'$ there is a $\lambda \geq f(\mu)$, $f(\mu')$ such that $f_\mu p_{f(\mu)\lambda} = q_{\mu\mu'} f_{\mu'} p_{f(\mu')\lambda}$. Two maps of systems $(f; f_\mu)$, $(f', f'_\mu) : \underline{X} \longrightarrow \underline{Y}$ are considered equivalent provided for each $\mu \in M$ there is a $\lambda \geq f(\mu)$, $f'(\mu)$ such that $f_\mu p_{f(\mu)\lambda} = f'_\mu p_{f(\mu)\lambda}$. Morphisms $\underline{f} : \underline{X} \to \underline{Y}$ in pro $(\mathcal{K})$ are equivalence classes of maps of systems $(f; f_\mu) : \underline{X} \to \underline{Y}$. If $\underline{g} : \underline{Y} \to \underline{Z} = (Z_\nu, r_{\nu\nu}, N)$ is given by $(g; g_\nu)$, then the composition $\underline{gf} : \underline{X} \to \underline{Z}$ is given by $(fg; g_\nu f_{g(\nu)})$. The identity $1_{\underline{X}} : \underline{X} \to \underline{X}$ is given by $(1_\Lambda; 1_{X_\lambda})$.

Generalizing the situation encountered in [14] and [8] Morita calls an inverse system $\underline{X} = (X_\lambda, p_{\lambda\lambda'}, \Lambda)$ in $\mathcal{W}$, i.e. an object of pro $(\mathcal{W})$, associated with a topological space X provided there exist homotopy classes of maps $p_\lambda : X \to X_\lambda$ such that $\lambda \leq \lambda'$ implies $p_\lambda = p_{\lambda\lambda'} p_{\lambda'}$ and the following two conditions hold for every $P \in Ob(\mathcal{W})$.

(i) For every homotopy class $m \in [X, P]$ there is a $\lambda \in \Lambda$ and a homotopy class $m_\lambda \in [X_\lambda, P]$ such that $m = m_\lambda p_\lambda$.

(ii) Whenever $m_\lambda p_\lambda = m'_\lambda p_\lambda$, m_λ, $m'_\lambda \in [X_\lambda, P]$, then there is a $\lambda' \geq \lambda$ such that $m_\lambda p_{\lambda\lambda'} = m'_\lambda p_{\lambda\lambda'}$.

In other words, the mapping

$$\text{Dir lim } ([X_\lambda, P], p_{\lambda\lambda' \#}, \Lambda) \longrightarrow [X, P]$$

induced by $(p_\lambda, \lambda \in \Lambda)$ is a bijection.

It is not difficult to see that there is a natural bijection between shape maps $X \longrightarrow Y$ and morphisms $\underline{f} : \underline{X} \rightarrow \underline{Y}$ in pro ($\mathcal{W}$), where $\underline{X}$ and $\underline{Y}$ are systems associated with X and Y respectively [16].

Every topological space X admits an associated system $\underline{X}$ namely the Čech system which consists of nerves of locally finite normal coverings [16]. It is however important to be able to use other associated systems as well. E.g., if X is the inverse limit of an inverse system $\underline{X}$ of compact ANR's in the category Top, then $\underline{X}$ is associated with X [14]. Also if X is embedded as a closed subset in an ANR for metric spaces, then the open neighborhoods of X form a system associated with X [16]. This is the reason why the Fox approach yields the same notion of shape for metric spaces.

In a similar way one can define the shape category of pairs of spaces, of pointed spaces or pointed pairs.

With every inverse system $\underline{X} = (X_\lambda, p_{\lambda\lambda'}, \Lambda)$ in $\mathcal{W}$ one can associate homology pro-groups. These are the inverse systems of groups $H_m(\underline{X}) = (H_m(X_\lambda), p_{\lambda\lambda' *}, \Lambda)$, hence, objects of pro (Grp). If $\underline{X}$ and $\underline{X}'$ are inverse systems in $\mathcal{W}$ associated with a space X, then $H_m(\underline{X})$ and $H_m(\underline{X}')$ are naturally isomorphic pro-groups, i.e. isomorphic objects of pro (Grp). Therefore, one can define homology pro-groups, of spaces X as homology pro-groups of associated systems $\underline{X}$ and they are determined up to a natural isomorphism. Clearly, isomorphic pro-groups have isomorphic inverse limits but the converse is not true. E.g., the pro-group

$$\mathbb{Z} \xleftarrow{2} \mathbb{Z} \xleftarrow{2} \mathbb{Z} \leftarrow \dots\dots$$

where 2 denotes mulitplication by 2 is not isomorphic to {o} although both have the inverse limit o. The inverse limit of the homology pro-group $H_m(\underline{X})$ is the usual Čech homology group $H_m(X)$. Homology pro-groups are finer invariants than homology groups. E.g., H_1 of the dyadic solenoid vanishes but the corresponding pro-group is non-trivial.

In a similar way one defines homotopy pro-groups $\pi_m(\underline{X,x})$ and their limits called shape groups. In shape theory the homotopy pro-groups play the role of homotopy groups in homotopy theory of CW-complexes.

For these reasons it is of interest to study the category pro (Grp) of pro-groups. This is a category with zero-objects, i.e. objects which are simultaneously initial and terminal. Such is the system $\underline{O} = \{O\}$ consisting only of one trivial group. In general the pro-group $\underline{G} = (G_\lambda, p_{\lambda\lambda'}, \Lambda)$ is a zero-object if and only if each $\lambda \in \Lambda$ admits a $\lambda' \geq \lambda$ such that $p_{\lambda\lambda'} = O$. In pro-groups there exist kernels and cokernels but pro (Grp) is not an exact category.

Nevertheless, one can speak of exact sequences of pro-groups. A sequence $G \xrightarrow{\underline{f}} H \xrightarrow{\underline{g}} K$ is exact at $\underline{H}$ provided $\underline{g}\underline{f} = 0$ and in the unique factorization $\underline{f} = \underline{i}\underline{f}'$, where $\underline{i}$ is the kernel of $\underline{g}$, the morphism $\underline{f}'$ is an epimorphism.

One can prove that for pointed pairs of spaces (X,A,x) the corresponding homology and homotopy sequences of pro-groups are always exact [11].

The author [12] and K.Morita [17] have proved independently that for pro-groups isomorphisms and bimorphisms coincide. In various situations it is important to be able to decide when is a morphism of pro-groups a monomorphism or an epimorphism. In [12] the following necessary and sufficient condition for $\underline{f} : \underline{G} \rightarrow \underline{H}$ generated by $(f,f_\mu) : (G_\lambda, p_{\lambda\lambda'}, \Lambda) \longrightarrow (H_\mu, q_{\mu\mu'}, M)$ are given:

(i) $\underline{f}$ is a monomorphism if and only if

$$(\forall\ \lambda \in \Lambda)(\exists\ \mu \in M)(\exists\ \lambda' \geq \lambda\ , f(\mu))$$

$$p_{\lambda\lambda'}(f_\mu p_{f(\mu)\lambda'})^{-1}(1) = 1\ .$$

(ii) $\underline{f}$ is an epimorphism if and only if

$$(\forall\ \mu \in M)(\forall\ \lambda \geq f(\mu))(\exists\ \mu' \geq \mu\ )$$

$$q_{\mu\mu'}(H_{\mu'}) \subseteq f_\mu\ p_{f(\mu)\lambda}\ (G_\lambda)\ .$$

One of the most important applications of pro-groups is the Whitehead theorem in shape theory:

Let $\underline{f} : (X,x_o) \rightarrow (Y,y_o)$ be a shape map of connected topological spaces having finite covering dimension.

If $\underline{f}$ induces an isomorphism of homotopy pro-groups $\pi_m \underline{(X,x)} \longrightarrow \pi_m \underline{(Y,y)}$ for all m, then $\underline{f}$ is a shape equivalence.

The theorem was first proved by M.Moszyńska for metric compacta [18]. Her proof was simplified and also extended to cover the case of topological spaces and shape maps generated by continuous maps by the author [11]. Finally, the general result was obtained by Morita [16].

The assumptions that X and Y be finite-dimensional cannot be omitted as shown by a counterexample due to J.Draper and J.Keesling [5]. In their example X and Y are infinite-dimensional metric continua and $\underline{f}$ is generated by a continuous map.

Recently D.A.Edwards and R.Geoghegan [6] have proved an infinite-dimensional Whitehead theorem for shape theory. Their result asserts that $\underline{f}$ is a shape equivalence provided it induces isomorphisms of homotopy pro-groups, X and Y are metric continua, Y has the shape of a CW-complex and X is movable. Movability is an important shape invariant notion introduced by K.Borsuk in [3] (also see [13]). An important corollary asserts that a map $f : (X,x_o) \rightarrow (Y,y_o)$ of metric continua such that sh $(f^{-1}(y))$ = sh (point), for every $y \in Y$, is a shape equivalence provided (X,x_o) is movable and (Y,y_o) has the shape of a CW-complex.

Institute of Mathematics
University of Zagreb / Zagreb, Yugoslavia

References:

[1] K. Borsuk: Concerning the notion of the shape of compacta. Proc. Intern. Symp. on Topology and its Applications. (Herceg-Novi 1968), Belgrade 1969, pp. 98-1o4.

[2] -------- : Concerning homotopy properties of compacta. Fund. Math. 62 (1968), 223-254.

[3] -------- : On movable compacta. Fund. Math. 66 (1969), 137-146.

[4] -------- : On homotopy properties of compact subsets of the Hilbert cube. Ann. Math. Studies 69 (1972),25-36.

[5] J. Draper and J. Keesling: An example concerning the Whitehead theorem in shape theory. To appear in Fund.Math.

[6] D.A. Edwards and R. Geoghegan: Infinite-dimensional Whitehead and Vietoris theorems in shape and pro-homotopy. To appear in Trans. Amer. Math. Soc.

[7] R. H. Fox: On shape. Fund. Math. 74 (1972), 47-71.

[8] W. Holsztyński: An extension and axiomatic characterization of Borsuk's theory of shape. Fund. Math. 7o (1971), 157-168.

[9] J.H. Le Van: Shape theory. Thesis, Univ. of Kentucky, Lexington, Kentucky, 1973.

[1o] S. Mardešić: Shapes for topological spaces. General Topology Appl. 3 (1973), 265-282.

[11] ---- : On the Whitehead theorem in shape theory I. To appear in Fund. Math.

[12] ---- : On the Whitehead theorem in shape theory II. To appear in Fund. Math.

[13] S. Mardešić and J. Segal: Movable compacta and ANR-systems. Bull. Acad. Polon. Sci. Sér. Sci. Math. Astronom. Phys. 18 (197o), 649-654.

[14] ----- : Shapes of compacta and ANR-systems. Fund. Math. 72 (1971), 41-59.

[15] ----- : Equivalence of the Borsuk and the ANR-system approach to shapes. Fund. Math. 72 (1971), 61-68.

[16] K. Morita: On shapes of topological spaces. Fund. Math. 86 (1975), 251-259.

[17] ----- : The Hurewicz and the Whitehead theorems in shape theory. Sc.Rep. Tokyo Kyoiku Daigaku, Sect. A 12 (1974), 246-258.

[18] M. Moszyńska: The Whitehead theorem in the theory of shapes. Fund. Math. 8o (1973), 221-263.

[19] C. Weber: La forme d'un espace topologique est une complétion. C.R. Acad. Sci. Paris, Sér. A-B 277 (1973), A 7-A 9 .

A note on the inverse mapping theorem of F. Berquier

P. Michor

We show that the notion of strict differentiability of [1], § IV is rather restrictive. In fact, we give a complete characterization of strictly differentiable mappings and use it to give a short proof of the main theorem of [1]. Notation is from [1], we only remark, that X is a finite dimensional C^0 manifold and C(R,X) is the space of continuous realvalued functions on X with the Whitney C^0 topology.

Theorem 1: Let $\Phi: C(R,X) \to C(R,X)$ be strictly differentiable at $\varphi_0 \in C(R,X)$. Then there exists an open neighbourhood V_0 of φ_0 in C(R,X) and a continuous function $f: \Omega \to R$, where Ω is a suitable open neighbourhood of the graph of φ_0 in $X\times R$ such that $\Phi(\varphi)(x) = f(x,\varphi(x))$, $x \in X$ for all $\varphi \in V_0$ and furthermore the map $f(x,.)$ is differentiable at $\varphi(x)$ for all $x \in X$ and
$(D\Phi(\varphi_0)h)(x) = df(x,.)(\varphi_0(x)).h(x)$, $x \in X$ for all $h \in C(R,X)$.
If Φ is furthermore differentiable in V_0 (cf. [1], § III) then $f(x,.)$ is differentiable in $\Omega \cap \{x\}\times R$ and $df(x,.)$ is continuous on each point of $\varphi_0(X)$.

Remark: The theorem says, that each strictly differentiable mapping $\Phi: C(R,X) \to C(R,X)$ looks locally like pushing forward sections of the trivial vector bundle $X\times R$ by a suitably differentiable fibre bundle homomorphism. Of course each such map is strictly differentiable, so we have obtained a complete characterization.

Proof: First we remark that the topology on C(R,X) can be described in the followig way: C(R,X) is a topological ring and sets of the form $V_\varepsilon = \{ g \in C(R,X) : |g(x)| < \varepsilon(x) , x \in X\}$ are a base of open neighbourhoods of 0 , where $\epsilon: X \to R$ is strictly positive and continuous.

Now by definition IV-1 of [1] we may write in a neighbourhood of φ_o $\Phi(g+h) - \Phi(g) = D\Phi(\varphi_o)h + R(g,h)$ where R satisfies the following condition: For each V_ϵ there are V_δ , V_λ such that $R(g,hk) \in h.V_\epsilon$ for all $g \in \varphi_o + V_\delta$, $h \in V_\lambda$ and $k \in C(R,X)$ with $|k(x)| \leq 1$, $x \in X$. Let $V_\epsilon = V_1$, $k = 1$, then there are V_δ , V_λ such that $R(g,h) \in hV_1$ for all $g \in \varphi_o + V_\delta$, $h \in V_\lambda$. Let $V_o = \varphi_o + (V_\delta \cap V_{\lambda/2})$.

We claim that if $\varphi_1, \varphi_2 \in V_o$ and $x \in X$ such that $\varphi_1(x) = \varphi_2(x)$ then $\Phi(\varphi_1)(x) = \Phi(\varphi_2)(x)$. This follows from the equation $\Phi(\varphi_1) - \Phi(\varphi_2) = D\Phi(\varphi_o)(\varphi_1 - \varphi_2) + R(\varphi_2, \varphi_1 - \varphi_2)$, since $[D\Phi(\varphi_o)(\varphi_1 - \varphi_2)](x) = (\varphi_1 - \varphi_2)(x).[D\Phi(\varphi_o)(1)](x) = 0$ and $R(\varphi_2,\varphi_1 - \varphi_2) \in (\varphi_1 - \varphi_2).V_1$, so $R(\varphi_2,\varphi_1 - \varphi_2)(x) = 0$.

If $\varphi \in C(R,X)$ denote the graph of φ by $X_\varphi = \{ (x,\varphi(x)): x \in X \}$. Let $\Omega = \bigcup \{ X_\varphi : \varphi \in V_o \}$. By the form of V_o it is clear that Ω is an open neighbourhood of X_{φ_o}. For $\varphi \in V_o$ define $f_\varphi: X_\varphi \to R$ by $f_\varphi(x,\varphi(x)) = \Phi(\varphi)(x)$. By the claim above we see that we have $f_\varphi | X_\varphi \cap X_\Psi = f_\Psi | X_\varphi \cap X_\Psi$ if φ and Ψ are in V_o, so we have got a mapping $f: \Omega \to R$, and $\Phi(\varphi)(x) = f(x,\varphi(x))$ for all $\varphi \in V_o$ and $x \in X$.

We show that f is continuous. If $(x_n,t_n) \to (x,t)$ in $\Omega \subseteq X{\times}R$ we may choose a sequence $\varphi_n \to \varphi$ in $C(R,X)$ such that $(x_n,\varphi_n(x_n)) = (x_n,t_n)$, $\varphi(x) = t$ (remembering that a sequence converges in the Whitney C^o topology iff it coincides with its limit off a compact set K of X after a while and converges uniformly on K). But then $\Phi(\varphi_n) \to \Phi(\varphi)$ uniformly, and $x_n \to x$, so $\Phi(\varphi_n)(x_n) = f(x_n,t_n) \to \Phi(\varphi)(x) = f(x,t)$.

Now we show that f is differentiable at each point of X_φ if Φ is differentiable at φ (strict differentiability implies differentiability, see [1]). We have $\Phi(\varphi+ h) - \Phi(\varphi) = D\Phi(\varphi)h + r_\varphi(h)$, where r_φ is a "small" mapping ([1], § III), i.e. for each V_ϵ there is V_δ such that $r_\varphi(h) \in h.V_\epsilon$ for all $h \in V_\delta$. Evaluating this equation at x we get $f(x,\varphi(x) + h(x)) - f(x,\varphi(x)) = [D\Phi(\varphi)(1)](x).h(x) + r_\varphi(h)(x)$. It is clear that the map $h(x) \to r_\varphi(h)(x)$ is $o(h(x))$ by the "smallness" of r_φ , so $f(x,.)$ is differentiable at $\varphi(x)$ and $[D\Phi(\varphi)h](x) = = df(x,.)(\varphi(x)).h(x)$.

It remains to show that $df(x,.)$ is continuous at each point of X_{φ_o} . This follows easily from Proposition IV-2 of [1] with the method we just applied to show that f is continuous. qed.

<u>Theorem 2</u>: Let $\Phi: C(R,X) \to C(R,X)$ be differentiable in a neighbourhood of $\varphi_o \in C(R,X)$ and strictly differentiable at φ_o and suppose that $D\Phi(\varphi_o)$ is surjective. Then there exists a neighbourhood V_o of φ_o and a neighbourhood W_o of $\Phi(\varphi_o)$ in $C(R,X)$ such that $\Phi: V_o \to W_o$ is a homeomorphism onto. Furthermore the map $\Phi^{-1}: W_o \to V_o$ is differentiable on W_o, strictly differentiable at $\Phi(\varphi_o)$ and for each $\varphi \in V_o$ we have $D(\Phi^{-1})(\Phi(\varphi)) = (D\Phi(\varphi))^{-1}$.

<u>Proof</u>: By theorem 1 we have that $\Phi(\varphi)(x) = f(x,\varphi(x))$ and $D\Phi(\varphi)(1)(x) = df(x,.)(\varphi(x))$. Since $D\Phi(\varphi_o)$ is surjective we conclude that $df(x,.)(\varphi_o(x)) \neq 0$ for all $x \in X$, and since $df(x,.)$ is continuous at $\varphi_o(x)$ it is $\neq 0$ on a neighbourhood of $\varphi_o(x)$ in R. Writing $f_x = f(x,.)$ we see that f_x^{-1} exists and is differentiable on some neighbourhood of $\Phi(\varphi_o)(x)$ in R by the ordinary inverse function theorem. So the map $(x,t) \to (x,f(x,t))$ is locally invetible at each point of the graph X_{φ_o} of φ_o ; one may construct a neighbourhood Ω of X_{φ_o} in $X\times R$ such that this map is invertible there (considering neighbourhoods $U_x \times V_{\varphi_o(x)}$ of $(x,\varphi_o(x))$ where $Id \times f$ is invertible and taking $\Omega = \bigcup_x U_x \times V_{\varphi_o(x)}$). Then $\Phi^{-1}(\Psi)(x) = f_x^{-1}(\Psi(x))$; all other claims of the theorem ase easily checked up. qed.

Remark: Theorem 2 is a little more general than the result in [1]. The method of proof is adapted from [2], 4.1 and 4.2 where we treated an anlogous smooth result.

References

[1] F. BERQUIER: Un theoreme d'inversion locale, to appear in the Proceedings of the Conference on Categorical Topology, Mannheim 1975.

[2] P.MICHOR: Manifolds of smooth maps, to appear.

P. Michor
Mathematisches Institut der Universität
Strudlhofgasse 4
A-1090 Wien, austria.

CARTESIAN CLOSED TOPOLOGICAL CATEGORIES

L.D. Nel

In section 1 we take stock of categories from general topology which admit straightforward axiomatic description and are cartesian closed. Several new ones have recently come to light, all of which are definable by filter axioms. Section 2 discusses topological categories, in the sense of Herrlich. The axioms, which blend initial completeness with simple smallness conditions, allow a rich theory including an efficient charaterization of cartesian closedness. Categories of spaces satisfying a separation axiom cannot form a topological category but may be included in a more general theory of initially structured categories. This is what section 3 is about. The next section discusses sufficient conditions for a reflective or coreflective subcategory of a cartesian closed topological category to inherit cartesian closedness. We conclude with a consideration of possiblities for the embedding of a given concrete category into a cartesian closed topological category.

Generally speaking our terminology will follow the book of Herrlich and Strecker [29]. Subcategory will mean full and isomorphism-closed subcategory. Recall that a category with finite products is called *cartesian closed* when for any object A the functor $A\times-$ has a right adjoint, denoted by $(-)^A$. The categories in which we are interested always have structured sets as objects and for them cartesian closedness means that for given spaces A,B,C there is always a function space B^A available, structured strongly enough to make the natural evaluation function $A\times B^A \to B$ a morphism and at the same time weakly enough to ensure that for any morphism $f:A\times C \to B$ the associated function $f^*:C \to B^A$ is also a morphism. For the definitions of the categories *Con*, *Lim*, *PsTop* and *Fil* see the appendix; for *PNear* and *SNear* see Herrlich [63].

1 Cartesian closed categories with simple axiomatic description

The lack of natural fuction space structures in *Top* makes it an awkward category for several theories such as homotopy theory and topological algebra. Steenrod [55] and MacLane [48] have advocated its replacement by the cartesian closed category *k-Haus* of compactly generated Hausdorff spaces (= k-spaces = Kelley spaces). Dubuc and Porta [20] demonstrated convincingly how topological algebra (particularly Gelfand duality theory) benefits from being cast in *k-Haus*. For related work see also Binz [7] where the cartesian closed category *Lim* is used and Franke [23] where an approach via abstract cartesian closed categories is studied. The advantage of a cartesian closed setting is already illustrated by the formation of function algebras with suitable structure. Whereas in *Top* the search for a suitable topology on an algebra of functions $A \to B$ would not always be successful, the availability of a categorically determined power object B^A ensures that the "right" topological structure for the function algebra is obtained by embedding into B^A.

In recent theories of infinite dimensional differential calculus *Top* has largely been replaced by *Lim* or *k-Haus* , see the papers by Frolicher and Seip at this conference and also Frolicher and Bucher [24], Keller [40], Machado [46], Seip [53]. For use in topology cartesian closed replacements for *Top* have been suggested by Spanier [54], Vogt [56], Wyler [61]. It is a pity that some of these suggested replacements of *Top* are awkward to describe axiomatically (e.g. *k-Haus*) while Spanier's Quasi-topological spaces have the smallness problems to be discussed later. So it seems of interest to list a few cartesian closed categories with simple axiomatic description and no attendant smallness problems: *Con* (Kent [42], Nel [49]), *Lim* (Bastiani [3], Cook and Fischer [14], Binz and Keller [8], Fischer [22], Kowalsky [44]and others), *PsTop* (Choquet [16], Machado [47], Nel [49]), *Fil* (Katetov [39], Robertson [52]).

Cartesian closed categories within the realm of nearness spaces were recently discovered and studied by Robertson [52] and Bentley, Herrlich and Robertson [5]. *Grill* is the subcategory of *SNear* formed by the objects whose near families are all contained in grills (recall that a family of subsets is a grill if all are non-empty and a union of two sets belongs to the family iff at least one of the two sets do). Now *Grill* is a cartesian closed coreflective subcategory of *SNear*. It is equivalent to the category *Fil* and contains the category of proximity spaces as a bireflective subcategory. It also contains suitably restricted convergence spaces as a coreflective subcategory, namely those that satisfy the following axiom:

R_0 If a filter F converges to x and y belongs to every member of F, then F converges to y.

This axiom by the way, reduces to the usual R_0 axiom (x is in every neighbourhood of y iff y is in every neighbourhood of x) when restricted to topological spaces. The category R_0Con thus defined is again cartesian closed and being bireflective in *Con* and coreflective in *Grill* it provides a link between convergence and nearness structures. In similar fashion one obtains two further cartesian closed categories R_0Lim and R_0PsT, bireflectively embedded in *Lim* and *PsTop* respectively and also bireflectively embedded in R_0Con.

R.M. Vogt [56] remarked that "many topologists dislike working with things that are not topological spaces". The nice properties of the above categories, in particular the simple form that the usual categorical constructions take in *Con*, *Lim* and *Grill* make it seem possible that Vogt's remark will become less true in future.

2 Topological categories and cartesian closedness

The categories *Con*, *Lim*, *PsTop*, *Top*, *SNear*, *Grill* along with a multitude of others share many categorical features. These can usefully be studied in terms of an abstract category satisfying certain axioms.

A is called a *topological category* if it comes equipped with a faithful functor $U:A \to Set$ such that

T1 *A* has initial structures for all sources to U*A* and

T2 for any set X the fibre $U^{-1}X$ has a representative set of objects and when X has cardinality 1 its fibre is represented by just one object.

The first axiom is a straightforward abstraction of the well-known existence of a smallest topology on a domain making a given source of functions into topological spaces continuous. Its fundamental role has been recognized and exploited by Bourbaki [9] and a host of others e.g. Antoine[1], Bentley [4], Brümmer [13], Hoffmann [30,31], Kamnitzer [37], Wischnewsky [57,58], Wyler [59]. The smallnes condition (T2) formulated by Herrlich [28] seems to be a very suitable companion for T1. It is simple to check in special cases, does not exclude any category of interest in general topology and yet is strong enough in conjunction with initiality to yield a rich theory. A striking example of how successfully T1 blends with T2 is the following result.

Theorem (Herrlich [28]) For a topological category *A* the following statements are equivalent:

1 *A* is cartesian closed

2 the functor A×- always preserves colimits

3 the functor A×- always preserves coproducts and quotients

4 the functor A×- always preserves final epi-sinks.

("always" means for any *A*-object A; a quotient is a final epimorphism)

Topological categories are stable under formation of bireflective and coreflective (automatically bicoreflective) subcategories. Examples of an apparently non-topological origin include the category of bornological spaces (bounded sets are axiomatized, see Hogbe-Nlend [32]) and of pre-ordered spaces, both of which are in fact also cartesian closed. Topological categories are (co-)complete with U preserving both limits and colimits and (co-) well-powered.

They automatically have final (i.e.coinitial) structures. For further properties see Herrlich [28] and the next section.

3 Initially structured categories

The category of Hausdorff topological spaces is not a topological category since it lacks initial structures for sources that are not point separating. Nevertheless this category has many features in common with topological categories. In view of the importance of categories formed by spaces satisfying a separation axiom it seems worthwhile to have a similar abstract theory for them. To this end we relax axiom T1 by demanding instead:

T1* A has initial structures for mono-sources to UA.

In the terminology of Herrlich [26] this means that U is an (epi, monosource)-topological functor. If we think of A-objects as structured sets it means that for any class of objects $(X_i,\alpha_i)_I$ and functions $f_i:X \to X_i$ the quotient set Q, obtained by collapsing points not separated by the f_i, has a smallest A-structure available for which all the induced functions $Q \to X_i$ are morphisms.

A significant portion of the theory of topological categories generalizes although some statements are complicated by the necessity of passage to a quotient set. It is no longer true that embeddings coincide with regular monomorphisms and also with extremal monomorphisms, but final epi-sinks do still coincide with extremal epi-sinks and quotients with extremal epis and also with regular epis. Well-poweredness, completeness and co-completeness are still with us but U no longer preserves co-limits. Factorization properties abound: every initially structured category is an (epi_U, embedding)-category, (epi, extremal mono)-category, (epi_U, initial monosource)-category, (quotient, mono)-category and a (final episink, mono)-category. Here epi_U denotes the class of e such that Ue is an epimorphism and (E,M)-category is used as in [29].

By using the above facts one can show that Herrlich's characterizations of cartesian closedness given in section 2 remains valid for initially structured categories.

Initially structured categories have stability under formation of subcategories that is better than that of topological categories. They are not only closed under coreflective subcategories (which incidently are characterized by being closed under formation of colimits or equivalently under final epi-sinks) but also under all epireflective subcategories. For the results of this section and further details see Nel [49].

4 Subcategories that inherit cartesian closedness

When is a subcategory $\mathcal{B}$ of a cartesian closed category $\mathcal{C}$ again cartesian closed? For reflective subcategories we have the following:

(a) (Day [18]) Suppose $\mathcal{B}$ is reflective in the monoidal category $\mathcal{C}$ and that the reflector $R:\mathcal{C} \to \mathcal{B}$ preserves finite products. Suppose also that $\mathcal{B}$ is dense in $\mathcal{C}$ (i.e. adequate in the sense of Isbell [38]). Then $\mathcal{B}$ inherits cartesian closedness from $\mathcal{C}$.

(b) (Robertson [52]) Suppose $\mathcal{C}$ is a topological category and $\mathcal{B}$ is a bireflective subcategory whose reflector R satisfies $R(B\times C) = B\times RC$ for all B in $\mathcal{B}$ and C in $\mathcal{C}$. Then $\mathcal{B}$ inherits cartesian closedness from $\mathcal{C}$.

(c) (Nel [49]) Suppose $\mathcal{C}$ is initially structured and $\mathcal{B}$ is a quotient-reflective subcategory. Then $\mathcal{B}$ inherits cartesian closedness from $\mathcal{C}$.

For coreflective subcategories there is a similar result:

(d) (Nel [49]) Suppose $\mathcal{C}$ is initially structured and $\mathcal{B}$ is a coreflective subcategory closed under finite products in $\mathcal{C}$. Then $\mathcal{B}$ inherits cartesian closedness from $\mathcal{C}$.

As a useful corollary of (d) we note that if $\mathcal{K}$ is any finitely productive subcategory of $\mathcal{C}$, then its coreflective hull inherits cartesian closedness from $\mathcal{C}$.

5 Embedding into cartesian closed topological categories

The categories *Con*, *Lim*, *PsTop* discussed in section 1 were apparently not intrduced with cartesian closedness in mind: in each case this property was discovered several years later. But there have been deliberate constructions to create cartesian closed categories for use in topology. The first of these was the category *Quasi-Top* introduces and studied by Spanier [54]. *Quasi-Top* does not satisfy axiom T2 and thus is not a topological category. In fact an object whose underlying set has more than one point may be undefinable in terms of sets. The same disadvantage is present in similar later constructions by Antoine [1] and Day [18].

Other constructions of cartesian closed categories were carried out within a given special category. Thus Vogt [56] and Wyler [61] studied the embedding of compact Hausdorff spaces into cartesian closed subcategories of *Top*. Antoine [1] and Machado [47] studied the embedding of *Top* into cartesian closed subcategories of *Lim*. In particular Machado showed that Antoine's epitopological spaces formed the smallest such category between *Top* and *Lim*. Bourdaud [10,62] obtained corresponding results for an embedding into *Lim* of pretopological spaces (see also Bourdaud's paper at this conference).

Embeddings of suitably restricted abstract categories into cartesian closed topological categories are being studied by H. Herrlich and myself. We conclude with a preliminary report about this. Suppose A to satisfy T2 and the following conditions: A has quotients, finite products preserved by U and in A the product of two quotients is a quotient. Let us call such A *preconvenient*. The category A^* is now constructed as follows. Its objects are pairs (ξ,X) where ξ is a set of pairs (A,a) such that $a:UA \to X$ is a 1-1 function subject to (1*) if UP is a singleton, then $(P,p)\varepsilon\xi$ for any $p:UP \to X$; (2*) if $(A,a)\varepsilon\xi$ and $a = c\circ Uq$ where c is 1-1 and $q:A \to C$ is a quotient, then $(C,c)\varepsilon\xi$. Morphisms from (ξ,X) to (η,Y) are functions $f:X \to Y$ such that for any $(A,a)\varepsilon\xi$ we have

$(B,b)\in\eta$ where (B,b) is the unique pair such that $f\circ a = b\circ Uq$ with $q:A \to B$ a quotient .

Then A^* is a cartesian closed topological category into which A can be embedded as a subcategory so that existing initial structures and powers are preserved. The coreflective hull of A is all of A^*. If A is already a topological category then it is bireflective in A^*.

The preconvenient categories to which this construction applies include *PNear*, *SNear*, the category of finite topological spaces, the category of *Top*-quotients of compact Hausdorff spaces.

By considering a concrete category A with embeddings and by using onto functions $a:X \to UA$ one can also construct an embedding $A \to A'$ where A' now turns out to be a preconvenient category containing A (if A is topological it is in fact coreflectively embedded). Thus any topological category can be fully embedded (in two steps) into a cartesian closed topological category. Unfortunately this embedding need not preserve initial structures.

However if some embedding of A into a cartesian closed topological category C is known to exist such that C coincides with the coreflective hull of A, then A is contained in a smallest cartesian closed topological subcategory B of C. In fact B can be constructed as the bireflective hull in C of all C-powers formed out of A-objects. Thus the embedding of A into B preserves initial structures and moreover it preserves powers. In the special case A = *Top*, C = *Lim* one obtains as a corollary Machado's result mentioned above.

Appendix. A *convergence space* [42] is a pair (X,q) where q is a function which assigns to each x in X a set qx of filters on X "convergent to x"; moreover the following conditions must hold: (F1) : every principal ultrafilter $\dot{x}$ is in qx; (F2) : if F is in qx and G refines F, then G is in qx; (C) : if F is in qx, then $F\wedge\dot{x}$ is in qx. A *limit space* [44], [22] is a convergence space in which (C) is strengthened to (L): if F,G are in qx, then $F\wedge G$ are in qx. A *pseudo-topological* *space* [16] is a limit space in which (L) is strenghtened to (PsT): if F is such that all its ultrafilter refinements are in qx, then F is in qx. Further strengthenings of this axiom lead to pretopological and topological spaces. By taking these spaces as objects and continuous (i.e. convergence preserving) functions as morphisms we obtain respectively the category *Con*, *Lim*, *PsTop*. If (F1) and (F2) are modified by axiomatizing only a family of convergent filters (not convergent to points) the two modified axioms give rise to the category *Fil* [39].

Bibliography

1 P. Antoine, Etude élémentaire des categories d'ensembles structurés, Bull. Soc. Math. Belge 18 (1966) 142-164 and 387-414.

2 P. Antoine, Notion de compacité et quasi-topologie, Cahiers de Top. et Geom. Diff. 14 (1973) 291-308.

3 A. Bastiani, Applications differentiables et variétés differentiables de dimension infinie, J. Analyse Math. 13 (1964) 1-114.

4 H.L. Bentley, T-categories and some representation theorems, Portugaliae Math. 32 (1973) 201-222.

5 H.L. Bentley, H. Herrlich and W.A. Robertson, Convenient categories for topologists, Comm. Math. Univ. Carolinae (to appear).

6 E. Binz, Bemerkungen zu limitierten Funktionenalgebren, Math. Ann 175 (1968) 169-184.

7 E. Binz, Continuous Convergence on C(X), Springer Lecture Notes in Math. 469 (1975).

8 E. Binz and H.H. Keller, Funktionenraume in der Kategorie der Limesraume, Ann. Acad. Sci. Fenn. Sec. AI 383 (1966) 1-21.

9 N. Bourbaki, Topologie Générale, Hermann et cie, Paris (1948).

10 G. Bourdaud, Structure d'Antoine associées aux semi-topologies et aux topologies, C.R. Acad. Sci. Paris Ser. A279 (1974) 591-594.

11 H. Breger, Die Kategorie der kompakt-erzeugten Raume als in Top coreflective Kategorie mit Exponentialgesetz, Diplomarbeit, Univ. Heidelberg (1971).

12 R. Brown, Function spaces and product topologies, Quart. J. Math. Oxford 15 (1964) 238-250.

13 G.C.L. Brummer, A categorical study of initiality in uniform topology, Ph.D. thesis, Univ. Cape Town 1971.

14 C.H. Cook and H.R. Fischer, On equicontinuity and uniform convergence, Math. Ann. 159 (1965) 94-104.

15 C.H. Cook and H.R. Fischer, Uniform convergence structures, Math. Ann. 173 (1967) 290-306.

16 G. Choquet, Convergences, Ann. Univ. Grenoble (i.e. Ann.Inst. Fourier) 23 (1947/48) 57-112.

17 D. Damerov, Die Kategorie der Kelley-Raume, Diplomarbeit, Freie Univ. Berlin (1969).

18 B. Day, A reflection theorm for closed categories, J. Pure Appl. Algebra 2 (1972) 1-11.

19 B. Day, An embedding theorem for closed categories, Category Seminar, Sydney (1972/73) Springer Lecture Notes in Math. 420 (1974).

20 E.J. Dubuc and H. Porta, Convenient categories of topological algebras and their duality theory, J. Pure Appl. Algebra 1 (1971) 281-316.

21 S. Eilenberg and G.M Kelly, Closed categories, Proc. Conf. on Categorical Algebra, La Jolla (1965) Springer, Berlin (1966) 421-562.

22 H.R. Fischer, Limesraume, Math. Ann 137 (1959) 269-303.

23 D. Franke, Funktionenalgebren in cartesisch abgeschlossenen Kategorien, Thesis, Free Univ. Berlin (1975).

24 A. Frolicher and W. Bucher, Calculus in vector spaces without norm, Springer Lecture Notes in Math. 30 (1966).

25 K.A. Hardie, Derived homotopy constructions, J. London Math. Soc. 35 (1960) 465-480.

26 H. Herrlich, Topological functors, Gen. Top. Appl. 4 (1974) 125-142.

27 H. Herrlich, Topological structures, Math. Centre Tract 25 (1974) 59-122.

28 H. Herrlich, Cartesian closed topological categories, Math. Colloq. Univ. Cape Town 9 (1974) 1-16..

29 H. Herrlich and G. E. Strecker, Category Theory, Allyn and Bacon, Boston (1973).

30 R. E. Hoffmann, Topological functors and factorizations (preprint)

31 R.E. Hoffmann, (E,M)-universally topological functors (preprint)

32 H. Hogbe-Nlend, Theorie des bornologies et applications, Springer Lecture Notes in Math. 213 (1971).

33 W. N. Hunsaker and P.L. Sharma, Proximity spaces and topological functors, Proc. Amer. Math. Soc.

34 M. Husek, S-categories, Comment. Math. Univ. Caroliniae 5 (1964) 37-46.

35 M. Husek, Categorical methods in topology, Proc. Symp. Gen. Topol. Appl. Prague (1966) 190-194.

36 M. Husek, Construction of special functors and its applications, Comment. Math. Univ. Carolinae 8 (1967) 555-556.

37 S.H. Kamnitzer, Protoreflections, relational algebras and topology, Thesis, Univ. Cape Town (1974).

38 J. Isbell, Adequate subcategories, Illinois J. Math. 4 (1960) 541-552.

39 M. Katetov, On continuity structures and spaces of mappings, Comment. Math. Univ. Carolinae 6 (1965) 257- 278.

40 H. Keller, Differential Calculus in Locally Convex Spaces, Springer Lecture Notes in Math. 417 (1974).

41 D.C. Kent, Convergence functions and their related toplogies, Fund. Math. 54 (1964) 125-133.

42 D.C. Kent, On convergence groups and convergence uniformities, Fund. Math.

60 (1967) 213-222.

43 D.C. Kent and G.D. Richartdson, Locally compact convergence spaces (preprint)

44 H.J. Kowalsky, Limesraume und Komplettierrung, Math. Nachr. 12 (1954) 301-340.

45 W.F. LaMartin, k-groups, Thesis, Tulane Univ. (1973).

46 A. Machado, Quasi-variétés complexed, Cahiers de Top. et Geom. Diff. 11 (1970) 231-279.

47 A. Machado, Espaces d'Antoine et pseudo-topologies, Cahiers de Top. et Geom. Diff. 14 (1973) 309-327.

48 S. MacLane, Categories for the Working Mathematician, Springer, New York (1971).

49 L.D Nel, Initially structured categories and cartesian closedness, Canad. J. Math. 27 (1975) 1361-1377.

50 H. Poppe, Compactness in general function spaces, VEB Deutscher Verlag der Wissenschaften, Berlin (1974)..

51 R. Pupier, Precompactologie et structures uniformes, Fund. Math. 83 (1974) 251-262.

52 W.A. Robertson, Convergence as a nearness concept, Thesis, Carleton Univ. (1975).

53 U. Seip, Kompakt erzeugte Vektorraume und Analysis, Springer Lecture Notes in Math. 273 (1972).

54 E. Spanier, Quasi-topologies, Duke Math. J. 30 (1963) 1-14.

55 N.E. Steenrod, Aconvenient category of topological spaces, Michigan Math. J. 14 (1967) 133-152.

56 R.M. Vogt, Convenient categories of topological spaces for homotopy theory, Archiv. Math. 22 (1971) 545-555.

57 M. Wischnewsky, Partielle Algebren in Initial-kategorien, Math. Zeitschr. 127 (1972) 17-28.

58 M. Wischnewsky, Generalized universal algebra in initialstructure categories, Algebra- Berichte (Uni-Druck, Munchen) 10 (1973) 1-35.

59 O. Wyler, On the categories of general topology and topological algebra, Archiv. Math. 22 (1971) 7-17.

60 O. Wyler, TOP Categories and categorical topology, Gen. Topol. Appl. 1 (1971) 17-28.

61 O. Wyler, Convenient categories for topology, Gen. Topol. Appl. 3 (1973) 225-242.

Closely related talks given at this conference:

62 G.Bourdaud, Some closed topological categories of convergence spaces.

63 H. Herrlich, Some topological theorems which fail to be true.

64 W.A. Robertson, Cartesian closed categories of Nearness structures.

65 O. Wyler, Are there topoi in topology?

Carleton University

Ottawa, Canada

This research was aided by NRC grant A5297.

EPIREFLECTIVE CATEGORIES
OF HAUSDORFF SPACES

Peter J. Nyikos, University of Illinois, Urbana, Ill.

0. Introduction. The study of epireflective categories of Hausdorff spaces is a natural outgrowth of a considerable body of topological theory. In fact, the Stone-Čech conpactification is the archetypal example of an epireflection, and one hardly needs to be reminded of its central role in such "classics" as [GJ]. I will be mentioning many more examples in the course of this talk, whose principal theme is the various methods, old and new, that can be used to obtain epireflective subcategories of the category T2 of Hausdorff spaces.

The methods employed fall naturally into two theories, to which Sections 1 and 2 of this talk are devoted. The first is based upon the pioneering work done by Mrówka, Engelking, and Herrlich on $\mathcal{E}$-compact and $\mathcal{E}$-regular spaces $[M_1]$, [EM], $[H_1]$. The second is of much more recent vintage and has to do with the various types of "disconnectedness" as defined by G. Preuss and most recently summarized and clarified by Arhangel'skii and Wiegandt [AW]. There is no hard and fast division between these theories, and Section 3 gives one way of bringing them together.

Here follow some of the definitions and results basic to this area of categorical topology.

0.1 DEFINITION. Let $\mathfrak{C}$ be a category. A subcategory $\mathcal{B}$ of $\mathfrak{C}$ is a reflective subcategory of $\mathfrak{C}$ if for each object $C \in \mathfrak{C}$ there exists an object $rC \in \mathcal{B}$ and a map $r_C : C \to rC$ with the following property: if B is an object of $\mathcal{B}_1$ then for each map $f : C \to B$ there exists a unique map $f^r : rC \to B$ in $\mathcal{B}$ such that $f^r \circ r_C = f_1$ i. e. such that the diagram at right commutes. The pair (r_C, rC) is called <u>the reflection of</u> C in $\mathcal{B}$.

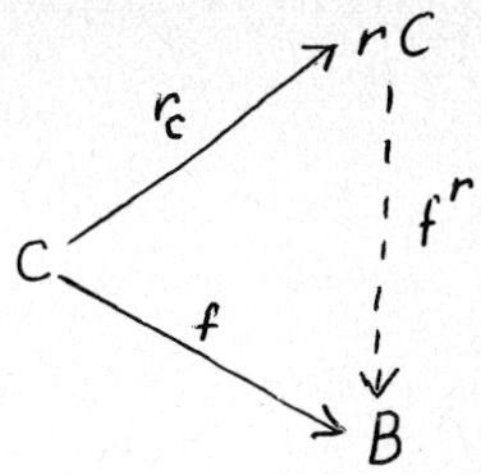

We can speak of "the" reflection of C in $\mathcal{B}$ because of the elementary categorical fact that universal objects are determined up to a unique isomorphism. In the case of topological spaces, the "isomorphisms" are the homeomorphisms.

0.2. NOTATION. The following categories will be symbolized as follows:

- TOP = the category of topological spaces and continuous maps between them.
- T2 = the category of Hausdorff spaces
- FT2 = the category of functionally Hausdorff spaces, i.e. spaces X such that for any two points $x_1, x_2 \in X$, there exists a continuous function $f : X \to \mathbb{R}$ such that $f(x_1) \neq f(x_2)$.
- $T3\frac{1}{2}$ = the category of Tychonoff (completely regular Hausdorff) spaces.
- D2 = the category of spaces whose quasicomponents are single points, where the quasicomponent of a point is defined to be the intersection of all clopen sets containing it.
- D3 = the category of zero-dimensional spaces, i. e. Hausdorff spaces with a base of clopen sets.
- CHSP = the category of compact Hausdorff spaces.

We will use the convention that all categories are <u>full</u> subcategories of TOP (that is, if two spaces are in a subcategory, so are all continuous functions between them) and that they are replete (that is, if X is in a subcategory, so is every space homeomorphic to X). Thus, by abuse of language, a class of spaces will be spoken of as a "category" even if there is no explicit mention of the maps between them. Mostly, we will be restricting our attention to Hausdorff spaces, but sometimes we will be forced to step outside and look at some non-Hausdorff construction.

0.3. DEFINITION. A reflective subcategory B of a category $\mathcal{C}$ is epireflective if the reflection map r_C is an epimorphism for all $C \in \mathcal{C}$; monoreflective if r_C is a monomorphism; extremally epireflective if r_C is an extremal epimorphism.

For the definitions of [extremal] epimorphism and monomorphism, see [H_2]. For the present, it is enough to know that the monomorphisms of all the above subcategories of TOP are the 1-1 maps [not just the embeddings, although in the case of CHSP the two concepts merge] and the extremal epimorphisms are the quotient onto maps. In TOP, the empimorphisms are the onto maps, while in all the other cases, they are the maps with dense range [H_2, pp. 114-S]; of course, in the case of CHSP, these two concepts coincide.

0.4. THEOREM. Let B be a subcategory of TOP [resp. T2] which includes a nonempty space.

(a) B is epireflective in TOP if, and only if, B is productive and hereditary.

(b) B is epireflective in T2 if, and only if, B is productive and closed hereditary.

(c) B is extremally epireflective if, and only if, B is productive, hereditary, and closed under the taking of finer topologies.

For a proof of (a) and (b), see [H_2]. The proof of (C) is similar, and involves writing the reflection map r_C as a composition of a quotient map and 1-1 map of the quotient space into rC.

Part (b) holds true of all the categories of spaces listed in 0.2, other than TOP, while part (C) holds not only for TOP and T2 but also for FT2, D_2, and CHSP. (Of course, in the case of CHSP, parts (b) and (c) say the same thing. In fact, every surjective map in CHSP is quotient, even closed.)

Theorem 0.4 shows that it is not enough to call as category "epireflective"; one must specify what it is epireflective in. For example, by (b) the category of compact Hausdorff spaces is epireflective in T2, but not in TOP. The same is true of the class of realcompact spaces.

On the other hand, the T_i - spaces for $i = 0,1,2,3$, and $3\frac{1}{2}$ are epireflective in TOP, and the first four are extremally epireflective as well. The categories D2 and FT2 are extremally epireflective in TOP (and T2 - there is no distinction here as long as the category is contained in T2) and the category D3 is epireflective in TOP.

As a result of Theorem 0.4, we know that the intersection of a class of epireflective subcategories of T2 is itself epireflective in T2. So, given a subcategory $\mathcal{E}$ of T2, one can speak of "the epireflective subcategory T2($\mathcal{E}$) generated by $\mathcal{E}$", viz. the intersection of all epireflective subcategories of T2 containing $\mathcal{E}$. Similarly, we define TOP($\mathcal{E}$). The same goes for extremally epireflective subcategories, and we will let Q($\mathcal{E}$) stand for the extremally epireflective subcategory (of TOP or T2, depending on whether $\mathcal{E} \subset$ T2) generated by $\mathcal{E}$.

One might expect, given Theorem 0.4, that (say) T2 ($\mathcal{E}$) can only be obtained by taking closed subspaces of products of spaces in $\mathcal{E}$, repeating the process on the new category, and so on *ad infinitum*. Actually, a bit of topological insight is enough to show that one step of this process is ample:

0.5. THEOREM. *Let* $\mathcal{E}$ *be a subcategory of* TOP.

(a) TOP ($\mathcal{E}$) *is the category of all subspaces of products of spaces in* $\mathcal{E}$.

(b) Q ($\mathcal{E}$) *is the category of all spaces which admit* 1-1 *maps into a product of spaces in* $\mathcal{E}$.

(c) *Let* $\mathcal{E} \subset$ T2. *Then* T2($\mathcal{E}$) *is the category of all closed subspaces of products of spaces in* $\mathcal{E}$.

A categorical proof of parts (a) and (c) may be found in [H_2], and one of (b) may be constructed along the same lines.

Of course, since T2 is extremally epireflective in TOP, Q ($\mathcal{E}$) $\subset$ T2 whenever $\mathcal{E} \subset$ T2. *A fortiori*, TOP ($\mathcal{E}$) $\subset$ T2 whenever $\mathcal{E} \subset$ T2. However, even in this case, TOP($\mathcal{E}$) usually contains T2($\mathcal{E}$) properly. For example, when $\mathcal{E} = \{\mathbb{R}\}$, TOP ($\mathcal{E}$) is the category of completely regular spaces, and T2($\mathcal{E}$) is the category of realcompact spaces. It is also important to note that T2($\mathcal{E}$) is defined only if $\mathcal{E} \subset$ T2.

Section 1. $\mathcal{E}$ - *regular*, $\mathcal{E}$ - *compact*, *and* $\mathcal{E}_c$-*compactlike spaces*.

With Theorem 0.5, we have already entered the realm of $\mathcal{E}$ - regular and $\mathcal{E}$ - compact spaces. As a class, these were first studied systematically by S. Mrówka and R. Engelking [M_1], [EM], [M_2]. The credit for fully realizing the categorical significance of $\mathcal{E}$-regular and $\mathcal{E}$- compact spaces goes to H. Herrlich [H_1], [H_2]. For example, it was he who dropped the insistence that $\mathcal{E}$ consist of a single space (or, equivalently, a set of spaces).

1.1. DEFINITION. Let $\mathcal{E}$ be a class of spaces. A space X is $\mathcal{E}$ - Hausdorff if for each pair of points $x, y \in X$, there exists $f : X \to E \in \mathcal{E}$ such that $f(x) \neq f(y)$. A space X is $\mathcal{E}$ - regular [Mrówka: $\mathcal{E}$ - completely regular] if it is $\mathcal{E}$ - Hausdorff and, for each closed set $A \subset X$, and each $x \in A$, there exists $f : X \to Y$, where Y is a finite product of spaces in $\mathcal{E}$, such that $f(x) \notin cl f(A)$.

1.2. THEOREM. Let X be a topological space, $\mathcal{E} \subset$ TOP.

(a) [M_1] X is $\mathcal{E}$ - regular if, and only if, it can be embedded as a subspace of a product of spaces in $\mathcal{E}$.

(b) X is $\mathcal{E}$ - Hausdorff if, and only if, it admits a 1-1 map into a product of spaces in $\mathcal{E}$.

In other words, $X \in$ TOP ($\mathcal{E}$) if, and only if, it is $\mathcal{E}$ - regular, and $X \in Q(\mathcal{E})$ if, and only if, X is $\mathcal{E}$ - Hausdorff. (Of course, $\mathcal{E}$ - Haudorff implies Hausdorff if, and only if, $\mathcal{E} \subset$ T2.)

The definition of $\mathcal{E}$ - compact spaces given in [EM] is a bit technical, and the most convenient definition is simply:

1.3. DEFINITION. A space is $\mathcal{E}$ - compact if it can be embedded as a closed subspace in a product of spaces in

1.4. NOTATION. We will let $\mathbb{I}$ stand for the closed unit interval [0,1] and $\mathbb{R}$ for the real line, both with the usual topology. The countably infinite discrete space will be identified with the natural numbers and denoted $\mathbb{N}$. We let 2 stand for the two-point discrete space.

Here is a table listing some of the best-known examples in this area:

TABLE 1

$\mathcal{E}$	T 2 ($\mathcal{E}$)	TOP ($\mathcal{E}$)	Q ($\mathcal{E}$)
$\{\mathbb{I}\}$	CHSP	T $3\frac{1}{2}$	FT2
$\{\mathbb{R}\}$	realcompact spaces	T $3\frac{1}{2}$	FT2
$\{\underline{2}\}$	D3 $\cap$ CHSP (=D2 $\cap$ CHSP)	D 3	D 2
$\{\mathbb{N}\}$	N - compact spaces	D 3	D 2

Several characterizations of the spaces in T2($\mathcal{E}$) for these examples may be found in $[H_1]$, $[H_2]$, $[H_4]$, and $[M_2]$. Also, necessary and sufficient conditions exist on $\mathcal{E}$ for TOP ($\mathcal{E}$) and T2 ($\mathcal{E}$) to be the above categories. Perhaps the most striking of these conditions is that, if $\mathcal{E} \subset$ CHSP, then CHSP = T2 ($\mathcal{E}$) iff $T3\frac{1}{2}$ = TOP($\mathcal{E}$) iff there exists in $\mathcal{E}$ a space which contains a copy of $\mathbb{I}$. [Proof: Of course, the latter condition implies the former; conversely, for CHSP to equal T2 ($\mathcal{E}$) or $T3\frac{1}{2}$ to equal TOP($\mathcal{E}$), some product of spaces in $\mathcal{E}$ must contain $\mathbb{I}$. There exists a projection map with respect to which the image of $\mathbb{I}$ is non-trivial, and this image contains a copy of $\mathbb{I}$ - in fact [HS] it is a Peano continuum, hence arcwise connected.] We will make use of similar results in Section 3.

Less familiar than the above examples is the case of $T2(\omega_\lambda)$, where ω_λ is the λ - th cardinal, looked upon as the space of all ordinals less than itself, with the order topology. If ω_λ is of countable cofinality, this is T2($\mathbb{N}$) again: on the one hand, ω_λ contains a closed, discrete, countably infinite space; on the other, ω_λ is strongly zero-dimensional and realcompact, hence $[H_1]$ it is $\mathbb{N}$ - compact. The situation in the case where ω_λ is of uncountable cofinality is different. Every such space is countably compact and not compact, hence not in T2 ($\mathbb{N}$). This is also true

[B], $[M_2]$ of any product of copies of ω_λ. More generally, as Blefko has shown [B], the categories $T2(\omega_\lambda)$ and $T2(\omega_\mu)$ are incomparable (with respect to containment) for any two regular cardinals ω_λ and ω_μ, including the case where one of them is $\omega_0 = \mathbb{N}$. Additional results of Blefko's concerning these spaces, and the open problems they leave, may be found in $[M_2]$.

In principle, Theorem 0.5 (c) solves the problem of characterizing the epireflective subcategories of T2, and (a) characterizes those which are apireflective in TOP as well. But this theorem alone can only be expected to yield essentially haphazard results such as the above. I will now describe one approach, worked out by S. Hong $[H_5]$ $[H_6]$ which makes the search for epireflective subcategories a bit more systematic. To introduce it, let me recall the following generalization of real compact spaces.

1.5. EXAMPLE. Let k be a cardinal number. Herrlich has referred to a Tychonoff space X as being k - compact if every Z - ultrafilter with the k - intersection property[1] is fixed. For example, the $\aleph_0$ - compact spaces are the compact Hausdorff spaces, while the $\aleph_1$ - compact spaces are the realcompact spaces. For each infinite cardinal k, the k- compact spaces form an epireflective subcategory of T2. Moreover, the category of k- compact spaces is distinct for distinct k : if $k < k^1$, then the category of k^1 - compact spaces contains the category of k - compact spaces properly. Specifically, Hušek has shown $[H_7]$ that the category of

---

[1]In this definition, the k-intersection property refers to a collection such that the intersection of fewer than k sets is nonempty; thus what is commonly called the "countable intersection property" is here referred to as the $\aleph_1$ - intersection property.

k- compact spaces is $T2(P_k)$ where $P_{\omega_0} = \amalg$; P_k where $k = \omega_{\nu+1}$ is $\amalg^{\omega_\nu} - \{p\}$ where p is any point; and P_k for a limit cardinal is $\prod \{P_t \mid t<k\}$.

It is a third characterization of these spaces which Hong utilized in his system.

1.6. DEFINITION. A subset A of a topological space X is k- closed if for every point $x \in X - A$, there exists a G_k - set disjoint from A and containing x. [A G_k- set is a set which is the intersection of fewer than k open sets.]

1.7. THEOREM [H_5] A space X is k- compact if, and only if, it is Tychonoff and k- closed in βX. For any Tychonoff space X, the reflection of X in the category of k- compact spaces is its k- closure in βX.

Now, given any epireflective subcategory $\mathcal{B}$ of T2 (in the case of k - compact spaces, CHSP plays the role of $\mathcal{B}$) one can define the category $\mathcal{B}_k$ of all spaces X for which the reflection map r_X to $\mathcal{B}$ is an embedding, such that the image $r_X(X)$ is k- closed in the reflection space rX.

1.8. THEOREM [H_5] For any epireflective subcategory $\mathcal{B}$ of T2 and any k, $\mathcal{B}_k$ is an epireflective subcategory of T2 containing $\mathcal{B}$.

Because spaces in $\mathcal{B}_k$ are embeddable in spaces of $\mathcal{B}$, it follows from Theorem 0.4 that $\mathcal{B}_k \subset$ TOP ($\mathcal{B}$). In fact, from Theorem 0.4 we obtain:

1.9. COROLLARY. Let $\mathcal{B}$ be an epireflective subcategory of T2. The reflection map r_X is an embedding if, and only if, $X \in$ TOP ($\mathcal{B}$).

Simple cardinality arguments show that every member of TOP($\mathcal{B}$) is in $\mathcal{B}_k$ for some k, so that (although foundationists shy away from such statements) one may informally say that TOP($\mathcal{B}$) is the union of the spaces in $\mathcal{B}_k$ for all k.

Now, obviously, if $\mathcal{B}$ is epireflective in TOP, then $\mathcal{B}_k = \mathcal{B}$ for all k. At the opposite extreme, if all the spaces in $\mathcal{B}$ are compact Hausdorff, and $\mathcal{B}$ contains a space with more than one point, then the categories $\mathcal{B}_k$ are distinct for all k. Indeed, the space D_k obtained by replacing $\amalg$ with $\underline{2}$ in Example 1.5 is in $\mathcal{B}_k$, but not in $\mathcal{B}_t$ for any $t < k$. See Example 1.18, below.

So even this system is enough to give us a wealth of epireflective subcategories of T2, one hierarchy for each epireflective subcategory of CHSP [and we will see in Section 3 how to obtain a number of such epireflective subcategories], the members in each hierarchy in one-to-one correspondence with the infinite cardinal numbers. Moreover, if $\mathcal{B}$ and $\mathcal{B}^1$ are contained in CHSP and $\mathcal{B} \neq \mathcal{B}^1$, then also $\mathcal{B}_k$ and $\mathcal{B}_k^1$ are distinct, as can be easily shown from:

1.10. LEMMA. _If_ $\mathcal{B}$ _and_ $\mathcal{B}^1$ _are epireflective subcategories of_ CHSP _and_ $\mathcal{B} \not\subset \mathcal{B}^1$, _then_ $\mathcal{B}_k \not\subset \mathcal{B}_k^1$ _for any_ k.

Proof. Suppose $\mathcal{B}_k$ were contained in $\mathcal{B}_k^1$. Then this would imply that TOP($\mathcal{B}_k$) is contained in TOP($\mathcal{B}_k^1$). Since TOP($\mathcal{B}_k$) = TOP($\mathcal{B}$), this would imply that every member of $\mathcal{B}$ can be embedded in a member of $\mathcal{B}^1$, and since each member of $\mathcal{B}$ is compact, we would have $\mathcal{B} \subset$ T2($\mathcal{B}^1$) = $\mathcal{B}^1$, contradiction.

But S. Hong has carried the system a step further. His generalization is motivated by the observation that one can form a new topology $\mathcal{T}_k$ on the space $(X, \mathcal{T})$ by using the G_k - sets as a base. Now $(X, \mathcal{T}_k)$ is the coreflection of $(X, \mathcal{T})$ in the category $\mathcal{P}_k$ of spaces in which the intersection of fewer than k open sets is open, and B_k is the category of all spaces in TOP(B) which are closed with respect to the $\mathcal{P}_k$ - coreflection of their B - reflection. This can be generalized to any coreflective subcategory of T2. The word "coreflective" is defined dually to "reflective":

1.11. DEFINITION. A subcategory $\mathcal{C}$ of a category $\mathcal{A}$ is coreflective if for each $X \in \mathcal{A}$ there exists $cX \in \mathcal{C}$ and a map $c_X : cX \to X$ such that for every map $f : C \to X$, where $C \in \mathcal{C}$, there exists a unique map $f_c \in \mathcal{C}$ such that the diagram at right commutes.

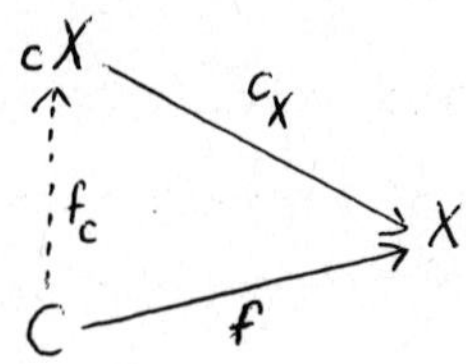

The following theorem is basic to the theory of topological coreflections:

1.12. THEOREM [H_2]. Let $\mathcal{C}$ be a coreflective subcategory of T2 or TOP. Then the coreflection map c_X is bijective.

In other words, cX can be thought of as X with a finer topology. H. Herrlich has given four axioms for a so-called idempotent closure operator on all topological (or Hausdorff) spaces simultaneously, in such a way that there is a 1-1 correspondence between coreflective subcategories of TOP [resp. T2] and idempotent closure operators on TOP [resp. T2]. Using this characterization, S. Hong showed [H_6]:

1.13. THEOREM. Let $\mathcal{C}$ be a coreflective subcategory of TOP, B an epireflective subcategory of T2. Let $B_{\mathcal{C}}$ be the category of spaces in TOP(B) which are closed in the $\mathcal{C}$ - coreflections of their B - reflection spaces. Then $B_{\mathcal{C}}$ is

<u>an epireflective subcategory of</u> T2.

The proof of Lemma 1.10 can be generalized to show the first part of:

1.14. THEOREM. <u>Let</u> $\mathcal{B}$ <u>and</u> $\mathcal{B}^1$ <u>be epireflective subcategories of</u> CHSP, $\mathcal{C}$ <u>a coreflective subcategory of</u> TOP.

(a) <u>if</u> $\mathcal{B}$ <u>and</u> $\mathcal{B}^1$ <u>are distinct, then</u> $\mathcal{B}_{\mathcal{C}}$ <u>and</u> $\mathcal{B}^1_{\mathcal{C}}$ <u>are distinct.</u>

(b) <u>if</u> $\mathcal{B} \subset \mathcal{B}^1$, <u>then</u> $\mathcal{B}_{\mathcal{C}} \subset \mathcal{B}^1_{\mathcal{C}}$.

(c) <u>if</u> $\mathcal{C} \subset \mathcal{C}'$, <u>then</u> $\mathcal{B}_{\mathcal{C}} \supset \mathcal{B}_{\mathcal{C}'}$.

The diagram below is the key to b). We assume $X \in \mathcal{B}_{\mathcal{C}}$. The maps f and h are the reflection maps to r^1X and rX respectively, and the unlabeled maps are the canonical reflection and coreflection maps. The map f induces f^{r^1} which in turn induces $(f^{r^1})_c$. The maps $h \circ g$ and $f \circ g$ are monomorphisms, hence so are $(h \circ g)_c$ and $(f \circ g)_c$

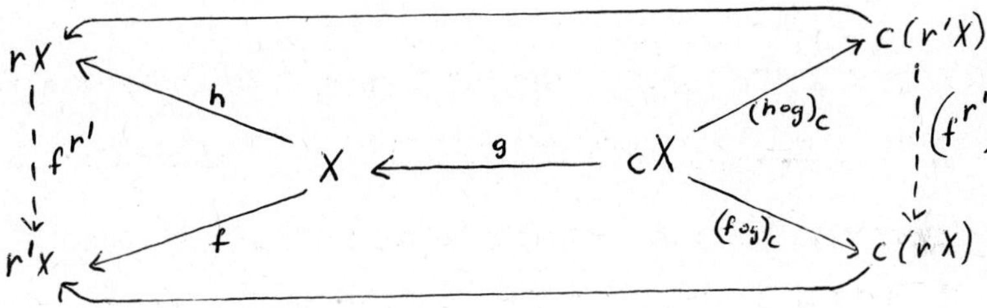

The map $(f^{r^1})_c$ maps the image of cX under $(h \circ g)_c$ bijectively onto the image of cX under $(f \circ g)_c$. Since "X is closed in the coreflection of its $\mathcal{B}$ - reflection" [more exactly, the image of cX under $(f \circ g)_c$ is closed in c(rX)] the same is true, by continuity of $(f^{r^1})_c$, of the coreflection of its $\mathcal{B}^1$ - reflection.

The proof of (c) is much simpler and relies upon the existence of the bijective map $c'X \to cX$ induced by $c'_X : c'X \to X$.

In contrast to Theorem 1.14, distinct $\mathcal{C}$ may give rise to the same $B_{\mathcal{C}}$.

1.15. EXAMPLE. Let $\mathcal{C}$ and $\mathcal{C}'$ be distinct coreflective subcategories of TOP, each of which contains CHSP. Then the $\mathcal{C}$ - coreflection of any compact Hausdorff space is that space itself, so $B_{\mathcal{C}} = B_{\mathcal{C}'} = B$.

However, the following problem seems to be open:

1.16. PROBLEM. Let $\mathcal{C}$ and $\mathcal{C}'$ be distinct coreflective subcategories of TOP, and let B be an epireflective subcategory of CHSP, contained neither in $\mathcal{C}$ nor $\mathcal{C}'$; are $B_{\mathcal{C}}$ and $B_{\mathcal{C}'}$ always distinct?

Once we step outside the realm of compact spaces, $B_{\mathcal{C}}$ and $B^1_{\mathcal{C}}$ may be identical. (A trivial example is where $\mathcal{C}$ is the category $\mathcal{P}_{\omega_1}$, B = CHSP, and B^1 is the category of real-compact spaces.) So it would be nice if every epireflective subcategory of $T3\frac{1}{2}$ (= TOP(CHSP)) were of the form $B_{\mathcal{C}}$ for some class B of compact spaces. Unfortunately, that is not true; for example, any epireflective subcategory which consists of totally disconnected Tychonoff spaces, some of which are not zero-dimensional, cannot be of this form. This is obviously true of the class of totally disconnected Tychonoff spaces, itself. Nevertheless, the order which these classes $B_{\mathcal{C}}$ bring to the study of epireflective subcategories is still considerable.

In closing this section, we have the following applications of a generalization of Theorem 1.14.

1.17. EXAMPLE. If E is a Tychonoff space with more than one point, then TOP ({E}) contains T2 ({E}) properly. Here is the reasoning. Since the two-point discrete space embeds in

E, $D3 \subset TOP(\{E\})$. Therefore the space D_k obtained by replacing Π with $\underline{2}$ in Example 1.5 is in $TOP(\{E\})$. However, E is t- compact for some cardinal t, whereas D_k is not t- compact for any $k > t$. So, while $T2(\{E\}) \subset (CHSP)_t$, D_k is not in $(CHSP)_t$.

That $T2(\{E\}) \subset (CHSP)_t$ follows from the generalization of (b) of Theorem 1.14 to epireflective subcategories $\mathcal{E}$ of T2; its proof, which was strictly categorical, goes through without change.

1.18 EXAMPLE. Let B be epireflective in CHSP. Let $\mathcal{D}$ denote the category of compact zero-dimensional spaces. If B contains a non-trivial space, then by Theorem 1.14, $\mathcal{D}_k \subset B_k \subset (CHSP)_k$ for all k. Thus, since D_k is in $\mathcal{D}_k$ (in fact $[H_7]$ we have $\mathcal{D}_k = T2(D_k))$, but is not t-compact for any $t < k$, it follows that $D_k \in B_k$, $D_k \notin B_t$ for any $t < k$.

Section 2. Extremally epireflective subcategories and disconectednesses.

A great broadening of horizons takes place when we move up to extremally epireflective subcategories of T2: the smallest nontrivial such category, D2, already spills out beyond the class of Tychonoff spaces. We could, of course, restrict our study to Tychonoff spaces or even compact Hausdorff spaces, considering only those spaces in e.g. D2 which are Tychonoff or compact Hausdorff. But in the latter case, as Preuss has pointed out $[P_2]$, we obtain nothing new: every epireflective subcategory of CHSP is extremally epireflective (in CHSP). On the other hand, the extremally epireflective subcategories of $T3\frac{1}{2}$ are just the intersections of $T3\frac{1}{2}$ with the extremally epireflective subcategories of TOP, and this is true also of the extremally epireflective subcategories of T2.

A special kind of extremally epireflective subcategory is a <u>disconnectedness</u>. To define it, it is probably best to begin with a continuously closed category, that is, one closed under the taking of continuous images of its members. (Of course, it makes a big difference which subcategory of TOP these images are allowed to lie in; for the moment we will use all of TOP itself.) Let $\mathcal{A}$ be a continuously closed subcategory of TOP; then $\mathcal{D}_{TOP}\mathcal{A}$ is the category $\{X \in TOP | X$ does not contain non-trivial $\mathcal{A}$- subspaces$\}$

Then $\mathcal{D}_{TOP}\mathcal{A}$ is a disconnectedness, and every TOP - disconnectedness is by definition a category of this sort.[1] We define T2 - disconnectedness, $T3\frac{1}{2}$ - disconnectedness, etc. anagolously.

2.1. EXAMPLE. Let $\mathcal{A}$ be the class of connected spaces; then $\mathcal{D}_{TOP}\mathcal{A}$ is the class of totally disconnected spaces.

2.2. EXAMPLE. A Peano continuum is defined as a continuous Hausdorff image of $\mathbb{I}$. The category $\mathcal{A}$ of Peano continua is continuously closed in T2, and (of course) in any of its subcategories. Since every Peano continuum is arcwise connected [HS], it follows that $\mathcal{D}_{T2}\mathcal{A}$ is the class of arcless spaces. Sometimes referred to as "arcwise totally disconnected," these are the spaces containing no subspace homeomorphic to a nondegenerate interval of $\mathbb{R}$.

This is one of the rare cases where $\mathcal{A}$ can be replaced by a single space in the definition of $\mathcal{D}_{T2}\mathcal{A}$ which we have chosen.

Since $T3\frac{1}{2} = TOP(\{\mathbb{R}\})$, an epireflective subcategory of T2 which is properly contained in $T3\frac{1}{2}$ is also contained in $\mathcal{D}_{T2}\mathcal{A}$.

---

1. In [P_1] and [AW] there is a more general definition, to include the case where $\mathcal{A}$ is not continuously closed: $\mathcal{D}_{TOP}\mathcal{A}$ for any subcategory $\mathcal{A}$ of TOP is the category of all spaces such that every map $f : X \to Y$, $X \in \mathcal{A}$, is constant. But while $\mathcal{D}_{TOP}$ is made more general, the classes of disconnectednesses do coincide with the disconnectednesses defined as in this paper.

The concept of a connectedness can be defined dually to that of a disconnectedness, but it is more natural to deviate slightly from strict duality. Let $\mathcal{B}$ be a hereditary class of spaces. The connectedness $\mathcal{C}_{TOP}\,\mathcal{B}$ is

$\{X \in TOP \mid X$ cannot be mapped continuously onto a non-trivial space in $\mathcal{B}\}$

A category of the form $\mathcal{C}_{TOP}\,\mathcal{B}$ for a hereditary $\mathcal{B}$ is called a connectedness.

The appropriatedness of the word "connectedness" is debatable since e.g. the class of arcwise connected spaces is not a connectedness, not even of the form $\mathcal{C}_{T2}\,\mathcal{B}$ or $\mathcal{C}_{CHSP}\,\mathcal{B}$. Consider the graph of $\sin \frac{1}{x}$ together with the portion of the y - axis between 1 and -1: this space is not arcwise connected, but any connectedness which contains $\mathbb{I}$ also contains this space (see Theorem 2.4). Nevertheless, this is the only logical dual to the concept of a disconnectedness, and it works together well with it. For instance, the two operators $\mathcal{D}$ and $\mathcal{C}$ form a Galois corespondence, so that $\mathcal{D}\mathcal{C}\mathcal{D}(\mathcal{A}) = \mathcal{D}(\mathcal{A})$ and $\mathcal{C}\mathcal{D}\mathcal{C}(\mathcal{B}) = \mathcal{C}(\mathcal{B})$ provided the same subscript (TOP,T2, etc.) is used throughout. Moreover, we have the following fundamental theorem, applying equally well to TOP-disconnectedness or T2-disconnectedness.

2.3. THEOREM. [AW, Theorem 3.7]. Let $\mathcal{B}$ be a disconnectedness, and let $\mathcal{A} = \mathcal{C}(\mathcal{B})$ be the connectedness dual to $\mathcal{B}$. Then $\mathcal{B}$ is a extremally epireflective subcategory of TOP. Moreover, the reflection map $f : X \to rX$ has the properties:

(i) $f^{-1}(y) \in \mathcal{A}$ for each $y \in rX$

(ii) If X' is a subspace of X belonging to $\mathcal{A}$, then X' is contained in $f^{-1}(y)$ for some $y \in rX$.

So the reflection map can be thought of as a breaking up of X into its "$\mathcal{A}$ - components", the sets $f^{-1}(f(x))$. Further clarification of the situation is provided by:

2.4. THEOREM. (a) Given a subcategory B of TOP [resp. T2], there is a smallest disconnectedness containing it, viz. $\mathcal{DC}(B^1)$, where B^1 is the category of all subspaces of spaces in B.

(b) Given any subcategory $\mathcal{A}$ of TOP [resp. T2], there is a smallest connectedness containing it, viz. $\mathcal{CD}(\mathcal{A}^1)$, where $\mathcal{A}^1$ is the category of all continuous images of spaces in $\mathcal{A}$.

Proof is immediate from the Galois correspondence.

2.5. THEOREM. [AW] A nonempty subcategory B of TOP is a TOP-disconnectedness if, and only if, the following conditions hold (where $X \in$ TOP)

(i) B is productive.
(ii) B is hereditary.
(iii) If $f : X \to Y$ is a continuous function such that $Y \in B$ and $f^{-1}(y) \in B$ for all $y \in Y$, then $X \in B$.

Note that if we replace "$f^{-1}(y) \in B$" by "$f^{-1}(y)$ is a singleton," we get a necessary and sufficient condition for "B is extremally epireflective", while if we drop (iii) altogether, we get the equivalent of "B is epireflective in TOP".

If we replace TOP by T2, or T$3\frac{1}{2}$, or CHSP, the three conditions are necessary for B to be a disconnectedness, but are they sufficient? In [AW] a negative answer for T2 is given, but the "counterexample" is faulty: it is claimed in [AW] that the category D2 satisfies (iii), but it does not.

2.6. EXAMPLE. Let T be the Tychonoff corkscrew as described in [SS, Examples 90 - 91]. The two extreme points, a^+ and a^-, are in the same quasicomponent of T. Let Y be the space which results from identifying a^+ and a^-, and let f be the quotient map from T to Y. Now $Y \in$ D2, and $f^{-1}(y)$ is in D2 for all $y \in Y$ (in fact, it is a singleton for all but one y, and a doubleton for that one!).

Not only is T not in D2, it is not even in FT2; on the other hand, T is totally diconnected (in fact, scattered) so that every subspace of T admits a map onto a two-point discrete space and thus a non-constant map to the real line. This answers Problem (a) of [AW] in the affirmative.

2.7. EXAMPLE. A Tychonoff space exhibiting the same phenomenon as T with respect to D2 can be constructed using Dowker's example M. This space is constructed by taking a collection $\{S_\alpha | \alpha < \omega_1\}$ of subsets of $\amalg$ such that $S_\alpha \subset S_\beta$ whenever $\alpha < \beta_1$ such that the complement of S_α is dense for all α, and such that $\bigcup\{S_\alpha | \alpha < \omega_1\} = \amalg$. Now let M be the subspace of $\amalg \times \omega_1$ consisting of the points $S_\alpha \times \{\alpha\}$ as α ranges over ω_1. It can be shown (cf. [GJ], exercise 16L, where M is referred to as Δ_1, and [E], pp. 254-255, where it is referred to as Y) that M is 0-dimensional and that, given any clopen subset A of M, either A or A^c must contain a cofinal subset of M. In other words, there exists α such that $\{S_\beta \times \beta | \beta > \alpha\}$ is a subset either of A or A^c. Thus if we enlarge M by "two points at ω_1" say $(0,\omega_1)$ and $(1,\omega_1)$, giving these the product topology neighborhoods, then the resulting space is still totally disconnected, but the two extra points are in the same quasicomponent. Now if we identify these two points, we obtain a space in D2. Thus D2 does not satisfy (iii) even within the category of Tychonoff spaces.

One can even obtain a metrizable example, e.g. the Knaster-Kuratowski "Cantor fan" (denoted Y* in [SS, Example 129]). Identifying the points within each quasicomponent gives a homeomorph of the Cantor set, and each quasicomponent is zero- diemnsional.

So, apparently, the following question is still open.

2.8 PROBLEM. Let $\mathcal{B}$ be a subcategory of T2, or $T3\frac{1}{2}$, or CHSP, satisfying the three consitions of Theorem 2.5 (with "closed hereditary" in (ii) in the case of CHSP). Can it ever happen that $\mathcal{B}$ is not a disconnectedness?

In the case of CHSP, we can ask a bolder question: is every epireflective subcategory a CHSP-disconnectedness? This splits up into Problem 2.8 on the one hand and a kind of left-fitting problem on the other:

2.9. PROBLEM. Let $\mathcal{B}$ be an epireflective subcategory of CHSP and let $f : X \to Y$ be a mapping with $X \in$ CHSP, $Y \in \mathcal{B}$. If $f^{-1}(y) \in \mathcal{B}$ for all $y \in Y$, is X in $\mathcal{B}$?

Whatever the answers to these last two problems, it is still interesting to study categories satisfying the three conditions of Theorem 2.5. Suppose, for example, we are given a subcategory $\mathcal{A}$ of T2 and asked, "what is the smallest subcategory $\mathcal{B}$ of T2 that contains $\mathcal{A}$ and satisfies those three conditions? Of course, $\mathcal{B}$ will contain TOP($\mathcal{A}$) and even Q($\mathcal{A}$) - - see the comment after Theorem 2.5 - and we can define, by transfinite induction, an ascending sequence extremally of epireflective subcategories of T2 which together will constitute $\mathcal{B}$.

To simplify matters, we can assume that $\mathcal{A}$ is extremally epireflective to begin with. There is no loss of generality, of course, and it allows us to use the label $\mathcal{A}_\alpha$ without danger of confusion with the categories $\mathcal{B}_k$ defined earlier.

2.10. LEMMA. *Let* $\mathcal{A}$ *and* $\mathcal{A}^1$ *be extremally epireflective in* T2, *with* r_X *the reflection map with respect to* $\mathcal{A}$ *for each* $X \in$ T2. *Let* $r(\mathcal{A}^1, \mathcal{A})$ *be the category whose objects are the spaces* $X \in$ T2 *such that* $r_X{}^{-1}(z) \in \mathcal{A}^1$ *for all* $z \in rX$. *Then* $r(\mathcal{A}^1, \mathcal{A})$ *is extremally epireflective.*

Moreover, $r(\mathcal{A}^1, \mathcal{A})$ is the category of spaces $X \in$ T2 which admit a map f into some member Y of $\mathcal{A}$, such that $f^{-1}(y) \in \mathcal{A}^1$ for all $y \in Y$.

Proof. The two descriptions of $r(\mathcal{A}^1, \mathcal{A})$ coincide: clearly the first implies the second; on the other hand, r_X will compose with f^r to produce f, so that each set of the form $r_X{}^{-1}(z)$ is contained (as a closed subset) in a set of the form $f^{-1}(y)$, hence is an object in $\mathcal{A}^1$.

By the second description, it is clear that $r(\mathcal{A}^1,\mathcal{A})$ is closed under subobjects in T2, so it remains only to show that $r(\mathcal{A}^1,\mathcal{A})$ is productive. Let $\{r_\gamma : X_\gamma \to Y_\gamma | \gamma \in \Gamma\}$ be a set of reflection maps, with $X_\gamma \in \mathcal{A}$ for all γ. The induced map $f : \Pi_\gamma X_\gamma \to \Pi_\gamma Y_\gamma (= Y)$ has the property that $f^{-1}(y)$ $(= \Pi_\gamma r_\gamma^{-1}(y_\gamma))$ is in $\mathcal{A}^1$ for all $y \in Y$, since $\mathcal{A}$ is productive. So by the second description, $r(\mathcal{A}^1,\mathcal{A})$ is productive.

Had we wanted to show merely that $r(\mathcal{A}^1,\mathcal{A})$ is epireflective in TOP, it would have been enough to assume that $\mathcal{A}^1$ is epireflective in TOP. The same goes for "epireflective in T2", mutatis mutandis, because the inverse image of each point is closed.

We can apply Lemma 2.10 inductively by starting with an epireflective subcategory $(\mathcal{A}, r)$ and letting $\mathcal{A}^1 = \mathcal{A}$. Then $(\mathcal{A}_1,r_1) = r(\mathcal{A},\mathcal{A})$ will be epireflective and contain $(\mathcal{A}, r)$. We seem to have a choice of ways to continue the induction: we could define $(\mathcal{A}_2, r_2)$ to be either $r(\mathcal{A}, \mathcal{A}_1)$ or $r(\mathcal{A}_1,\mathcal{A})$. Actually, the two are the same, but the proof requires a good deal of machinery to state with suitable generality. For purposes of characterizing the category $\mathcal{B}$ for which we are aiming, we can (fortunately) skirt the whole issue by instead defining $(\mathcal{A}_2, r_2)$ as $r(\mathcal{A}_1,\mathcal{A}_1)$. Clearly, $\mathcal{A}_2$ contains "both" of the epireflective subcategories just named.

Supposing $(\mathcal{A}_\alpha, r_\alpha)$ to be defined, let $\mathcal{A}_{\alpha+1}$ be $r(\mathcal{A}_\alpha, \mathcal{A}_\alpha)$. If α is a limit ordinal, let $\mathcal{A}_\alpha$ be the category of all spaces which admit a 1 - 1 map into a product of spaces from any of the $\mathcal{A}_\beta$ with $\beta < \alpha$.

It is clear that the category $\mathcal{B}$ of all spaces belonging to some $\mathcal{A}_\alpha$ is productive, hereditary, and inversely closed under maps $f : X \to Y$ with point - inverses in $\mathcal{B}$. [The last named condition can be satisfied e.g. by letting $\mathcal{A}_\alpha$ be the supremum of the $\mathcal{A}_\beta$ to which the point-inverses and the space Y belong; then the domain X will be in $\mathcal{A}_{\alpha+1}$.]

This whole procedure is susceptible to a wide categorical generalization. In particular, the proof of Lemma 2.10 can be modified for complete categories in which the extremally epireflective subcategories are those closed under products and subobjects - - provided we can give meaning to the concept of "inverse image". But this is best handled in a separate paper.

Section 3. Generalization of a famous problem.

Given a Tychonoff space X belonging to a disconnectedness, it is natural to ask whether βX also belongs. In this section I will handle some conditions on the disconnectedness for which the answer is affirmative if X is realcompact. The general idea behind the conditions is that the closed subsets of $\beta X - X$ are "pathological" for a realcompact space X.

First, we will look at an example where the answer is very much in the negative: the category of totally disconnected spaces. We define the following classes of spaces, each containing the one before:

- $\mathcal{L}$: realcompact X such that βX is totally disconnected.
- $\mathcal{N}$: N - compact spaces
- $\mathcal{O}$: zero-dimensional realcompact spaces
- $\mathcal{R}$: realcompact spaces, every compact subspace of which is totally disconnected.

It is easy to see that $\mathcal{O}$ and $\mathcal{R}$ are distinct: the Cantor fan, for example, is in $\mathcal{R}$ but not $\mathcal{O}$. A famous problem that stood for twelve years was whether $\mathcal{O} = \mathcal{N}$. I solved it in 1970 by showing that Prabir Roy's space Δ is not N-compact [N_y]. As for the relationship between $\mathcal{S}$ and $\mathcal{N}$, S. Mrówka settled it in 1972 by describing a space in $\mathcal{N}$ which is not in $\mathcal{S}$ [M_3].

These classes all have categorical interpretations in terms of the class $\mathcal{B}$ of totally disconnected Hausdorff spaces. In fact, if $\mathcal{B}$ is an epireflective subcategory of T2, we can define:

$$\mathcal{S}_1(\mathcal{B}) = \{X \mid X \text{ realcompact, } \beta X \in \mathcal{B}\}$$

$$\mathcal{N}_1(\mathcal{B}) = (\mathcal{B} \cap \text{CHSP})_{\aleph_1}$$

$$\mathcal{O}_1(\mathcal{B}) = \{X \mid X \text{ realcompact, } X \in \text{TOP } \mathcal{B}\}$$

$$\mathcal{R}_1(\mathcal{B}) = \{X \mid X \text{ realcompact, every subspace of } X \text{ closed in } \beta X \text{ is in } \mathcal{B}\}$$

From now on, we will assume $\mathcal{B} \subset$ CHSP, so that the "$\cap$ CHSP" in the description of $\mathcal{N}_1(\mathcal{B})$ becomes redundant, and we can say

$$\mathcal{R}_1(\mathcal{B}) = \{X \mid X \text{ realcompact, every compact subspace of } X \text{ is in } \mathcal{B}\}$$

If $\mathcal{B} = \mathcal{D}_{\text{CHSP}}(\mathcal{A})$ for some continuously closed $\mathcal{A}$, then this is just the category of realcompact spaces belonging to $\mathcal{D}_{\text{T2}}(\mathcal{A})$. In any case, since $\mathcal{B}$ is closed hereditary, we have $\mathcal{O}_1(\mathcal{B}) \subset \mathcal{R}_1(\mathcal{B})$. Next, since the class of realcompact spaces is CHSP_k where $k = \aleph_1$, it follows from Theorem 1.14(b) that $\mathcal{N}_1(\mathcal{B}) \subset \mathcal{O}_1(\mathcal{B})$. Finally, if X is realcompact, then X is $\mathcal{T}_k$ - closed in βX ($k = \aleph_1$), so that if $\beta X \in \mathcal{B}$, then $X \in \mathcal{N}_1(\mathcal{B})$. Summing up,

$$\mathcal{S}_1(\mathcal{B}) \subset \mathcal{N}_1(\mathcal{B}) \subset \mathcal{O}_1(\mathcal{B}) \subset \mathcal{R}_1(\mathcal{B})$$

and the case where $\mathcal{B}$ is the class of totally disconnected compact Hausdorff spaces is an example where all containments are proper. We will now give some example where the four categories coalesce.

In each case, $\mathcal{B}$ will be of the form $\mathcal{D}_{CHSP}\mathcal{A}$, raising afresh the question of whether every epireflective subcategory of CHSP is a disconnectedness.

3.1. EXAMPLE. Let $\mathcal{B}$ be the class of arcless compact Hausdorff spaces (see Example 2.2) To show $\mathcal{S}_1(\mathcal{B}) = \mathcal{R}_1(\mathcal{B})$ --- so that all four classes are equal - - it is enough to show that if a realcompact space X contains no arcs, then βX contains no arcs either. This will follow immediately from Theorem 9.11 of [GJ], which I will restate as follows.

3.2. LEMMA. <u>Let</u> A <u>be a closed subset of</u> βX. <u>If</u> A <u>does not contain a copy of</u> $\beta\mathbb{N}$, <u>then</u> $A \cap (\beta X - \upsilon X)$ <u>is closed and discrete in the relative topology of</u> $\beta X - \upsilon X$.

And, since no arc contains $\beta\mathbb{N}$, it follows that if Y is realcompact and βY contains an arc, all but a discrete subspace of that arc must lie inside Y, and so Y itself contains an arc. Equality between the four classes follows immediately.

Here is a related result which is even easier to apply.

3.3. THEOREM. <u>Let</u> A <u>be a closed subset of</u> βX. <u>If</u> A <u>does not contain a copy of</u> $\beta\mathbb{N}$, <u>then the points of</u> A <u>lying outside</u> υX <u>are isolated in the relative topology of</u> A .

<u>Proof</u>. By Lemma 3.2, any point x of A which is not in υX has a closed neighborhood N such that $N \cap A$ is contained in υX, except for x itself. Since x is not in υX, there is a zero-set Z containing x which misses X[H_6]. Let $E = (N \cap A)\setminus\{x\}$. If x is not isolated in A, then E and Z fulfill the conditions of Lemma 9.4 in [GJ], and so E contains a C* - embedded (in βX) copy of $\mathbb{N}$. In other words, the closure of E in βX contains a copy of $\beta\mathbb{N}$. But this contradicts the assumption that A (which

contains E) is closed in βX.

3.4. COROLLARY. Let A be a closed, dense-in-itself subspace of βX. If A does not contain a copy of $\beta\mathbb{N}$, then $A \subset \upsilon X$.

I will refer to a compact, connected Hausdorff space as a "continuum".

3.5. COROLLARY. Let A be a continuum in βX. If A does not contain a copy of $\beta\mathbb{N}$, then $A \subset \upsilon X$.

For example, if A is metrizable, or first countable, or even sequential; or if A is orderable; or hereditarily separable; or of cardinal less than $2^{\mathfrak{c}}$, then $A \subset \upsilon X$.

3.6. COROLLARY. Let X be a realcompact space, and let A be a continuum contained in βX. If A does not contain a copy of $\beta\mathbb{N}$, then $A \subset X$.

And so we come to:

3.7 THEOREM. Let $\mathcal{A}$ be a class of continua, none of which contains a copy of $\beta\mathbb{N}$, and let $B = \mathcal{D}_{CHSP}\mathcal{A}$. Then

$$\mathcal{S}_1(B) = \mathcal{N}_1(B) = \mathcal{O}_1(B) = \mathcal{R}_1(B) .$$

Proof. If $\mathcal{A}^1$ is the class of all continuous images of spaces in $\mathcal{A}$, then $B = \mathcal{D}_{CHSP}(\mathcal{A}^1)$. Suppose some member Y of $\mathcal{A}^1$ contains a copy of $\beta\mathbb{N}$. Let $f : X \to Y$ be a map with $X \in \mathcal{A}$. Then, by projectivity of $\beta\mathbb{N}$ [H_4], there is a copy of $\beta\mathbb{N}$ in X mapping 1-1 onto the copy in Y, contradicting the hypothesis on $\mathcal{A}$. Therefore, if X is a realcompact space, any nontrivial member of $\mathcal{A}^1$ contained in βX must be contained in X, and X cannot be in $\mathcal{R}_1(B)$. In other words, if X is not in $\mathcal{S}_1(B)$, then X is not in $\mathcal{R}_1(B)$ either.

3.8. EXAMPLE. Let $\mathcal{A}_0$ be the class of all continua containing no copy of $\beta\mathbb{N}$. From the proof of Theorem 3.7, $\mathcal{A}_0$ is continuously closed. Let $B_0 = \mathcal{D}_{CHSP}\mathcal{A}_0$. Clearly, B_0 is the class of all compact Hausdorff spaces, every nontrivial subcontinuum of which contains a copy of $\beta\mathbb{N}$. An example of a connected space belonging to this class is $\beta\,\mathbb{R}^+ \setminus \mathbb{R}^+$[GJ].

$\mathcal{A}_0$ is obviously the largest class of continua (and hence B_0 the smallest epireflective subcategory of CHSP) to which Theorem 3.7 can be applied directly. For any disconnectedness containing B_0, the four categories defined above will coalesce. Of course, the <u>largest</u> disconnectedness to which 3.7 can be applied is the class ATDC of arcless compact Hausdorff spaces, since any epireflective subcategory of CHSP which contains a space outside ATDC is all of CHSP.

Here are some intermediate examples.

3.9. EXAMPLE. Let $\mathcal{A}$ be the class of orderable compact Hausdorff spaces. Since $\beta\mathbb{N}$ cannot be embedded in an orderable space (in fact, it is not even hereditary normal), $\mathcal{A}$ satisfies the hypotheses of Theorem 3.7.

Characterizing the continuous images of spaces in $\mathcal{A}$ is an unsolved problem. They must all be locally connected, since the members of $\mathcal{A}$ are locally connected and compact: every closed, continuous image of a locally connected space is locally connected [HS]. But we lack a convenient necessary and sufficient condition; certainly we have nothing like the Hahn-Mazurkiewicz Theorem, which characterizes the continuous images of $\mathbb{I}$, the Peano continua, as those continua which are locally connected and metrizable.

3.10. EXAMPLE. For each infinite cardinal k we can construct an analogue $\mathbb{I}_k$ of $\mathbb{I}$ as follows. Let β be the first

ordinal of cardinality k . Points of II_k are transfinite sequences of the form $(b_\alpha)_{\alpha < \beta}$ where each b_α is either 0 or 1,
* and* for each $\alpha < \beta$ there exists $\gamma > \alpha$ such that $b_\gamma = 0$. Give II_k the lexicographical order topology. Think of the members of II_k as transfinite binary decimals, and it is not hard to show that II_k is Dedekind complete. [Let B be a subset of II_k . For each ordinal α, say that B eventually agrees up to α if there exists a member b of B such that all members of B greater than b agree with b up to and including α. The only complication occurs if there exists α such that B does not eventually agree up to α. In that case, take the smallest such ordinal; the supremum of B will have a 0 at α and all ordinals beyond.] Thus, II_k is connected and locally compact. Since it contains a greatest and least element, it is compact.

It is easy to show that $_{\mathrm{CHSP}}\mathrm{II}_k$ contains $_{\mathrm{CHSP}}\mathrm{II}_\ell$ properly whenever $k > \ell$.

3.11 EXAMPLE. Let $\mathcal{A}$ be the class of all perfectly normal continua. Since a regular space is perfectly normal and Lindelöf if, and only if, it is hereditary Lindelöf, it follows that $\mathcal{A}$ coincides with the class of hereditarily Lindelöf continua. Thus $\mathcal{A}$ is continuously closed, and no member of $\mathcal{A}$ contains a copy of $\beta\mathbb{N}$. In fact, every nontrivial member of $\mathcal{A}$ has the cardinality of the continuum, being countable on the one hand and connected on the other.

The disconnectedness of all compact Hausdorff spaces containing no metric continua contains $\mathcal{D}_{\mathrm{CHSP}}\mathcal{A}$; but is the containment proper? Or is it true that every perfectly normal continuum must contain a nontrivial metric continuum?

We can ask similar questions about other easily describable classes $\mathcal{A}$. For example, must every sequential continum contain a nontrivial first countable subcontinuum, or even a metrizable subcontinuum? The big problem seems not the finding of different $\mathcal{A}$

* see Errata at the end of the paper!

to plug into Theorem 3.7, but the determining of whether the resulting disconnectednesses coincide.

What about classes not covered by Theorem 3.7? We have, for example, the class $\mathcal{A}$ of locally connected continua. It is possible for a locally connected continuum to contain a copy of $\beta\mathbb{N}$ - for example, $\mathbb{I}^{\mathfrak{c}}$ where $\mathfrak{c}$ is the cardinality of the continuum. But we can apply Corollary 3.5 to show that $\mathbb{I}^{\mathfrak{c}}$, if it occurs in βX, must already be in X if X is realcompact: every point of $\mathbb{I}^{\mathfrak{c}}$ is contained in an arc A, which must lie completely inside X. In line with this, we may ask:

3.12. PROBLEM. Let X be a locally connected continuum. Does X have a locally connected subcontinuum which does not contain a copy of $\beta\mathbb{N}$?

If the answer is always "yes", then the four classes coincide for $\mathcal{B} = \mathcal{D}_{CHSP}\mathcal{A}$ ($\mathcal{A}$ = all locally connected continua) also. In fact, Theorem 3.7 could be broadened to include all continuously closed classes $\mathcal{A}$ such that each member of $\mathcal{A}$ contains a subcontinuum with no $\beta\mathbb{N}$ - isomorphic subspace.

If we pass from realcompact spaces to $\aleph_2$ - compact spaces, there is no possibility of extending the results in this section. Say we let

$$\mathcal{S}_2(\mathcal{B}) = \{X \mid X \text{ is } \aleph_2 \text{ - compact and } \beta X \in \mathcal{B}\}$$

and define $\mathcal{N}_2(\mathcal{B})$ to be $\mathcal{B}_k$ with $k = \aleph_2$. Now $\mathcal{N}_2(\underline{2})$ already contains Dowker's example M, because every free clopen ultrafilter on M has at most the $\aleph_1$ - intersection property [see H_6, Lemma 2.8]. Hence by Theorem 1.14, $\mathcal{N}_2(\mathcal{B})$ contains M for all nontrivial $\mathcal{B}$. On the other hand, βM contains a copy of $\mathbb{I}$, so that the only $\mathcal{B}$ for which $M \in \mathcal{S}_2(\mathcal{B})$ is CHSP itself.

ERRATA

(i) for each $\alpha < \beta$ there exists $\gamma > \alpha$ such that $b_\gamma = 0$

(ii) there exists a limit ordinal δ such that $b_\alpha=1$ for all $\alpha \geq \delta$, and for each $\alpha < \delta$, there exists $\gamma > \alpha$ such that $b_\gamma=0$

(iii) $b_\alpha=1$ for all $\alpha < \beta$.

BIBLIOGRAPHY

[AW] A. V. Arhangel'skii and Wiegandt, Connectednesses and disconnectednesses in topology, Gen. Top. Appl. 5(1975)9-33.

[B] R. Blefko, Doctoral dissertation. University Park, Pennsylvania 1965.

[E] R. Engelking, Outline of General Topology. Amsterdam, North Holland, 1968.

[EM] R. Engelking and S. Mrówka, On E-compact spaces, Bull. Acad. Pol. Sci. Math. Astr. Phys., 6 (1958) 429-436.

[GJ] L. Gillman and M. Jerison, Rings of Continuous Functions. Princeton, Van Nostrand, 1960.

[H_1] H. Herrlich, $\mathcal{E}$- kompakte Räume, Math. Z. 96 (1967) 228-255.

[H_2] H. Herrlich, Topologische Reflexionen und Coreflexionen. New York, Springer-Verlag, 1968.

[H_3] H. Herrlich, Limit - operators and topological coreflections, AMS Transactions 146 (1969) 203-210.

[H_4] H. Herrlich, Categorical topology, Gen. Top. Appl. 1 (1972) 1-15.

[H_5] S. S. Hong, Limit-operators and reflective subcategories, in: TOPO 72 - General Topology and its applications (New York, Springer-Verlag, 1974).

[H_6] S. S. Hong, On k- compactlike spaces and reflective subcategories, Gen. Top. Appl. 3 (1973) 319-330.

[H_7] M. Hušek, The Class of k- compact spaces is simple, Math. Z., 110 (1969) 123-126.

[HS] D. W. Hall and G. L. Spencer, Elementary Topology. Wiley, New York, 1955.

[M_1] S. Mrówka, On universal spaces, Bull. Acad. Pol. Sci. 4(1956) 479-481.

[M_2] S. Mrówka, Further results on E-compact spaces I, Acta Math. 120 (1968) 161-185.

[M_3] S. Mrówka, Recent results on E-compact spaces, in: TOP 72 - General Topology and its Applications (New York, Springer-Verlag, 1974).

[N_y] P. Nyikos, Prabir Roy's space Δ is not N-compact, Gen. Top. Appl. 3 (1973) 197-210.

[P_1] G. Preuss, Trennung und Zusammenhang, Monatsh. Math. 74 (1970) 70-87.

[P_2] G. Preuss, A categorical generalization of completely Hausdorff spaces, in: General Topology and its Relations to Modern Analysis and Algebra IV. New York, Academic Press, 1972.

[SS] L. A. Steen and J. A. Seebach, Counterexamples in Topology. New York, Holt, Rinehart and Winston, 1970.

CATEGORICAL PROBLEMS IN MINIMAL SPACES

BY

JACK R. PORTER

Abstract. A space X with a topological property P is called minimal P if X has no strictly coarser topology with property P and is called P-closed if X is a closed set in every space with property P that contains X as a subspace. This paper surveys, from a categorical viewpoint, a number of results recently obtained in minimal P and P-closed spaces where P includes the properties of regular Hausdorff, extremally disconnected Hausdorff, and the separation axioms $S(\alpha)$ for each ordinal $\alpha > 0$. Particular attention is focused on some of the categorical problems in these areas.

CATEGORICAL PROBLEMS IN MINIMAL SPACES

BY

JACK R. PORTER[1]

For a topological property P, a minimal P space is a set X with a topology that is minimal among the partially ordered set of all P-topologies on X. If P is a property that implies Hausdorff, then a compact P space in minimal P. Closely associated with minimal P spaces is the property of P-closed--a P-space X is P-closed if X is a closed set in every P-space containing X as a subspace. For a large number of properties P (cf.[BPS]), a minimal P space is P-closed.

The theory of categorical topology has touched the area of minimal P and P-closed spaces, as many other areas of general topology. This paper surveys a number of categorically related results recently obtained in the minimal P and P-closed spaces where P is the properties of regular Hausdorff, extremally disconnected Hausdorff, and the separation axioms $S(\alpha)$ for each ordinal $\alpha > 0$ ($S(\alpha)$ defined below); emphasis is placed on the problems in these areas. This survey paper is a continuation, in spirit, of [BPS, S1].

Let TOP (resp. HAUS) denote the category of spaces (resp. Hausdorff spaces) and continuous functions. The full subcategory of HAUS of

---

1 The research of the author was partially supported by the University of Kansas General Research Fund.

regular (includes Hausdorff) spaces is denoted by REG. For each ordinal $\alpha > 0$, a space X is said to be $R(\alpha)$ (resp. $U(\alpha)$) [PV1] if for every pair of <u>distinct points</u> $x, y \in X$, there are subfamilies $\{F_\beta : \beta < \alpha\}$ and $\{G_\beta : \beta < \alpha\}$ of open neighborhoods of x and y, respectively, such that $F_o \cap G_o = \emptyset$ (resp. $cl_X F_o \cap cl_X G_o = \emptyset$) and for $\gamma + 1 < \alpha$, $cl_X G_{\gamma+1} \subseteq G_\gamma$ and $cl_X F_{\gamma+1} \subseteq F_\gamma$. For $\alpha \geq \omega$ (the first infinite ordinal), it is easy to verify that $R(\alpha)$ and $U(\alpha)$ are equivalent concepts. For notational convenience, the symbols $R(\alpha)$ and $U(\alpha)$ for $\alpha \geq \omega$ are replaced by a single symbol $S(\alpha)$; for $n \in \mathbb{N}$, $R(n)$ is replaced by $S(2n-1)$ and $U(n)$ by $S(2n)$. Thus, a space is Hausdorff (resp. Urysohn) if and only if it is $S(1)$ (resp. $S(2)$). The full subcategory of HAUS of $S(\alpha)$ spaces is denoted by $S(\alpha)$.

The notation $\mathcal{A} \subset \mathcal{B}$ is used to denote that the category $\mathcal{A}$ is a subcategory of the category $\mathcal{B}$; by the largest subcategory of $\mathcal{B}$ with a certain property, we are referring to this inclusion. As usual, a <u>replete</u> subcategory means an isomorphism-closed subcategory.

A subset A of a topological space X is <u>regular-open</u> if $\text{int}(\text{cl}A) = A$. The family of regular-open sets form an open basis for a topology, coarser than the original topology; X with this new topology is denoted by X_s and called the <u>semiregularization</u> of X. A space X is said to be <u>semiregular</u> if $X = X_s$. Let SR denote the full subcategory of HAUS of semiregular spaces.

A function $f: X \to Y$ where X and Y are spaces is <u>θ-continuous</u> [F] if for each $x \in X$ and open set U containing $f(x)$, there is open set V containing x such that $f(\text{cl}V) \subseteq \text{cl}U$. The proof of the following facts are straight-forward.

(1.1) If $f: X \to Y$ is θ-continuous where X is a space and Y is a regular space, then f is continuous.

(1.2) For a space X, the identity function $X_s \to X$ is θ-continuous.

2. <u>P = Hausdorff.</u> The term "Hausdorff-closed" is shortened to "H-closed". There are two well-known methods to densely embed a Hausdorff space X in an H-closed space--(1) using the Katětov extension [K] (denoted as κX) and (2) using the Fomin extension [F] (denoted as σX). The full subcategory of HAUS of H-closed spaces is denoted by HC. Three situations motivate the first problem.

(I) In 1968, Herrlich and Strecker [HS1] showed that HAUS is the epireflective hull of HC, however, they did find a subcategory, denoted $HAUS^*$, of HAUS such that $HC \subset HAUS^*$, $ob(HAUS^*) = ob(HAUS)$, and HC is epireflective in $HAUS^*$ with κ as the epireflector. In 1971, Harris [Ha1, Ha2] proved there is a largest subcategory, denoted as pHAUS, of HAUS such that $HC \subset pHAUS$ and HC is epireflective in pHAUS with κ as the epireflector. Thus, $HAUS^* \subset pHAUS$ and $ob(pHAUS) = ob(HAUS)$.

This situation does not fall under the scope of the usual theory of epireflection; the point of failure is best understood with the following characterization theorem of the theory.

<u>Theorem 2.1.</u> [HS2, Th. 37.2] If $\mathcal{A}$ is a full, replete subcategory of a complete, well-powered, co-(well-powered) category $\mathcal{B}$, then $\mathcal{A}$ is epireflective in $\mathcal{B}$ if and only if $\mathcal{A}$ is closed under the formation

of products and extremal subobjects in $\mathcal{B}$.

Now HC is a full, replete, epireflective subcategory of pHAUS and HC is closed under the formation of products and under the formation of extremal subobjects in pHAUS. Also, pHAUS is well-powered and co-(well-powered). By Theorem 23.8 in [HS2], completeness is equivalent to being closed under the formation of products and finite intersections. But pHAUS is neither closed under the formation of products as noted in Example 2.2 nor under finite intersections as noted in Example 2.3.

Example 2.2. Let C denote the Cantor space, i.e., $C = \Pi\{X_n : n \in \mathbb{N}\}$ where X_n is $\{0,2\}$ with the discrete topology. Since HC is a full subcategory of HAUS and $X_n \in ob(HC)$ for each $n \in \mathbb{N}$, it follows that C is categorical product of $\{X_n : n \in \mathbb{N}\}$ in HC and, thus, is the only candidate for the categorical product of $\{X_n : n \in \mathbb{N}\}$ in pHAUS. Let $\pi_n : C \to X_n$ be the usual coordinate projection function. It is straightforward to show that C contains a countable subspace $X = \{a_i, b_{ij} : i,j \in \mathbb{N}\}$ such that (1) there is a family $\{U_i : i \in \mathbb{N}\}$ of pairwise disjoint open sets such that $a_i \in U_i$, (2) $\{b_{ij} : j \in \mathbb{N}\} \subseteq U_i$ for $i \in \mathbb{N}$, and (3) for each $i \in \mathbb{N}$, the sequence $\{b_{ij} : j \in \mathbb{N}\}$ converges to a_i. Let $f: \mathbb{N} \to X$ be a bijection. For each $n \in \mathbb{N}$, $\pi_n \circ f: \mathbb{N} \to X_n$ is a p-map (recall that an open cover of a space Y is a p-cover if there is a finite subfamily whose union is dense and a continuous function $f: Y \to Z$ is a p-map if for each p-cover $\mathcal{U}$ of

$Z, f^{-1}(\mathcal{U}) = \{f^{-1}(U): U \in \mathcal{U}\}$ is a p-cover of Y) since each open cover of X_n is finite. Now, $V = X \setminus \{a_i: i \in \mathbb{N}\}$ is open subset of X and $\mathcal{V} = \{U_i: i \in \mathbb{N}\} \cup \{V\}$ is p-cover of X. Now $f^{-1}(\mathcal{V})$ has no finite subcover. But the only p-cover of a discrete space is one with a finite subcover. Thus, $f^{-1}(\mathcal{V})$ is not a p-cover and f is not a p-map. So, C is not the categorical product of $\{X_n: n \in \mathbb{N}\}$ in pHAUS. This shows that pHAUS is not closed under the formation of products.

Example 2.3. In this example, pHAUS is shown not to be closed under finite intersections with a space $X \in ob(HC)$. Let $X = \{(1/n, 1/m): n, m, -m \in \mathbb{N}\} \cup \{1/n, 0): n \in \mathbb{N}\} \cup \{(0,1), (0,-1)\}$ where $X \setminus \{(0,1), (0,-1)\}$ has the usual subspace topology inherited from the plane and a basic neighborhood of $(0,1)$ (resp., $(0,-1)$) is a set containing $\{(0,1)\} \cup \{(1/n, 1/m): n \geq k, m \in \mathbb{N}\}$ (resp. $\{(0,-1)\} \cup \{(1/n, -1/m): n \geq k, m \in \mathbb{N}\}$) for some $k \in \mathbb{N}$, cf. Ex. 3.14 in [BPS]. The space X is H-closed. Let A_1 (resp. A_{-1}) be the subspace $\{(0,1)\} \cup \{(1/n, 1/m): n, m \in \mathbb{N}\} \cup \{(1/n, 0): n \in \mathbb{N}\}$ (resp. $\{(0,-1)\} \cup \{(1/n, -1/m): n, m \in \mathbb{N}\} \cup \{(1/n, 0): n \in \mathbb{N}\}$). Both subspaces A_1 and A_{-1} are H-closed and the inclusion functions $i_1: A_1 \to X$ and $i_{-1}: A_{-1} \to X$ are p-maps. Let $A = A_1 \cap A_{-1}$ and $i: A \to X$ be the inclusion function. It is straightforward to show that if the subobjects (A_1, i_1) and (A_{-1}, i_{-1}) have an intersection in pHAUS it must be (A, i). But (A, i) is not a subobject in pHAUS as i is not a p-map for (1) by Theorem D in [Hal], i is a p-map if and only if $cl_X A$ is H-closed and (2) $cl_X A = A$ is an infinite discrete subspace and not H-closed.

One of the inherent flaws of the class ob(HC) is not being closed hereditary; in fact, by results in [L,SW], every Hausdorff space is a closed subspace of some H-closed space. However, the class ob(HC) is regularly closed hereditary (cf., Prop. 3 in [HS1]), i.e., if $X \in ob(HC)$ and $A = cl_X int_X A \subseteq X$, then $A \in ob(HC)$.

(II) The existence of a largest subcategory of HAUS in which HC is an epireflective subcategory with κ as the epireflection occurs in a more general setting as noted by the following theorem.

Theorem 2.4. [Pol] Let $\mathcal{B} \subseteq$ HAUS and $\mathcal{A}$ be a full, replete subcategory of HAUS. Let $F: ob(\mathcal{B}) \to ob(\mathcal{A})$ such that for each $X \in ob(\mathcal{B})$, FX is a topological extension of X and for $X \in ob(\mathcal{A})$, FX = X. For $X \in ob(\mathcal{B})$, let $F_X: X \to FX$ denote the inclusion function. Then there is a largest subcategory $\mathcal{C}$ of HAUS such that $\mathcal{A}$ is an epireflective subcategory of $\mathcal{C}$ with F as the epireflection. Also, $ob(\mathcal{C}) = ob(\mathcal{B})$.

Let σHAUS denote the largest subcategory of HAUS such that HC is an epireflective subcategory of σHAUS with σ as the HC-epireflection. An old problem has been to characterize the morphisms of σHAUS which is equivalent to characterizing for X,Y in ob(HAUS), those continuous functions $f: X \to Y$ that have a continuous extension $\sigma f: \sigma X \to \sigma Y$; D. Harris communicated to me, at the Memphis Topology Conference during March 1975, that he has solved this problem.

In [Pol], it is shown that σHAUS $\subset$ pHAUS, i.e., if a continuous function between Hausdorff spaces has a continuous extension to their Fomin extensions, then the continuous function has a continuous extension to their Katětov extensions. Similar to the category pHAUS, the extremal subobjects of HC in σHAUS are the H-closed subspaces and σHAUS is not closed under finite intersections (same spaces and argument in Example 2.3) and is not closed under formations of products (use the same spaces as in Example 2.2 and the following fact to show that π_n is a morphism in σHAUS). Thus, the category σHAUS, like the category pHAUS, is not complete and, hence, does not satisfy the hypothesis of Theorem 2.4.

Proposition 2.5. A continuous function $f: X \to Y$ where X is Hausdorff and Y is compact Hausdorff has a continuous extension $F: \sigma X \to Y$.

Proof. By Theorem 2.1 in [K], f has a continuous extension $g: \kappa X \to Y$. By Theorem 8 in [F], there is a θ-continuous function $h: \sigma X \to \kappa X$ such that $h(x) = x$ for $x \in X$. Let $F = h \circ g$. Thus, F is an extension of f and is θ-continuous since the composition of θ-continuous functions is θ-continuous. By 1.1, F is continuous.

(III) Banaschewski proved [B] that a Hausdorff space has a minimal Hausdorff extension if and only if it is semiregular. Each semiregular Hausdorff space X has a largest (in a partition sense, cf. [PV2,3]) minimal Hausdorff extension, denoted as μX and called the Banaschewski-Fomin-Shanin (abbreviated to BFS) minimal Hausdorff extension of X. Let MH denote the full subcategory of HAUS of minimal Hausdorff spaces. By Theorem 2.4, there is a largest subcategory of SR, denoted μSR, such that MH is an epireflective subcategory of μSR with μ as the epireflection. The category MH is closed under the formation of products [O], but the class ob(MH) is not regularly closed hereditary (the regularly closed subspace A_1 of the minimal Hausdorff space X in Example 2.3 is not minimal Hausdorff). The extremal subobjects of MH in μSR are unknown. D. Harris communicated to me at the Memphis Topology Conference that he has characterized the morphisms of μSR, i.e., those continuous functions between objects in SR that can be extended to their BFS minimal Hausdorff extensions. The category μSR is not closed under the formation of products (use the same spaces as in Example 2.2 and the following fact to show that π_n is a morphism in μSR for each $n \in \mathbb{N}$) and is not closed under finite intersections by the following example. So, the category μSR, like the categories pHAUS and σHAUS is not complete and does not satisfy the hypothesis of Theorem 2.1.

(2.6) Let $X \in ob(SR)$ and Y be a compact Hausdorff space. A continuous function $f: X \to Y$ has a continuous extension $F: \mu X \to Y$.

Proof. Use the same proof as in 2.5 with his modification: Let $h:(\kappa X)_s \to \kappa X$ be the identity function and note that h is θ-continuous by 1.2 and the fact that $\mu X = (\kappa X)_s$ by Theorem 5.5 in [PT].

Example 2.7. Let X be the space in Example 2.3 and $X^* = \{x^*: x \in X\}$ be a copy of the space X. Let Y be the topological sum of X and X^* with the points $(1/n,0)$ and $(1/n,0)^*$ identified for each $n \in \mathbb{N}$. The spaces X, X^*, and Y are minimal Hausdorff. Let $i: X \to Y$ and $i^*: X^* \to Y$ be the inclusion functions. Let $A = i(X) \cap i^*(X^*)$ and $j: A \to Y$ be the inclusion function. Now, i and i^* are morphisms in MH, and it is straightforward to show that if (X,i) and (X^*,i^*) have an intersection, it must be (A,j). However, A is infinite discrete, non-H-closed, closed subspace of Y. So, if j has a continuous extension from μA to $\mu Y(=Y)$, then $cl_Y A(=A)$ would be H-closed. So, j is not a morphism in μSR and μSR is not closed under finite intersections.

In situations I, II, and III, we have three cases of epireflections (κ,σ, and μ) not falling under the scope of the usual theory of epireflections.

Problem A. Generalize the theory of epireflection to cover the three epireflections in situations I, II, and III.

It must be remarked that G. Strecker, at this conference, has extended the theory of epireflections to cover the epireflection κ in situation I. The possibility still exists that Strecker's extended theory of epireflections may also cover the epireflections σ and μ in situations II and III.

Also, it must be remarked that the existence of the largest subcategory μSR of SR on which MH is epireflective with μ as the epireflection solves part of problem 5 in [Hel] and a part of the problem on page 308 in [HS1]. The solution to the other part of the problem in [HS1] uses this variation of Theorem 2.4.

Theorem 2.4′. Let $\mathcal{B} \subseteq$ TOP and $\mathcal{A}$ be a full, replete subcategory of TOP. Let $F: ob(\mathcal{B}) \to ob(\mathcal{A})$ such that for each $X \in ob(\mathcal{B})$, there is a continuous function $F_X: X \to FX$ with this property: for each $f \in M_{\mathcal{B}}(X,Y)$, there is at most one morphism $g \in M_{\mathcal{A}}(FX,FY)$ such that

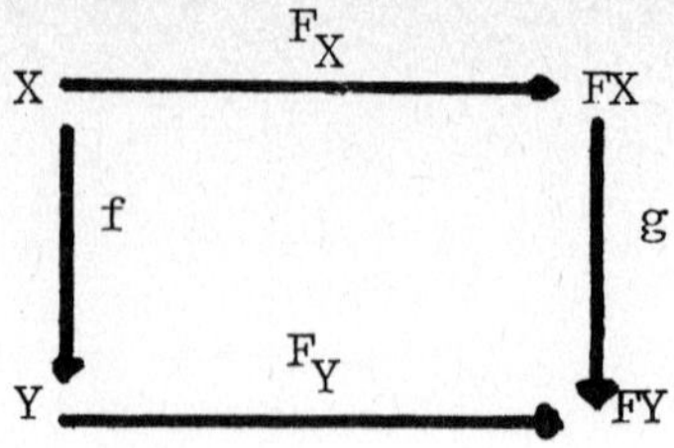

commutes. If $\mathcal{A} \subseteq \mathcal{B}$ and $FX = X$ for each $X \in ob(\mathcal{A})$, then there is a largest subcategory $\mathcal{C}$ of $\mathcal{B}$ on which F is an $\mathcal{A}$-epireflection. Also, $ob(\mathcal{C}) = ob(\mathcal{B})$ and $\mathcal{A} \subseteq \mathcal{C}$.

Proof. The proof is similar to the proof of Theorem 2.4 in [Pol].

To apply Theorem 2.4′ to second part of the problem on page 308 in [HS1], let $\mathcal{B}$ = HAUS, $\mathcal{A}$ = MH, and for $X \in ob(\text{HAUS})$, define $FX = \mu(X_s)$. So, $F_X\colon X \to \mu(X_s)$ is the inclusion function (not necessarily an embedding). By Theorem 2.4′ , there is a largest subcategory, denoted μHAUS, on which MH is an epireflective subcategory with $\mu(\ )_s$ as the epireflection.

Our next problem is also motivated by three situations.

(IV) In 1930, Tychonoff [T] proved that every Hausdorff space can be embedded (not necessarily densely) in an H-closed space. Actually, he showed for an open basis $\mathcal{B}$ of a Hausdorff space X, X can be embedded in a decomposition of $\Pi_{\mathcal{B}} I$ with a special topology where I is the unit interval with the usual topology.

(V) Frolik and Liu [FL] proved that every H-closed space can be embedded as a maximal separated subspace of its closure in the product of the unit intervals with the upper semicontinuity topology.

(VI) Parovičenko [Pa] proved that if $\{Y_\alpha : \alpha \in A\}$ is a set of H-closed extensions of a Hausdorff space X and $e: X \to \Pi\{Y_\alpha : \alpha \in A\}$ is the embedding map defined by $e(x)(\alpha) = x$, then $cl(e(X))$ is an H-closed extension of X which is the supremum of $\{Y_\alpha : \alpha \in A\}$ relative to the usual partial ordering defined between extensions.

A popular construction of the Stone-Cech compactification of a Tychonoff space is taking the closure of an embedding into a product of unit intervals. Situations IV, V, and VI motivate our next problem by presenting some evidence that a variation of the embedding into a product may be possible for the Katětov, Fomin, and BFS extensions.

Problem B. For $X \in ob(HAUS)$ (resp. $X \in ob(SR)$), construct, hopefully along the lines of (V), κX or σX (resp. μX) in terms of products.

S. Salbany has communicated to me, at this conference, that he has a method of constructing the Katětov extension κX of a Hausdorff space along the lines of the Frolik-Liu technique.

3. $P = S(\alpha)$ for $\alpha > 0$. Katětov [K] showed that a space is minimal $S(1)$ if and only if it is $S(1)$-closed and semiregular. The corresponding fact for $S(\alpha)$ spaces, $\alpha > 1$, is not only false but no reasonable substitute has been found for the property of semiregular (cf. Ex. 4.8 in [BPS]). In fact, for limit ordinals α, minimal $S(\alpha)$ spaces are regular [PV1]. Even though it is false that the class of $S(2)$-closed [He2] and the class of minimal $S(2)$ [S2] spaces are closed under the formation of products, it is true [PV1] that every $S(\alpha)$ space can be densely embedded in a $S(\alpha)$-closed space. For $\alpha > 0$, let $S(\alpha)C$ (resp. $MS(\alpha)$) be the full subcategory of HAUS of $S(\alpha)$-closed (resp. minimal

S(α)) spaces.

(3.1) There is nontrivial subcategory A of S(α) such that S(α)C is epireflective in A and ob(A) = ob(S(α)).

Proof. Since every S(α) space can be densely embedded in a S(α)-closed space [PV1], then by using the axiom of choice for a class of nonempty sets, for each X ∈ ob(S(α)), assign a S(α)-closed extension FX of X. By Theorem 2.4, there is a largest subcategory A of S(α) on which S(α)C is epireflective with F as the epireflection. Also, ob(S(α)) = ob(A).

For X ∈ ob(HAUS), let $s_X: X \to X_s$ be the identity function. By Theorem 2.4′, there is a largest subcategory A of HAUS on which SR is epireflective with s as the epireflection; actually, in this case s is a monoreflection. Also, s is a surjection from ob(S(1)C) onto ob(MS(1)) [K]; however, in general, a S(2)-clsoed space may have no coarser minimal S(2) topology (cf. [Po2]). It is still unknown for α > 1 (the α = 2 case is Problem 16(b) in [BPS]) if an S(α)-closed space with at least one coarser minimal S(α) topology has only one such topology; in the case α = 1, this is true. Let S(α)C′ be the full subcategory of S(α)C whose objects have at least one coarser minimal S(α) topology. For each X ∈ ob(S(α)C′), using the axiom of choice for a class of nonempty sets, let X_m be X with some coarser minimal S(α) topology. Let $m_X: X \to X_m$ be the identity function; m_X has the uniqueness property of Theorem 2.4′. An application of Theorem 2.4′ yields the following fact.

(3.2) For $\alpha > 0$, there is a largest subcategory A of $S(\alpha)C'$ on which $MS(\alpha)$ is a monoreflective subcategory with m as the monoreflection. Also, $ob(A) = ob(S(\alpha)C')$.

We conclude this section by noting that two of the unsolved problems (Problems 7 and 16(b)) in [PBS] for S(2) spaces should be extended to $S(\alpha)$ spaces for $\alpha > 1$.

Problem C.

(1) [BPS, Prob. 7]. For $\alpha > 1$, find a necessary and sufficient condition for a $S(\alpha)$ space to be embedded (densely embedded) in a minimal $S(\alpha)$ space.

(2) [BPS, Prob. 16(b)]. For $\alpha > 1$, prove or disprove that each object in $S(\alpha)C'$ has at most one coarser minimal $S(\alpha)$ topology.

4. P = regular and $S(\omega)$. Recall that our definition of regularity includes Hausdorff. The relationship between $S(\omega)$ and regularity is given by this theorem.

Theorem 4.1. [PV1]

(a) A regular space is $S(\omega)$, but the converse is false in general.

(b) A space is regular-closed if and only if it is regular and $S(\omega)$-closed.

(c) A space is minimal regular if and only if it is minimal $S(\omega)$.

Even though it is false that every regular space has a regular-closed extension [He3], it is true that every $S(\omega)$ space has $S(\omega)$-closed extension [PV1]. The category REG is an epireflective subcategory of HAUS [He4, Th. 2.12]; denote the REG-epireflector by r. By 3.1(e) of [PV1] if X is a $S(\omega)$-closed space, then rX is regular-closed. We are quite interested in those $S(\omega)$-closed spaces with coarser regular-closed topologies, and, of course, these are precisely the $S(\omega)$-closed spaces on which r is one-to-one.

Problem D. Internally characterize those $S(\omega)$-closed spaces on which the REG-reflector is a monomorphism.

By tracing the construction of the REG-reflector r, it is straightforward to show for $X \in ob(TOP)$, $r: X \to rX$ is a monomorphism if and only if for every pair of distinct points $x, y \in X$, there is $Y \in ob(REG)$ and continuous function $f: X \to Y$ such that $f(x) \neq f(y)$. This type of external characterization does not seem to be useful in identifying those $S(\omega)$-closed spaces with coarser regular-closed topologies.

The full subcategory of REG of objects possessing regular-closed extensions is denoted as RC-REG. D. Harris [Ha3] has developed a proximity theory (called RC-proximities) that generate, in a bijective manner, the set of regular-closed extensions for each $X \in ob(RC\text{-}REG)$. So, a topological space X is a object in RC-REG if and only if the topology of X has a compatible RC-proximity. This characterization of objects in RC-REG reveals a global nature of such spaces; however, an internal characterization of such spaces might provide valuable insight into the internal structure of regular spaces.

Problem E. Find an internal characterization of those regular spaces possessing regular-closed extensions.

Since every regular space can be densely embedded in a $S(\omega)$-closed extension, then a solution to Problem D might lead to a solution to

Problem E.

5. P = Extremally disconnect Hausdorff. In this section, the property of extremally disconnected Hausdorff (resp. extremally disconnected Tychonoff) is abbreviated to EDH (resp. EDT). In [PW], an EDH space is shown to be EDH-closed if and only if it is H-closed. If X is EDH space, then X_s is EDT space [PW, 2.1]. So, a minimal EDH space is EDT, and is seeking a characterization of minimal EDH spaces, it suffices to restrict ourselves to the class of EDT spaces.

Theorem 5.1. [PW] An EDT space X is not minimal EDH if and only if there is a nonempty clopen subset $B \subseteq X$ and a continuous injection $B \to cl_{\beta X}(X \setminus B) \setminus X$.

In the case that X is a locally compact, EDT space, the "continuous injection" can be replaced by "embedding" into $\beta X \setminus X$. The role of the Stone-Cech compactification in Theorem 4.1 hints of the possibility of a categorical theory that would characterize minimal P spaces where P is EDH plus, hopefully, many other properties. An intermediate step in this direction would be a solution to the next problem.

Problem F. [PW] Find an internal characterization of minimal EDH spaces.

Since the absolute (or projective cover) of a Hausdorff space is EDT, it is natural to inquire which Hausdorff spaces give rise to absolutes that are minimal EDH. The absolute of an H-closed space is compact EDH and, hence, minimal EDH; however, there are minimal regular spaces whose absolutes are not minimal EDH. The complete solution is provided by the next theorem which extends Theorem 5.1.

Theorem 5.2. [PW] Let X be a Hausdorff space and EX the absolute of X. EX is not minimal EDH if and only if there is a nonempty regularly closed subset $A \subseteq X$ (i.e., $A = cl_X int_X(A)$) and a continuous injection $EA \to cl_{\sigma X}(X \setminus A) \setminus X$ (recall that σX is the Fomin H-closed extension of X).

There is some evidence that a minimal EDH space is pseudocompact, e.g., a separable minimal EDH space is countably compact. An affirmative answer to the conjecture that a minimal EDH space is pseudocompact should be useful in obtaining a categorical theory that characterizes minimal EDH spaces.

Problem G. [PW] Prove or disprove that minimal EDH spaces are pseudocompact.

In conclusion, we remark that the regular-closed extension problem (Problem E) and the problem of finding a categorical theory that characterizes minimal EDH spaces are global problems (in the latter case, note the reference to a nonempty clopen set in Theorem 5.1). Much of the theory of minimal spaces accomplished during the 1960's was of a local nature, i.e., certain properties had to hold at each point. It is now clear that many of the unsolved problems remaining in the area of minimal spaces are of a global nature and seem to be ripe for an attack by global machinery, such as, categorical theory.

Department of Mathematics
University of Kansas
Lawrence, Kansas 66044
U.S.A.

REFERENCES

B B. Banaschewski, Über Hausdorffsch-minimale Erweiterungen von Räumen, Arch. Math. 12 (1961), 355-365.

BPS M. P. Berri, J. R. Porter, and R. M. Stephenson, Jr., A survey of minimal topological spaces, General Topology and its Relations to Modern Analysis and Algebra III, Proc. Kanpur Top. Conf. 1968, Acad. Press, New York, 1970, 93-114.

F S. Fomin, Extensions of topological spaces, Ann. of Math. 44 (1943), 471-480.

FL Z. Frolík and C. T. Liu, An embedding characterization of almost realcompact spaces, Proc. Amer. Math. Soc. 32 (1972), 294-298.

Ha1 D. Harris, Katětov extension as a functor, Math. Ann. 193 (1971), 171-175.

Ha2 ________, Structures in topology, Mem. Amer. Math. Soc. 115 (1971).

Ha3 ________, Regular-closed spaces and proximities, Pacific J. Math 34 (1970), 675-685.

He1 H. Herrlich, On the concept of reflections in general topology, Contributions to Extension Theory of Topological Structures, VEB Deutscher Verlag der Wissenschaften, Berlin (1969), 105-114.

He2 ________, Regular-closed, Urysohn-closed and completely Hausdorff-closed spaces, Proc. Amer. Math. Soc. 26 (1970), 695-698.

He3 ________, T_ν-Abgeschlossenheit und T_ν-Minimalität, Math. Z. 88 (1965), 285-294.

He4 ________, Categorical topology, Gen. Top. and its Appl. 1 (1971), 1-15.

HS1 ________, and G. E. Strecker, H-closed spaces and reflective subcategories, Math. Annalen 177 (1968), 302-309.

HS2 ________________, Category Theory, Allyn and Bacon, Boston, 400 pp.

K M. Katětov, Über H-abgeschlossene und bikompakt Räume, Časopis Pěst. Mat., 69 (1940), 36-49.

L C. T. Liu, Absolutely closed spaces, Trans. Amer. Math. Soc. 130 (1968), 86-104.

O F. Obreanu, Spatii Separate Minimale, An. Acad. Repub., Pop. Romîne, Sect. Sti. Mat. Fiz. Chem. Ser. A 3(1950), 325-349.

Pa I. I. Parovičenko, On suprema of families of H-closed extensions of Hausdorff spaces, Soviet Math. Kokl. 11 (1970), 1114-1118.

Po1 J. R. Porter, Extension function and subcategories of HAUS, Canad. Math. Bull. 18 (4) (1975), 587-590.

Po2 ________, Not all semiregular Urysohn-closed spaces are Katětov-Urysohn, Proc. Amer. Math. Soc. 25 (1970), 518-520.

PT ________ and J. D. Thomas, On H-closed and minimal Hausdorff spaces, Trans. Amer. Math. Soc. 138 (1969), 159-170.

PV1 ________ and C. Votaw, S(α) spaces and regular Hausdorff extensions, Pacific J. Math. 45 (1973), 327-345.

PV2 ________________, H-closed extension I, Gen. Top. and its Appl. 3 (1973), 211-224.

PV3 ________________, H-closed extension II, Trans. Amer. Math. Soc. 202 (1975), 193-209.

PW ________ and R. G. Woods, Minimal extremally disconnected Hausdorff spaces, submitted.

S1 R. M. Stephenson, Jr., Some unsolved problems concerning P-minimal and P-closed spaces, Proc. Memphis Top. Conf. 1975, to appear.

S2 ________, Products of minimal Urysohn spaces, Duke Math. J. 38 (1971), 703-707.

SW G. E. Strecker and E. Wattel, On semiregular and minimal Hausdorff embeddings, Proc. Kon. Ned. Akad. v. Wet. A70 (1967), 234-237.

T A. Tychonoff, Über die topologische Erweiterung von Räumen, Math. Ann., 102 (1930), 544-561.

SOME OUTSTANDING PROBLEMS IN TOPOLOGY
AND THE V-PROCESS

by

M. Rajagopalan

(Dedicated to Swami Gnanananda and Professor M. Venkataraman)

INTRODUCTION.

We mention some well-known problems in topology:

(a) Is a product of sequentially compact, T_2 spaces countably compact in general?

(b) Is it true that a completely regular, T_2, scattered, countably compact space X cannot be mapped continuously onto the closed interval $[0,1]$?

(c) Is every completely regular, T_2, scattered space 0-dimensional?

(d) Does every scattered completely regular, T_2 space admit a scattered compactification?

All these problems have been raised by well-known mathematicians and have proven to be hard problems. The problem (a) was raised by C. Scarborough and A. H. Stone [9]; (b) by P. Nyikos and J. J. Schäffer [5]; (c) by Z. Semadeni [10] and (d) by R. Telgarsky [11] and Z. Semadeni [10].

Though the above problems seem to be unrelated to each other at first sight all can be answered in the negative by constructing suitable counterexamples and all these examples can be obtained by a single method which we call V-PROCESS (which is a short form for VENKATARAMAN PROCESS). Some of them can be answered under weak set-theoretic axioms as well. Under suitable axiom of set theory this V-process yields very strong examples also. For example, using the Jensen's axiom called $\diamondsuit$

and the V-process we can show the existence of a hereditarily separable, perfectly normal, locally compact, locally countable, first countable, normal, sequentially compact, scattered space which is not Lindelöf.

Many persons have asked the author to explain this v-process clearly. In this paper we explain the V-process. Then, we also produce a family of sequentially compact spaces whose product is not countably compact using the axiom ⊛ (mentioned below) and ZFC only. Thus we give a negative solution to the problem (a) of C. Scarborough and A. H. Stone under fairly weak set-theoretic axioms.

NOTATIONS AND DEFINITIONS:

All spaces are assumed to be T_2. N denotes the discrete space of integers > 0. If X is a completely regular space then βX denotes the Stone-Čech compactification of X. Let Y be a topological space, $A \subset Y$ a subset of Y and π a partition of Y. Then we use int A to denote the interior of A and Y/π to denote both the quotient set and quotient space and $|Y|$ the cardinality of Y. R denotes the set of real numbers and $|R|$ is denoted by c. We put $\aleph_0$ for $|N|$.

ZFC denotes the Zermelo-Fraenkel axioms of set theory together with the Axiom of Choice. (CH) denotes the continuum hypothesis which states that there is no cardinal number λ so that $\aleph_0 < \lambda < c$. (MA) denotes the Martin's Axiom which states:

(MA) denotes the Martin's Axiom which states: Let X be a compact Hausdorff space in which every pairwise disjoint family F of non-empty open sets is countable. Then X cannot be expressed as a union $\bigcup_{A \in G} A$ of a family G of closed sets A so that int $A = \phi$ for all $A \in G$ and $|G| = c$.

(Axiom ⊛ is the following statement: Let J be a set so that $|J| < c$. Let 0_α be a compact open set of $\beta N-N$ for each $\alpha \in J$. Let the family $\{0_\alpha \mid \alpha \in J\}$ have the finite intersection property. Then $\bigcap_{\alpha \in J} 0_\alpha$ has non-empty interior in $\beta N-N$.

We use Λ to denote the initial ordinal of cardinal c. We use Ω to denote the first uncountable ordinal. We may not use (CH) in every one of our assertions. So Λ may not be equal to Ω. We note that given a countable set $\alpha_1 < \alpha_2 < \cdots < \alpha_n < \cdots$ of ordinals so that $\alpha_n < \Lambda$ for all $n \in N$ there is an ordinal $\gamma < \Lambda$ so that $\alpha_n < \gamma$ for all $n \in N$.

In the sequel, we generally use Greek letters like $\alpha, \beta, \gamma, \ldots$ to denote ordinals $< \Lambda$ and English letters like m, n etc. to denote the elements of N. It is known that (CH) implies (MA) and (MA) implies the axiom ⊛. (See [7] for this). We follow [1] and [12] for notions in General Topology, extremal disconnectedness and βN. We follow [2] and [4] for properties of scattered spaces. We note that a topological space is scattered if every non-empty subspace has a relative isolated point in it. In [4] it is shown that a compact, T_2 space is scattered if and only if it cannot be mapped continuously onto $[0,1]$.

SECTION 1: A Description of V-Process.

DEFINITION 1.1. Let $A_1, A_2, \ldots, A_n, \ldots$ be a sequence of subsets of βN. The growth of the sequence (A_n) is defined to be the set $\overline{\bigcup_{n=1}^{\infty} A_n} - \bigcup_{n=1}^{\infty} A_n$.

DEFINITION 1.2. Let P be a property of closed subsets of βN such that $\{n\}$ has P for all $n \in N$. P is called a good property if the following hold:

(a) If F is a closed subset of βN which does not have P, then there is a subset $A \subset F$ which is relatively open as well as closed in F and $A \neq \phi$ and A has P.

(b) βN cannot be expressed as the union of a family $\mathcal{F}$ of closed sets with property P and also with $|\mathcal{F}| < c$.

DEFINITION 1.3. Let P be a good property of closed subsets of βN. Let Y be a dense open subset of βN and π a partition of Y by closed subsets of βN. The pair (Y,π) is said to satisfy the condition (V_p) if the following hold:

(i) $\frac{Y}{\pi}$ is locally compact, T_2 and first countable.

(ii) Each member of π has P.

(iii) $|\frac{Y}{\pi}| < c$

(iv) Given $A \in \pi$ we have that there is a compact open subset M of βN so that M is saturated under π and $A \subset M \subset Y$,

(v) $\{n\} \in \pi$ for all $n \in N$.

Now we proceed to describe the V-process which is used to solve the problems (a), (b), (c) and (d) mentioned in the introduction.

STEP I OF THE V-PROCESS.

The step I of the V-process begins by choosing a suitable good property P_0 of closed subsets of βN to attack the problem concerned.

STEP II OF THE V-PROCESS.

This step consists of proving the following lemma.

LEMMA 1.4. *Let* P_0 *be the given good property of Step* I. *Let* Y *be a dense open set of* βN *and* π *a partition of* Y *by compact subsets of* βN. *Let* (Y,π) *satisfy the condition* $\textcircled{V}_{P_0}$. *Let* $A_1, A_2, \ldots, A_n, \ldots$ *be a sequence of distinct members of* π *so that* $Y \cap A = \phi$ *where* A *is the growth of the sequence* (A_n). *Let* $A \neq \phi$ *and closed. Then there exists an open set* Y_0 *and a partition* π_0 *of* Y_0 *by compact sets so that the following hold*:

(a) (Y_0, π_0) *satisfies the condition* $\textcircled{V}_{P_0}$.

(b) $Y_0 \supset Y$ *and* $\pi_0 \supset \pi$.

(c) $Y_0 \cap A \neq \phi$.

STEP III OF THE V-PROCESS.

This consists in proving the following lemma.

LEMMA 1.5. *Let* γ *be a limit ordinal which is* $< \Lambda$. *Let* P_0 *be as in Step* I *above. Suppose that* Y_α *is a dense open subset of* βN *and* π_α *is a partition of* Y_α *by compact subset of* βN *and* (Y_α, π_α) *satisfies the condition* $\textcircled{V}_{P_0}$ *for all* $\alpha < \gamma$. *Let* $Y_\alpha \supset Y_\beta$ *and* $\pi_\alpha \supset \pi_\beta$ *for all ordinals* $\alpha, \beta < \gamma$ *so that* $\alpha \geq \beta$. *Then the pair* (Y,π) *satisfies the condition* $\textcircled{V}_{P_0}$ *where*

$$Y = \bigcup_{\alpha<\gamma} Y_\alpha \quad \text{and} \quad \pi = \bigcup_{\alpha<\gamma} \pi_\alpha .$$

We remark that this lemma is easy to prove.

STEP IV OF THE V-PROCESS.

We now prove the following lemma.

LEMMA 1.6. *Let* P_0 *be as in Step* I *above. Let* (Y,π) *be as in Lemma* 1.4. *Let* F *be a family of closed subsets of* βN *so that the following hold*:

(a) $|F| < c$.

(b) <u>If</u> $A \in F$ <u>and</u> $A \neq \phi$ <u>then there is a sequence</u> $A_1, A_2, \ldots, A_n, \ldots$ <u>of distinct members of</u> π <u>so that</u> A <u>is the growth of the sequence</u> (A_n).

<u>Then there is an open set</u> Y_0 <u>of</u> βN <u>and a partition</u> π_0 <u>of</u> Y_0 <u>by compact subsets of</u> βN <u>so that the following hold</u>:

(i) $Y_0 \supset Y$ <u>and</u> $\pi_0 \supset \pi$.

(ii) (Y_0, π_0) <u>satisfies</u> (V_{P_0}).

(iii) $Y_0 \cap A \neq \phi$ <u>if</u> $A \in F$ <u>and</u> $A \neq \phi$.

<u>Proof</u>: The proof of this lemma is obtained by well-ordering F and using transfinite induction. To be more precise, let the non-empty sets in F be written as $F_1, F_2, \ldots, F_\alpha, \ldots$ where the suffix α is understood to be an ordinal $< \Lambda$. If $Y \cap F_1 \neq \phi$, put $Y_1 = Y$ and $\pi_1 = \pi$. If $Y \cap F_1 = \phi$, then replace Y, π, A of Lemma 1.4 by Y, π, F_1 of this lemma and get (Y_1, π_1) so that $Y_1 \cap F_1 \neq \phi$ and (Y_1, π_1) satisfies the conclusions (a) and (b) of Lemma 1.4. Now assume that γ is a successor ordinal and $\gamma = \beta+1$ and $\beta < \Lambda$ and we have defined (Y_β, π_β). If $Y_\beta \cap F_\beta \neq \phi$ then put $Y_{\beta+2} = Y_\beta$ and $\pi_{\beta+1} = \pi_\beta$. If $Y_\beta \cap F_\beta = \phi$ then replace Y, π, A of Lemma 1.4 by Y_β, π_β, F_β and get $(Y_{\beta+1}, \pi_{\beta+1})$ so that this pair satisfies the conclusions (a), (b) and (c) of Lemma 1.4 when the obvious substitutions are made in those statements.

Now suppose γ is a limit ordinal and we have defined (Y_α, π_α) for all $\alpha < \gamma$ and $\gamma < \Lambda$. Put $Y_{\gamma-} = \bigcup_{\alpha<\gamma} Y_\alpha$ and $\pi_{\gamma-} = \bigcup_{\alpha<\gamma} \pi_\alpha$. If $Y_{\gamma-} \cap F_\gamma \neq \phi$ then put $Y_\gamma = Y_{\gamma-}$ and $\pi_\gamma = \pi_{\gamma-}$. If $Y_{\gamma-} \cap F_\gamma = \phi$ then use Lemma 1.5 to observe that $(Y_{\gamma-}, \pi_{\gamma-})$ satisfies the hypotheses of Lemma 1.4. Then replace Y, π, A of that lemma by $Y_{\gamma-}$, $\pi_{\gamma-}$, F_γ and get an (Y_γ, π_γ) so that this pair satisfies the conclusions of Lemma 1.4 with the obvious substitutions made in them. Thus by transfinite induction we get a well ordered sequence of pairs (Y_α, π_α). Finally put $Y_0 = \bigcup Y_\alpha$ and $\pi_0 = \bigcup \pi_\alpha$. Then this (Y_0, π_0) will be the required pair.

STEP (V) OF THE V-PROCESS.

This is essentially the last step in the V-process. This will construct an open set Y of βN and a partition π of Y by compact sets of βN. The construction will be such that Y/π will always be a scattered, locally compact, first countable, non-compact, T_2, space that is countably compact. Hence the quotient space Y/π will always be sequentially compact, too. Let us give the details of this step which is done by transfinite induction using all ordinals $\alpha < \Lambda$.

CONSTRUCTION 1.7. Put $Y_1 = N$ and $\pi_1 = \{\{n\} \mid n \in N\}$. Put G_1 to be collection of all non-empty sets A so that A is the growth of a sequence $A_1, A_2, \ldots, A_n, \ldots$ of distinct elements of π and A is compact. Well order G_1 as $A_{11}, A_{12}, \ldots, A_{1\alpha}, \ldots$ where $\alpha < \Lambda$. Now suppose that γ is a successor ordinal and $\gamma = \beta+1$ and $\beta < \Lambda$ and we have defined (Y_β, π_β) and $A_{\alpha\delta}$ for all ordinals α and δ so that $1 \leq \alpha \leq \beta$ and $1 \leq \delta < \Lambda$. Put $F = \{A_{\alpha\delta} \mid 1 \leq \alpha \leq \beta,\ 1 \leq \delta \leq \beta\}$. Then replace (Y, π, F) of Lemma 1.6 by Y_β, π_β, F above. Then get $(Y_{\beta+1}, \pi_{\beta+1})$ so that this pair satisfies the conclusions (a), (b) of Lemma 1.6 with the obvious substitutions made in those statements. Let $G_{\beta+1}$ be the set of all non-empty closed sets A of βN so that A is the growth of a distinct sequence $(A_1, A_2, \ldots, A_n, \ldots)$ of elements of $\pi_{\beta+1}$. Write $G_{\beta+1}$ as a well ordered sequence $A_{(\beta+1)1}, A_{(\beta+1)2}, \ldots, A_{(\beta+1)\alpha}, \ldots$ where $\alpha < \Lambda$.

Now suppose that γ is a limit ordinal $< \Lambda$ and that we have defined Y_α, π_α, $A_{\alpha\delta}$ for all ordinals $\alpha < \gamma$ and $\delta < \Lambda$. Put $Y_{\gamma-} = \bigcup_{\alpha<\gamma} Y_\alpha$ and $\pi_{\gamma-} = \bigcup_{\alpha<\gamma} \pi_\alpha$. Put $F_{\gamma-}$ to be the collection $\{A_{\alpha\delta} \mid 1 \leq \alpha < \gamma$ and $1 \leq \delta < \gamma\}$. Now use Lemma 1.6 taking $Y_{\gamma-}$, $\pi_{\gamma-}$, $F_{\gamma-}$ in the place of Y, π, F of that lemma. We get (Y_γ, π_γ) so that $Y_\gamma \cap A_{\alpha\delta} \neq \phi$ for all α, δ so that $1 \leq \alpha < \gamma$ and $1 \leq \delta < \gamma$. Now put G_γ to be the collection of all non-empty closed sets A which are growths of a distinct sequence of members of π_γ. Well order G_γ as $A_{\gamma 1}, A_{\gamma 2}, \ldots A_{\gamma\delta}, \ldots$ for $1 \leq \delta < \Lambda$. Thus the transfinite induction gives open sets Y_α and partition π_α of Y_α for all $\alpha < \Lambda$. Finally put $Y_{\Lambda-} = \bigcup_{\alpha<\Lambda} Y_\alpha$ and $\pi_{\Lambda-} = \bigcup_{\alpha<\Lambda} \pi_\alpha$. This $Y_{\Lambda-}$ and $\pi_{\Lambda-}$ is the goal of the Step (V) of the V-process. If we put Y for $Y_{\Lambda-}$ and π for $\pi_{\Lambda-}$ for short, then the quotient space $X = \frac{Y}{\pi}$ is the required counter-example.

<u>THEOREM 1.8.</u> *The space X obtained in Construction 1.7 at the end of the V-process described above, is locally compact, T_2, first countable, separable and scattered.*

<u>Proof</u>: Let Y_α, π_α be as in Construction 1.7. Put $Y_\alpha/\pi_\alpha = X_\alpha$ for $\alpha < \Lambda$. Then X is an ascending union of the subspaces X_α. Since $\pi \supset \pi_\alpha$ and Y_α is open in βN and saturated under π, we have that X_α is open in X for all $\alpha < \Lambda$. Since (Y_α, π_α) satisfies condition (V_{P_0}) for all $\alpha < \Lambda$ we get that X is locally compact, T_2, first countable. Let $q : Y_{\Lambda-} \to X$ be the quotient map. Then $q(N)$ is dense in X and hence X is separable. Now X_α is locally compact and T_2 and $|X_\alpha| < c$ for all $\alpha < \Lambda$. So the one-point compactification $X_\alpha \cup \{\infty\}$ of

X_α cannot be mapped continuously onto $[0,1]$. So $X_\alpha \cup \{\infty\}$ and hence X_α is scattered for all $\alpha < \Lambda$. So if $F \subset X$ is a non-empty set and $F \cap X_{\alpha_0} \neq \phi$ for some $\alpha_0 < \Lambda$ then $F \cap X_{\alpha_0}$ has a point x_0 which is isolated in $F \cap X_{\alpha_0}$. Then $\{x_0\}$ is open in F because $F \cap X_{\alpha_0}$ is open in F. Thus X is scattered.

<u>THEOREM 1.9. Let X be the space constructed at the end of the V-process in Construction 1.7. Then X is sequentially compact.</u>

<u>Proof:</u> Let Y_α, π_α, $Y_{\Lambda-}$, $\pi_{\Lambda-}$ be all as in Construction 1.7. Put $q : Y_{\Lambda-} \to X$ to be the natural quotient map and $q(Y_\alpha) = Y_\alpha | \pi_\alpha = X_\alpha$ for all $\alpha < \Lambda$. Now it is enough if we show that X is countably compact because X is first countable by Theorem 1.8. For this it is enough to show that if $\{a_1, a_2, \ldots, a_n \ldots\}$ is an infinite set of distinct elements in X and $\{a_n\}$ is isolated in $\{a_1, a_2, \ldots, a_n, \ldots\}$ for all $n \in N$ then $\{a_1, a_2, \ldots a_n, \ldots\}$ has a cluster point in X. So we take such a subset $\{a_1, a_2, \ldots, a_n, \ldots\}$ in X. Put $q^{-1}(a_n) = F_n$ for all $n \in N$ and $M = \bigcup_{n=1}^{\infty} F_n$. Now M cannot be compact because $q(M) = \{a_1, a_2, \ldots a_n, \ldots\}$ cannot be compact. Moreover F_n is open relative to M for $n \in N$. So the growth $F = \overline{\bigcup_{n=1}^{\infty} F_n} - \bigcup_{n=1}^{\infty} F_n$ of the sequence (F_n) is non-empty and closed in βN. So there are ordinals $\alpha, \delta < \Lambda$ so that $F = A_{\alpha\delta}$. If γ is an ordinal $< \Lambda$ so that $\gamma > \alpha$ and δ then $Y_\gamma \cap A_{\alpha\delta} \neq \phi$. So $Y_{\Lambda-} \cap A_{\alpha\delta} \neq \phi$. Let $y_0 \in Y_{\Lambda-} \cap A_{\alpha\delta}$. Then $q(y_0)$ is a cluster point of the set $\{a_1, a_2, \ldots a_n, \ldots\}$. Thus we get the theorem.

<u>THEOREM 1.10. The space X of Construction 1.7 is not compact.</u>

Proof: Keep the notations of Theorem 1.9. Since Y_α is a dense open set of βN and $|Y_\alpha / \pi_\alpha| < c$ and p_0 is a good property of closed sets of βN and every member of π_α has p_0, we have that $Y_\alpha \neq \beta N$ for all $\alpha < \Lambda$. So $X_\alpha \neq X$ for all $\alpha < \Lambda$. Then the family $\{X_\alpha \mid \alpha<\Lambda\}$ is an open cover of X from which a finite subcover cannot be extracted.

REMARK 1.11. Whatever be the axioms of set theory we use, if we can prove Lemma 1.4 under those axioms then the remaining steps in the construction of the space X of the V-process follow and that final space X is always a locally compact, T_2, first countable, scattered, separable, sequentially compact, non-compact space. The role of the set-theoretic axioms we use is only to prove Lemma 1.4 for a particular property P_0 and the space X that we finally obtain by the V-process will be the required example.

In the next section we show how we can use the V-process to solve the problem (a) of C. Scarborough and A. H. Stone of the introduction. We solve it using only the axiom ⊛ beyond ZFC.

SECTION 2: A Solution to a Problem of C. Scarborough and A. H. Stone.

C. Scarborough and A. H. Stone [9] asked whether the product of sequentially compact spaces is always countably compact. Rajagopalan and R. G. Woods [8] showed the answer to be negative to this problem using continuum hypothesis. They used the V-process. J. Vaughn announced in a private communication that he also solved the problem of C. Scarborough and A. H. Stone above in the negative by using [CH]. Eric van Douwen informed the author that he also got a solution to the same problem under the axiom BF and ZFC. Neither the solution of Eric van Douwen nor that of J. Vaughn appeared yet in print. We produce a family of sequentially compact, locally compact, T_2 spaces whose product is not countably compact.

In this section we use only the axiom ⊛ which is weaker than both Martin's Axiom and continuum hypothesis.

DEFINITION 2.1. Let x be a given element of $\beta N-N$. We say that a closed set $F \subset \beta N$ has property P_x if $x \notin F$.

REMARK 2.2. It is clearly seen that P_x is a good property. Now we try to prove Lemma 1.4 for this P_x using only ⊛ and ZFC.

<u>THEOREM 2.3</u>. <u>Let Y be a dense subset of βN and π a partition of Y by compact sets of βN. Let (Y,π) satisfy the condition $\textcircled{V_{P_x}}$. Let $A_1, A_2, \ldots, A_n, \ldots$ be a sequence of distinct members of π so that the growth $A = \overline{\bigcup_{n=1}^{\infty} A_n} - \bigcup_{n=1}^{\infty} A_n$ of the sequence (A_n) is closed and non-empty. Let $A \cap Y = \phi$. Then there exists an open set Y_0 of βN and a partition π_0 of Y_0 by compact sets of βN so that $Y_0 \supset Y$ and $\pi_0 \supset \pi$ and $Y_0 \cap A \neq \phi$.</u>

<u>Proof</u>: Given $n \in N$ choose a compact open set W_n of βN so that $A_n \subset W_n \subset Y$ and W_n is saturated under π. (This is possible because (Y,π) satisfies the condition $\textcircled{V_{P_x}}$.) Put

$M_n = W_n - \bigcup_{i=1}^{n-1} W_i$ for all $n \in N$ and $n > 1$ and $M_1 = W_1$. Now

$\bigcup_{n=1}^{\infty} A_n \subset \bigcup_{n=1}^{\infty} M_n \subset Y$ and $A = \overline{\bigcup_{n=1}^{\infty} A_n} - \bigcup_{n=1}^{\infty} A_n$ is non-empty and disjoint with Y. So $\bigcup_{n=1}^{\infty} A_n$ cannot be compact. So there is an increasing sequence $n_1 < n_2 < \ldots < n_k < \ldots$ of integers so that $M_{n_k} \supset A_{r_k}$ for some integer r_k for all $k \in N$. Let E be the set of even integers. Let $G_1 = \bigcup_{k \in E} M_{n_k}$ and $G_2 = \bigcup_{k \in N-E} M_{n_k}$. Then G_1, G_2 are disjoint open sets of βN. Since βN is extremally disconnected x can belong to atmost one of the sets $\overline{G}_1$ or $\overline{G}_2$ (See [12,1]). So there is a subsequence $(M_{n_{k_i}})$ of (M_{n_k}) so that $x \notin \overline{\bigcup_{i=1}^{\infty} M_{n_{k_i}}}$. Put $D_i = M_{n_{k_i}}$ and $C_i = A_{r_{k_i}}$ for $i \in N$. Then $\bigcup_{i=1}^{\infty} C_i$ is not compact because $C_i \subset D_i$ for $i \in N$ and $\{D_i \mid i \in N\}$ is a pairwise disjoint collection of open sets of βN. So $C = \overline{\bigcup_{i=1}^{\infty} C_i} - \bigcup_{i=1}^{\infty} C_i \neq \phi$ and $C \subset A$. By assumption $\frac{Y}{\pi}$ has cardinality $< c$. So $\frac{D_k}{\pi}$ is a compact, T_2 space of cardinality less than c and hence totally disconnected for all $k \in N$. Since $\frac{Y}{\pi}$ is first countable by assumption, it follows that given $k \in N$ there is a descending sequence $V_{k1} \supset V_{k2} \supset \ldots \supset V_{kn} \supset \ldots$ of compact open sets V_{kn} of βN so that $V_{k1} = D_k$ and $\bigcap_{n=1}^{\infty} V_{kn} = C_k$ and V_{kn} is saturated under π for all $n \in N$. We put $\bigcup_{k=1}^{\infty} D_k = D$. We note that $C \subset \overline{D}$ and $x \notin \overline{D}$ and $\overline{D}$ is a compact open subset of βN. Now let $F \in \pi$ and $F \cap D = \phi$. Then using the normality of βN we get a compact open set V_F of βN so that $C \subset U_F$ and $F \cap U_F = \phi$. Let $\mathcal{F}$ be the family $\{U_F \mid F \in \pi$ and $F \cap D = \phi\}$

$\bigcup \{\bar{D}-D_n \mid n\varepsilon N\}$. Then $C \subset A$ for all $A \varepsilon F$. So the axiom ⊛ gives us that there is a set $G \subset D-Y$ which is non-empty and compact and open relative to $\beta N-N$. Then there is a compact open set M of βN so that $M \cap (\beta N-N) = G$. Then $M \cap D_{\ell_n} \neq \phi$ for an infinite set $\ell_1, \ell_2, \ldots, \ell_n, \ldots$ of integers ℓ_n. Then given $n \varepsilon N$ there is an integer s_n so that $M \cap V_{\ell_n s_n} \neq \phi$. Put

$R = \bigcup_{n=1}^{\infty} V_{\ell_n s_n}$. Then $(\bar{R}-R) \cap Y = \phi$ and $(\bar{R}-R) \cap C \neq \phi$ and $\bar{R} \cap Y = R$ is saturated under π and $\bar{R}-R$ has the property P_X. Also $\bar{R}$ is compact open in βN. Put $Y_0 = Y \cup \bar{R}$ and $\pi_0 = \pi \cup \{\bar{R}-R\}$. Then (Y_0, π_0) is the required pair of the theorem.

LEMMA 2.4. *Let X be a topological space. Then X is countably compact if and only if given a 1-1 map $f : N \to X$ from N into X there is an element $p \varepsilon \beta N-N$ and a continuous extension $g : N \cup \{p\} \to X$ of f.*

Proof: Let us assume that given a 1-1 map $f : N \to X$ there is an element $p \varepsilon \beta N-N$ and a continuous extension $g : N \cup \{p\} \to X$ of f. Let $\{a_1, a_2, \ldots, a_n, \ldots\}$ be an infinite subset of X. Let $f_0 : N \to X$ be the map given by $f_0(n) = a_n$ for $n \varepsilon N$. Then there is a $p \varepsilon \beta N-N$ and a continuous extension $g_0 : N \cup \{p\} \to X$ of f_0 to $N \cup \{p\}$. Then $g_0(p)$ is a cluster point of $\{a_1, a_2, \ldots, a_n, \ldots\}$ in X. Conversely, let us assume that X is countably compact. Let $f : N \to X$ be a 1-1 function from N into X. Then $f(N)$ is an infinite subset of X. Since X is countably compact, $f(N)$ has a cluster point 'ℓ'. Let us now consider two cases:

Case (i): Let $\ell \notin f(N)$. Let $\mathcal{V}$ be the filter of all neighborhoods of ℓ in X. Let $\mathcal{F}$ be the collection of all sets of the form $f^{-1}(V \cap f(N))$ where $V \in \mathcal{V}$. Then $\mathcal{F}$ is a filter on N and extends to an ultrafilter $\mathcal{U}$ on N. Then there is a unique element $p \in \beta N - N$ such that $p \in \bigcap_{A \in \mathcal{U}} \overline{A}$ (see [1,12]). Now put

$g : N \cup \{p\} \to X$ as

$$g(x) = \begin{cases} f(x) & \text{if} \quad x \in N \\ \ell & \text{if} \quad x = p \end{cases}$$

Then this $g : N \cup \{p\} \to X$ is a continuous extension of f to $N \cup \{p\}$.

Case (ii): Let $\ell \in f(N)$. Without loss of generality we can take $\ell = f(1)$. Then $\ell \in \overline{f(N-\{1\})}$ and $\ell \notin f(N-\{1\})$. So, by Case (i) above we have an element $p \in \beta N - N$ and a continuous extension $g : (N-\{1\}) \cup \{p\} \to X$ of $f / N-\{1\}$. Then put

$$\chi : N \cup \{p\} \to X \text{ as}$$

$$\chi(x) = \begin{cases} f(x) & \text{if} \quad x \in N \\ g(p) & \text{if} \quad x = p \end{cases}$$

This gives a continuous extension of f to $N \cup \{p\}$. Thus we get the result.

LEMMA 2.5. Let $p \in \beta N - N$ be given. Let X_p be the space constructed by the Construction 1.8 of the V-process using the property P_p of Definition 2.1. Let $q_p : N \to X_p$ be the restriction of the natural quotient map $q : Y_{\Lambda-} \to X_p$ of Theorem 1.9. Then q_p is a 1-1 map of N into X_p which cannot be extended to a continuous map $g : N \cup \{p\} \to X_p$.

Proof: Let $X_p \cup \{\infty\}$ be the one-point compactification of X_p. Then the map $\tilde{q}_p : \beta N \to X_p \cup \{\infty\}$ given by

$$\tilde{q}_p(x) = \begin{cases} q(x) & \text{if } x \in Y_{\Lambda-} \\ \infty & \text{if } x \notin Y_{\Lambda-} \end{cases}$$

is easily seen to be a continuous map from βN onto $X_p \cup \{\infty\}$. So the function $\tilde{g} : (N \cup \{p\}) \to X_p \cup \{\infty\}$ which is the restriction of $\tilde{q}_p$ to $N \cup \{p\}$ is continuous and maps p on ∞. Since N is dense in $N \cup \{p\}$ there can be only one continuous extension of f from $N \cup \{p\} \to X \cup \{\infty\}$ and $\tilde{g}$ is already one such. So there cannot be a continuous extension $g : N \cup \{p\} \to X$ of f because in that case $g(p) \in X$ and hence g and $\tilde{g}$ will be two continuous extensions of f from $N \cup \{p\}$ to $X_p \cup \{\infty\}$. This gives the lemma.

THEOREM 2.6. *Assuming the Axiom ⊛ and ZFC there is a family of locally compact, first countable, T_2, scattered, sequentually compact, seperable spaces X whose product is not countably compact.*

Proof: Given $p \in \beta N\text{-}N$ let X_p be the space as in Lemma 2.5. Then the collection $\{X_p \mid p \in \beta N\text{-}N\}$ is a collection of spaces as in the statement of Theorem 2.6. Let $q_p : N \to X_p$ be the map as in Lemma 2.5 for each $p \in \beta N\text{-}N$. Let $H = \prod_{p \in \beta N\text{-}N} X_p$. We claim that H is not countably compact. For let $\Pi q_p : N \to H$ be the map that so that the pth coordinate of $(\Pi q_p)(n)$ is $q_p(n)$ for all $p \in \beta N\text{-}N$ and $n \in N$. Then an easy application of Lemma 2.5 gives that there is no 'p_0' $\in \beta N\text{-}N$ so that Πq_p extends to a continuous function $g : N \cup \{p_0\} \to H$. So Lemma 2.4 gives that H is not countably compact. Thus we have the theorem.

REMARK 2.7. To make the paper short we are not giving the good properties here which are used to solve the problems (b), (c), (d) of the introduction. The interested reader can look into [3] for details about the solution to (b) and into [6] for details about the problems (c) and (d). In [3] and [6] CH is used. Jerry Vaughn has shown in a private communication that a method due to ostazewski can be used to solve the problem (a) of Scarborough and Stone using CH. The same method of ostazewski and an axiom different from ⊛ was used by Eric van Douwen (in a private communication) to solve the same problem (a).

Finally we raise the following problems:

OPEN PROBLEMS:

1. Can any of the spaces X_p obtained in Lemma 2.5 be made normal? How about if we use either [CH] or [MA]?
2. Can any of the spaces X_p obtained in Lemma 2.5 be made hereditarily separable if we assume [CH]?
3. Can the family of spaces X_p of Lemma 2.5 be so constructed that X_p is not homeomorphic to X_a if $p, a \in \beta N-N$ and $p \neq a$?
4. Is there a nicely described procedure to generate good properties of closed subsets of βN? In particular, is there a nice way to find all the good properties?
5. Is it possible to find $p \in \beta N-N$ and X_p as in Lemma 2.5 so that $\beta(X_p)$ is the one-point compactification of X_p? Is this possible under stronger set-theoretic axioms like $\lozenge$ or [CH]?

REFERENCES

[1] DUGUNDJI, R.; Topology, Allyn and Bacon, Boston (1966).

[2] KANNAN, V. and M. RAJAGOPALAN; On scattered spaces, Proc. Amer. Math. Soc., 43 (1974), 402-408.

[3] KANNAN, V. and M. RAJAGOPALAN; Scattered spaces II, Ill. J. Math (To appear).

[4] MROWKA, S., M. RAJAGOPALAN, and T. SOUNDARARAJAN; A characterisation of compact scattered spaces through chain limits (Chain compact spaces), TOPO 72, General Topology and Its Applications, Second Pittsburg International Conference, Dec (1972), Springer-Verlag, Berlin (1974), 288-297.

[5] NYIKOS, P. and J. J. SCHÄFFER; Flat spaces of continuous functions, Stud. Math., 42 (1972), 221-229.

[6] RAJAGOPALAN, M.; Scattered spaces III, J. Ind. Math. Soc. (To appear).

[7] RUDIN, M. E.; Lecture notes on set-theoretic topology, CBMS Series, No. 23, AMS, Providence, R.I. (1975).

[8] RAJAGOPALAN, M. and R. GRANT WOODS; Products of sequentially compact spaces, (To appear).

[9] SCARBOROUGH, C. and A. H. STONE; Products of nearly compact spaces, Trans. Amer. Math. Soc., 124 (1966), 131-147.

[10] SEMADENI, Z.; Sur les ensembles clairesemés, Rozprawy Math., 19 (1959), 1-39.

[11] TELGARSKY, R.; C-scattered spaces and paracompact spaces, Fund. Math., 73 (1971), 59-74.

[12] VAIDYNATHASWAMY, R.; Set topology, Chelsea Publishing Co., New York (1960).

[13] VAN DOUWEN, ERIC; First countable regular spaces and ω^{ω}, (To appear).

[14] VAUGHN, J. E.; Products of perfectly normal sequentially compact spaces, (To appear).

Nearness and Metrization

H.C. Reichel, Vienna

0. Introduction:

From an heuristic point of view the general ideas of "nearness" and "distance functions" in topology seem to be strongly connected. This paper deals with some part of these concepts; it tries to compare distance-functions which satisfy all the usual "metric" axioms, with several "topological structures", especially with the concept of nearness-spaces recently developped by Horst Herrlich (viz. § 3) and, since then, studied by several authors. - Real, as well as non-numerical, distance-functions have been studied under several aspects since the beginning of general topology. Especially those (generalized) distance-functions taking their values in an ordered group yield a satisfactory theory. (Compare the bibliography of this paper and [22] e.g.). - Working with such vector - (or, more generally, group-) valued metrics on X, one trivially uses only the positive cone of this group G. Thus one may try to replace G by a partially (or totally) ordered semigroup S from the very beginning of the theory. This is also motivated by several different problems in metrization theory and the theory of general "convergence structures". (For example, think of statistical metric spaces or the interesting question to what extent "countability" inherent in metrization theorems can be generalized using the well-ordering of natural numbers instead of their cardinality). - But using semigroups S, the S-valued metrics d_S need not induce uniform structures (even if <u>all</u> metric axioms are satisfied included the triangular inequality), neither must the convergence structure induced by d_S on X be compatible with a topology on X. (For a detailed study, see a paper of Reichel and Ruppert [30] and §1 of this paper). The "reason" for this disadvantage is that a totally ordered abelian group is a topological group in its order-topology which need not be true for totally ordered abelian semigroups S. In §1 we study uniform structures generated by "semigroup-valued metrics" d_S. Here, d_S induces a (semi)uniform structure on X in the sense of E. Cech and A. Weil, respectively, from which a topologically equivalent one can be "rediscovered" again. Surprisingly, the latter is not true if we use structures in the sense of Tukey and Isbell (see example 3.1. and §3).-

§2 indicates some facts showing to what extent the topological theory of metrizable spaces becomes more general if we admit also (non-numerical) "semigroup-valued" metrics and quasimetrics. (Compare [21], [31] and especially [30]).

The situation becomes more transparent if we consider **nearness**-spaces (in the sense of H. Herrlich) which provide an incisive frame for studying general metrization problems (§ 3). Since metrizability of nearness spaces clearly is connected with the study of nearness-structures having linearly ordered bases, § 3 can be interpreted as a general study of such structures. The definition of metrizability of N-spaces given by H. Herrlich [7] and Hunsaker and Sharma [10], is rich in meaning only for <u>uniform</u> nearness spaces with linearly ordered bases (as they have been investigated in many papers); so we supplement the theory by another definition (compatible with the old one for uniform N-spaces) and propose several problems in this context. Here, the crucial point is to find the correct generalization of the concept of the diameter of bounded subsets of S-metrizable spaces. (Note, that there need not exist suprema of bounded subsets in S.) (See page 15).We conclude by characterizing regular, semiuniform,uniform and paracompact nearness-spaces with linearly ordered bases. At the same time the latter yields conditions for topological nearness spaces to be also uniform, and thus includes Herrlichs characterization of Nagata-spaces. As a byproduct we generalize a theorem of Atsuji, [1], by characterizing those uniform spaces X with linearly ordered bases where all continuous mappings from X into any other uniform space are uniformly continuous. And finally, ω_μ-compactness of a completely regular space X is discovered as a necessary condition for the fine uniformity (Isbell [11]) on X to possess a linearly ordered base.

§ 1. <u>Semigroup metrics and uniform structures</u>

In this paper we consider only commutative semigroups (S,+) with identity 0. (S;<) is a *partially ordered* (p.o.) *semigroup* iff (S;<) is a partially ordered set and $a < b$ implies $a + c < b + c$ for all $a,b,c \in S$. Moreover we assume $a < 0$ or $a > 0$ for $a \neq 0$ and, if a has an inverse -a, $-a < 0$ iff $a > 0$. A *totally ordered* (t.o.) *semigroup* is a p.o. semigroup where $a < b$ or $a > b$ if $a \neq b$.

S^+ denotes the set of all $s > 0$, $s \in S$. S is said to be of character ω_μ iff there exists a decreasing ω_μ-sequence converging to 0 in the order topology of S. (Here ω_μ denotes the μth infinite cardinal number. The power of ω_μ is denoted by $\aleph_\mu$. - Iff τ is a topology on S, (S,τ) is called a *topological semigroup* iff the binary operation is a continuous function on the product space $S \times S$. -

- An *S-metric* on a set X is a function $d_S: X \times X \to S^+$ (i.e. $d_S(x,y) \geq 0$, for all $x,y \in X$) such that

(i) $d(x,x) = 0$	for all $x \in X$
(ii) $d(x,y) = 0$	implies $x = y$
(iii) $d(x,y) = d(y,x)$	for all $x,y \in X$
(iv) $d(x,y) \leq d(x,z) + d(z,y)$	for all $x,y,z \in X$

(X,d_S) is called an S-metric space or a space metrized by the semigroup S. A special case are the (usual) metric spaces $(X,d_{\mathbb{R}})$ which are $\mathbb{R}$-metric in our notation.

- Various "topological structures" compatible with S-metric spaces, as well as their relations to the structure of the linearly ordered semigroup S will be studied in a paper by H.C. Reichel and W. Ruppert [30] appearing in "Monatshefte für Mathematik" 1976/77. Concerning nearness-structures compatible with S-metrics see § 3.

For a p.o.-semigroup-metric space (X,d_S) it is possible to define "topological" structures in several ways. R. de Marr and Fleischer, [4] for example, study p.o. semigroups S such that, for every nonempty $M \subset S$, inf M exists; and they associate with d_S the following convergence-structure:

a net $\{x_\alpha / x_\alpha \in X, \alpha \in I\}$ is said to converge to $x \in X$ iff $\limsup d(x_\alpha, x) = 0$; [4]. Nevertheless, there need not be a topology on X for this convergence:

let S be the semigroup of all non-negative bounded real-valued measurable functions on [0,1], p.o. by $f < g$ a.e.; and let X be the set of all real-valued measurable functions on [0,1], then

$$d(f,g) = |f-g| \cdot (1-|f-g|)^{-1}$$

is an S-metric on X which - in the above mentioned sense - yields convergence almost everywhere for sequences (f_n). But there is no topology for this convergence.-

An other possibility is to study the entourages induced on $X \times X$ by d_S in the sense of A. Weil. Obviously, every p.o.-semigroup-metric d_S induces a semi-uniform structure $\mathfrak{U}_{d_S}$ by letting

$$U_\varepsilon = \{(x,y)/d_S(x,y) < \varepsilon\}, \quad \varepsilon \in S^+,$$

be a subbase for $\mathfrak{U}_{d_S}$. Hereby, as usual, a semiuniform structure $\mathfrak{U}$ on X is defined analogously to a diagonal-uniformity on X with the exception of the "uniform axiom":

$\forall\ U \in \mathfrak{U}\ \exists\ V \in \mathfrak{U}$ such that $V \circ V \subset \mathfrak{U}$.

For details see e.g. the book of Cech [2]. - By a theorem of G.K. Kallisch [13], $\mathfrak{U}_{d_S}$ is a uniformity on S if S is a p.o. vector group and, conversely, any uniform space $(X,\mathfrak{U})$ can be induced by a p.o.-vector-group-metric on X in this way.

For arbitrary semigroups S, the theory shows remarkable differences. So, for example, the semiuniformity $\mathfrak{U}_{d_S}$ need not be a uniformity, even if S is totally ordered. See example 1.5. below. (Theorem 1.1. shows the "reason" for this disadvantage). - A semiuniformity $\mathfrak{U}$ on X is called S-metrizable iff there is a p.o. semigroup S such that $\mathfrak{U} = \mathfrak{U}_{d_S}$. The question arises to characterize those semigroups S which yield uniform structures:

Theorem 1.1.: The following assertions are equivalent:

(i) $\mathfrak{U}$ is a separated uniform structure with a linearly ordered base $\mathfrak{B}$ ($B_1 < B_2$ iff $B_1 \supset B_2$; $B_{1,2} \in \mathfrak{B}$)

(ii) $\mathfrak{U}$ is G-metrizable by a t.o. (abelian) group G

(iii) $\mathfrak{U}$ is S-metrizable by a t.o. semigroup S such that S, with the order topology τ_o, becomes a topological semigroup

(iv) $\mathfrak{U}$ is S-metrizable by a totally ordered semigroup S such that the function $s \to s+s, s \in S$, is continuous at 0 with respect to the order topology on S.

Proof: The equivalence of (i) and (ii) was shown explicitely by F.W. Stevenson and W.J. Thron in [26]. To prove (ii) $\Rightarrow$ (iii) we only have to show that every t.o. group G is a topological group with respect to its order topology τ_o (which is not true for semigroups as we shall see). Without loss of generality, let G be abelian (compare also P. Nyikos and H.C. Reichel [21]):

Inversion is continuous which follows directly from the definitions. To show for each neighbourhood W of 0 the existence of a neighbourhood V with $V + V \subset W$, proceed as follows: take $s \in G$, $s > 0$, such that the interval $(-s,s) \subset W$, pick t such that $t < s$, and let $\{s_i / i < \alpha\}$ be a net converging to 0 from above. (If no such t or no such net exists, then G is discrete and we are done).

Now $\{t + s_i / i < \alpha\}$ converges to t and we need only pick s_i such that $s_i < t$ and $t + s_i < s$, whence $s_i + s_i < t + s_i < s$. Thus $V = (-s_i, s_i)$ satisfies $V + V \subset W$.

Since (iii) $\Rightarrow$ (iv) is trivial, let us show (iv) $\Rightarrow$ (i):
Let S be a t.o. semigroup such that $s \to s+s$, $s \in S$, is continuous at 0 with respect to τ_o. If S^+ has a smallest element a, the S-metric d_S

obviously induces the discrete uniformity on X. Otherwise, let $\{\epsilon_\tau / \tau < \omega_\mu\}$ be a cofinal decreasing well-ordered net converging to 0 with respect to τ_0. Then, for every ϵ_τ, there exists an index τ_1 such that $\epsilon_\sigma + \epsilon_\sigma < \epsilon_\tau$ if $\sigma > \tau_1$. Similarly, start from $\tau_1 + 1$ and construct τ_2 such that $\epsilon_\sigma + \epsilon_\sigma < \epsilon_{\tau_1 + 1}$ for all $\sigma > \tau_2$. Then, for all $\sigma > \tau_2$, $\epsilon_\sigma + \epsilon_\sigma + \epsilon_\sigma + \epsilon_\sigma < \epsilon_{\tau_1 + 1} + \epsilon_{\tau_1 + 1} < \epsilon_\tau$.
Now the system of all

$$U_{\epsilon_\tau} = \{(x,y)/d_S(x,y) < \epsilon_\tau\}, \tau < \omega_\mu,$$

is a linearly ordered base for a separated uniformity $\mathfrak{U}$ on X, since, for every $\tau < \omega_\mu$, $U_{\tau_2 + 1} \circ U_{\tau_2 + 1} \subset U_\tau$. ∎

There is another assertion equivalent to (i) - (iv), showing that the theory of non-archimedean (n.-a.) semigroup-metrics deserves an extra study. An S-metric d_S is n.-a. iff S is totally ordered and d_S satisfies the "strong" triangular inequality:

$$d_S(x,y) \leq \max\ (d_S(x,z), d_S(z,y)),$$

for all $x,y,z \in X$.

(N.-a. group-metrics play an important rôle in general valuation theory (Krull [15], Schilling [23])).

Theorem 1.2.: $(X, \mathfrak{U})$ is a separated *uniform* space with a linearly ordered base $\mathfrak{B}$ if and only if $\mathfrak{U}$ can be metrized either by a suitable *$\mathbb{R}$-metric* or by a *non-archimedean semigroup-metric* d_S.

Proof: Let S be a t.o. semigroup and d_S a n.-a. S-metric on X. For $\epsilon \in S^+$, let $B_\epsilon(x) = \{y/d(x,y) < \epsilon\}$ be the ϵ-ball with centre $x \in X$; then the covering $\mathfrak{B}_\epsilon = \{B_\epsilon(x)/x \in X\}$ essentially is a partition of X, because, by double application of the strong triangular inequality, two such balls either have empty intersection or they are identical. Now the coverings $\mathfrak{B}_\varepsilon, \varepsilon \in S^+$, can serve as a base for a (Tukey-) uniformity $\mathfrak{U}$ on X which is equivalent to $\mathfrak{U}_{d_S}$. (Clearly, each $\mathfrak{B}_\varepsilon$ is a star-refinement of itself). Conversely, we sharpen the result of Stevenson and Thron, mentioned before. (Compare also the paper of P. Nyikos and H.C. Reichel [21] on ω_μ-metric spaces). If $\mathfrak{B}$ is countable we are done; so let $\mathfrak{B} = \{B_i / i \in I\}$ be a linearly ordered base for $\mathfrak{U}$ such that $B_i < B_j$ iff $B_i \supset B_j$. By a construction of A. Hayes [6], to each $B_i \in \mathfrak{B}$ we can associate a sequence (B_n), n=1,2,...,

$$B_1 = B_i, \quad B_{n+1} \circ B_{n+1} \subset B_n,$$

such that

$$\bigcap_{n=1}^{\infty} B_n = B^{(i)}$$

has the property

$$B^{(i)} \circ B^{(i)} = B^{(i)}, \text{ and } \mathfrak{B}' = \{B^{(i)} / i \in I\}$$

is a linearly ordered base of $\mathfrak{U}$. The uniform cover

$$\mathfrak{C}_i = \{B^{(i)}(x) / x \in X\}, \text{ where } B^{(i)}(x) = \{y / (x,y) \in B^{(i)}\},$$

is a partition of X and we obtain a n.-a. group-metric by a construction which - although different - is essentially equivalent to the proof of the above mentioned theorem of Stevenson and Thron: Let $S = \prod \mathbb{Z}_i$ $(i \in I)$ where $\mathbb{Z}_i$ is the group of the integers, S being ordered lexicographically, and set $d_i(x,y) = 0$ iff x and y belong to the same set in $\mathfrak{C}_i$, $d_i(x,y) = 1$ otherwise. Now

$$d_S(x,y) := (d_i(x,y))_{i \in I}$$

is the desired non-archimedean metric inducing $\mathfrak{U}$.

With the theorem of Stevenson and Thron in mind (and Kallisch's result) one could conjecture a similar characterization of <u>semi-uniform</u> spaces $(X,\mathfrak{U})$ using p.o. and t.o. semigroups, respectively. Indeed, one can show an analogue (theorem 1.4.), but the proof has to be completely different, since both theorems mentioned above, indispensably use the fact that every uniform space is unimorphic with a subspace of a product of $\mathbb{R}$-metric spaces, for which of course there is no analogue in the theory of semi-uniform spaces. (There would be an analogue only if we did not require the distance functions to satisfy the triangular inequality).

Lemma 1.3.: A semiuniform space $(X,\mathfrak{U})$ induces a T_1-topology if and only if $\mathfrak{U}$ is "separated", i.e.:

$$\bigcap U_i \ (U_i \in \mathfrak{U}) = \Delta, \text{ the diagonal of } X \times X.$$

Remark: The straight forward proof is omitted, however, remark that a separated uniform space induces a T_2-topology, but to prove the converse you indispensably have to use the "uniform-axiom" or Tukeys equivalent "star-refinement axiom" either. Example 1.5. presents a separated semiuniform space which induces a non-Hausdorff topology.

Theorem 1.4.: A separated semiuniform space $(X,\mathfrak{U})$ has a linearly ordered base if and only if $\mathfrak{U}$ can be metrized by a t.o. semigroup S; *) i.e.: $\mathfrak{U} = \mathfrak{U}_{d_S}$ for the semigroup-metric d_S.

Proof: We only have to show S-metrizability of $(X,\mathfrak{U})$ where $\mathfrak{U}$ is a semiuniformity having a linearly ordered base. Let $\mathfrak{B}$ denote such a base of least power.

---

*) Without loss of generality, we can restrict ourselves to semigroups S being two-sided canvellative. (Of course, if moreover S is holoidal (see [Ljapin; Semigroups, AMS Translations 1963], i.e. "naturally ordered" (see [Fuchs; Partially ordered algebraic systems, Pergamon Oxford 1963]), then S is the positive cone of some t.o. group G. However, the latter would be a rather strong additional condition.

$\mathfrak{B} = \{B_i / i < \omega_\mu\}$ for some ordinal ω_μ. Here $i < j$ need not be equivalent with $B_i \supset B_j$. But, if, for every $k < \omega_\mu$, we let $V_k = \cap\, B_i$ $(i \leq k)$, we obtain a well ordered base $\mathfrak{B} = \{V_i / i < \omega_\mu\}$ for $\mathfrak{U}$ such that $i < j$ is equivalent with $V_i \supset V_j$. Indeed, $\mathfrak{B}$ is a base for $\mathfrak{U}$, since otherwise there would be a $k < \omega_\mu$ such that V_k is not refined by any B_i. But $\mathfrak{B}$ is linearly ordered so that in this case, for every $i > k$, there is a $j < k$ such that $B_i \supset B_j$. Therefore, the system $\{B_j / j < k\}$ would be a linearly ordered base for $\mathfrak{U}$ with a power less than ω_μ, which yields a contradiction.
- Now let us construct the t.o. semigroup S by which $(X, \mathfrak{U})$ can be metrized (S will be used in 1.5. again):
Let M be a set of arbitrary elements x_i $(i < \omega_\mu)$ and set $x_i < x_j$ iff $i > j$. Now let S be the free abelian semigroup over the set M. which, of course, can be visualized as the semigroup of all formal "linear combinations" $\Sigma\, m_i x_i$ (m_i a natural number or 0) where only finitely many $m_i \neq 0$. Identify the empty word with $0 \in S$. Further, for every pair $a = \Sigma\, m_i x_i$, $b = \Sigma\, n_j x_j$ with $\Sigma\, m_i \neq \Sigma\, n_j$ let $a < b$ iff $\Sigma\, m_i < \Sigma\, n_j$. If $a \neq b$ and $\Sigma\, m_i = \Sigma\, n_j$, pick the minimum index k such that $m_k \neq n_k$ and set $a < b$ iff $m_k < n_k$. For those $a = \Sigma\, m_i x_i$ where $\Sigma\, m_i = 1$, this order coincides with the given order of the "basis set" $\{x_i / i < j\}$. It is easy to show that $(S, <)$ is a totally ordered (and therefore cancellative semigroup: $a < b \Rightarrow a + c < b + c$, for all $a, b, c \in S$, is proved by direct computation. - Now let $d_S(x,y) = x_k$ iff $(x,y) \in V_i$ for all $i < k$ and $(x,y) \notin V_k$. Moreover, let $d_S(x,y) = 0$ iff $(x,y) \in V_i$ for all $i < \omega_\mu$. Thus we obtain an S-valued metric d_S on X, since the triangular inequality is satisfied for all $x, y, z \in X$ by the special order of S. And obviously, this S-metric induces $\mathfrak{U}$, i.e.: $\mathfrak{U} = \mathfrak{U}_{d_S}$. Note that in our semigroup S, $\{x_i / i < \omega_\mu\}$ is a net converging to 0 in the order topology of S, but $\lim\, (x_i + x_i) = x_1 \neq 0$. - (For our metric d_S, every set consisting of elements $d_S(x,y)$ has a greatest lower bound). -

We supplement theorem 1.4. by an example of a semiuniform space $(X,\mathfrak{U})$ with a linearly ordered base for which there is no equivalent uniform structure. Note that, by theorem 1.2., no such semiuniformity can be compatible with a non-archimedean metric d_S.

Example 1.5.: Let S be as in the proof of the proceeding theorem, and let $X = M$, metrized by $d_S(x,y) = \min\{x,y\}$ if $x \neq y$; $x,y \in X$; and $d_S(x,x) = 0$ for all $x \in X$.-Because of the special ordering of S, d_S satisfies the triangular inequality, and, as usual, the system $\mathfrak{B}$ of all $U_\epsilon = \{(x,y)/d_S(x,y) < \epsilon\}$, $\epsilon \in S^+$, is a base of a separated semiuniformity $\mathfrak{U}$ on X. But certainly, there is no uniform structure $\mathfrak{U}'$ equivalent with $\mathfrak{U}$, since, for every $\epsilon \in S^+$, $U_\epsilon \circ U_\epsilon = X \times X$ (of course, for every ϵ and every (x,y) there is a $z < \epsilon$ so that (x,z) and $(z,y) \in U_\epsilon$, yielding $(x,y) \in U_\epsilon \circ U_\epsilon$). Thus no entourage U_η can have a "uniform refinement" U_ϵ. - (The semigroup metric $d_S(x,y) = \max\{x,y\}$, $x \neq y$, and $d_S(x,x) = 0$, for all $x \in X$, would induce a uniform structure on X inducing the discrete topology).

§ 2. Semigroup metrics and topologies:

With similar methods as used in theorem 1.4. we can prove a theorem concerning S-metrizability of toplogical spaces; hereby a topology τ on X is metrizable by a p.o. (t.o.) semigroup S iff

$$\mathfrak{B} = \{B_\epsilon(x)/x \in X,\ \epsilon \in S^+\}$$

is a base for τ. Since we want to exclude trivial cases we assume that vor every $\epsilon > 0$, there exists an η with $0 < \eta < \epsilon$. Then for every t.o. semigroup S, let $\{\epsilon_i/i < \omega_\mu\}$, $\epsilon_i \neq 0$, be a well-ordered net of minimal power that converges to 0 with respect to the order topology of S. Then, if the system $\mathfrak{B}$ of all balls $\{B_{\varepsilon_i}(x) \;/\; i<\omega_\mu \;;\; x \in X\}$ is a base for a topology τ on X, then τ will be a T_1-topology such that every non-isolated point has a well-ordered local base $\{U_i(x)/i<\omega_\mu\}$ $U_i \subset U_j$ iff $i > j$. (Note that $\mathfrak{B}$ is a base for a topology on X if S is (the positive cone of) an abelian group. If S is only a totally-ordered abelian semigroup with minimal element $0 \in S$ the system of all d_S-balls need not form a base for a topology on X (see [30]). On the other hand side, example 1.5. shows readily that $\mathfrak{B}$ could be a base for a topology τ on X even if addition in S is not continous at $0 \in S$). -

Conversely, every T_1-space (X,τ) such that every non-isolated point $p \in S$ has a totally-ordered neighbourhood base of open sets and of of minimal cardinality ω_μ, can be metrized by a distance-function d_S over a t.o. semigroup S, satisfying (i), (ii) and (iv), i.e. the triangular inequality (page 3). By several authors such distance-functions are called W-(Wilson)-quasimetrics, see e.g. [29]. The proof of this assertion is modelled by the proof of theorem 1.4.: construct the t.o. semigroup S in the same manner as in this proof (1.4.), and for all $p,q \in X$, let $d_S(p,q) = x_i \in S$ iff $q \in U_j(p)$ for all $j < i$ and $q \notin U_i(p)$; and $d_S(p,q) = 0$ iff $q \in U_i$ for all $i < \omega_\mu$. Since, by the construction of S, the triangular inequality is satisfied for all $u,p,q \in X$, we obtain the desired distance-function that generates the topology τ. - If p is isolated let $d_S(p,q)=x_1$ $\forall q \in X$. Omitting all details which are similar to the proof of theorem 1.4., we obtain

<u>Theorem 2.1.:</u> A T_1-topology τ on a set X can be generated by a W - quasimetric d_S taking its values in a totally-ordered abelian semigroup S of character ω_μ, if and only if every non-isolated point $p \in X$ has a well-ordered open neighbourhoodbase of minimal cardinality ω_μ $\{U_i(p) \;/\; i < \omega_\mu\}$ $U_i \subset U_j$ iff $i>j$.

Remark 2.2.: Note that by this theorem, every first-countable space (X,τ) can be quasimetrized by a t.o. semigroup S of countable character. This, of course, does not mean that (X,τ) is quasimetrizable by $\mathbb{R}$ as it would follow if S were a group, e.g. (see § 1). On the other side, W.A. Wilson [29] showed that every second countable T_1-space is $\mathbb{R}$-quasimetrizable. In [22a] H. Ribeiro has shown that (X,τ) is $\mathbb{R}$-quasimetrizable iff every $p \in X$ has a countable neighbourhoodbase $\{U_i(p)/i \in \mathbb{N}\}$ such that for $q \in U_{i+1}(p)$, $U_{i+1}(q) \subset U_i(p)$, $i = 1,2,\ldots$

Other conditions for $\mathbb{R}$-quasimetrizability of (X,τ) can be found in [24] or [29] e.g. -

Similary we can ask under what conditions in Theorem 2.1., S can be chosen as a t.o. group of character ω_μ. This question will be answered fully in the above mentioned paper by H.C. Reichel and W. Ruppert.[30] This paper contains also many examples and applications of the theory of S-quasimetrics.

- In [20] P. Nyikos characterized spaces of the type considered in our theorem 2.1. by distance functions of a different type, by so called protometrics. Compare also his concept of m-semimetrics ([20a]).

Let us go back to semigroup-metrics as defined in § 1. An $\mathbb{R}$-metrizable space is always paracompact and - as it is well known (see J. Juhász [12] or A. Hayes [6]) - the same is true for spaces which can be metrized by t.o. groups. (Compare also [21]). However, working with semigroup-metrics we obtain from theorem 1.1. and the preceeding assertion:

Proposition 2.3.: Spaces (X,τ) which are metrizable by t.o. semigroups S need not be paracompact. However, (X,τ) is paracompact if the function $s \to s+s$, $s \in S$, is continuous at $0 \in S$ with respect to the order topology on S (or, more specially, if S is a topological semigroup with respect to the order-topology on S). Note that S-metrizable spaces are only T_1 in general, therefore it is convenient to define a space to be paracompact if it is T_1 and every open covering has a locally-finite open refinement.

<u>Proof</u>: We only have to present a non-paracompact space (X,d_S) metrized by a t.o. semigroup. Therefore, consider example 1.5., and let $\mathfrak{D}$ be an open covering of X. Let $O \in \mathfrak{D}$ and, for $x \in O$, pick a ball $B_\epsilon(x) \subset O$. Note that for every $\eta > 0$ and every $y \in X$, $B_\eta(y) \cap B_\epsilon(x) \neq \emptyset$. So $\mathfrak{D}$ cannot have a star-refinement. Moreover, (X,τ) is not T_2 although all "metric axioms" (page 3) are satisfied; neither is (X,τ) T_3, since a set $A \subset X$ is closed if and only if there is an $\epsilon \in X$ such that $\epsilon < a$ for all $a \in A$. - Concerning paracompactness in the realm of nearness-spaces, see § 3, theorem 3.9.

§ 3. Metrizability of nearness-structures:

Nearness spaces have been defined by H. Herrlich ([7], [8], [9]) as a unifying concept for the theories of topological spaces, uniform spaces, proximity spaces, contiguity spaces, merotopic spaces and other well known "topological" structures. - (Compare also [19]).

H. Herrlich and several other authors (see the bibliography in [7]) developped a rich theory of nearness-spaces ("N-spaces") and their applications. So for example, N-spaces provide an incisive and adequate method for studying extensions of spaces. - In the following we study some metrization problems connected with nearness-spaces. For detailed definitions, we refer to [7]. (See also a forthcoming paper of the author on metrizability of nearness-spaces). - Usually, N-spaces (X,ξ) are defined as an axiomatization of the concept of "nearness of arbitrary collections of sets", i.e.:

ξ is a collection of subsets of the power set PX of a set X described by some system of axioms. ξ is then the collection of those systems $\mathfrak{A}$ of sets $A \subset X$ which are "near". In a topological R_0-space X, for example, a nearness structure ξ is defined by $\mathfrak{A} \in \xi \Leftrightarrow \bigcap cl\ (A) \neq \emptyset,\ A \in \mathfrak{A}$.

For our purposes, it seems to be more useful to use another, equivalent, definition (see [7]): a system μ of coverings of X is called a nearness structure on X if

(i) $\mathfrak{A} \in \mu$ und $\mathfrak{A} < \mathfrak{B}$ ($\mathfrak{A}$ is a refinement of $\mathfrak{B}$) implies $\mathfrak{B} \in \mu$.
(ii) $\emptyset \neq \mu \neq P(PX)$.
(iii) if $\mathfrak{A}, \mathfrak{B} \in \mu$ then $(\mathfrak{A} \wedge \mathfrak{B}) \in \mu$
(iv) if $\mathfrak{A} \in \mu$ then $\{\mathrm{int}_\mu A \mid A \in \mathfrak{A}\} \in \mu$ where $\mathrm{int}_\mu A = \{x \in X \mid \{A, X - \{x\}\} \in \mu\}$.

$\varkappa \subset \mu$ is a base of μ iff, for every $\mathfrak{B} \in \mu$, there is a $\mathfrak{A} \in \varkappa$ such that $\mathfrak{A} < \mathfrak{B}$. $\varkappa$ is a subbase of μ iff $\{\mathfrak{A}_1 \wedge \ldots \wedge \mathfrak{A}_n \mid \mathfrak{A}_i \in \varkappa\}$ is a base of μ. (For an equivalent definition involving the ξ-structures see [7] and [10] - μ is a seminearness-structure iff it satisfies (i) - (iii). -

If μ is the system of all uniform covers of a uniform space $(X, \mathfrak{U})$, μ is a nearness-structure. Conversely, the category of those nearness-spaces (X, μ) which can be derived from a uniform structure $\mathfrak{U}$ on X, is called U-Near ("uniform" N-spaces). U-Near is isomorphic with Unif and a full bireflective subcategory of Near. The morphisms of Unif and Near are the uniformly continuous mappings and the nearness-preserving maps [8], respectively. By Tukeys theory, μ is a uniform nearness-structure iff every $\mathfrak{A} \in \mu$ has a star refinement $\mathfrak{B} \in \mu$.

Analogously, for every topological R_0-space (X, τ), the system μ of all open covers is a base of a nearness-structure on X. And the full subcategory T-Near of Near consisting of those N-spaces which are derived from topological spaces forms a bicoreflective subcategory of Near. - For more details, see [7] e.g.

By a definition of Hunsaker and Sharma [10] (see also Herrlich [7], a nearness space (X, ξ) is metrizable iff there exists a metric d on X such that $\mathfrak{A} \in \xi$ iff, for any positive real number ϵ there exists a point $x \in X$ with the ball $B_\epsilon(y) \in \sec \mathfrak{A}$, where $\sec \mathfrak{A} = \{B \subset X \mid \forall A \in \mathfrak{A}\ A \cap B \neq \emptyset\}$. Equivalently, (X, μ) is metrizable iff there is a metric d on X such that $\mathfrak{B}_\epsilon = \{B_\epsilon(x)/x \in X\}$, $\epsilon > 0$ yields a base of μ. It is easy to show that μ is metrizable iff μ is uniform and has a countable base. Naturally, for arbitrary (semi)nearness spaces we can define metrizability over (t.o.) groups S in an analogous manner, and a reformulation of § 1, shows that metrizability by t.o. groups characterizes exactly the category of uniform nearness-spaces with linearly ordered bases (ordered with respect to refinement). However, as we shall see, the theory of arbitrary nearness-spaces with linearly ordered bases seems to be more difficult. - In the following, an in § 1, every semigroup is abelian and has an identity-element 0. - To prevent trivialities, the zeroelement should not be isolated from above with respect to the order.

Example 3.1.: Analogous to the theory of (semi) uniform spaces, for any semigroup metric d_S on X, let μ_R be the (semi) nearness structure on X induced by d_S by taking $\mathfrak{B}_\varepsilon = \{B_\varepsilon(x) \mid x \in X\}$, $\varepsilon \in S^+$, $B_\varepsilon(x) = \{y \mid d_S(x,y) < \varepsilon\}$, as a (sub) base of μ_R. Now consider $\mu_R(d_S)$ induced by the semigroup-metric d_S constructed in example 1.5. $\{\mathfrak{B}_\varepsilon \mid \varepsilon \in S^+\}$ is then a base for the set of all "uniform coverings" of the semiuniform space (X, u_{d_S}) considered in § 1, and certainly, u_{d_S} is not a uniform structure.

Nevertheless, μ_R is a uniform nearness-structure: every $\mathfrak{B}_\delta$ is trivially a star-refinement of every $\mathfrak{B}_\varepsilon$, since every $\mathfrak{B}_\varepsilon$ contains the whole space X, ($B_\varepsilon(x) = X$ if $\varepsilon > x$!). More exactly, μ_R is the trivial nearness-structure on X which has {X} as a base. In other words: the metric d_S cannot be "rediscovered" from the nearness-structure μ_R induced by it. The situation is completely different if we study another (semi-) nearness-structure μ_D induced by d_S, letting $\mathfrak{A}_\varepsilon = \{A \subset X \mid \text{diam } A = \varepsilon\}$, $\varepsilon \in S^+$, be a (sub-) base for μ_D. (Notice the following problem in defining "diam A" which will be discussed in all details in a forthcoming paper of the author on metrizability of nearness-spaces by linearly ordered abelian (semi)groups: if the cofinality of S at $0 \in S$ is $> \omega_0$ inf and sup of bounded subsets of S need not exist in general. Therefore, in order to define "diam A", we choose a net $\{x_i / i \in I\}$ with wellordered index-set I converging to $0 \in S$ with respect to the order topology of S and with minimal cardinality (i.e. the well-defined cofinality of S at $0 \in S$), and define a set $A \subset X$ to be bounded if $d(x,y) < x_1$, for all $x,y \in A$. For a bounded set A consider $i = \min \{j \in I \;/\; \exists\; x,y \in A$ such that $d(x,y) \geq x_j\}$ and define

$$\text{diam } A = \begin{cases} x_{i-1} & \text{if } i \text{ is not a limit ordinal, and} \\ x_i & \text{if } i \text{ is a limit ordinal.} \end{cases}$$

Then we can show that $d(u,v) \leq \text{diam } A$, for all $u,v \in A$. (Moreover, $d(u,v) < \text{diam } A$ in the first case which appears if A is bounded and $\nexists$ sup $\{d(x,y) \;/\; x,y \in A\}$ in S, e.g.).- For the theory of S-metrizability of nearness-spaces, it does not matter that the terms "bounded" and "diam A" is defined only with respect to a 0-cofinal net $\{x_i / i \in I\}$ as chosen above. For more details see the paper mentioned before.

Note that in our example, a ball $B_\epsilon(x)$ either is equal to X (if $x<\epsilon$) or, for $x \geq \epsilon$ and $\epsilon = x_i \in X$, diam $B_\epsilon(x) = x_{i+1} < \epsilon$! Besides, for every $A \subset X$ with diam $A \leq \epsilon$, $\epsilon \neq 0$, there exists $x \in X$ and ϵ', $x \geq \epsilon' > \epsilon$ such that the ball $B_{\epsilon'}(x)$ covers A and diam $B_{\epsilon'}(x) \leq \epsilon$. Thus the system of all $\mathfrak{C}_\epsilon = \{B_\epsilon(x) \ / \ x \geq \epsilon\}$, $\epsilon \in S^+$, is a base for $\mu_D(d_S)$. Letting $\epsilon = x_i$ $(i < \omega_\mu)$, we see that μ_D is a nearness-structure with a linearly ordered base, and μ_D is <u>not</u> uniform, since the $\mathfrak{C}_\epsilon$-star of every $B_\epsilon(x) \in \mathfrak{C}_\epsilon$ is the whole space X! So no $\mathfrak{C}_\epsilon$ can have a star-refinement here. (Neither is μ_D a regular nearness-structure as it can be deduced straight forward from the definitions in [7],§11).- Compare theorem 3.8. - A more detailed study of μ_D will be found in the forthcoming paper of the author cited above.

<u>Remark 3.2.:</u> The advantage of considering μ_D instead of μ_R is that - at least in our example - the metric d_S (or, generally, an "equivalent" one) can be "rediscovered" from $\mu_D(d_S)$ whereas μ_R in our example is the trivial nearness-structure:
for $x,y \in X$, let $d(x,y) = x_i \in S^+$ iff, for all $\alpha < i$, there is one set in $\mathfrak{C}_{x_\alpha}$ containing both, x and y, and there is no set in $\mathfrak{C}_{x_i}$ containing x and y. if no no such x_i exists, set $d(x,y) = 0$. This procedure yields the S-metric we started with. (Compare remark 3.5.).-

Given two semigroup-metrics d_S, d_S' on X, call them R-equivalent (D-equivalent) if $\mu_R(d_S) = \mu_R(d_S')$, and $\mu_D(d_S) = \mu_D(d_S')$ respectively. Consider again example 3.1. and let d be the trivial pseudometric on X: $d(x,y) = 0$ for all $x,y \in X$; then d_S and d are R-equivalent whereas they are <u>not</u> D-equivalent.

This could be another motivation to prefer the structures μ_D instead of μ_R in studying the nearness-structures induced by semigroup-metrics on X.

Resumé 3.3.: Let S be a totally-ordered semigroup, then every S-metric d_S on X induces the (semi) nearness-structures μ_R and μ_D. In general $\mu_R \neq \mu_D$; however, if S is a totally-ordered abelian group (or a semigroup where the function $s \to s + s$, $s \in S^+$, is continous at $0 \in S$ with respect to the order-topology of S), then $\mu_D = \mu_R$ by the considerations in § 1.
Another condition assuring $\mu_D = \mu_R$ will be given in theorem 3.4.-
Summarizing, the following definition of metrizabilitiy of nearness-spaces seems to be the most adequate one. By the comments above, it coincides with Hunsaker and Sharma's definition if we consider only real metrics.

Definition: Let S be a totally-ordered abelian semigroup as above, then a nearness-space (X,μ) is metrizable by the S-metric d_S, $d_S: X \times X \to S$, iff $\mu = \mu_D(d_S)$, where the latter is described in example 3.1. -
Equivalently, we can describe nearness-spaces by their merotopic structure γ (see [7]); then γ induced by d_S on X is equal to $\{ \mathfrak{A} \subset PX \;/\;$ for every $\epsilon \in S^+$, there is an $A \in \mathfrak{A}$ such that diam $A < \epsilon \}$. The covering structure μ_γ on X belonging to this γ coincides with $\mu_d(d_S)$.

Theorem 3.4.: If $\mu_D(d_S)$ is a uniform nearness-structure then

$$\mu_D(d_S) = \mu_R(d_S).$$

Proof: Let $\mathfrak{C}_\varepsilon \in \mu_D$ and let $\eta \in S^+$ such that $\mathfrak{C}_\eta$ is a star-refinement of $\mathfrak{C}_\varepsilon$. Now let $A \in \mathfrak{C}_\eta$, then, for every $x \in A$ and $\sigma < \eta$, the ball $K_\sigma(x)$ is a subset of the star $St(A, \mathfrak{C}_\eta)$, since for every $y \in K_\sigma(x)$, there is a set B such that $x, y \in B$ and diam $B \leq \sigma < \eta$. therefore $B \in \mathfrak{C}_\eta$ and $y \in B \subset St\ (A, \mathfrak{C}_\eta)$. Thus the covering $\mathfrak{B}_\sigma$ consisting of all σ-balls $B_\sigma(x)$, $x \in X$, is a refinement of $\mathfrak{C}_\varepsilon$; so μ_R is finer than μ_D. Conversely, every covering $\mathfrak{B}_\varepsilon$ is refined by $\mathfrak{C}_\eta$ if $\eta < \varepsilon$; so μ_D is finer than μ_R, hence $\mu_R = \mu_D$.

Remark 3.5.: If (X,μ) is a uniform nearness-space with a linearly ordered base $\varkappa$ then, by the preceeding text, there is a t.o group S and an S-metric d_S on X such that $\mu = \mu_D = \mu_R(d_S)$. If μ is an arbitrary nearness-structure with a well-ordered base $\varkappa$ of least power $\aleph_\mu$, we can immitate the technique of remark 3.2. using the t.o. semigroup S with generators $\{x_i / i < \omega_\mu\}$ as it was constructed in the proof of theorem 1.4. Then we obtain an S-metric d_S on X such that μ is finer than $\mu_D(d_S)$ and this structure being finer than $\mu_R(d_S)$. Clearly, for every $B \in \mathfrak{B}_i \in \varkappa$ we have diam $B \leq x_i$; therefore every covering $\{A / \text{diam } A < x_i\}$ of X is refined by $\mathfrak{B}_{i+1} \in \varkappa$.

- Let (X,μ) be the N_1-nearness-structure of example 3.1., then, by remark 3.2., $\mu = \mu_D(d_S) \neq \mu_R(d_S)$. So it seems interesting to deal with the following problems:

Problem 1: For an arbitrary N_1-space (X, μ) with a linearly ordered base $\varkappa$ let d_S be the S-metric constructed in (3.2.). Hereby S is the semigroup constructed as indicated in theorem 1.4. Under which conditions is $\mu = \mu_D(d_S)$? More generally, find necessary and sufficient conditions assuring $\mu = \mu_D(d_S)$ for a given N_1-space (X,μ) and a suitable S-metric d_S.

Problem 2: If S is p.o. semigroup and d_S an S-metric on X, we can derive prenearness-structures μ_R and μ_D as before but using subbases instead of basis. Under what conditions is μ_R or μ_D a nearness-structure, which nearness-structures can be described in this way by S-metrics?-(Added in proof: a forthcoming paper of the author is concerned with partial solutions of 1,2).

Another problem arises from the following observation:
let (X,μ) be any N_1-space with a (linearly ordered) base $\varkappa = \{\mathfrak{B}_i / i \in I\}$; then the system $\mathfrak{U}_i = \bigcup (B \times B)$, $B \in \mathfrak{B}_i$ - $i \in I$ - is a base of a separated semiuniformity $\mathfrak{U}_\mu$ on X (in the sense of Cech-Weil; § 1). Conversely, given any separated semiuniformity $\mathfrak{U}$ on X, the system of all uniform coverings $\mathfrak{C}_i = \{\mathfrak{U}_i(x) / x \in X\}$ where $\mathfrak{U}_i(x) = \{y \mid (x,y) \in \mathfrak{U}_i\}$, $\mathfrak{U}_i \in \mathfrak{U}$, is a base for a N_1-structure $\mu_{\mathfrak{U}}$ on X. Example 3.1. shows that, for a given N-structure μ, μ need not be equivalent with $\mu_{\mathfrak{U}_\mu}$:
in example 3.1., $\mu_{\mathfrak{U}_{(\mu_D)}} = \mu_R(d_S)$ is the trivial nearness-structure on X whereas μ_D is not trivial!

Of course, μ is a uniform N-structure iff $\mu = \mu_{\mathfrak{U}_\mu}$, since - by a straightforward arguement - μ is uniform if every μ-cover $\mathfrak{A}$ has a barycentric refinement $\mathfrak{B}$.

Call an N-structure μ on X semiuniform iff $\mu = \mathfrak{U}_\mu$ for a suitable semiuniformity $\mathfrak{U}$ on X, then all uniform N-spaces are also semiuniform; and we ask for

(i) a characterization of all semiuniform N-spaces within the realm of the theory of nearness-structures; and

(ii) a characterization of those semiuniform N-structures which have a linearly ordered base. (See also theorem 1.4.). - -

Regular and paracompact N-spaces with linearly ordered bases:

As we have seen the concept of metrizability of nearness-structures μ is definitely rich (and unique) in meaning if we restrict ourselves to metrics d_S over t.o. groups S (or, equivalently by theorem 1.1., to t.o. semigroups S which are topological semigroups in their order topology). Then the category of metrizable N-spaces in this sense is isomorphic with the category of uniform N-spaces with a linearly ordered base. Hereby these categories are equipped with the usual morphisms. - More generally, let us now study regular N-spaces (for the exact definition see [7], § 11, or [8]). These structures are equivalent with K. Moritas regular T-uniformities, [17], called "semi-uniformities" by A.K. Steiner and E.F. Steiner in [25]. - Note that every uniform N-space is regular.

Definition: A nearness-space (X,μ) is submetrizable by an ordered abelian group G iff there is a G-valued metric d_G on X such that the N-structure $\mu_R(d_G)$ induced by d_G is finer than μ (i.e.: $\mu_R(d_G) \geq \mu$).

Moreover, two N-structures μ_1 and μ_2 on X are topologically equivalent iff, for their topological coreflections μ_1^T, μ_2^T, we have $\mu_1^T = \mu_2^T$. Of course, μ_1 and μ_2 are topologically equivalent iff they induce the same topology on X: $\tau_{\mu_1} = \tau_{\mu_2}$, where - for any N-structure μ - τ_μ is induced by the system of all stars $st(x,\mathfrak{A})$, $x \in X$, $\mathfrak{A} \in \mu$ as a base for τ_μ. (For regular N-spaces (X,μ), the system of all sets $A \in \mathfrak{A}$, $\mathfrak{A} \in \mu$, is also a base for τ_μ).

<u>**Theorem 3.8.**</u>: Every *regular* nearness space (X,μ) with a linearly ordered base $\varkappa$ is submetrizable by a t.o. group G. More exactly:

For every regular N-space (X,μ) with a linearly ordered base $\varkappa$ there is a topologically equivalent <u>uniform</u> N-structure μ^* on X with a linearly ordered base $\mathfrak{B}$ which is finer than μ. ($\mathfrak{B}$ and $\varkappa$ are of the same cofinality).

<u>Remark</u>: In [25] we find an example of a non-uniform regular nearness-space with a countable base. - By theorem 3.8. every such N-space is submetrizable by a realvalued metric $d_{\mathbb{R}}$ on X.

<u>Proof</u>: We have to show that there is a uniform nearness-structure μ^* with a linearly ordered base $\varkappa^*$ such that μ^* is finer than μ, and which is topologically equivalent with μ. This will be done by applying a theorem which generalizes a metrization theorem of A.H. Frink to group-valued metrics (P. Nyikos and H.C. Reichel [21], theorem 4).

Let (X,μ) be a regular N-space and $\varkappa$ a linearly ordered base of μ, say of least power $\omega_\mu (\mu \geq 0)$. Then by the same method as used in theorem 1.4., we can assume $\varkappa = \{\mathfrak{A}_i / i < \omega_\mu\}$ to be well ordered by refinement. By regularity of μ, for every $i < \omega_\mu$, there is a $j > i$ such that $\mathfrak{A}_j$ locally star-refines $\mathfrak{A}_i$ (i.e.: for each $A \in \mathfrak{A}_j$ there is a covering $\mathfrak{A}_k \in \varkappa$, $k = k(A)$, and a set $B \in \mathfrak{A}_i$ such that star $(A,\mathfrak{A}_k) \subset B$). So, for every $x \in X$, the sets $W_i(x) = st(x,\mathfrak{A}_i)$, $i < \omega_\mu$, form a nested neighbourhoodbase for a topology τ on X compatible with μ (all $A \in \mathfrak{A}_i$, $i < \omega_\mu$, are open in τ); by the same techniques as used in § 1 for semiuniform structures, the set $\bigcap W_i(x)$, $i \leq k < \omega_\mu$, is open with respect to τ. Moreover, for every i and x, there is a $j = j(i,x)$ such that $W_j(x) \cap W_j(y) \neq \emptyset$ implies $W_j(y) \subset W_i(x)$. But these are exactly the sufficient conditions which - by the above mentioned generalization of Frinks theorem - assure the existence of a uniform structure μ^*, with a linearly ordered base $\varkappa^*$, which is compatible with the topology τ, and which is finer than μ. Finally apply the theorem of Stevenson and Thron (see our theorem 1.1.).

(For the contable case, i.e. $\omega_\mu = \omega_o$, compare A.K. Steiner and E.F. Steiner [25], theorem 4.10. - For $\omega_\mu > \omega_o$, note that the proof of the generalization of Frinks theorem is essentially different to the countable case, for details see the above mentioned paper of Nyikos and Reichel [21]). - Now, if there is no countable base $\varkappa^*$ for the nearness-structure μ^*, we know from § 1 (e.g. from theorem 1.2.) that μ^* has a base consisting of partitions $\mathfrak{B}_i$ of X ($i \in I$). μ^* is finer than μ, thus (X, τ_μ) is zerodimensional: ind X = Ind X = dim $X = 0$, even stronger: τ_μ (the topology induced by μ) has a base of rank 1 (it is a non-archimedean topological space). See [22].

Ba combining theorem 3.7., the remark before theorem 3.7., and the theorem of Stevenson and Thron (viz. our theorem 1.1.) we obtain:

Corollary 3.8.: A topology τ on a set X is metrizable by a totally-ordered abelian group G (of character ω_μ) if and only if it is induced by a <u>regular</u> N-structure μ on X with a linearly ordered base (of cofinality ω_μ). From this result follows that a topology τ induced by a regular N-structure μ with a linearly ordered base is paracompact, therefore the N-structure $\mu_\tau = \mu^T$ induced by τ is a uniform (and topological, hence paracompact) N-structure on X.-

Summarizing we obtain:

Corollary 3.8.a.: If μ is a regular nearness-structure on a set X with a linearly ordered base $\varkappa$, then its topological coreflection μ^T is a uniform (and hence paracompact) nearness-structure on X. -
(Note that, of course, μ^T need not have a linearly ordered base itself).

Topological N-spaces and metrizability:

Now let us consider topological nearness-spaces with linearly ordered bases.
A topological nearness-space is paracompact iff it is uniform (Herrlich [7]). Sharpening our results we study now paracompact N-spaces (X,μ) with linearly ordered bases. - (X,μ) is ω_μ-contigual ([7], 18.11.) iff for every $\mathfrak{A} \in \mu$ there exists a subset $\mathfrak{B}$ of $\mathfrak{A}$ with less than ω_μ elements such that $\mathfrak{B} \in \mu$. - As usual, a topological space (X,τ) is called ω_μ-compact (or "initially ω_μ-compact; *)) iff every open cover has a subcover consisting of fewer than ω_μ sets. A topological N-space is ω_μ-contigual iff it is ω_μ-compact.

Theorem 3.9.: Let (X,μ) be a uniform nearness-space with a linearly ordered base $\varkappa$ of least power $\aleph_\mu$ and let τ be the topology on X induced by μ. Let X have no isolated points, then μ is topological (hence paracompact) iff (X,τ) is ω_μ-compact (equivalently, iff (X,μ^T) is ω_μ-contigual, where μ^T denotes the topological coreflection of μ).

*) For more details about ω_μ-compact spaces, see e.g.[R.Sikorski:Remarks on spaces of high power; Fund.Math. 37(1950),125-136] or [26] or D. Harris: Transfinite metrics, sequences and top. spaces Fund.M.73(1972) 137-142.
-ω_0-compact spaces are exactly the compact ones.-

<u>Remark</u>: In [7], H. Herrlich defined an N-space (X,ξ) to be a Nagata space iff (X,ξ) is metrizable and topological. Theorem 3.9. can be interpreted as a generalization of proposition 19.5. of [7] where Herrlich characterized the class of Nagata spaces by using results of N. Atsuji, M. Katetov, J. Nagata and A.H. Stone.

<u>Proof</u>: Let (X,μ) be a uniform N-space with a linearly ordered base $\varkappa$ of least power $\aleph_\mu$, then μ either is metrizable by a real ($\mathbb{R}$-) valued metric d (iff $\aleph_\mu = \aleph_o$) or μ has a well-ordered base σ consisting of partitions $\mathfrak{B}_\alpha$ of X, $\alpha < \omega_\mu$, (see the proof of theorem 1.2.). In this case, all sets $B \in \mathfrak{B}_\alpha$ are closed and open with respect to τ. - Suppose now, that (X,τ) is ω_μ-compact and let $\mathfrak{C} = \{O_i / i \leq k < \omega_\mu\}$ be an arbitrary open cover of X. If μ is metrizable by an ($\mathbb{R}$-valued) metric d, i.e. iff $\omega_\mu = \omega_o$, chose a Lebesgue-number $\delta' = 2\delta \in \mathbb{R}$ of $\mathfrak{C}$ such that the covering of X consisting of all balls $B_\delta(x) = \{y | d(x,y) < \delta\}$, $x \in X$, is finer than $\mathfrak{C}$. Thus we conclude that every open cover of X is refined by some $\mathfrak{B} \in \mu$ which implies that μ is topological, hence paracompact.

If $\omega_\mu > \omega_o$, consider the base σ of μ described above and note that the system of all sets $B \in \mathfrak{B}_\alpha$, $\alpha < \omega_\mu$, is a clopen base for the topology τ. Suppose now that there is no $\mathfrak{B}_\alpha \in \sigma$ refining a given open cover $\mathfrak{C}$, then, for every α, pick a $B_\alpha \in \mathfrak{B}_\alpha$ such that $B_\alpha \not\subset O_i$ for every $i \leq k$. Moreover, for every $\alpha < \omega_\mu$, pick $x_\alpha \in B_\alpha$, and obtain a ω_μ-sequence $\{x_\alpha\}$ in X. Since (X,τ) is ω_μ-compact, $\{x_\alpha\}$ has a cluster point x, [26], theorem 1.1. and $x \in O_i$ for some i.

Since O_i is open there is a $\beta < \omega_\mu$ and a set $B \in \mathfrak{B}_\beta$ such that $x \in B \subset O_i$. But all $\mathfrak{B}_\beta$ are clopen partitions of X, and x is a cluster point of $\{x_\alpha\}$, so there must be an index α such that our B_α is a subset of B. And this yields a contradiction to $B_\alpha \not\subset O_i$ for all $\alpha < \omega_\mu$ and all $O_i \in \mathfrak{C}$.

Therefore, concluding similarly to the countable case, every open cover of (X,τ) is refined by some $\mathfrak{B}_\alpha \in \sigma$; thus μ is topological, hence paracompact.

Conversely, let (X,μ) be a paracompact topological nearness space with a base $\varkappa = \{\mathfrak{B}_\alpha / \alpha < \omega_\mu\}$ well-ordered by refinement. Since (X,τ) is paracompact we can assume that all $\mathfrak{B}_\alpha$ are locally finite open coverings of X; if moreover $\omega_\mu > \omega_o$, $\mathfrak{B}_\alpha$ can be visualized as clopen partitions of X (see § 1).- Suppose (X',τ) is not ω_μ-compact then there is a set $Z = \{z_\alpha / \alpha < \omega_\mu\}$ such that, for every α, there is an open neighbourhood $V(z_\alpha)$ which does not contain any other point of Z. Moreover, chose $V(z_\alpha)$ "small enough" such that $V(z_\alpha) \subsetneq \cap B$ $(z_\alpha \in B \in \mathfrak{B}_\alpha)$, then the system
$(X \setminus Z) \cup \{V(z_\alpha) / \alpha < \omega_\mu\}$ is an open covering of X which cannot be refined by any $\mathfrak{B}_\alpha$, hence (X,μ) cannot be topological, which is a contradiction.

From this theorem we obtain a corollary which generalizes a theorem of N. Atsuji about metric spaces:

<u>Corollary 3.10.</u>: Let $(X,\mathfrak{U})$ be a uniform space with a linearly ordered base of least power ω_μ, and let (X,τ) have no isolated points, then the following is equivalent:

(i) every continuous mapping f from $(X,\mathfrak{U})$ into any other uniform space Y is uniformly continuous.

(ii) (X,τ) is ω_μ-compact. (τ is the topology induced by $\mathfrak{U}$.

<u>Proof</u>: (ii) $\Rightarrow$ (i): follows from the proceeding theorem: for every uniform covering $\mathfrak{B}$ of Y, $f^{-1}(\mathfrak{B})$ is an open covering of X, hence refineable by a uniform covering $\mathfrak{W}$ of $(X,\mathfrak{U})$. - To prove (i) $\Rightarrow$ (ii), remember that (X,τ) is paracompact if $\mathfrak{U}$ has a linearly ordered base (see e.g. Hayes [6] or Juhász [12]).

Therefore the "fine" uniformity $\mathfrak{B}$ of X consists of all open coverings of X, since X is fully normal by A.H. Stones theorem. Thus id: $(X,\mathfrak{U}) \to (X,\mathfrak{B})$, as a continuous mapping, is uniformly contiuous by the assumption of our theorem, and therefore, every open cover of X is refineable by an uniform cover of $(X,\mathfrak{U})$. - The rest of the theorem is a consequence of theorem 3.9.

<u>Remark</u>: The above mentioned result of N. Atsuji follows from this corollary by letting $\omega_\mu = \omega_0$. In a similar manner we could generalize the other parts of Atsujis theorem on metric spaces in [1]: [H.C. Reichel: "On a theorem of N. Atsuji";] to appear. (This paper studies also the case, where (X,τ) has isolated points).

Finally, by the same paracompactness-argument as it was used in the last proof, we derive another corollary from theorem 3.9.:

<u>Corollary 3.11.</u>: If the fine uniformity $\mathfrak{U}$ of an arbitrary completely regular topological space (X,τ) has a linearly ordered base of least power $\aleph_\mu$, and if X has no isolated points, then X must be ω_μ-compact.

-In [7], § 4, H. Herrlich remarks the completely symmetric relation between uniform and topological structures viewed in the realm of nearness-structures: a uniform space (X,μ) is "topologizable" iff there is a topological N-structure η on X such that its uniform (bi-)reflection η, on X is equivalent with u. -

Problem: Are there any non-trivial sufficient conditions for topologizability of a uniform space with a linearly ordered base? -

- Die Grundidee zu dieser Arbeit faßte ich während eines Aufenthaltes bei Prof. Dr. Horst Herrlich, Bremen. Ihm, wie auch Herrn Dr. W. Ruppert danke ich für wertvolle Hinweise und Diskussionen.

B I B L I O G R A P H Y :

[1] Atsuji, M.: Uniform continuity of continuous functions of metric spaces; Pacific J. Math. 8 (1958) 11 - 16.

[2] Čech, E.: Topological spaces; Z. Frolik and M. Katětov (eds.); Prague 1966.

[3] Császár, A.: Grundlagen der allgemeinen Topologie; Akadémicei Kiado, Budapest 1963.

[4] De Marr, R. and Fleischer, I.: Metric spaces over partially ordered semi-groups; CMUC 7 (1966) 501 - 508.

[5] Engelking, R.: Outline of General Topology; North-Holland, Amsterdam 1968

[5a] Fletcher, P. and Lindgren, W.F.: Transitive quasi-uniformities; J. Math. Anal. Appl. 39 (1972) 397 - 405.

[6] Hayes, A.: Uniform spaces with linearly ordered bases are paracompact; Proc. Cambridge Phil. Soc. 74 (1973) 67 - 68.

[7] Herrlich, H.: Topological Structures; P.C. Baayen (ed.); Math. Centre Tracts 52 (1974) 59 - 122.

[8] Herrlich, H.: A concept of nearness; J.Gen.Top.Appl.5 (1974) 191 - 212.

[9] Herrlich, H.: Some topological theorems which fail to be true; preprint 1975. (Int.Conf.on Categorical Top., Mannheim 1975)

[10] Hunsaker, W.N. and Sharma, P.L.: Nearness structures compatible with a topological space; preprint 1974.

[11] Isbell, J.R.: Uniform spaces;
Amer. Math. Soc. Math. Surveys 12 (1964).

[12] Juhász, I.: Untersuchungen über ω_μ-metrisierbare Räume;
Ann. Univ. Sci. Sect. Math., Budapest, 8 (1965) 129 - 145.

[13] Kalisch, G.K.: On uniform spaces and topological algebra;
Bull. Amer. Math. Soc. 52 (1946) 936 - 939.

[14] Katětov, M.: On continuity structures and spaces of mappings;
CMUC 6 (1956) 257 - 278.

[15] Krull, W.: Allgemeine Berwertungstheorie;
J. Reine u. Angew. Math. 167 (1932) 160 - 196.

[16] Mammuzić, Z.: Introduction to General Topology;
Noordhoff, Groningen 1963.

[17] Morita, K.: On the simple extension of a space with respect to a uniformity I - IV;
Proc. Japan Acad. 27 (1951) 65 - 72, 130 - 137, 166 - 171, 632 - 636.

[18] Nagata, J.: On the uniform topology of bicompactifications;
J. Inst. Pol. Osaka City Univ. 1 (1950) 28 - 38.

[18a] Nagata, J.: Modern General Topology;
North-Holland, Amsterdam 1968.

[19] Naimpally, S.A.: Reflective functors via nearness; Fund. Math. 85(1974), 245-255.

[20] Nyikos, P.: Some surprising base properties in topology;
Studies in Topology (Proc. Conf. Univ. North Carolina, Charlotte N.C., 1974; 427 - 450;
Academic Press, New York 1974.

[20a] Nyikos, P.: On the product of suborderable spaces;
Preprint 1974.

[21] Nyikos, P. and Reichel, H.C.: On uniform spaces with linearly ordered bases II; (ca. 16 pg) to appear in Fund. Math.

[22] Reichel, H.C.: Some results on uniform spaces with linearly ordered bases; (ca. 26 pg) to appear in Fund. Math.

[22a] Ribeiro, H.: Sur les éspaces a metrique faible; Portugaliae Math. 4 (1943), 21 - 40.

[23] Schilling, O.F.G.: General theory of valuations; Math. Surveys IV, Amer. Math. Soc. 1950.

[24] Sion M. and Zelner G.: On quasi-metrizability; Canad. J. 19 (1967) 1243 - 1249.

[25] Steiner A.K. and Steiner E.F.: On semi-uniformities; Fund. Math. 83 (1973) 47 - 58.

[26] Stevenson F.W. and Thron W.J.: Results on ω_{μ}-metric spaces; Fund. Math. 65 (1969) 317 - 324.

[27] Stone A.H.: Universal spaces for some metrizable uniformities; Quart. J. Math. 11 (1960) 105 - 115.

[28] Weil, A.: Sur les éspaces á structure uniforme et sur la topologie géneral; Paris 1937.

[29] Wilson, W.A.: On quasi-metric spaces; Amer. J. Math. 53 (1931) 675 - 684.

[30] Reichel, H.C. and Ruppert, W.: Über Metrisierbarkeit durch Distanzfunktionen mit Werten in angeordneten Halbgruppen (to appear in "Monatshefte f. Math.").

[31] Reichel, H.C.: A characterization of metrizable spaces(to appear).

Address of the author: Mathematisches Institut der Universität Wien
A-1090 Wien, Strudlhofgasse 4; Austria.

REFLECTIVE SUBCATEGORIES AND CLOSURE OPERATORS

by Sergio Salbany

Introduction

The following familiar examples illustrate the problem of associating a closure operator with a reflective subcategory.

Example 1 The Stone-Čech compactification βX of a Tychonoff space X is a compact Hausdorff extension of X, in which X is dense and such that every continuous map into a compact Hausdorff space C, has a continuous extension to $\beta X \to C$.

In other words, the category of compact Hausdorff spaces is a reflective subcategory of the category of Tychonoff spaces Ty.

In fact, it is a reflection in the category of topological spaces, and, as such, it may be regarded as the composite of

(i) Initial reflection - Top $\to$ CR, which assigns to each topological space X the weak topology induced by its continuous maps into $I = [0,1]$ with its usual topology.

(ii) Separating reflection - CR $\to$ Ty, which identifies the points in a completely regular space

X which are indistinguishable in the context of the discussion - points x and y whose images under continuous maps $f : X \to I$ are the same. For a given X the Tychonoff reflection is simply the image of X in the product $I^{C(X,I)}$, where $C(X,I)$ denotes the set of continuous maps from X to I.

(iii) The Stone-Čech reflection in Ty.

An alternative way of regarding βX, which has been the basis of many generalizations of compactness (e.g. [3],[7]), is based on the fact that the product $I^{C(X,I)}$ is compact Hausdorff and that closed subspaces of compact spaces are compact. If e is the product map $X \to I^{C(X,I)}$, then $\beta X = \overline{e[X]}$, where $\overline{}$ denotes the closure in the product topology on $I^{C(X,I)}$.

Example 2 If $\mathbb{R}$ = real line with usual topology replaces I in the above example, the resulting reflection is the Hewitt-realcompactification υX. We re-emphasize that

$$\upsilon X = \overline{e[X]} \subset \mathbb{R}^{C(X,\mathbb{R})}$$

where $\overline{}$ is the closure in the product topology.

Example 3 If $e : X \to I^{C(X,I)}$ and Q denotes the Q-closure operator of Mrowka ($x \in Q(A) \iff$ every G_δ-neighbourhood of x intersects A), then $\upsilon X = Q[e[X]]$.

Example 4 Every topological space is initial with respect to its continuous maps into the Sierpinski two point space $D = \{0,1\}$ with only one non-trivial open set $\{0\}$. Let us compare this situation with that of example 1 .

(i) The Initial reflector is the identity reflector.

(ii) The Separating reflector is the T_0-reflection.

(iii) The analogue of the Stone-Čech reflector in T_0 is very interesting. A tempting analogue is $\beta_0 X = \overline{e[X]} \subset D^{C(X,D)}$, where D has the topology specified above, e is the product map $e : X \to D^{C(X,D)}$ and $\overline{}$ denotes closure in the product topology.

This procedure was claimed to yield a space $\beta_0 X$ with the extension property for continuous maps from X to compact Hausdorff spaces by Nielsen and Sloyer [13] , but their claim was disproved with a simple observation by Salbany and Brümmer [14] .

When Baron characterized epimorphisms in T_0 [1], the proper closure to use became apparent. The reflector $X \rightsquigarrow$ b-closure-$(e[X]) \subset D^{C(X,D)}$ was studied by Skula [16] , Nel [12] . The b-closure of a set A in a topological space X consists of all points x such that $\mathrm{cl}\, x \cap V \cap A \neq \phi$, for all neighbourhoods V of x ($\mathrm{cl}_X x$ = closure of $\{x\}$).

2. The closure associated with a reflector

2.1 The closure operator

If $\underline{C}$ is a reflective subcategory with reflector R and reflection map $\eta_X : X \to R[X]$, then any map into $C \in \underline{C}$ $f : X \to C$ has a unique extension $\overline{f} : R[X] \to C$ such that $\overline{f} \circ \eta_X = f$. Thus, for any two maps $h,g : R[X] \to C$ such that $h \circ \eta_X = g \circ \eta_X$, $h = g$. Let $K(h,g)$ denote the coincidence set of h and g, consisting of all points x such that $h(x) = g(x)$. The uniqueness requirement is expressed by saying that $K(h,g) \supset \eta_X[X]$ $\Rightarrow K(h,g) = R[X]$.

Motivated by these considerations and by the account of Lambek and Rattray of the Fakir construction ([11]) that we shall discuss in a later section, we propose the following definition.

Definition Let $\underline{C}$ be a class of spaces. Given a space X and $A \subset X$, put

$$[A] = \cap\{K(f,g) \mid K(f,g) \supset A,\ f,g : X \to C,\ C \in \underline{C}\}$$

Proposition (i) $A \subset [A]$.

(ii) $A \subset B \Rightarrow [A] \subset [B]$.

(iii) $[[A]] = [A]$.

Proof (i) is clear, since each $K(f,g)$ in the definition of $[A]$ contains the set A.

(ii) Each $K(f,g)$ in the definition of $[B]$ contains B, hence contains A, hence is in the class of coincidence sets which determine $[A]$. Hence $[A] \subset [B]$.

(iii) By (i) $[A] \subset \big[[A]\big]$. To prove $\big[[A]\big] \subset [A]$, let $K(f,g)$ be such that $K(f,g) \supset [A]$ Then $K(f,g) \supset \big[[A]\big]$. Hence, the sets $K(f,g)$ in the class determining $[A]$ are in the class determining $\big[[A]\big]$. Hence $\big[[A]\big] \subset [A]$, as required.

<u>Note</u> similar considerations establish the formula $[A] \cup [B] \subset [A \cup B]$. From this formula it follows that $\big[[A] \cup [B]\big] = [A \cup B]$.

Although I believe that examples where $[A \cup B] \neq [A] \cup [B]$ exist in profusion, I have not been able to find one.

Thus [ ] is an operator whose induced neighbourhood structure is a Neighbourhood space (Espace à Voisinages) in the sense of M. Fréchet ([5]) (V is defined to be a neighbourhood of x if $x \notin [X - V]$). Such expansive, monotone and idempotent operators have been rediscovered and extensively used by P.C. Hammer ([6]). These generalized closures should be contrasted with generalized closures in the sense of Čech ([2]), where the operator is no longer required to be idempotent but is expansive and preserves finite unions (and is, consequently, monotone).

We show that [ ] induced by mappings into the Sierpinski dyad D is a Kuratowski closure.

<u>Proposition</u> The operator [ ] induced by mappings into the Sierpinski dyad is a Kuratowski closure.

<u>Proof</u> By our remarks above, it is sufficient to prove that $[A \cup B] \subset [A] \cup [B]$. Suppose $x \notin [A]$ and $x \notin [B]$. Then there are functions f_1, f_2, g_1, g_2: $X \to D$ such that $K(f_1,f_2) \supset A$, $K(g_1,g_2) \supset B$ and $f_1(x) \neq f_2(x)$, $g_1(x) \neq g_2(x)$. Let $h_1 = f_1 g_1 \vee f_2 g_2$ $h_2 = f_2 g_1 \vee f_1 g_2$ (where $(f \vee g)(x) = \sup\{f(x), g(x)\}$). Then $K(h_1,h_2) \supset A \cup B$ and $h_1(x) \neq h_2(x)$. The proof is complete.

<u>Note</u> The choice of h_1, h_2 is based on the following formal identities: $K(f_1,f_2) \cup K(g_1,g_2) = Z(f_1-f_2) \cup$ $\cup Z(g_1-g_2) = Z(f_1-f_2)(g_1-g_2) = Z[(f_1g_1+f_2g_2) - (f_1g_2+f_2g_1)]$ $= K(f_1g_1+f_2g_2, f_1g_2+f_2g_1)$.

2.2 <u>Comparing [ ]-closure and closure</u>

The following two propositions state sufficient conditions for the neighbourhood structure $T([\ ])$ to be comparable with the topology T on a given set.

To state one of the relationships one has to consider the initial category $In(\underline{A})$ determined by a class of spaces $\underline{A}$. This is a reflective subcategory of Top whose objects are topological spaces which are initial with respect to their mappings into objects of $\underline{A}$.

Proposition The following statements are equivalent.

(1) $T \subseteq T([\ ])$ for all X in $In(\underline{A})$.

(2) $T \subseteq T([\ ])$ for all finite products of objects in $\underline{A}$.

Proof (1) ⇒ (2) since all finite products of objects in $\underline{A}$ are in $In(A)$. (2) ⇒ (1). Suppose $x \notin \overline{A}$, then there is a finite product $B = A_1 \times A_2 \times \ldots \times A_n$ of objects A_i in $\underline{A}$ and a map $f : X \to B$ and an open set V in B such that $f(x) \in V$ and $f^{\leftarrow}[V] \cap A = \phi$. By assumption (2), there are functions $h_1, h_2 : B \to A_0$, $A_0 \in \underline{A}$ such that $x \notin K(h_1, h_2)$ and $X - V \subset K(h_1, h_2)$. Then $K(h_1 \circ f, h_2 \circ f) \supset \overline{A}$ and $x \notin K(h_1 \circ f, h_2 \circ f)$. Hence $\overline{A}$ is $T([\ ])$-closed. It follows that $T \subseteq T([\ ])$.

Corollary Suppose $\underline{A}$ is closed under finite products and such that if X is in $In(\underline{A})$ and $A \subset X$ is a closed set not containing x, then there are functions $f_1, f_2 : X \to A_0$, $A_0 \in \underline{A}$, such that $x \notin K(f_1, f_2)$, $K(f_1, f_2) \supset A$. Then $T \subseteq T([\ ])$.

Note The class of spaces considered in the example is simply the analogue of the class of completely regular spaces : points and disjoint closed sets can be separated by a $[0,1]$-valued continuous function, 0 on the set and 1 at the point.

Proposition The following are equivalent :

(1) $T([\ ]) \subset T$ for all X .

(2) $\underline{A}$ consists only of Hausdorff spaces.

Proof (1) $\Rightarrow$ (2) . Recall that (X,T) is a Hausdorff space iff the diagonal Δ is closed in $(X\times X, T\times T)$. Now, let X be in $\underline{A}$. $\Delta = k(\pi_1,\pi_2)$, where π_1, π_2 are the projection maps from $X\times X$ to X . Now $K(\pi_1,\pi_2)$ is $T([\ ])$-closed, hence T-closed. Hence X is a Hausdorff space.

(2) $\Rightarrow$ (1) . Note the following implications $T([\ ]) \subset T \Leftrightarrow K(f,g)$ is T-closed $\Leftrightarrow (f\times g)^{\leftarrow}[\Delta]$ is T-closed. Hence (2) $\Rightarrow$ (1) .

Combining the above criteria, we have the following.

Proposition Suppose $\underline{A}$ is a class of spaces closed under finite products and consisting only of Hausdorff spaces. Then the following are equivalent .

(1) $T = T([\ ])$ for all objects in $In(\underline{A})$.

(2) $T = T([\ ])$ for all objects in $\underline{A}$.

We now identify the [ ]-closure in the examples in the introduction.

<u>Example 1</u> The class $\underline{A}$ consists of products of copies of I , hence all the objects in $\underline{A}$ are Hausdorff and completely regular. By the corollary to our proposition and the comparison criterion, it follows that T([ ]) = T .

<u>Example 2</u> The class $\underline{A}$ consists of products of copies of $\mathbb{R}$. As in example 1, T([ ]) = T .

<u>Example 3</u> The class $\underline{A}$ consists of products of copies of the Sierpinski dyad D . By the corollary above, $T \subset T([\ ])$; and since all spaces involved in $\underline{A}$ are not Hausdorff we have, in fact, $T \subsetneqq T([\ ])$.

We have already shown that this neighbourhood structure T([ ]) is in fact a topology. We prove that this topology coincides with the b-topology ([16]) (also called front topology by L.Nel ([12])).

Proposition [ ]-closure = b-closure.

Proof Suppose $x \notin [A]$, then there are $f,g : X \to D$ such that $A \subset K(f,g)$, $f(x) \neq g(x)$. Assume $f(x) = 0$ and $g(x) = 1$. Then $x \in f^{\leftarrow}[0] \cap g^{\leftarrow}[1] \subset X-A$. Now $f^{\leftarrow}[0]$ is open, $cl\,\{x\} \subset g^{\leftarrow}[1]$, hence $x \notin$ b-closure (A). Similarly, if $g(x) = 0$, $f(x) = 1$. Conversely, suppose $x \notin$ b-closure (A). Then there is V open, such that $cl\{x\} \cap V \cap A = \phi$. Let $f : X \to D$ be $f = 0$ on V, $f = 1$ off V. Let $g : X \to D$ be $g = 1$ on $cl\{x\}$, $g = 0$ off $cl\{x\}$. Then $A \subset K(g,f.g)$ and $x \notin K(g,f.g)$. Hence $x \notin [A]$. Proof is complete.

Note The above argument simply reproves Baron's characterization of epimorphisms in T_0 as being maps onto b-dense subspaces ([1]) : Essentially because T_0 spaces are subspaces of products of D, we have: $A \xrightarrow{f} B$ is epi in T_0 iff $f[A]$ is T([ ])-dense in B.

We now discuss a related example involving bitopological spaces (X,P,Q).

Example 5 A bitopological space (X,P,Q) is pairwise completely regular if for every x and disjoint P-closed set F, there is a bicontinuous map $f : (X,P,Q) \to (I,u,l,)$ such that $f(x) = 0$

and $f(x) = 1$ on F, and for every x and disjoint Q-closed set G, there is $g : (X,P,Q) \to (I,l,u)$ such that $g(x) = 0$ and $g(x) = 1$ on G([15]). It is shown in ([15]) that these are precisely the bitopological spaces which are initial with respect to their mappings into (I,u,l). (In $\mathbb{R}$ and I, the non-trivial u-open sets are of the form $(-\infty,a)$ and the non-trivial l-open sets of the form (b,∞).)

There have been many proposals in the literature for what a "compact object" ought to be in the category of bitopological spaces. In ([15]) a case has been made for calling a bitopological space (X,P,Q) compact if the supremum topology $P \vee Q$ is compact (not necessarily Hausdorff). The present discussion also shows how naturally these spaces arise : in looking for an analogue of the Stone-Čech reflector for bitopological spaces, one is led to consider the embedding $e : X \to I^{C(X,I)}$, where $I = ([0,1],u,l)$ and $C(X,I)$ denotes the set of bicontinuous functions $f : (X,P,Q) \to (I,u,l)$; now, the Stone-Čech bitopological compactification $\beta_2 X$ "ought to be" $[e(X)]$, where $[\ ]$ denotes $T([\ ])$-closure as defined previously via coincidence sets. We show that the topology $T([\ ])$ induced in (X,P,Q) is simply $P \vee Q$, confirming that it is natural to regard pairwise compact spaces as $P \vee Q$-closed subspaces of products of copies of (I,u,l), in other words, as $P \vee Q$-compact spaces.

<u>Proposition</u> Let $[\ ]$ be the operator induced by mappings $f : (X,P,Q) \to (I,u,l)$. Then $T([\ ]) = P\vee Q$.

<u>Proof</u> $K(f,g)$ is $P\vee Q$-closed for any maps $f,g : (X,P,Q) \to (I,u,l)$, as $f,g : (X,P\vee Q) \to (I,u\vee l)$ and $u \vee l$ is the usual topology on I . Thus $T([\ ]) \subseteq P\vee Q$. Conversely, suppose that A is $P\vee Q$-closed. To show that A is $T([\ ])$-closed, we assume $x \notin A$ and exhibit $h_1,h_2 : (X,P,Q) \to (I,u,l)$ such that $x \notin K(h_1,h_2)$ and $A \subset K(h_1,h_2)$. Since A is $P\vee Q$-closed and $x \notin A$, there is a P-open set P and a Q-open set Q such that $x \in P \cap Q \subset X-A$. Let $f : (X,P,Q) \to (I,u,l)$, $g : (X,P,Q) \to (I,u,l)$ be such that $f(x) = 0$, $g(x) = 1$, $f = 1$ off P , $g = 0$ off Q . Then $x \notin K(g.f,g)$, $A \subset K(g.f,g)$.

<u>Note</u> As remarked earlier in the discussion of example 4, this proposition essentially proves that epis in the category of separated pairwise completely regular spaces are maps onto $P\vee Q$-dense subspaces ([15]) . The proof given above is a simplification of the characterization of epis given in [15] .

Examples 4 and 5 above are intimately related. Suppose (X,T) is a topological space. Let T^* be the topology on X for which points x have minimal neighbourhoods: T-closure $\{x\}$. The bitopological space (X,T,T^*) is pairwise completely regular. Moreover, it is pairwise separated if and only if X is a T_0 - space. The correspondence $(X,T) \to (X,T,T^*)$ has been studied in [15]. It is interesting to note that $T \vee T^*$ is simply the b-topology. Thus, the characterization of epis in T_0 follows from the characterization of epis in pairwise Tychonoff spaces. However, in general, $\beta_2 X \supsetneqq bX$ as a set of points, where β_2 is the bitopological Stone-Čech extension reflector and b is the pc - reflector which takes X to the b - closure of its image in the canonical product of copies of D .

One further comment will finalize this discussion.

<u>Example 6</u> Let $\underline{A}$ be the category of uniform spaces (not necessarily separated). Let M be a metric space and U the metrizable space of uniformly continuous functions $f : M \to I$, $I = [0,1]$. U is an injective in $\underline{A}$ [9]. Lambek and Rattray [11] raise the question of describing the associated reflective subcategory. Based on the preceding work we can provide the first step towards an answer: The reflective subcategory consists of all uniform spaces which are isomorphic to <u>closed</u> subspaces of products of copies of U .

2.3 Characterizing reflections by coincidence kernels

This section raises a question which we have been

unable to settle even within situations which are fairly algebraic, such as those involved with semi-continuous functions $(X,T) \to (\mathbb{R},u)$ or with bicontinuous functions $(X,P,Q) \to (\mathbb{R},u,l)$.

The Stone-Čech compactification can be characterized as being a compact Hausdorff extension C of X , in which X is densely embedded and such that

$$cl_C[Z(f) \cap Z(g)] = cl_C Z(f) \cap cl_C Z(g) .$$

The problem is to find an analogous characterization which is more categorial. The obvious substitutes for the zero sets are the coincidence sets. We conjecture that the proper substitute for cl_C ought to be the [ ] - operator discussed in this note.

3. Lambek-Rattray localization

We conclude our discussion with a brief description of the work by Lambek and Rattray in so far as it is pertinent to this note.

3.1 Let $\underline{A}$ be a complete category and I a fixed object in $\underline{A}$. The object I determines functors

$$\underline{A} \xrightarrow{(-,I)} \underline{Ens}^{op} \xrightarrow{I^{(-)}} \underline{A}$$

where $(-,I)$ is a left adjoint of $I^{(-)}$. Thus, the composition $S = I^{(-,I)}$ is a part of a triple, (S,η,μ) on $\underline{A}$.

The Fakir construction [4], associates with S another triple (Q,η_1,μ_1) where $Q : \underline{A} \to \underline{A}$ is the equalizer

$$Q \xrightarrow{\kappa} S \underset{S\eta}{\overset{\eta S}{\rightrightarrows}} S^2 .$$

Let FixQ denote the full sub-category of all objects A such that $\eta_1(A) : A \to QA$ is an isomorphism. The question arises - when is Q a reflector ? This is answered by the Theorem in the Lambek-Rattray paper which we shall quote for completeness.

<u>Theorem</u> The following statements are equivalent.

(a) I is injective with regard to $\kappa(A)$, for each A in $\underline{A}$.

(b) Q is idempotent, i.e. becomes a reflector $Q : \underline{A} \to \mathrm{Fix}Q$.

(c) FixQ is the limit closure of I.

We add one more equivalent condition to the theorem which we have found useful in discussing topological examples.

Let us say that $A \xrightarrow{f} B$ is $\underline{C}$-epi (where $\underline{C}$ is a class of spaces) if $A \xrightarrow{f} B \xrightarrow{g} C = A \xrightarrow{f} B \xrightarrow{h} C$ $\Rightarrow g = h$, whenever C is in $\underline{C}$.

The equivalent condition that we mentioned is then

(a') $A \xrightarrow{\eta_1(A)} Q(A)$ is I-injective for every object A in $\underline{A}$.

3.2 Finally, the connection between Q and our discussion of the closure associated with a reflection:

$$Q(A) = \cap\{K(f,g) \supset \eta(A) \mid f,g : S(A) \to I\} = [\eta(A)] .$$

It was only after proving this that we realized that this is the content of Lambek-Rattray's lemma 1, that $\kappa(A) : QA \to SA$ is the joint equalizer of all pairs of mappings $SA \rightrightarrows I$ which coequalize $\eta(A) : A \to SA$.

I wish to thank the organisers of the Conference for their invitation and financial support and also the Council for Scientific and Industrial Research (C.S.I.R.) and the University of Cape Town for their financial support. I also wish to express my thanks, through Prof. K.Hardie, to the T.R.G. for providing an opportunity for the discussion of these ideas.

References

[1] S. Baron - Note on epi in T_o , Canadian Mathematical Bull. 11, 503-504, 1968.

[2] E. Čech - Topological spaces. Academia Prague, 1966.

[3] R. Engelking and S. Mrowka - On E -compact spaces, Bull. Acad. Polon. Sci., 6, 429-436, (1958).

[4] S. Fakir - Monade idempotente associée à une monade, C. R. Acad. Sci., Paris 270, A99-A101, 1970.

[5] M. Fréchet - Espaces Abstraits, Paris, 1928.

[6] P.C. Hammer - Extended Topology: Continuity I, Portugaliae Math. 23, 79-93 (1964).

[7] H. Herrlich - 𝔊 - kompakte - Räume, Math Z. 96, 228-255, (1967).

[8] H. Herrlich - Topologische Reflexionen und Coreflexionen. Lecture Notes 78. Berlin-Heidelberg - New York Springer 1968.

[9] J.R. Isbell - Uniform Spaces, Amer. Math. Soc. Surveys No.12, Providence, R.I., 1964

[10] J.F. Kennison - Reflective functors in general topology and elsewhere. Trans. Amer. Math. Soc. 118, 303-315 (1965).

[11] J. Lambek and B.A. Rattray - Localization at injectives in complete categories, Proc. Amer. Math. Soc. (41), 1-9, 1973.

[12] L.D. Nel and R.G. Wilson - Epi-reflections in the category of T_o spaces. Fund. Math. 75, 69-74 (1972).

[13] R. Nielsen and C. Sloyer - On embedding in quasi-cubes. Amer. Math. Monthly 75, 514-515 (1968).

[14] S. Salbany and G.C.L. Brümmer - Pathology of Upper Stone-cech compactifications, Amer. Math. Monthly 78, 1971.

[15] S. Salbany - Bitopological spaces, compactifications and completions. Math. Monographs, University of Cape Town, 1974.

[16] L. Skula - On a reflective subcategory of the category of topological spaces, Trans. Amer. Math. Soc. 142, 137-141 (1969).

COMPACTNESS THEOREMS

by

M. Schroder

ABSTRACT: Theorems due to E. Binz, and to G.D. Richardson and D.C. Kent, saying that in certain categories of convergence spaces compact spaces are topological, both have a common extension. A convenient proof involves a "convolution" of convergence structures.

INTRODUCTION

Lately several theorems have appeared, each saying that in a specified category of convergence spaces, compact spaces are topological. To start with, in 1968 E. Binz proved that the category of c-embedded convergence spaces has this property. Then in 1972, as a by-product of their work on compactifications, G.D. Richardson and D.C. Kent did the same for principal T_3 spaces. A.C. Cochran and R.B. Trail [2] and C.H. Cook [3, Theorem 2] obtained this result as well, using other methods.

Here it is shown first that the "principal" condition can be relaxed to "pseudotopological". Together with [5], [7] or [9], which all characterise c-embedded spaces internally, this yields an easier proof of Binz's theorem. Second, the method used here allows one to weaken the "T_3" condition as well.

§1 TERMINOLOGY

Throughout this note, X is a non-empty set, $\underline{\underline{X}}$ its power set and $F(X)$ the set of all proper filters on X. As usual, $\underline{\underline{X}}$ is regarded as an improper filter.

Mainly because the term "convergence space" has recently become ambiguous, it is convenient to give the following definitions. A subset H of $F(X)$ is called a *segment* if for any ψ in H every filter finer than ψ also belongs to H, or a *filter-ideal* if in addition $\psi \cap \chi$ belongs to H whenever both ψ and χ do. (Set notation is used throughout, with exactly its usual set-theoretic meaning; for example, if ψ and χ are filters on X, then ψ is finer than χ iff $\psi \supseteq \chi$.)

Some authors use the term "convergence structure" for a map $\gamma : X \to \underline{\underline{F(X)}}$ under which each γ_x is a filter-ideal containing $\dot{x}$, the ultrafilter at x. Others call this a "limit structure", and use "convergence structure" for those maps γ in which each γ_x is merely a segment containing $\dot{x}$. Here, the latter idea is referred to simply as a "structure", the term "convergence structure" being reserved for the filter-ideal version. Similarly, the (convergence) space X_γ is the set X together with the (convergence) structure γ.

A tool used often is the multiplication of structures described in [10]. Since it is not widely known, its definition and some of its properties are sketched below.

As in [10], a map $\phi : X \to F(X)$ is called a *selection* (on X); it is called a γ-*selection* if γ is a structure and $\phi_x \in \gamma_x$ for all x. Any selection can be condensed onto a set or a filter as follows: if ϕ is a selection, P a subset of X, and ψ a filter on X, then the equations

$$\phi \cdot P = \cap\{\phi_x : x \in P\} \quad \text{and}$$

$$\phi \cdot \psi = \cup\{\phi \cdot P : P \in \psi\}$$

both define filters on X (except that $\phi\cdot\emptyset = \underline{\underline{X}}$ is improper). Now if θ is a selection as well, the filters $\theta\cdot\phi_x$ clearly define a selection $\theta\cdot\phi$. It is easy to verify that the filters $\theta\cdot(\phi\cdot\psi)$ and $(\theta\cdot\phi)\cdot\psi$ coincide, and as a result, that multiplication of selections is associative.

Now let γ and δ be structures on X. By defining $\gamma\cdot\delta_x$ to be the set of all filters χ such that $\chi \supseteq \phi\cdot\psi$, for some γ-selection ϕ and some ψ in δ_x, one obtains a structure $\gamma\cdot\delta$, coarser than both γ and δ. Moreover, if both γ and δ are (principal) convergence structures then so is $\gamma\cdot\delta$. The resulting multiplication or convolution of structures is not very well understood. None-the-less, it is known that

(i) the discrete topology is the identity element,

(ii) $\cdot$ is not always commutative,

(iii) for any structures γ, δ, η on X, $\gamma\cdot(\delta\cdot\eta) = (\gamma\cdot\delta)\cdot\eta$ if either γ or δ is principal (however, I do not know if the associative law holds in general), and

(iv) a convergence structure γ is diagonal [6] iff $\gamma = \gamma\cdot\gamma$. (O. Wyler's recent treatise [11] provides a natural framework for diagonal structures, and probably $\cdot$ as well.)

Perhaps the most significant fact is that composition of adherence operators reflects this multiplication; that is, $a_{\gamma\cdot\delta} = a_\delta a_\gamma$ for any structures γ and δ on X. These facts and others dealt with in [10] were proved for convergence structures, but the proofs given there still go through almost unchanged.

Using this fact, one can characterise those convergence structures γ whose adherence operators are idempotent.

(1) a_γ is idempotent (meaning that $a_\gamma = a_\gamma a_\gamma$) iff $\gamma\cdot\gamma$ is finer than $\pi\gamma$, the principal modification of γ.

Proof: Suppose first that $\gamma\cdot\gamma$ is finer than $\pi\gamma$. Then for each subset A of X,

$$\begin{aligned} a_\gamma(A) &\subsetneq a_\gamma(a_\gamma(A)) \\ &= a_{\gamma\cdot\gamma}(A), \quad \text{as noted above} \\ &\subsetneq a_{\pi\gamma}(A) = a_\gamma(A). \end{aligned}$$

Thus $a_\gamma = a_\gamma a_\gamma$. Conversely, if a_γ is idempotent, then

$$a_{\gamma\cdot\gamma} = a_\gamma a_\gamma = a_\gamma = a_{\pi\gamma} .$$

Hence $\gamma\cdot\gamma$ is finer than $\pi\gamma$, since $\pi\gamma$ is the coarsest structure whose adherence operator is exactly a_γ. #

This is probably a convenient place to point out that the processes of solidification [9] and Choquet modification [3] are just different ways of achieving the same end, as D.C. Kent told me in a letter. To be precise, let H be a non-empty segment in F(X). A function T associating with each ψ in H a member $T(\psi)$ of ψ is called an H-cover. Now the set of all H-covers defines a set σH of filters: $\chi \in \sigma H$ iff for each H-cover T, a finite subset $\{\psi_1, \ldots, \psi_k\}$ of H can be found, with

$$T(\psi_1) \cup \ldots \cup T(\psi_k) \in \chi .$$

It is an easy exercise in set theory to prove

(2) $\chi \in \sigma H$ iff every ultrafilter finer than χ belongs to H (that is, iff χ belongs to the Choquet modification of H). #

Here too, ambiguous terminology has arisen. The difficulty is avoided by referring solely to solidifications and solidity, while using lemma (2) above to translate, when necessary.

The solidification $\sigma\gamma$ of a structure γ is obtained by using this construction pointwise. Clearly $\sigma\gamma$ is coarser than γ, but finer than $\pi\gamma$.

Lemma (2) has the following corollary, whose proof is a similar exercise in set theory.

(3) Let a be an adherence operator on X and H a segment, and define $a(H)$ to be the segment generated by the set $\{a(\psi) : \psi \in H\}$. Then

$$a(H) \subseteq a(\sigma H) \subseteq \sigma a(H) ,$$

and in particular, the same ultrafilters belong to all three segments.#

Finally, §1 ends with a corollary of lemma (1) which turns out to be surprisingly useful later.

(4) Suppose that γ is a structure satisfying

(i) $\sigma\gamma = \pi\gamma$ and

(ii) $\gamma\cdot\gamma$ is finer than $\pi\gamma$.

Then $\sigma\gamma$ is topological.

Proof: Any principal structure whose adherence operator is idempotent is topological.#

§2 COMPACTNESS

Throughout, compactness means topological compactness, that is, X_γ is a compact space iff every ultrafilter on X belongs to some γ_x. In the proof of his result, C.H. Cook showed in effect that if an ultrafilter belongs to $\gamma\cdot\gamma_x$ then it already belongs to γ_x. Equivalently, $\gamma\cdot\gamma$ is finer than $\sigma\gamma$. Under the same conditions (namely that X_γ is compact and T_3), a similar method shows that any ultrafilter in $\pi\gamma_x$ belongs to γ_x, or in other words, that $\pi\gamma = \sigma\gamma$. Applying lemma (4), one sees that X_γ is nearly topological, in the sense that its solidification is topological. The main purpose of this section is to derive results which can be similarly coupled to lemma (4).

From now on, let γ be a fixed structure on X, and a its adherence operator. First it is well known that if ψ belongs to γ_x then $\dot{x}$ is finer than $a(\psi)$. But as a is the adherence operator of $\pi\gamma$ as well, this proves

(5) $\quad \psi \in \pi\gamma_x \Longrightarrow \dot{x} \supseteq a(\psi)$. #

(In fact, lemma (5) has a partial converse: namely, if ψ is an ultrafilter such that $\dot{x}$ is finer than $a(\psi)$, then it belongs to $\pi\gamma_x$.)

Together with (4), the next two lemmas form a powerful tool for studying a wide range of compact spaces.

(6) $\quad \gamma_x \cap a(\gamma_y) = \emptyset \Longrightarrow \pi\gamma_x \cap \gamma_y = \emptyset$.

Proof: Suppose that ψ belongs to both $\pi\gamma_x$ and γ_y . By (5), $\dot{x}$ is finer than $a(\psi)$, and so $\dot{x}$ belongs to both $a(\gamma_y)$ and γ_x . #

(7) $\quad \gamma_x \cap a(\gamma_y) = \emptyset \Longrightarrow \gamma\cdot\gamma_x \cap \gamma_y = \emptyset$.

Proof: Assume that γ_x and $a(\gamma_y)$ are disjoint, but that χ belongs to both $\gamma\cdot\gamma_x$ and γ_y. By definition, there is a γ-selection ϕ and a filter ψ in γ_x, with $\chi \supseteq \phi\cdot\psi$. The first assumption shows that there can be no proper filter finer than both ψ and $a(\chi)$. This means that sets P in ψ and K in χ can be found, such that $P \cap a(K) = \emptyset$. Thus each $z \in P$ lies outside $a(K)$, and in particular, $K' \in \phi_z$ since $\phi_z \in \gamma_z$. In other words,

$$K' \in \phi\cdot P \subsetneq \phi\cdot\psi \subsetneq \chi .$$

However, this completes the proof, by contradicting the fact that K belongs to χ.#

The first application is in the category of solid T_3 spaces, and yields a theorem which could also be proved using the method of [8] almost without change.

Theorem 1: Any compact solid T_3 space is a compact Hausdorff topological space.

Proof: Take such a space X_γ. First, let ψ be an ultrafilter belonging to $\gamma\cdot\gamma_x$. By compactness, $\psi \in \gamma_y$ for some y. However since X_γ is T_3, if $x \neq z$ then

$$\gamma_x \cap a(\gamma_z) = \gamma_x \cap \gamma_z = \emptyset ,$$

and so $\gamma\cdot\gamma_x \cap \gamma_z = \emptyset$, by lemma (7). Thus $x = y$, showing that $\gamma\cdot\gamma$ is finer than $\sigma\gamma$. Similarly, one can show using lemma (6) that $\pi\gamma$ is finer than $\sigma\gamma$. Thus as $X_\gamma = X_{\sigma\gamma}$ by assumption, X_γ is topological by (4).#

This theorem clearly extends those of [2], [3] and [8], since principal spaces are solid. Moreover, it can be used to attack [1, Satz 9] as well: this states that for any c-embedded space X_γ, the following conditions are equivalent -

(i) X_γ is compact,

(ii) X_γ is compact Hausdorff and topological, and

(iii) $C_c(X_\gamma)$ is a Banach algebra.

Binz's proof of "(i) $\Longrightarrow$ (ii)" used the path "(i) $\Longrightarrow$ (iii) $\Longrightarrow$ (ii)"; a slightly more direct proof is now possible, as outlined below.

By [5, Theorem 2.4], [7, Satz 4] or [9, Theorem 3.6], every c-embedded space is solid, Hausdorff and w-regular, and in particular, T_3. Thus theorem 1 applies to any compact c-embedded space.

An alternative proof uses the fact that c-embedded spaces are solid and functionally Hausdorff. Thus by [3, Theorem 8], if X_γ is compact and c-embedded, γ can be identified with the compact Hausdorff weak topology on X.

§3 RELAXATION

To extend theorem 1 to cope with spaces which are not necessarily solid, Hausdorff or regular, one must unfortunately give a list of regularity axioms (R) and symmetry axioms (S). To complete the picture and avoid ambiguity, a few separation axioms (T) are given as well. Some of these are filter versions of corresponding axioms in topology, while others are suggested by lemmas (6) and (7). In this list, X_γ is a space, with x and y ranging over X.

$R_0 : \dot{y} \notin \gamma_x \Longrightarrow \dot{x} \notin \gamma_y$

$R_1 : \dot{y} \notin \gamma_x \Longrightarrow \gamma_x \cap \gamma_y = \emptyset$

$R_{1.1} : \dot{y} \notin \gamma_x \Longrightarrow \pi\gamma_x \cap \gamma_y = \emptyset$

$R_{1.2} : \dot{y} \notin \gamma_x \Longrightarrow \gamma_x \cap a(\gamma_y) = \emptyset$

$R_{1.5} : \dot{y} \notin \gamma_x \Longrightarrow a(\gamma_x) \cap a(\gamma_y) = \emptyset$

$R_2 : a(\gamma_x) = \gamma_x$, for all x in X

$S_0 : \dot{y} \in \gamma_x \Longrightarrow \gamma_y \subseteq \gamma_x$

$S_1 : \dot{y} \in \gamma_x \Longrightarrow \gamma_y = \gamma_x$

$T_0 : \dot{y} \in \gamma_x$ and $\dot{x} \in \gamma_y \Longrightarrow x = y$

$T_1 : \dot{y} \in \gamma_x \Longrightarrow x = y$

$T_2 : \gamma_x \cap \gamma_y = \emptyset$ unless $x = y$

$T_3 : T_0$ and R_2

The following facts will be needed, some of which are well known, and all of which are easily proved either directly, or with the help of lemmas (3), (5) and (6).

(i) $R_2 \Longrightarrow R_{1.5} \Longrightarrow R_{1.2} \Longrightarrow R_{1.1} \Longrightarrow R_1 \Longrightarrow R_0$.

(ii) Even in the context of topological spaces, $R_{1.5}$ is strictly weaker than R_2; for example, it is well known that a functionally Hausdorff topological space may fail to be R_2, but it is $R_{1.5}$.

(iii) R_0 and $T_0 \Longleftrightarrow T_1$.

(iv) R_1 and $T_0 \Longleftrightarrow T_2$.

(v) R_0 and $S_0 \Longleftrightarrow S_1$.

(vi) The properties (R), (S) and (T) are all preserved under solidification.

(vii) A space is $R_{1.5}$ if it is k-regular.

(The idea of k-regularity arose in [2], where it was shown that a space is k-regular iff its principal modification is R_2.)

Now by putting all these ideas together in the obvious way, one obtains the following results, which significantly extend their earlier counterparts.

Theorem 2 : The solidification of any compact $R_{1.2}$ and S_0 space is a completely regular (not necessarily Hausdorff) topological space.

Proof: Let X_γ be a compact space which is $R_{1.2}$ and S_0. First, suppose ψ is an ultrafilter in $\pi\gamma_x$. By compactness, ψ belongs to γ_y for some y.

Thus by (6), γ_x and $a(\gamma_y)$ are not disjoint, and hence $\dot{y} \in \gamma_x$ $(R_{1.2})$. Consequently, $\psi \in \gamma_x$ (S_0). Thus $\pi\gamma_x \subseteq \sigma\gamma_x$, or in other words, $\pi\gamma = \sigma\gamma$. The second step is similar, with (7) showing that the other condition of (4) is satisfied as well.

Thus $X_{\sigma\gamma}$ is topological: more, it is R_2. By (3), for any y the segments $a(\gamma_y)$ and $a(\sigma\gamma_y)$ share the same ultrafilters. So take an ultrafilter ψ in $a(\gamma_y)$. Again by compactness, it belongs to some γ_x, and so $\dot{y} \in \gamma_x$ $(R_{1.2})$. Thus as X_γ is S_1, ψ belongs to γ_y. It follows that $a(\sigma\gamma_y) \subseteq \sigma\gamma_y$, as desired.

In short, $X_{\sigma\gamma}$ is a compact R_2 topological space, and hence completely regular [4, page 138 Theorem 5, and page 110 Theorem 7]. #

Corollary 1 : The solidification of any compact T_3 space is a compact Hausdorff topological space.

Corollary 2 : Any compact solid space which is $R_{1.2}$ and S_0 is a completely regular topological space.

Corollary 3 : The solidification of a compact k-regular S_0 space is a completely regular topological space.

Corollary 4 : Any compact solid k-regular T_0 space is a compact Hausdorff topological space.

References:

[1] E. Binz "Kompakte Limesräume und limitierte Funktionenalgebren" Comm. Math. Helv. 43 (1968), 195-203.

[2] A.C. Cochran and R.B. Trail "Regularity and complete regularity for convergence spaces" in Lecture Notes in Mathematics 375, 64-70 Springer (Berlin) 1974.

[3] C.H. Cook "Compact Pseudo-Convergences" Math. Ann. 202 (1973), 193-202.

[4] S.A. Gaal "Point Set Topology" Academic Press (New York, London) 1964.

[5] D.C. Kent, K. McKennon, G.D. Richardson and M. Schroder "Continuous convergence in C(X)" Pac. J. Math 52 (1974), 457-465.

[6] H.J. Kowalsky "Limesräume und Komplettierung" Math. Nachr. 11 (1954), 143-186.

[7] B. Müller "L_c- und c-einbettbare Limesräume" To appear.

[8] G.D. Richardson and D.C. Kent "Regular compactifications of convergence spaces" Proc. A.M.S. 31 (1972), 571-573.

[9] M. Schroder "Solid Convergence Spaces" Bull. Austral. Math. Soc. 8 (1973), 443-459.

[10] M. Schroder "Adherence operators and a way of multiplying convergence structures" Mathematics Preprint 29, (1975) University of Waikato.

[11] O. Wyler "Filter space monads, regularity, completions" in Lecture Notes in Mathematics 378, 591-637 Springer (Berlin) 1974.

Mathematics Department,
University of Waikato,
Hamilton, New Zealand.

Differential Calculus and Cartesian Closedness

by

Ulrich Seip

Cartesian closed categories play an important rôle in many aspects of mathematics. They appear in modern forms of Algebraic Geometry, in Logic, in Topology. This leads quite naturally to the idea of trying to use this notion also for Differential Calculus.

The first attempts in this direction were undertaken by A.Bastiani [1], and then by A.Froelicher-W.Bucher [4]. They all used the notion of limit spaces to generalize calculus in order to be able to form function spaces of differentiable maps between limit vector spaces. In retrospect one may say that A.Bastiani used a "good" definition of continuous differentiability but the category she chose did not allow the desired cartesian closedness property. A.Froelicher and W.Bucher took cartesian closedness exactly as their goal. But because of their "bad" definition of continuous differentiability, their special types of limit vector spaces became increasingly complicated.

To me it seemed always desirable to establish differential calculus in a pure topological setting. This because questions of continuity and differentiability are local questions and topological spaces have by definition a local structure. So the question was, whether or not a differential calculus can be established in a topological setting so as to obtain cartesian closedness in the infinitely often differentiable case.

For a long time it even seemed impossible to establish anything like cartesian closedness for continuous maps - not to speak of differentiable ones. But in 1963 Gabriel and Zisman proved [6] that the category $\mathcal{CG}$ of compactly generated hausdorff spaces is cartesian closed - and this is a full subcategory of topological spaces.

Starting from this category one is quite naturally led to the investigation of the category $\mathcal{CGV}$ of compactly generated vector spaces and continuous linear maps. Observing that the Hahn-Banach theorem is the tool for proving the so-called mean value theorem of differential calculus, it becomes clear that not all compactly generated vector spaces can be used for a differential calculus if one wants the basic well known theorems of calculus to hold (all of them are consequences of the mean value theorem). But there is a nice complete and cocomplete full subcategory of $\mathcal{CGV}$ which is in a one-to-one correspondence with a full subcategory of convex vector spaces. The objects of this category $\mathcal{CGV}^{\#}$ are suitable for a differential calculus, but they do still not allow to prove such an important fact as the existence of a primitive map for a given continuous map $\alpha:\mathbb{R}\to E$. To obtain this, one has to impose sequential completeness on the compactly generated vector spaces under consideration, and this leads to the category $\widehat{\mathcal{CGV}}$ of convenient compactly generated vector spaces.

The objects of this category then form the base for our differential calculus. Our notion of (continuous) differentiability is the simplest one possible: We say that a map $\alpha:E\supset U\to F$ is (continuously) differentiable if it has a Gâteaux derivative $D\alpha:U\to L(E,F)$ which is continuous (Gâteaux derivative stands for directional derivative in every direction). Since $\mathcal{CG}$ is cartesian closed, the convenient vector spaces do not only behave well for continuous linear maps but also for continuous maps. From this follows immediately that the notions of "weak" and "strong" differentiability coincide in our setting. And because banach and fréchet vector spaces are also convenient vector spaces, our calculus becomes a generalisation of the well known fréchet calculus.

The main theorem of our differential calculus for convenient vector spaces states that the category of convenient vector spaces and smooth (infinitely differentiable) maps is cartesian closed.

The article is divided into 5 sections:

Section 1: Consists mainly of the Gabriel-Zisman theorem for $\mathcal{CG}$ and the Kelley theorem which says that for X compactly generated and U complete, the compact-uniform topology CU(X,U) for continuous maps from X to the topological space underlying U is again complete.

Section 2: The general theory of compactly generated vector spaces and continuous linear maps. The main theorem states that the categories $\mathcal{CGV}^{\#}$ and $\widehat{\mathcal{CGV}}$ are complete, cocomplete, additive, have an internal functor L and a tensor product functor $\otimes$.

Section 3: Developes the differential calculus and gives the proofs of the basic theorems like: Differentiability implies continuity, functoriality of the tangent operator, differentiable maps into a product and from a finite product, convergence theorem of differentiable maps, symmetry of higher derivatives, existence of primitive maps. Poincaré lemma and Stoke theorem are left to the reader.

Section 4: Introduces convenient vector space structures for differentiable and smooth function spaces. The highlight is theorem 4.9, stating that the category of convenient real vector spaces and smooth maps is cartesian closed.

Section 5: Contains other results and problems. Mainly an attempt is made to extend theorem 4.9 to the case of smooth manifolds.

1. <u>Topological Background</u> We denote by $\mathcal{H}$ the category of hausdorff spaces and by $\mathcal{CG}$ the full subcategory of compactly generated spaces. We remind the reader that a hausdorff space is called compactly generated if it carries the final topology with respect to the inclusions of its compact subspaces. To each hausdorff space X we associate a compactly generated space CG(X) with the same points as X, by requiring that CG(X) carries the final topology with respect to the inclusions of the compact subspaces of X. Hence the identity function $1: CG(X) \to X$ becomes continuous and CG becomes a coreflector $CG: \mathcal{H} \to \mathcal{CG}$, not changing the functions with respect to the underlying sets. Since $\mathcal{H}$ is complete and cocomplete, it follows that $\mathcal{CG}$ is complete and cocomplete.

In order to avoid notational difficulties, we shall denote by the symbol $\times$ the usual topological product (with respect to $\mathcal{H}$), whereas the symbol $\sqcap$ stands for the product with respect to $\mathcal{CG}$ calculated as $\sqcap = CG \circ \times$.

From elementary topology we recall that the compact-open topology on function spaces of continuous maps defines an internal functor $CO: \mathcal{H}^{op} \times \mathcal{H} \to \mathcal{H}$. Hence $CG \circ CO = C: \mathcal{CG}^{op} \times \mathcal{CG} \to \mathcal{CG}$ is an internal functor for $\mathcal{CG}$.

<u>Proposition 1.1.</u> Let X be compactly generated and S a subset of X. If S is open or closed in X, the subspace topology on S is compactly generated.

<u>Proof.</u> Clearly a hausdorff space is compactly generated iff it carries the final topology with respect to the inclusions of its locally compact subspaces. If S is open or closed in X, the intersection subspaces $S \cap K$ are locally compact for every compact subset K of X. The proposition follows.

<u>Lemma 1.2.</u> Let X be compactly generated, Y hausdorff. Then the evaluation map $\varepsilon: CG \circ CO(X,Y) \sqcap X \to Y$, defined by $(\alpha, x) \mapsto \alpha(x)$, is continuous.

<u>Proof.</u> The domain of ε is compactly generated. Hence it suffices to prove continuity on compact subspaces. Every compact subset being contained in the product of its projections, we may restrict ourselves to prove the continuity of ε on compact subspaces of type $K \times L$, where K is compact in $CG \circ CO(X,Y)$ and L is compact in X. If $(\alpha, x) \in K \times L$ and $\alpha(x) \in U$ with U open in Y, we obtain in $L \cap \alpha^{-1} U$ a relative-open neighborhood of x in L. L being compact, there exists a relative-open neighborhood V of x in L such that the L-closure $\overline{V}$ is contained in $L \cap \alpha^{-1} U$. Hence $(\alpha, x) \in [(\overline{V}, U) \cap K] \times V$ and ε maps this relative-open neighborhood of (α, x) into U.

<u>Lemma 1.3.</u> Let X and Y be compactly generated. Then the map $\eta: X \to C(Y, X \sqcap Y)$, defined by $\eta(x): y \mapsto (x,y)$, is continuous.

<u>Proof.</u> The domain of η is compactly generated and the category $\mathcal{CG}$ is

coreflective in $\mathcal{H}$. Hence it suffices to show that $\eta: X \to CO(Y, X \sqcap Y)$ is continuous on compact subspaces K of X. If (L,U) denotes a subbasis open subset of $CO(Y, X\sqcap Y)$ with L compact in Y and U open in $X \sqcap Y$, we clearly have $[\eta^{-1}\complement(L,U)] \cap K = proj_X[(K\times L)\cap \complement U]$, whence continuity of η on K follows.

From the lemmas we obtain the theorems:

<u>Theorem 1.4</u> (Gabriel-Zisman). The category $\mathcal{CG}$ is complete, cocomplete, and cartesian closed with $C: \mathcal{CG}^{op}\times\mathcal{CG}\to\mathcal{CG}$ as internal functor. For each object X the natural transformations $\eta_X: 1_{\mathcal{CG}} \to C(X, -\sqcap X)$ and $\varepsilon_X: C(X,-)\sqcap X \to 1_{\mathcal{CG}}$ are the unit and counit of the cartesian adjointness. Hence a function $\alpha: X \to C(Y,Z)$ is continuous iff the corresponding function $\hat{\alpha} = \varepsilon\circ(\alpha\sqcap 1): X\sqcap Y \to Z$, defined by $\hat{\alpha}(x,y) = \alpha x(y)$, is continuous.

<u>Theorem 1.5</u> (Steenrod). The following diagram commutes:

$$\begin{array}{ccc} \mathcal{CG}^{op}\times\mathcal{H} & \xrightarrow{CO} & \mathcal{H} \\ \downarrow{\scriptstyle 1\times CG} & & \downarrow{\scriptstyle CG} \\ \mathcal{CG}^{op}\times\mathcal{CG} & \xrightarrow{C} & \mathcal{CG} \end{array}$$

Theorem 1.4 is an immediate consequence of the lemmas. To prove theorem 1.5, it suffices to prove commutativity on objects. Clearly $C(X, CG\ Y)$ and $CG\circ CO(X,Y)$ have the same underlying sets. The identity function $1: CG\ Y \to Y$ being continuous, we obtain immediately the continuity of the identity function $1: C(X, CG\ Y) \to CG\circ CO(X,Y)$.- By lemma 1.2 the evaluation $\varepsilon: CG\circ CO(X,Y)\sqcap X \to Y$ is continuous, whence by coreflectiveness this is also true for the evaluation $\varepsilon: CG\circ CO(X,Y)\sqcap X \to CG\ Y$. By theorem 1.4 the continuity of $1: CG\circ CO(X,Y) \to C(X, CG\ Y)$ follows.

We shall now exhibit relations between the category $\mathcal{CG}$ and categories of uniform spaces. First we note that there is an adjoint relation between the complete and cocomplete categories $\mathcal{H}$ and $\mathcal{HU}$, the latter denoting the category of hausdorff uniform spaces and uniformly continuous maps. This, because the usual topologizing functor $H: \mathcal{HU} \to \mathcal{H}$ clearly preserves limits and the solution set condition for application of the adjoint functor theorem evidently can be satisfied. We shall denote by $HU: \mathcal{H} \to \mathcal{HU}$ the coadjoint to H.

Now observe that for a hausdorff space X and a hausdorff uniform space U the function space $CO(X, H\ U)$ is uniformizable as follows: If K is compact in X and V an entourage of U, we define (K,V) by $(K,V) = \{(\alpha_1,\alpha_2)\in CO(X,H\ U)\times CO(X,H\ U) \mid (\alpha_1 x, \alpha_2 x)\in V \text{ for all } x\in K\}$. These sets (K,V) then form a subbase of a uniformity on the set of continuous maps from X to H(U). The resulting uniform space will be denoted by $CU(X,U)$. Then CU becomes a functor $CU: \mathcal{H}^{op}\times\mathcal{HU} \to \mathcal{HU}$ in the obvious way and the following proposition follows:

Proposition 1.6. The category $\mathcal{HU}$ of hausdorff uniform spaces is complete and cocomplete. The topologizing functor $H:\mathcal{HU}\to\mathcal{H}$ has a coadjoint functor $HU:\mathcal{H}\to\mathcal{HU}$. Furthermore, the compact-uniform functor $CU:\mathcal{H}^{op}\times\mathcal{HU}\to\mathcal{HU}$ renders the diagram

$$\begin{array}{ccc} \mathcal{H}^{op}\times\mathcal{HU} & \xrightarrow{CU} & \mathcal{HU} \\ \downarrow{\scriptstyle 1\times H} & & \downarrow{\scriptstyle H} \\ \mathcal{H}^{op}\times\mathcal{H} & \xrightarrow{CO} & \mathcal{H} \end{array}$$

commutative.

The category $\mathcal{HU}$ contains two important full reflective subcategories: The category $\mathcal{H}\hat{\mathcal{U}}$ of sequentially complete hausdorff uniform spaces, and the category $\widehat{\mathcal{HU}}$ of complete hausdorff uniform spaces. The reflectors $\frown:\mathcal{HU}\to\mathcal{H}\hat{\mathcal{U}}$ and $\wedge:\mathcal{HU}\to\widehat{\mathcal{HU}}$ can be constructed by using the adjoint functor theorem. The fact that a dense subspace D of a hausdorff space X implies card $X\leqslant 2^{2^{\text{card } D}}$ may be used to construct solution sets. The explicit construction of the completion $\hat{U}$ of a hausdorff uniform space U in terms of minimal cauchy filters can be found in [2]. Clearly U may be considered a uniform subspace of $\hat{U}$ and $\hat{U}$ as the intersection of all sequentially complete subspaces of $\hat{U}$ containing U.

Now again the category $\mathcal{CG}$ enters the considerations. Kelley proved that if X is compactly generated and U is complete uniform, then $CU(X,U)$ is again complete (whereas this is not true for arbitrary hausdorff spaces X).

Theorem 1.7 (Kelley). The functor $CU:\mathcal{H}^{op}\times\mathcal{HU}\to\mathcal{HU}$ factors by restriction to $\mathcal{CG}^{op}\times\mathcal{H}\hat{\mathcal{U}}$ through $\mathcal{H}\hat{\mathcal{U}}$, and by restriction to $\mathcal{CG}^{op}\times\widehat{\mathcal{HU}}$ through $\widehat{\mathcal{HU}}$. Hence the following diagram commutes:

$$\begin{array}{ccc} \mathcal{CG}^{op}\times\mathcal{H}\hat{\mathcal{U}} & \xrightarrow{CU} & \mathcal{H}\hat{\mathcal{U}} \\ \downarrow & & \downarrow \\ \mathcal{CG}^{op}\times\widehat{\mathcal{HU}} & \xrightarrow{CU} & \widehat{\mathcal{HU}} \\ \downarrow & & \downarrow \\ \mathcal{CG}^{op}\times\mathcal{HU} & \xrightarrow{CU} & \mathcal{HU} \\ {\scriptstyle 1\times(CG\circ H)}\downarrow & & \downarrow{\scriptstyle CG\circ H} \\ \mathcal{CG}^{op}\times\mathcal{CG} & \xrightarrow{C} & \mathcal{CG} \end{array}$$

Proof. Let X be compactly generated, U hausdorff uniform, and $\mathcal{F}$ be a given cauchy filter on $CU(X,U)$. Then there exists for any subbasis entourage (K,V) of $CU(X,U)$ a set $F\in\mathcal{F}$ with $F\times F\subset(K,V)$. If $K=\{x\}$, we obtain $F(x)\times F(x)\subset V$, whence for every $x\in X$ the image $\varepsilon_x(\mathcal{F})$ of $\mathcal{F}$ under evaluation $\varepsilon_x:CU(X,U)\to U$ at x is a cauchy filter on U. To prove the theorem we may therefore assume that the function $\lambda:X\to H(U)$, defined by $\lambda:x\mapsto\lim\varepsilon_x(\mathcal{F})$, exists. Let us prove that this function is continuous: Since X is compactly generated, it suffices to prove continuity of λ on compact subsets K of X. So let $\mathcal{G}$ be a convergent filter on K with $\lim\mathcal{G}=x_o$. Then there exists to any continuous map $\alpha:X\to H(U)$ and any entourage V of U a set $G_\alpha\in\mathcal{G}$ such that $(\alpha x,\alpha x_o)\in V$ for all $x\in G_\alpha$.

If V is any symmetric entourage of U, we choose first a set $F\in\mathcal{F}$ with $F\times F\subset(K,V)$, and then we select for a fixed $\alpha\in F$ a $G_\alpha\in\mathcal{G}$ with the property mentioned before. Hence we have for all $x\in G_\alpha$ that $(\lambda(x),\lambda(x_0))\in V^5$. Because every entourage W contains a symmetric entourage V with $V^5\subset W$, the continuity of $\lambda:X\to H(U)$ follows.- To complete the proof of the theorem, we observe that for a symmetric entourage V of U and a set $F\in\mathcal{F}$ with $F\times F\subset(K,V)$ we have $(\lambda x,\alpha x)\in V^2$ for all $\alpha\in F$ and all $x\in K$. This, because by definition of λ exists for every $x\in K$ a map $\alpha_x\in F$ such that $(\lambda x,\alpha_x x)\in V$. It follows $F\subset(K,V)[\lambda]$, whence $\lambda=\lim\mathcal{F}$.

We end this short discussion by exhibiting the adjoint relationships between the various categories $\mathcal{CG}$, $\mathcal{H}$, $\mathcal{HU}$, $\mathcal{H}\hat{\mathcal{U}}$, $\widehat{\mathcal{HU}}$ in form of a diagram.

<u>Proposition 1.8</u>. The categories $\mathcal{CG}$, $\mathcal{H}$, $\mathcal{HU}$, $\mathcal{H}\hat{\mathcal{U}}$, $\widehat{\mathcal{HU}}$ are all complete and cocomplete. They are related to each other according to the following diagram:

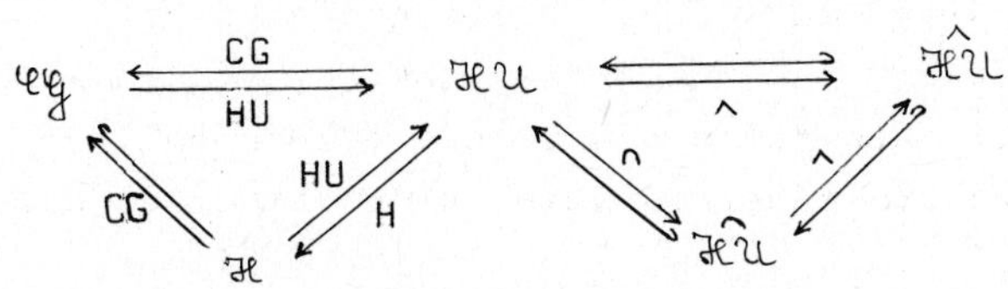

In this diagram commute the inner and the outer triangles (we have shortened the notation $CG\circ H$ to CG). Moreover, every pair of functors in opposite direction is an adjoint pair with the "outer" functor adjoint to its "inner" counterpart.

2. <u>Compactly Generated Real or Complex Vector Spaces</u> We begin by reviewing the main properties of locally convex hausdorff topological vector spaces (short: convex vector spaces) over the field $\mathbb{F}$, with $\mathbb{F}$ either denoting the real numbers $\mathbb{R}$ or the complex numbers $\mathbb{C}$. The category of these spaces with the continuous linear maps as arrows will be denoted by $\mathcal{LCV}$.

We observe the existence of the functor $CO:\mathcal{H}^{op}\times\mathcal{LCV}\to\mathcal{LCV}$, defined as follows: If X is a hausdorff space and M a convex vector space, the underlying vector space structure of $CO(X,M)$ is obtained by pointwise addition and scalar multiplication of continuous maps from X to M and the convex topology on this vector space is the compact-open topology. On arrows CO is defined by composition. Analogously we obtain the functor $LCO:\mathcal{LCV}^{op}\times\mathcal{LCV}\to\mathcal{LCV}$ with $LCO(M,N)$ the vector space of continuous linear maps from M to N equipped with the compact-open topology. Finally we see that we have for any $n\in\mathbb{N}$ corresponding functors $L^nCO:(\overset{n}{\times}\mathcal{LCV})^{op}\times\mathcal{LCV}\to\mathcal{LCV}$, where $L^nCO(M_1,\dots,M_n;N)$ is the vector space of n-linear continuous maps $\lambda: \times_{i=1}^{n} M_i \to N$ equipped with the compact-open topology.

There are two outstanding full reflective subcategories of $\mathcal{LCV}$: The first is the category $\mathcal{L}\widehat{\mathcal{CV}}$ of sequentially complete, the second is the category $\mathcal{L}\hat{\mathcal{CV}}$ of complete convex vector spaces. The reflectors $\wedge: \mathcal{LCV} \to \mathcal{L}\widehat{\mathcal{CV}}$ and $\wedge: \mathcal{LCV} \to \mathcal{L}\hat{\mathcal{CV}}$ are constructed as for hausdorff uniform spaces. Using Kelley's theorem 1.7, we obtain immediately:

Theorem 2.1. The functor $CO: \mathcal{H}^{op} \times \mathcal{LCV} \to \mathcal{LCV}$ factors by restriction to $\mathcal{CG}^{op} \times \mathcal{L}\widehat{\mathcal{CV}}$ through $\mathcal{L}\widehat{\mathcal{CV}}$, and by restriction to $\mathcal{CG}^{op} \times \mathcal{L}\hat{\mathcal{CV}}$ through $\mathcal{L}\hat{\mathcal{CV}}$. Hence the following diagram commutes:

$$\begin{array}{ccc}
\mathcal{CG}^{op} \times \mathcal{L}\hat{\mathcal{CV}} & \xrightarrow{\ CO\ } & \mathcal{L}\hat{\mathcal{CV}} \\
\downarrow & & \downarrow \\
\mathcal{CG}^{op} \times \mathcal{L}\widehat{\mathcal{CV}} & \xrightarrow{\ CO\ } & \mathcal{L}\widehat{\mathcal{CV}} \\
\downarrow & & \downarrow \\
\mathcal{CG}^{op} \times \mathcal{LCV} & \xrightarrow{\ CO\ } & \mathcal{LCV} \\
\downarrow & & \| \\
\mathcal{H}^{op} \times \mathcal{LCV} & \xrightarrow{\ CO\ } & \mathcal{LCV}
\end{array}$$

The main theorem for $\mathcal{LCV}$ is the theorem of Hahn-Banach. Together with the other relevant properties of $\mathcal{LCV}$ one has:

Theorem 2.2 (Hahn-Banach etc.). The categories $\mathcal{LCV}, \mathcal{L}\widehat{\mathcal{CV}}, \mathcal{L}\hat{\mathcal{CV}}$ are complete, cocomplete, additive. The ground field $\mathbb{F}$ is a generator and a cogenerator for each of these categories. More precise: If K is any closed convex subset of a convex vector space M and if $x \notin K$, then there exists a continuous linear map $\lambda: M \to \mathbb{F}$ such that $\lambda(x) \notin \overline{\lambda K}$. Further: For every $n \in \mathbb{N}$, the compact-open topology gives a functor $L^n CO: (\overset{n}{\times} \mathcal{LCV})^{op} \times \mathcal{LCV} \to \mathcal{LCV}$ with $L^n CO(M_1, \dots, M_n; N)$ the convex vector space of n-linear continuous maps with the compact-open topology.

The proof of this theorem is standard and can be found in any good textbook on topological vector spaces [9].

What happens to a convex vector space M if we apply the functor CG to its topological structure? Since CG preserves underlying sets and products, we may - and shall - consider CG(M) as the same vector space with addition and scalar multiplication now continuous with respect to the compactly generated product. Since the ground field $\mathbb{F}$ is locally compact, it is already itself a compactly generated vector space (as it is a convex one).

More generally, we say that a vector space E over $\mathbb{F}$ and equipped with a compactly generated topology, is a compactly generated vector space if addition and scalar multiplication are continuous maps with respect to the $\mathcal{CG}$-product $\sqcap$. It is easy to see that $E \sqcap \mathbb{F} = E \times \mathbb{F}$ since the latter already is compactly generated because $\mathbb{F}$ is locally compact. We denote by $\mathcal{CGV}$ the category of compactly generated vector spaces and continuous linear maps.

Evidently we have a functor CG: $\mathcal{LCV} \to \mathcal{CGV}$, defined on objects as described before and not changing the linear maps underlying arrows of $\mathcal{LCV}$. We denote by $\mathcal{CGV}^{\#}$ the full subcategory of $\mathcal{CGV}$ generated by all compactly generated vector spaces of type CG(M) with M any convex vector space. Further we denote by $\overset{\frown}{\mathcal{CGV}}$ the full subcategory of $\mathcal{CGV}^{\#}$ generated by all compactly generated vector spaces of type CG(M) with M any sequentially complete convex vector space. Clearly these definitions suggest to define $\widehat{\mathcal{CGV}}$ as the full subcategory with objects of type CG(M) where M is complete. But we shall see that this category is troublesome.

Now we define a functor LC: $\mathcal{CGV}^{\#} \to \mathcal{LCV}$ as follows: If E is in $\mathcal{CGV}^{\#}$, LC(E) has the same underlying vector space as E and carries the topology generated by the convex open subsets of E. One proves easily that addition and scalar multiplication remain continuous for LC(E) and the ordinary topological product. Since E=CG(M) for some convex vector space M, the topology of LC(E), being finer than the topology of M, is hausdorff.

We prove now that the restriction of LC to $\overset{\frown}{\mathcal{CGV}}$ factors through $\overset{\frown}{\mathcal{CGV}}$: For this, let (x_n) be a cauchy sequence in LC(E), where E=CG(M) with M sequentially complete. Since the topology of LC(E) is finer than the topology of M, the sequence (x_n) is also a cauchy sequence in M, hence convergent to some vector x_o in M. The set $\{x_n | n \in \mathbb{N}\} \cup \{x_o\}$ is then compact in M and hence compact in E=CG(M). Consequently (x_n) also converges in E to x_o. Because the topology of E is finer than the topology of LC(E), we finally see that the sequence (x_n) converges to x_o in LC(E).

If we try to do the same for $\widehat{\mathcal{CGV}}$, we clearly run into problems. The ensuing difficulties can not simply be circumvented by applying afterwards the completion functor $\wedge: \mathcal{LCV} \to \widehat{\mathcal{LCV}}$, because this functor generally enlarges the underlying vector spaces.

<u>Lemma 2.3</u>. The following diagram commutes:

$$\begin{array}{ccccc} \overset{\frown}{\mathcal{LCV}} & \xrightarrow{CG} & \overset{\frown}{\mathcal{CGV}} & \xrightarrow{LC} & \overset{\frown}{\mathcal{LCV}} \\ \downarrow & & \updownarrow & & \downarrow \\ \mathcal{LCV} & \xrightarrow{CG} & \mathcal{CGV}^{\#} & \xrightarrow{LC} & \mathcal{LCV} \end{array}$$

Moreover, the functor CG is left inverse and adjoint to the functor LC.

<u>Proof</u>. We are only left with the proof that CG is left inverse and adjoint to LC. If E=CG(M), we have already seen that LC(E) has a topology finer than M. Hence CG∘LC(E) has a topology finer than CG(M)=E. On the other hand, the topology of E is certainly finer than the topology of LC(E). Hence also finer than the topology of CG∘LC(E). It follows that CG is left inverse to LC.- To prove adjointness, we show

that $\mathcal{LCV}(LC\ E,M)=\mathcal{CGV}^{\#}(E,CG\ M)$ for any E in $\mathcal{CGV}^{\#}$ and M in $\mathcal{LCV}$: Let $\lambda\in\mathcal{LCV}(LC\ E,M)$. Applying CG and observing that CG is left inverse to LC we get $\lambda\in\mathcal{CGV}^{\#}(E,CG\ M)$. Conversely let $\lambda\in\mathcal{CGV}^{\#}(E,CG\ M)$. Applying LC we get $\lambda\in\mathcal{LCV}(LC\ E,LC\circ CG\ M)$. Since the topology of $LC\circ CG(M)$ is finer than the topology of M, we obtain $\lambda\in\mathcal{LCV}(LC\ E,M)$ as desired.

We now introduce an internal functor $L:\mathcal{CGV}^{op}\times\mathcal{CGV}\to\mathcal{CGV}$ by defining L(E,F) as the vector space of continuous linear maps from E to F, equipped with the subspace topology of C(E,F), where C(E,F) denotes the vector space of continuous maps from E to F with the cartesian closed compactly generated function space topology described in § 1. Since the subspace L(E,F) is evidently closed in C(E,F), we know from proposition 1.1 that the topology of L(E,F) is compactly generated. From the cartesian closedness of $\mathcal{CG}$ we deduce immediately that C(E,F) is a compactly generated vector space. Hence L(E,F) is a compactly generated vector space.

In an analogous way we get a functor $C:\mathcal{CG}^{op}\times\mathcal{CGV}\to\mathcal{CGV}$.

<u>Lemma 2.4</u>. The restriction of the internal functor $L:\mathcal{CGV}^{op}\times\mathcal{CGV}\to\mathcal{CGV}$ to $\mathcal{CGV}^{\#op}\times\mathcal{CGV}^{\#}$ factors through $\mathcal{CGV}^{\#}$, and the restriction of L to $\widehat{\mathcal{CGV}}^{op}\times\widehat{\mathcal{CGV}}$ factors through $\widehat{\mathcal{CGV}}$. Moreover, the following diagram commutes:

$$\begin{array}{ccc} \mathcal{CGV}^{\#op}\times\mathcal{CGV}^{\#} & \xrightarrow{\quad L\quad} & \mathcal{CGV}^{\#} \\ \Big\downarrow{\scriptstyle LC^{op}\times LC} & & \Big\uparrow{\scriptstyle CG} \\ \mathcal{LCV}^{op}\times\mathcal{LCV} & \xrightarrow{\quad LCO\quad} & \mathcal{LCV} \end{array}$$

Analogous statements hold in the multilinear cases.

<u>Proof</u>. First we prove commutativity of the diagram: Since CG preserves initial structures, we have $L(E,F)=CG\circ LCO(E,F)$. Since $CG\circ LC(F)=F$, we get from Steenrod's theorem 1.5 that $CG\circ LCO(E,F)=CG\circ LCO(E,LC\ F)$. Since $CG\circ LC(E)=E$, we get from lemma 2.3 that $LCO(E,LC\ F)=LCO(LC\ E,LC\ F)$. Hence $CG\circ LCO(E,LC\ F)=CG\circ LCO(LC\ E,LC\ F)$. So the diagram commutes, and this also shows that the restriction of L to $\mathcal{CGV}^{\#op}\times\mathcal{CGV}^{\#}$ factors through $\mathcal{CGV}^{\#}$.- If F belongs to $\widehat{\mathcal{CGV}}$, we know that CO(E,LC F) is sequentially complete by theorem 2.1. Hence LCO(E,LC F) is sequentially complete because it is a closed subspace of CO(E,LC F). Hence L(E,F) is of the form CG(M) with M=LCO(E,LC F) sequentially complete.

This sets the stage for proving the main theorem for the categories $\mathcal{CGV}^{\#}$ and $\widehat{\mathcal{CGV}}$:

<u>Theorem 2.5</u>. The categories $\mathcal{CGV}^{\#}$ and $\widehat{\mathcal{CGV}}$ are complete, cocomplete, additive. They are symmetric multiplicatively closed with the ground field as unit. The internal closing functor is L and the multiplicative functor $\otimes$ satisfies $L(E\otimes F,G)\approx L^2(E,F;G)$ with a natural isomorphism. The ground field is a generator and cogenerator for these categories.

Proof. Since CG is left inverse and adjoint to LC, completeness, cocompleteness, additivity follow from theorem 2.2. The same argument shows that the ground field is a generator and cogenerator. The functors $L^n(E_1,\dots,E_n;-)$ are limit preserving for every $n\in\mathbb{N}$ since $C(\prod_{i=1}^{n} E_i,-)$ is limit preserving for $\mathcal{CG}$ and the topology for limits is evidently the one obtained in $\mathcal{CG}$ after application of the forgetful functor. Hence $L^2(E,F;-)$ has a coadjoint functor $T_{(E,F)}$ by the special adjoint functor theorem and we obtain $E\otimes F$ as $T_{(E,F)}(\mathbb{F})$. Clearly $E\otimes F\approx F\otimes E$, $\mathbb{F}\otimes E\approx E\approx E\otimes\mathbb{F}$, $L(\mathbb{F},E)\approx E$, $L(E,L(F,G))\approx L(E\otimes F,G)\approx L^2(E,F;G)$ with natural isomorphisms in all variables.

We define the dual E^* of a compactly generated vector space by $E^*=L(E,\mathbb{F})$. The identity $1:E^*\to E^*$ gives us by adjointness the continuous linear evaluation map $\varepsilon:E^*\otimes E\to\mathbb{F}$, and using the commutativity of tensor products and then again adjointness we obtain the continuous linear map $\sigma:E\to E^{**}$ with $\sigma x:\lambda\mapsto\lambda(x)$. If E is in $\mathcal{CGV}^{\#}$, the map σ is clearly injective because $L(\mathbb{F},E)\approx E$ and $\mathbb{F}$ is a cogenerator for $\mathcal{CGV}^{\#}$. We say that E is embedded in its double dual space if E carries the initial compactly generated topology with respect to $\sigma:E\rightarrowtail E^{**}$.

Theorem 2.6 (Frölicher-Jarchow). Every compactly generated vector space E of $\mathcal{CGV}^{\#}$ is embedded in its double dual. Hence this holds also for the category $\widehat{\mathcal{CG}}\mathcal{V}$.

Proof. We have only to show that the σ-induced compactly generated topology on the vector space underlying E is finer than the topology of E. For convenience, denote by E' the vector space of continuous linear maps from E to $\mathbb{F}$ and put $E^{\vee}=LCO(LC\ E,\mathbb{F})$. By lemma 2.4 we have $E^*=CG(E^{\vee})$. Now consider the injective linear function $\sigma:LC(E)\rightarrowtail LCO(LC\circ CG\ E^{\vee},\mathbb{F})$ and denote by E_σ the σ-induced convex vector space structure on the vector space underlying LC(E). If U is any closed circled convex zero neighborhood of LC(E), the polar U^o of U is defined as $U^o=\{\lambda\in E'\mid |\lambda x|\leq 1 \text{ for all } x\in U\}$. Since U is closed, circled and convex, we have $U=\{x\in LC(E)\mid |\lambda x|\leq 1 \text{ for all } \lambda\in U^o\}=$ $=\sigma^{-1}(U^o,|f|\leq 1)$. Hence U is a zero neighborhood in E_σ if U^o is compact in $LC\circ CG(E^{\vee})$. Since U^o obviously is an equicontinuous subset of E', the topology of U^o is the same whether one considers U^o as a subspace of $E^{\vee}$ or as a subspace of E' with the topology of pointwise convergence. But as a subspace of the latter, U^o clearly is compact. Hence U^o is compact in $E^{\vee}$, whence compact in $LC\circ CG(E^{\vee})$. Since the closed circled convex zero neighborhoods of LC(E) form a zero neighborhood base, it follows that the topology of E_σ is finer than the topology of LC(E). Since the functor CG preserves initial topologies, $CG(E_\sigma)$ has the compactly generated topology induced by $\sigma:E\rightarrowtail E^{**}$, because $E^{**}=CG\circ LCO(LC\circ CG\ E^{\vee},\mathbb{F})$ by lemma 2.4. Hence this topology is finer than the topology of $CG\circ LC(E)=E$.

As has been already mentioned, we have an evident functor $C:\mathcal{CGV}^{op}\times\mathcal{CGV}\to\mathcal{CGV}$. Denoting by $\mathcal{CGV}_{cont}$ the category of compactly generated vector spaces with the continuous maps as arrows, we get also evident functors $C:\mathcal{CGV}^{op}\times\mathcal{CGV}_{cont}\to\mathcal{CGV}_{cont}$ and $C:\mathcal{CGV}^{op}_{cont}\times\mathcal{CGV}_{cont}\to\mathcal{CGV}_{cont}$. Replacing $\mathcal{CGV}$ by $\mathcal{CGV}^{\#}$ or $\widehat{\mathcal{CGV}}$, and replacing $\mathcal{CGV}_{cont}$ by $\mathcal{CGV}^{\#}_{cont}$ or $\widehat{\mathcal{CGV}}_{cont}$ we obtain restrictions of all these functors to the respective full subcategories and all the restricted functors factor through the corresponding subcategories. In case of $\mathcal{CGV}^{\#}$ and $\mathcal{CGV}^{\#}_{cont}$ this follows from Steenrod's theorem 1.5, and for $\widehat{\mathcal{CGV}}$ and $\widehat{\mathcal{CGV}}_{cont}$ this follows then from Kelley's theorem 1.7. And because $CG\circ LC(E)=E$ in these cases, we have moreover that $C(X,E)=CG\circ CO(X,LC\ E)$.

We obtain especially:

Theorem 2.7. The category $\widehat{\mathcal{CGV}}_{cont}$ has arbitrary products which are calculated as in $\widehat{\mathcal{CGV}}$. It is cartesian closed with the internal functor $C:\widehat{\mathcal{CGV}}^{op}_{cont}\times\widehat{\mathcal{CGV}}_{cont}\to\widehat{\mathcal{CGV}}_{cont}$. The natural homeomorphism $\sigma:C(E,C(F,G))\xrightarrow{\simeq}C(E\sqcap F,G)$ is linear. The ground field is a generator and a cogenerator for $\widehat{\mathcal{CGV}}_{cont}$.

3. Differential Calculus for Convenient Vector Spaces over the Reals

From now on we shall fix the ground field to be the real numbers $\mathbb{R}$. We shall call a compactly generated real vector space E a convenient vector space if it is of type CG(M) where M is a sequentially complete convex real vector space. Put differently: A compactly generated real vector space is called convenient if it belongs to $\widehat{\mathcal{CGV}}$.

Definition 3.1. Let E and F be convenient vector spaces. Let U be open in E and V open in F, and let $\alpha:E\supset U\to V\subset F$ be a given function (defined on U with values in V). Then we call α differentiable (on U), if there exists a continuous map $\beta:E\supset U\to L(E,F)$ such that for every fixed $(x,y)\in U\sqcap E$ we have: $\lim\limits_{0\neq t\to 0}\frac{1}{t}\{\alpha(x+ty)-\alpha(x)\}=\beta x(y)$.

Proposition 3.2. If $\alpha:E\supset U\to V\subset F$ is differentiable, the map $\beta:U\to L(E,F)$ with $\beta x(y)=\lim\limits_{0\neq t\to 0}\frac{1}{t}\{\alpha(x+ty)-\alpha(x)\}$ is unique.

This is clear since F is hausdorff.

Definition 3.3. Let $\alpha:E\supset U\to V\subset F$ be differentiable. Then we define:

(i) The derivative $D\alpha:U\to L(E,F)$ by $D\alpha x(y)=\lim\limits_{0\neq t\to 0}\frac{1}{t}\{\alpha(x+ty)-\alpha(x)\}$

(ii) The differential $d\alpha:U\sqcap E\to F$ by $d\alpha(x,y)=\lim\limits_{0\neq t\to 0}\frac{1}{t}\{\alpha(x+ty)-\alpha(x)\}$

(iii) The tangent $T\alpha:U\sqcap E\to V\sqcap F$ by $T\alpha(x,y)=(\alpha(x),\ \lim\limits_{0\neq t\to 0}\frac{1}{t}\{\alpha(x+ty)-\alpha(x)\})$.

In each of these cases $(x,y)\in U\sqcap E$ is considered arbitrary but fixed, so that the limit on the right side of the equations exists.

In case that the domain U of a differentiable map α is an open subset of $\mathbb{R}$, we denote by $\alpha':\mathbb{R}\supset U\to F$ the continuous map defined by $\alpha'(r)=D\alpha r(1)$. We shall also use the notation $\frac{d}{dr}\alpha$ for α'.

We have:

<u>Proposition 3.4</u>. If $\alpha:\mathbb{R}\supset U\to V\subset F$ is differentiable, then α is continuous.

<u>Proposition 3.5</u>. If $\alpha:\mathbb{R}\supset U\to V\subset F$ is differentiable and $\lambda:F\to\mathbb{R}$ is a continuous linear map, then $\lambda\circ\alpha:\mathbb{R}\supset U\to\mathbb{R}$ is differentiable and $(\lambda\circ\alpha)'=\lambda\circ\alpha'$.

These two propositions are simple consequences of the definitions.

Now we shall state and prove the central theorem for differential calculus, often misleadingly called the mean value theorem. From this theorem all the important theorems of differential calculus follow.

<u>Lemma 3.6</u> (The fundamental lemma of differential calculus). Let $I=[0,1]$ be the closed unit interval in $\mathbb{R}$, let E be a convenient vector space, and let K be a closed convex subset of $LC(E)$. If $\alpha:\mathbb{R}\supset I\to E$ is a continuous map such that its restriction to the open interval $(0,1)$ is differentiable and if $\alpha'(r)\in K$ for all $r\in(0,1)$, then $\alpha(1)-\alpha(0)\in K$.

<u>Proof</u>. If $E=\mathbb{R}$ there exists an $r_0\in(0,1)$ such that $\alpha'(r_0)=\alpha(1)-\alpha(0)$ as everyone knows. Hence the lemma holds for $E=\mathbb{R}$.- Assume that the lemma does not hold for some $\alpha:I\to E$, t.i. $\alpha(1)-\alpha(0)\notin K$. By Hahn-Banach exists a continuous linear map $\lambda:LC(E)\to\mathbb{R}$ such that $\lambda(\alpha(1)-\alpha(0))\notin\overline{\lambda K}$. Since $CG\circ LC(E)=E$, the linear map λ is also continuous as a map $\lambda:E\to\mathbb{R}$. By proposition 3.5 we have $(\lambda\circ\alpha)'=\lambda\circ\alpha'$. Hence $(\lambda\circ\alpha)'(r)\in\overline{\lambda K}$ for all $r\in(0,1)$. Since the lemma holds for $\mathbb{R}$ we must have $\lambda\circ\alpha(1)-\lambda\circ\alpha(0)\in\overline{\lambda K}$. Contradiction!

<u>The fundamental lemma at work</u>. First we introduce a useful notation for differentiable maps: If $\alpha:E\supset U\to V\subset F$ is differentiable, we define $\Theta R\alpha:E\sqcap E\sqcap\mathbb{R}\supset\{(x,y,t)\mid x\in U \text{ and } x+ty\in U\}\to F$ by $\Theta R\alpha(x,y,0)=0$ and for $t\neq 0$ we put $\Theta R\alpha(x,y,t)=\frac{1}{t}\{\alpha(x+ty)-\alpha(x)\}-d\alpha(x,y)$. We observe that the domain of $\Theta R\alpha$ is open in $E\sqcap E\sqcap\mathbb{R}$. We shall prove that $\Theta R\alpha$ is continuous: First observe that for any $(x,y)\in U\sqcap E$ and any real numbers s and t with $x+sty\in U$ the formula $\Theta R\alpha(x,sy,t)=s\Theta R\alpha(x,y,st)$ holds. Hence the function $\Theta R\alpha(x,sy,t)$ is differentiable with respect to s for any fixed (x,y,t), and the derivative is given by $\frac{d}{ds}\Theta R\alpha(x,sy,t)=d\alpha(x+tsy,y)-d\alpha(x,y)\underset{\text{def}}{=}\varphi(x,y,t,s)$. This function $\varphi:E\sqcap E\sqcap\mathbb{R}\sqcap\mathbb{R}\supset\{(x,y,t,s)\mid x\in U \text{ and } x+tsy\in U\}\to F$ is continuous, has open domain, and satisfies $\varphi(x,y,t,0)=0$. Therefore: If $(x_0,y_0)\in U\sqcap E$ is fixed and K is any closed convex zero neighborhood in $LC(F)$, there exist positive real numbers δ and ε and a neighborhood $N_{(x_0,y_0)}$ of (x_0,y_0) in $U\sqcap E$, such that $\varphi(x,y,t,s)\in K$ whenever $|t|<\delta$, $|s|<\varepsilon$, $(x,y)\in N_{(x_0,y_0)}$. From the form of φ we infer that this implies $\varphi(x,y,t,s)\in K$ for $|t|<\delta\varepsilon$, $|s|\leq 1$, $(x,y)\in N_{(x_0,y_0)}$. By proposition 3.4, $\Theta R\alpha(x,sy,t)$ is continuous in s for $0\leq s\leq 1$ for any fixed t with $|t|<\delta\varepsilon$ and any fixed $(x,y)\in N_{(x_0,y_0)}$. Applying the fundamental lemma we get $\Theta R\alpha(x,y,t)-\Theta R\alpha(x,0,t)=\Theta R\alpha(x,y,t)\in K$ for all $(x,y)\in N_{(x_0,y_0)}$ and $|t|<\delta\varepsilon$. Since $F=CG\circ LC(F)$ we see that $\Theta R\alpha$ is continuous at all points $(x,y,0)$.- Define $N_{y_0}=\{y\mid (x_0,y)\in N_{(x_0,y_0)}\}$ and $Z=\{ty\mid y\in N_{y_0} \text{ and } |t|<\delta\varepsilon\}$. Hence Z is a zero neighborhood in E and for any $z\in Z$ we have $x_0+z\in U$ and

$\Theta R\alpha(x_0,z,1)=\alpha(x_0+z)-\alpha(x_0)-d\alpha(x_0,z)\in K$. This proves that $R\alpha: U\sqcap E\supset\{(x,y)\mid x\in U \text{ and } x+y\in U\}\to F$, defined by $R\alpha(x,y)=\alpha(x+y)-\alpha(x)-d\alpha(x,y)$, is continuous at all points $(x,0)$. Since $d\alpha: U\sqcap E\to F$ is continuous, we obtain the continuity of $\alpha: E\supset U\to V\subset F$ which in turn clearly implies the continuity of $\Theta R\alpha$ at all points (x,y,t) with $t\neq 0$, $x\in U$, $x+ty\in U$.

Thus we have proved:

<u>Theorem 3.7</u>. If a map $\alpha: E\supset U\to V\subset F$ is differentiable, then α is continuous.

<u>Theorem 3.8</u>. For a map $\alpha: E\supset U\to V\subset F$ are equivalent:

(i) α is differentiable

(ii) there exists a continuous map $\beta: U\sqcap E\to F$ which is linear in the second variable, such that the map $\Theta R\alpha: E\sqcap E\sqcap \mathbb{R}\supset\{(x,y,t)\mid x\in U \wedge x+ty\in U\}\to F$, defined by $\Theta R\alpha(x,y,t)=\frac{1}{t}\{\alpha(x+ty)-\alpha(x)\}-\beta(x,y)$ for $t\neq 0$ and $\Theta R\alpha(x,y,0)=0$, is continuous.

Since we work in a cartesian closed setting for continuous maps, it is clear that the continuity of the maps $D\alpha: U\to L(E,F)$, $d\alpha: U\sqcap E\to F$, $T\alpha: U\sqcap E\to V\sqcap F$ imply each other for any differentiable function $\alpha: E\supset U\to V\subset F$. Thus, so-called "weak" and "strong" differentiability are the same notions in our context. We shall make extensive use of this fact.

Next we prove compositions of differentiable maps to be differentiable.

<u>Proposition 3.9</u>. Let $\alpha: E\supset U\to V\subset F$ and $\beta: F\supset V\to W\subset G$ be differentiable. Then the composite map $\beta\circ\alpha: E\supset U\to W\subset G$ is differentiable.

<u>Proof</u>. Consider the map $\Theta R(\beta\circ\alpha): E\sqcap E\sqcap\mathbb{R}\supset\{(x,y,t)\mid x\in U \text{ and } x+ty\in U\}\to G$, defined by $\Theta R(\beta\circ\alpha)(x,y,t)=\frac{1}{t}\{\beta\circ\alpha(x+ty)-\beta\circ\alpha(x)\}-d\beta(\alpha x,d\alpha(x,y))$ for $t\neq 0$ and $\Theta R(\beta\circ\alpha)(x,y,0)=0$. Then we have:

$\Theta R(\beta\circ\alpha)(x,y,t)=\Theta R\beta(\alpha x,\Theta R\alpha(x,y,t)+d\alpha(x,y),t)+d\beta(\alpha x,\Theta R\alpha(x,y,t))$. By theorem 3.8 this map $\Theta R(\beta\circ\alpha)$ is continuous. Differentiability of $\beta\circ\alpha$ follows since $d\beta(\alpha x,d\alpha(x,y))$ is linear in y.

<u>Theorem 3.10</u> (The chain rules). Let $\alpha: E\supset U\to V\subset F$ and $\beta: F\supset V\to W\subset G$ be differentiable. Then the following formulae hold for the composite map:

(i) $D(\beta\circ\alpha)=\mathrm{comp}\circ\{D\alpha,D\beta\circ\alpha\}: E\supset U\to L(E,F)\sqcap L(F,G)\xrightarrow{\mathrm{comp}} L(E,G)$

(ii) $d(\beta\circ\alpha)=d\beta\circ(\alpha\sqcap d\alpha)\circ(\Delta_U\sqcap 1_E): E\sqcap E\supset U\sqcap E\to U\sqcap U\sqcap E\to V\sqcap F\to G$

(iii) $T(\beta\circ\alpha)=T\beta\circ T\alpha$.

This is clear. The fact that only the tangent of differentiable mappings behaves functorially is the main reason why this explicit form of differentiation is in most cases more useful than the derivative or the differential. Observe that $T(1_U)=1_{U\sqcap E}$ for the identity map $1_U: E\supset U\to U\subset E$. We put $TU=U\sqcap E$ and call it the tangent space of $U\subset E$.

In passing we note that constant maps, translations, continuous linear and multilinear maps obviously are differentiable. Moreover, pointwise sums and pointwise scalar multiplication with respect to differentiable maps give differentiable maps whenever defined.

Theorem 3.11. Let $\alpha: E\supset U\to V\subset F$ be differentiable. If U is connected in E, the following statements are equivalent:

(i) α is a constant map

(ii) $D\alpha=0$

Proof. (i)⇒(ii) is trivial. (ii)⇒(i): Let $x_0\in U$. Then $S=\{x\in U\,|\,\alpha x=\alpha x_0\}$ is clearly closed in U. For any $x\in S$ choose a radial open neighborhood $R(x)\subset U$. Since $\frac{d}{ds}\Theta R\alpha(x,sy,t)=0$ for any $y\in R(x)-x$, $s\in I$, $t\in I$, we see that $\Theta R\alpha(x,y,t)\in K$ for every $t\in[0,1]$ and any closed convex zero neighborhood K in $LC(F)$. Hence $\Theta R\alpha(x,y,1)=\alpha(x+y)-\alpha(x)\in K$ for all closed convex zero neighborhoods K in $LC(F)$. Hence $\alpha(x+y)=\alpha(x)$, whence the set S is open in U. The theorem follows.

Theorem 3.12 (Differentiable maps into a product). Let $\alpha: E\supset U\to V\subset \prod_{\iota\in I} F_\iota$ be a map into a product space. Then the following statements are equivalent:

(i) α is differentiable

(ii) for every $\iota\in I$ the map $pr_\iota\circ\alpha: E\supset U\to V_\iota=pr_\iota(V)\subset F_\iota$ is differentiable.

This is clear. To prove an analogous statement for maps from a product, we have to restrict ourselves to the case of finite products.

Definition 3.13. Let $\alpha: E_1\sqcap E_2\supset U\to V\subset F$ be a function, where U is open in the convenient product vector space $E_1\sqcap E_2$ and V open in the convenient vector space F. We say that α is partially differentiable with respect to the first variable, if there exists a continuous map $\beta_1: U\to L(E_1,F)$ such that for any fixed $(x,y_1)\in U\sqcap E_1$ we have:
$\lim_{0\neq t\to 0}\frac{1}{t}\{\alpha(x+t(y_1,0))-\alpha(x)\}=\beta_1 x(y_1)$. – In an analogous way one defines partial differentiability with respect to the second variable and extends the definition to the case of any finite number of convenient factors E_i for $i\geqslant 2$. – Finally, we say that an $\alpha: \prod_{i=1}^{n} E_i\supset U\to V\subset F$ is partially differentiable, if it is partially differentiable with respect to all variables i with $1\leqslant i\leqslant n$.

Proposition 3.14. If $\alpha: \prod_{i=1}^{n} E_i\supset U\to V\subset F$ is partially differentiable with respect to the j-th variable, the corresponding $\beta_j: U\to L(E_j,F)$ is unique.

This is clear. Hence we define:

Definition 3.15. Let $\alpha: \prod_{i=1}^{n} E_i\supset U\to V\subset F$ be partially differentiable with respect to the j-th variable. Then we define:

(i) the j-th partial derivative $D_j\alpha: \prod_{i=1}^{n} E_i\supset U\to L(E_j,F)$ by
$D_j\alpha x(y_j)=\lim_{0\neq t\to 0}\frac{1}{t}\{\alpha(x+t(0,\dots,0,y_j,0,\dots,0))-\alpha(x)\}$

(ii) the j-th partial differential $d_j\alpha: (\prod_{i=1}^{n} E_i)\sqcap E_j\supset U\sqcap E_j\to F$ by
$d_j\alpha(x,y_j)=D_j\alpha x(y_j)$

(iii) the j-th partial tangent $T_j\alpha: (\prod_{i=1}^{n} E_i)\sqcap E_j\supset U\sqcap E_j\to V\sqcap F\subset F\sqcap F$ by
$T_j\alpha(x,y_j)=(\alpha(x),d_j\alpha(x,y_j))$

The limits are always calculated by fixed $(x,y_j)\in U\sqcap E_j$.

Theorem 3.16 (Differentiable maps from a finite product). Let $\alpha: \prod_{i=1}^{n} E_i \supset U \to V \subset F$ be a map out of a finite product space. Then the following statements are equivalent:

(i) α is differentiable

(ii) α is partially differentiable.

Proof. The implication (i)$\Rightarrow$(ii) is trivial. To see that (ii)$\Rightarrow$(i) it clearly suffices to consider the case of a product $E_1 \sqcap E_2$ of two factors. For fixed $(x,y_1,y_2) \in U \sqcap (E_1 \sqcap E_2)$ and $t \neq 0$ we consider the function $\theta R\alpha(x,y_1,y_2,t) = \frac{1}{t}\{\alpha(x+t(y_1,y_2))-\alpha(x)\} - d_1\alpha(x,y_1) - d_2\alpha(x,y_2) =$
$= \frac{1}{t}\{\alpha(x+t(y_1,y_2))-\alpha(x+t(0,y_2))\} - d_1\alpha(x+t(0,y_2),y_1) +$
$+ \frac{1}{t}\{\alpha(x+t(0\ ,y_2))-\alpha(x)\} - d_2\alpha(x,y_2) \quad + \quad d_1\alpha(x+t(0,y_2),y_1) - d_1\alpha(x,y_1) =$
$\theta R_1\alpha(x+t(0,y_2),y_1,t) + \theta R_2\alpha(x,y_2,t) + d_1\alpha(x+t(0,y_2),y_1) - d_1\alpha(x,y_1)$. So we are left to prove that $\lim_{0 \neq t \to 0} \theta R_1\alpha(x+t(0,y_2),y_1,t) = 0$. To see this, consider the map $\theta R_1\alpha(x+t(0,y_2),sy_1,t)$ for $t \neq 0$ and $\theta R_1\alpha(x,sy_1,0)=0$. This map is differentiable with respect to s, and we obtain:
$\frac{d}{ds}\theta R_1\alpha(x+t(0,y_2),sy_1,t) = d_1\alpha(x+t(0,y_2)+st(y_1,0),y_1) - d_1\alpha(x+t(0,y_2),y_1) \underset{def}{=}$
$= \varphi(x,y_1,y_2,t,s)$. The map φ is a continuous map with open domain and satisfying $\varphi(x,y_1,y_2,t,0)=0$. Hence: For any closed convex zero neighborhood K of LC(F) exist positive real numbers δ and ε such that $\varphi(x,y_1,y_2,t,s) \in K$ whenever $|t|<\delta$ and $|s|<\varepsilon$. From the form of φ we infer that this implies $\varphi(x,y_1,y_2,t,s) \in K$ whenever $|t| < \mathrm{Min}\{\delta,\delta\varepsilon\}$ and $|s| \leqslant 1$. Application of the fundamental lemma gives $\theta R_1\alpha(x+t(0,y_2),y_1,t) \in K$ for $|t| < \mathrm{Min}\{\delta,\delta\varepsilon\}$. Hence $\lim_{0 \neq t \to 0} \theta R_1\alpha(x+t(0,y_2),y_1,t)=0$.

As a corollary we obtain that our notion of differentiability coincides in the finite dimensional case with the usual notion of continuous differentiability.

Next we prove the decisive lemma which will allow us later the forming of convenient function spaces of differentiable mappings.

Lemma 3.17 (Convergence theorem for sequences of differentiable maps). Let E and F be convenient vector spaces, and let U be open in E. Further suppose that (α_n) is a sequence of differentiable maps $\alpha_n: E \supset U \to F$. Then the following holds: If (α_n) is a cauchy sequence in CO(U,LC F), and if $(d\alpha_n)$ is a cauchy sequence in CO(U$\sqcap$E,LC F), then the limit functions $\lim(\alpha_n) = \alpha: U \to F$ and $\lim(d\alpha_n) = \beta: U \sqcap E \to F$ exist and are continuous. Moreover, α is differentiable and $d\alpha = \beta$.

Proof. Since CG$\circ$LC(F)=F, we obtain from Kelley's theorem 1.7 the existence and the continuity of α and β. Clearly, β is linear in the second variable. - If $(x,y) \in U \sqcap E$ is fixed, there exists an $\varepsilon > 0$ such that $x+ty \in U$ for all $|t| < \varepsilon$. Hence $\theta R\alpha_n(x,y,t): \mathbb{R} \supset (-\varepsilon,\varepsilon) \to F$ is defined and continuous for all $n \in \mathbb{N}$. Because $\theta R\alpha_n$ is continuous, we get continuity of $\varphi_n: \mathbb{R} \sqcap \mathbb{R} \supset (-\varepsilon,\varepsilon) \sqcap I \to F$ with $\varphi_n(t,s) = \theta R\alpha_n(x,sy,t)$. We have $\frac{d}{ds}\varphi_n(t,s) = d\alpha_n(x+tsy,y) - d\alpha_n(x,y)$. Since the $d\alpha_n$ form a cauchy sequence, we see that the sequence $(\frac{d}{ds}\varphi_n)\in CO((-\varepsilon,\varepsilon)\sqcap I,LC\ F)$ is also cauchy. Therefore, if C is compact in $(-\varepsilon,\varepsilon)$ and K is any closed convex zero neighborhood of LC(F), there exists a natural number N such that for all m>N and n>N we have $(\frac{d}{ds}\varphi_m-\frac{d}{ds}\varphi_n)(C\sqcap I)\subset K$. Application of the fundamental lemma yields $(\Theta R\alpha_m-\Theta R\alpha_n)(x,y,t)\in K$ for all $t\in C$. This proves that $(\Theta R\alpha_n(x,y,-))$ is a cauchy sequence in $CO((-\varepsilon,\varepsilon),LC\ F)$. By Kelley's theorem there exists a continuous $\rho:(-\varepsilon,\varepsilon)\to LC(F)$ with $\rho=\lim(\Theta R\alpha_n(x,y,-))$, and since $CG\circ LC(F)=F$ we may consider ρ as a continuous map $\rho:(-\varepsilon,\varepsilon)\to F$. For any fixed $t\neq 0$ we have $\rho(t)=\lim_{n\to\infty}\Theta R\alpha_n(t)=\lim_{n\to\infty}\frac{1}{t}\{\alpha_n(x+ty)-\alpha_n(x)\}-d\alpha_n(x,y)=$ $=\frac{1}{t}\{\alpha(x+ty)-\alpha(x)\}-\beta(x,y)$ and obviously we have $\rho(0)=0$. Hence it follows: $\lim_{0\neq t\to 0}\frac{1}{t}\{\alpha(x+ty)-\alpha(x)\}=\beta(x,y)$ which proves differentiability of α with differential $d\alpha=\beta$.

Higher orders of differentiability are introduced inductively with respect to the derivative operator D or the tangent operator T. So we have for example in the case of a 2-times differentiable map $\alpha:E\supset U\to V\subset F$ the first derivative $D\alpha:U\to L(E,F)$ and differentiating this map we obtain the second derivative $D^2\alpha:U\to L(E\otimes E,F)$. These derivatives are related to the second tangent $T^2\alpha:U\sqcap E\sqcap E\sqcap E\to V\sqcap F\sqcap F\sqcap F$ by the formula $T^2\alpha(x,y,z_1,z_2)=(\alpha x,D\alpha x(y),D\alpha x(z_1),D^2\alpha x(y\otimes z_1)+D\alpha x(y_2))$. It is clear that the notion of higher order differentiability is the same whether defined inductively from D or from T. We have already defined the tangent space of an open subset U of a convenient vector space E as $TU=U\sqcap E$. We now define the tangent space of order n of U inductively by $T^nU=T(T^{n-1}U)$ for n>1. With this notation we have for the i-th tangent of an n-times differentiable map $\alpha:U\to V$ that $T^i\alpha:T^iU\to T^iV$ where $1\leqslant i\leqslant n$.

<u>Theorem 3.18</u> (Functoriality of the tangent operator). Let $\alpha:E\supset U\to V\subset F$ and $\beta:F\supset V\to W\subset G$ be n-times differentiable. Then the composite map $\beta\circ\alpha:U\to W$ is n-times differentiable, and we obtain for the i-th tangent $T^i(\beta\circ\alpha):T^iU\to T^iW$ that $T^i(\beta\circ\alpha)=T^i\beta\circ T^i\alpha$ for $1\leqslant i\leqslant n$.

We note in passing that constant maps, translations, continuous linear and multilinear maps are n-times differentiable for arbitrary $n\in\mathbb{N}$.

The most important theorem on higher derivatives is:

<u>Theorem 3.19</u> (Symmetry of higher derivatives). If $\alpha:E\supset U\to V\subset F$ is n-times differentiable, then for each $x\in U$ the i-th derivative $D^i\alpha x:\overset{i}{\otimes}E\to F$ is totally symmetric for every $i\in(1,\dots,n)$.

<u>Proof</u>. For i=1 there is nothing to prove. Assume i=2: Fix $(x,y_1,y_2)\in U\sqcap E\sqcap E$ and consider - whenever defined - the map $\varphi(t,s)=\Theta R\alpha(x,sy_1+y_2,t)-\Theta R\alpha(x,sy_1,t)-\Theta R\alpha(x,sy_2+y_1,t)+\Theta R\alpha(x,sy_2,t)$. Clearly we can find a $\delta>0$ such that $\varphi(t,s)$ is defined for $0\leqslant s\leqslant 1$ and $|t|<\delta$. Differentiating with respect to s gives us:

$\frac{d}{ds}\varphi(t,s)=t\{D^2\alpha x(y_2\otimes y_1-y_1\otimes y_2)+[\Theta RD\alpha(x,sy_1+y_2,t)-\Theta RD\alpha(x,sy_1,t)](y_1)+$ $-[\Theta RD\alpha(x,sy_2+y_1,t)-\Theta RD\alpha(x,sy_2,t)](y_2)\}=t\{D^2\alpha x(y_2\otimes y_1-y_1\otimes y_2)+\psi(t,s)\}$ where ψ is continuous and satisfies $\psi(0,s)=0$. Since $[0,1]$ is compact, there exists for any given closed convex zero neighborhood K of LC(F) an $\varepsilon>0$ such that $\psi(t,s)\in K$ for $0\leqslant s\leqslant 1$ and $|t|<\varepsilon$. Let us assume that $D^2\alpha x(y_2\otimes y_1-y_1\otimes y_2)\neq 0$. Then select a closed convex zero neighborhood K such that $D^2\alpha x(y_1\otimes y_2-y_2\otimes y_1)\notin K$ and choose $\varepsilon>0$ such that $\psi(t,s)\in K$ for $0\leqslant s\leqslant 1$ and $|t|<\varepsilon$. Applying the fundamental lemma we get that $0=\varphi(t,1)-\varphi(t,0)\in t\{D^2\alpha x(y_2\otimes y_1-y_1\otimes y_2)+K\}$ for $|t|<\varepsilon$. Contradiction. - By the inductive definition of higher differentiability it is now clear that $D^i\alpha x$ is also totally symmetric for $i>2$.

Finally we prove the existence of a primitive function for a given continuous map $\alpha:\mathbb{R}\supset U\to E$. Here a map $\beta:\mathbb{R}\supset U\to E$ is called a primitive map for α if $\beta'=\alpha$.

Lemma 3.20. Let E be a convenient vector space and $\alpha:\mathbb{R}\supset[0,1]=I\to E$ a continuous map. Then there exists a continuous map $\beta:\mathbb{R}\supset I\to E$ which is differentiable on $(0,1)$ with $\beta'=\alpha$ on the open interval $(0,1)$.

Proof. We subdivide I into 2^n parts of equal length and define $\alpha_n:I\to E$ by $\alpha_n(t)=\alpha(2^{-n}i)+(2^nt-i)\{\alpha(2^{-n}(i+1))-\alpha(2^{-n}i)\}$ for $i\leqslant 2^nt\leqslant i+1$ and $i\in\{0,1,\dots,2^{n-1}\}$. Further we define $\beta_n:I\to E$ by $\beta_n(t)=2^{-1}(t-2^{-n}i)\{2\alpha(2^{-n}i)+(2^nt-i)\{\alpha(2^{-n}(i+1))-\alpha(2^{-n}i)\}\}+$ $+2^{-n-1}\sum_{j=0}^{i-1}\{\alpha(2^{-n}j)+\alpha(2^{-n}(j+1))\}$ for $i\leqslant 2^nt\leqslant i+1$ and $i\in\{0,1,\dots,2^{n-1}\}$. Obviously β_n is differentiable on $(0,1)$ with $\beta_n'=\alpha_n$. Moreover $\beta_n(0)=0$. Let us consider now the sequences (α_n) and (β_n) in CO(I,LC E). Since I is compact and α is continuous, we can find a natural number N to any closed convex zero neighborhood K of LC(E) such that $\alpha(t)-\alpha(t')\in K$ whenever $|t-t'|\leqslant 2^{-N}$. Hence $\alpha_n(t)-\alpha(t)\in IK+K$ whenever $n>N$. Consequently the sequence (α_n) is a cauchy sequence in CO(I,LC E) convergent to α. Now consider the maps $\varphi_{m,n}(t,s)=\beta_m(ts)-\beta_n(ts)$. These continuous maps satisfy $\varphi_{m,n}(t,0)=0$ and are differentiable for any fixed t with respect to s. We have $\frac{d}{ds}\varphi_{m,n}(t,s)=t\{\alpha_m(ts)-\alpha_n(ts)\}$. Since the (α_n) form a cauchy sequence there exists for any closed convex zero neighborhood K a natural number N such that $(\alpha_m-\alpha_n)(I)\subset K$ whenever $m>N$ and $n>N$. Then we obtain $\frac{d}{ds}\varphi_{m,n}(t,s)\in tK$ whence application of the fundamental lemma yields $\varphi_{m,n}(t,1)=(\beta_m-\beta_n)(t)\in tK$ whenever $m>N$ and $n>N$. From this we conclude that the sequence (β_n) is a cauchy sequence in CO(I,LC E). Lemma 3.17 now tells us that the continuous map $\beta=\lim(\beta_n):I\to E$ is differentiable on $(0,1)$ with $\beta'=\alpha$.

Theorem 3.31 (Existence of primitive maps). Let E be a convenient vector space, let U be open in $\mathbb{R}$, and let $\alpha:\mathbb{R}\supset U\to E$ be a continuous map. Then there exists a primitive map $\beta:\mathbb{R}\supset U\to E$ for α. If U is connected (t.i. an open interval), the difference of any two primitive maps for α is constant.

Proof. The existence of a primitive map follows directly from lemma 3.30. The difference of any two such maps in the connected case must be constant by theorem 3.11.

4. Convenient Function Spaces for Differentiable and Smooth Maps

Let E and F be convenient vector spaces and let U be open in E. If α and β are n-times differentiable maps from U to F, and if r is any real number, the pointwise sum $\alpha+\beta$ and the pointwise scalar product $r\alpha$ are obviously again n-times differentiable maps from U to F and we have $D^n(r\alpha+\beta)=rD^n\alpha+D^n\beta$ and $T^n(r\alpha+\beta)=rT^n\alpha+T^n\beta$. Hence the n-times differentiable maps from U to F form a vector space denoted by $D^n(U,F)$. We extend this notion to the case n=0 by defining $D^0(U,F)$ as the vector space of continuous maps from U to F. Since differentiability implies continuity, and higher differentiability is defined inductively, we have linear inclusions $i_{n,m}:D^n(U,F)\hookrightarrow D^m(U,F)$ whenever $0\leq m\leq n$. Hence the intersection $\bigcap_{n=0}^{\infty} D^n(U,F)$ is a vector space, denoted by $D^\infty(U,F)$ and called the vector space of smooth maps from U to F. Clearly $D^\infty(U,F)=\lim D^n(U,F)$ in the category of real vector spaces. Hence we have for every $n\in\mathbb{N}$ linear inclusions $i_n:D^\infty(U,F)\hookrightarrow D^n(U,F)$ satisfying the universal property associated with a limit.

We shall now turn the vector spaces $D^n(U,F)$ and $D^\infty(U,F)$ into convenient vector spaces as follows: For n=0 we provide $D^0(U,F)$ with the convenient structure given by the functor $C:\mathcal{CG}^{op}\times\widehat{\mathcal{CGV}}_{cont}\to\widehat{\mathcal{CGV}}_{cont}$ described at the end of section 2. The resulting convenient vector space will be henceforth denoted by $C^0(U,F)$. For n=1 we consider the linear injective function $T_\#:D^1(U,F)\rightarrowtail C^0(TU,TF)$, defined by $\alpha\mapsto T\alpha$. Now we induce on $D^1(U,F)$ the initial compactly generated topology with respect to $T_\#$. This topology turns $D^1(U,F)$ into a compactly generated vector space. Let us prove that this is even a convenient vector space: We know that $C^0(TU,TF)$ is a convenient vector space. If $(\alpha_n)\in D^1(U,F)$ is a sequence such that the sequence $(T\alpha_n)\in C0(TU,TF)$ is cauchy, we have by lemma 3.17 that the sequence (α_n) converges to a differentiable map α with $T\alpha=\lim(T\alpha_n)\in C0(TU,TF)$. It follows that $D^1(U,F)$ with the initial compactly generated topology induced by $T_\#$ is a convenient vector space. This vector space will be denoted by $C^1(U,F)$. Now we proceed inductively by setting $C^n(U,F)=C^1(C^{n-1}(U,F))$ for n>1. Since the linear inclusion $i_{1,0}:C^1(U,F)\hookrightarrow C^0(U,F)$ is continuous, we see that all linear inclusions $i_{n,m}:C^n(U,F)\hookrightarrow C^m(U,F)$ are continuous whenever $0\leq m\leq n$.

Consequently, since $\widehat{\mathcal{CGV}}$ is complete, the limit for the diagram $\{i_{n,m}:C^n(U,F)\hookrightarrow C^m(U,F)\mid n,m\in\mathbb{N} \text{ and } 0\leq m\leq n\}$ exists in $\widehat{\mathcal{CGV}}$, and we may and shall select $D(U,F)$ equipped with the corresponding compactly generated limit topology to define the convenient vector space $S(U,F)$ of smooth maps from U to F. We observe that by construction the linear inclusions $i_n: S(U,F) \hookrightarrow C^n(U,F)$ are continuous. Since we may identify $S(U,F)$ as a subspace of the product space $\prod_{n=0}^{\infty} C^n(U,F)$ in $\mathcal{C}\widehat{\mathcal{G}}\mathcal{V}$, we get from theorem 3.12 that a map $\alpha: G \supset W \to S(U,F)$ is differentiable iff the maps $i_n \circ \alpha: W \to C^n(U,F)$ are differentiable for all $n \in \mathbb{N}$. Since theorem 3.12 clearly extends to n-times differentiable maps into a product, the construction of $S(U,F)$ shows as well that a map $\alpha: G \supset W \to S(U,F)$ is smooth iff the maps $i_n \circ \alpha: W \to C^n(U,F)$ are smooth for all $n \in \mathbb{N}$.

We have thus proved the theorem:

<u>Theorem 4.1</u>. Let E, F, G be convenient vector spaces, let U be open in E and W open in G. Then we have for each $n \in \mathbb{N}$ a convenient vector space $C^n(U,F)$ with elements the n-times differentiable maps from U to F such that

(i) $C^0(U,F) = C(U,F)$ in the notation of section 2

(ii) $C^n(U,F)$ has the initial compactly generated topology induced by the continuous linear injective map $T^n_{\#}: C^n(U,F) \rightarrowtail C^0(T^nU, T^nF)$ which is defined by $T^n_{\#}(\alpha) = T^n\alpha$.

Hence for all $0 \leq j \leq n$ the linear injections $T^j_{\#}: C^n(U,F) \rightarrowtail C^{n-j}(T^jU, T^jF)$ are continuous and $C^n(U,F)$ has the initial compactly generated topology with respect to all of these maps.

Further the linear inclusions $i_{n,m}: C^n(U,F) \hookrightarrow C^m(U,F)$ are continuous whenever $0 \leq m \leq n$.

Since the functor $C(U,-): \mathcal{C}\widehat{\mathcal{G}}\mathcal{V}_{cont} \to \mathcal{C}\widehat{\mathcal{G}}\mathcal{V}_{cont}$ is compatible with arbitrary products and by the validity of theorem 3.12 the obvious linear map $\iota: C^n(U, \prod_{i\in I} F_i) \twoheadrightarrow \prod_{i\in I} C^n(U, F_i)$ is a homeomorphism.

A map $\alpha: G \supset W \to C^n(U,F)$ is j-times differentiable iff the map $T^n_{\#} \circ \alpha: G \supset W \to C^0(T^nU, T^nF)$ is j-times differentiable.

The convenient vector space $S(U,F)$ of smooth maps from U to F is the intersection of the convenient vector spaces $C^n(U,F)$ for all $n \in \mathbb{N}$ and is equipped with the corresponding compactly generated limit topology.

Therefore we have continuous linear inclusions $i_n: S(U,F) \hookrightarrow C^n(U,F)$ for every $n \in \mathbb{N}$ and continuous linear injective maps $T^n_{\#}: S(U,F) \rightarrowtail S(T^nU, T^nF)$.

A map $\alpha: G \supset W \to S(U,F)$ is n-times differentiable (or smooth) iff for every $n \in \mathbb{N}$ the map $i_n \circ \alpha = \alpha: G \supset W \to C^n(U,F)$ is n-times differentiable (or smooth).

Finally we have for any product $\prod_{i\in I} F_i$ the obvious linear homeomorphism $\iota: S(U, \prod_{i\in I} F_i) \twoheadrightarrow \prod_{i\in I} S(U, F_i)$.

End of theorem 4.1.

We note especially that constant maps, translations, continuous linear and multilinear maps are smooth.

Lemma 4.2. Let E, F, G be convenient vector spaces, let U be open in E, and let $\beta:F\to G$ be an n-times differentiable map where $n\geq 1$. Then β induces a differentiable map $\beta_*:C^n(U,F)\to C^{n-1}(U,G)$ defined by $\beta_*(\alpha)=\beta\circ\alpha$.

Proof. First we show continuity of β_*: This is immediate from the commutativity of the diagram

$$\begin{array}{ccc} C^n(U,F) & \xrightarrow{T^n_\#} & C^0(T^nU,T^nF) \\ \downarrow \beta_* & & \downarrow (T^n\beta)_* \\ C^n(U,G) & \xrightarrow{T^n_\#} & C^0(T^nU,T^nG) \\ \downarrow i_{n,n-1} & & \\ C^{n-1}(U,G) & & \end{array}$$

(with $\beta_*: C^n(U,F)\to C^{n-1}(U,G)$ the composite on the left).

Next we define $T\beta_*:TC^n(U,F)\to TC^{n-1}(U,G)$ by commutativity of the upper side of the diagram

$$\begin{array}{ccc} TC^n(U,F)=C^n(U,F)\sqcap C^n(U,F) & \xrightarrow{T\beta_*} & TC^{n-1}(U,G)=C^{n-1}(U,G)\sqcap C^{n-1}(U,G) \\ \cong\downarrow \iota^{-1} & & \cong\uparrow \iota \\ C^n(U,TF) & \xrightarrow{(T\beta)_*} & C^{n-1}(U,TG) \\ \downarrow i_{n,n-1} & & \\ C^{n-1}(U,TF) & & \downarrow T^{n-1}_\# \\ \downarrow T^{n-1}_\# & & \\ C^0(T^{n-1}U,T^nF) & \xrightarrow{(T^n\beta)_*} & C^0(T^{n-1}U,T^nG) \end{array}$$

The lower side of this diagram clearly commutes. Hence $T\beta_*$ - so defined - is continuous. The differential $d\beta_*:TC^n(U,F)\to C^{n-1}(U,G)$, corresponding to $T\beta_*$, is then given by $d\beta_*(\alpha,\varphi)(x)=d\beta(\alpha x,\varphi x)$, whence linear in the second variable. So we are left to prove that for fixed $(\alpha,\varphi)\in TC^n(U,F)$ the map $\Theta R\beta_*(\alpha,\varphi,-):\mathbb{R}\to C^{n-1}(U,G)$, defined by $\Theta R\beta_*(\alpha,\varphi,t)=\frac{1}{t}\{\beta_*(\alpha+t\varphi)-\beta_*(\alpha)\}-d\beta_*(\alpha,\varphi)$ for $t\neq 0$ and $\Theta R\beta_*(\alpha,\varphi,0)=0$ is continuous for $t=0$. This is by definition of $C^{n-1}(U,G)$ equivalent to continuity of $T^{n-1}_\#\circ\Theta R\beta_*(\alpha,\varphi,-)$ and this is by cartesian closedness of C^0 with respect to continuous maps equivalent to continuity of the associated explicit map $\Phi:T^{n-1}U\sqcap\mathbb{R}\to T^{n-1}G$ with $\Phi(\xi,t)=[T^{n-1}_\#\circ\Theta R\beta_*(\alpha,\varphi,-)](\xi,t)$. Since T^{n-1} defines and is defined by all the D^i for $0\leq i\leq n-1$, we have to show that for fixed $t\in\mathbb{R}$ the map $\Theta R\beta_*(\alpha,\varphi,t):E\supset U\to G$ is (n-1)-times differentiable and the maps $D^i\Theta R\beta_*(\alpha,\varphi,t):U\to L(\overset{i}{\otimes}E,G)$ are continuous in (x,t) with $D^i\Theta R\beta_*(\alpha,\varphi,0)=0$ for $0\leq i\leq n-1$. An easy computation gives us $D^i\Theta R\beta_*(\alpha,\varphi,t)(x)(x_1\otimes\ldots\otimes x_i)=$

$$=\Theta RD^i\beta(\alpha x,\varphi x,t)(d[\alpha+t\varphi](x,x_1)\otimes\ldots\otimes d[\alpha+t\varphi](x,x_i))+$$
$$+tD^{i+1}\beta(\alpha x)(d[\alpha+t\varphi](x,x_1)\otimes\ldots\otimes d[\alpha+t\varphi](x,x_{i-1})\otimes d\varphi(x,x_i)\otimes\varphi(x))+$$
$$\sum_{k=1}^{i-1}\Theta RD^{i-1}\beta(\alpha x,\varphi x,t)(d[\alpha+t\varphi](x,x_1)\otimes\ldots\otimes d[\alpha+t\varphi](x,x_{k-1})\otimes d^2[\alpha+t\varphi](x,x_k\otimes x_i)\otimes$$
$$\otimes d[\alpha+t\varphi](x,x_{k+1})\otimes\ldots\otimes d[\alpha+t\varphi](x,x_i))=$$
$$=\Theta RD^i\beta(\alpha x,\varphi x,t)(d[\alpha+t\varphi](x,x_1)\otimes\ldots\otimes d[\alpha+t\varphi](x,x_i))+F_i((x,t),x_1,\ldots,x_i).$$

For $i=0$ we have $\Theta R\beta_*(\alpha,\varphi,t)(x)=\Theta R\beta(\alpha x,\varphi x,t)$. Hence $F_0=0$. By induction we see from our formula that $\Theta R\beta_*(\alpha,\varphi,t):U\to G$ is (n-1)-times differentiable. By another induction we see that the F_i are $(n-1-i)$-times differentiable, are continuous in $((x,t),x_1,\dots,x_i)$ and satisfy $F_i((x,0),x_1,\dots,x_i)=0$. This proves the assertion.

Corollary 4.3. Let E, F, G be convenient vector spaces, let U be open in E and $\beta:F\to G$ an n-times differentiable map. Then β induces i-times differentiable maps $\beta_*:C^n(U,F)\to C^{n-i}(U,G)$ for all $i\in(0,1,\dots,n)$ which are defined by $\beta_*(\alpha)=\beta\circ\alpha$.

Proof. For i=0 this is clear by commutativity of the diagram

$$\begin{array}{ccc} C^n(U,F) & \xrightarrow{T^n_\#} & C^o(T^nU,T^nF) \\ \downarrow{\scriptstyle\beta_*} & & \downarrow{\scriptstyle(T^n\beta)_*} \\ C^n(U,G) & \xrightarrow{T^n_\#} & C^o(T^nU,T^nG) \end{array}$$

For i=1 this has been proved in lemma 4.2. Suppose the corollary true for j-1 with $1\leqslant j-1<n$ and let $\beta:F\to G$ be j-times differentiable. Then $\beta_*:C^n(U,F)\to C^{n-j+1}(U,G)$ is (j-1)-times differentiable by hypothesis, whence $\beta_*=i_{n-j+1,n-j}\circ\beta_*:C^n(U,F)\to C^{n-j}(U,G)$ is (j-1)-times differentiable. The commutative diagram

$$T^{j-1}\beta_*\left[\begin{array}{ccccc} T^{j-1}C^n(U,F) & \xrightarrow[\approx]{\iota^{-1}} & C^n(U,T^{j-1}F) & \xrightarrow{i_{n,n-j+1}} & C^{n-j+1}(U,T^{j-1}F) \\ \downarrow{\scriptstyle T^{j-1}\beta_*} & & \downarrow{\scriptstyle(T^{j-1}\beta)_*} & & \downarrow{\scriptstyle(T^{j-1}\beta)_*} \\ T^{j-1}C^{n-j+1}(U,G) & \xrightarrow[\approx]{\iota^{-1}} & C^{n-j+1}(U,T^{j-1}G) & \xrightarrow{i_{n-j+1,n-j}} & C^{n-j}(U,T^{j-1}G) \\ \downarrow{\scriptstyle Ti_{n-j+1,n-j}} & & \downarrow{\scriptstyle(T^{j-1}1_G)_*} & & \| \\ T^{j-1}C^{n-j}(U,G) & \xrightarrow[\approx]{} & C^{n-j}(U,T^{j-1}G) & = & \end{array}\right.$$

shows $T^{j-1}\beta_*$ differentiable, since by lemma 4.2 the far right map $(T^{j-1}\beta)_*:C^{n-j+1}(U,T^{j-1}F)\to C^{n-j}(U,T^{j-1}G)$ is differentiable.

Lemma 4.4. Let E, F, G be convenient vector spaces, let U be open in E and V open in F, and let $\alpha:E\supset U\to V\subset F$ be an n-times differentiable map. Then α induces a continuous linear map $\alpha^*:C^n(V,G)\to C^n(U,G)$, defined by $\alpha^*(\beta)=\beta\circ\alpha$.

Proof. Clearly α^* is linear. The continuity follows from the commutative diagram

$$\begin{array}{ccc} C^n(V,G) & \xrightarrow{T^n_\#} & C^o(T^nV,T^nG) \\ \downarrow{\scriptstyle\alpha^*} & & \downarrow{\scriptstyle(T^n\alpha)^*} \\ C^n(U,G) & \xrightarrow{T^n_\#} & C^o(T^nU,T^nG) \end{array}$$

Lemma 4.5. Let E, F be convenient vector spaces and let U be open in E. If $n\geqslant 1$, the evaluation map $\varepsilon:C^n(U,F)\sqcap U\to F$, defined by $\varepsilon(\alpha,x)=\alpha x$, is differentiable.

Proof. Continuity of evaluation is clear from the commutative diagram

$$\begin{array}{ccc} C^n(U,F)\sqcap U & \xrightarrow{i_{n,o}\sqcap 1_U} & C^o(U,F)\sqcap U \\ \downarrow{\scriptstyle\varepsilon} & & \downarrow{\scriptstyle\varepsilon} \\ F & = & F \end{array}$$

and cartesian closedness with respect to C^o. Next we define maps

$liso: T(C^n(U,F)\sqcap U) \to TC^n(U,F)\sqcap TU$ by $liso(\alpha,x,\varphi,y)=((\alpha,\varphi),(x,y))$, $pr_j: TC^n(U,F) \to C^n(U,F)$ by $pr_1(\alpha,\varphi)=\alpha$ and $pr_2(\alpha,\varphi)=\varphi$, $\sigma_j: TF \to TF$ by $\sigma_1(y,z)=(y,z)$ and $\sigma_2(y,z)=(0,y)$. Obviously all these maps are continuous linear maps. From these maps we obtain by composition for $j=1,2$ the maps $\varrho_j=\sigma_j\circ\varepsilon\circ(T_{\#}\sqcap 1_{TU})\circ(pr_j\sqcap 1_{TU})\circ liso$ from $T(C^n(U,F)\sqcap U)$ to TF. Since we know that $\varepsilon: C^{n-1}(TU,TF)\sqcap TU \to TF$ is continuous, we deduce that $\varrho_1+\varrho_2$ is continuous. We shall prove that $T\varepsilon=\varrho_1+\varrho_2$: To see this, we observe that the corresponding differential $d\varepsilon$ satisfies $d\varepsilon((\alpha,x),(\varphi,y))=d\alpha(x,y)+\varphi(x)$, whence $d\varepsilon$ is linear in the second variable. For fixed $((\alpha,x),(\varphi,y))\in T(C^n(U,F)\sqcap U)$ and $t\neq 0$ we have $\Theta R\varepsilon((\alpha,x),(\varphi,y),t)=\frac{1}{t}\{\varepsilon(\alpha+t\varphi,x+ty)-\varepsilon(\alpha,x)\}-d\varepsilon((\alpha,x),(\varphi,y))=$ $=\Theta R(\alpha+t\varphi)(x,y,t)+td\varphi(x,y)$. Hence $\lim_{0\neq t\to 0}\Theta R\varepsilon((\alpha,x),(\varphi,y),t)=0$.

Corollary 4.6. Let E, F be convenient vector spaces and let U be open in E. Then the evaluation map $\varepsilon: C^n(U,F)\sqcap U \to F$ is n-times differentiable.

Proof. For $n=0$ we have continuity of ε by cartesian closedness with respect to continuous maps. For $n=1$ we have differentiability by lemma 4.5. Suppose the corollary true for $n=k\geq 1$. Then we have for $n=k+1$ differentiability by lemma 4.5 with derivative $T\varepsilon=\varrho_1+\varrho_2$. The definition of the ϱ_j involved - apart from $\varepsilon: C^k(TU,TF)\sqcap TU \to TF$ - only continuous linear maps. By hypothesis the evaluation map $\varepsilon: C^k(TU,TF)\sqcap TU \to TF$ is k-times differentiable. Hence $T\varepsilon$ is k-times differentiable.

Lemma 4.7. Let E, F be convenient vector spaces, let U be open in E and V open in F. Then the insertion map $\eta: E\supset U \to C^n(F\supset V, E\sqcap F)$, defined by $\eta x: y \mapsto (x,y)$, is i-times differentiable for every $i\in\mathbb{N}$.

Proof. We have $\eta=\kappa+\lambda$ where $\kappa: U \to C^n(V,E\sqcap F)$ is defined by $\kappa x: y\mapsto(0,y)$, and where $\lambda: U \to C^n(V,E\sqcap F)$ is defined by $\lambda x: y\mapsto(x,0)$. Hence η is the sum of the constant map κ and the linear map λ. We are left to prove the continuity of λ which is equivalent to the continuity of $T^n_{\#}\circ\lambda: U \to C^0(T^nV, T^n(E\sqcap F))$. Since $T^n_{\#}\circ\lambda(x)(y_1,\dots,y_{2^n})=((x,0),(0,0),\dots,(0,0))$ this is the case.

Definition 4.8. We denote by $\widehat{\mathfrak{Cgv}}_{smooth}$ the category with objects the convenient vector spaces and arrows the smooth maps.

We have the following fundamental theorem:

Theorem 4.9 (The fundamental theorem for smooth maps). The category $\widehat{\mathfrak{Cgv}}_{smooth}$ contains the category $\widehat{\mathfrak{Cgv}}$ and is contained in the category $\mathfrak{Cgv}_{cont}$. All these categories have the same objects and the same (arbitrary) products. The category $\widehat{\mathfrak{Cgv}}_{smooth}$ is cartesian closed with $S: \widehat{\mathfrak{Cgv}}^{op}_{smooth}\times\widehat{\mathfrak{Cgv}}_{smooth} \to \widehat{\mathfrak{Cgv}}_{smooth}$ as internal functor. This functor S is defined on objects as the limit of the diagram $\{C^n(E,F) \xrightarrow{in_{n,n-1}} C^{n-1}(E,F) \mid n\in\mathbb{N}\}$ in $\widehat{\mathfrak{Cgv}}$, where C^0 is the cartesian internal functor for $\widehat{\mathfrak{Cgv}}_{cont}$ and $C^n(E,F)$ has the initial compactly generated topology induced by the linear injective map

$T^n : C^n(E,F) \rightarrowtail C^0(T^nE, T^nF)$. For arrows (t.i. smooth maps) the functor S is defined by composition.
The unit for cartesian closedness is given by the smooth insertion maps $\eta : E \to S(F, E\sqcap F)$ with $\eta x : y \mapsto (x,y)$. The counit for cartesian closedness is given by the smooth evaluation maps $\varepsilon : S(E,F)\sqcap E \to F$ with $\varepsilon(\alpha, x) = \alpha x$. A map $\alpha : E \to S(F,G)$ is smooth iff the corresponding map $\hat{\alpha} = \varepsilon \circ (\alpha \sqcap 1) : E \sqcap F \to G$, defined by $\hat{\alpha}(x,y) = \alpha x(y)$, is smooth. The natural diffeomorphism $\sigma : S(E, S(F,G)) \overset{\approx}{\rightarrowtail\!\!\twoheadrightarrow} S(E\sqcap F, G)$ is linear.
The ground field $\mathbb{R}$ is a generator and a cogenerator for $\widehat{\mathfrak{ECV}}_{smooth}$.
The tangent functor $T : \widehat{\mathfrak{ECV}}_{smooth} \to \widehat{\mathfrak{ECV}}_{smooth}$ with $T : E \mapsto TE = E\sqcap E$ and $T : \alpha \mapsto T\alpha$ with $T\alpha(x,y) = (\alpha x, d\alpha(x,y))$ is linear.
Finally the diagram

$$\begin{array}{ccc} \widehat{\mathfrak{ECV}}^{op}_{smooth} \times \widehat{\mathfrak{ECV}}_{smooth} & \xrightarrow{\ S\ } & \widehat{\mathfrak{ECV}}_{smooth} \\ \downarrow{\scriptstyle 1\times T} & & \downarrow{\scriptstyle T} \\ \widehat{\mathfrak{ECV}}^{op}_{smooth} \times \widehat{\mathfrak{ECV}}_{smooth} & \xrightarrow{\ S\ } & \widehat{\mathfrak{ECV}}_{smooth} \end{array}$$

commutes up to a smooth natural isomorphism.

<u>Proof</u>. 1. Functoriality of S: Let $\alpha : E \to F$ be smooth. Then the following diagram commutes for all $n \in \mathbb{N}$:

$$\begin{array}{ccc} S(F,G) & \xrightarrow{\ i_n\ } & C^n(F,G) \\ {\scriptstyle S(\alpha,G)=\alpha^*}\downarrow & & \downarrow{\scriptstyle \alpha^*} \\ S(E,G) & \xrightarrow{\ i_n\ } & C^n(E,G) \end{array}$$

From lemma 4.4 we know that $\alpha^* : C^n(F,G) \to C^n(E,G)$ is smooth for all $n \in \mathbb{N}$. Hence by the limit definition of S we see that $\alpha^* : S(F,G) \to S(E,G)$ is smooth. - Let $\beta : F \to G$ be smooth. Then the following diagram commutes for all $n \in \mathbb{N}$ and $k \in \mathbb{N}$:

$$\begin{array}{ccc} S(E,F) & \xrightarrow{\ i_{n+k}\ } & C^{n+k}(E,F) \\ {\scriptstyle S(E,\beta)=\beta_*}\downarrow & & \downarrow{\scriptstyle \beta_*} \\ S(E,G) & \xrightarrow{\ i_n\ } & C^n(E,G) \end{array}$$

The map $\beta_* : C^{n+k}(E,F) \to C^n(E,G)$ is k-times differentiable by corollary 4.3. Hence by the limit definition of S we see that $\beta_* : S(E,F) \to S(E,G)$ is k-times differentiable. Since k is arbitrary we see that $\beta_* : S(E,F) \to S(E,G)$ is smooth..
2. Smoothness of the unit and counit: The insertions $\eta : E \to C^n(F, E\sqcap F)$ are smooth for all $n \in \mathbb{N}$ by lemma 4.7. Hence $\eta : E \to S(F, E\sqcap F)$ is smooth by the limit definition of S. - The evaluations $\varepsilon : C^n(E,F)\sqcap E \to F$ are n-times differentiable for every $n \in \mathbb{N}$ by corollary 4.6. Since the diagram

$$\begin{array}{ccc} S(E,F)\sqcap E & \xrightarrow{\ \varepsilon\ } & F \\ {\scriptstyle i_n \sqcap 1}\downarrow & & \| \\ C^n(E,F)\sqcap E & \xrightarrow{\ \varepsilon\ } & F \end{array}$$

commutes, we see that $\varepsilon : S(E,F)\sqcap E \to F$ is smooth.

This proves that $\widehat{\mathfrak{CGV}}_{smooth}$ is cartesian closed with S as internal functor. All other statements are evident.

We close with the following important remark: If we define the category ${}_{open}\widehat{\mathfrak{CGV}}_{smooth}$ as the category of open subsets of convenient vector spaces and smooth maps, we obtain the more general functor $S: {}_{open}\widehat{\mathfrak{CGV}}^{op}_{smooth} \times \widehat{\mathfrak{CGV}}_{smooth} \to \widehat{\mathfrak{CGV}}_{smooth}$. We see as before that insertion $\eta: E \supset U \to S(F \supset V, E \sqcap F)$ and evaluation $\varepsilon: S(E \supset U, F) \sqcap U \to F$ are smooth maps, that a map $\alpha: E \supset U \to S(F \supset V, G)$ is smooth iff the corresponding map $\hat{\alpha} = \varepsilon \circ (\alpha \sqcap 1): E \sqcap F \supset U \sqcap V \to G$ is smooth, and that we have a smooth linear diffeomorphism $\sigma: S(E \supset U, S(F \supset V, G)) \xrightarrow{\approx} S(E \sqcap F \supset U \sqcap V, G)$.

5. Other Results and Problems Arising in this Context It is fairly simple to establish the main theorems of differential forms in this setting. The Poincaré lemma and the Stoke theorem are easily seen to hold. - The most troublesome question coming up concerns the inverse function theorem. But this is unavoidable as we see from the following example: Let $\exp: S(\mathbb{R},\mathbb{R}) \to S(\mathbb{R},\mathbb{R})$ denote the exponential map defined by $\exp(\alpha)(t) = e^{\alpha(t)}$. Clearly $D\exp(0) = 1_{S(\mathbb{R},\mathbb{R})}$. Hence $D\exp(0)$ is invertible, and the usual inverse function theorem would imply the invertibility of exp in a neighborhood of the zero map. But because on the one side $\exp(\alpha): \mathbb{R} \to \mathbb{R}$ only takes positive values (whatever $\alpha: \mathbb{R} \to \mathbb{R}$ one starts with), and since on the other side there exist smooth maps also assuming negative values in every neighborhood of $\exp(0): \mathbb{R} \to \mathbb{R}$, we see that even the invertibility of exp near the zero map is impossible. Therefore one has to impose additional conditions then are necessary in the banach space case, in order to obtain an inverse function theorem in our generalized setting. - Another interesting point concerns differential equations in the new setting. By proving suitable fixed point theorems for convenient vector spaces one obtains the necessary foundation for existence and uniqueness theorems for ordinary differential equations. These fixed point theorems clearly involve bounded subsets. More details can be found in [10]. Until now I have not established a Frobenius theorem for partial differential equations.

Finally I shall touch upon some questions concerning the general theory of differentiable manifolds. Instead of modelling differentiable manifolds on hilbert or banach spaces [8], we shall now consider differentiable manifolds modelled on convenient vector spaces, where atlases consist of differentiably compatible charts in the sense of our generalized differential calculus. We shall then speak of convenient differentiable manifolds. In the same spirit we define convenient differentiable vector bundles. What one would like to obtain is the cartesian closedness of the category $\mathcal{M}_{smooth}$ of convenient smooth manifolds and smooth maps. There are several obstructions: The first consists in defining charts for the set $S(M,N)$ of smooth maps from one convenient smooth manifold to another. It is clear however, on what convenient vector spaces the "manifold" $S(M,N)$ has to be modelled: Near a smooth map $\alpha:M\to N$ the corresponding vector space has to be the vector space $S_\alpha(\tau_N)$ of smooth sections over α in the tangent bundle $\tau_N:TN\to N$. More formally: $S_\alpha(\tau_N)=\{\sigma:M\to TN|\tau_N\circ\sigma=\alpha,\ \sigma \text{ smooth}\}$. This fact was noticed by Eells [3]. The convenient vector space topology of $S_\alpha(\tau_N)$ is now obtained as follows: For every $i\in\mathbb{N}$ we have $T^i_\#:S_\alpha(\tau_N)\rightarrowtail C(T^iM,T^{i+1}N)$ and $LC\circ C(T^iM,T^{i+1}N)$ clearly is a sequentially complete hausdorff convex vector space. Hence $CG\circ LC\circ C(T^iM,T^{i+1}N)$ is a convenient vector space and we take for $S_\alpha(\tau_N)$ the convenient compactly generated initial topology with respect to the linear injective map $\{T^i\}:S_\alpha(\tau_N)\rightarrowtail\prod_{i=0}^{\infty}CG\circ LC\circ C(T^iM,T^{i+1}N)$. So we know where the charts have to be situated, but we lack the corresponding chart maps.

To obtain chart maps I recall the following construction, possible for every smooth hilbert manifold N [8]: There exists the so-called exponential map $\exp:TN\supset U\to N$, defined on a neighborhood U of the zero section $0:N\to TN$, which together with the projection $\tau_N:TN\to N$ gives a smooth diffeomorphism $\{\tau_N|U,\ \exp\}:U\rightarrowtail\!\!\!\twoheadrightarrow V\subset N\times N$ with an open neighborhood V of the diagonal in $N\times N$. Applying now lemma II, 7.4 in [7] we can select a smaller neighborhood U of the zero section in TN which is fiberwise smoothly diffeomorphic to TN itself. These two results together imply in the case of smooth hilbert manifolds the existence of a smooth map $\varepsilon:TN\to N$ satisfying the following two conditions: (i) the composition $\varepsilon\circ 0$ of ε with the zero section is the identity map on N, and (ii) the smooth map $\{\tau_N,\varepsilon\}:TN\to N\times N$ establishes a smooth diffeomorphism between TN and its image in $N\times N$ where this image is an open neighborhood of the diagonal in $N\times N$.

In the general case we take now this situation as the definition of what we call a smooth addition for a convenient smooth manifold.

Definition 5.1. Let N be a convenient smooth manifold. Then we say that a smooth map $\varepsilon_N:TN\to N$ is a smooth addition for N if the following holds: (i) If $0_N:N\to TN$ is the zero section, then $\varepsilon_N\circ 0_N=1_N:N\to N$, and (ii) the map $\{\tau_N,\varepsilon_N\}:TN\to N\times N$ is a smooth diffeomorphism between TN and its image in $N\times N$ with this image open in $N\times N$.

Assume that N is a convenient smooth manifold with a smooth addition $\varepsilon_N:TN\to N$. Then we obtain for every $x\in N$ by restriction of ε_N to the fiber $\tau_N^{-1}(x)$ over x a chart $\varepsilon_x=\varepsilon_N|\tau^{-1}(x):\tau^{-1}(x)\rightarrowtail\!\!\!\twoheadrightarrow U_x\subset N$ near x, and these charts form a smooth atlas which defines the (original) smooth structure for N.

Now we consider $S(M,N)$ where N is a convenient smooth manifold with a smooth addition $\varepsilon_N : TN \to N$ and M is any convenient smooth manifold. Then we obtain for every smooth $\alpha \in S(M,N)$ by composition with ε_N a map $\varepsilon_\alpha : S_\alpha(\tau_N) \to S(M,N)$ defined by $\varepsilon_\alpha(\sigma) = \varepsilon \circ \sigma$. Clearly these maps ε_α are one-to-one for every α, whence they are the natural candidates for defining (induced) smooth charts for $S(M,N)$. The only problem is their smooth compatibility: In general they are only compatible if M is a compact smooth manifold. But in this case the smooth structure of $S(M,N)$ does obviously not depend on the choice of the smooth addition ε_N for N.

Denoting by ${}_{comp}\mathcal{M}_{smooth}$ the category of compact smooth manifolds and smooth maps (which coincides with the category of compact convenient smooth manifolds) and denoting by ${}_{add}\mathcal{M}_{smooth}$ the category of convenient smooth manifolds with a smooth addition, we obtain the following theorem:

Theorem 5.2. There exists a functor $S : {}_{comp}\mathcal{M}^{op}_{smooth} \times {}_{add}\mathcal{M}_{smooth} \to {}_{add}\mathcal{M}_{smooth}$ given on objects by $S(M,N)$ with smooth charts as above, and defined by composition for smooth maps. This functor has the property that whenever M and N are compact smooth manifolds and P is a convenient smooth manifold with a smooth addition, then the smooth manifolds $S(M,S(N,P))$ and $S(M \sqcap N,P)$ are naturally diffeomorphic by the usual correspondence of $\alpha : M \to S(N,P)$ with $\hat{\alpha} : M \sqcap N \to P$ where $\hat{\alpha}(x,y) = \alpha x(y)$.

The proof of theorem 5.2 is straight forward, using of course the generalisation of theorem 4.9 at the end of section 4 to show the smoothness of induced maps. To obtain a smooth addition for $S(M,N)$ one establishes first a natural smooth diffeomorphism $\delta : TS(M,N) \overset{\approx}{\rightarrowtail\!\!\!\twoheadrightarrow} S(M,TN)$ analogous to the one in theorem 4.9. The smooth addition $\varepsilon_N : TN \to N$ then induces the smooth map $(\varepsilon_N)_* : S(M,TN) \to S(M,N)$ and one gets in $(\varepsilon_N)_* \circ \delta : TS(M,N) \to S(M,N)$ the desired smooth addition for $S(M,N)$.

It rests to state that we did not exactly obtain what we wanted: Namely a cartesian closed category of convenient smooth manifolds. The compactness condition on M for forming $S(M,N)$ as a convenient smooth manifold should be removed. But for doing this, the topology on the section spaces seems to be too coarse. May be that there exists another generalisation of differential calculus which works as good in the manifold case as the one developed here works in the vector space case.

References

[1] A.Bastiani: Applications différentiables et variétés différentiables de dimension infinie, J.d'Anal.Math.13 (1964), 1-114

[2] N.Bourbaki: Topologie générale, Hermann, Paris

[3] J.Eells : A setting for global analysis, Bull.Am.Math.Soc.72 (1966), 751-807

[4] A.Frölicher-W.Bucher: Calculus in vector spaces without norm, SLN 30, Springer, Berlin (1966)

[5] A.Frölicher-H.Jarchow: Zur Dualitätstheorie kompakt erzeugter und lokal konvexer Vektorräume, Comm.Math.Helv.47 (1972), 289-310

[6] P.Gabriel-M.Zisman: Fondement de la topologie simpliciale, Séminaire homotopique, Université de Strasbourg (1963/64)

[7] M.Golubitsky-V.Guillemin: Stable mappings and their singularities, GTM 14, Springer, New York-Heidelberg-Berlin (1973)

[8] S.Lang : Differential manifolds, Addison-Wesley, Reading (1972)

[9] H.H.Schäfer:Topological vector spaces, GTM 3, Springer, New York-Heidelberg-Berlin (1971)

[10] U.Seip : Kompakt erzeugte Vektorräume und Analysis, SLN 273, Springer, Berlin-Heidelberg-New York (1972)

Ulrich Seip
Fachbereich Mathematik
Universität Konstanz
Postfach 7733
D 775 KONSTANZ
Deutschland

Ulrich Seip
Instituto de Matemática e Estatística
Universidade de São Paulo
Cx. Postal 20.570 (Ag. Iguatemi)
BR 01451 SÃO PAULO
Brasil

PERFECT SOURCES

by

G. E. Strecker

Abstract:

A survey is given of various approaches to suitable categorical analogues of perfect maps. The notions of α-perfect source and $\mathcal{R}$-strongly perfect source are defined, and are shown to be ideally suited to factorization theory and theorems demonstrating the existence and construction of epireflective hulls in a quite general setting. A characterization is given of when the two types of perfectness coincide, and suggestions for further study are provided.

§1. Introduction

It is not surprising that soon after finding appropriate categorical analogues for the important topological entities of homeomorphisms, injective and surjective mappings, embeddings, closed embeddings, dense maps, and quotient maps, topologists would focus attention on the problem of obtaining a suitable analogue for the important class of perfect mappings.

The significance of these mappings stems from the fact that even though they need be neither injective nor surjective, they come close to being both structure preserving as well as structure reflecting, and are very closely related to the important notion of compactness. Consequently, as a class, perfect maps have quite nice characteristics and many topological properties are preserved or inversely preserved by them.

A topological map $f : X \longrightarrow Y$ is said to be perfect if and only if it is continuous, closed, and has compact point-inverses. Bourbaki $[B_2]$ has shown that this is equivalent to:

(*) for each space Z, $f \times 1_Z : X \times Z \longrightarrow Y \times Z$ is closed.

In [HI] Henriksen and Isbell showed that for a mapping $f : X \longrightarrow Y$ between completely regular spaces:

(**) f is perfect if and only if $\beta f[\beta X \setminus X] \subset \beta Y \setminus Y$.

Each of the above characterizations is subject to a categorical generalization. For the first, however, one needs the existence of finite products and, more importantly, the notion of "closed mapping" and so afortiori the notion of "closed sets" in objects of the category in question. Such a generalization has been obtained by Manes [M_1]. The assumption is made that one is dealing with a category of structured sets which reasonably creates finite products and for which each object is assigned a family of "closed" subsets of its underlying set that behave appropriately (e.g. they are finitely intersective and each morphism inversely preserves them). He defines a perfect map as the obvious analogue of (*), calls an object compact iff the terminal map is perfect and calls it Hausdorff iff its diagonal is closed. Some quite general proofs of standard topological results are obtained; e.g.

(1) perfect maps form a subcategory.

(2) an object is compact iff each projection parallel to it is closed.

(3) compactness if finitely productive, closed hereditary, preserved by surjective maps and inversely preserved by perfect maps.

(4) an object is Hausdorff iff the graph of any map with it as codomain is closed.

(5) any map with compact domain and Hausdorff codomain is perfect.

The obvious analogue of (**) would be in the setting of a category $\mathcal{C}$ having a given epireflective full subcategory, $\mathcal{R}$ (and reflection maps $r_A: A \to rA$).

A morphism $f:A \to B$ is called $\mathcal{R}$- strongly perfect iff

$$\begin{array}{ccc} A & \xrightarrow{r_A} & rA \\ f\downarrow & & \downarrow rf \\ B & \xrightarrow[r_B]{} & rB \end{array}$$

is a pullback square.

Franklin $[F_3]$, Hager $[H_1]$, Tsai [T], and Błaszczyk and Mioduszewski [BM] have each used the above idea to generalize or obtain analogues of perfect mappings in various restricted settings. In particular, Franklin and Hager deal only with the categories of Tychonoff spaces and uniform spaces and their reflection maps are assumed to be embeddings, Tsai only considers subcategories of Hausdorff spaces, and Blaszczyk and Mioduszewski's setting is within the category of Hausdorff spaces and mappings which can be extended to the Katětov H-closed extensions $[K_1]$ of their domains and codomains. Whereas in [T] and [BM] the main thrust is to obtain topological (internal) characterizations of the $\mathcal{R}$-strongly perfect morphisms and their relationship to the question of extendibility, in $[F_3]$ and $[H_1]$ some easily abstracted, quite general proofs of characteristics of $\mathcal{R}$-strongly perfect maps are obtained; namely it is shown that $\mathcal{R}$-strongly perfect maps:

(1) form a subcategory

(2) are arbitrarily productive

(3) are closed under the formation of pullbacks;
e.g., projections parallel to $\mathcal{R}$-factors.

In this setting Franklin and Hager have also obtained generalizations of Franklin's

$[F_2]$ and Herrlich and van der Slot's $[HS_1]$ theorems dealing with left-fitting hulls of various topological properties. In this connection we should also mention a very nice generalization of Franklin's result due to Nel $[N_3]$. Also see $[HS_3, 37C]$.

Since the characterizations (*) nad (**) are the same for maps between completely regular spaces it seems natural to inquire as to the disparity of their generalizations. Herrlich $[H_3]$ has shown that they are different even in the category $\mathcal{C}$ of Hausdorff spaces, where $\mathcal{K}$ is the subcategory of compact spaces. Namely, there are closed embeddings that are not strongly perfect. In $[H_2]$ Herrlich has suggested and later ($[H_3]$) more fully analyzed a third general approach to perfect maps that agrees with the analogues to (*) and (**) for completely regular spaces and lies strictly between them when applied to Hausdorff spaces. Nakagawa $[N_2]$ has also investigated an approach that for suitably nice categories yields the same classes of perfect maps. Herrlich's approach, which is closely related to the right bicategorical structures of Kennison $[K_4]$, can be described as follows:

For any class of objects ω in a category $\mathcal{C}$, call $e: A \to B$ an ω-<u>extendible epimorphsim</u> iff e is an epimorphism and whenever $K \in \omega$ and $f: A \to K$, there exists some h such that $f = h \circ e$. A morphism f is called ω-perfect iff whenever

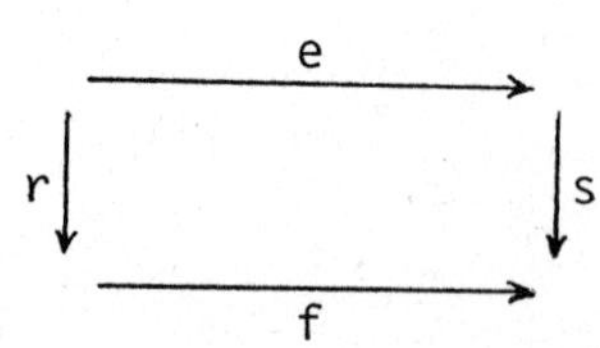

is a commutative square with e an ω-extendible epimorphism, there exists some d such that $r = d \circ e$ and $s = f \circ d$.

In [H_2] the classes of epimorphisms that are the ω-extendible epimorphisms for some ω are characterized, the ω-perfect morphisms are shown to be:

(1) closed under composition,

(2) closed under (multiple) pullbacks,

(3) closed under products,

(4) left cancellative, and

(5) a superclass of the class of all strong monomorphisms.

The relationship between the ω-perfect morphisms, epireflective hulls, and factorizations is investigated and a characterization for those classes of morphisms that are the ω-perfect morphisms for some ω is called for. Suitable such characterizations of perfect morphisms have since been given in [S_2] and [S_3]. In these papers and in [H_3], [N_2], and [N_3] numerous related results and refinements and improvements in the theory have been obtained and many examples given.

In the next section we shall make some further refinements, a major one being to extend the theory to perfect sources, rather than perfect morphisms. It has been brought to our attention that an approach to source diagonalizations and factorizations similar to that given below will appear in [P].

§2. Definitions

We will assume throughout that we have a given category $\mathcal{C}$ and that all epireflective subcategories are both full and isomorphism-closed.

A source is a pair (A,F) where A is a $\mathcal{C}$-object and F is a class of morphisms each with domain A. Such a source will sometimes simply be denoted by F. If (A,F) is a source and for each $f : A \to B_f$ in F, (B_f, G_f) is a source, then $(G_f) \circ F$ will denote the source (A,H) where $H = \{g \circ f \mid f \in F \text{ and } g \in G_f\}$. In the special case where (A,F) is a source and $g : B \to A$ is a morphism, $F \circ g$ is the source $(B, (f \circ g)_{f \in F})$.

If β is a family of sources, then β_1 denotes the family of all singleton sources in β. Each such source $(B, \{f\})$ will be identified with its single morphism f.

A family of sources, β, is called <u>basic</u> provided that:

(B1) β_1 contains all isomorphisms in $\mathcal{C}$.

(B2) β is closed under composition; i.e., whenever F is a source in β and (G_f) is a family of sources in β such that $(G_f) \circ F$ exists, $(G_f) \circ F$ must be in β.

(B3) β is closed under the formation of joint pullbacks; i.e., if $(L, K \cup \{h\})$ is the limit of the diagram

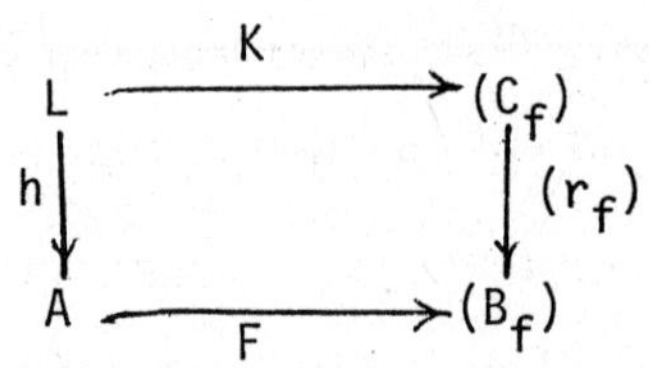

where $(A,F) \in \beta$, then $(L,K) \in \beta$.

(B4) β is closed under the formation of multiple pullbacks; i.e., whenever $(F_i)_I$ is a family of sources belonging to β with the property that for each $i, j \in I$ the class of all codomains of F_i is the same as the class of all codomains of F_j, then if (L,K) is their limit, it follows that $(F_i)\circ K$ belongs to β.

A class of morphisms is called <u>cobasic</u> provided that it, considered as a class of singleton sinks, is a cobasic class of sinks; i.e., satisfies the conditions dual to (B1)-(B4) above.

Let α be any class of morphisms. A source F is called α-<u>lower diagonalizable</u> provided that whenever e and r are morphisms with $e \in \alpha$ and G is a source such that $G\circ e = F\circ r$, there exists a morphism d such that the diagram

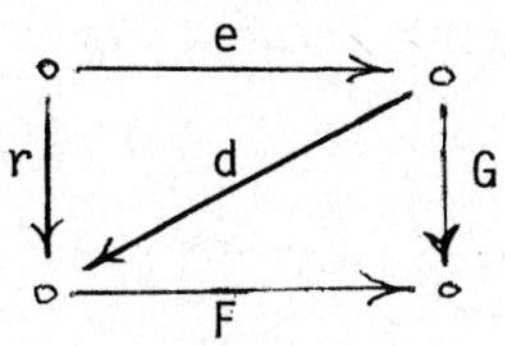

commutes. The class of all α-lower diagonalizable sources will be denoted by $\Lambda(\alpha)$. Similarly if β is a family of sources, a morphism e is called β-<u>upper diagonalizable</u>, and we write $e \in \Upsilon(\beta)$, iff whenever r is a morhpism and F and G are sources with $F \in \beta$ such that $G\circ e = F\circ r$ there exists some d such that the above diagram commutes. If α is a class of epimorphisms, then a member of $\Lambda(\alpha)$ is called an α-<u>perfect source</u> and a member of $\Lambda(\alpha)_1$ is called an α-<u>perfect morphism</u>. β is called a <u>class of perfect sources</u> if $\beta = \Lambda(\alpha)$ for some class of epimorphisms α.

If ω is any class of objects, then we say that $f : A \to B$ is ω-<u>extendible</u> provided that for each $D \in \omega$ and

$g : A \to D$ there exists some $h : B \to D$ such that $g = h \circ f$. The class of ω-extendible epimorphisms will be denoted by $\chi(\omega)$.

If β is a family of sources, then $\Delta(\beta)$ will denote the class of all objects A having the property that each source with domain A belongs to β .

A source F is called a mono-source iff whenever $F \circ f = F \circ g$ it follows that $f = g$. The class of all mono-sources that are also epi-perfect is called the class of all strong mono-sources (cf. [K_3])

A family of sources is said to be:

left-cancellative

iff whenever F is a source and (G_f) is a family of sources such that $(G_f) \circ F$ exists and belongs to β , then F must belong to β .

fundamental iff β :

(F1) is basic,

(F2) contains all strong mono-sources, and

(F3) is left-cancellative.

closed under the formation of products

iff whenever for each $i \in I$, (A_i, F_i) is a source in β for which the induced map $\langle F_i \rangle$ from A_i to the product of the codomains of F_i exists, and such that $(\Pi A_i, \pi_i)$ exists, then the source $(\Pi A_i, (\langle F_i \rangle) \circ (\pi_i))$ belongs to β .

For a given class, α, of epimorphisms, $\mathcal{C}$ is said to be:

α-compatible

iff every α-source in $\mathcal{C}$ has a cointersection, and each two-element source in $\mathcal{C}$

having at least one member in α has a pushout.

an α-<u>perfect category</u>

provided that each source F in $\mathcal{C}$ has an essentially unique factorization $F = G \bullet e$, with $e \in \alpha$ and G α-perfect.

If $\mathcal{R}$ is an epireflective subcategory of $\mathcal{C}$ with reflection morphisms $r_A : A \longrightarrow rA$, then a source (A,F) is called an $\mathcal{R}$-<u>strongly perfect source</u> iff the source $(A, F \cup \{r_A\})$ is the limit of the diagram

$$\begin{array}{ccc} & & rA \\ & & \Big\downarrow{\scriptstyle rf} \\ B_f & \xrightarrow{\;r_{B_f}\;} & rB_f \end{array}$$

where for each $f \in F$, $f : A \to B_f$.

§3 Results

<u>Theorem 1</u>

For any class of sources, β, the following are equivalent:

(1) β is a class of perfect sources.

(2) $\beta = \Lambda(\Upsilon\beta \cap \alpha)$, for some class of epimorphisms, α.

(3) $\beta = \Lambda(\Upsilon\beta \cap \mathit{epi})$, where epi is the class of all epimorphisms of $\mathcal{C}$.

(4) $\beta = \Lambda(\delta)$ for some cobasic family of epimorphisms, δ.

Theorem 2

For any class α of morphisms:

(1) $\alpha \cap \Lambda(\alpha)_1$ is contained in the class of all isomorphisms.

(2) $\Lambda(\alpha)$ is basic and closed under the formation of products.

Theorem 3

For any class α of epimorphisms;

(1) $\Lambda(\alpha)$ is fundamental, and $\Lambda(\alpha)_1$ is closed under the formation of inverse limits of inverse spectra; i.e., if each spectrum map is in $\Lambda(\alpha)_1$, then each member of the inverse limit source will be in $\Lambda(\alpha)_1$.

(2) if $\mathcal{C}$ is α-perfect, then every class ω of objects in $\mathcal{C}$ has an α-reflective hull, $R_\alpha(\omega)$. Furthermore, for each object A the $R_\alpha(\omega)$-reflection morphism is the first factor of the $(\alpha, \Lambda(\alpha))$ factorization of the source consisting of all morphisms with domain A and codomain belonging to ω.

(3) if $\mathcal{C}$ is α-compatible, then the following are equivalent:

(a) α is cobasic,

(b) $\mathcal{C}$ is α-perfect and α is closed under composition with isomorphisms,

(c) $\alpha = \Upsilon\Lambda(\alpha)$;

if, in addition, α has the property that whenever $f \circ g \in \alpha$ and g is an epimorphism, it follows that $g \in \alpha$, then (a), (b) and (c) are equivalent to:

(d) $\alpha = \chi(\omega)$ for some class of objects ω

(e) $\alpha = \chi\, \Delta\, \Lambda(\alpha)$.

(4) if ω is any class of objects for which $(\chi(\omega) \cap \alpha)$ is cobasic and $\mathcal{C}$ is $(\chi(\omega) \cap \alpha)$-compatible, then ω has an α-reflective hull in $\mathcal{C}$ given by: $R_\alpha(\omega) = \Delta\Lambda(\chi(\omega) \cap \alpha)$.

Thus under these conditions $\omega = \Delta\Lambda(\chi(\omega) \cap \alpha)$ if and only if ω is the object class of a full α-reflective subcategory of $\mathcal{C}$.

Theorem 4

Let ω be any class of objects, α be any class of epimorphisms, and $\mathcal{R}$ be α-reflective in $\mathcal{C}$. Then

(1) each $\mathcal{R}$-strongly perfect source is $(\chi(\mathcal{R}) \cap \alpha)$-perfect.

(2) the following are equivalent:

(a) The $\mathcal{R}$-strongly perfect sources and the $(\chi(\mathcal{R}) \cap \alpha)$-perfect sources coincide and $\mathcal{C}$ is an $(\chi(\mathcal{R}) \cap \alpha)$-perfect category.

(b) For each source $(A, (f_i))$ in $\mathcal{C}$ and each diagram

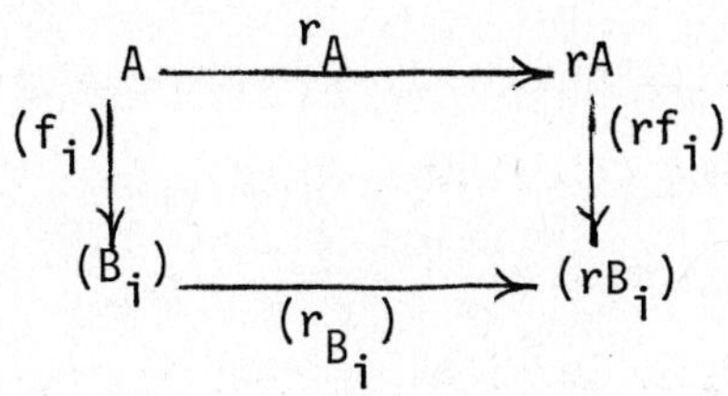

the lower corners have a limit $(L, (k_i), k_A)$ and the induced morphism, h, belongs to α.

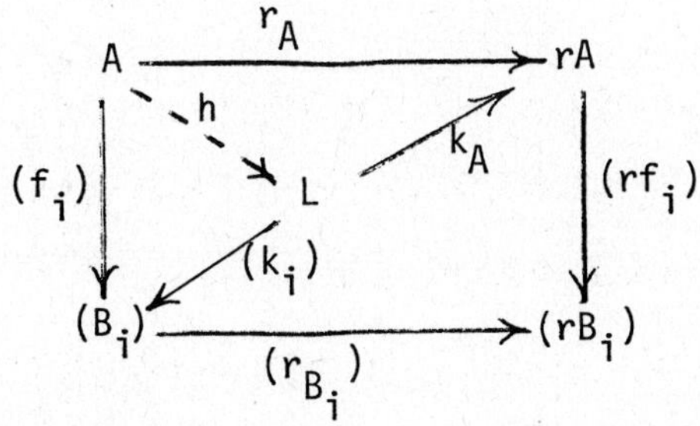

§4 Consequences and Applications

Since singleton perfect sources are perfect morphisms, Theorems 1,

2, and 3 clearly generalize many of the main results of [H_2], [H_3], [N_2], [S_2] and [S_3]. Note that we have defined perfectness relative to classes of epimorphisms, rather than to classes of objects as was done in the papers cited above. In particular for any class of objects, ω, our $\Lambda(\chi(\omega))_1$ is precisely the ω-perfect morphisms of [H_2], [H_3], [N_2], [S_2] and [S_3]. This alteration has allowed us to obtain as special cases of our theorems many of the results of Nel [N_3], who defined perfect morphisms relative to development classes of morphisms, thereby broadening the scope of the theory. The general results of §3 also shed new light on factorizations of the "monotone quotient-light" variety and their attendant "totally disconnected" reflections (cf. [S_4]). Note that in Theorem 2 there is no restriction on the class of morphisms α; in particular members of α need not be epimorphisms. Also there is no assumption of uniqueness of the diagonal morphism in the definition of Λ. Thus Theorem 2 yields results e.g., for closed embeddings since in the category of topological spaces they are $\Lambda_1(\alpha)$ where α = all dense maps; and for (amnestic) topological functors since, by the recently announced result of Brümmer and Hoffman, in the category of all categories and faithful functors, they are $\Lambda(\alpha)$, where α = all full faithful functors.

Besides specializations to the general theories mentioned above, there are also interesting new specializations all the way down to classical topological results. For example Theorem 3(1), for the case where $\mathcal{C}$ is the completely regular spaces and α is the compact extendible epimorphisms yields Morita's result [M_2] that every inverse limit of an inverse spectrum of metrizable spaces and perfect bonding maps is a paracompact M-space.

It should be mentioned that a special case of our general notion of factoring sources so that the second factor is a perfect source has in essence been considered by Whyburn [W_1], [W_2], Cain [C_1], [C_2], and others. In the Whyburn-Cain setting one has a continuous function (between Hausdorff spaces) $f:A \to B$ and a compactification $\widetilde{A}$ of A. An attempt is then made to obtain a perfect mapping $f^*:A^* \to B$ with A a dense subset of A^*, $f^*|A = f$ and a mapping $h:A^* \to A$ whose restriction to A is the inclusion of A into $\widetilde{A}$. Looking at this from a "source" standpoint one has the given source $(A, (f,e))$, where $e:A \to \widetilde{A}$. If one then takes the (epi, strong monosource)-factorization in the category of Hausdorff spaces, $(A^*, (f^*,h))$ is immediately obtained as the second factor. Using this approach many of the theorems involving "mapping compactifications" become quite easy to prove. For example the two main results of [C_2] reduce to trivialities. Also using the "factorization of two-element sources" approach, the result of Nagata [N_1] that a space of weight m is a paracompact M-space if and only if it is homeomorphic with a closed subspace of the product of a metrizable space and $[0,1]^m$, is easily shown and the construction of Morita's paracompactification μX of a given M-space [M_2] and the Katětov-Viglino absolute closure of a Hausdorff mapping [V] are readily obtained. For more details see [DS].

Theorem 4 above has some significance in that it characterizes the situation when two of the major categorical approaches to perfectness coincide, and thus gives perhaps some more insight as to why they yield the same classes of "compact-perfect" mappings (resp. sources) in the category of completely regular spaces, but different classes in the category of Hausdorff spaces.

§5. Areas for Further Study

The theory of perfect maps of Manes [M_1] which was described at the beginning is the most recent of the three "perfect map" theories and is one which it seems deserves further exploitation. An earlier paper of Brown [B_3] essentially shows that the Manes approach "works" for the category of topological spaces and sequentially continuous functions. Here "closed" means "sequentially closed" [F_4], "compact" translates to "sequentially compact" and "Hausdorff" becomes "space with unique sequential limits." Since sequential compactness is finitely productive, but not arbitrarily so, this shows that the Manes approach is appropriate in situations where the α-perfect or $\mathcal{R}$-strongly perfect approaches are not. Also it is apparent that with some slight tightening of the Manes axioms (e.g., making closed sets arbitrarily intersective, and having the canonical inclusion of a closed subset of a structured set be not only perfect but also an optimal lift) one could obtain appropriate general versions of many more classical results relating perfectness, compactness, Hausdorff, and various factorizations. Thus, in the full subcategory of Hausdorff objects, each dense morphism would be an epimorphism, each regular monomorphism would be a closed embedding, and unique (dense, closed embedding dyad)-factorizations would occur. (cf. [HS_2] and [DS]).

Not long after Gleason's [G] fundamental discovery concerning onto-projectives and projective covers in the category of compact Hausdorff spaces, it became clear that to extend the results to wider topological categories the appropriate mappings for projectivity to be relative to, should be the perfect onto maps (see e.g., [F_1] and [S_1]). Banaschewski [B] has obtained quite nice general results that extend many earlier ones in the area. Here, again, projectivity is considered relative to the perfect onto mappings between Hausdorff spaces. It seems as though a fruitful area to investigate would be that of projectives in still more general settings using the appropriate categorical analogues of perfectness given above.

We should also mention that since the approaches to perfectness mentioned in this paper are categorical, they are all subject to dualization; i.e., coperfect morphisms, coperfect sinks, monocoreflective hulls, etc.

Although some interesting examples and applications in this dual realm are presented in [H_3] and [S_3], it appears to be an area which historically has been neglected and which deserves further study and development. In this connection it is interesting to observe that sequential spaces yield a nice example for Manes-perfectness and at the same time a fundamental example of a monocoreflective hull.

Finally, it appears that many useful results could be obtained by a determination of internal characterizations of various categorical perfect sources and morphisms in the recently developed, quite important and convenient categories of nearness spaces, seminearness spaces and grills [BHR], [H_4], [H_5], [H_6], and [K_2].

REFERENCES

[B_1] B. Banaschewski, Projective covers in categories of topological spaces and topological algebras. General Topology and Its Relations to Modern Analysis and Algebra, III (Proc. Conf. Kanpur, 1968), pp. 63-91. Academia, Prague, 1971.

[B_2] N. Bourbaki, General Topology, Part 1, Addison-Wesley, Reading, Mass., 1966.

[B_3] R. Brown, On sequentially proper maps and a sequential compactification, J. London Math. Soc. (2), 7 (1973), 515-522.

[BHR] H. L. Bentley, H. Herrlich, and W. A. Robertson, Convenient categories for topologists, preprint.

[BM] A. Błaszczyk and J. Mioduzewski, On factorization of maps through τX, Colloq. Math. 23 (1971), 45-52.

[C_1] G. L. Cain, Jr., Extensions and compactifications of mappings, Math. Ann. 191 (1971), 333-336.

[C_2] __________, Metrizable mapping compactifications, General Topology and Appl., 2 (1972), 271-275.

[DS] F. A. Delahan and G. E. Strecker, Graphic extensions of mappings, preprint.

[F_1] J. Flachsmeyer, Topologische Projektivräume, Math. Nachr. 26 (1963) 57-66.

[F_2] S. P. Franklin, On epi-reflective hulls, General Topology and Appl. 1 (1971), 29-31.

[F_3] __________, On epi-reflective hulls II, Notes for Meerut Univ. Summer Inst. on Topology, 1971.

[F_4] __________, Spaces in which sequences suffice, Fund. Math. 57 (1965), 107-115.

[G] A. M. Gleason, Projective topological spaces, Illinois J. Math. 2 (1958), 482-489.

[H_1] A. W. Hager, Perfect maps and epireflective hulls, Canad. J. Math. 27 (1974), 11-24.

[H_2] H. Herrlich, A generalization of perfect maps, General Topology and Its Relations to Modern Analysis and Algebra, III (Proc. Third Prague Topological Sympos. (1971), pp. 187-191. Academia, Prague, 1972.

[H_3] __________, Perfect subcategories and factorizations, Topics in Topology (Proc. Colloq. Kaszthely, 1972), pp. 387-403. Colloq. Math. Soc. János Bolyai, Vol 8, North Holland, Amsterdam, 1974.

[H_4] __________, A concept of nearness, General Topology and Appl. 5 (1974), 191-212.

[H_5] __________, Topological structures, Math. Centre Tract 52 (1974), 59-122.

[H_6] __________, Some topological theorems which fail to be true, preprint.

[HI] M. Henriksen and J. R. Isbell, Some properties of compactifications, Duke Math. J., 25 (1958), 83-106.

[HS_1] H. Herrlich and J. van der Slot, Properties which are closely related to compactness, Indag. Math. 29 (1967), 524-529.

[HS_2] H. Herrlich and G. E. Strecker, Coreflective subcategories in general topology, Fund. Math. 73 (1972), 199-218.

[HS_3] ______________________________, Category Theory, Allyn and Bacon, Boston, 1973.

[K_1] M. Katětov, Über H-abgeschlossene und bikompakte Räume Časopis Pěst, Mat. Fys. 69 (1940), 36-49.

[K_2] __________, On continuity structures and spaces of mappings, Comment. Math. Univ. Carolinae 6 (1965), 257-278.

[K_3] G. M. Kelley, Monomorphisms, epimorphisms, and pull-backs, J. Austral. Math. Soc. 9 (1969), 124-142.

[K_4] J. F. Kennison, Full reflective subcategories and generalized covering spaces, Illinois J. Math. 12 (1968), 353-365.

[M_1] E. G. Manes, Compact Hausdorff objects, General Topology and Appl. 4 (1974), 341-360.

[M_2] K. Morita, Topological completions and M-spaces, Sci. Rep. Tokyo Kyoiku Daigaku 10 , No. 271 (1970), 271-288.

[N_1] J. Nagata, A note on M-spaces and topologically complete spaces, Proc. Japan Acad. 45 (1969), 541-543.

[N_2] R. Nakagawa, Classes of morphisms and reflections, preprint.

[N_3] L. D. Nel, Development classes: an approach to perfectness, reflectivesness and extension problems, TOPO 72-General Topology and its Applications (Proc. Second Pittsburgh Internat. Conf.) pp. 322-340. Lecture Notes in Math., Vol. 378, Springer, Berlin 1974.

[P] D. Pumplün, Kategorien, to appear.

[S_1] D. P. Strauss, Extremally disconnected spaces, Proc. Amer. Math. Soc. 18 (1967), 305-309.

[S_2] G. E. Strecker, Epireflection operators vs. perfect morphisms and closed classes of epimorphisms, Bull. Austral. Math. Soc. 7 (1972), 359-366.

[S_3] ____________, On characterizations of perfect morphisms and epireflective hulls, TOPO 72-General Topology and its Applications (Proc. Second Pittsburgh Internat. Conf.) pp. 468-500. Lecture Notes in Math., Vol. 378, Springer, Berlin 1974.

[S_4] ____________, Component properties and factorizations, Topological Structures, Math. Centre Tract 52 (1974), 123-140.

[T] J. H. Tsai, On a generalization of perfect maps, Pacific J. Math. 46 (1973), 275-282.

[V] G. Viglino, Extensions of functions and spaces, Trans. Amer. Math. Soc. 179 (1973), 61-69.

[W_1] G. T. Whyburn, Compactifications of mappings, Math. Ann. 166 (1966), 168-174.

[W_2] ____________, Dynamic topology, Amer. Math. Monthly 77 (1970), 556-570.

Department of Mathematics
Kansas State University
Manhattan, Kansas 66506

Espaces fonctionnels et structures syntopogènes.

par Daniel Tanré

Les structures syntopogènes ont été introduites par Csaszar[4] afin de pouvoir considérer les applications continues entre espaces topologiques, uniformes entre espaces uniformes, les morphismes usuels entre espaces de proximité comme des cas particuliers d'applications T-continues entre espaces syntopogènes. Cependant, la catégorie des structures syntopogènes définies dans (4) n'étant pas proprement fibrée[8], ce que nous appellerons ici structure syntopogène est une classe d'équivalence de structures syntopogènes au sens de (4). Ceci coïncide d'ailleurs avec la définition de T-espaces préidempotents donnée par Hacque (7). La différence de présentation entre les deux notions est la suivante: $A < B$ dans (4) signifie $B \in \rho(A)$ dans (7). Nous notons $\mathcal{V}$ la catégorie des T-espaces et $\mathcal{S}$ la sous-catégorie pleine de $\mathcal{V}$ formée des espaces syntopogènes. Nous montrons que $\mathcal{V}$ est une catégorie topologique proprement fibrée[8] et que $\mathcal{S}$ est stable pour la formation de structures initiales dans $\mathcal{V}$.

Nous abordons ensuite l'essentiel de notre étude: les espaces fonctionnels de structures syntopogènes. Si $\mathbb{Y}$ est un espace syntopogène sur Y et X un ensemble, nous mettons sur l'ensemble $M(Y,X)$ des applications de X vers Y une structure d'espace syntopogène qui se réduit à la structure de la convergence uniforme si Y est un espace uniforme.

Si $\mathbb{X}$ et $\mathbb{Y}$ sont des espaces syntopogènes sur X et Y respectivement, nous définissons les parties T-équicontinues de $M(Y,X)$, de sorte que, si $\mathbb{X}$ et $\mathbb{Y}$ sont des espaces topologiques (resp. des espaces uniformes, resp. si $\mathbb{X}$ est un espace topologique et $\mathbb{Y}$ un espace uniforme) nous retrouvions la définition de l'égale continuité[10] de Kelley (resp. équicontinuité uniforme[2], resp. équicontinuité). Enfin, nous relions ces deux notions en les appliquant dans le cas particulier d'applications partielles déduites d'une application f de $X \times K$ vers Y, où K est un ensemble.

Dans ce contexte, on ne peut que regretter la lourdeur des notations et la technicité des démonstrations. Aussi, pour terminer, nous définissons un foncteur Hom interne à la catégorie des N-espaces[8], foncteur qui coïncide avec la convergence simple. Ici, les démonstrations se déduisent, dans leur grande majorité, des résultats catégoriques obtenus sur les N-espaces.

1- Les catégories des T-espaces et des structures syntopogènes:

Soit E un ensemble, $\mathcal{P}(E)$ l'ensemble des parties de E et $\mathcal{F}(E)$ l'ensemble des filtres sur E. Nous noterons $\emptyset$ l'ensemble vide.

<u>Définition 1-1</u>: Une application ρ de $P(E)$ dans $P^2(E)$ définit un *ordre semi-topogène*(4) sur E si les axiomes suivants sont vérifiés:

(ST_1) $\emptyset \in \rho(\emptyset)$, $E \in \rho(E)$;

(ST_2) $B \subset B'$, $A \subset A'$ et $B \in \rho(A)$ entraînent $B' \in \rho(A')$;

(ST_3) $B \in \rho(A)$ entraîne $A \subset B$.

<u>Définition 1-2</u>: Une application ρ de $P(E)$ dans $P^2(E)$ définit un *ordre topogène* (4) sur E (ou (E,ρ) est un T-*espace simple*(7)) si ρ est un ordre semi-topogène vérifiant l'axiome supplémentaire:

(Q) $B \in \rho(A)$ et $B' \in \rho(A')$ entraînent $B \cap B' \in \rho(A \cap A')$ et $B \cup B' \in \rho(A \cup A')$.

Nous pouvons exprimer la définition d'ordre topogène et de T-espace simple sous la forme plus légère suivante(7),(11) :

ρ est une application de $P(E)$ dans $F(E)$ vérifiant:

(T_1) $\emptyset \in \rho(\emptyset)$;

(T_2) $B \in \rho(A)$ entraîne $A \subset B$;

(T_3) $\rho(A \cup B) = \rho(A) \cap \rho(B)$.

L'ensemble des ordres topogènes sur E est ordonné par la relation $\rho_1 \subset \rho_2$, ρ_2 est plus fine que ρ_1. Si ρ_1 et ρ_2 sont deux ordres topogènes, on définit un nouvel ordre topogène $\rho = \rho_1 \circ \rho_2$ en posant, pour tout $A \subset E$, $\rho(A) = \bigcup_{B \in \rho_2(A)} \rho_1(B)$.

A un ordre topogène ρ correspond canoniquement une relation d'éloignement $\bar{\delta}$ caractérisée par $A \,\bar{\delta}\, B$ ssi il existe $C \in \rho(B)$ tel que $C \cap A = \emptyset$.

Un ordre topogène ρ sur E est dit:

ponctuel si $\rho(A) = \bigcap_{x \in A} \rho(x)$, pour tout $A \in P(E)$;

parfait s'il est ponctuel et si les filtres $\rho(A)$ sont principaux;

symétrique si $\bar{\delta}$ est symétrique;

idempotent si $\rho^2 = \rho$.

<u>Remarque</u>: La terminologie ci-dessus est celle de (7); dans (4) ponctuel est remplacé par parfait et parfait par biparfait. De plus, dans (4), la relation ρ est dite symétrique si $B \in \rho(A)$ implique $E-A \in \rho(E-B)$. On constate immédiatement que les deux définitions d'ordre topogène symétrique coïncident.

<u>Définition 1-3</u>: Soit (E,ρ) et (E',ρ') deux T-espaces simples. Une application f de E vers E' définit une *application* T-*continue*(7) de (E,ρ) vers (E',ρ') si, pour tout $B \in P(E)$, l'image réciproque par f de tout ρ'-voisinage de $f(B)$ est un ρ-voisinage de B, i.e. $f^{-1}\rho'(f(B)) \subset \rho(B)$.

Introduisons maintenant les structures étudiées dans (4) et (7): les T-espaces et les structures syntopogènes.

<u>Définition 1-4</u>: Une T-*base* $\mathcal{B}$ sur un ensemble E est un ensemble filtrant à droite d'ordres topogènes sur E, i.e.

(S_1) pour tout $(\rho_1,\rho_2) \in \mathcal{B}\times\mathcal{B}$, il existe $\rho \in \mathcal{B}$ tel que $\rho_1 \subset \rho$ et $\rho_2 \subset \rho$.

L'ensemble des T-bases sur E est préordonné par la relation suivante: $\mathcal{B} < \mathcal{B}'$ ssi, pour tout $\rho \in \mathcal{B}$, il existe $\rho' \in \mathcal{B}'$ tel que $\rho \subset \rho'$. Cette relation de préordre détermine une relation d'équivalence et une relation d'ordre sur l'ensemble quotient.

<u>Définition 1-5</u>: Un T-*espace* $\mathbb{E} = (E,[\rho_i])$ est un ensemble E muni d'une classe d'équivalence $[\rho_i]$ de T-bases sur E.

Par abus de notation, une classe d'équivalence sera souvent désignée par l'un de ses éléments $\mathcal{B}$, les ordres ρ_i de $\mathcal{B}$ étant appelés ordres génériques de $\mathbb{E}$. Un T-espace a une propriété (par exemple symétrique) si les ordres génériques d'une de ses T-bases la possèdent.

<u>Définition 1-6</u>: Un *espace syntopogène*(4) (ou T-*espace préidempotent*(7)) est un T-espace $\mathbb{E} = (E,[\rho_i])$ vérifiant:

(S_2) pour tout $\rho_i \in [\rho_i]$, il existe $\rho_j \in [\rho_i]$ tel que $\rho_i \subset \rho_j^2$.

<u>Définition 1-7</u>: Soit $\mathbb{E} = (E,[\rho_i])$ et $\mathbb{E}' = (E',[\rho_j'])$ deux T-espaces. Une application f de $\mathbb{E}$ dans $\mathbb{E}'$ est T-*continue* si, pour tout ordre générique ρ_j' de $\mathbb{E}'$, il existe un ordre générique ρ_i de $\mathbb{E}$ tel que l'application f soit T-continue du T-espace simple (E,ρ_i) dans le T-espace simple (E',ρ_j').

Parmi les T-espaces, on distingue les espaces prétopologiques(3), topologiques, de proximité(5), uniformes. Pour plus de détails sur ces résultats, le lecteur consultera (4) et (7).

Nous allons maintenant établir l'existence de structures initiales dans $\mathcal{V}$. Cette construction n'est faite ni dans (4) ni dans (7), on peut cependant déduire cette existence de résultats de (4) pour la catégorie $\mathcal{S}$ ($\mathcal{S}$ est à produits et à structures images réciproques) des espaces syntopogènes. Nous noterons $\mathcal{V}$ la catégorie des applications T-continues entre T-espaces dont les ensembles sous-jacents appartiennent à un univers $\mathcal{U}$; $\mathcal{S}$ est une sous-catégorie pleine de $\mathcal{V}$.

<u>Proposition 1-8</u>: *La catégorie $\mathcal{V}$ est à structures initiales.*

→ Soit, pour tout $\alpha \in J$, où J appartient à l'univers $\mathcal{U}$, un T-espace $\mathbb{Y}_\alpha = (Y_\alpha,(\eta_\beta^\alpha)_{\beta\in I_\alpha})$. Soit E un ensemble, élément de $\mathcal{U}$, et f_α une application de E dans Y_α, pour tout $\alpha \in J$. Soit J' une partie finie de J; pour tout $\alpha \in J'$, on choisit $\gamma(\alpha) \in I_\alpha$. Notons γ ce choix. On pose, pour tout $B \subset E$;

$C \in \rho^{J',\gamma}(B)$ ssi, pour tout $\alpha \in J'$, il existe $C_\alpha \in \eta^\alpha_{\gamma(\alpha)}(f_\alpha(B))$ tels que

$\bigcap_{\alpha\in J'} f_\alpha^{-1}(C_\alpha) \subset C$.

L'ensemble $\rho^{J',\gamma}(B)$ étant saturé par induction et filtrant à gauche, $\rho^{J',\gamma}$ est une application de $\mathcal{P}(E)$ dans $\mathcal{F}(E)$. L'axiome (T_1) est aussi trivialement vérifié par $\rho^{J',\gamma}$. Pour (T_2), soit $C \in \rho^{J',\gamma}(B)$, il existe $C_\alpha \in \eta^\alpha_{\gamma(\alpha)}(f_\alpha(B))$ tels que $\bigcap_{\alpha\in J'} f_\alpha^{-1}(C_\alpha) \subset C$. De $f_\alpha(B) \subset C_\alpha$, on déduit $B \subset \bigcap_{\alpha\in J'} f_\alpha^{-1}(C_\alpha) \subset C$. Pour l'axiome (T_3), nous remarquons d'abord que $B \subset B'$ entraîne $\rho^{J',\gamma}(B') \subset \rho^{J',\gamma}(B)$. On a ainsi, pour deux parties B et B' quelconques de E:
$\rho^{J',\gamma}(B \cup B') \subset \rho^{J',\gamma}(B) \cap \rho^{J',\gamma}(B')$. Soit $C \in \rho^{J',\gamma}(B) \cap \rho^{J',\gamma}(B')$, il existe $C_\alpha \in \eta^\alpha_{\gamma(\alpha)}(f_\alpha(B))$ et $C'_\alpha \in \eta^\alpha_{\gamma(\alpha)}(f_\alpha(B'))$ tels que $\bigcap_{\alpha\in J'} f_\alpha^{-1}(C_\alpha) \subset C$ et $\bigcap_{\alpha\in J'} f_\alpha^{-1}(C'_\alpha) \subset C$. Or $C_\alpha \cup C'_\alpha \in \eta^\alpha_{\gamma(\alpha)}(f_\alpha(B \cup B'))$ et $\bigcap_{\alpha\in J'} f_\alpha^{-1}(C_\alpha \cup C'_\alpha) \subset C$ entraînent $C \in \rho^{J',\gamma}(B \cup B')$.
On a ainsi montré que $\rho^{J',\gamma}$ est un ordre topogène sur E.

Faisons parcourir à J' l'ensemble des parties finies de J et à γ l'ensemble des choix possibles; $\mathbb{E} = (E,[\rho^{J',\gamma}])$ est un T-espace, c'est-à-dire vérifie l'axiome (S_1). En effet, soit $\rho^{J',\gamma'}$ et $\rho^{J'',\gamma''}$ deux ordres topogènes génériques de E; on pose $K = J' \cup J''$ et on définit un choix γ de la façon suivante:
si $\alpha \in J'$ et $\alpha \notin J''$, on prend $\eta^\alpha_{\gamma(\alpha)} = \eta^\alpha_{\gamma'(\alpha)}$; si $\alpha \notin J'$ et $\alpha \in J''$, on prend $\eta^\alpha_{\gamma(\alpha)} = \eta^\alpha_{\gamma''(\alpha)}$; si $\alpha \in J'$ et $\alpha \in J''$, on sait qu'il existe $\eta^\alpha_{\gamma(\alpha)}$ tel que $\eta^\alpha_{\gamma'(\alpha)} \subset \eta^\alpha_{\gamma(\alpha)}$ et $\eta^\alpha_{\gamma''(\alpha)} \subset \eta^\alpha_{\gamma(\alpha)}$. On a alors:
$\rho^{J',\gamma'} \subset \rho^{K,\gamma}$ et $\rho^{J'',\gamma''} \subset \rho^{K,\gamma}$.

Il reste à montrer que $\mathbb{E}$ est la structure initiale pour la famille $(\mathcal{Y}_\alpha, f_\alpha)_{\alpha\in J}$. Soit $\mathcal{Z} = (Z,[z_j])$ un T-espace et g une application de Z dans E; on pose $g_\alpha = f_\alpha \circ g$. L'application f_α est T-continue de $\mathbb{E}$ vers $\mathcal{Y}_\alpha$. En effet, soit η^α_β un ordre générique de $\mathcal{Y}_\alpha$, on pose $J' = \{\alpha\}$ et $\gamma(\alpha) = \beta$; si $C \in \eta^\alpha_\beta(f_\alpha(B))$, alors $f_\alpha^{-1}(C)$ appartient à $\rho^{J',\gamma}(B)$, par définition. La T-continuité de l'application g de $\mathcal{Z}$ vers $\mathbb{E}$ entraîne donc celle de g_α de $\mathcal{Y}_\alpha$ vers $\mathcal{Z}$.
Réciproquement, supposons que g_α soit T-continue de $\mathcal{Y}_\alpha$ vers $\mathcal{Z}$. Soit $\rho^{J',\gamma}$ un ordre générique de $\mathbb{E}$ et soit $(\eta^\alpha_{\gamma(\alpha)})_{\alpha\in J'}$ la famille définissant $\rho^{J',\gamma}$. Par hypothèse, pour tout $\alpha \in J'$, il existe un ordre générique z^α_j de $\mathcal{Z}$ tel que g_α soit T-continue de z^α_j vers $\eta^\alpha_{\gamma(\alpha)}$. D'après (S_1), l'ensemble J' étant fini, il existe $z^{J'}_j$ tel que $z^\alpha_j \subset z^{J'}_j$, pour tout $\alpha \in J'$. On vérifie aisément que g est T-continue de $(Z,z^{J'}_j)$ vers $(E,\rho^{J',\gamma})$. ←

La catégorie $\mathcal{V}$ est donc une catégorie topologique proprement fibrée au sens de (8).

Proposition 1-9: *Si tous les ordres topogènes considérés dans la démonstration précédente sont symétriques, les ordres topogènes* $\rho^{J',\gamma}$ *construits sont aussi symétriques.*

→ Soit $\rho^{J',\gamma}$ défini par la famille $(\eta^{\alpha}_{\gamma(\alpha)})_{\alpha\in J'}$. Soit $C \in \rho^{J',\gamma}(B)$, il existe $C_\alpha \in \eta^{\alpha}_{\gamma(\alpha)}(f_\alpha(B))$ tels que $\bigcap_{\alpha\in J'} f_\alpha^{-1}(C_\alpha) \subset C$. On obtient alors, pour tout $\alpha \in J'$: $Y_\alpha - f_\alpha(B) \in \eta^{\alpha}_{\gamma(\alpha)}(Y_\alpha - C_\alpha)$ et $E-B \in \rho^{J',\gamma}(E-f_\alpha^{-1}(C_\alpha))$. L'ensemble J' étant fini, E-B appartient à $\bigcap_{\alpha\in J'} \rho^{J',\gamma}(E-f_\alpha^{-1}(C_\alpha)) = \rho^{J',\gamma}(E-\bigcap_{\alpha\in J'} f_\alpha^{-1}(C_\alpha))$. On en déduit: $E-B \in \rho^{J',\gamma}(E-C)$. ←

Corollaire 1-10: *La sous-catégorie pleine* V_s *de* V, *dont les objets sont les T-espaces symétriques, est fermée pour la construction de structures initiales dans* V.

Proposition 1-11: *La sous-catégorie pleine* S *des espaces syntopogènes est fermée pour la construction de structures initiales dans* V.

→ On suppose donc tous les $\mathcal{Y}_\alpha$ syntopogènes. Soit $\rho^{J',\gamma'}$ défini par la famille $(\eta^{\alpha}_{\gamma'(\alpha)})_{\alpha\in J'}$. Pour tout $\alpha \in J'$, il existe $\eta^{\alpha}_{\gamma(\alpha)}$ tel que $\eta^{\alpha}_{\gamma'(\alpha)} \subset \overset{2}{\eta}{}^{\alpha}_{\gamma(\alpha)}$. Soit $\rho^{J',\gamma}$ l'ordre topogène sur E associé à $(\eta^{\alpha}_{\gamma(\alpha)})_{\alpha\in J'}$ et montrons: $\rho^{J',\gamma'} \subset \overset{2}{\rho}{}^{J',\gamma}$.

Soit $C \in \rho^{J',\gamma'}(B)$, il existe $C_\alpha \in \eta^{\alpha}_{\gamma'(\alpha)}(f_\alpha(B))$ tel que $\bigcap_{\alpha\in J'} f_\alpha^{-1}(C_\alpha) \subset C$. De $C_\alpha \in \overset{2}{\eta}{}^{\alpha}_{\gamma(\alpha)}(f_\alpha(B))$, on déduit l'existence d'un D_α tel que $C_\alpha \in \eta^{\alpha}_{\gamma(\alpha)}(D_\alpha)$ et $D_\alpha \in \eta^{\alpha}_{\gamma(\alpha)}(f_\alpha(B))$.

Posons $D = \bigcap_{\alpha\in J'} f_\alpha^{-1}(D_\alpha)$, on a $D \in \rho^{J',\gamma}(B)$ par construction et on montre, en utilisant la continuité de f_α: $C \in \rho^{J',\gamma}(D)$. Il en résulte $C \in \overset{2}{\rho}{}^{J',\gamma}(B)$. ←

Remarques: 1) Si l'ensemble J est infini, la structure initiale d'une famille de structures simples n'est pas en général simple. Il s'ensuit en particulier que la catégorie T des topologies n'est pas fermée pour la construction de produits dans V.

2) La structure initiale d'une famille de structures ponctuelles n'est généralement pas ponctuelle. Cependant, la ponctualité est conservée par structure image réciproque. On en déduit (énoncé dans (*4*) pour S): la structure image réciproque dans V d'un espace topologique est un espace topologique.

3) Ce qui précède et (*8*) nous donnent l'existence de structures finales et de limites dans V et dans S.

4) Pour terminer, remarquons que la réunion des T-bases d'une même classe d'équivalence pour < est encore une T-base de cette classe. C'est cet élément maximal que nous prenons comme représentant, dans la plupart des démonstrations.

2- <u>Espaces fonctionnels: structure de la convergence totale:</u>

Nous nous placerons ici dans la catégorie S des espaces syntopogènes.

Soit $\mathcal{Y} = (Y,[\eta_i])$ un espace syntopogène et X un ensemble. Notons $M(Y,X)$ l'ensemble des applications de X dans Y. Soit η et η' des ordres génériques de $\mathcal{Y}$ tels que $\eta \subset \overset{2}{\eta}'$; nous définissons β sur $P(M(Y,X))$ par:

$B \in \beta(A)$ ssi, pour tout $C \subset Y$, il existe $U_C \in \eta(C)$ et il existe $V_C \in \eta'(C)$ tels que $U_C \in \eta'(V_C)$ et $\bigcap_{C\subset Y} \{g \mid t \in A^{-1}(V_C) \Rightarrow g(t) \in U_C\} \subset B$.

L'expression $t \in A^{-1}(V_C)$ signifie $f(t) \in V_C$, pour tout $f \in A$. Si $A = \emptyset$, on pose $\emptyset \in \eta(\emptyset)$.

<u>Proposition 2-1</u>: *Pour tout couple* (η,η') *tel que* $\eta \subset \overset{2}{\eta}'$, *l'application* β *définit un ordre semi-topogène sur* $M(Y,X)$.

$\rightarrow$ (ST_1) et (ST_3) sont bien entendus vérifiés.

Pour (ST_2), remarquons que, par définition, si B est inclus dans B' et si B appartient à $\beta(A)$, on a: $B' \in \beta(A)$. Soit $A \subset A'$ et $B \in \beta(A')$, alors, pour tout $C \subset Y$, il existe $U_C \in \eta(C)$ et $V_C \in \eta'(C)$ tels que:

$U_C \in \eta'(V_C)$ et $\bigcap_{C\subset Y} \{g \mid t \in A'^{-1}(V_C) \Rightarrow g(t) \in U_C\} \subset B$.

Notons: $\quad H = \bigcap_{C\subset Y} \{g \mid t \in A^{-1}(V_C) \Rightarrow g(t) \in U_C\}$,

$\quad H' = \bigcap_{C\subset Y} \{g \mid t \in A'^{-1}(V_C) \Rightarrow g(t) \in U_C\}$;

on a: $H \subset H'$, d'où: $B \in \beta(A)$ et $\beta(A') \subset \beta(A)$. $\leftarrow$

Notons β^q l'ordre topogène associé[(4)] à β.

<u>Proposition 2-2</u>: *En faisant varier dans* $[\eta_i]\times[\eta_i]$ *le couple* (η,η') *tel que* $\eta \subset \overset{2}{\eta}'$, *les applications* β^q *associées définissent une structure syntopogène sur* $M(Y,X)$, *notée* $M(\mathcal{Y},X)$.

$\rightarrow$ Etablissons d'abord (S_1). Soit β associé à (η,η') et β_1 associé à (η_1,η_1'); il existe η_2 tel que $\eta \subset \eta_2$ et $\eta_1 \subset \eta_2$; il existe η_2' tel que $\eta' \subset \eta_2'$ et $\eta_1' \subset \eta_2'$; il existe η'' tel que $\eta_2 \subset \overset{2}{\eta}''$ et il existe η_3 tel que $\eta_2' \subset \eta_3$ et $\eta'' \subset \eta_3$. Notons β_2 l'application associée à (η,η_3); on a $\beta \subset \beta_2$ et $\beta_1 \subset \beta_2$. On déduit de ceci[(4)]: $\beta^q \subset \beta_2^q$ et $\beta_1^q \subset \beta_2^q$.

Montrons maintenant l'axiome de préidempotence (S_2). Soit β l'ordre semi-topogène associé à (η,η'); il existe η'' tel que $\eta' \subset \overset{2}{\eta}''$. Notons β' l'ordre semi-topogène associé à $(\overset{2}{\eta}'',\eta'')$; on a: $\eta \subset \overset{2}{\eta}' \subset \eta' \subset \overset{2}{\eta}'' \subset \eta''$.

Soit $B \in \beta(A)$, alors pour tout $C \subset Y$, il existe $U_C \in \eta(C)$ et il existe $V_C \in \eta'(C)$ tels que $U_C \in \eta'(V_C)$ et $\bigcap_{C\subset Y} \{g \mid t \in A^{-1}(V_C) \Rightarrow g(t) \in U_C\} \subset B$.

De $\eta' \subset \overset{2}{\eta}''$ et $U_C \in \eta'(V_C)$, on déduit l'existence d'un V_C' tel que $U_C \in \eta''(V_C')$ et $V_C' \in \eta''(V_C)$.

On a obtenu: pour tout $C \subset Y$, il existe $V_C' \in \overset{2}{\eta}''(C)$ et $V_C \in \eta''(C)$ tels que

$V'_C \in \eta''(V_C)$. Posons $B' = \bigcap_{C \subset Y} \{g \mid t \in A^{-1}(V_C) \Rightarrow g(t) \in V'_C\}$.
On a $B' \subset B$ et $B' \in \beta'(A)$. Posons maintenant:
$B'' = \bigcap_{C \subset Y} \{g \mid t \in B'^{-1}(V'_C) \Rightarrow g(t) \in U_C\}$. Par définition, B" appartient à $\beta'(B')$.
Montrons $B'' \subset B$. Soit $g_o \in B''$ et $t \in A^{-1}(V_C)$. Par définition de B', $g(t) \in V'_C$, pour tout $g \in B'$, donc $t \in B'^{-1}(V'_C)$. Il s'ensuit $g_o(t) \in U_C$ car $g_o \in B''$. Ceci est vrai pour tout $C \subset Y$, donc $g_o \in B$ et $B \in \beta'(B')$.
En résumé, nous avons montré: soit β un ordre semi-topogène déduit de (η,η'), il existe β' tel que: pour tout $A \subset M(Y,X)$, tout $B \in \beta(A)$, il existe $B' \subset M(Y,X)$ tel que $B \in \beta'(B')$ et $B' \in \beta'(A)$, d'où $\beta \subset \overset{2}{\beta}'$.
La démonstration d'une propriété semblable pour le topogénisé β^q de β se fait simplement en utilisant une caractérisation du topogénisé mise en évidence dans (*4*). On obtient ainsi $\beta^q \subset \overset{2}{\beta}'^q$. ←

On peut ainsi définir un foncteur $M(\mathcal{Y},-)$ de M^* dans S, où M^* est la duale de la catégorie des applications.

Si $\mathcal{Y}$ est ponctuel, l'espace fonctionnel $M(\mathcal{Y},X)$ n'est pas nécessairement ponctuel; mais l'inconvénient est mineur car on peut considérer l'espace ponctuel $M(\mathcal{Y},X)^p$ associé pour avoir des espaces fonctionnels dans la sous-catégorie pleine S_p des espaces syntopogènes ponctuels. Par contre, si $\mathcal{Y}$ est défini par un seul ordre η, l'espace fonctionnel construit ci-dessus est simple. Si $\mathcal{Y}$ est topologique, il suffit de prendre $M(\mathcal{Y},X)^p$ pour obtenir un espace topologique. Nous allons maintenant simplifier la construction de l'espace fonctionnel ponctuel $M(\mathcal{Y},X)^p$.
Pour cela, supposons les ordres η_i de $[\eta_i]$ tous ponctuels et soit (η,η') tels que $\eta \subset \overset{2}{\eta}'$. Pour tout $f \in M(Y,X)$, notons $\gamma(f)$ le filtre engendré par les intersections finies des parties B telles que:
pour tout $y \in Y$, il existe $U_y \in \eta(y)$ et $V_y \in \eta'(y)$ tels que $U_y \in \eta'(V_y)$ et
$B = \bigcap_{y \in Y} \{g \mid f(t) \in V_y \Rightarrow g(t) \in U_y\}$.
On pose $\gamma(A) = \bigcap_{f \in A} \gamma(f)$.

<u>Proposition 2-3</u>: *Avec les notations précédentes, on a* $M(\mathcal{Y},X)^p = (M(Y,X),[\gamma])$.

→ Par construction, γ est un ordre topogène ponctuel plus fin que β, on en déduit(*4*) : $\beta^{qp} \subset \gamma$. En explicitant les définitions de γ et β , on obtient: $B \in \gamma(f)$ implique $B \in \beta^q(f)$. Il s'ensuit: $\gamma \subset \beta^{qp}$. ←

<u>Proposition 2-4</u>: *Si* $\mathcal{Y}$ *est un espace uniforme et* X *un ensemble,* $M(\mathcal{Y},X)^p$ *est la structure de la convergence uniforme.*

→ Soit (η,η') des ordres génériques de $\mathcal{Y}$ tels que $\eta \in \overset{2}{\eta}'$. Par hypothèse, η (resp. η') est canoniquement associé à un entourage W (resp.W') de Y, tels que $\overset{2}{W}' \subset W$. Notons γ l'application associée à (η,η'). Soit
$\langle W(f)\rangle = \{g \mid (f(x),g(x)) \in W$, pour tout $x \in X\}$.

Soit $B \in \gamma(f)$ défini par $U_y = W(y)$ et U'_y; on a: $W'(y) \subset U'_y$, $W'(U'_y) \subset W(y)$ et $B = \bigcap_{y \in Y} \{g \mid f(t) \in U'_y \Rightarrow g(t) \in W(y)\}$.

Si on choisit $y = f(x)$, de $f(x) \in U'_{f(x)}$, on déduit $g(x) \in W(f(x))$, ceci pour tout $x \in X$ et tout $g \in B$, d'où $B \subset \langle W(f) \rangle$.

Soit maintenant $g \in \langle W'(f) \rangle$ et t tel que $f(t) \in U'_y$. De $(g(t),f(t)) \in W'$, on déduit: $g(t) \in W'(U'_y)$ et $g(t) \in W(y)$, d'où: $\langle W'(f) \rangle \subset B$.

Notons $\hat{\eta}$ (resp. $\hat{\eta}'$) l'ordre topogène engendré par η (resp. η') sur $M(Y,X)$, i.e.: $\hat{\eta}(f)$ est le filtre des surensembles de $\langle W(f) \rangle$. On déduit de ce qui précède: $\hat{\eta} \subset \gamma \subset \hat{\eta}'$.

En notant $[\hat{\eta}]$ la structure syntopogène de la convergence uniforme et $[\gamma]$ celle définie précédemment, on a $[\hat{\eta}] < [\gamma]$ et $[\gamma] < [\hat{\eta}]$, pour la relation $<$ définie dans le premier paragraphe. Ainsi, $[\gamma]$ est la structure de la convergence uniforme. ←

3- T-équicontinuité:

Soit $\mathcal{X} = (X,[\rho_i])$ et $\mathcal{Y} = (Y,[\eta_i])$ deux espaces syntopogènes; notons $M(Y,X)$ l'ensemble des applications de X dans Y.

Définition 3-1: Une partie H de $M(Y,X)$ est T-*équicontinue* si, pour tout couple (η,η') d'ordres génériques de $\mathcal{Y}$ tels que $\eta \subset \overset{2}{\eta'}$, il existe un ordre générique ρ de $\mathcal{X}$ tel que:

pour tout $C \subset Y$, tout $A \subset X$, tout $U \in \eta(C)$, il existe $V \in \eta'(C)$ et $U' \in \rho(A)$ tels que $U \in \eta'(V)$ et: $f \in H$ et $f(A) \subset V$ impliquent $f(U') \subset U$.

Proposition 3-2: *Si $\mathcal{X}$ et $\mathcal{Y}$ sont des espaces topologiques, la définition précédente coïncide avec celle d'égale continuité de Kelley*[10].

Proposition 3-3: *1) Si $\mathcal{X}$ et $\mathcal{Y}$ sont des espaces uniformes, la définition 3-1 est celle d'équicontinuité uniforme*[2].

2) *Si $\mathcal{X}$ est un espace topologique et $\mathcal{Y}$ un espace uniforme, la* T-*équicontinuité est l'équicontinuité*[2].

→ 1) Soit $\mathcal{X}$ et $\mathcal{Y}$ deux espaces uniformes. Soit H une partie de $M(Y,X)$ vérifiant la définition 3-1. Montrons que H est uniformément équicontinue. Soit W un entourage de Y, il existe W' tel que $\overset{2}{W'} \subset W$. Par hypothèse, il existe un entourage V' de X tel que:

pour tout $y = f(x_o)$, pour tout $x_o \in X$, pour tout $U = W(y)$, il existe V tel que $W'(f(x_o)) \subset V$ et $W'(V) \subset W(y)$, et il existe U' tel que $V'(x_o) \subset U'$, vérifiant: $f \in H$ et $f(x_o) \in V$ impliquent $f(U') \subset U$.

Or $f(x_o) \in W'(f(x_o)) \subset V$; donc, pour tout $f \in H$, on a $f(U') \subset U$ et par suite $f(V'(x_o)) \subset U$, i.e.: $(x,x_o) \in V'$ implique $(f(x),f(x_o)) \in W$.

On a ainsi montré: pour tout entourage W, il existe un entourage V' tel que

$(x',x) \in V$ entraîne $(f(x'),f(x)) \in W$.
Réciproquement, soit H une partie uniformément équicontinue de $M(Y,X)$; montrons qu'elle est T-équicontinue. Soit (W,W') un couple d'entourages de Y tels que $\overset{2}{W'} \subset W$. Soit V l'entourage de X correspondant à W', c'est-à-dire vérifiant: $(x',x) \in V$ implique $(f(x'),f(x)) \in W'$, pour tout $f \in H$.
Notons η (resp.η', resp.ρ) les ordres topogènes associés canoniquement à W (resp.W', resp.V). Les ordres topogènes considérés sont ponctuels, nous raisonnerons donc sur les éléments de Y et de X.
Soit $y \in Y$ et $x \in X$, soit $U \in \eta(y)$, i.e. $W(y) \subset U$; choisissons $U' = V(x) \in \rho(x)$ et $\overline{V} = W'(y) \in \eta'(y)$. On a: $W'(V) \subset W(y) \subset U$. Soit $f \in H$ tel que $f(x) \in W'(y)$. Si $x' \in V(x)$, par hypothèse, $(f(x'),f(x))$ est un élément de W'. On obtient ainsi: $(f(x'),y) = (f(x'),f(x)) \circ (f(x),y) \in W' \circ W' \subset W$.
Il s'ensuit: $f(V(x)) \subset W(y) \subset U$.

2) La démonstration est semblable à la précédente. ←

Nous allons maintenant relier les deux définitions posées: celle d'espaces fonctionnels et celle de T-équicontinuité.

Soit K un ensemble, $\mathscr{Y}$ et $\mathscr{X}$ deux espaces syntopogènes. Soit f une application de $K \times X$ dans Y; on note H l'ensemble des applications partielles $f(t,-)$ de X dans Y, où t parcourt K.

<u>Proposition 3-4</u>: *L'application* $\phi: x \mapsto f(-,x)$ *de* X *dans* $M(Y,K)$ *est* T-*continue de* $\mathscr{X}$ *vers* $M(\mathscr{Y},K)$ *ssi* H *est une partie* T-*équicontinue de* $M(Y,X)$.

→ Explicitons chaque définition.
H est T-équicontinue: pour tout couple (η,η') tel que $\eta \subset \overset{2}{\eta'}$, il existe ρ, ordre générique de $\mathscr{X}$, tel que: pour tout $C \subset Y$, tout $A \subset X$, tout $U \in \eta(C)$, il existe $V \in \eta'(C)$ avec $U \in \eta'(V)$ et il existe $U' \in \rho(A)$ vérifiant: $f(t,A) \subset V$ implique $f(t,U') \subset U$.
ϕ est T-continue de $\mathscr{X}$ vers $M(\mathscr{Y},K)$: pour tout ordre générique β^q de $M(\mathscr{Y},X)$, il existe un ordre générique ρ de $\mathscr{X}$ tel que ϕ soit continue de ρ vers β^q. L'ordre ρ étant topogène, la T-continuité de ϕ, de ρ vers β^q, est équivalente à la T-continuité de ρ vers β. Par définition de β, l'expression précédente devient: pour tout couple (η,η') tel que $\eta \subset \overset{2}{\eta'}$, il existe ρ, ordre générique de $\mathscr{X}$, tel que: pour tout $B \in \beta(\phi(A))$, il existe $U' \in \rho(A)$ tel que $\phi(U') \subset B$. En explicitant la définition de β, on constate que les deux assertions sont équivalentes. ←

<u>Appendice</u>: <u>Convergence simple dans les N-espaces</u>.

Nous allons mettre une structure de foncteur Hom interne sur la catégorie des N-espaces[8] (que nous traduirons par espaces de rapprochement), structure présentant une comparaison avec l'analyse usuelle. Nous ne rappellerons pas ici les définitions et propriétés des N-espaces, le lecteur se reportera à (8).

Soit E un ensemble et (E',ξ) un espace de prérapprochement. Soit $A \subset P(M(E',E))$, on pose: $A \in \eta$ ssi, pour tout $x \in E$, $\{A(x) \mid A \in A\} \in \xi$.

Proposition: $(M(E',E),\eta)$ *est un espace de prérapprochement.*

C'est l'espace produit de E copies de (E',ξ).

Désormais, nous supposerons que ξ est un N-espace et nous noterons η_q l'espace Q-proche associé à η (8).

Proposition: $(M(E',E),\eta_q)$ *est un N-espace.*

Soit (E',ξ) un N-espace uniforme (i.e. provenant d'une uniformité W sur E'). Soit η la structure définie précédemment.

Proposition: $(M(E',E),\eta)$ *est le N-espace uniforme de la convergence simple.*

Soit α une structure de N-espace sur E, nous noterons η' la structure induite par η sur l'ensemble Hom$((E',\xi),(E,\alpha))$ des N-applications de (E,α) vers (E',ξ). Soit (E'',β) un N-espace et f une N-application de (E',ξ) vers (E'',β); notons η'' la structure de N-espace sur Hom$((E'',\beta),(E,\alpha))$.

Alors, Hom$(f,(E,\alpha))$ est une N-application de (Hom$((E',\xi),(E,\alpha)),\eta'$) vers (Hom$((E'',\beta),(E,\alpha)),\eta''$) qui à g associe $f \circ g$. De même, Hom$((E,\alpha),f)$ est une N-application de (Hom$((E,\alpha),(E'',\beta)),\eta''$) vers (Hom$((E,\alpha),(E',\xi)),\eta'$) associant $h \circ f$ à f.

Ainsi, la structure de convergence simple définit un foncteur Hom interne à la catégorie des N-espaces.

Remarque: Dans ce paragraphe, les démonstrations se déduisent dans leur presque totalité des propriétés catégoriques établies dans (8).

Cette étude nous conduit aux question suivantes: définition d'une σ-convergence et existence d'un théorème d'Ascoli dans le cadre des structures syntopogènes et des N-espaces. Nous aborderons celles-ci dans un prochain travail.

-:-:-:-:-:-:-:-

(1) A.*Bastiani: Topologie générale, Cours multigraphié, Amiens 1973.*

(2) N.*Bourbaki: Topologie générale, Hermann, Paris.*

(3) G.*Choquet: Convergences, Ann. Univ. Grenoble, nouvelle série, 23, 1947.*

(4) A.*Csaszar: Foundations of general topology, Macmillan, New-York 1963.*

(5) V.A.*Efremovic: Geometry of proximity, Mat. Sbornik, 31 (73), 1952.*

(6) C.*Ehresmann: Catégories et structures, Dunod, Paris 1965.*

(7) M.*Hacque: Les T-espaces et leurs applications, Cahiers de Top. et Géom. Diff., IX-3, 1968.*

(8) H.*Herrlich: Topological structures, Mathematical Centre Tracts, 52, 1974.*

(9) J.R.*Isbell: Uniform spaces, Amer. Math. Soc. Math. Surveys, 12, 1964.*

(10) J.L.*Kelley: General Topology, Van Nostrand, 1955.*

(11) D.*Tanré: Sur les T-espaces simples, Esquisses math. 18, Paris, 1972.*

FILTERS AND UNIFORMITIES IN GENERAL CATEGORIES

S.J.R. VORSTER†

§ 1 INTRODUCTION

In [1] B. Eckmann and P.J. Hilton translated the group axioms into categorical language and studied the resulting concept of a group-object in the context of general categories. In this paper a similar approach will be followed by using uniform spaces instead of groups.

Arbitrary sets equipped with uniformities are replaced by arbitrary objects supplied with "uni-structures", which yields the concept of uni-objects in general categories. Uniformly continuous maps translate into categorical language without any difficulty. Furthermore, the ideas of filters on sets, groups, etc. are unified by introducing objectfilters on objects of categories. Analogues of some basic convergence properties of filters on uniform spaces are proved in a more general context, viz. for objectfilters on uni-objects in categories. The theory of uniform spaces will thus be made available and may possibly become useful in areas of mathematics other than general topology.

§ 2 PRELIMINARIES

Any notations and concepts which are used but not defined in this paper, will have the meaning assigned to them by H. Herrlich and G.E. Strecker [4].

---

† This research was made possible by a grant from the South African C.S.I.R.

Let $\underline{C}$ be a well-powered category with $A \in \mathrm{Ob}(\underline{C})$ an arbitrary object of $\underline{C}$ and $\psi: A \to B$ any morphism of $\underline{C}$. The following notations will be used (where all limits mentioned are assumed to exist):

E for a terminal object in $\underline{C}$; $\underline{E}$ for the class of all epimorphisms (or extremal epimorphisms) in $\underline{C}$; $\underline{M}$ for the class of all monomorphisms (or extremal monomorphisms) in $\underline{C}$; $\mathrm{Sub}(A)$ (resp. $\underline{M}(A)$) for the set of all subobjects (resp. all $\underline{M}$-subobjects) of A; $(X,f) \leqslant (Y,g)$ if there exists a morphism $h: X \to Y$ such that $f = gh$, and $(X,f) \approx (Y,g)$ if both $(X,f) \leqslant (Y,g)$ and $(Y,g) \leqslant (X,f)$ hold, where $(X,f), (Y,g) \in \mathrm{Sub}(A)$; $\bigcap_{i=1}^{n} (X_i, f_i)$ for the intersection of any finite family of subobjects $(X_i, f_i) \in \mathrm{Sub}(A)$, $i = 1,2,\cdots,n$.

If

Diagram 1

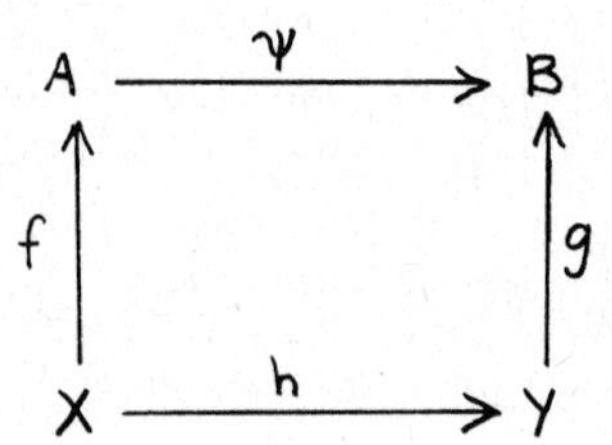

with $(X,f) \in \mathrm{Sub}(A)$ and $(Y,g) \in \mathrm{Sub}(B)$, is a pullback square (resp. $(\underline{E},\underline{M})$-factorization of ψf, i.e. $\psi f = gh$ with $h \in \underline{E}$ and $g \in \underline{M}$) then we shall write $\psi^{-1}(Y,g) = (X,f)$ (resp. $\psi(X,f) = (Y,g)$).

Let I be an arbitrary set (resp. $I = \{1,2\}$). The product of a family $(A_i)_{i \in I}$, $A_i \in \mathrm{Ob}(\underline{C})$ will be denoted by $(\Pi A_i, p_i)$ (resp. $(A_1 \times A_2, p_1, p_2)$) or simply by ΠA_i (resp. $A_1 \times A_2$) and, if $(\psi_i: B \to A_i)_{i \in I}$ is given $\langle \psi_i \rangle : B \to \Pi A_i$ (resp. $\langle \psi_1, \psi_2 \rangle : B \to A_1 \times A_2$) will denote the unique morphism such that $p_i \langle \psi_i \rangle = \psi_i$ for each $i \in I$ (resp. $p_1 \langle \psi_1, \psi_2 \rangle = \psi_1$ and $p_2 \langle \psi_1, \psi_2 \rangle = \psi_2$). For a given family $(\psi_i: A_i \to B_i)_{i \in I}$ we shall write

$\Pi\psi_i = \langle \psi_i p_i \rangle : \Pi A_i \to \Pi B_i$ (resp. $\psi_1 \times \psi_2 = \langle \psi_1 p_1, \psi_2 p_2 \rangle : A_1 \times A_2 \to B_1 \times B_2$).

The following results will frequently be used:

(2.1) Proposition If $\underline{C}$ is a category with finite products, then

(i) $\langle \psi_1, \psi_2 \rangle \phi = \langle \psi_1 \phi, \psi_2 \phi \rangle$, (ii) $(\phi_1 \times \phi_2)\langle \psi_1, \psi_2 \rangle = \langle \phi_1 \psi_1, \phi_2 \psi_2 \rangle$,

(iii) $(\phi_1 \times \phi_2)(\psi_1 \times \psi_2) = \phi_1 \psi_1 \times \phi_2 \psi_2$,

hold whenever both sides of each equality make sense.

The commutator morphism $k = k_{AB}: A \times B \to B \times A$ defined by $k = \langle p_2, p_1 \rangle$ will play an important role. Clearly, k is an isomorphism with inverse $k^{-1} = k$. One easily verifies

(2.2) Proposition If $\underline{C}$ is a category with finite products, then

(i) $k\langle \psi_1, \psi_2 \rangle = \langle \psi_2, \psi_1 \rangle$, and (ii) $k(\psi \times \phi) = (\phi \times \psi)k$

hold whenever both sides of each equality make sense.

(2.3) Proposition If $\underline{C}$ is a well-powered category with finite products and if $A \in Ob(\underline{C})$ and $(X,f),(Y,g) \in Sub(A \times A)$, then

(i) $(X,f) \leqslant (Y,g)$ if and only if $(X,kf) \leqslant (Y,kg)$,

(ii) $(X,kf) \leqslant (Y,g)$ if and only if $(X,f) \leqslant (Y,kg)$.

§ 3 OBJECTFILTERS IN CATEGORIES

In this section the concept of filters on objects of general categories will be introduced, which will unify the concepts of filters on sets, groups and objects of other categories.

(3.1) Definition Let $\underline{C}$ be any category with a terminal object E.

(i) An $A \in Ob(\underline{C})$ is a non-empty object if and only if there exists an $m: E \to A$ (which is necessarily a monomorphism).

(ii) A subobject (X,f) of an $A \in Ob(\underline{C})$ is a non-empty subobject of

A if and only if there exists an m: $E \to A$ such that $(E,m) \leqslant (X,f)$.

(3.2) Examples In all concrete categories having the "singletons" as terminal objects (for example, Set, Top, Grp, etc.) non-empty objects and subobjects correspond precisely to objects of these categories having non-empty underlying sets.

(3.3) Definition Let $\underline{C}$ be a well-powered $(\underline{E},\underline{M})$ category with finite intersections and a terminal object. A non-empty subset $\underline{F}$ of $\underline{M}(A)$, $A \in Ob(\underline{C})$, is an objectfilter on A if and only if

(OF.1) $(X,f),(Y,g) \in \underline{F}$ implies $(X,f) \cap (Y,g) \in \underline{F}$,

(OF.2) $(X,f) \in \underline{F}$ and $(X,f) \leqslant (Y,g)$ imply $(Y,g) \in \underline{F}$,

(OF.3) $(X,f) \in F$ implies (X,f) is a non-empty subobject of A.

If $\underline{F}$ and $\underline{G}$ are objectfilters on A such that $(X,f) \in \underline{F}$ implies $(X,f) \in \underline{G}$ then $\underline{F}$ is said to be contained in $\underline{G}$ and we write $\underline{F} \subset \underline{G}$.

(3.4) Examples (i) In Set objectfilters correspond precisely to filters on sets.

(ii) In Grp (resp. Ab) an objectfilter on a group (resp. abelian group) G is a non-empty family of subgroups of G such that $\underline{F}$ is closed under finite intersections and any subgroup of G which contains an element of $\underline{F}$ also belongs to $\underline{F}$. (It is known that an objectfilter on an abelian group gives rise to a topology on G such that G becomes a topological group.)

(iii) In Top an objectfilter on a topological space X is a non-empty family $\underline{F}$ of non-empty subspaces of X which is closed under finite intersections and such that any subspace of X which contains an element of $\underline{F}$ also belongs to $\underline{F}$.

(3.5) Definition Let $\underline{C}$ be a well-powered $(\underline{E},\underline{M})$ category with finite intersections and a terminal object E. A non-empty subset $\underline{B}$ of $\underline{M}(A)$, $A \in Ob(\underline{C})$, is a basis for an objectfilter on A if and only if

(OFB.1) $(X,f) \in \underline{B}$ implies that (X,f) is a non-empty subobject of A.

(OFB.2) $(X,f),(Y,g) \in \underline{B}$ implies that there exists a $(Z,h) \in \underline{B}$ such that $(Z,h) \leqslant (X,f) \cap (Y,g)$.

(3.6) Proposition Let $\underline{C}$ be a well-powered $(\underline{E},\underline{M})$ category with finite intersections and a terminal object. If A is a non-empty object of $\underline{C}$ and if a subset $\underline{B}$ of $\underline{M}(A)$ is a basis for an objectfilter on A, then

$$(\underline{B}) = \{(X,f) \in \text{Sub}(A) \;/\; (Y,g) \leqslant (X,f) \quad \text{for some} \quad (Y,g) \in \underline{B}\}$$

is an objectfilter on A. $(\underline{B})$ is said to be the objectfilter generated by $\underline{B}$.

(3.7) Definition If $\underline{C}$ is a well-powered category, a morphism $\psi\colon A \to B$ of $\underline{C}$ is said to preserve terminality if, for any $(E,m) \in \text{Sub}(A)$, E a terminal object in $\underline{C}$, it holds that $\psi(E,m) = (E',m') \in \text{Sub}(B)$, where E' is also a terminal object in $\underline{C}$.

(3.8) Theorem Let $\underline{C}$ be a well-powered $(\underline{E},\underline{M})$ category with finite intersections and a terminal object E. Suppose every $\psi\colon A \to B$ of $\underline{C}$ preserves terminality. For any objectfilter $\underline{F}$ on A the set $\underline{B} = \{\psi(X,f)/(X,f) \in \underline{F}\}$ forms a basis for an objectfilter (denoted by $\psi(\underline{F})$) on B.

Proof We have to verify (OFB.1) and (OFB.2). Firstly, let $(Y,g) \in \underline{B}$, i.e. $(Y,g) = \psi(X,f)$ for some $(X,f) \in \underline{F}$. Now, since (X,f) is a non-empty subobject of A there exists an $m\colon E \to A$ such that $(E,m) \leqslant (X,f)$, i.e. there is an $f'\colon E \to X$ satisfying $ff' = m$. Let $\psi(E,m) = (E',m')$ and consider

Diagram 2

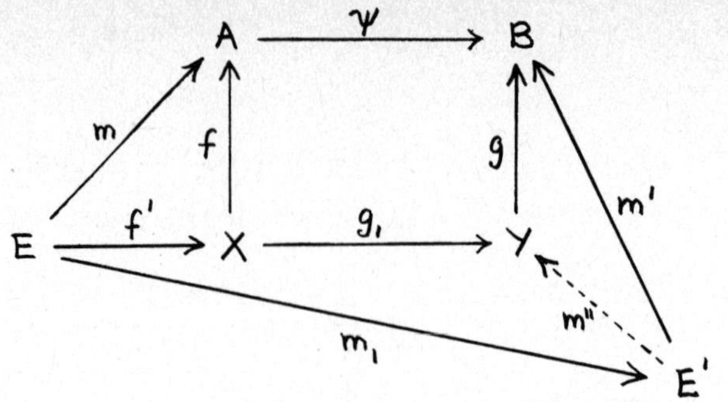

where $g_1, m_1 \in \underline{E}$ and $g, m' \in \underline{M}$. Then we have $m'm_1 = \psi m = \psi ff' = gg_1 f'$ and hence it follows from the $(\underline{E},\underline{M})$-diagonalization property ([4], 33.2 and 33.3) that there exists an $m'': E' \to Y$ such that $m' = gm''$, i.e. $(E',m') \leqslant (Y,g)$ and since ψ preserves terminality (Y,g) is a non-empty subobject of $\underline{B}$. Thus (OFB.1) is satisfied.

Secondly, one easily verifies that $\underline{B}$ also satisfies (OFB.2), by making use of the $(\underline{E},\underline{M})$-diagonalization property ([4],33.2) and the definition of intersection.

(3.9) Theorem Let $\underline{C}$ be a well-powered $(\underline{E},\underline{M})$ category with finite intersections and products. Let $\underline{F}_i$ be an objectfilter on $A_i \in Ob(\underline{C})$ for each $i \in I$, I an arbitrary set. For any family $(X_i,f_i)_{i\in I}$, $(X_i,f_i) \in \underline{F}_i$ for each $i \in I$, it holds that $p_j(\Pi X_i, \Pi f_i) \approx (X_j,f_j)$ for every $j \in I$, where $p_j: \Pi A_i \to A_j$ denotes the j-th projection morphism.

Proof Clearly, $\underline{C}$ is finitely complete (by [4],23.7) and hence has a terminal object E. Consider any fixed $j \in I$. It will be shown that $p_j \in \underline{E}$. Now, since $(X_i,f_i) \in \underline{F}_i$ is a non-empty subobject of A_i for each $i \in I$, there exists an $m_i: E \to A_i$ such that $(E,m_i) \leqslant (X_i,f_i)$, i.e. there is an $m_i': E \to X_i$ such that

Diagram 3

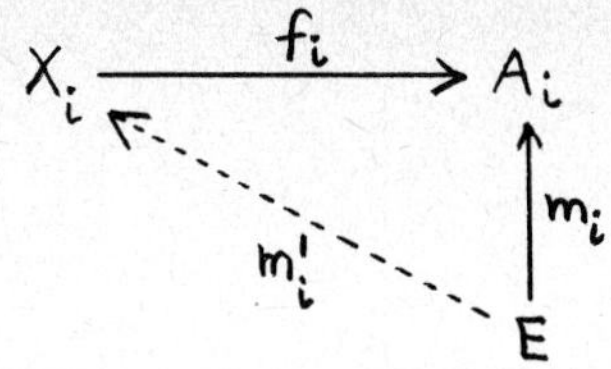

commutes. For each $i \in I$, $i \neq j$, let

$h_i: X_j \to X_i = X_j \overset{e_j}{\to} E \overset{m_j'}{\to} X_i$ where $Hom(X_j, E) = \{e_j\}$ and let $h_j = 1: X_j \to X_j$. Then, by the definition of product there exists $\langle h_i \rangle : X_j \to \Pi X_i$ such that $p_j \langle h_i \rangle = 1$, i.e. p_j is a retraction and therefore also an epimorphism. In addition, p_j is an extremal epimorphism since $p_j = mf$, m a monomorphism, implies that $1 = p_j \langle h_i \rangle = mf \langle h_i \rangle$ so that m is also a retraction (See [4],6.7). Hence $p_j \in \underline{E}$.

Now, let $p_j(\Pi X_i, \Pi f_i) = (Y_j, g_j)$, then in the

Diagram 4

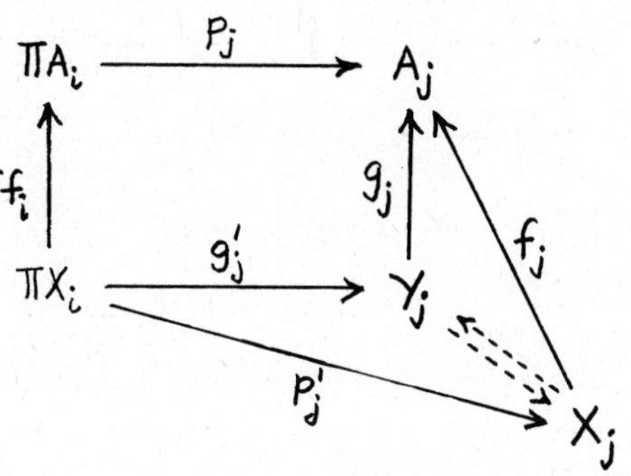

$p_j \Pi f_i = g_j g_j' = f_j p_j'$, where $g_j', p_j' \in \underline{E}$ and $g_j, f_j \in \underline{M}$, and hence it follows from the $(\underline{E}, \underline{M})$-diagonalization property that $p_j(\Pi X_i, \Pi f_i) = (Y_j, g_j) \approx (X_j^{\cdot}, f_j)$.

(3.10) Theorem Let $\underline{C}$ be a well-powered, finitely complete $(\underline{E}, \underline{M})$ category with products. Let $(A_i)_{i \in I}$ be an arbitrary family of non-empty objects of $\underline{C}$ and let $\underline{F}_i$ be an objectfilter on A_i for each $i \in I$. Then

(i) $\underline{B} = \{(\Pi X_i, \Pi f_i)/(X_i, f_i) \in \underline{F}_i$ for each $i \in I$ and there exists a finite set J such that $(X_i, f_i) \approx (A_i, 1)$ for all $i \in I-J\}$ is a basis for an objectfilter on $A = \Pi A_i$.

(ii) $p_j(\underline{F}) = \underline{F}_j$ for each $j \in I$, where $\underline{F} = (\underline{B})$ is the objectfilter generated by $\underline{B}$.

Proof (i) One easily verifies that $\underline{B}$ satisfies (OFB.1) and (OFB.2).
(ii) Firstly, let $(X,f) \in p_j(\underline{F})$, i.e. there exists $(Y,g) = (\Pi Y_i, \Pi g_i) \in \underline{B}$ such that $p_j(Y,g) \leqslant (X,f)$. It follows from (3.9) that $(Y_j, g_j) \leqslant p_j(Y,g) \leqslant (X,f)$ where $(Y_j, g_j) \in \underline{F}_j$ so that $(X,f) \in \underline{F}_j$.

Conversely, suppose $(X,f) \in \underline{F}_j$ and let $(Y,g) = (\Pi X_i, \Pi f_i)$, where $(X_i, f_i) = (A_i, 1)$ for each $i \neq j$ and $(X_j, f_j) = (X,f)$. Then, by (3.9) $p_j(Y,g) \leqslant (X,f)$ and since $(Y,g) \in \underline{F} = (\underline{B})$ (because $(A_i, 1) \in \underline{F}_i$ for each $i \neq j$) it follows that $(X,f) \in p_j(\underline{F})$. Thus $p_j(\underline{F}) = \underline{F}_j$, where $\underline{F} = (\underline{B})$.

§ 4 UNI-OBJECTS AND UNI-MORPHISMS

In order to generalize the definition of uniform spaces to abstract categories, the idea of the "composition of relations" will be replaced by a more general concept.

(4.1) Definition Let $\underline{C}$ be a well-powered category with finite products and consider an $A \in Ob(\underline{C})$ with product $(A \times A, p_1, p_2)$. The

Diagram 5

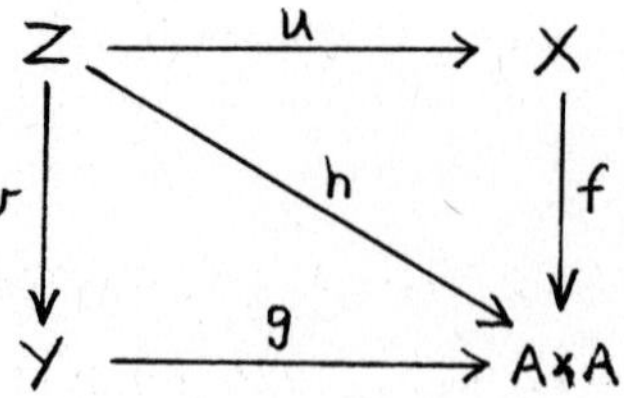

with $(X,f),(Y,g),(Z,h) \in \mathrm{Sub}(A \times A)$ is said to (p_1,p_2)-commute (or alternatively, (h,u,v) (p_1,p_2)-commutes with (f,g)) iff $p_1 h = p_1 fu$, $p_2 h = p_2 gv$ and $p_2 fu = p_1 gv$.

(4.2) Definition Under the conditions of (4.1), diagram 5 is called a (p_1,p_2)-pullback iff it (p_1,p_2)-commutes and for any other (p_1,p_2)-commutative square

Diagram 6

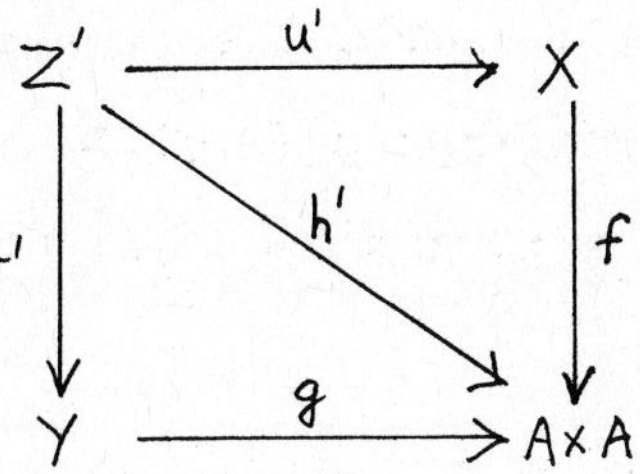

with $(Z',h') \in \mathrm{Sub}(A \times A)$, there exists an $h'': Z' \to Z$ such that $h' = hh''$.

Diagram 7

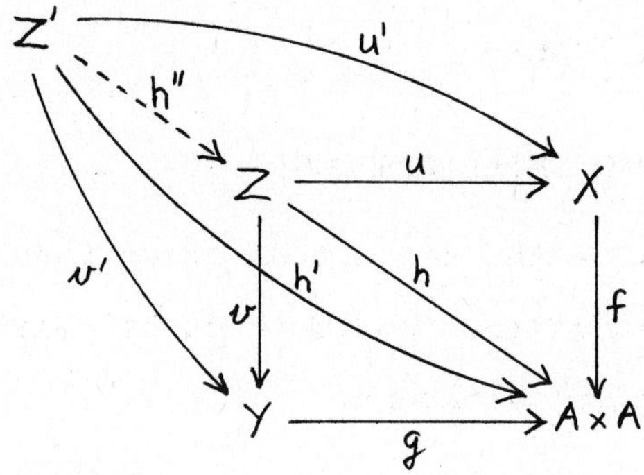

If no confusion is possible we shall simply call (Z,h) a (p_1,p_2)-pullback of (X,f) and (Y,g) (resp., of (X,f) if $(X,f) = (Y,g)$) and write $(Z,h) = (X,f) * (Y,g)$.

(4.3) Examples (i) Let $\underline{C}$ be any well-powered category with finite products. For any $A \in \mathrm{Ob}(\underline{C})$ and each $(X,f) \in \mathrm{Sub}(A)$, with products $(A \times A, p_1, p_2)$ and $(X \times X, p_1', p_2')$ respectively.

Diagram 8

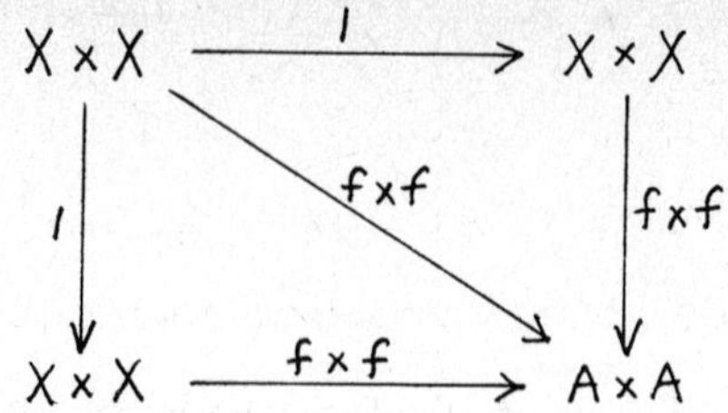

is a (p_1, p_2)-pullback square, for suppose

Diagram 9

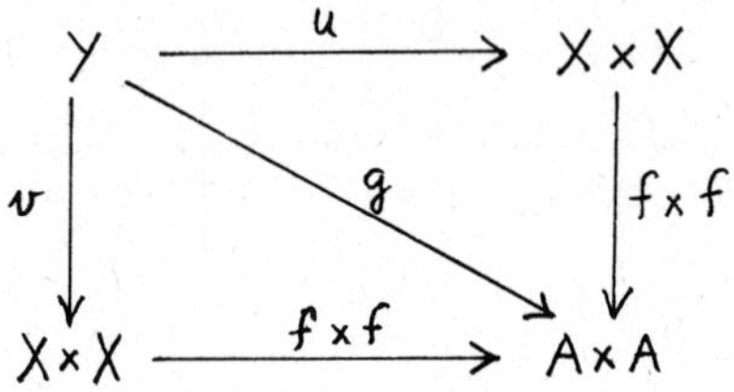

(p_1, p_2)-commutes, where $(Y,g) \in \mathrm{Sub}(A \times A)$, then by using (2.1) $h = \langle p_1' u, p_2' v\rangle : Y \to X \times X$ satisfies

$$(f \times f)h = \langle fp_1' u, fp_2' v\rangle = \langle p_1 (f \times f)u, p_2 (f \times f)v\rangle = \langle p_1 g, p_2 g\rangle = g.$$

(ii) Let $\underline{C}$ = $\underline{\mathrm{Set}}$ and consider any set A with Cartesian product $(A \times A, p_1, p_2)$ and arbitrary subsets X and Y of $A \times A$ with inclusion maps $i_X: X \to A \times A$ and $i_Y: Y \to A \times A$ respectively. If $Z = Y \circ X = \{(x,y) \in A \times A/$ there exists a $z \in A$ such that $(x,z) \in X$ and $(z,y) \in Y\}$ with inclusion $i_Z: Z \to A \times A$, then one easily verifies that

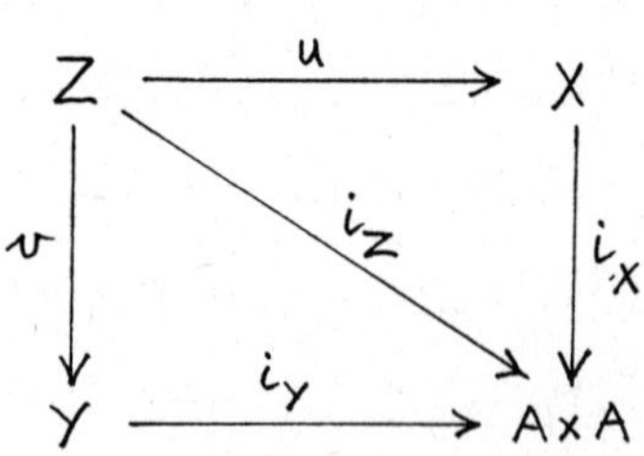

Diagram 10

is a (p_1, p_2)-pullback square. The maps u and v are defined by $u(x,y) = (x, z_{xy})$ and $v(x,y) = (z_{xy}, y)$ for every $(x,y) \in Z$, where for each $(x,y) \in Z$ a $z_{xy} \in \{z \in A/(x,z) \in X, (z,y) \in Y\}$ has been selected by the Axiom of Choice.

(iii) (Due to R.J. Wille) Let $\underline{C} = \underline{Vect}_R$ be the category of all vector spaces over the field R of real numbers. For any vector space A with subspace X of $A \times A$, we consider the subspace

$$X \circ X = \{(x,y) \mid \exists z \in A \text{ such that } (x,z),(z,y) \in X\}$$

of $A \times A$. We shall show that there exist linear maps u,v such that

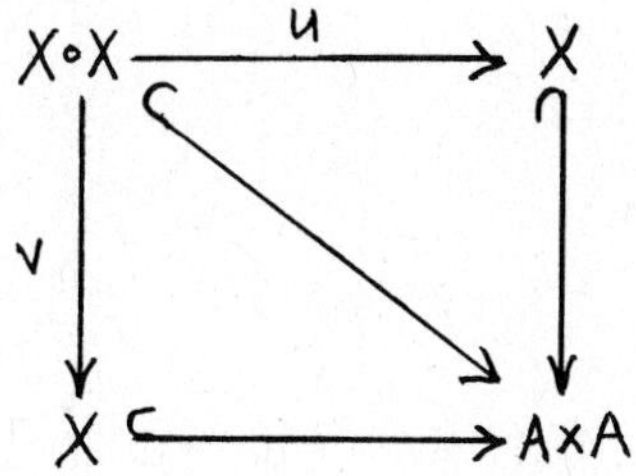

is a (p_1, p_2)-pullback diagram.

By using the Axiom of Choice, with each $(x,y) \in X \circ X$ we associate a $z(x,y) \in Z(x,y) = \{z \in A \mid (x,z) \in X, (z,y) \in X\}$ in the following way: With $(x,y)_0 = (0,0) \in X \circ X$ we associate $z(0,0) = 0 \in Z(0,0)$. Since $X' = X \circ X - \{(0,0)\}$ is well-ordered it has a least element $(x,y)_1$ for which we choose a $z(x,y)_1 \in Z(x,y)_1$. With $r(x,y)_1$, $r \in R$, we associate $rz(x,y)_1$. For the least element $(x,y)_2$ of $X' - \{r(x,y) \mid r \in R\}$ select a $z(x,y)_2 \in Z(x,y)_2$ and with $s(x,y)_2$, $s \in R$, associate $sz(x,y)_2$. With $r(x,y)_1 + s(x,y)_2$, $r,s \in R$, associate $rz(x,y)_1 + sz(x,y)_2$. For the least element $(x,y)_3$ of $X' - \{r(x,y)_1 + s(x,y)_2 \mid r,s \in R\}$ choose a $z(x,y)_3 \in Z(x,y)_3$. Associate $tz(x,y)_3$ with $t(x,y)_3$, $t \in R$. For all linear combinations of $(x,y)_i$, $i = 1,2,3$, the associated $z(x,y)_i \in Z(x,y)_i$ is determined. Continue this process indefinitely.

One easily checks that the above diagram with u,v defined by $u(x,y) = (x,z(x,y))$ and $v(x,y) = (z(x,y),y)$ for all $(x,y) \in X \circ X$, is a (p_1,p_2)-pullback.

If $\underline{C}$ is a well-powered $(\underline{E},\underline{M})$ category with finite products and $A \in Ob(\underline{C})$ we shall write

$\underline{M}^*(A\times A) = \{(X,f) \in \underline{M}(A\times A)/(X,f) * (X,f) \text{ exists}\}$, which is non-empty since $(A\times A,1) \in \underline{M}^*(A\times A)$.

(4.4) Proposition Let $\underline{C}$ be a well-powered $(\underline{E},\underline{M})$ category with finite products. Consider any $A \in Ob(\underline{C})$ and $(X,f),(Y,g),(Z,h) \in Sub(A\times A)$ such that $(X,f) \leqslant (Y,g)$.

(i) If (h,u,v) (p_1,p_2)-commutes with (f,f) then it also (p_1,p_2)-commutes with (g,g).

(ii) If $(X,f),(Y,g) \in \underline{M}^*(A\times A)$ then $(X,f) * (X,f) \leqslant (Y,g) * (Y,g)$.

(iii) If $(X,f),(Y,g) \in \underline{M}^*(A\times A)$ and $(X,f) \approx (Y,g)$ then

$$(X,f) * (X,f) \approx (Y,g) * (Y,g).$$

(4.5) Proposition Let $\underline{C}$ be a well-powered $(\underline{E},\underline{M})$ category with finite products and let $A \in Ob(\underline{C})$.

(i) $(X,f) \in \underline{M}^*(A\times A)$ iff $(X,kf) \in \underline{M}^*(A\times A)$.

(ii) If $(X,f) \in \underline{M}^*(A\times A)$ and $(X,f) \approx (Y,g)$ then $(X,f) * (Y,g)$ exists and is equivalent to $(X,f) * (X,f)$.

It is now possible to generalize the definition of uniform spaces.

(4.6) Definition Let $\underline{C}$ be a well-powered, finitely complete $(\underline{E},\underline{M})$ category. A pair (A,α), $A \in Ob(\underline{C})$ and α a subset of $\underline{M}(A\times A)$, is a uni-object in $\underline{C}$ iff

(U0.1) $(X,f) \in \alpha$ implies $(A,\Delta) \leqslant (X,f)$, where $\Delta = \langle 1,1\rangle : A \to A\times A$,

(U0.2) $(X,f) \in \alpha$ implies $(X,kf) \in \alpha$,

(U0.3) $(X,f) \in \alpha$ implies that there exists a $(Y,g) \in \underline{M}^*(A \times A)$ such that $(Y,g) \in \alpha$ and $(Y,g) * (Y,g) \leqslant (X,f)$,

(U0.4) $(X,f),(Y,g) \in \alpha$ implies $(X,f) \cap (Y,g) \in \alpha$,

(U.05) $(X,f) \in \alpha$ and $(X,f) \leqslant (Y,g)$ imply $(Y,g) \in \alpha$.

If (A,α) is a uni-object, α will be called a uni-structure on A.

(4.7) Examples (i) If $\underline{C} = \underline{\text{Set}}$ and if (p_1, p_2)-pullbacks are interpreted as in example (4.3)(ii), uni-objects correspond precisely to uniform spaces.

(ii) Let $\underline{C} = \underline{\text{Vect}}_R$ and consider any real vector space A. Clearly, any family α of subspaces of $A \times A$ satisfying the following properties is a uni-structure on A: (a) $\Delta \subset X$ for any $X \in \alpha$;

(b) $X \in \alpha$ implies $X' = \{(y,x) \mid (x,y) \in X\} \in \alpha$;

(c) $X, Y \in \alpha$ implies $X \cap Y \in \alpha$;

(d) $X \in \alpha$, $X \leqslant Y$ imply $Y \in \alpha$;

(e) For any $X \in \alpha$ there exist a $Y \in \alpha$ such that $Y \circ Y \leqslant X$ (see Example 4.3 (ii)).

The reader is urged to find some more examples by interpreting the definitions of (p_1, p_2)-pullbacks and uni-objects in other categories.

(4.8) Proposition Let $\underline{C}$ be a well-powered, finitely complete $(\underline{E},\underline{M})$ category and consider any uni-object (A,α) in $\underline{C}$. For any $(X,f) \in \alpha$ there exists a $(Y,g) \in \underline{M}^*(A \times A)$ such that $(Y,g) \in \alpha$, $(Y,g) \approx (Y,kg)$ and $(Y,g) * (Y,g) \leqslant (X,f)$.

The definition of uniformly continuous maps between uniform spaces can now also be generalized without difficulty.

(4.9) Definition Let $\underline{C}$ be a well-powered, finitely complete $(\underline{E},\underline{M})$ category and let (A,α) and (B,β) be uni-objects in $\underline{C}$.

A morphism $\psi: A \to B$ of $\underline{C}$ is a uni-morphism iff $(\psi \times \psi)^{-1}(X,f) \in \alpha$ for each $(X,f) \in \beta$.

(4.10) Proposition Let $\underline{C}$ be a well-powered, finitely complete $(\underline{E},\underline{M})$ category. The class of all uni-objects in $\underline{C}$ together with all uni-morphisms in $\underline{C}$ form a subcategory $\underline{Uni}$ of $\underline{C}$.

§ 5 OBJECTFILTERS ON UNI-OBJECTS

In this section it will be shown that analogues of the results that, in a uniform space every convergent filter is a Cauchy filter and every Cauchy filter converges to each of its adherence points, also hold in general categories.

(5.1) Definition Let $\underline{C}$ be a well-powered, finitely complete $(\underline{E},\underline{M})$ category (with terminal object E). Let (A,α) be a uni-object in $\underline{C}$ and $\underline{F}$ an objectfilter on A.

(i) The objectfilter $\underline{F}$ converges to an $m: E \to A$ on (A,α) (and m is called a limitmorphism of $\underline{F}$) iff for any $(X,f) \in \alpha$ there exists a $(Y,g) \in \underline{F}$ such that $(Y \times E, g \times m) \leqslant (X,f)$.

(ii) $\underline{F}$ is a Cauchy objectfilter on (A,α) iff for any $(X,f) \in \alpha$ there exists a $(Y,g) \in \underline{F}$ such that $(Y \times Y, g \times g) \leqslant (X,f)$.

(iii) An $m: E \to A$ is adherent to an objectfilter $\underline{F}$ on (A,α) iff there exists an objectfilter $\underline{G}$ on A such that $\underline{F} \subset \underline{G}$ and $\underline{G}$ converges to $m: E \to A$ on (A,α).

(5.2) Remarks (i) Note that in a uniform space $(X,\underline{U})$ (5.1)(i) means that a filter $\underline{F}$ converges to $x \in X$ iff given $u \in \underline{U}$ there exists an $F \in \underline{F}$ such that $F \times \{x\} \subseteq u$. This is a generalization of sequential convergence in metric spaces for, it is easily seen that the Fréchet filter associated with a sequence $\{x_n\}$ converges to x iff $\{x_n\}$ converges to x.

(ii) Definitions (5.1)(ii) and (iii) are obvious generalizations of the usual definition of Cauchy filters on uniform spaces and a well-known result respectively.

(5.3) Theorem Let $\underline{C}$ be a well-powered, finitely complete $(\underline{E},\underline{M})$ category. If $\underline{F}$ is an objectfilter on A which converges to $m: E \to A$ on a uni-object (A,α) then $\underline{F}$ is a Cauchy objectfilter on (A,α).

Proof Consider any $(X,f) \in \alpha$. By (4.8) there exists a $(Y,g) \in \underline{M}^*(A \times A)$ such that $(Y,g) \in \alpha, (Y,g) \approx (Y,kg)$ and $(Y,g) * (Y,g) \leqslant (X,f)$. Hence $(Y,g) * (Y,kg) \approx (Y,g) * (Y,g)$ by (4.5)(ii). Now, since $\underline{F}$ converges to $m: E \to A$ on (A,α), given $(Y,g) \in \alpha$ there exists a $(Z,h) \in \underline{F}$ such that $(Z \times E, h \times m) \leqslant (Y,g)$, i.e. there is an $h': Z \times E \to Y$ such that $h'g = h \times m$.

Now, consider the products $(A \times A, p_1, p_2)$ and $(Z \times Z, p_1', p_2')$ and let $Hom(Z,E) = \{e\}$. It will be shown that the

Diagram 11

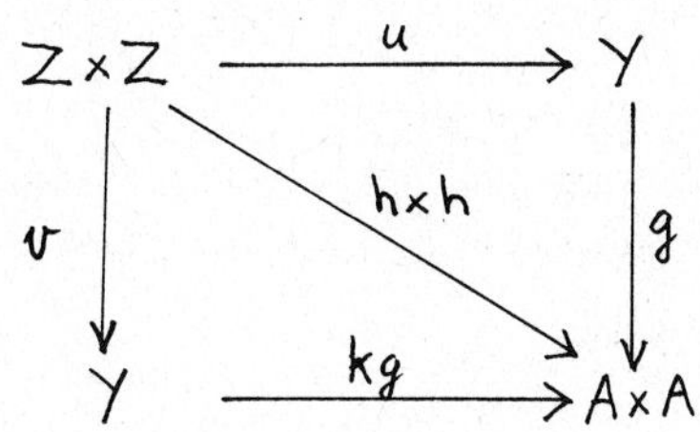

(p_1, p_2)-commutes, where $u = h'\langle 1,e\rangle p_1'$ and $v = h'\langle 1,e\rangle p_2'$.

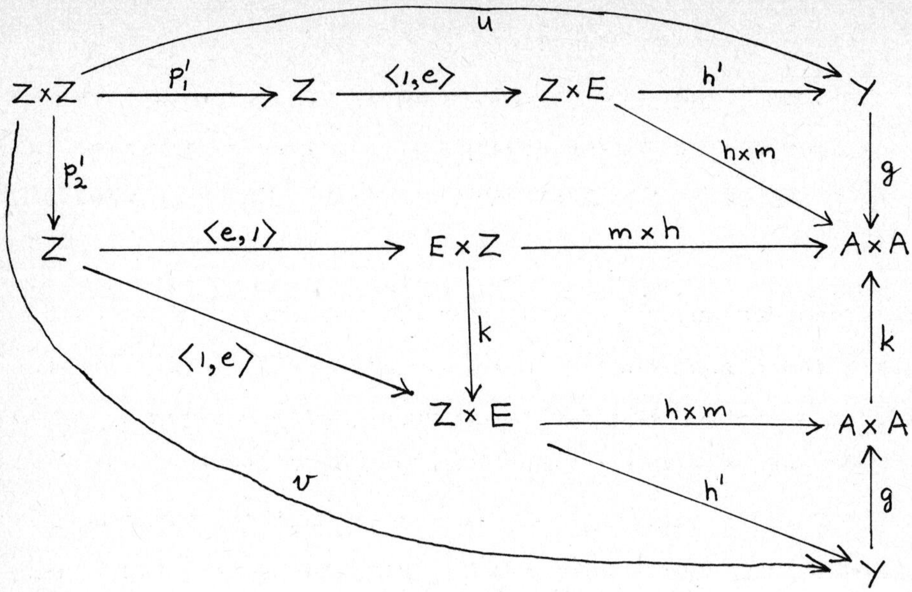

Diagram 12

By using (2.1) and (2.2), we have

$$gu = gh'\langle 1,e\rangle p_1' = (h\times m)\langle 1,e\rangle p_1' = \langle h, me\rangle p_1' = \langle hp_1', mep_1'\rangle$$

and

$$kgv = kgh'\langle 1,e\rangle p_2' = k(h\times m)\langle 1,e\rangle p_2' = (m\times h)k\langle 1,e\rangle p_2'$$

$$= (m\times h)\langle e,1\rangle p_2' = \langle me, h\rangle p_2' = \langle mep_2', hp_2'\rangle .$$

Hence, $p_1 gu = hp_1' = p_1(h\times h)$ and $p_2 kgv = hp_2' = p_2(h\times h)$. Furthermore, since E is terminal it holds that $ep_1' = e' = ep_2'$, where we have put $\mathrm{Hom}(Z\times Z, E) = \{e'\}$, so that $p_2 gu = mep_1' = me' =$ $= mep_2' = p_1 kgv$.

Thus diagram 11 (p_1, p_2)-commutes, so that

$$(Z\times Z, h\times h) \leqslant (Y,g) * (Y,kg) \approx (Y,g) * (Y,g) \leqslant (X,f).$$

Hence, given any $(X,f) \in \alpha$ there exists a $(Z,h) \in \underline{F}$ such that

$(Z \times Z, h \times h) \leqslant (X,f)$, i.e. $\underline{F}$ is a Cauchy objectfilter on (A,α).

(5.4) Theorem Let $\underline{C}$ be a well-powered, finitely complete $(\underline{E},\underline{M})$ category. If $\underline{F}$ is a Cauchy objectfilter on a uni-object (A,α) in $\underline{C}$ and if $m\colon E \to A$ is adherent to $\underline{F}$, then $\underline{F}$ converges to $m\colon E \to A$.

Proof Consider any $(X,f) \in \alpha$. By (U0.4) there exists an $(X',f') \in \underline{M}^*(A \times A)$ such that $(X',f') \in \alpha$ and $(X',f') * (X',f') \leqslant (X,f)$. Since $\underline{F}$ is a Cauchy objectfilter, given $(X',f') \in \alpha$ there exists a $(Y,g) \in \underline{F}$ such that $(Y \times Y, g \times g) \leqslant (X',f')$, i.e. there is a $g'\colon Y \times Y \to X'$ such that $f'g' = g \times g$. Furthermore, since $m : E \to A$ is adherent to $\underline{F}$ there exists an objectfilter $\underline{G}$ on A such that $\underline{F} \subset \underline{G}$ and $\underline{G}$ converges to $m\colon E \to A$ on (A,α). Hence, given $(X',f') \in \alpha$ there exists a $(Z,h) \in \underline{G}$ such that $(Z \times E, h \times m) \leqslant (X',f')$, i.e. there is a $g'' : Z \times E \to X'$ such that $f'g'' = h \times m$. Since $\underline{F} \subset \underline{G}$ it holds that $(Y,g) \in \underline{G}$ and therefore $(Z',h') = (Y,g) \cap (Z,h) \in \underline{G}$, which implies that there exist $g_1 : Z' \to Y$ and $h_1 : Z' \to Z$ such that the

Diagram 13

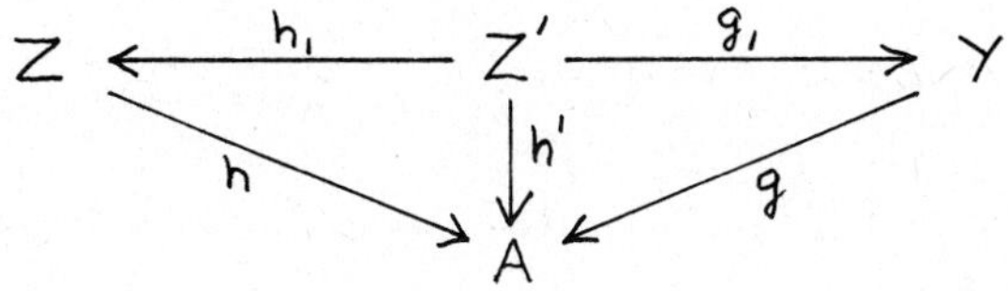

commutes. Since $(Z',h') \in \underline{G}$ it follows from (OF.3) that there exists an $m'\colon E \to A$ such that $(E,m') \leqslant (Z',h')$, i.e. there is an $h_2 : E \to Z'$ such that $h'h_2 = m'$.

It will now be shown that

Diagram 14

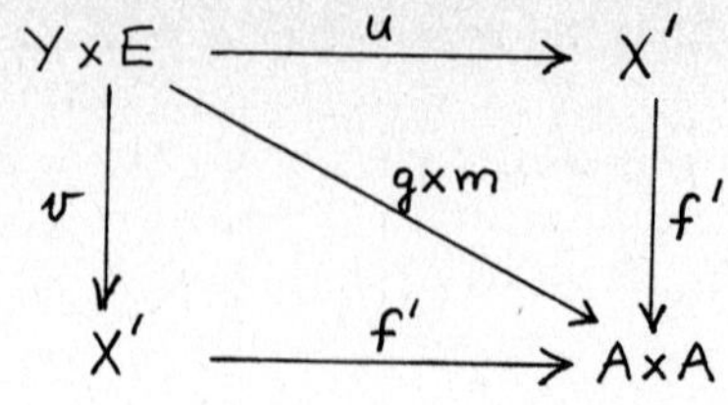

(p_1, p_2)-commutes, where $u = g'(1 \times g_1 h_2)$ and $v = g''(h_1 h_2 e \times 1)$.

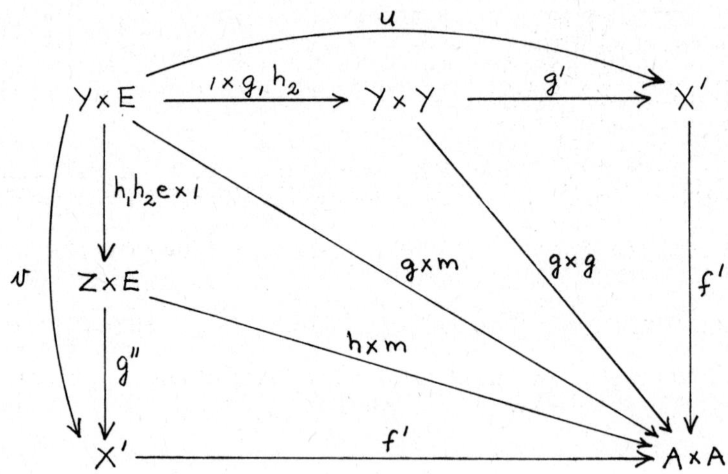

Diagram 15

We have

$$f'u = f'g'(1 \times g_1 h_2) = (g \times g)(1 \times g_1 h_2) = g \times gg_1 h_2 = g \times h'h_2 =$$

$$= g \times h'h_2 = g \times m'$$

and

$$f'v = f'g''(h_1 h_2 e \times 1) = (h \times m)(h_1 h_2 e \times 1) = hh_1 h_2 e \times m =$$

$$= h'h_2 e \times m = m'e \times m.$$

Hence, $p_1 f'u = p_1(g \times m') = gp_1 = p_1(g \times m)$, $p_2 f'v = p_2(m'e \times m) =$ $= mp_2 = p_2(g \times m)$ and $p_2 f'u = m'p_2' = m'ep_1' = p_1 f'v$, where

$p_1': Y \times E \to Y$ and $p_2': Y \times E \to E$ are projection morphisms.

Thus diagram 14 (p_1, p_2)-commutes, so that

$(Y \times E, g \times m) \leqslant (X', f') * (X', f') \leqslant (X, f)$. Hence, given $(X,f) \in \alpha$ there exists a $(Y,g) \in \underline{F}$ such that $(Y \times E, g \times m) \leqslant (X,f)$, i.e. $\underline{F}$ converges to $m: E \to A$ on (A, α).

REFERENCES

[1] B. ECKMANN and P.J. HILTON. Group-like structures in general categories I. Math. Ann. 145 (1962), 227 - 255

[2] M. FRÉCHET. Sur quelques points du calcul fonctionnel. Rend. Palermo, XXII (1906), 1 - 74

[3] M. FRÉCHET. Les ensembles abstraits et le calcul fonctionnel. Rend. Palermo. XXX (1910), 1 - 26

[4] H. HERRLICH and G.E. STRECKER. Category Theory. Allyn and Bacon, Inc., 1973

[5] G. PREUSS. Allgemeine Topologie. Springer Verlag, 1972

[6] A. WEIL. Sur les espaces à structure uniforme et sur la topologie générale. Hermann, 1938

CATEGORIES OF TOPOLOGICAL TRANSFORMATION GROUPS

J. de Vries

1. INTRODUCTION

The theory of topological transformation groups (ttg's) forms a fascinating and comprehensive realm in the world of mathematics, bordering on the domains of abstract harmonic analysis, ergodic theory, geometry, differential equations and topology. In this talk I cannot give you even a flavour of the subject. Instead, I would like to discuss certain *categories of ttg's*. I shall use category theory in a rather "naive" way. Some categories of ttg's will be defined and investigated. In this context, the categories TOPGRP (all topological groups and continuous homomorphisms) and TOP (all topological spaces and continuous functions) will be regarded as "known", and most questions will be reduced to questions about these categories. In the attempt to do so, some "classical" problems and techniques appear in a natural way. This shows that these problems are interesting, not only because they turned out to be so in the development of the subject (by "accident"), but also from the more "intrinsic" point of view of category theory. However, the solutions of these problems have been given independently of category theory. On the other hand, attempts to place certain problems and their solutions into a categorical setting can be very illuminating, and may require the definition of interesting new categories. This will be illustrated when we consider the problem of embedding arbitrary ttg's in linear ttg's.

Let me first recall some definitions. A *topological transformation group* (ttg) is a system $\langle G,X,\pi\rangle$ in which G is a topological group (the *phase group*), X is a topological space (the *phase space*) and π (the *action* of G on X) is a continuous function, π: $G \times X \to X$, such that

$$\pi(e,x) = x;\ \pi(s,\pi(t,x)) = \pi(st,x)$$

for every $x \in X$ and $s, t \in G$ (e denotes the unit of any group under consideration). A ttg with phase group G is often called *a G-space*. If π is an action of G on X, continuous mappings $\pi^t: X \to X$ and $\pi_x: G \to X$ can be defined by

$$\pi^t x := \pi(t,x) =: \pi_x t$$

for $t \in G$ and $x \in X$. Plainly, each π^t is an autohomeomorphism of X, and the mapping $\bar{\pi}: t \mapsto \pi^t$ is a homomorphism of the underlying group of G into the full homeomorphism group H(X) of X. The closure of the group $\bar{\pi}[G]$ in X^X is a semigroup (with composition of mappings as multiplication), called the *enveloping semigroup* of $\langle G,X,\pi\rangle$. For $x \in X$, the subset $\pi_x[G]$ of X is called the *orbit* of x under G. The orbits form a partition of X. The corresponding quotient space and quotient mapping are denoted by X/C_π and $c_\pi: X \to X/C_\pi$, respectively.

We give here a few examples of ttg's . In all cases, G denotes a topological group, and $\lambda: G \times G \to G$ its multiplication.

(i) For every topological space X, define $\mu_X^G := \lambda \times 1_X: G \times G \times X \to G \times X$. Then $\langle G, G\times X, \mu_X^G\rangle$ is a ttg.

(ii) If H is a subgroup of G then G acts on the space G/H of left cosets by means of an action π, defined by $\pi^t(sH) := tsH$, $s,t\in G$.

(iii) If Y is a topological space then a mapping $\tilde{\rho}: G \times C_c(G,Y) \to C_c(G,Y)$ can be defined by $\tilde{\rho}^t f(s) := f(st)$, $s,t\in G$, $f\in C(G,Y)$. If G is locally compact, then $\tilde{\rho}$ is continuous, and $\langle G, C_c(G,Y), \tilde{\rho}\rangle$ is a ttg.

(v) Similarly, a ttg $\langle G, L^p(G), \tilde{\rho}\rangle$ can be defined if G is locally compact and $1 \le p < \infty$. Here $L^p(G)$ is the usual space of measurable functions whose p-th power is integrable with respect to the right Haar measure on G.

(vi) If f is a sufficiently nice $\mathbb{R}^n$-valued function on an open domain Ω in $\mathbb{R}^n$, then the autonomous differential equation $\dot{x} = f(x)$ defines an action of $\mathbb{R}$ on Ω such that the orbits are just the solution curves. (Due to this example, actions of $\mathbb{R}$ on arbitrary spaces are often called *flows*.)

We cannot go into details here about the application of ttg's. Originating from differential geometry (the work of S. LIE on "continuous groups") there is the theory of Lie groups and their actions, including work of HILBERT, BROUWER, CARTAN and WEYL, to mention only a few names from the classical period. Introductions into these areas can be found in [29] or [9]. For applications in harmonic analysis, see for instance [40]. Related is the theory of fibre bundles.

Ttg's are also studied in Topological Dynamics. This field of research grew from classical dynamics and the qualitative theory of differential equations in an attempt to prove theorems about stability, recurrence, asymptoticity, etc. by purely topological means, whenever possible. The most notable early work was by H. POINCARÉ and G.D. BIRKHOFF. A large body of results for flows which are of interest for classical dynamics has been developed since that time, without reference to the fact that the flows arise from differential equations. Later, results were extended to general ttg's. A landmark in this development towards abstraction is the book [19]; a more recent introduction is [16]. Here the link between ttg's and dynamics is not so clear for a non-specialist. More closely related to differential equations are books like [8] or [21]. Also in some books on differential equations one can find results on flows. See for instance [31] or [25]*). These theories are "local", in the sense that questions are asked like "what does the ω-limit set look like?"; "what happens in the neighbourhood of a fixed point?"; etc. Related are the "global" theories of SMALE and others, where the object of study is vector fields on manifolds. For an introduction, see [2], and for applications, [1].

In the development of the theory of ttg's an important role has been played by the quotient mapping $c_\pi : X \to X/C_\pi$ for a ttg $\langle G,X,\pi\rangle$. Let me formulate here two related problems:

(i) If G is compact, then c_π is a perfect mapping, and there exists a nice relationship between the topological properties of X and X/C_π. In this context, also the normalized Haar measure of G can be used.

*) For flows in the plane, see also [7].

For which ttg's with a non-compact phase group does there exist such a nice relationship? Paracompactness of X/C_π turns out to be of particular interest.

(ii) A (global) *continuous cross-section* of a ttg $\langle G,X,\pi\rangle$ is a pair (S,τ) with $S \subseteq X$ and $\tau\colon X \to G$ a continuous function such that, for every $x \in X$, $\tau(x)$ is the *unique* element of G for which $\pi(\tau(x),x) \in S$. It is easily seen that $\langle G,X,\pi\rangle$ has a continuous cross-section iff it is isomorphic as a G-space with $\langle G,G\times Y,\mu_Y^G\rangle$ for some space Y. In that case, Y, S and X/C_π are homeomorphic. The question of which ttg's have such a global continuous cross-section is important, not only in abstract theories, but also in the study of flows ("which flows are *parallelizable?*").

These two problems are also related to general questions in Topological Dynamics, dealing with the structure of orbit closures. See e.g. [22]. Concerning the relationship between (i) and (ii), we confine ourselves to the remark that in order to prove that certain ttg's have a global continuous cross-section one usually shows first the existence of local continuous cross-sections; then, using nice properties of X/C_π and G, these are pasted together to a global one. Cf. for instance [34], [27], [22] and the references given there.

A number of solutions of the following problem use also the existence of local cross-sections for certain ttg's. The problem is

(iii) Which ttg's $\langle G,X,\pi\rangle$ can be embedded in a topological vector space V, or even in a Hilbert space, in such a way that the π^t's become restrictions of invertible linear operators ρ^t, $t \in G$, such that $\langle G,V,\rho\rangle$ is a ttg.

For flows, see for example (the proof of) BEBUTOV's theorem and generalizations thereof in [30], [26] or [23]. For actions of Lie groups G and embeddings in Hilbert G-spaces using the method of local cross-sections (or, more general, of *slices*), see [33] and the references given there. We shall return to problem (iii) in the last part of this paper. First, I shall indicate why problems (i) and (ii) are also interesting from a categorical point of view.

2. THE CATEGORY TTG

2.1. A *morphism of ttg's* from $\langle G,X,\pi\rangle$ to $\langle H,Y,\sigma\rangle$ is a morphism $(\psi,f)\colon (G,X) \to (H,Y)$ in the category TOPGRP × TOP for which the diagram

$$\begin{array}{ccc} G \times X & \xrightarrow{\ \pi\ } & X \\ {\scriptstyle \psi\times f}\big\downarrow & & \big\downarrow{\scriptstyle f} \\ H \times Y & \xrightarrow[\ \sigma\ ]{} & Y \end{array}$$

commutes.*) *Notation*: $\langle\psi,f\rangle\colon \langle G,X,\pi\rangle \to \langle H,Y,\sigma\rangle$. Here, f will be called a *ψ-equivariant* mapping; an 1_G-equivariant mapping will just be called equivariant.

Let TTG denote the category having the class of all ttg's as its object class (also ttg's with an empty phase space are admitted). The morphisms in TTG are the above defined morphisms of ttg's, with coordinatewise composition.

2.2. Important for the investigation of the category TTG are the following forgetful functors, whose obvious definitions we leave to the reader:

$$K\colon \mathrm{TTG} \to \mathrm{TOPGRP} \times \mathrm{TOP};$$
$$G\colon \mathrm{TTG} \to \mathrm{TOPGRP};$$
$$S\colon \mathrm{TTG} \to \mathrm{TOP}.$$

These functors forget all about actions, so they cannot be expected to reveal much about the "internal" structure of ttg's. In this respect, the following functor may be expected to be more useful:

$$S_1\colon \mathrm{TTG} \to \mathrm{TOP}.$$

It is defined in the following way. For an object $\langle G,X,\pi\rangle$ in TTG, set $S_1\langle G,X,\pi\rangle := X/C_\pi$, the orbit space of $\langle G,X,\pi\rangle$. If $\langle\psi,f\rangle\colon \langle G,X,\pi\rangle \to \langle H,Y,\sigma\rangle$

*) The products here are ordinary cartesian products, i.e. products in the category TOP. In this context, we shall consider TOPGRP just as a subcategory of TOP, and we shall always suppress the corresponding inclusion functor.

is a morphism in TTG, then f maps each orbit of X *into* an orbit of Y, hence there is a unique continuous function $f'\colon X/C_\pi \to Y/C_\sigma$ such that $f' \circ c_\pi = c_\sigma \circ f$. Now set $S_1\langle\psi,f\rangle := f'$.

2.3. THEOREM. *The functor* $K\colon \mathrm{TTG} \to \mathrm{TOPGRP} \times \mathrm{TOP}$ *is monadic. Consequently,* TTG *is complete, and* K *preserves and reflects all limits and all monomorphisms.*

PROOF. Let $C := \mathrm{TOPGRP} \times \mathrm{TOP}$, and define a functor $H\colon C \to C$ by means of the assignments

$$H\colon \begin{cases} (G,X) \mapsto (G,G\times X) & \text{on objects;} \\ (\psi,f) \mapsto (\psi,\psi\times f) & \text{on morphisms.} \end{cases}$$

Some straightforward arguments show that by

$$\eta_{(G,X)} := (1_G,\eta_X^G) \text{ and } \mu_{(G,X)} := (1_G,\mu_X^G),$$

(G,X) any object in C, two natural transformations

$$\eta\colon I_C \to H \text{ and } \mu\colon H^2 \to H$$

are defined. Here $\eta_X^G(x) := (e,x)$ and $\mu_X^G(s,(t,x)) := (st,x)$ for $s, t \in G$ and $x \in X$. It is easily verified that the triple (H,η,μ) satisfies the definition of a monad (cf.[28], Chap.VI). The *algebras* over this monad are easily seen to be the systems $((G,X), (\psi,\pi))$ with (G,X) an object in C, $\psi = 1_G$, and $\pi\colon G \times X \to X$ a morphism in TOP making the diagrams

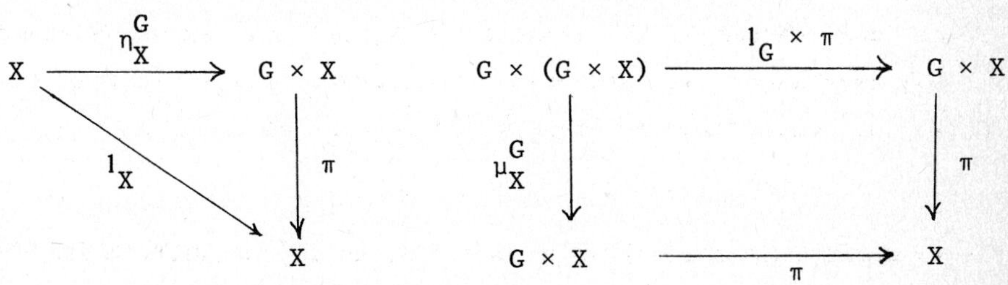

commutative, i.e. π is an action of G on X. So the algebras over (H,η,μ) can unambiguously be identified with objects in TTG. In doing so, the morphisms between such algebras become morphisms in TTG, and the category of all algebras over (H,η,μ) turns out to be isomorphic (can be identified with) TTG. In making this identification, K corresponds to the forgetful functor of this category of algebras to C; this is equivalent to saying that K is monadic.

Now the remaining statements in the theorem are a direct consequence of the general theory of monads (cf.[28], Chap.VI). □

2.4. COROLLARY. *The functor* K: TTG → TOPGRP × TOP *has a left adjoint* F, *defined by the rules*

$$F: \begin{cases} (G,X) \mapsto \langle G, G\times X, \mu_X^G\rangle & \textit{on objects;} \\ (\psi,f) \mapsto \langle \psi, \psi\times f\rangle & \textit{on morphisms.} \end{cases}$$

PROOF. Either by the theory of monads, using the identification of TTG with the category of algebras over (H,η,μ) as indicated in the proof of 2.3, or by a direct argument, showing that for every object (G,X) in TOPGRP × TOP the arrow

$$\eta_{(G,X)} : (G,X) \longrightarrow (G,G\times X)$$

has the desired universal property. □

2.5. The unit of the adjunction of F and K is the natural transformation η (cf. also the proof of 2.4); the counit is given by the arrows

$$\xi_{\langle G,X,\pi\rangle} : \langle G,G\times X,\mu_X^G\rangle \xrightarrow{\langle 1_G,\pi\rangle} \langle G,X,\pi\rangle$$

in TTG. Therefore, we may call the objects $\langle G,G\times X,\mu_X^G\rangle$ in TTG *free ttg's* (compare [24], p.231). This terminology can cause some confusion, because usually a ttg $\langle G,X,\pi\rangle$ is called free if $\pi^t x = x$ for some $x \in X$ implies $t = e$; we shall use the term *strongly effective* for this notion. It is obvious that a free ttg is strongly effective, but the converse is not generally true; a well-known class of counterexamples is provided by groups G which are subgroups of topological groups X such that the quotient mapping of H onto the space of right cosets does not admit a continuous section (let G act on X by left translations). *The free ttg's are plainly just the ttg's which have a continuous global cross-section.*

There is yet another way in which we arrive at the need of characterizing ttg's with continuous global cross-sections. Indeed, as in any adjoint situation, we have not only a monad, but also a comonad which is defined by the adjunction (F,K,η,ξ). Thus, in TTG, we have the comonad $(FK,\xi,F\eta K)$.

2.6. THEOREM. *The coalgebras for the comonad* $(FK,\xi,F\eta K)$ *in* TTG *are the systems* $(\langle G,X,\pi\rangle,(S,u))$ *with* (S,u) *a continuous cross-section of* $\langle G,X,\pi\rangle$. *The morphisms of coalgebras from* $(\langle G,X,\pi\rangle,(S,u))$ *to* $(\langle H,Y,\sigma\rangle,(T,v))$ *are the morphisms* $\langle\psi,f\rangle: \langle G,X,\pi\rangle \to \langle H,Y,\sigma\rangle$ in TTG *with* $f[S] \subseteq T$.

PROOF. Straightforward. For details, cf. [39], p. 92. □

2.7. As is well-known, comonads give rise to a cohomology theory (see e.g. [28], Chap.VII,§6). It would be interesting to investigate how this can be used with respect to the above mentioned comonad, (if not in TTG, then restricted to a suitable subcategory).

2.8. We shall show now that the category TTG is cocomplete, but that the functor K does not have nice preservation properties for colimits. In view of Beck's theorem (cf.[28], p.147) and theorem 2.3 above, certain coequalizers are preserved by K. An example where K does *not* preserve the coequalizer will be given in 2.13 below. The bad behaviour of K with respect to colimits is due to the functor S. Indeed, we have the following results:

2.9. LEMMA. *The functor* G: TTG $\to$ TOPGRP *has a right adjoint. Hence it preserves all colimits and all epimorphisms.* □

2.10. LEMMA. *The functor* S_1: TTG $\to$ TOP *has a right adjoint. Hence it preserves all colimits and all epimorphisms.* □

2.11. COROLLARY. *The functor* K: TTG $\to$ TOPGRP $\times$ TOP *preserves and reflects all epimorphisms.*

PROOF. Reflection: K is faithful.
Preservation: If $<\psi,f>: <G,X,\pi> \to <H,Y,\sigma>$ is epic in TTG, then by 2.9, ψ is a surjection. Hence f maps each orbit of X *onto* some orbit in Y. By 2.10, $S_1 f$ maps X/C_π onto Y/C_σ. Combining these results it follows easily that f is a surjection. □

2.12. The category TOPGRP × TOP is well-powered and co-(well-powered). Since the functor K preserves all monomorphisms and all epimorphisms, *the category* TTG *is well-powered and co-(well-powered)* as well. In addition, TTG is complete. Hence theorem 23.11 in [24] implies that *the category* TTG *has all coequalizers.*

2.13. EXAMPLE. Let $<G,Y,\sigma>$ be a ttg. Call an equivalence relation X in Y *invariant* if, considered as a subset of $Y \times Y$, it is invariant under the coordinate-wise action π of G on $Y \times Y$. In order to obtain an example which shows that the functor K (or rather, the functor S) does not preserve co-equalizers, it is sufficient to construct a ttg $<G,Y,\sigma>$ and an invariant equivalence relation X in Y such that there exists *no* continuous action of G on Y/X making the quotient mapping $g: Y \to Y/X$ equivariant.

To this end, take $G := \mathbb{Q}$, the additive group of the rationals, $Y := \mathbb{Q} \times ([0,1] \times \mathbb{N})$, and $X := \Delta_{\mathbb{Q}} \times R$, where $\Delta_{\mathbb{Q}}$ is the diagonal in $\mathbb{Q} \times \mathbb{Q}$ and R is the equivalence relation in $[0,1] \times \mathbb{N}$ obtained by identifying all points $(0,n)$, $n \in \mathbb{N}$, with each other. If we consider the action $\mu^{\mathbb{Q}}_{[0,1]\times\mathbb{N}}$ of $\mathbb{Q}$ on $Y = \mathbb{Q} \times ([0,1] \times \mathbb{N})$, then the equivalence relation X in Y is invariant, and there is *only one candidate* ζ for an action of $\mathbb{Q}$ on Y/X which makes the quotient mapping $g: Y \to Y/X$ equivariant. Now continuity of $\zeta: \mathbb{Q} \times (Y/X) \to Y/X$ can easily be seen to imply the equality of the following two topologies on Y/X: (i) the quotient topology induced by g and (ii) the product topology obtained by identifying Y/X with $\mathbb{Q} \times (([0,1] \times \mathbb{N})/R)$, where $([0,1]\times\mathbb{N})/R$ has its usual quotient topology. It can be seen, however, that these two topologies do not coincide. For details, cf. [39], 3.4.4.

2.14. The easiest way to prove that TTG is cocomplete is by invoking theorem 23.13 in [24]. First, we recall some definitions. An *epi-sink* in a category X is a family $\{f_i: X_i \to X\}_{i\in I}$ of morphisms in X such that for every pair of morphisms g, h: $X \to Y$ in X the condition $gf_i = hf_i$ for all $i \in I$ implies $g = h$. The category X is called *strongly co-(well-powered)* if for every set-indexed family $\{X_i : i \in I\}$ of objects there is at most a set of objects X in X for which there exists an epi-sink $\{f_i : X_i \to X\}_{i \in I}$. The theorem referred to above reads as follows:

If the category X *is complete and well-powered, then the following are equivalent:*

(i) X *is strongly co-(well-powered);*

(ii) X *is cocomplete and co-(well-powered).*

Observe that the categories TOPGRP and TOP are complete and well-powered, and that they satisfy condition (ii). Hence these categories are strongly co-(well-powered). We shall use this in proving the following

2.15. LEMMA. *The category* TTG *is strongly co-(well-powered).*

PROOF. Let $\{\langle\psi_i, f_i\rangle : \langle G_i, X_i, \pi_i\rangle \to \langle G, X, \pi\rangle\}_{i \in I}$ be a set-indexed epi-sink in TTG. Since left adjoints preserve epi-sinks, if follows from 2.9 that $\{\psi_i : G_i \to G\}_{i \in I}$ is an epi-sink in TOPGRP. This allows G only to be taken from a set of possible topological groups. Similarly, 2.10 implies that $\{S_1 f_i : X_i/C_{\pi_i} \to X/C_\pi\}_{i \in I}$ is an epi-sink in TOP, leaving for X/C_π only a set of possibilities. Plainly, $\mathrm{card}(X) \leq \mathrm{card}(G).\mathrm{card}\,(X/C_\pi)$, hence there is at most a set of possibilities for X. Finally, for each G and each X there is only a set of actions of G on X. So there is at most a set of objects in TTG from which $\langle G, X, \pi\rangle$ can be taken. □

2.16. THEOREM. *The category* TTG *is cocomplete.*

PROOF. Clear from 2.14 and 2.15. □

2.17. In [39], a different proof of the cocompleteness of TTG is given, using a technique which is a generalization of the construction of a "canonical" extension of the action of a subgroup to an action of the whole group. It is also related to the construction of "induced representations". Both techniques are very important in the theory of actions of compact groups. See [9] and [40]. The methods, used in [39] also indicate how examples can be constructed which show that K does not preserve all coproducts.

3. SUBCATEGORIES OF TTG

3.1. Let A and B denote subcategories of TOPGRP and TOP, respectively, and set X: = $K^{\leftarrow}[A \times B]$. Then X is a subcategory of TTG. The restrictions and corestrictions of the functors K,G and S will be denoted by the same symbols; so we have K: X → A × B, G: X → A and S: X → B.

3.2. If one wants to show that K: X → A × B is monadic using the same methods as in 2.3, one has to require, among others, that G × X is an object in B for every object (G,X) in A × B. This condition appears to be rather harmless at first sight. However, a large portion of Topological Dynamics deals with actions of discrete groups on compact Hausdorff spaces (cf.[16]). So one might try to apply category theory in this field by taking B: = COMP (the category of all compact Hausdorff spaces) and A a category having discrete groups as objects. Then the above condition is only fulfilled if the objects of A are all finite. For Topological Dynamics, the restriction to finite groups is unacceptable. For other parts of the theory of ttg's, actions of finite groups on compact spaces is very important: it is one of the corner stones of the general theory of actions of compact Lie groups (cf.[29], p.222).

Although monadicity of the functor K: X → A × B may be unattractive in view of practical purposes, K and X do have nice properties under rather mild conditions. The proofs of the following propositions can be found in [39], section 4.

3.3. PROPOSITION. *Suppose that the inclusion functor of* B *into* TOP *preserves all limits. Then the functor* K: X → A × B *creates all limits. Hence, if* A *and* B *are complete, then so is* X, *and all limits and monomorphisms are preserved and reflected by* K. □

3.4. PROPOSITION. *Suppose* B *is a full subcategory of* TOP. *If either* A ⊆ B *or* B *is productive and closed hereditary, then* K *preserves and reflects monomorphisms.* □

3.5. If one wants to show that X is cocomplete or that K: $X \to A \times B$ preserves epimorphisms, then the proofs given for TTG cannot be adapted to the present situation, unless the restricted functor S_1: $X \to$ TOP actually sends X into B. So this brings us directly to the first problem, mentioned in section 1. In addition, something must be known about the epimorphisms in A.

Although the question of which nice properties of the phase space of a ttg are inherited by the orbit space (and under what circumstances!) is very interesting, the following proposition avoids this problem (for details, cf.[39], section 4).

3.6. PROPOSITION. *Suppose that the following conditions are fulfilled:*

(i) *Epimorphisms in* A *have dense ranges;*

(ii) B *is a full subcategory of* HAUS, *having a terminal object;*

(iii) *If* Y *is an object in* B *and* A *is a closed subset of* Y, *then* $Y \cup_A Y$ *is in* B.

Then the functor K: $X \to A \times B$ *preserves and reflects epimorphisms.* □

3.7. The conditions (ii) and (iii) are rather mild. Yet the above proposition is of restricted applicability, because of condition (i). Indeed, it is still unknown to me whether the category HAUSGRP satisfies condition (i). It is known, however, that COMPGRP does! Of course, the preceding proposition can also be applied to the very important case (which we neglected untill now) that the category A has only one object, a fixed topological group G, and only one morphism, namely 1_G. In that case, the category X will be denoted B^G (the category of all G-spaces with phase space in B).

3.8. I want to make now a few remarks concerning reflective subcategories of TTG. I shall restrict myself to the question of the "preservation of reflections" by the functor K for only one particular case. In view of the following lemma, it is the functor S which causes difficulties.

3.9. LEMMA. *Suppose that* X: = $\overleftarrow{K}[A \times B]$ *is a reflective subcategory of* TTG *and that* B *has a final object. Then* A *is a reflective subcategory of* TOPGRP. *In addition, if* $\langle G,X,\pi\rangle$ *is an object in* TTG *and*

$$<\psi,f> : <G,X,\pi> \to <H,Y,\sigma>$$

is its reflection into X, *then* ψ: $G \to H$ *is a reflection of* G *into* A.

PROOF. This is a consequence of the fact that under rather mild conditions a functor having a right adjoint (c.q. the functor G) preserves reflections. For a different proof, see [39], p. 133. □

3.10. It is not difficult to show that $K^{\leftarrow}$[COMPGRP × COMP] is a reflective subcategory of TTG. So by the preceding lemma, the reflection of an object $<G,X,\pi>$ of TTG into this subcategory has the form

$$<\alpha_G,f> : <G,X,\pi> \to <G^c,Y,\sigma>,$$

where α_G: $G \to G^c$ is the *Bohr compactification* of G. In general, f: X → Y is not the reflection of X into COMP. In fact, there are examples which show that Y can be a one-point space even if X is a non-trivial compact Hausdorff space. The problem whether Y is trivial or not has been important in Topological Dynamics, and is related to many interesting questions. This follows from the following observation (cf.[39], p.141):

Using the above notation with X *a compact Hausdorff space, the morphism*

$$<1_G,f> : <G,X,\pi> \to <G,Y,\sigma^{\alpha}>$$

where $\sigma^{\alpha}(t,y) := \sigma(\alpha(t),y)$, $(t,y) \in G \times Y$, *is just the maximal equicontinuous factor of* $<G,X,\pi>$.

For definitions and results concerning maximal equicontinuous factors, cf. [17], [18] and [35].

3.11. Using the notation explained in 3.7, it is not difficult to show that $COMP^G$ is a reflective subcategory of TOP^G, for any topological group G. Let the reflection of the G-space $<G,X,\pi>$ into $COMP^G$ be denoted by

$$<1_G,k> : <G,X,\pi> \to <G,Z,\zeta>.$$

If G is discrete, $k: X \to Z$ is just the reflection of X into COMP. There are examples which show that for non-discrete groups (e.g. $G=\mathbb{R}$) $k: X \to Z$ may be not the reflection of X into COMP. See [11]; it can also be shown that $\langle G,G,\lambda\rangle$ gives such an example, provided $RUC^*(G) \neq C^*(G)$. For the inequality $RUC^*(G) \neq C^*(G)$, cf. [13]. The reflection of $\langle G,G,\lambda\rangle$ into $COMP^G$ plays an important role in Topological Dynamics; there it is called the *greatest* (or *maximal*) G-*ambit*. See [10] and the references given there.

I do not know whether the mapping $k: X \to Z$ is a topological embedding if X is a Tychonoff space. I have some partial results, including the cases that X is locally compact Hausdorff or that $\langle G,X,\pi\rangle$ has the form $\langle G,(G/H)\times Y,\sigma\rangle$ with $\sigma^t(sH,y) := (tsH,y)$, $t, s \in G$, $y \in Y$, where H is a closed subgroup of G. See [37].

4. THE CATEGORY TTG_* AND GENERALIZATIONS

4.1. The objects of TTG_* are the same as the objects of TTG, viz. the ttg's. The categories differ from each other with respect to their morphisms. We shall first give a brief motivation for the definition of the morphisms in TTG_*. The idea stems from the following problem: Given a ttg $\langle G,X,\pi\rangle$, does there exist a ttg $\langle H,Y,\sigma\rangle$ such that

(i) Y is a topological vector space ;

(ii) Each σ^t, $t \in H$, is an invertible continuous linear operator on Y*);

(iii) X can be embedded in Y as an invariant subset in such a way that

$$\overline{\pi}[G] = \{\sigma^t|_X : t \in H\}.$$

If $\langle G,X,\pi\rangle$ is effective (i.e. $\pi^t \neq \pi^s$ if $t \neq s$), then it follows from (iii) that we obtain a homomorphism $\psi: H \to G$ such that for every $t \in H$ the following diagram commutes

$$\begin{array}{ccc} X & \xrightarrow{\pi^{\psi(t)}} & X \\ {\scriptstyle f}\downarrow & & \downarrow{\scriptstyle f} \\ Y & \xrightarrow{\sigma^t} & Y \end{array} \qquad\qquad \begin{array}{c} G \\ \uparrow{\scriptstyle \psi} \\ H \end{array}$$

*) Such a ttg $\langle H,Y,\sigma\rangle$ will be called *a linear ttg*.

Here f: $X \to Y$ is the embedding mapping of X into Y.

For more details about the above mentioned linearization problem we refer to section 5. At this point we are only interested in the diagram which expresses the relationship between ψ and f. We shall use it in the following definition:

4.2. The object class of TTG_* is the class of all ttg's. A *morphism in* TTG_* *from* $\langle G,X,\pi\rangle$ *to* $\langle H,Y,\sigma\rangle$ is a morphism (ψ^{op},f): $(G,X) \to (H,Y)$ in the category $TOPGRP^{op} \times TOP$ such that for every $t \in H$ the diagram in 4.1 commutes (now f is not necessarily an embedding). *Notation:* $\langle\psi^{op},f\rangle$: $\langle G,X,\pi\rangle \to \langle H,Y,\sigma\rangle$. The composition of the morphisms in TTG_* is defined coordinatewise.

4.3. The obvious forgetful functor from TTG_* to $TOPGRP^{op} \times TOP$ will be denoted by K_*. *It can be shown that this functor preserves all colimits.* Using this, it is fairly easy to construct an example which shows that *the category* TTG_* *is not complete* (the example is related to the one in 2.13). It can also be shown that K_* *preserves all monomorphisms.* In particular it follows that TTG_* is well-powered. We shall see in 4.7 below that TTG_* has a coseparator. According to Theorem 23.14 in [24], a complete, well-powered category having a coseparator is cocomplete. *It follows, that* TTG_* *is not complete.*

4.4. We want to say something more about the existence of coseparators in TTG_*, also because such objects are related to the general embedding problem, mentioned in 4.1. The notation will be similar to the notation in section 3, except for some obvious modifications. Thus, A and B are subcategories of TOPGRP and TOP, respectively, and $X_* := K_*^{\leftarrow}[A^{op} \times B]$. Moreover, for any object (G,X) in $TOPGRP^{op} \times TOP$, the evaluation mapping $f \mapsto f(e)$: $C_c(G,X) \to X$ will be denoted by δ_X^G.

4.5. LEMMA. *For every object* (G,X) *in* $TOPGRP^{op} \times TOP$ *with* G *a locally compact Hausdorff group, the pair* $(\langle G,C_c(G,X),\tilde{\rho}\rangle$, $(1_G^{op},\delta_X^G))$ *is a co-universal arrow for* (G,X) *with respect to the functor* K_*.

PROOF. Consider the following diagram:

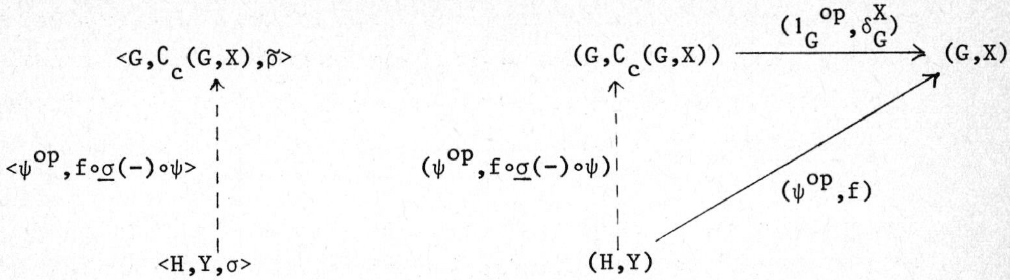

Here for every $y \in Y$, $\underline{\sigma}(y) := \sigma_y : H \to Y$, so that $f\circ\underline{\sigma}(-)\circ\psi : y \mapsto f\circ\sigma_y\circ\psi :$ $G \to X$. Observe, that local compactness of G is needed to ensure that $\tilde{\rho} : G \times C_c(G,X) \to C_c(G,X)$ is continuous. □

4.6. COROLLARY. *If* A *has a generator* G *which is locally compact Hausdorff and if* B *has a coseparator* X *such that* $C_c(G,X)$ *is an object in* B, *then* $\langle G, C_c(G,X), \tilde{\rho}\rangle$ *is a coseparator in* X_*.

PROOF. Apply the dual of Prop. 31.11 of [24]. □

4.7. EXAMPLES.

(i) Let E be the indiscrete 2-point space. Then the ttg $\langle \mathbb{Z}, E^{\mathbb{Z}}, \tilde{\rho}\rangle$ is a coseparator in TTG_*.

(ii) $\langle \mathbb{Z}, [0,1]^{\mathbb{Z}}, \tilde{\rho}\rangle$ is a coseparator for the full subcategories of TTG_*, determined by all ttg's with a Tychonoff, resp. with a compact T_2, phase space.

(iii) If G is a fixed locally compact Hausdorff group, then $\langle G, C_c(G,[0,1], \tilde{\rho}\rangle$ is a coseparator for the full subcategory of TOP^G defined by all Tychonoff G-spaces (not for $COMP^G$, unless G is discrete).

4.8. In the remainder of this section, A shall denote the full subcategory of TOPGRP, defined by all *locally compact Hausdorff groups*, and $X_* := K_*^{\leftarrow}[A^{op} \times TOP]$ (so we take $B := TOP$). In this case, it follows immediately from 4.5, that the functor $K_* : X_* \to A^{op} \times TOP$ has a right adjoint. This follows also from our next theorem:

4.9. THEOREM. *The functor* $K_*: X_* \to A^{op} \times TOP$ *is comonadic. Consequently,* X_* *is a finitely cocomplete (for so is* $A^{op} \times TOP$*), and all existing colimits and all epimorphisms are preserved and reflected by* K_*.

PROOF. Details will be published elsewhere. We only mention that in the proof essential use is made of the canonical homeomorphisms

$$C_c(G\times G,X) \simeq C_c(G,C_c(G,X)),$$

$$C_c(G\times X,X) \simeq C_c(X,C_c(G,X)),$$

G any locally compact Hausdorff space and X an arbitrary topological space. □

4.10. If G is a fixed locally compact Hausdorff topological group, then the category TOP^G can be considered as a subcategory of TTG_*. Similar to 4.9, one shows that TOP^G may be considered as a category of coalgebras over a comonad in TOP. On the other hand, similar to 2.3, TOP^G may be considered as a category of algebras over a monad in TOP. The monad and the comonad considered here are nicely related: it can be shown that they are adjoint to each other according to the definition given in [15]. Although this seems to be known, I could find no references to this fact in the literature.

4.11. One might conclude from the above remarks that ttg's with locally compact phase groups are the nice objects which deserve further study. Although this conclusion is true as far as it concerns the applications, from a categorical point of view there is a much nicer class of objects. Indeed, the homeomorphisms used in the proof of 4.9 are an indication of the fact that we should work in the cartesian closed category KR of all k-spaces. The proper objects are the systems $[G,X,\pi]$ where G is a k-group (i.e. a group G with a k-topology making the mapping $(s,t) \mapsto st^{-1}: G \otimes G \to G$ continuous), X is a k-space and $\pi: G \otimes G \to X$ is a continuous mapping satisfying the usual equations (here $\otimes$ denotes the product in the category KR: the k-refinement of the cartesian product). With the class of these k-ttg's as object class, one can form the categories k-TTG and k-TTG_* (morphisms similar to TTG and TTG_*, respectively). The study of these categories is initiated in [39].

5. LINEARIZATION OF ACTIONS

5.1. The general problem which we described in 4.1 has a trivial solution if we work in the category TTG_*. Indeed, lemma 5.2 below shows that the only condition which must be imposed on a ttg $\langle G,X,\pi\rangle$ in order that the action can be linearized, is that X can be embedded in a topological vector space. If we restrict ourselves to Hausdorff topological vector spaces, this means exactly that X is a Tychonoff space. In the following lemma, G_d is the group G with the discrete topology, and $\iota\colon G_d \to G$ is the identity mapping.

5.2. LEMMA. *If* $\langle G,X,\pi\rangle$ *is a ttg with* X *a Tychonoff space then there exists a morphism* $\langle \iota^{op}, f\rangle\colon \langle G,X,\pi\rangle \to \langle G_d,V,\sigma\rangle$ *in* TTG_* *such that* V *is a topological vector space,* σ *is a linear action, and* $f\colon X \to V$ *is a topological embedding.*

PROOF. There exists a topological embedding $g\colon X \to \mathbb{R}^\kappa$ for some cardinal number κ. Moreover, the mapping $\underline{\pi}\colon x \mapsto \pi_x\colon X \to C_c(G,X)$ is a topological embedding. Hence $f := C_c(G,g)\circ\underline{\pi}$ is a topological embedding of X into the topological vector space $C_c(G,\mathbb{R}^\kappa) =: V$. Plainly, $\langle G_d, C_c(G,\mathbb{R}^\kappa), \tilde{\rho}\rangle$ is a ttg, and $\langle \iota^{op}, f\rangle$ is a morphism in TTG_*. □

5.3. In order to make the problem more interesting, the following extra conditions will be imposed. First, a G-space $\langle G,X,\pi\rangle$ should be linearized in a G-space rather than in a G_d-space. Second, if X is a metric space, then $\langle G,X,\pi\rangle$ should be linearized in a *Hilbert* G-*space*. And finally, a large class of G-spaces should be linearized simultaneously in one and the same linear G-space.

The proof of the following theorem is a modification of the proof of lemma 5.2 above. Observe, that the apparent relationship with lemma 4.5 has a categorical background (cf. [24], Prop. 19.6).

5.4. THEOREM. *Let* G *be a locally compact Hausdorff group and let* κ *be a cardinal number. Then every* G-*space* $\langle G,X,\pi\rangle$ *with* X *a Tychonoff space of weight* $\leq \kappa$ *can equivariantly be embedded in the linear* G-*space* $\langle G, C_c(G,\mathbb{R})^\kappa, \tilde{\rho}^\kappa\rangle$. □

Using similar ideas, in combination with results from [5] and [32], the following theorem can be proved. Here $H(\kappa)$ is the Hilbert sum of κ copies of the Hilbert space $L^2(G)$, and the action $\sigma(\kappa)$ induces on each copy of $L^2(G)$ a "weighted" right translation. For a proof, see [36].

5.5. THEOREM. *Let* G *be a sigma-compact locally compact Hausdorff group and let* κ *be a cardinal number. Then there exists a linear* G*-space* $\langle G,H(\kappa),\sigma(\kappa)\rangle$ *in which every* G*-space* $\langle G,X,\pi\rangle$ *with* X *a metric space of weight* $\leq \kappa$ *can be equivariantly embedded.* □

5.6. Let G be as in 5.5. and assume, for convenience, that G is infinite. The weight of the Hilbert space $H(\kappa)$ equals $\kappa . w(G)$, where $w(G)$ is the weight of G (for compact groups this is well-known; the proof for the non-compact case can be found in [36]). If we take $\kappa = w(G)$, then $H(\kappa)$ is isomorphic to $L^2(G)$. So there is an action σ of G on $L^2(G)$ such that $\langle G,L^2(G),\sigma\rangle$ is a linear ttg in which every metric G-space $\langle G,X,\pi\rangle$ with $w(X) \leq w(G)$ can be equivariantly embedded. No explicit description of σ can be given in this case. However, there is an action τ of G on $L^2(G\times G)$ which can easily be described explicitly, such that $\langle G,L^2(G\times G),\tau\rangle$ is a linear ttg in which every metric G-space $\langle G,X,\pi\rangle$ can equivariantly be embedded, provided $w(X) \leq L(G)$, the Lindelöf degree of G). The proof is highly non-categorical; see [38].

5.7. One of the most notable early results on linearizations of actions is BEBUTOV's theorem. See [26] and also [23]. These results have applications in the theory of differential equations. Also related to differential equations are the results in [12]. In these three papers, only actions of $\mathbb{R}$ are linearized. Actions of Lie groups are considered, among others, in [33]. See also [34] and the references given there. Linearizations in Hilbert spaces of actions of more general locally compact groups appear in [3] and [5], where earlier work of COPELAND and DE GROOT ([14],[20]) was generalized. More information about the history of these results can be found in [3] and in [6].

REFERENCES.

[1] ABRAHAM, R. & J.E. MARSDEN, *Foundations of mechanics*, Benjamin, New York, 1967.

[2] ABRAHAM, R. & J. ROBBIN, *Transversal mappings and flows*, Benjamin, New York, 1967.

[3] BAAYEN, P.C., *Universal morphisms*, Mathematical Centre Tracts no. 9, Mathematisch Centrum, Amsterdam, 1964.

[4] BAAYEN, P.C., Topological linearization of locally compact transformation groups, *Report no. 2, Wiskundig Seminarium, Vrije Universiteit*, Amsterdam, 1967.

[5] BAAYEN, P.C. & J. DE GROOT, Linearization of locally compact transformation groups in Hilbert space, *Math. Systems Theory* 2(1968), 363-379.

[6] BAAYEN, P.C. & M.A. MAURICE, Johannes de Groot 1914-1972, *General Topology and Appl.* 3(1973), 3-32.

[7] BECK, A., *Continuous flows in the plane*, Springer-Verlag, Berlin, Heidelberg, New York, 1974.

[8] BHATIA, N.P. & G.P. SZEGÖ, *Stability theory of dynamical systems*, Springer-Verlag, Berlin, Heidelberg, New York, 1970.

[9] BREDON, G.E., *Introduction to compact transformation groups*, Academic Press, New York, 1972.

[10] BROOK, R.B., A construction of the greatest ambit, *Math. Systems Theory* 4(1970), 243-248.

[11] CARLSON, D.H., Extensions of dynamical systems via prolongations, *Funkcial. Ekvac.* 14(1971), 35-46.

[12] CARLSON, D.H., Universal dynamical systems, *Math. Systems Theory* 6(1972), 90-95.

[13] COMFORT W.W. & K.A. ROSS, Pseudocompactness and uniform continuity in topological groups, *Pacific J. Math.* 16(1966), 483-496.

[14] COPELAND Jr., A.H. & J. DE GROOT, Linearization of a homeomorphism, *Math. Annalen* 144 (1961), 80-92.

[15] EILENBERG, S. & J.C. MOORE, Adjoint functors and triples, *Illinois J. Math.* 9(1965), 381-398.

[16] ELLIS, R., *Lectures on topological dynamics*, Benjamin, New York, 1969.

[17] ELLIS, R. & W.H. GOTTSCHALK, Homomorphisms of transformation groups, *Trans. Amer. Math. Soc.* 94(1960), 258-271.

[18] ELLIS, R. & H. KEYNES, A characterization of the equicontinuous structure relation, *Trans. Amer. Math. Soc.* 161(1971), 171-183.

[19] GOTTSCHALK, W.H. & G.A. HEDLUND, *Topological dynamics*, Amer. Math. Soc. Colloquium Publications, Vol. 36, Providence, R.I., 1955.

[20] GROOT, J. DE, Linearization of mappings, in *General topology and its relation to modern analysis and algebra*, Proc. 1961 Prague Symposium, Prague, 1972, p.191-193.

[21] HAJEK, O., *Dynamical systems in the plane*, Academic Press, New York, 1968.

[22] HAJEK, O., Parallelizability revisited, *Proc. Amer. Math. Soc.* 27(1971), 77-84.

[23] HAJEK, O., Representations of dynamical systems, *Funkcial. Ekvac.* 14(1971), 25-34.

[24] HERRLICH, H. & G.E. STRECKER, *Category theory*, Allyn and Bacon Inc., Boston, 1973.

[25] HIRSCH, M.W. & S. SMALE, *Differential equations, dynamical systems, and linear algebra*, Academic Press, New York, 1974.

[26] KAKUTANI, S., *A proof of Bebutov's theorem*, J. Differential Equations 4(1968), 194-201.

[27] KOSZUL, J.L., *Lectures on groups of transformations*, Tata Institute of Fundamental Research, Bombay, 1965.

[28] MACLANE, S., *Categories for the working mathematician*, Springer-Verlag, Berlin, Heidelberg, New York, 1971.

[29] MONTGOMERY, D. & L. ZIPPIN, *Topological transformation groups*, Interscience, New York, 1955.

[30] NEMYCKIĬ, V.V., Topological problems in the theory of dynamical systems, *Uspehi Mat. Nauk.* 4(1949), no. 6(34), 91-153 (English translation in: AMS Translation Series 1, Vol. 5, p.414-497).

[31] NEMYCKIĬ, V.V. & V.V. STEPANOV, *Qualitative theory of differential equations*, Princeton University Press, Princeton, N.J., 1960.

[32] PAALMAN - DE MIRANDA, A.B., A note on W-groups, *Math. Systems Theory* 5(1971), 168-171.

[33] PALAIS, R.S., Slices and equivariant embeddings, in: A. BOREL et al., *Seminar on transformation groups*, Annals of Mathematics Studies 46, Princeton University Press, Princeton, N.J., 1960, p.101-115.

[34] PALAIS, R.S., On the existence of slices for actions of non-compact Lie groups, *Ann. of Math.* 73(1961), 295-323.

[35] PELEG, R., Weak disjointness of transformation groups, *Proc. Amer. Math. Soc.* 33(1972), 165-170.

[36] VRIES, J. DE, A note on topological linearization of locally compact transformation groups in Hilbert space, *Math. Systems Theory* 6(1972), 49-59.

[37] VRIES, J. DE, Can every Tychonoff G-space equivariantly be embedded in a compact Hausdorff G-space? *Math. Centrum, Amsterdam, Afd. Zuivere Wisk.*, ZW 36, 1975.

[38] VRIES, J. DE, A universal topological transformation group in $L^2(G\times G)$, *Math. Systems Theory* 9(1975), 46-50.

[39] VRIES, J. DE, *Topological transformation groups* I *(a categorical approach)*, Mathematical Centre Tracts, no. 65, Mathematisch Centrum, Amsterdam, 1975.

[40] WALLACH, N.R., *Harmonic analysis on homogeneous spaces*, Marcel Dekker, Inc., New York, 1973.

ON MONOIDAL CLOSED TOPOLOGICAL CATEGORIES I

Manfred B. Wischnewsky

The aim of this paper initiated by L.D. Nel's talk at the Conference on Categorical Topology at Mannheim (1975) is to give two different characterizations of monoidal closed topological categories. Both of these characterizations include as special instance the Herrlich-Nel results on cartesian closed (relative) topological categories* over the base category Sets ([11], [13]). The main tools in this paper are a generalization of the Special Adjoint Functor Theorem - the Relative Special Adjoint Functor Theorem - and the Dubuc-Tholen theory of Adjoint Triangles. As corollaries we obtain a characterization of monoidal closed (relative) Top-categories over wellbounded (= locally bounded) categories and over cocomplete, wellpowered and cowellpowered monoidal closed categories.

It is assumed that the reader is familiar with the notation and the content of [12]. All other notions which are used in this paper are briefly recalled.

§ 1 Factorizations and Generators, Review.

Let us first recall some of the basic notions and propositions on factorizations and generators (see e.g. [9], [16], [18a]). Let $\underline{A}$ be a category. For two morphisms $e : A \longrightarrow B$ and $m : C \longrightarrow D$ we write $e \downarrow m$ if every commutative diagram $ge = mf$ can be made commutative by a unique morphism $w : B \longrightarrow C$. If $P \subset Mor(\underline{A})$ then let $P^{\uparrow} := \{e: e\downarrow m$ for all $m \in P\}$ and $P^{\downarrow} := \{m: e\downarrow m$ for all $e \in P\}$. Let $E, M \subset Mor\underline{A}$. Then (E,M) is called a prefactorization in $\underline{A}$ if $E^{\downarrow} = M$ and $M^{\uparrow} = E$. A prefactorization (E,M) is called a factorization if every morphism f in $\underline{A}$ is of the form $f = me$ with $e \in E$ and $m \in M$. A factorization (E,M) is called proper if every $e \in E$ is an

*) Relative topological category ≡ E- reflective subcategory of an (absolute) topological category.

epimorphism and every $m \in M$ is a monomorphism. We say that a category $\underline{A}$ has a M-factorization for a class M of $\underline{A}$-morphisms if $\underline{A}$ has a $(M^{\uparrow}, M)$-factorization.

(1.1) LEMMA. *Let (E,M) be a prefactorization in* $\underline{A}$ *and consider the following assertions:*

(a) *Every* E *is an epimorphism.*

(b) $pq \in M$ *implies* $q \in M$.

(c) *Every equalizer is in* M.

(d) *Every section is in* M.

Then

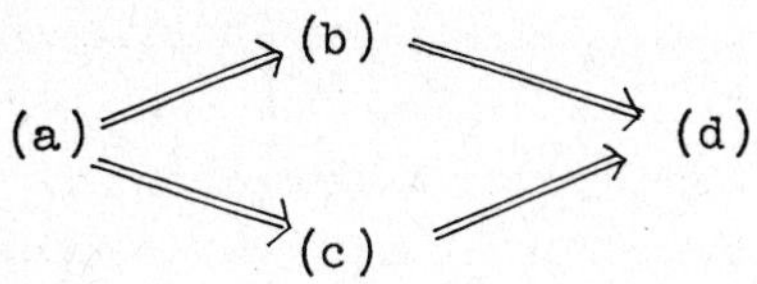

while $(d) \Longrightarrow (a)$ *if* $\underline{A}$ *admits either finite products or weak cokernel pairs.*

(1.2) LEMMA. *Let (E,M) be a prefactorization. Then*

(a) M *contains all isomorphisms and is closed under composition.*

(b) *Every pullback of a* M *is a* M.

(c) *The fibred product of* $m_i : A_i \longrightarrow B$, $i \in I$, *is a* M *if each* m_i *is a* M.

(d) *If* mn *is a* M *so is* n *provided* m *is either a* M *or a monomorphism.*

In the sequel let $\underline{A}$ be a category with a proper (E,M)-factorization. By a subobject of $A \in \underline{A}$ I mean a morphism $i : U \longrightarrow A$ in M. $\underline{A}$ is M-wellpowered if for each $A \in \underline{A}$ the class of subobjects of A is small up to isomorphisms. (Dually E-cowellpowered). If the $u_i : U_i \longrightarrow A$, $i \in I$, are M-subobjects of A and if the monomorphisms u_i admit an intersection then the intersection is again an M-subobject of A. An E-generator of $\underline{A}$ is a small full subcategory $\underline{G}$ of $\underline{A}$ such that for each $A \in \underline{A}$ the family of all morphisms $G \longrightarrow A$ with domain $G \in \underline{G}$ is in E. We say that a family

$q_i : B_i \longrightarrow A$, $i \in I$, is in E if whenever we have $m : C \longrightarrow D$ in M and morphisms $u_i : B_i \longrightarrow C$ and $f : A \longrightarrow D$ such that for each $i \in I$ $fq_i = mu_i$, then there is an A-morphism $w : A \longrightarrow C$ with $f = mw$ (Freyd-Kelly [9]). Dually one defines M-cogenerator.

(1.3) LEMMA Let A be a category with a proper (E,M)-factorization. If A has an E-generator and admits finite intersections of subobjects, it is M-wellpowered. ([9])

(1.4) PROPOSITION ([9]). Let A admit coproducts or be finitely complete. Consider the following statements:

(a) G is a small dense subcategory of A.

(b) G is a generator with respect to a proper (E,M)-factorization.

(c) Whenever $f \neq g : A \rightrightarrows B$ there is a $G \in G$ and an $h : G \longrightarrow A$ such that $fh \neq gh$.

Then (a) $\Longrightarrow$ (b) $\Longrightarrow$ (c). Moreover (c) $\Longrightarrow$ (b) if the factorization (E,M) is the (epi, ext mono)-factorization.

§ 2 The Relative Special Adjoint Functor Theorem (RSAFT).

The classical Special Adjoint Functor Theorem (SAFT) - a powerful categorical tool - replaces the solution set condition of the Adjoint Functor Theorem by the condition that the domain category has a cogenerator and is wellpowered. Typical applications of the SAFT are Watt's Theorem (see e.g. [14]) or the existence of the Čech - Stone - Compactification. If one wants to apply the dual of the SAFT then one often has the situation that the domain category has a generator but is unfortunately not cowellpowered. Hence the SAFT can not be applied. In this paragraph I'll prove a useful generalization of the SAFT - the Relative Special Adjoint Functor Theorem (RSAFT) by assuming that all data are relative to a proper (E,M)-factorization. The proofs are straightforward and hence only sketched or omitted.

(2.1) THEOREM (Relative Special Initial Object Theorem).
Let A be a complete category with a proper (E,M)-factorization. Assume that A has a M-cogenerator Q and that every class of M-subobjects of an object A in A has an intersection. Then A has an initial object.

Proof: Let $Q_o := \prod_{Q \in \underline{Q}} Q$ be the product of all objects in Q and let I be the intersection of all M-subobjects of Q_o. Then I is an initial object in A. The proof is the same as in the classical case (see[12]V §8 Theorem 1) if one takes into consideration LEMMA 1.1 and LEMMA 1.2.

Let $X \in \underline{X}$ be an object in an arbitrary category X, $G : \underline{A} \longrightarrow \underline{X}$ be a functor and denote by $Q : (X{\downarrow}G) \longrightarrow \underline{A}$ the projection functor from the comma category $(X{\downarrow}G)$ to A.

(2.2) LEMMA. *Let A be a category with a proper (E,M)-factorization. Then $(X{\downarrow}G)$ has a proper factorization (E_G, M_G) which is preserved by Q (i.e. Q creates $Q^{-1}M$ - factorizations from M - factorizations). where $E_G := Q^{-1}E$ and $M_G := Q^{-1}M$. Furthermore if A is M-wellpowered (E- cowellpowered) then $(X{\downarrow}G)$ is M_G - wellpowered (E_G - cowellpowered).*

Proof: Straightforward, if one takes into consideration that equality of morphisms in $(X{\downarrow}G)$ means equality as morphisms in A.

(2.3) THEOREM (Relative Special Adjoint Functor Theorem).
Let A be a complete category with a proper (E,M)-factorization, a M - cogenerator and with the property that every class of M-subobjects has an intersection. Then a functor $G : \underline{A} \longrightarrow \underline{X}$ has a left adjoint if and only if G preserves all limits and all intersections of classes

of M-morphisms.

Proof: We have to show that each category $(X\downarrow G)$ has an initial object i.e. a G- universal morphism. Since G preserves limits and $\underline{A}$ is complete and has a proper (E,M)-factorization, each category $(X\downarrow G)$, $X \in \underline{X}$, is complete and has a proper factorization (E_G, M_G).It is easy to see that the subcategory of $(X\downarrow G)$ consisting of all objects $k : X \longrightarrow GQ$, $Q \in \underline{Q}$, is a M_G- cogenerator. Then continue in the same vein as in [12] V §8 Theorem 2 .

(2.4) *COROLLARY. Let A be a complete category with a proper (E,M)-factorization. If A is M-wellpowered and has a M- cogenerator then a functor G : A ⟶ X has a left adjoint if and only if G preserves limits. In particular every continuous functor G : A ⟶ Sets is representable.*

(2.5) *COROLLARY. Let A be a category with a proper (E,M)-factorization. If A is complete and M-wellpowered and has a M- cogenerator then A is also cocomplete.*

§ 3 Wellbounded Categories

Wellbounded categories - a generalization of locally presentable categories - play an important role in the theory of Categorical Universal Algebra as P. Freyd and M. Kelly showed in the fundamental paper on "Categories of continuous functors I " ([9]).

Let $\underline{K}$ be a complete and cocomplete category with a proper (E,M)-factorization. A M-subobject $m : U \longrightarrow K$ of a $\underline{K}$-object K is the M-union of M-subobjects $u_i : U_i \longrightarrow K$,$i \in I$, if $u_i \leq m$ for all $i \in I$ and if for every i $u_i = mf_i$ implies that the family of K-morphisms f_i , $i \in I$, is in E. The union of a family $u_i : U_i \longrightarrow K$, $i \in I$, is denoted by $\bigcup_{i\in I} U_i$. Let r be a regular cardinal. An ordered set I is r- directed if every subset of I of cardinality r has an upper bound in I. An r- directed family of M-subobjects $u_i : U_i \longrightarrow K$, $i \in I$, is a family of M-subobjects u_i where I is r- directed and where $u_i \leq u_j$ whenever $i \leq j$. Then the M- union $\bigcup U_i$ of the M- subobjects u_i is called an r-directed union. An object $K \in \underline{K}$ is said to be bounded for a regular cardinal r if any morphism from K into an r- directed union $\bigcup U_i$ factors through some U_i. $K \in \underline{K}$ is bounded if K is bounded for a regular cardinal. The category $\underline{K}$ is bounded if each object K in $\underline{K}$ is bounded.

(3.1) DEFINITION. A bicomplete category $\underline{K}$ with a proper (E,M)-factorization is wellbounded if it is bounded, E- cowellpowered and possesses an E- generator.

(3.2) EXAMPLES.

(1) Every locally presentable category in the sense of Gabriel-Ulmer is wellbounded (Freyd-Kelly[9]), as for example the categories of sets, groups, rings, Lie-algebras, sheaves over Sets, Grothendieck- categories with generators, or the dual of the category of compact spaces.

(2) Let $\underline{K}$ be a Top-category over a wellbounded category. Then $\underline{K}$ is again wellbounded (Wischnewsky [19]). So for instance the categories of topological, measurable, uniform, compactly generated or limit spaces, groups, rings,... are wellbounded.

(3) Let $\underline{U}$ be an E_K - reflective subcategory of a wellbounded category $\underline{K}$. Then $\underline{U}$ is wellbounded (Wischnewsky[19]). So for instance all epireflective subcategories of a Top-category over Sets are wellbounded as the categories of T_0-, T_1-, T_2-, T_3- spaces, of zero dimensional or completely regular spaces.

(4) The categories of coalgebras, bialgebras, Hopf-algebras or formal groups are wellbounded (Röhrl- Wischnewsky [17]).

§ 4 Monoidal Closed Topological Categories Over Wellbounded Categories

Recall that a monoidal category $\underline{V} = \langle \underline{V}, \square, E, \alpha\, \lambda\, \rho \rangle$ consists of a category $\underline{V}$, a bifunctor $\square : \underline{V} \times \underline{V} \longrightarrow \underline{V}$, an object $E \in \underline{V}$ and three natural isomorphisms

$$\alpha : A \square (B \square C) \cong (A \square B) \square C$$

$$\lambda : E \square A \cong A$$

$$\rho : A \square E \cong A$$

satisfying the usual coherence axioms (Mac Lane [12]).

A monoidal category $\underline{V}$ is called symmetrical if there is a functorial and coherent isomorphism

$$\gamma : A \square B \cong B \square A .$$

A closed category or monoidal closed category is a symmetrical monoidal category such that for any $V \in \underline{V}$ the functor $\underline{V} \xrightarrow{- \square V} \underline{V}$ has a right adjoint $\underline{V} \xrightarrow{(-)^V} \underline{V}$. I refer for examples to [1], [2], [3], [5] [6], [7], [8], [11], [13].

Let $T : \underline{K} \longrightarrow \underline{L}$ be a fibresmall functor. T is called topological or an initialstructure functor if for all (small) categories $\underline{D}$ and all functors $I : \underline{D} \longrightarrow \underline{K}$ the canonically induced functor between the comma-categories $(\Delta \downarrow I)$ and $(\Delta \downarrow TI)$

$$T_I : (\Delta \downarrow I) \longrightarrow (\Delta \downarrow TI)$$

$$\langle K, \varphi : \Delta K \longrightarrow I \rangle \longmapsto \langle TK, T\varphi : \Delta TK \longrightarrow TI \rangle$$

has a right adjoint with the identity as counit. The cones in the image category of the right adjoint right-inverse are called T- initial cones.

(4.1) REMARKS: The above definition allows at once some important generalizations by simply restricting the classes of admissible index-categories $\underline{D}$ or of admissible functors I or of admissible cones.

(1) If one restricts oneself to small categories $\underline{D}$ with cardOb($\underline{D}$) $\leq \alpha$ where α is a fixed regular cardinal then one obtains the α-restricted Top- categories. For instance the category of pseudo-metric spaces is an $\aleph_0$- Top- category (Čech[4]).

(2) Take any class M of cones in thecategories of type ($\Delta\downarrow$TI). A M-Top functor T : $\underline{K} \longrightarrow \underline{L}$ is a functor which generates only INS- cones of M- cones. The most important examples are the relative topological functors in the sense of H.Herrlich ([10]) which correspond to E- reflective subcategories of Top- categories. More general every M-Top-category $\underline{U}$ over a category $\underline{L}$ can be embedded "initialstructure compatibly" into a Top- category over the category $\underline{L}$ (Wischnewsky; to be published elsewhere).

(4.2) THEOREM. Let K be a cocomplete category with a proper (E,M)-factorization. If K has an E- generator and is E- cowellpowered then for any (symmetrical) monoidal category $\langle K, \square, E, \alpha, \lambda, \rho \rangle$ over K there are equivalent:

(i) K is monoidal closed.

(ii) For any $K \in \underline{K}$ the functor $- \square K : \underline{K} \longrightarrow \underline{K}$ preserves colimits.

(iii) (a) For any $K \in \underline{K}$ the functor - K preserves coproducts.

(b) The tensor product of any two regular epimorphisms is a regular epimorphism.

Proof: One has only to prove (ii)$\Longrightarrow$(i). But this is obvious from the Relative Special Adjoint Functor Theorem (§2 THEOREM 2.3).

We obtain now immediately as a corollary the following theorem which shows that

(1) the base category Sets in Herrlich-Nel's results can be replaced

by an arbitrary wellbounded category, and that

(2) the characterization remains the same if we replace the cartesian product by an arbitrary tensor product .

(4.3) THEOREM. Let K be a relative topological category over a well-bounded category L i.e. an E- reflective subcategory of a Top- category over L. Then for any (symmetrical) monoidal category $\langle K, \Box, E, \alpha, \lambda, \rho\rangle$ over K there are equivalent:

(i) K is monoidal closed.

(ii) For any $K \in K$ the functor $- \Box K$ preserves colimits.

(iii) (a) For any $K \in K$ the functor $- \Box K$ preserves coproducts.

(b) The tensor product of any two regular epimorphisms is a regular epimorphism.

If one of these conditions is fulfilled then we obtain for all $K, K' \in K$

$$\underline{K}(K, K') \cong \underline{K}(E, (K')^K).$$

In particular if $T : K \longrightarrow$ Sets is relative topological over Sets and E is the "free object" over a one point set then

$$\underline{K}(K, K') \cong T((K')^K).$$

Proof: Any relative topological category over a wellbounded category L is again wellbounded.

§ 5 Monoidal Closed Topological Categories And Adjoint Triangles

In this paragragh I consider a slightly different situation. I assume that the base category L is already monoidal closed (as for example the category of sets or the category of R- modules) and that the Top-functor from a monoidal Top-category K over L is strict i.e. preserves the monoidal structure (see e.g. [12]). More exactly let $\underline{K} = \langle \underline{K}, \Box, E, \alpha, \lambda\rho\rangle$ and $\underline{L} = \langle \underline{L}, \Box, E, \alpha\ \lambda\ \rho\rangle$ be monoidal categories over K resp. over L.

Let T : K ⟶ L be a Top-functor which is moreover strict monoidal . In this case K is called a strict monoidal Top-category over L. In the sequel I will apply W.Tholen's generalizations ([18b]) of Dubuc's results on adjoint triangles to the following Dubuc triangle:

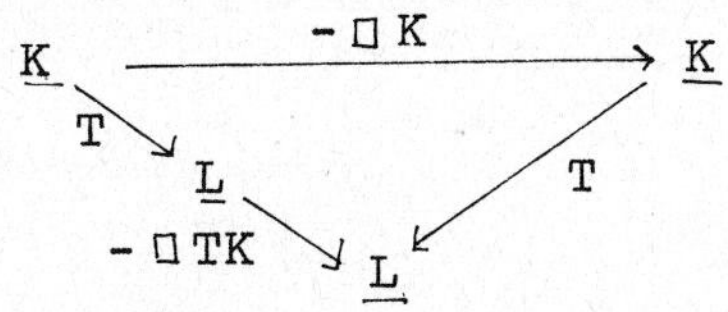

Tholen's Adjoint Functor Theorem for Dubuc-triangles applied to the above situation delivers at once the following theorem.

(5.1) THEOREM (Monoidal Closedness Theorem for Topological Categories). Let K be a strict monoidal Top-category over a monoidal closed category L. If K has coproducts and a proper (E,M)-factorization and if for any K ∈ K and e ∈ E the K-morphism e □ K is T-final and if finally K is M-wellpowered then K is monoidal closed if and only if for any K ∈ K the functor - □ K preserves coproducts.

This THEOREM has now some important corollaries.

(5.2) COROLLARY. Let K be a strict monoidal Top-category over a monoidal closed category L. Let L have coproducts, let L be wellpowered and let every morphism in L factorize through a regular epimorphism and a monomorphism. Then there are equivalent:

(i) K is monoidal closed.

(ii) For any K ∈ K the functor - □ K preserves coproducts and regular epimorphisms.

(5.3) COROLLARY. Let K be a strict monoidal Top-category over a monoidal closed category L. Let L be cocomplete, wellpowered and cowellpowered. Then K is monoidal closed if and only if the functors - □ K preserve colimits.

(5.4) COROLLARY. Let K be a Top-category over a wellpowered, cowellpowered and cocomplete cartesian closed category L. Then K is cartesian closed if and only if all functors $- \times K$, $K \in \underline{K}$, preserve colimits.

This COROLLARY contains again Herrlich's result as a special instance.

REFERENCES

1 BASTIANI,A.,EHRESMANN,C.: Categories of sketched structures. Cahier Topo. Geo. diff. XIII,2, 105 - 214 (1972)

2 BENTLEY,HERRLICH,ROBERTSON : Convenient categories for topologists Preprint (1975).

3 BINZ,E.KELLER : Funktionenräume in der Kategorie der Limesräume. Ann. Acad. Sci. Fenn. Sec. AI 383 ,1 - 21 (1966).

4 Čech,E.: Topological spaces. Prague 1966.

5 DAY,B.: A reflection theorem for closed categories. J. of Pure and Applied Algebra 2 (1972) 1 - 11.

6 DUBUC,E.,PORTA,H.: Convenient categories of topological algebras their duality theory. J. Pure Appl. Algebra 1, 281 - 316 (1970).

7 EILENBERG,S.,KELLY,G.M.: Closed Categories. Proc. Conf. on Categorical Algebra. La Jolla,Springer,Berlin,Heidelberg,New York 1966.

8 FRANKE,D.: Funktionenalgebren in kartesisch abgeschlossenen Kategorien. thesis.Free Univ. Berlin 1975.

9 FREYD,P.,KELLY,G.M.: Categories of continuous functors I. J. of Pure a. Appl. Alg. 2, 169 - 191 (1972).

10 HERRLICH,H.: Topological functors.General Topology Appl. 4,125-145 (1974).

11 " : Cartesian closed topological categories.Math. Colloqu. Univ. Cape Town IX, 1 - 16 (1974).

12 MAC LANE,S.: Categories for the working mathematician. Springer, Berlin,Heidelberg, New York 1971.

13 NEL,L.D.: Recent results on cartesian closed topological categories. (this volume).

14 PAREIGIS,B.:Categories and functors.Academic Press New York 1970.

15 PUMPLÜN,D.,THOLEN,W.:Covollständigkeit vollständiger Kategorien Manuscripta math. 11, 127 - 140 (1974).

16 RINGEL,C.M.: Diagonalisierungspaare I (II) .Math.Z. 117. 248 - 266 (1970) (Math.Z. 122, 10 - 32 (1971).)

17 RÖHRL,H.,WISCHNEWSKY,M.B.: Universal Algebra over Hopf-algebras. Algebra-Bericht Nr.26 ,1 - 35,(1974).

17a SCHUBERT,H.: Categories,Springer,Berlin,Heidelberg,New York 1974.

18a THOLEN,W.: Relative Bildzerlegungen und algebraische Kategorien. thesis.Münster 1974.

18b " : Adjungierte Dreiecke,Colimites und Kan-Erweiterungen. to appear in Math. Ann.

19 WISCHNEWSKY,M.B.: On the boundedness of topological categories. Manuscripta math. 12, 205 - 215 (1974).

20 WYLER,O.: Convenient categories for topology.Gen. Topo. Appl. 3 , 225 - 241 (1973).

M.B.Wischnewsky

Mathematisches Institut

der Universität

8 München 2, Theresienstr. 39

W-Germany

ON TOPOLOGICAL ALGEBRAS RELATIVE TO FULL AND FAITHFUL DENSE FUNCTORS

Manfred B. Wischnewsky

In this paper I consider algebras in the sense of Diers ([5]) in topological categories ([3], [8], [10], [11], [12], [14], [15], [18], [19], [20], [24], [25]). This notion of an algebra generalizes at the same time the notion of an algebra for a monad and the notion of a sheaf on a Grothendieck-topology. The underlying algebraic theories are defined in a natural way with respect to full and faithful dense functors. The results given here combine and generalize in an easy way some of the basic theorems of the theory of Topological Universal Algebra whose fundamental results and techniques can be found for instance in the author's survey article ([23]). The proofs are only sketched or even omitted since only standard techniques of Categorical Topology are used.

§ 1 Algebraic Theories And Algebraic Categories

In this paragraph I'll recall Diers' notion of an algebraic theory resp. category ([5]) which generalizes at the same time the notion of an algebra for a monad and the notion of an algebra on an esquisse in the sense of Ehresmann (see e.g. [1]) resp. of a sheaf on a Grothendieck-topology. One uses the notion of relative adjoint functors ([17]) Let $J : \underline{A}_o \longrightarrow \underline{A}$ be a functor, $A \in \underline{A}$ an object, and $J_A : (J\downarrow A) \longrightarrow \underline{A}_o$ the forgetful functor from the comma category $(J\downarrow A)$ to $\underline{A}_o$ and $\varphi_A(J) : JJ_A \longrightarrow \Delta A$ the canonical transformation $(X,f) \longmapsto f$. J is dense at A ([7], [17], [27]) if $\varphi_A(J)$ is a colimit cone. If J is dense at all $A \in \underline{A}$ then J is called dense.

Let now $J : \underline{A}_o \longrightarrow \underline{A}$ be a full and faithful dense functor.

A J-theory (Linton [13]) is a pair $(\underline{T}, T)$ consisting of a category $\underline{T}$, having the same objects as $\underline{A}_o$, and a functor $T : \underline{A}_o^{op} \longrightarrow \underline{T}$ which induces the identity on the objects.

(1.1) DEFINITION (Diers). A J-theory $(\underline{T},T)$ is said to be algebraic if the functor $T^{op} : \underline{A}_o \longrightarrow \underline{T}^{op}$ has a J-right adjoint in the sense of Ulmer ([17]).

The category Th(J) of all algebraic J-theories is the full subcategory of $(\underline{A}_o^{op} \downarrow \mathrm{Cat})$ having as objects the J-algebraic functors. The category $\underline{A}^T$ of all T-algebras is defined by the following pullback:

$$\begin{array}{ccc} \underline{A}^T & \longrightarrow & [\underline{T},\mathrm{Sets}] \\ {\scriptstyle U^T}\downarrow & \text{pullback} & \downarrow{\scriptstyle [T,\mathrm{Sets}]} \\ \underline{A} & \xrightarrow[\underline{A}(J(-),\cdot)]{} & [\underline{A}_o^{op},\mathrm{Sets}] \end{array}$$

If $(\underline{T},T)$ is an algebraic J-theory then the functor $U^T : \underline{A}^T \longrightarrow \underline{A}$ has a J-left adjoint. Furthermore U^T creates limits and J-absolute colimits. A J-absolute colimit in $\underline{A}$ is a cocone $\varphi : D \longrightarrow \Delta A$ over $\underline{A}$ such that for any A_o in $\underline{A}_o$ the cocone $\underline{A}(JA_o,\varphi) : \underline{A}(JA_o,D-) \longrightarrow \Delta\underline{A}(JA_o,A)$ is a colimit in Sets.

(1.2) REMARK. The above notion of an algebraic J-theory can be generalized in the following way:

Let $\underline{A}$ be a category with a factorization (E,M) and with a (not necessarily small) E-generator $\underline{G}$. By Definition $\underline{G}$ is a subcategory $J : \underline{G} \longrightarrow \underline{A}$. Then a J-theory $(\underline{T},T)$ (in the sense of Linton) is G-algebraic if the functor $T^{op} : \underline{G} \longrightarrow \underline{T}^{op}$ has a J-right adjoint.

(1.3) DEFINITION. A T-algebra with values in a category $\underline{K}$ is a functor $P : \underline{T} \longrightarrow \underline{K}$ such that the functor $P^{op}T^{op}$ has a J-right adjoint.
The category $\mathrm{Alg}(\underline{T},\underline{K})$ of all T-algebras in $\underline{K}$ is the full subcategory of $[\underline{T},\underline{K}]$ having as objects the T-algebras in $\underline{K}$.

(1.4) EXAMPLES of algebraic J-theories resp. algebraic J-categories ([5]).

1) Algebraic theories in the sense of Lawvere (see e.g. [28]).

Let Cardfin be the full subcategory in Sets of all finite cardinals and let J : Cardfin $\longrightarrow$ Sets be the inclusion. A functor F : Cardfin $\longrightarrow \underline{K}$ has a J-right adjoint if and only if F preserves finite sums. Hence each algebraic J-theory is a theory in the sense of Lawvere and conversely. The category Alg($\underline{T},\underline{K}$) is the category of all T-algebras in $\underline{K}$ in the sense of Lawvere.

2) Algebraic theories in the sense of Benabou ([2]).

Let I be a set and denote by $\text{Cardfin}^{(I)}$ the full subcategory of Sets^I having as objects all I-families (X_i, $i \in I$) for which $X_i \in$ Cardfin and $\{i \in I, X_i \neq \emptyset\}$ is finite and denote by J : $\text{Cardfin}^{(I)} \longrightarrow \text{Sets}^I$ the inclusion. A functor F : $\text{Cardfin}^{(I)} \longrightarrow \underline{K}$ has a J-right adjoint if and only if it commutes with finite sums. An algebraic theory with respect J is an algebraic theory in the sense of Benabou and conversely ([2]). The category Alg($\underline{T},\underline{K}$) is the category of all T-algebras in $\underline{K}$ in the sense of Benabou.

3) Eilenberg-Moore Categories.

Let (T,η,μ) be a monad on $\underline{A}$. Denote by $\underline{A}_T$ the corresponding Kleisli-category and by $F_T : \underline{A} \longrightarrow \underline{A}_T$ the free functor. Then $(\underline{A}_T^{op}, F_T^{op})$ is an algebraic theory with respect to the identity $id_{\underline{A}}$ on $\underline{A}$. The category $\underline{A}^T$ is isomprphic to the Eilenberg-Moore category induced by the monad (T,η,μ).

4) Esquisse in the sense of Ehresmann ([1]) resp. algebraic theories in the sense of Gabriel-Ulmer ([7]).

Let $(\underline{C},\Sigma)$ be a small category $\underline{C}$ together with a class Σ of colimits. Then there exists in a canonical way ([5] 6.5) a full and faithful dense functor J : $\underline{A}_o \longrightarrow [\underline{C}^{op}, \text{Sets}]$ and an algebraic J-theory $(\underline{C}^{op}, C)$ such that a functor S : $\underline{C}^{op} \longrightarrow \underline{K}$ is a C-algebra if and only if S is a $(\underline{C},\Sigma)$-sheaf.

Let $F : \underline{A}_o \longrightarrow \underline{X}$ be a J-left adjoint of $U : \underline{X} \longrightarrow \underline{A}$. Let $F_1 : \underline{A}_o \longrightarrow \underline{C}$ and $F_2 : \underline{C} \longrightarrow \underline{X}$ be respectively the full coimage and the full image of F. Then F_1 is J-left adjoint to UF_2. $(\underline{C}^{op}, F_1^{op})$ is an algebraic J-theory. It is said to be generated by F. Furthermore there exists a canonical comparison functor $C : \underline{X} \longrightarrow \underline{A}^T$ such that $U^T C = U$. The pair $(\underline{X}, U)$ is called J-algebraic if the functor C is an equivalence. With these notations I can give now Diers' characterization of algebraic categories.

(1.5) THEOREM (Diers). A pair (X,U) is J-algebraic if and only if U has the following properties:

- (i) U has a J-left adjoint.
- (ii) U reflects isomorphisms and
- (iii) U reflects J-absolute colimits.

§ 2 Topological Algebras Relative To Full And Faithful Dense Functors

Let $J : \underline{A}_o \longrightarrow \underline{A}$ be a full and faithful dense functor. Furthermore we assume that $\underline{A}$ is complete in order not to complicate the following presentation. Let $(\underline{T}, T)$ be an algebraic J-theory.

(2.1) LEMMA. The inclusion functor $E : Alg(\underline{T}, \underline{K}) \longrightarrow [\underline{T}, \underline{K}]$ creates limits.

Proof: Let $D : \underline{D} \longrightarrow Alg(\underline{T}, \underline{K})$ be a diagram and let $\Delta(\lim ED) \longrightarrow ED$ be a limit cone of ED in $[T,K]$. Let S_d be a J-right adjoint of $D(d)^{op}T^{op}$ where $d \in \underline{D}$. Then D induces a functor $D^* : \underline{D} \longrightarrow [\underline{K}^{op}, \underline{A}]$ by the assignment $d \mapsto S_d$. Let S be a limit of D^* in $[\underline{K}^{op}, \underline{A}]$. Then we obtain:

$$\underline{A}(JA_o, SK) = \underline{A}(JA_o, \lim S_d(K)) \cong \lim \underline{A}(JA_o, S_d(K)) \cong \lim \underline{K}(K, D(d)TA_o) \cong$$
$$\cong \underline{K}(K, \lim D(d)TA_o) .$$

Hence S is a J-right adjoint of $(\lim ED)^{op}T^{op}$ i.e. $\lim ED$ is a T-algebra in $\underline{K}$.

(2.2) LEMMA. Let $F : \underline{K} \longrightarrow \underline{L}$ be a right adjoint functor. Then the induced functor $[\underline{T}, F] : [\underline{T}, \underline{K}] \longrightarrow [\underline{T}, \underline{L}] : A \longmapsto FA$
factors through $\mathrm{Alg}(\underline{T}, \underline{K})$ and $\mathrm{Alg}(\underline{T}, \underline{L})$.

Proof: Let A be a T-algebra in $\underline{K}$ and let $S : \underline{K}^{op} \longrightarrow \underline{A}$ be a J-right adjoint of $A^{op}T^{op}$. Let D be a left adjoint of F. Then

$$\underline{L}(L, FAT(A_o)) \cong \underline{K}(DL, AT(A_o)) \cong \underline{A}(JA_o, SD(L)) \text{ for all } L \in \underline{L} \text{ and } A_o \in \underline{A}_o$$

Hence SD^{op} is a J-right adjoint of $F^{op}A^{op}T^{op}$.

In the sequel let us furthermore assume that the base categories of topological categories are complete.

(2.3) THEOREM. Let $(\underline{T}, T)$ be an algebraic J-theory and $F : \underline{K} \longrightarrow \underline{L}$ be a Top-functor. Then the functor

$$\mathrm{Alg}(\underline{T}, F) : \mathrm{Alg}(\underline{T}, \underline{K}) \longrightarrow \mathrm{Alg}(\underline{T}, \underline{L}) : A \longmapsto FA$$

is again a Top-functor provided $\mathrm{Alg}(\underline{T}, F)$ is fibresmall.

Proof: Let I be the right adjoint right inverse functor of F. Let A be a T-algebra in $\underline{L}$ with J-right adjoint S. Then IA is a T-algebra in $\underline{K}$ with J-right adjoint SF^{op}. One can show easily that the assignment $A \longmapsto IA$ defines a right adjoint right inverse of $\mathrm{Alg}(\underline{T}, F)$. Since $\mathrm{Alg}(\underline{T}, \underline{K})$ and $\mathrm{Alg}(\underline{T}, \underline{L})$ are complete (LEMMA 2.1) and $\mathrm{Alg}(\underline{T}, F)$ preserves obviously limits we obtain the above theorem by applying Hoffmann's characterization of Top-functors ([10]).

(2.4) REMARK. The condition that $\mathrm{Alg}(\underline{T}, F)$ is fibresmall is fulfilled in all examples given in § 1, in particular if $\underline{A}_o$ is small or $\mathrm{Alg}(\underline{T}, L)$ is an Eilenberg-Moore category. For the rest of this paper we will always assume that $\mathrm{Alg}(\underline{T}, F)$ is fibresmall.

Let $F : \underline{K} \longrightarrow \underline{L}$ and $F' : \underline{K} \longrightarrow \underline{L}$ be Top-functors and $M : \underline{K} \longrightarrow \underline{K}'$ and $N : \underline{L} \longrightarrow \underline{L}'$ be arbitrary functors. Recall that $(M,N) : (\underline{K},F) \longrightarrow (\underline{K}',F')$ is initial continuous or an initial morphism if (I1) $F'M = NF$ and (I2) for every INS-cone $\Delta K \longrightarrow T$, $T \in [\underline{D},\underline{K}]$, the cone $\Delta MK \longrightarrow MT$ is again an INS-cone. If $\underline{L}$ and $\underline{L}'$ are complete then (M,N) is initial continuous if and only if M preserves limits and codiscrete objects ([23] THEOREM 1.9)

(2.5) THEOREM. *Let $(\underline{T},T)$ be a small algebraic J-theory and let $F : \underline{K} \rightarrow \underline{L}$ be a Top-functor. Then the pair of inclusion functors*

$$E_{\underline{K}} : Alg(\underline{T},\underline{K}) \longrightarrow [\underline{T},\underline{K}] \quad \text{and}$$
$$E_{\underline{L}} : Alg(\underline{T},\underline{L}) \longrightarrow [\underline{T},\underline{L}] \quad \text{is initial continuous.}$$

Proof: Clear from the above characterization.

(2.6) COROLLARY. *Notation as above.*

$E_{\underline{K}}$ is adjoint if and only if $E_{\underline{L}}$ is adjoint.

Proof: One has only to prove that $E_{\underline{K}}$ is adjoint provided $E_{\underline{L}}$ is adjoint. But this is trivial since $E_{\underline{K}}$ preserves limits and since $Alg(\underline{T},F)^{-1}R_{\underline{L}}T,FA$ is obviously a solution set for $A \in [\underline{T},\underline{K}]$ where $R_{\underline{L}}$ denotes a coadjoint of $E_{\underline{L}}$.

(2.7) REMARKS.

1) The above COROLLARY follows also immediately from Wyler's taut lifting theorem ([24]) by using THEOREM 2.5.
2) If $Alg(\underline{T},\underline{L})$ is for example wellpowered and cowellpowered then the above COROLLARY can also be derived from the following Dubuc-triangle (see e.g. [26] Thm 28.12 or [16]):

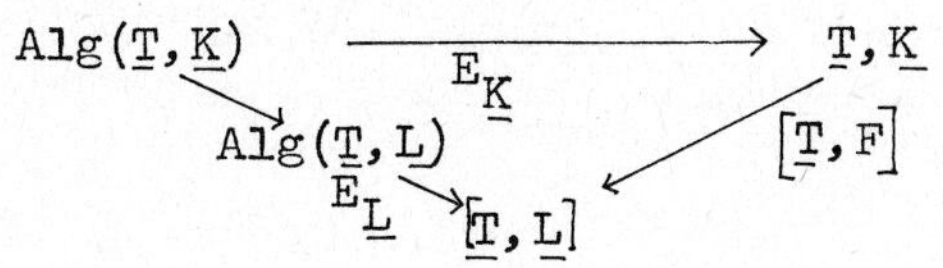

For instance all categories of continuous functors with values in a locally presentable category $\underline{L}$ or with values in the dual of a locally presentable category fulfill this assumption.

A standard example for the above THEOREM resp. COROLLARY is given by the following

(2.8) COROLLARY (Wischnewsky [20], [23]). *Let* $(\underline{C},\Sigma)$ *be an esquisse in the sense of Ehresmann resp. in the sense of Gabriel-Ulmer and let L be a locally presentable category. Let K be an arbitrary Top-category over L. Then we obtain the following assertions:*

(1) *The inclusion functor* $Alg(\underline{C},\underline{K}) \longrightarrow [\underline{C},\underline{K}]$ *is reflective.*

(2) *The inclusion functor* $Coalg(\underline{C},\underline{K}) \longrightarrow [\underline{C},\underline{K}^{op}]^{op}$ *is coreflective.*

(3) *The categories* $Alg(\underline{C},\underline{K})$ *and* $Coalg(\underline{C},\underline{K})$ *are complete, cocomplete, wellpowered and cowellpowered and have a generator.*

(2.9) THEOREM. *Let* $(\underline{T},T)$ *be an algebraic J-theory. Then the pair of evaluation functors*

$$U_C^K : Alg(\underline{T},\underline{K}) \longrightarrow \underline{K} : A \longmapsto AC \quad \text{and}$$
$$U_C^L : Alg(\underline{T},\underline{L}) \longrightarrow \underline{L} : A \longmapsto AC \quad \text{where } C \in \underline{T} \text{ is}$$

initial continuous.

(2.10) COROLLARY. *Notation as above.*

U_C^K *is adjoint if and only if* U_C^L *is adjoint.*

(2.11) COROLLARY. *Let* $F : \underline{K} \longrightarrow \underline{L}$ *be a Top-functor and* (T,η,μ) *be a monad over L. Then the underlying functor*

$$U : Alg(\underline{T},\underline{K}) \longrightarrow \underline{K} \quad \text{is monadic.}$$

Denote by $Top(\underline{L})$ the category of all topological functors over $\underline{L}$

and all initial continuous morphisms between Top-categories over $\underline{L}$ as morphisms. Then we obtain

(2.12) *THEOREM. Let* $(\underline{T},T)$ *be an algebraic* J-*theory. Then there exists a functor* $\mathrm{Alg}(\underline{T},-) : \mathrm{Top}(\underline{L}) \longrightarrow \mathrm{Top}(\mathrm{Alg}(\underline{T},\underline{L}))$

$$F : \underline{K} \longrightarrow \underline{L} \qquad \mathrm{Alg}(\underline{T},F) : \mathrm{Alg}(\underline{T},\underline{K}) \longrightarrow \mathrm{Alg}(\underline{T},\underline{L})$$

$$\begin{array}{ccc} \underline{K} & \xrightarrow{H} & \underline{K}' \\ & {\scriptstyle F}\searrow \quad \swarrow{\scriptstyle F'} & \\ & \underline{L} & \end{array} \qquad \begin{array}{ccc} \mathrm{Alg}(\underline{T},\underline{K}) & \xrightarrow{\mathrm{Alg}(\underline{T},H)} & \mathrm{Alg}(\underline{T},\underline{K}') \\ & \searrow \quad \swarrow & \\ & \mathrm{Alg}(\underline{T},\underline{L}) & \end{array}$$

where $\mathrm{Alg}(\underline{T},H)$ *is defined by* $A \longmapsto HA$ (Take into consideration that if S is a J-right adjoint of $A^{op}T^{op}$ then SR^{op} is a J-right adjoint of $H^{op}A^{op}T^{op}$ where R is a coadjoint of the initial continuous functor H).

Proof: Easy calculation.

(2.13) *COROLLARY. Let* $H : \underline{K} \longrightarrow \underline{K}'$ *be initial continuous over* $\underline{L}$. *Then the functor* $\mathrm{Alg}(\underline{T},H)$ *has a left adjoint.*

A standard example is given by the initial continuous functor

uniform spaces $\longrightarrow$ topological spaces.

(2.14) *DEFINITION.* Let $\underline{A}_o$ be a category together with a class Σ of co-limits. A full and faithful dense functor $J : \underline{A}_o \longrightarrow \underline{A}$ is said to be Σ-*dense* if for every category $\underline{K}$ and for every functor $F : \underline{A}_o \longrightarrow \underline{K}$ holds: F has a J-right adjoint if and only if F preserves the colimits in Σ.

Examples of Σ-dense functors can be found in (1.4) 1, 2 .

(2.15) *THEOREM. Let* $J : \underline{A}_o \longrightarrow \underline{A}$ *be a full and faithful* Σ-*dense functor and* $(\underline{T},T)$ *an algebraic* J-*theory. Then* $(\underline{T},T)$ *induces a functor*

$$\mathrm{Top(CAT)} \longrightarrow \mathrm{Top(CAT)}$$

$$\begin{array}{ccc} \underline{K} & \xrightarrow{F} & \underline{L} \\ M\downarrow & \text{init. cont.} & \downarrow N \\ \underline{K}' & \xrightarrow{F'} & \underline{L}' \end{array} \quad \longmapsto \quad \begin{array}{ccc} \mathrm{Alg}(\underline{T},\underline{K}) & \longrightarrow & \mathrm{Alg}(\underline{T},\underline{L}) \\ \mathrm{Alg}(\underline{T},M)\downarrow & & \mathrm{Alg}(\underline{T},N)\downarrow \\ \mathrm{Alg}(\underline{T},\underline{K}') & \longrightarrow & \mathrm{Alg}(\underline{T},\underline{L}') \end{array}$$

In particular Alg($\underline{T}$,M) *is adjoint if and only if* Alg($\underline{T}$,N) *is adjoint.*

Let H : ($\underline{T}'$,T') $\longrightarrow$ ($\underline{T}$,T) be a morphism between algebraic J-theories. Then H induces a functor Alg(H,$\underline{K}$) : Alg($\underline{T}$,$\underline{K}$) $\longrightarrow$ Alg($\underline{T}'$,$\underline{K}$) : A $\longmapsto$ HA . Alg(H,$\underline{K}$) is called a *J-algebraic functor*. In the same vein as THEOREM 5.2 in [21] one can prove the following

(2.16) *THEOREM. Let* F : $\underline{K} \longrightarrow$ L *be a Top-functor and let* H : ($\underline{T}'$,T') $\longrightarrow$ ($\underline{T}$,T) *be a morphism of J-theories. Then we obtain the following statements:*

(1) *The pair of functors* (Alg(H,$\underline{K}$),Alg(H,$\underline{L}$)) *is initial continuous.*

(2) Alg(H,$\underline{K}$) *is adjoint if and only if* Alg(H,$\underline{L}$) *is adjoint.*

(2.17) *Final Observation.*

In the same vein as for instance in [21] and [23] we can now study T-algebras in reflective or coreflective subcategories of Top-categories. One obtains similar results. Hence one can state the following METATHEOREM. Replace theory in ([16], [18], [19], [20], [21], [23], [24]) by algebraic J-theory then you will get the same results for algebras over Top-categories.

REFERENCES

1 BASTIANI,A.,EHRESMANN,C.: Categories of sketched structures. Cahier Topo. Geo. diff. XIII,2, 105 - 214 (1972).

2 BENABOU,I.: Structures algébriques dans les catégories. Cahier Topo. Geo. diff. X,1, 1 - 126 (1968).

3 BRÜMMER,G.C.L,: A categorical study of initiality. Thesis. Cape Town (1971).

4 DIERS,Y.: Type de densité d'une sous-catégorie pleine. Preprint Université de Lille 1975.

5 " : Foncteur pleinement fidèle dense classant les algèbres. Preprint. Université de Lille. 1975.

6 ERTEL,H.G.:Algebrenkategorien mit Stetigkeit in gewissen Variablen familien, thesis, Univ. Düsseldorf, 1972.

7 GABRIEL,P.,ULMER,F.: Lokal präsentierbare Kategorien. LN 221, Springer, Berlin, Heidelberg, New York (1971).

8 HERRLICH,H.: Topological functors. General Topology and Appl., 4 (1974).

9 " : Cartesian Closed Topological Categories. Math. Coll. Univ. Capetown,9 (1974).

10 HOFFMANN,R.E.: Die kategorielle Auffassung der Initial- und Final-topologie.thesis, Univ. Bochum 1972.

11 HUSEK,M.: S-categories. Comm. Math. Univ. Carol. 5 (1964).

12 KENNISON,J.F.: Reflective functors in general topology and elsewhere. Trans. Amer. Math. Soc. 118,303 - 315 (1965).

13 LINTON,F.W.: An outline of functorial semantic, LN 80,Springer 1968

14 ROBERTS,J.E.: A characterization of initial functors. J.Algebra 8 , 181 - 193,(1968).

15 TAYLOR,J.C.: Weak families of maps. Canad. Math. Bull. 8,77-95,(1968)

16 THOLEN,W.: Relative Bildzerlegungen und algebraische Kategorien, thesis, Univ. Münster,1974.

17 ULMER,F.: Properties of dense and relative adjoint functors. J.Algebra 8 , 77 - 95 (1968).

18 WISCHNEWSKY,M.B.: Algebren und Coalgebren in Initial- und Gabriel-kategorien. Diplomarbeit, Univ. München 1971.

19 " :Partielle Algebren in Initialkategorien,Math.Z. 127, 83 - 91 (1972).

20 " : Generalized Universal Algebra in Initialstructure categories. Algebra-Berichte Nr. 10 (1973) 1 - 35.

21 " : On regular topological algebras over arbitrary base categories. Algebra- Berichte Nr. 16 (1973) 1 -36.

22 " : On the boundedness of topological categories, Manuscripta math. 12, 205 - 215 (1974).

23 " : Aspects of Universal Algebra in Initialstructure categories. Cahier Topo. Geo. diff. X V, 1 - 27 (1974).

24 WYLER, O.: On the categories of general topology and topological algebra. Arch. d. Math., 22/1, 7 - 17 (1971).

25 " : Top categories and categorical topology. General topology and its applications 1, 17 - 28 (1971).

Books on Category Theory

26 HERRLICH, H., STRECKER, G.E.: Category theory, Allyn and Bacon, Boston, 1973.

27 MAC LANE, S.: Categories for the working mathematician. Springer, Berlin, Heidelberg, New York 1971.

28 PAREIGIS, B.: Categories and Functors. Academic Press, New York (1970)

29 EHRESMANN, C.: Catégories et structures. Dunod, Paris, (1965).

30 SCHUBERT, H.: Categories. Springer, Berlin, Heidelberg, New York (1973)

M.B. Wischnewsky
Mathematisches Institut
der Universität
8 München 2
Theresienstr. 39
W - Germany

ARE THERE TOPOI IN TOPOLOGY ?

Oswald Wyler
Department of Mathematics
Carnegie-Mellon University
Pittsburgh, PA 15213

ABSTRACT. The straight answer is no. Topoi are too set-like to occur as categories of sets with topological structure. However, if $\underline{\underline{A}}$ is a category of sets with structure, and if $\underline{\underline{A}}$ has enough substructures, then $\underline{\underline{A}}$ has a full and dense embedding into a complete quasitopos of sets with structure. There is a minimal embedding of this type; it embeds e.g. topological spaces into the quasitopos of Choquet spaces. Quasitopoy are still very set-like. They are cartesian closed, and all colimits in a quasitopos are preserved by pullbacks. Thus quasitopoi are in a sense ultra-convenient categories for topologists. Quasi-topoi inherit many properties from topoi, For example, the theory of geometric morphisms of topoi remains valid, almost without changes, for quasitopoi.

ARE THERE TOPOI IN TOPOLOGY ?

Oswald Wyler

Introduction

Topoi were introduced in SGA 4 [21] as categories of set-valued sheaves. Grothendieck stated in the introduction of SGA 4 that topologists should be concerned with the topos of sheaves instead of the underlying topological space, but this advice was not followed. Lawvere and Tierney recognized the set-like and logical properties of topoi, and they introduced elementary topoi as categories with these properties. Tierney [20], Kock and Wraith [13], and Freyd [9] gave introductions to elementary topoi. The latest and simplest version of the axioms for an elementary topos will be found in § 1 of this paper.

For topological purposes, topoi are too set-like. They can serve as base categories for non-standard topology, as in L. Stout's thesis [19], but this seems to be their only use. On the other hand, there has been an intensive search for topological categories more set-like -- or more convenient as Steenrod [18] called them -- than topological spaces; see Herrlich [10] for a survey of this search. Convenient categories should be at least cartesian closed; B. Day [6] suggested that all categories $\underline{\underline{T}}/A$, for objects A of a convenient category $\underline{\underline{T}}$, should be cartesian closed. A category $\underline{\underline{T}}$ with finite limits and this property is called span-closed.

We go one step further. We show that the span-closed categories which occur in topology are in fact quasitopoi and thus very set-like indeed. Quasitopoi were introduced by J. Penon [16] as a generalization of elementary topoi. The generalization is broad enough to allow topological examples, but not too broad so that quasitopoi retain many useful properties of topoi. Thus quasitopoi are useful and convenient for topologists, and we obtain a quasi-affirmative answer to the title question of this paper by studying quasitopoi in topology.

We begin in § 1 by defining topoi and quasitopoi and stating some of their basic properties. § 2 describes categories of P-sieves for a set-valued functor P . These categories were invented by P. Antoine [1], [2], as cartesian closed completions of concrete categories. Day [6] showed that categories of P-sieves are span-closed; we show that P-sieves form a quasitopos if P allows enough subobject inclusions. Thus every topological category with enough subspaces can be densely embedded into a quasitopos.

Quasitopoi of P-sieves are quite large; thus we devote §§ 3 and 4 mainly to the construction of smaller quasitopoi from a given quasitopos. In § 3, we describe the general theory of geometric morphisms of quasitopoi; this is essentially a generalization of the corresponding theory for topoi. In § 4, we apply the results of § 3 to categories of P-sieves. This generalizes the results of B. Day [6] on closed-span categories of limit spaces. Our main result is that every concrete category with enough subobjects has a minimal quasitopos extension, resulting from a canonical Grothendieck topology. For topological spaces, this minimal extension has been known in two forms: it is the category of pseudotopological or Choquet spaces [4], and also M. Schröder's category of solid convergence spaces [17]. The observation that these are the same category seems to be new.

In presenting our theory of quasitopoi in topology, we suppress most of the proofs. Some of the proofs are quite involved, but only a few new ideas seem to be required. Thus the interested reader may be able to supply the proofs, using the existing literature on topoi as a guide. I plan to describe the theory with more details, and with full proofs, in a set of lecture notes.

1. Topoi and quasitopoi

1.1. An elementary topos can be defined as a category $\underline{\underline{E}}$ with finite limits and with powerset objects. $\underline{\underline{E}}$ has finite limits if $\underline{\underline{E}}$ has pullbacks and a terminal object; powerset objects represent relations in $\underline{\underline{E}}$ as follows.

1.2. We define a relation $(u,v) : A \rightharpoondown B$ in $\underline{\underline{E}}$ as a "span" or pair of morphisms $A \xrightarrow{u} X \xrightarrow{v} B$ with common domain, and with $\langle u,v \rangle : X \longrightarrow A \times B$ monomorphic. We call two relations (u,v) and (u',v') equivalent and write $(u,v) \simeq (u',v')$ if $u' = u\,x$ and $v' = v\,x$ for an isomorphism x, but we do not identify equivalent relations. We say that (u,v) is a partial morphism if u is a monomorphism.

For $f : A \longrightarrow B$ in $\underline{\underline{E}}$ and a relation $(u,v) : B \rightharpoondown C$ in $\underline{\underline{E}}$, we define a composition $(u,v) \circ f$ as a relation

$$(u,v) \circ f = (u',\ v\,f') : A \rightharpoondown C ,$$

where $\begin{array}{ccc} \cdot & \xrightarrow{f'} & \cdot \\ {\scriptstyle u'}\downarrow & & \downarrow{\scriptstyle u} \\ \cdot & \xrightarrow{f} & \cdot \end{array}$ is a pullback square in $\underline{\underline{E}}$.

Now a powerset object for an object A of $\underline{\underline{E}}$ is given by an object $P\,A$

and a relation $\in_A : PA \longrightarrow A$, with the universal property that every relation $(u,v) : X \longrightarrow A$ with codomain A has exactly one factorization $(u,v) \simeq \in_A \circ f$ with $f : X \longrightarrow PA$ in $\underline{\underline{E}}$. The morphism f thus obtained is called the <u>characteristic morphism</u> of the relation (u,v) .

<u>1.3</u>. The categories of sets and of finite sets are elementary topoi, with PA the set of all subsets of A , and with $\in_A$ given by all pairs (X,x) such that $x \in X$ and $X \subset A$. Categories of set-valued sheaves are also topoi.

<u>1.4</u>. We note some basic properties of a topos $\underline{\underline{E}}$.

<u>1.4.1</u>. Every monomorphism of $\underline{\underline{E}}$ is an equalizer, and every epimorphism a coequalizer.

<u>1.4.2</u>. $\underline{\underline{E}}$ has finite colimits.

<u>1.4.3</u>. $\underline{\underline{E}}$ is cartesian closed.

<u>1.4.4</u>. Partial morphisms in $\underline{\underline{E}}$ can be represented (see 1.6).

<u>1.5</u>. In a topological situation, or in a lattice regarded as a category, not every monomorphism is an equalizer, and not every epimorphism a coequalizer, but the remainder of 1.4 and other properties of topoi may still be valid. This led J. Penon [16] to define quasitopoi.

We recall first that a monomorphism m is called <u>strong</u> [12] or <u>strict</u> [11] if for every commutative square $m\,u = v\,e$ with e epimorphic, there is a morphism t such that $u = t\,e$ and $v = m\,t$. Strong epimorphisms are defined dually. Strong monomorphisms are closed under composition and pullbacks, and every equalizer is a strong monomorphism. We say that a partial morphism (m,f) is <u>strong</u> if m is a strong monomorphism.

1.6. We define a quasitopos as a category $\underline{\underline{E}}$ with the following properties.

QT 1. $\underline{\underline{E}}$ has finite limits and colimits.

QT 2. $\underline{\underline{E}}$ is cartesian closed.

QT 3. Strong partial morphisms of $\underline{\underline{E}}$ are represented.

The last statement means that for every object A of $\underline{\underline{E}}$ there is a strong partial morphism $\zeta_A : \tilde{A} \rightharpoonup A$ such that every strong partial morphism $(m,f) : X \rightharpoonup A$ factors $(m,f) \simeq \zeta_A \circ \varphi$ for a unique morphism $\varphi : X \longrightarrow A$ in $\underline{\underline{E}}$. It follows that $\zeta_A \simeq (\vartheta_A, \mathrm{id}\, A)$, with $\zeta_A \circ \vartheta_A \simeq (\mathrm{id}\, A, \mathrm{id}\, A)$.

1.7. Topoi clearly are quasitopoi, and a quasitopos $\underline{\underline{E}}$ is a topos if every monomorphism of $\underline{\underline{E}}$ is strong. The limit spaces of Kowalsky [14] and Fischer [8] form a quasitopos. Heyting algebras, also called relatively pseudocomplemented lattices [3] are quasitopoi. We shall obtain other examples.

1.8. The "fundamental theorem for topoi" is valid for quasitopoi. This means that every pullback functor $f^* : \underline{\underline{E}}/B \longrightarrow \underline{\underline{E}}/A$, for $f : A \longrightarrow B$ in $\underline{\underline{E}}$, has not only the usual left adjoint $\Sigma_f : \underline{\underline{E}}/A \longrightarrow \underline{\underline{E}}/B$, given by $\Sigma_f u = f u$ for an object $u : X \longrightarrow A$ of $\underline{\underline{E}}/A$, but also a right adjoint Π_f.

As Day [6] has shown, right adjoints for all pullback functors f^* exist if and only if every category $\underline{\underline{E}}/A$ is cartesian closed.

1.9. We list some additional basic properties of a quasitopos $\underline{\underline{E}}$.

1.9.1. Every strong monomorphism of $\underline{\underline{E}}$ is an equalizer, and every strong epimorphism a coequalizer.

1.9.2. Every morphism f of $\underline{\underline{E}}$ has a factorization $f = m u e$ with e a strong epimorphism, m a strong monomorphism, and u epimorphic and monomorphic. Pullbacks in $\underline{\underline{E}}$ preserve this factorization.

1.9.3. Strong relations in $\underline{\underline{E}}$, i.e. relations (u,v) with $\langle u,v\rangle$ a strong monomorphism, are representable.

1.9.4. Strong equivalence relations in $\underline{\underline{E}}$ are kernel pairs of their characteristic morphisms.

1.9.5. Pullbacks preserve colimit cones in $\underline{\underline{E}}$.

2. Quasitopoi of sieves

We consider in this section a concrete category $\underline{\underline{A}}$, i.e. $\underline{\underline{A}}$ is equipped with a faithful functor $P : \underline{\underline{A}} \longrightarrow \underline{\underline{Ens}}$ to the category of sets. We assume for convenience that P has skeletal fibres, i.e. if $P\,u$ is an identity mapping for an isomorphism u of $\underline{\underline{A}}$, then u is an identity morphism. By the usual abus de langage, we often use the same symbol for a morphism f of $\underline{\underline{A}}$ and its underlying mapping $P\,f$.

2.1. We define a P-sieve on a set E as a pair (Φ, E) consisting of E and a subfunctor Φ of the contravariant functor $\underline{\underline{Ens}}(P-, E) : \underline{\underline{A}}^{op} \longrightarrow \underline{\underline{Ens}}$. In this situation, Φ is given by assigning to every object X of $\underline{\underline{A}}$ a set ΦX of mappings $u : P\,X \longrightarrow E$, with the property that $v \cdot P\,f$ always is in ΦX for $f : X \longrightarrow Y$ in $\underline{\underline{A}}$ and v in ΦY . With built-in abus de langage, we define a morphism $h : (\Phi, E) \longrightarrow (\Psi, F)$ of P-sieves as a mapping $h : E \longrightarrow F$ such that always $h\,u \in \Psi X$ for u in ΦX .

This defines a category of P-sieves which we denote by $\underline{\underline{A}}^{cr}$ (for "crible"), and putting $P^{cr}(\Phi, E) = E$ defines a forgetful functor $P^{cr} : \underline{\underline{A}}^{cr} \longrightarrow \underline{\underline{Ens}}$. Composition of morphisms in $\underline{\underline{A}}^{cr}$ is composition of the underlying mappings.

2.2. The functor $P^{cr} : \underline{\underline{A}}^{cr} \longrightarrow \underline{\underline{Ens}}$ admits all possible initial and final structures. Thus $\underline{\underline{A}}^{cr}$ is a top category over sets, in the sense of [22] and [23], except that fibres of P^{cr} may be large. The large fibres do not matter, however, since P^{cr} admits initial and final structures for all admissible families of data, large or small.

For a family of P-sieves (Φ_i, E_i) and of mappings $f_i : E \longrightarrow E_i$, the initial structure (Φ, E) for P^{cr} has ΦX consisting of all mappings $u : P X \longrightarrow E$ such that every $f_i u$ is in the corresponding $\Phi_i X$. Final structures are obtained dually, with "every" replaced by "some".

2.3. For an object A of $\underline{\underline{A}}$, we define a P-sieve $\mathcal{Y} A$ on $P A$ by letting $(\mathcal{Y} A) X$ be the set of all $P f$ for $f : X \longrightarrow A$ in $\underline{\underline{A}}$. Putting $\mathcal{Y} g = P g : \mathcal{Y} A \longrightarrow \mathcal{Y} B$ for $g : A \longrightarrow B$ in $\underline{\underline{A}}$ then defines a functor $\mathcal{Y} : \underline{\underline{A}} \longrightarrow \underline{\underline{A}}^{cr}$, with $P = P^{cr} \mathcal{Y}$. By abus de langage, we may use P to identify subsieves of $\mathcal{Y} A$ with subfunctors of $\underline{\underline{A}}(-, A)$, i.e. with sieves on A in the usual sense of the word. Antoine [1], [2], who introduced P-sieves, proved that $\mathcal{Y}$ preserves all initial structures which P admits, and he obtained the following Yoneda lemma.

2.4. PROPOSITION. <u>For a</u> P-<u>sieve</u> (Φ, E), <u>a mapping</u> $u : P A \longrightarrow E$ <u>is in</u> ΦA <u>if and only if</u> $u : \mathcal{Y} A \longrightarrow (\Phi, E)$ <u>in</u> $\underline{\underline{A}}^{cr}$. <u>In particular</u>, $\mathcal{Y} : \underline{\underline{A}} \longrightarrow \underline{\underline{A}}^{cr}$ <u>is a full embedding</u>.

2.5. For P-sieves (Φ, E) and (Ψ, F), we construct a P-sieve $[\Phi, \Psi]$ on the set F^E by letting $[\Phi, \Psi] X$ consist of all mappings $\hat{u} : P X \longrightarrow F^E$ such that the corresponding mapping $u : P X \times E \longrightarrow F$ is a morphism $u : \mathcal{Y} X \times (\Phi, E) \longrightarrow (\Psi, F)$ of P-sieves. By 2.4, this is the only way to construct a cartesian closed structure for P-sieves, and it works. In fact, $\underline{\underline{A}}^{cr}$ is not only cartesian closed but span-closed.

We often can say more. We say that $\underline{\underline{A}}$ <u>has</u> P-<u>inclusions</u> [24] if P admits an initial structure for every subset inclusion $E \subset PA$.

$\underline{\underline{2.6}}$. THEOREM. <u>Let</u> $P : \underline{\underline{A}} \longrightarrow \underline{\underline{Ens}}$ <u>be a faithful functor</u>. <u>If</u> $\underline{\underline{A}}$ <u>has</u> P-<u>inclusions, then the category</u> $\underline{\underline{A}}^{cr}$ <u>of</u> P-<u>sieves is a quasitopos</u>.

<u>Proof</u>. $\underline{\underline{A}}^{cr}$ has limits and colimits, and we have seen that $\underline{\underline{A}}^{cr}$ is cartesian closed. Strong monomorphisms $m : (\Phi, E) \longrightarrow (\Psi, F)$ of $\underline{\underline{A}}^{cr}$ are injective mappings with Φ the initial structure for m and Ψ, i.e. $u : PX \longrightarrow E$ is in ΦX iff $m\,u \in \Psi X$.

For sets, $\widetilde{E}$ is E with one point added, and ϑ_E is the inclusion. For P-sieves, we claim that $\vartheta_{\Phi,E} = \vartheta_E : (\Phi, E) \longrightarrow (\widetilde{\Phi}, \widetilde{E})$, with $\widetilde{\Phi}$ constructed as follows. For $u : PX \longrightarrow E$, we construct a pullback square

$$\begin{array}{ccc} E' & \xrightarrow{u'} & E \\ \downarrow & & \downarrow \vartheta_E \\ PX & \xrightarrow{u} & \widetilde{E} \end{array}$$

of sets, with a set inclusion at left. If X' is the initial structure for P and this inclusion, then we put $u \in \widetilde{\Phi} X$ iff $u' \in \Phi X'$.

$\underline{\underline{2.7}}$. We say that $\underline{\underline{A}}$ <u>has constant morphisms</u> if $\underline{\underline{A}}$ satisfies the following two conditions. (i) $\underline{\underline{A}}$ has a terminal object A_1 with PA_1 a singleton, and every constant mapping $f : PA_1 \longrightarrow PX$ lifts to a morphism $f : A_1 \longrightarrow X$ of $\underline{\underline{A}}$. (ii) $\underline{\underline{A}}$ has an object A_o with PA_o empty, and an object A_o with this property is an initial object of $\underline{\underline{A}}$. The categories occurring in topology usually satisfy these two conditions.

We obtain a category of P-sieves with constant morphisms in two steps. We denote by $\underline{\underline{A}}^{ci}$ the full subcategory of $\underline{\underline{A}}^{cr}$ with those P-sieves (Φ, E) as objects for which the unique $u : PA \longrightarrow E$ is in ΦA if PA is empty, and by $\underline{\underline{A}}^{c}$ the full subcategory of $\underline{\underline{A}}^{ci}$ with the P-sieves (Φ, E) as objects for which every ΦX contains all constant mappings $PX \longrightarrow E$. $\underline{\underline{A}}^{c}$ has constant morphisms, and we have a Yoneda embedding $\mathcal{Y} : \underline{\underline{A}} \longrightarrow \underline{\underline{A}}^{c}$, with $P^{c}\mathcal{Y} = P$ for the forgetful functor $P^{c} : \underline{\underline{A}}^{c} \longrightarrow \underline{\underline{\mathrm{Ens}}}$, if $\underline{\underline{A}}$ has constant morphisms.

2.8. PROPOSITION. *If* $\underline{\underline{A}}^{cr}$ *is a quasitopos, then* $\underline{\underline{A}}^{c}$ *is a quasitopos.*

Proof. The reflector $\underline{\underline{A}}^{cr} \longrightarrow \underline{\underline{A}}^{ci}$ adds the unique $u : PA \longrightarrow E$ to ΦA if necessary, for PA empty. This functor preserves limits; thus $\underline{\underline{A}}^{ci}$ is a quasitopos by 3.7 below.

If (Φ, E) is an object of $\underline{\underline{A}}^{ci}$, let E' be the set of all $x \in E$ such that the constant mapping $u : PX \longrightarrow E$ with range $\{x\}$ is in ΦX, for every object X of $\underline{\underline{A}}$ with PX not empty, and let (Φ', E') be the initial P-sieve for Φ and the inclusion $E' \longrightarrow E$. This is an object of $\underline{\underline{A}}^{c}$, and if $h : (\Psi, F) \quad (\Phi, E)$ with (Ψ, F) an object of $\underline{\underline{A}}^{c}$, then h factors through E' and hence through (Φ', E'). Thus $\underline{\underline{A}}^{c}$ is coreflective in $\underline{\underline{A}}^{ci}$, and isomorphic to the category of coalgebras for an idempotent comonad. As this comonad is exact, $\underline{\underline{A}}^{c}$ is a quasitopos, by 3.1 below, if $\underline{\underline{A}}^{ci}$ is one.

2.9. REMARKS. (i) We note that P^{cr} preserves the full quasitopos structure. P^{c} does not. (ii) If $\underline{\underline{A}}$ has constant morphisms, and if PA and E are both not empty, then a P-sieve (Φ, E) is an object of $\underline{\underline{A}}^{c}$ iff the mappings $u : PA \longrightarrow E$ in ΦA are collectively surjective. (iii) If $\underline{\underline{A}}$ has constant morphisms, then the functor $P : \underline{\underline{A}} \longrightarrow \underline{\underline{\mathrm{Ens}}}$ preserves limits.

3. Geometric morphisms of quasitopoi

Geometric morphisms of quasitopoi are defined in the same way as for topoi: they are adjunctions $f^* \dashv f_* : \underline{\underline{E}} \longrightarrow \underline{\underline{F}}$ such that f^* is left exact, i.e. preserves finite limits. It follows that f^* preserves monomorphisms and strong monomorphisms. We consider in this section only the geometric morphisms used to construct quasitopoi of coalgebras and of sheaves.

3.1. A comonad (G, δ, ψ) on a category $\underline{\underline{E}}$ is called <u>left exact</u> if the functor G preserves finite limits.

THEOREM. <u>For a left exact comonad</u> (G, δ, ψ) <u>on a quasitopos</u> $\underline{\underline{E}}$, <u>the category</u> $\underline{\underline{E}}_G$ <u>of coalgebras is a quasitopos</u>.

<u>Proof</u>. We indicate only the construction of $(A,\alpha)^{\sim}$ for a G-coalgebra. We have a strong partial morphism $(G\vartheta_A, \delta_A) : G\tilde{A} \longrightarrow A$, and hence a pullback

$$\begin{array}{ccc} GA & \xrightarrow{\delta_A} & A \\ {\scriptstyle G\vartheta_A}\downarrow & & \downarrow{\scriptstyle \vartheta_A} \\ G\tilde{A} & \xrightarrow{\rho_A} & \tilde{A} \end{array}$$

in $\underline{\underline{E}}$, with

$$G\rho_A \cdot \psi_{\tilde{A}} \cdot G\vartheta_A = G\rho_A \cdot GG\vartheta_A \cdot \psi_A$$
$$= G\vartheta_A \cdot G\delta_A \cdot \psi_A = G\vartheta_A \,..$$

The forgetful functor $U_G : \underline{\underline{E}}_G \quad \underline{\underline{E}}$ creates finite limits since G is left exact, Thus we have an equalizer fork

$$(A,\alpha)^{\sim} \xrightarrow{\;e\;} F_G\tilde{A} \underset{G\rho_A \cdot \psi_{\tilde{A}}}{\overset{id}{\rightrightarrows}} F_G\tilde{A}$$

of coalgebras, with $G\vartheta_A \cdot \alpha = e\,\vartheta_{A,\alpha}$ for $\vartheta_{A,\alpha} : (A,\alpha) \longrightarrow (A,\alpha)^{\sim}$.

3.2. We define a topology of a quasitopos $\underline{\underline{E}}$ as a natural closure operator γ for monomorphisms of $\underline{\underline{E}}$. Thus we require the following.

(i) m and γm have the same codomain, and $\gamma m \leq \gamma m'$ if $m \leq m'$.

(ii) $m \leq \gamma m$, and $\gamma\gamma m \simeq \gamma m$.

(iii) $f^*(\gamma m) \simeq \gamma(f^* m)$ if m and f have the same codomain.

We recall that $f^* m$ is the pullback of m by f . Axioms (i) - (iii) suffice for topoi; for quasitopoi we need an additional axiom.

(iv) γm is strong if m is strong.

3.3. A monomorphism m is called closed for a topology γ if $\gamma m \simeq m$, and m is called dense if γm is an isomorphism. We note some elementary properties of closed and dense monomorphisms.

3.3.1. If $m = m_1 m'$, then m' is dense iff $m_1 \leq \gamma m$.

3.3.2. If $f^* m$ is defined, then $f^* m$ is dense if m is dense, and $f^* m$ is closed if m is closed.

3.3.3. The composition of dense monomorphisms is dense, and the composition of closed monomorphisms is closed.

3.3.4. $\gamma(m_1 \cap m_2) \simeq \gamma m_1 \cap \gamma m_2$ if m_1 and m_2 have the same codomain.

3.3.5. If $m_1 u = v m'$ with m_1 closed and m' dense, then $u = t m'$ and $m_1 t = v$ for a unique morphism t of $\underline{\underline{E}}$.

3.4. An object S of $\underline{\underline{E}}$ is called separated for a topology γ of $\underline{\underline{E}}$ if $\underline{\underline{E}}(d,S)$ is injective for every dense monomorphism d , and an object F is called a sheaf for γ if $\underline{\underline{E}}(d,F)$ is bijective for every dense monomorphism d .

If F is a sheaf for γ, then we define a closure operator for strong partial morphisms $(m,f) : A \rightharpoondown F$. If $m = \gamma m \cdot d$, then $f = \bar{f} d$ for a unique $\bar{f}$, and we put $\gamma(m,f) = (\gamma m, \bar{f}) : A \rightharpoondown F$. This closure operator is idempotent and natural in A. Thus

$$\gamma((m,f) \circ h) \simeq \gamma(m,f) \circ h$$

if $(m,f) \circ h$ is defined. It follows that

$$(m,f) \simeq (\vartheta_F, \mathrm{id}) \circ \varphi \implies \gamma(m,f) \simeq (\vartheta_F, \mathrm{id}) \circ j_F \varphi ,$$

for $j_F : \tilde{F} \longrightarrow \tilde{F}$ given by $\gamma(\vartheta_F, \mathrm{id}) \simeq (\vartheta_F, \mathrm{id}) \circ j_F$.

<u>3.5</u>. THEOREM. <u>If</u> γ <u>is a topology of a quasitopos</u> $\underline{\underline{E}}$, <u>then separated objects and sheaves for</u> γ <u>define full reflective subcategories</u> $\mathrm{Sep}_\gamma \underline{\underline{E}}$ <u>and</u> $\mathrm{Sh}_\gamma \underline{\underline{E}}$ <u>of</u> $\underline{\underline{E}}$. <u>The category</u> $\mathrm{Sh}_\gamma \underline{\underline{E}}$ <u>of sheaves for</u> γ <u>is a quasitopos, and the reflector</u> $\underline{\underline{E}} \longrightarrow \mathrm{Sh}_\gamma \underline{\underline{E}}$ <u>preserves finite limits.</u>

<u>Proof</u>. In order to obtain QT 3 for sheaves, we construct an equalizer

$$\tilde{F}_j \xrightarrow{\ e\ } \tilde{F} \underset{\mathrm{id}}{\overset{j_F}{\rightrightarrows}} \tilde{F}$$

if F is a sheaf. This furnishes a <u>sheaf</u> $\tilde{F}_j$. Since $j_F \vartheta_F = \vartheta_F$ by the construction of j_F, we have $\vartheta_F = e\, \vartheta_F^*$ for a strong monomorphism ϑ_F^*, and $(\vartheta_F^*, \mathrm{id}) : \tilde{F}_j \rightharpoondown F$ represents strong partial morphisms in $\mathrm{Sh}_\gamma \underline{\underline{E}}$.

<u>3.6</u>. Strictly full reflective subcategories of $\underline{\underline{E}}$ can be characterized by idempotent monads $(T, \eta, \mathrm{id}\, T)$ on $\underline{\underline{E}}$, with $T\,T = T$. An object F of $\underline{\underline{E}}$ is in the reflective subcategory determined by T iff η_F is an isomorphism. If $\underline{\underline{E}}$ is a quasitopos and T obtained from a category $\mathrm{Sh}_\gamma \underline{\underline{E}}$ of sheaves for a topology γ, then T is left exact, i.e. T preserves finite limits.

Conversely, if T is left exact, then putting $\gamma_T m \simeq \bar{m}$ for a pullback

$$\begin{array}{ccc} \cdot & \xrightarrow{\bar{m}} & A \\ \downarrow & & \downarrow \eta_A \\ \cdot & \xrightarrow{T\,m} & T\,A \end{array}$$

in $\underline{\underline{E}}$ defines a topology γ_T of $\underline{\underline{E}}$. An object F of $\underline{\underline{E}}$ is a sheaf for γ_T if and only if η_F is isomorphic in $\underline{\underline{E}}$, and we have the following result.

3.7. THEOREM. *If $\underline{\underline{E}}$ is a quasitopos and $\underline{\underline{F}}$ a reflective full subcategory of $\underline{\underline{E}}$ with left exact reflector, then $\underline{\underline{F}}$ is a quasitopos.*

3.8. It remains to compare γ with γ_T if γ is a topology of a quasitopos $\underline{\underline{E}}$ and (T, η, id) the left exact idempotent monad on $\underline{\underline{E}}$ obtained from the category $\mathrm{Sh}_\gamma \underline{\underline{E}}$ of sheaves for γ.

We note first that γ and γ_T produce the same separated objects and the same sheaves. In 3.6, $T\,m$ and hence $\bar{m}$ are closed for γ. Thus $\gamma m \leq \gamma_T m$, and γ_T is coarser than γ. If m is strong, then $\gamma_T m \simeq \gamma m$ by the usual argument for topoi, and it is easy to see that $\gamma_T m \simeq \gamma m$ if the codomain of m is separated for γ. We do not know whether always $\gamma_T m \simeq \gamma m$, so that γ_T and γ are equivalent, or whether it is possible to obtain the same quasitopos of sheaves from two topologies which are not equivalent.

3.9. In a quasitopos $\underline{\underline{E}}$, every monomorphism has a factorization $m = m_1 u$ with m_1 a strong monomorphism and u epimorphic and monomorphic. Pullbacks preserve this factorization; thus putting $\gamma m \simeq m_1$ defines a topology of $\underline{\underline{E}}$. Closed monomorphisms for γ are strong; thus $\mathrm{Sh}_\gamma \underline{\underline{E}}$ is a topos. This is the topos of coarse objects of $\underline{\underline{E}}$, obtained by Penon [16] in a different way.

4. Quasitopoi in topology

All categories in this section are assumed to be concrete, with constant morphisms (2.7).

4.1. If a commutative triangle of faithful functors

$$\begin{array}{ccc} \mathbf{A} & \xrightarrow{\;G\;} & \mathbf{B} \\ & {\scriptstyle P}\searrow \quad \swarrow{\scriptstyle Q} & \\ & \mathbf{Ens} & \end{array}$$

is given, then we call G a dense embedding of $\mathbf{A}$ into $\mathbf{B}$ if G is full, and every object B of $\mathbf{B}$ has the final structure for the functor Q and all morphisms $u : G\,X \longrightarrow B$ in $\mathbf{B}$.

In this situation, the morphisms $u : G\,X \longrightarrow B$ form a colimit cone in $\mathbf{B}$. Thus a dense embedding is a dense functor as defined e.g. in [15].

4.2. PROPOSITION. A dense embedding $G : \mathbf{A} \longrightarrow \mathbf{B}$ preserves initial structures and limits. If $\mathbf{A}$ is complete, then G has a left adjoint left inverse which preserves underlying sets.

4.3. PROPOSITION. $\mathcal{Y} : \mathbf{A} \longrightarrow \mathbf{A}^c$ is a dense embedding. If $G : \mathbf{A} \longrightarrow \mathbf{B}$ is a dense embedding, then $\mathcal{Y} = G^c\,G$ for a dense embedding $G^c : \mathbf{B} \longrightarrow \mathbf{A}^c$.

Proofs. The proof of the first part of 4.2 is straightforward; the second part follows from [22; 6.3].

$\mathcal{Y}$ is a dense embedding by the definitions and 2.4, used for $\mathcal{Y} : \mathbf{A} \longrightarrow \mathbf{A}^c$. If $G : \mathbf{A} \longrightarrow \mathbf{B}$ is given as in 4.1, then we let $G^c\,B$ be the P-sieve over $Q\,B$ with $u : P\,X \longrightarrow Q\,B$ in $(G^c\,B)\,X$ iff $u : G\,X \longrightarrow B$ in $\mathbf{B}$. This defines a functor $G^c : \underline{\underline{B}} \longrightarrow \underline{\underline{A}}^c$, and one sees easily that G^c is a dense embedding such that $G^c G = \mathcal{Y}$.

We need a special case of a theorem of Day [5].

4.4. PROPOSITION. If $G : \underline{\underline{A}} \longrightarrow \underline{\underline{B}}$ is a dense embedding with left adjoint $H : \underline{\underline{B}} \longrightarrow \underline{\underline{A}}$, and if $\underline{\underline{B}}$ is cartesian closed, then $\underline{\underline{A}}$ is cartesian closed if and only if H preserves finite products.

4.5. A dense embedding $G : \underline{\underline{A}} \longrightarrow \underline{\underline{B}}$ induces a dense embedding $G_A : \underline{\underline{A}}/A \longrightarrow \underline{\underline{B}}/GA$ for every object A of $\underline{\underline{A}}$. A left adjoint H of G induces a left adjoint H_A for every functor G_A , and 4.4 is valid for these adjunctions. If H preserves finite limits, then every H_A preserves finite products.

Combining this information with 4.3, we see that we obtain dense embeddings of $\underline{\underline{A}}$ into complete span-closed categories by looking for full reflective subcategories of $\underline{\underline{A}}^c$ with left exact reflectors which preserve underlying sets. By 3.7, these subcategories are quasitopoi if $\underline{\underline{A}}^c$ is a quasitopos, and we are looking for topologies γ of $\underline{\underline{A}}^c$ with a reflector $\underline{\underline{A}}^c \longrightarrow Sh_\gamma \underline{\underline{A}}^c$ which preserves underlying sets, and such that every object $\mathcal{Y}A$ is a sheaf for γ .

The associated sheaf functor $\underline{\underline{A}}^c \longrightarrow Sh_\gamma \underline{\underline{A}}^c$ preserves underlying sets only if every dense monomorphism for γ is bijective at the set level, Conversely, if this is the case, then every object of $\underline{\underline{A}}^c$ is separated for γ , hence densely embedded into a sheaf, and thus an associated sheaf functor which preserves underlying sets exists.

4.6. Let $P : \underline{\underline{A}} \longrightarrow \underline{\underline{C}}$ be the forgetful functor. If $P^c \mu$ is bijective for every γ-dense monomorphism μ of $\underline{\underline{A}}^c$, then the topology γ of $\underline{\underline{A}}^c$ is equivalent to a unique topology $\check{\gamma}$ of $\underline{\underline{A}}^c$ such that $P^c(\check{\gamma}\mu) = P^c\mu$ for every monomorphism μ of $\underline{\underline{A}}^c$. We say that $\check{\gamma}$ with this property is a P-topology of $\underline{\underline{A}}^c$. Since every object of $\underline{\underline{A}}^c$ is separated for a P-topology, we have by 3.8 a bijection between P-topologies of $\underline{\underline{A}}^c$ and the corresponding categories of sheaves. These are the strictly full reflective subcategories of $\underline{\underline{A}}^c$ with a reflector which preserves finite limits and underlying sets.

Embeddings $(\Phi, PA) \longrightarrow yA$ in $\underline{\underline{A}}^c$ correspond bijectively to subfunctors of the functor $\underline{\underline{A}}(-, A)$, i.e. to sieves on A in the usual sense, if we allow only those sieves which contain all constant morphisms with codomain A. With the same restriction on sieves, we obtain a bijection between P-topologies of $\underline{\underline{A}}^c$ and Grothendieck topologies of $\underline{\underline{A}}$ as follows.

If a P-topology γ of $\underline{\underline{A}}^c$ is given, then we denote by $J_\gamma A$ the class of all sieves on an object A of $\underline{\underline{A}}$ for which the corresponding $(\Phi, PA) \longrightarrow yA$ is dense for γ. This defines a Grothendieck topology J_γ of $\underline{\underline{A}}$. If J is a Grothendieck topology of $\underline{\underline{A}}$, and if $\mu = m : (\Phi, E) \longrightarrow (\Psi, F)$ is a monomorphism of $\underline{\underline{A}}^c$, then we put $\gamma_J \mu = m : (\overline{\Phi}, E) \longrightarrow (\Psi, F)$, with $u : PA \longrightarrow E$ in $\overline{\Phi}A$ iff $m\,u \in \Psi A$ and $u^*\Phi \in JA$, where $(u^*\Phi)X$ consists of all $f : X \longrightarrow A$ in $\underline{\underline{A}}$ such that $u \cdot Pf$ is in ΦX. This defines a P-topology γ_J on $\underline{\underline{A}}^c$. One verifies easily that the correspondences $\gamma \longmapsto J_\gamma$ and $J \longmapsto \gamma_J$ are inverse bijections.

4.7. If R is a sieve on an object A of $\underline{\underline{A}}$ and Φ a P-sieve on a set E, both with constants, then we put $R \perp \Phi$ if a mapping $u : PA \longrightarrow E$ is in ΦA whenever $u \cdot Pf$ is in ΦX for every $f : X \longrightarrow A$ in a set RX.

If γ is a P-topology of $\underline{\underline{A}}^c$ and J the corresponding Grothendieck topology of $\underline{\underline{A}}$, then the sheaves for γ are the P-sieves orthogonal to J for the relation $\perp$ just defined, and J is the orthogonal complement of the class of all γ-sheaves, for $\perp$. The fact that $R \perp \Phi$ for R in J A and every sheaf (Φ, E) for γ can be expressed by saying that $\mathcal{Y}A$ has the final structure for the morphisms $f : X \longrightarrow A$ in R and the category of γ-sheaves.

4.8. We say that a sieve R on an object A of $\underline{\underline{A}}$ is a quotient sieve if A has the final structure for P and all morphisms $f : X \longrightarrow A$ in a set R X . With J and γ as in 4.7, we have the following corollary of 4.7.

PROPOSITION. Every object $\mathcal{Y}A$ of $\underline{\underline{A}}^c$ is a sheaf for γ if and only if every sieve in J is a quotient sieve.

With set inclusion as order relation, Grothendieck topologies of $\underline{\underline{A}}$ form a complete lattice. There is thus a largest Grothendieck topology of $\underline{\underline{A}}$ which consists of quotient sieves; this is the canonical topology of $\underline{\underline{A}}$. The corresponding category of sheaves in $\underline{\underline{A}}^c$ is the smallest complete quasitopos into which $\underline{\underline{A}}$ can be densely embedded -- provided that $\underline{\underline{A}}^c$ is a quasitopos. By 2.8 and 2.6, this is the case if $\underline{\underline{A}}$ has P-inclusions.

4.9. Let Top denote the category of topological spaces and Lim the category of limit spaces. The embedding Top $\longrightarrow$ Lim is dense, and Lim is a quasitopos [16]. Thus we can identify Lim with the category of sheaves for a P-topology of Top^c , by a functor G^c . If we do this, then the category of sheaves for the canonical topology of Top becomes a category of limit spaces.

Theorem 1 of [7] can easily be generalized from quotient maps to quotient sieves; thus we can describe the canonical topology of Top as follows.

THEOREM. For a quotient sieve R on a topological space Y , the following three conditions are logically equivalent.

(i) R *is in* J Y *for the canonical Grothendieck topology of* Top .

(ii) *If an ultrafilter* G *on* Y *converges to* y *in* Y , *then there are always* $f : X \longrightarrow Y$ *in* R , *an ultrafilter* F *on* X *and* x *in* X *such that* F *converges to* x *in* X , *and* $f(F) = G$ *and* $f(x) = y$.

(iii) *If we assign to every* $f : X \longrightarrow Y$ *in* R *an open cover* $(U_{f,i})$ *of* $f^{-1}(y)$ *in* X , *for a point* y *of* Y , *then there is always a finite union of sets* $f(U_{f,i})$ *which is a neighborhood of* y *in* Y .

4.10. With the notation of 4.7, condition (ii) in the Theorem above says that $R \perp (E,q)$ for every Choquet space (E,q) , and condition (iii) says that $R \perp (E,q)$ if (E,q) is a solid convergence space [17], i.e. a limit space which satisfies the following axiom L 3'.

L 3'. If $(F_i)_{i \in I}$ is a family of filters on E converging to x in E , and if G is a filter on E such that for every family of sets $A_i \in F_i$, one for every $i \in I$, a finite union of sets A_i belongs to G , then G q x .

Solid limit spaces and Choquet spaces both define strictly full reflective subcategories of Lim , with reflectors which preserve underlying sets and finite limits. Thus we have the following result from 4.7 and 4.9.

THEOREM. *Solid limit spaces and Choquet spaces define the same strictly full subcategory of* Lim . *This category is a quasitopos, isomorphic to the category of sheaves in* Top^c *for the canonical topology of* Top .

R e f e r e n c e s

1. Antoine, P., Extension minimale de la catégorie des espaces topologiques. C. R. Acad. Sc. Paris 262 (1966), sér. A, 1389 - 1392.

2. Antoine, P., Etude élémentaire des catégories d'ensembles structurés. Bull. Soc. Math. Belgique 18 (1966), 142 - 166 and 387 - 414.

3. Birkhoff, G., Lattice Theory. Rev. Ed. Providence, 1948.

4. Choquet, G., Convergences. Ann. Univ. Grenoble, Sect. Sc. Math. Phys. (N.S.) 23 (1948), 57 - 112.

5. Day, B., A reflection theorem for closed categories. J. Pure Appl. Alg. 2 (1972), 1 - 11.

6. Day, B., Limit spaces and closed span categories. Category Seminar, Sidney 1972/73, pp. 65 - 74. Lecture Notes in Math. 420 (1974).

7. Day, B.J., and G.M. Kelly, On topological quotient maps. Proc. Camb. Phil. Soc. 67 (1970), 553 - 558.

8. Fischer, H.R., Limesräume. Math. Annalen 137 (1959), 269 - 303.

9. Freyd, P., Aspects of topoi. Bull. Austral. Math. Soc. 7 (1972), 1 - 76.

10. Herrlich, H., Cartesian closed topological categories. Preprint (1974).

11. Jurchescu, A., and M. Lascu, Morphisme stricte, categorii cantoriene, functori de completare. Studii Cerc. Mat. 18 (1966), 219 - 234.

12. Kelly, G.M., Monomorphisms, epimorphisms, and pullbacks. J. Austral. Math. Soc. 9 (1969), 124 - 142.

13. Kock, A., and G.C. Wraith, Elementary toposes. Lecture Note Series, no. 30. Matematisk Institut, Aarhus Universitet, 1971.

14. Kowalsky, H.-J., Limesräume und Komplettierung. Math. Nachr. 12 (1954), 301 - 340.

15. MacLane, S., Categories for the Working Mathematician. Springer, 1971.

16. Penon, J., Quasi-topos. C. R. Acad. Sc. Paris 276 (1973), Sér. A, 237 - 240.

17. Schroder, M., Solid convergence spaces. Bull. Austral. Math. Soc. 8 (1973), 443 - 459.

18. Steenrod, N.E., A convenient category of topological spaces. Michigan Math. J. 14 (1967), 133 - 152.

19. Stout, L.N., General topology in an elementary topos. Ph.D. Thesis, University of Illinois, 1974.

20. Tierney, M., Axiomatic sheaf theory: some constructions and applications. C.I.M.E. III Ciclo 1971, Varenna 12-21 Settembre. Cremonese, Roma, 1973.

21. Verdier, J.L., SGA 4/I, Théorie des topos et cohomologie étale des schémas, Exposés I - IV. 2me.éd. Lecture Notes in Math. 269 (1972).

22. Wyler, O., On the categories of general topology and topological algebra. Arch. Math. 22 (1971), 7 - 17.

23. Wyler, O., Top categories and categorical topology. Gen. Top. Appl. 1 (1971), 17 - 28.

24. Wyler, O., Quotient maps. Gen. Top. Appl. 3 (1973), 149 - 160.

25. Wyler, O., Convenient categories for topology. Gen. Top. Appl. 3 (1973), 225 - 242.

Vol. 399: Functional Analysis and its Applications. Proceedings 1973. Edited by H. G. Garnir, K. R. Unni and J. H. Williamson. II, 584 pages. 1974.

Vol. 400: A Crash Course on Kleinian Groups. Proceedings 1974. Edited by L. Bers and I. Kra. VII, 130 pages. 1974.

Vol. 401: M. F. Atiyah, Elliptic Operators and Compact Groups. V, 93 pages. 1974.

Vol. 402: M. Waldschmidt, Nombres Transcendants. VIII, 277 pages. 1974.

Vol. 403: Combinatorial Mathematics. Proceedings 1972. Edited by D. A. Holton. VIII, 148 pages. 1974.

Vol. 404: Théorie du Potentiel et Analyse Harmonique. Edité par J. Faraut. V, 245 pages. 1974.

Vol. 405: K. J. Devlin and H. Johnsbråten, The Souslin Problem. VIII, 132 pages. 1974.

Vol. 406: Graphs and Combinatorics. Proceedings 1973. Edited by R. A. Bari and F. Harary. VIII, 355 pages. 1974.

Vol. 407: P. Berthelot, Cohomologie Cristalline des Schémas de Caracteristique p > o. II, 604 pages. 1974.

Vol. 408: J. Wermer, Potential Theory. VIII, 146 pages. 1974.

Vol. 409: Fonctions de Plusieurs Variables Complexes, Séminaire François Norguet 1970–1973. XIII, 612 pages. 1974.

Vol. 410: Séminaire Pierre Lelong (Analyse) Année 1972–1973. VI, 181 pages. 1974.

Vol. 411: Hypergraph Seminar. Ohio State University, 1972. Edited by C. Berge and D. Ray-Chaudhuri. IX, 287 pages. 1974.

Vol. 412: Classification of Algebraic Varieties and Compact Complex Manifolds. Proceedings 1974. Edited by H. Popp. V, 333 pages. 1974.

Vol. 413: M. Bruneau, Variation Totale d'une Fonction. XIV, 332 pages. 1974.

Vol. 414: T. Kambayashi, M. Miyanishi and M. Takeuchi, Unipotent Algebraic Groups. VI, 165 pages. 1974.

Vol. 415: Ordinary and Partial Differential Equations. Proceedings 1974. XVII, 447 pages. 1974.

Vol. 416: M. E. Taylor, Pseudo Differential Operators. IV, 155 pages. 1974.

Vol. 417: H. H. Keller, Differential Calculus in Locally Convex Spaces. XVI, 131 pages. 1974.

Vol. 418: Localization in Group Theory and Homotopy Theory and Related Topics. Battelle Seattle 1974 Seminar. Edited by P. J. Hilton. VI, 172 pages 1974.

Vol. 419: Topics in Analysis. Proceedings 1970. Edited by O. E. Lehto, I. S. Louhivaara, and R. H. Nevanlinna. XIII, 392 pages. 1974.

Vol. 420: Category Seminar. Proceedings 1972/73. Edited by G. M. Kelly. VI, 375 pages. 1974.

Vol. 421: V. Poénaru, Groupes Discrets. VI, 216 pages. 1974.

Vol. 422: J.-M. Lemaire, Algèbres Connexes et Homologie des Espaces de Lacets. XIV, 133 pages. 1974.

Vol. 423: S. S. Abhyankar and A. M. Sathaye, Geometric Theory of Algebraic Space Curves. XIV, 302 pages. 1974.

Vol. 424: L. Weiss and J. Wolfowitz, Maximum Probability Estimators and Related Topics. V, 106 pages. 1974.

Vol. 425: P. R. Chernoff and J. E. Marsden, Properties of Infinite Dimensional Hamiltonian Systems. IV, 160 pages. 1974.

Vol. 426: M. L. Silverstein, Symmetric Markov Processes. X, 287 pages. 1974.

Vol. 427: H. Omori, Infinite Dimensional Lie Transformation Groups. XII, 149 pages. 1974.

Vol. 428: Algebraic and Geometrical Methods in Topology, Proceedings 1973. Edited by L. F. McAuley. XI, 280 pages. 1974.

Vol. 429: L. Cohn, Analytic Theory of the Harish-Chandra C-Function. III, 154 pages. 1974.

Vol. 430: Constructive and Computational Methods for Differential and Integral Equations. Proceedings 1974. Edited by D. L. Colton and R. P. Gilbert. VII, 476 pages. 1974.

Vol. 431: Séminaire Bourbaki – vol. 1973/74. Exposés 436–452. IV, 347 pages. 1975.

Vol. 432: R. P. Pflug, Holomorphiegebiete, pseudokonvexe Gebiete und das Levi-Problem. VI, 210 Seiten. 1975.

Vol. 433: W. G. Faris, Self-Adjoint Operators. VII, 115 pages. 1975.

Vol. 434: P. Brenner, V. Thomée, and L. B. Wahlbin, Besov Spaces and Applications to Difference Methods for Initial Value Problems. II, 154 pages. 1975.

Vol. 435: C. F. Dunkl and D. E. Ramirez, Representations of Commutative Semitopological Semigroups. VI, 181 pages. 1975.

Vol. 436: L. Auslander and R. Tolimieri, Abelian Harmonic Analysis, Theta Functions and Function Algebras on a Nilmanifold. V, 99 pages. 1975.

Vol. 437: D. W. Masser, Elliptic Functions and Transcendence. XIV, 143 pages. 1975.

Vol. 438: Geometric Topology. Proceedings 1974. Edited by L. C. Glaser and T. B. Rushing. X, 459 pages. 1975.

Vol. 439: K. Ueno, Classification Theory of Algebraic Varieties and Compact Complex Spaces. XIX, 278 pages. 1975

Vol. 440: R. K. Getoor, Markov Processes: Ray Processes and Right Processes. V, 118 pages. 1975.

Vol. 441: N. Jacobson, PI-Algebras. An Introduction. V, 115 pages. 1975.

Vol. 442: C. H. Wilcox, Scattering Theory for the d'Alembert Equation in Exterior Domains. III, 184 pages. 1975.

Vol. 443: M. Lazard, Commutative Formal Groups. II, 236 pages. 1975.

Vol. 444: F. van Oystaeyen, Prime Spectra in Non-Commutative Algebra. V, 128 pages. 1975.

Vol. 445: Model Theory and Topoi. Edited by F. W. Lawvere, C. Maurer, and G. C. Wraith. III, 354 pages. 1975.

Vol. 446: Partial Differential Equations and Related Topics. Proceedings 1974. Edited by J. A. Goldstein. IV, 389 pages. 1975.

Vol. 447: S. Toledo, Tableau Systems for First Order Number Theory and Certain Higher Order Theories. III, 339 pages. 1975.

Vol. 448: Spectral Theory and Differential Equations. Proceedings 1974. Edited by W. N. Everitt. XII, 321 pages. 1975.

Vol. 449: Hyperfunctions and Theoretical Physics. Proceedings 1973. Edited by F. Pham. IV, 218 pages. 1975.

Vol. 450: Algebra and Logic. Proceedings 1974. Edited by J. N. Crossley. VIII, 307 pages. 1975.

Vol. 451: Probabilistic Methods in Differential Equations. Proceedings 1974. Edited by M. A. Pinsky. VII, 162 pages. 1975.

Vol. 452: Combinatorial Mathematics III. Proceedings 1974. Edited by Anne Penfold Street and W. D. Wallis. IX, 233 pages. 1975.

Vol. 453: Logic Colloquium. Symposium on Logic Held at Boston, 1972–73. Edited by R. Parikh. IV, 251 pages. 1975.

Vol. 454: J. Hirschfeld and W. H. Wheeler, Forcing, Arithmetic, Division Rings. VII, 266 pages. 1975.

Vol. 455: H. Kraft, Kommutative algebraische Gruppen und Ringe. III, 163 Seiten. 1975.

Vol. 456: R. M. Fossum, P. A. Griffith, and I. Reiten, Trivial Extensions of Abelian Categories. Homological Algebra of Trivial Extensions of Abelian Categories with Applications to Ring Theory. XI, 122 pages. 1975.

Vol. 457: Fractional Calculus and 1974. Edited by B. Ross. VI, 381 page

Vol. 458: P. Walters, Ergodic Theory – Introductory Lectures. VI, 198 pages. 1975.

Vol. 459: Fourier Integral Operators and Partial Differential Equations. Proceedings 1974. Edited by J. Chazarain. VI, 372 pages. 1975.

Vol. 460: O. Loos, Jordan Pairs. XVI, 218 pages. 1975.

Vol. 461: Computational Mechanics. Proceedings 1974. Edited by J. T. Oden. VII, 328 pages. 1975.

Vol. 462: P. Gérardin, Construction de Séries Discrètes p-adiques. »Sur les séries discrètes non ramifiées des groupes réductifs déployés p-adiques«. III, 180 pages. 1975.

Vol. 463: H.-H. Kuo, Gaussian Measures in Banach Spaces. VI, 224 pages. 1975.

Vol. 464: C. Rockland, Hypoellipticity and Eigenvalue Asymptotics. III, 171 pages. 1975.

Vol. 465: Séminaire de Probabilités IX. Proceedings 1973/74. Edité par P. A. Meyer. IV, 589 pages. 1975.

Vol. 466: Non-Commutative Harmonic Analysis. Proceedings 1974. Edited by J. Carmona, J. Dixmier and M. Vergne. VI, 231 pages. 1975.

Vol. 467: M. R. Essén, The Cos $\pi\lambda$ Theorem. With a paper by Christer Borell. VII, 112 pages. 1975.

Vol. 468: Dynamical Systems – Warwick 1974. Proceedings 1973/74. Edited by A. Manning. X, 405 pages. 1975.

Vol. 469: E. Binz, Continuous Convergence on C(X). IX, 140 pages. 1975.

Vol. 470: R. Bowen, Equilibrium States and the Ergodic Theory of Anosov Diffeomorphisms. III, 108 pages. 1975.

Vol. 471: R. S. Hamilton, Harmonic Maps of Manifolds with Boundary. III, 168 pages. 1975.

Vol. 472: Probability-Winter School. Proceedings 1975. Edited by Z. Ciesielski, K. Urbanik, and W. A. Woyczyński. VI, 283 pages. 1975.

Vol. 473: D. Burghelea, R. Lashof, and M. Rothenberg, Groups of Automorphisms of Manifolds. (with an appendix by E. Pedersen) VII, 156 pages. 1975.

Vol. 474: Séminaire Pierre Lelong (Analyse) Année 1973/74. Edité par P. Lelong. VI, 182 pages. 1975.

Vol. 475: Répartition Modulo 1. Actes du Colloque de Marseille-Luminy, 4 au 7 Juin 1974. Edité par G. Rauzy. V, 258 pages. 1975. 1975.

Vol. 476: Modular Functions of One Variable IV. Proceedings 1972. Edited by B. J. Birch and W. Kuyk. V, 151 pages. 1975.

Vol. 477: Optimization and Optimal Control. Proceedings 1974. Edited by R. Bulirsch, W. Oettli, and J. Stoer. VII, 294 pages. 1975.

Vol. 478: G. Schober, Univalent Functions – Selected Topics. V, 200 pages. 1975.

Vol. 479: S. D. Fisher and J. W. Jerome, Minimum Norm Extremals in Function Spaces. With Applications to Classical and Modern Analysis. VIII, 209 pages. 1975.

Vol. 480: X. M. Fernique, J. P. Conze et J. Gani, Ecole d'Eté de Probabilités de Saint-Flour IV-1974. Edité par P.-L. Hennequin. XI, 293 pages. 1975.

Vol. 481: M. de Guzmán, Differentiation of Integrals in R^n. XII, 226 pages. 1975.

Vol. 482: Fonctions de Plusieurs Variables Complexes II. Séminaire François Norguet 1974–1975. IX, 367 pages. 1975.

Vol. 483: R. D. M. Accola, Riemann Surfaces, Theta Functions, and Abelian Automorphisms Groups. III, 105 pages. 1975.

Vol. 484: Differential Topology and Geometry. Proceedings 1974. Edited by G. P. Joubert, R. P. Moussu, and R. H. Roussarie. IX, 287 pages. 1975.

etry of Banach Spaces – Selected Topics.

Vol. 486: S. Stratila and D. Voiculescu, Representations of AF-Algebras and of the Group U (∞). IX, 169 pages. 1975.

Vol. 487: H. M. Reimann und T. Rychener, Funktionen beschränkter mittlerer Oszillation. VI, 141 Seiten. 1975.

Vol. 488: Representations of Algebras, Ottawa 1974. Proceedings 1974. Edited by V. Dlab and P. Gabriel. XII, 378 pages. 1975.

Vol. 489: J. Bair and R. Fourneau, Etude Géométrique des Espaces Vectoriels. Une Introduction. VII, 185 pages. 1975.

Vol. 490: The Geometry of Metric and Linear Spaces. Proceedings 1974. Edited by L. M. Kelly. X, 244 pages. 1975.

Vol. 491: K. A. Broughan, Invariants for Real-Generated Uniform Topological and Algebraic Categories. X, 197 pages. 1975.

Vol. 492: Infinitary Logic: In Memoriam Carol Karp. Edited by D. W. Kueker. VI, 206 pages. 1975.

Vol. 493: F. W. Kamber and P. Tondeur, Foliated Bundles and Characteristic Classes. XIII, 208 pages. 1975.

Vol. 494: A Cornea and G. Licea. Order and Potential Resolvent Families of Kernels. IV, 154 pages. 1975.

Vol. 495: A. Kerber, Representations of Permutation Groups II. V, 175 pages. 1975.

Vol. 496: L. H. Hodgkin and V. P. Snaith, Topics in K-Theory. Two Independent Contributions. III, 294 pages. 1975.

Vol. 497: Analyse Harmonique sur les Groupes de Lie. Proceedings 1973–75. Edité par P. Eymard et al. VI, 710 pages. 1975.

Vol. 498: Model Theory and Algebra. A Memorial Tribute to Abraham Robinson. Edited by D. H. Saracino and V. B. Weispfenning. X, 463 pages. 1975.

Vol. 499: Logic Conference, Kiel 1974. Proceedings. Edited by G. H. Müller, A. Oberschelp, and K. Potthoff. V, 651 pages 1975.

Vol. 500: Proof Theory Symposion, Kiel 1974. Proceedings. Edited by J. Diller and G. H. Müller. VIII, 383 pages. 1975.

Vol. 501: Spline Functions, Karlsruhe 1975. Proceedings. Edited by K. Böhmer, G. Meinardus, and W. Schempp. VI, 421 pages. 1976.

Vol. 502: János Galambos, Representations of Real Numbers by Infinite Series. VI, 146 pages. 1976.

Vol. 503: Applications of Methods of Functional Analysis to Problems in Mechanics. Proceedings 1975. Edited by P. Germain and B. Nayroles. XIX, 531 pages. 1976.

Vol. 504: S. Lang and H. F. Trotter, Frobenius Distributions in GL_2-Extensions. III, 274 pages. 1976.

Vol. 505: Advances in Complex Function Theory. Proceedings 1973/74. Edited by W. E. Kirwan and L. Zalcman. VIII, 203 pages. 1976.

Vol. 506: Numerical Analysis, Dundee 1975. Proceedings. Edited by G. A. Watson. X, 201 pages. 1976.

Vol. 507: M. C. Reed, Abstract Non-Linear Wave Equations. VI, 128 pages. 1976.

Vol. 508: E. Seneta, Regularly Varying Functions. V, 112 pages. 1976.

Vol. 509: D. E. Blair, Contact Manifolds in Riemannian Geometry. VI, 146 pages. 1976.

Vol. 510: V. Poènaru, Singularités C^∞ en Présence de Symétrie. V, 174 pages. 1976.

Vol. 511: Séminaire de Probabilités X. Proceedings 1974/75. Edité par P. A. Meyer. VI, 593 pages. 1976.

Vol. 512: Spaces of Analytic Functions, Kristiansand, Norway 1975. Proceedings. Edited by O. B. Bekken, B. K. Øksendal, and A. Stray. VIII, 204 pages. 1976.

Vol. 513: R. B. Warfield, Jr. Nilpotent Groups. VIII, 115 pages. 1976.

Vol. 514: Séminaire Bourbaki vol. 1974/75. Exposés 453 – 470. IV, 276 pages. 1976.

Vol. 515: Bäcklund Transformations. Nashville, Tennessee 1974. Proceedings. Edited by R. M. Miura. VIII, 295 pages. 1976.